BIO-ENERGY
FROM
WASTEWATERS

Authors

Dr. S.N. Kaul obtained his bachelor and master degrees from IIT, Kanpur and was awarded Ph.D. degree from the University of New Castle Upon Tyne, UK. Dr. Kaul joined CSIR-National Environmental Engineering Research Institute [NEERI] Nagpur in 1970 and worked for over three decades. Dr. Kaul superannuated as Acting Director and Director Grade scientist from NEERI and was also associated with MIT college of Engineering, Pune as its Principal. Dr. Kaul has published over 350 Papers in National and International journals beside and authorship of over 30 books on Environmental Science and Engineering, 8 Patents and 150 Conference papers. Dr. Kaul has designed and operated several waste treatment plants in India. Dr. Kaul has supervised over a dozen of Ph.D. and 40 M.Tech. students. In addition, Dr. Kaul was responsible for organizing several National and International Conferences and has traveled extensively in India and aboard concerning various R&D programs. Dr. Kaul has received several National and International awards and is expert member of various important National committees on Environmental Science and Engineering.

Dr. D.R. Saini obtained his B.Tech. and M.Tech. from HBTI, Kanpur and Ph.D. (Chem. Eng.) from IIT-Delhi. Presently working as Principal Scientist at NCL, Pune. Dr. Saini has Coauthored three books, six chapters and published seventy papers in international journals. Dr. Saini has guided 3 Ph.D. and few M.Tech. students.

Professor Yogesh Chandra Sharma, born at Aligarh, India obtained Ph.D. in Applied Chemistry 1991 and Doctor of Science (D.Sc.) in 2010. Dr. Sharma has published large number of research articles in international professional journals. Dr. Sharma is referee of nearly four dozens of international professional journals including journals from Royal Society of Chemistry (RSC), American Chemical Society (ACS), American Society of Civil Engineers (ASCE), and journals from other reputed publishers. Dr. Sharma is also Editor in- Chief of the International Journal of Pollution and Solutions (CIP, USA). His current interests are synthesis, characterization and applications of nano-adsorbents, bio-fuel and catalysts. Currently Professor Sharma is teaching at the Institute of Technology, Banaras Hindu University, Varanasi.

Er. Parteek Kaul has obtained his B Tech and M Tech from NIT, Nagpur. He has worked in several multinational industries in India and abroad and has number of patents and publications to his credit.

BIO-ENERGY
FROM
WASTEWATERS

By :
Prof. (Dr.) S.N. Kaul
Dr. D.R. Saini
Prof. (Dr.) Y.C. Sharma
Er. Prateek Kaul

2014
DAYA PUBLISHING HOUSE®
A Division of
ASTRAL INTERNATIONAL (P) LTD.
NEW DELHI – 1100 002

ISBN 9789351301103

Published by : **Daya Publishing House®**
A Division of
Astral International Pvt. Ltd.
– ISO 9001:2008 Certified Company –
4760-61/23, Ansari Road, Darya Ganj
New Delhi-110 002
Ph. 011-43549197, 23278134
E-mail: info@astralint.com
Website: www.astralint.com

Laser Typesetting : **Classic Computer Services**, Delhi - 110 035

Printed at : **Replika Press Pvt. Ltd.**

PRINTED IN INDIA

Dedicated to the Sacred Memory

of

Sri Sri Sarda Maa

and

Swami Vivekananda

Preface

Very few books are available on biogas (Bio-energy) recovery from wastewaters. An attempt has been made to consolidate the relevant materials on this subject including kinetics and future R and D. The book will be useful addition on biogas recovery and may be used by students, teachers, researchers and practicing engineers in India.

Methane recovery from wastewaters through anaerobic treatment is potentially one of the most attractive methods of solving the twin problems of energy production and pollution control in a cost effective manner. Attempts towards the satisfaction of energy needs of mankind in the biosphere warrants substantial external energy subsidies. Therefore, shift in energy-mix towards renewable is now recognized as one of the necessary measures for meeting the energy needs of the future.

The anaerobic process is in many ways ideal for waste treatment. It has several significant advantages over other available methods and is almost certainly assured of increased usage in the future. Anaerobic treatment is currently employed at most municipal treatment and also in some industrial sectors. However, in spite of the present significance and large future potential of this process, it has not yet enjoyed the favourable reputation it truly deserves. The primary obstacle has been a lack of fundamental understanding of the process, required both to explain and control the occasional upsets which may occur, and to extend successfully this process of the treatment of wide variety of Industrial wastewaters.

The authors have extensive and intensive experiences in the field of Wastewater Treatment and the present book will be useful addition in this area. In addition, it narrates the potential of biogas recovery from various types of wastewaters.

The authors acknowledge the co-operation and assistance of all persons involved directly or indirectly in the compilation of this book, and are sure that it will be well received by all.

The trust reposed by Daya Publishing House, a division of Astral International (P) Ltd., New Delhi, is thankfully acknowledge by the authors.

Prof. (Dr.) S.N. Kaul
Dr. D.R. Saini
Prof. (Dr.) Y.C. Sharma
Er. Prateek Kaul

Contents

CHAPTER 1

ANAEROBIC TECHNOLOGIES FOR WASTEWATER TREATMENT

1.0 Introduction

Methane recovery from wastewaters through anaerobic treatment is potentially one of the most attractive methods of solving the twin problems of energy production and pollution control in a cost effective manner. Attempts towards the satisfaction of energy needs of mankind in the biosphere warrants substantial external energy subsidies. Therefore, shift in energy-mix towards renewable is now recognised as one of the necessary measures for meeting the energy needs of the future.

The anaerobic process is in many ways ideal for waste treatment. It has several significant advantages over other available methods and is almost certainly assured of increased usage in the future. Anaerobic treatment is currently employed at most municipal treatment and also in some industrial sectors. However, inspite of the present significance and large future potential of this process, it has not yet enjoyed the favourable reputation it truly deserves. The primary obstacle has been a lack of fundamental understanding of the process, required both to explain and control the occasional upsets which may occur, and to extend successfully this process to the treatment of wide variety of industrial wastewaters.

An increasing realization of the potential of anaerobic treatment is evident from the reporting each year of large number of recent investigations on this process. Already significant advances have been made extending the process so that it can be used successfully on many more organic wastewaters.

In the fifties and sixties aerobic processes were very popular in biological treatment which generally required energy for transferring oxygen into the waste. The scenario changed significantly as a result of environmental debate and the increase in energy prices both culminating in seventies. Rense and energy conservation became the current topics of research interest and anaerobic processes quickly emerged with new potential. Further, increasingly stringent pollution control regulations coupled with the rising energy costs of aerobic treatment systems in the early seventies greatly stimulated interest in anaerobic treatment as an energy saving waste treatment technology. This interest led to the development of range of reactor designs suitable for the treatment of low, medium and high strength wastewaters.

1.1 Need for Anaerobic Treatment

The advantages of anaerobic treatment can best be indicated by comparing this process with aerobic treatment. In aerobic treatment, as represented by activated sludge and trickling filter process, the wastewater is mixed with large quantities of micro-organism and air. Micro-organisms must the organic waste as food and use oxygen in the air to burn a portion of this food to carbon dioxide and water for energy. Since these organisms obtain much energy from this oxidation, their growth is rapid and a large portion of organic waste is converted into new cells. The portion converted to cells is not actually stabilized but is simple changed in form. Although these cells can be removed from the waste stream, the biological sludge they produce still presents a significant disposal problem.

In anaerobic treatment, the waste is also mixed with large quantities of micro-organisms, but here air is excluded. Under these conditions bacteria grow which are capable of converting the organic waste to carbon dioxide and methane gas. Unlike aerobic oxidation, the anaerobic conversion to methane gas yields relatively little energy to the micro-organisms. Thus their rate of growth is slow and only a small portion of the waste is converted to new cells, the major portion of the degradable waste being converted to methane gas. Such conversion of methane gas represents waste stabilization since this gas is insoluble and escapes from the waste stream where it can be collected and burned to carbon dioxide, water and heat.

As much as 80 to 90 per cent of the degradable organic portion of a waste can be stabilized in anaerobic treatment by conversion to methane gas, even in highly loaded systems. This is in contrast to aeration system, where only about 50 per cent of the waste is actually stabilized even at conventional loading.

The other advantages of anaerobic are shown in Table 1.1. Since only a small portion of the waste is converted to cells, the problem of disposal of excess sludge is greatly minimized. Also, the requirements for nutrients, nitrogen and phosphorus are proportionately reduced. This is especially important in treatment of industrial wastewaters which lack these materials. The sludge produced is quite stable and will not present nuisance problems.

Since anaerobic treatment does not require oxygen, treatment rates are not limited by oxygen transfer rates. The anaerobic treatment process does have some disadvantages which may limit the use of this process for certain industrial wastes. The major disadvantage is that relatively high temperatures are required for optimum operation. Dilute wastewater may not produce sufficient methane gas for waste heating and this may represent a major limitation. However, treatment costs will still be on a lower side considering energy requirements for same strength of wastewater by aerobic methods. This suggests a need for increased research activity on low temperature anaerobic treatment. There are indications that much lower temperatures can be used if the systems are adequately designed.

Another disadvantage of anaerobic treatment is related to the slow rate of growth of methane producing bacteria. Because of it, longer periods of time are required for starting the process. This slow rate of growth also limits the rate at which the process can adjust to changing waste loads, temperature or other environmental conditions. However, this has been offset by application of anaerobic fixed film reactors. Table 1.2 describes some of the disadvantages of the anaerobic process.

The advantages of anaerobic treatment are quite significant, while the disadvantages are relatively few. The advantages normally outweigh the disadvantages for more concentrated wastes, with COD greater than 10,000 mg/L. For less concentrated wastes, the disadvantages become more important and may limit the use of this process except for treatment purposes without recovering methane.

The most recent and significant advances in anaerobic digestion are related to the technology's ability to accommodate relatively high rates of organic loading. Industries are also interested in using anaerobic digestion for biodestruction of organic materials that are not removed in conventional aerobic treatment. As application of anaerobic technology to various process streams increase, more successes are inevitable, resulting in industries that are more economically competitive because of their more judicious use of natural resources. Industries adopting anaerobic digestion seem to fall into the following categories:

- Industries scheduled to expand their production process capacity and where facilities are already at capacity.
- Industries that discharge to public owned treatment works whose surcharge for treatment has increased substantially.
- Industries that have relatively monotonous, highly concentrated waste streams contributing a major portion of total waste load.
- Industries in areas where extremely high costs make conventional aerobic digestion too expensive.

Keeping the views expressed above, it is quite evident that there is urgent need for application of anaerobic processes for treatment and recovery of useful by product wherever possible. In addition recent advances in both the fundamental understanding of the anaerobic processes and engineering application of this process have taken place. These new developments show a great deal of promise in overcoming many of the limitations associated with treatment of both municipal and industrial wastewaters.

Table 1.1 : Advantages of Anaerobic Process

Sr. No.	*Advantages*
1.	A high degree of waste stabilization is possible even at high organic loading rates
2.	Low production of stabilized excess sludge
3.	Low nutrient requirements
4.	No Oxygen requirements
5.	The excess sludge has good dewatering characteristics
6.	Production of a useful end product in the form of methane gas
7.	Well adapted sludge can be preserved unfed for a period of one year or more without any possible deterioration
8.	May be less sensitive to toxic compounds than aerobic processes
9.	Low-strength wastewater can be treated but resources recovery may not be profitable

Table 1.2 : Disadvantages of Anaerobic Process

Sr. No.	*Disadvantages*
1.	Slow growth rate of methane producing bacteria (Methanogens)
2.	Long solids retention time, requiring a larger volume of reactor, however, this has been offset quite considerably by the application of fixed film technologies
3.	Need for auxiliary heating to maintain digesters at optimum temperature for growth of the essential bacteria
4.	The sensitive nature of the methanogens
5.	The general feeling of un-reliability associated with the process
6.	Start up of process is slightly longer than aerobic system
7.	Anaerobic digestion is essentially a pretreatment process especially for medium and high strength wastewaters
8.	Little practical experience has been gained with the application of the process to the direct treatment of wastewater.

1.2 Historical Developments of Anaerobic Systems

As early as the eighteenth century the formation of methane from anaerobic decomposition of organic deposits was known and in the middle of ninteenth century the involvement of bacteria in this decomposition became clear. However, it was just one century ago (1881) when anaerobic treatment was reported to be a useful method for reducing the mass and putrescible nature of suspended organic material removed from municipal wastewaters. In recent years interest in anaerobic treatment has expanded considerably because the methane by product can serve as a useful fuel to offset a growing demand for energy.

Alongwith this increase in interest, several newer processes have evolved that offer promise for more economical anaerobic treatment, and for the generation of methane fuel from a wider range of agricultural, industrial and municipal organic residue.

France is credited with having made the first significant contribution towards anaerobic treatment of wastewaters suspended solids. In December 1881 and January 1882 there appeared a description of an air tight Chamber developed by M. Louis Moyras[1] called 'Mouras Automatic Scavenger' in which suspended organic material in wastewater was liquified. This was different from the water tight cesspools then commonly used, as little attention or no attention was paid to trapping their inlets or outlets to exclude air.

Other studies on liquefaction of wastewater solids in the absence of air were conducted by Scott-Moncrieff[2] in about 1890 and in 1891, he constructed in England a tank with empty space below and a bed of stones above. This first application resembles the anaerobic filter. The value of this method was confirmed through studies by Houston[2] in 1892 and 1893 who reported an apparent great decrease in volume of sludge to be handled.

Donald Cameron of Exeter, England, in 1895 constructed a tank similar to Mouras automatic scavenger for the preliminary treatment of screened combined wastewater averaging about 270 m^3/d and termed this system as septic tank. A similar system was designed by A.N. Talbot for Urbana, Illinois in 1894 and for Champaign, Illinois in 1897[3].

In 1897, waste disposal tanks at a Leper colony in Matunga, Bombay are reported[4] to have also been equipped with gas collectors and the gas used to drive gas engines. The effluent from septic tanks often was black and offensive and still contained undigestible material which clogged contact beds often used for further subsequent chambers. Clark was first to recognise the solution to this problem in 1899, when he stated that the sludge should be permitted to ferment by itself in a separate tank[5].

In 1904, William O. Travis put into operation a new two stage process in which the suspended material was separated from wastewater and allowed to pass into a separate hydrolyzing chamber[2,5]. He also hung baffles in the chambers to attract fine non-depositable suspended solids.

Construction of Travis tank was begun at Emscher in 1905, but was modified by Karl Imhoff[2,5] when he took over as the sewerage engineer of Emscher Drainage Board. While the Imhoff tank with its separate digestion had clear advantages over the septic tank, it was not without problems. It was tall, and the digestion tank had to be connected immediately with the sedimentation tank. In 1927, the Ruhrverband installed in the clarification plant at Essen-Rellinghausen, the first sludge-heating apparatus in a separate digestion tank[6]. Collection of gas and heating equipment were conducted by Emschergenossenschaft in 1914 and in 1923 the gas was collected on a large scale and delivered to municipal gas system at the Essen-Rellinghausen plant[6]. As early as 1927, the Ruhrverband utilized the bludge gas in Iserlohn and then in Essen-Rellinghausen to generate power for a biological treatment plant and used the cooling water from motors for heating of the digestion tanks. In addition in 1930s many cities in Germany added compressing plant to store the gas in steel cylinders for use as a motor fuel[7].

In 1927, Rudolf[8] demonstrated that the total quality of gas produced from a given amount of sludge was independent of temperature of treatment, but the rate of digestion increased with increasing temperature. Fair and Moor[9] demonstrated in the early 1930s that there were two temperature optima for anaerobic treatment one the mesophilic range between 28-33°C and the other thermophilic range between 55-60°C.

1.2.1 Process Chemistry and Microbiology

According to Barker[10], the Italian physicist Alessandro Volta in 1976 showed that combustible air was formed from the sediments in lakes, ponds, streams and concluded it was derived from plant material in the sediments. Bunsen and Hoppe-Seyler[4,10] confirmed this observation and noted further that the gas production was particularly energetic during summer months. In 1856, Reiset found methane liberated from decomposing manure piles and proposed this process by studied to help the decomposition of organic material in general[4]. Bechamp[10] is credited with giving the first indication in 1868 that the methane was formed by a microbiological process. Popoff in 1875 reported on natural evolution of methane from stagnant ponds, hay, gum arabic, glucose, certain fatty acids, salts and numerous other products and indicated that 40°C was the optimum temperature for methane gas production.

In 1890 Van Senus was the first to attribute the anaerobic decomposition of material like cellulose to the joint activities of several micro-organisms. By 1927, Castellane[4] and others reported on extensive studies of symbiotic relationships in methane production. Omelianski's[4,10] classic studies on methane fermentation of cellulose were reported in 1890s. He isolated hydrogen, acetic and butyric acids and also reported the formation of methane from hydrogen and carbon dioxide. Sohngen intimated that fermentation of complex materials proceeded through oxidation-reduction reactions to form hydrogen, CO_2 and acetic acid[11].

In 1930, Buswell *et al.*[5] showed the importance of two phases in digestion process and credited Thum and Rechle in 1914 with recognising an initial acid phase and later alkaline one. Imhoff earlier termed these two phases as 'putrial' and 'odourless' but by 1916 adopted the expression 'acid digestion' and 'methane digestion'[11].

Buswell and his colleagues at Illinois Division of State Water Survey[3,5] demonstrated the anaerobic treatability of a wide range of industrial and agricultural residues in addition to municipal wastewater and showed the importance of volatile fatty acids as intermediate in the process. Of significance was the demonstration that the following stoichiometric equation was applicable for methane fermentation of substrate in general[4].

$$C_nH_aO_b + (n - a/4 - b/2)H_2O = (n/2 - a/8 + b/4)\,CO_2 + (n/2 + a/8 - b/4)CH_4 \qquad (1.1)$$

In 1948, Buswell, *et al.*[12] using ^{14}C tracers proved that methane formation from acetate did occur through CO_2 reduction and subsequently Stadtman *et al.*[13] and Pine *et al.*[14] conducted experiments that added further verification to the decarboxylation hypothesis.

Jeris[15] using radiotracers showed that 70 per cent of methane resulted from overall fermentation of most organic compounds and mixtures of compounds came from acetate which was formed as an intermediate. Barkers extensive studies[16] led to the reported isolation of organisms, Methanobacterium omelianski which oxidized ethanol to acetate and methane. Hungate developed techniques[17] which resulted in the isolation of second bacteria capable of converting CO_2 and H_2 to CH_4. It may be mentioned that it was not possible to isolate bacteria capable of converting propionate, butyrate or higher fatty acid salts to acetate and methane. A major break occured in 1967 when Byrant *et al.*[18] reported that original M. omlianski culture contained two bacterial species, not one. It was clearly demonstrated that one converted ethanol to acetate and hydrogen and other converted H_2 and CO_2 to CH_4. Thus, it became clear that the complete oxidation of a simple ethanol to CO_2 and CH_4 would require three types of bacterial species and is shown in Table 1.3.

Byrant *et al.* also demonstrated that a similar synthrophic association existed in the oxidation of butyrate[19] and propionate[20]. In order for energy to be available to the organisms, oxidising propionic acid and hydrogen the partial pressure of H_2 cannot exceed 10^{-6} atmospheres[21]. At this low pressure, the energy available to the hydrogen oxidising bacteria is reduced considerably from what it would be at partial pressures near one atmosphere. This results in much lower bacterial yield per mole of hydrogen gas oxidized, as confirmed by overall growth yields measured by Speece[22] and Lawrence[23] for complete methane fermentation of propionate and other fatty acids as well as by thermodynamic predictions[24].

Table 1.3 : Presence of Three Species for Requirement of Anaerobic Reactions

Sr.No.	*Species*	*Reaction*	*ΔG°, kcal*[24]
1.	Species – 1	$C_2H_5OH + H_2O = CH_3COO^- + H^+ + 2H_2$	1.42
2.	Species – 2	$2H_2 + 1/2\ CO_2 = 1/2\ CH_4 + H_2O$	– 15.63
3.	Species – 3	$CH_3COO^- + H^+ = CH_4 + CO_2$	– 6.77
	Net	$C_2H_5OH = 3/2\ CH_4 + 1/2\ CO_2$	– 20.98

* Standard free energy at pH-7 and 25°C

It would appear that our practical consideration of processes should somehow incorporate the three phase concept. However, the practical consequence of these findings have yet to be demonstrated. Further research is needed in this area.

1.2.2 Recent Process Developments

During the 1950s two developments were of practical significance, one was the use of mixing in digesters and the other was development of anaerobic contact process.

Prior to 1950, most separate digesters treating municipal wastewater sludges did not employ mechanical mixing. This resulted in separation of solids from the liquid forming a thickened sludge at the bottom of the tank and a floating scum layers at the top. Mixing helped to reduce the scum formation tendency to a great extent and also enhanced the rate of digestion by bringing bacteria and wastes more closely together, the value of this high rate digestion was demonstrated by Morgan[25] and Torpey[26].

Stander[27-29] recognised the value of maintaining a large population of bacteria in methane producing reactor (Clarigester-reactor). Schroepfer *et al.*[30] developed a similar concept in 1950.

Table 1.4 : Developments in Anaerobic Digestion*

Sr.No.	*Investigator*	*Process Description*
1.	M.Louis Mouras (1881)	Mouras-Automatic Scavenger
2.	W.D. Scot-Moncrieff (1880) England	The first application of an anaerobic filter
3.	Donald Cameron (1883) England	Septic Tank
4.	At Matunga, Bombay (1897)	Waste disposal tanks at leper colony with gas collectors.
5.	Harry W. Chark (1899) USA	Sludge was formed in a separate tank
6.	William O. Travis (1904)	Travis Tank with hydrolysing chamber
7.	Karl Imhoff (1905)	Modification of Travis Tank
8	Germany (1927)	The 1st sludge heating apparatus in a separate digestion tank was set up. The collected gas was delivered to municipal gas system
9.	Fair and Moore (1930)	Importance of seeding and pH control
10.	Morgan and Torpey (1950)	Mixing in digesters and development of high rate digestion
11.	Stander (1950)	Development of Clarigester and anaerobic baffled reactor based on RBC concepts
12.	Young and McCarty (1972)	Anaerobic Filter
13.	Lettinga *et al.* (1979)	UASB
14.	Switzenbum and Jewell (1980)	Developed the further concept of anaerobic filter to fixed film reactors

*Source - Khanna, P. Brainstorming Session on Environmental Biotechnology, Vol. 1, p. (67-68) (April 12, 1989), Department of Biotechnology, N.Delhi and NEERI, Nagpur (India).

Similar to aerobic trickling filter, anaerobic filter was developed by Young *etal.*[31]. This concept was extended further in the development of anaerobic attached film reactor by Switzenbaum *et al.*[32].

Lettinga, *et al.*[33] developed a process similar to Stander by incorporating a different mode of separation of gas and suspended solids (UASB).

Table 1.5 : Effect and Relevance of Biofilms of Various Rate Processes*

Sr. No.	*Effect*	*Specific Process and Result*
1.	Heat transfer reduction	Biofilm formation on condenser tubes and cooling tower fill material. ENERGY LOSSES.
2.	Increase in fluid frictional resistance	Biofilm formation in water and waste water conduits as well as condenser and heat exchange tubes. Causes increased power consumption for pumped systems or reduced capacity in gravity systems. ENERGY LOSSES.
		Biofilm formation on ship hulls causing increased fuel consumption. ENERGY LOSSES.
3.	Mass transfer and chemical transformations	Accelerated corrosion due to processes in the lower layers of the biofilm. Results in Material Deterioration in metal condenser tubes, sewage conduits, and cooling tower fill.
		Biofilm formation on remote sensors, submarine periscopes, sight glasses, etc. causing REDUCED EFFECTIVENESS.
		Detachment of micro-organisms from biofilms in cooling towers. Releases PATHOGENIC ORGANISMS (e.g., Legionella in aerosols).
		Biofilm formation and detachment in drinking water distribution systems. Changes WATER QUALITY in distribution system..
		Biofilm formation on teeth. Causes DENTAL PLAQUE AND CARIES.
		Attachment of microbial cells to animal tissue. Causes DISEASE of lungs, intestinal tract, and uninary tract. Extraction and oxidation of organic and inorganic compounds from water and wastewater (e.g. rotating biological contactors, biologically-aided carbon adsorption and benthal stream activity.) REDUCED POLLUTANT LOAD.
		Biofilm formation in industrial production processes REDUCES PRODUCT QUALITY.
		Immobilized organisms or community of organisms for conducting SPECIFIC CHEMICAL TRANSFORMATIONS.
		Fouling biofilm accumulation REDUCES EFFECTIVENESS of ion exchange and membrane processes used for high quality water treatment.

*Source - Characklis, W.G. and Zelver, N. Dynamics of Biofilm Processes : Methods; Water Research, p. 1207-1216 (1982).

Rotating biological reactors have also been used for anaerobic treatment of wastewaters and offer some promise[34]. Stander developed an anaerobic baffled reactor based on the above concept. Table 1.4 summarises the important events for the development of different types of anaerobic reactor systems over the past ten decades. The effects and relevance of biofilms on various processes are summarised in Table 1.5.

REFERENCES

1. Moigno, A.F. Mouras Automatic Scavenger, Cosmos, 662(1881) and 97 (1882), as reviews in the Minutes of the Proceedings of the Institution of Engineers, XLVIII, p. 350 (1882), London.
2. Matcalf, L and Eddy, H.P. American Sewerage Practice, III, Disposal of Sewage, 1st Edition (1915), McGraw Hill Book Company, Inc. New York.
3. Buswell, A.M. Studies on Two Stages Sludge Digestion, 1928/29, State Water Survey, Bulletin No. 29, State of Illinois, Urbana, Illinois.
4. Buswell A.M. Anaerobic Fermentations, State Water Survey, Bulletin No. 32 (1938), State of Illinois, Urbana, Illinois.
5. Buswell A.M. and Neave, S.L. Laboratory Studies of Sludge Digestion, State Water Survey, Bulletin No. 30 (1930), State of Illinois, Urbana, Illinois.
6. Imhoff, K. Sedimentation and Digestion in Germany Modern Sewage Disposal p.47 (1938) Editor Pearse L, Lancaster Press, Lancaster, Penn.
7. Hyde, C.G A review of progress in Sewage Treatment During the past Fifty years in USA, Modern Sewage Disposal P-1(1938), Editor Pearse L, Lancaster Press, Lancaster, Penn.
8. Rudolf, W. Effect of Temperature on Sewage Sludge Digestion. Industrial and Engg. Chemistry 19, p. 241(1927).
9. Fair, G.M. and Moor, M. Time and Rate of Sludge Digestion and their Variations with Temperature, Sewage Works Journal, 6, p. 3 (1934).
10. Barker, H.A. Biological Formation of Methane, Bacterial Fermentations, p. 1 (1956) John Wiley & Sons, Inc. New York.
11. Barker H.A. On Biochemistry of the Methane Fermentation Archiv. Fun Mikobologic. 7, p. 404 (1936).
12. Buswell A.M. and Sollo, F.W. The Mechanisms of the Methane Fermentation, American Chemical Society Journal 70, p1778 (1948).
13. Stadtman, R.C. and Barker, H.A. Studies on Methane Fermentation. VII - Tracer Experiments an Mechanisms of Methane Formation, Archives of Biochemistry, 21, p. 256 (1949).
14. Pine, M.J. and Barker, J.A. Studies an Methane Fermentation XII. The Pathway of Hydrogen in the Acetate Fermentation, Journal of Biochemistry 71, p. 644 (1956).
15. Jerris, J.S. and Mc Carty, P.L. Biochemistry of Methane Fermentation using C - 14 Tracer. H. Wat Poll. Cont. Fed. 37, p. 178 (1965).
16. Barker, H.A. Studies upon the Methane Fermentation IV. The Isolation and Culture of Methanobacterium-Omelianski, Antonie, V. Leeuwenhoek, 6 p. 261 (1940).

17. Hungate R.E. The Anaerobic Mesophellic Cellulotylic Bacteria, Bact, Rev 14. p. 1 (1950).
18. Bryant, M.P., Wolin, E.A. Wolin, M.J. and Wolfe, R.S. Methanobacillus omelianski, a Symbiotic Association of two Species of Bacteria, Archiv for Mikrobiologie. 59. p20(1967).
19. Mclnerney, M.J. Anaerobic Bacterium that Degrades Fatty Acid in Synthetic Association with Methanogens, Bact. Rev. 122, p. 129 (1979).
20. Boone, D.R. Propionate-Degrading Bacterium, Syntrophobacter Wolinii Sp. nov. gen. nov, from Methanogenic Ecosystem, Applied and Envn. Microbiology 40, p626(1981)
21. Thauer, R.K. Energy Conservation in Chemotrophic Anaerobic Bacteria, Bacteriol. Rev, 41, p100(1977).
22. Speece, R.E. and McCarty, P.L. Nutrient Requirements and Biological Solids Accumulation in Anaerobic Digestion, Advances in Water Poll. Res. 2 p. 305 (1964), Pergamon Press, London.
23. Lawrence, A. W. and McCarty, P. L. Kinetics of Methane Fermentation in Anaerobic Treatment J. Wat. Poll. Cont. Fed. 41, R1 (1969).
24. McCarty, P.L. Energetics and Bacterial Growth, Organic Compounds in Aquatic Environment p495 (1971) Ed. Faust, S.D., Dekkar Inc. New York.
25. Morgan, P.F. Studies on Accelerated Digestion of Sewage and Industrial Wastes 26, p462(1954).
26. Torpey, W.N. Loading to Failure of a Pilot Plant High Rate Digester, Sewage and Industrial Waste 27, RI(1955).
27. Standar, G.J. Effluents from Fermentation Industries, The Institutes of Sewage Purification, pt-4, p438(1956).
28. Standar G.J and Synidar, R. Effluents from Fermentation Industries. The Institute of Sewage Purification, pt-4p 447 (1950).
29. Standar, G.J. Water Pollution Research - A Key to Wastewater Management J.Wat Poll. Cont. Fed. 38 p. 774 (1966).
30. Schroepfer. G.J. and Fullman W.J. The Anaerobic Contact Process as Applied to Packing House Wastes, Sewage and Industrial Wastes, 27 p. 644 (1955).
31. Young, J.C. and McCarty, P.L. The Anaerobic Filter for Waste Treatment. J. Wat. Poll. Cont. Fed 41, R160(1969).
32. Switzenbaum M.S. and Jewel, W.J. Anaerobic Attached Film Expanded Bed Reactor Treatment, J. Wat. Poll Conts Fed 52, p1953(1980).
33. Lettinga, G. and van Velsen, A.F.M. Feasibility of Upflow Anaerobic Sludge Blanket (UASB) - Process, Proc. Nat. Conf. on Envn. Engg., p35(1979) Amer. Soc. Civil Engrs. New York.
34. Friedman, A.A and Young, K.S., New Observations with Anaerobic Fixed Film Reactors, Workshop on Anaerobic Fillers (1980) Orlando, Florida.

CHAPTER 2

PRESENT SCENARIO OF ANAEROBIC SYSTEMS IN INDIA

2.0 Introduction

The assessment of current and future power scenario clearly brings out the existing supply demand gap as also the perception that unless a different strategy are adopted the situation could become more difficult over a period of time. In such a situation, biomass has tremendous potential to meet a significant part of country's energy requirement by way of using various conversion technologies *viz.*, methanation, combustion and gasification for thermal energy, captive power generation, lift irrigation and rural electrification in a decentralised manner.

According to a country-wide study conducted by National Productivity Council, the total annual availability of agro and agro-industrial residues was observed to be about 370 million tons in the year 1985-86. Out of this huge quantity, 50 million tons of agro-industrial residues in the form of cotton stalks, sugarcane trash, tur stalks, maize cobs, tapioca stalks, coconut, cashew nuts, arecanut trunk, jute sticks and oil seed stalks are suitable and also available for power generation. Utilization of 50 per cent of this quantity would ensure power generation of 3000 MW; 50 million tons of agro-industrial residues in the form of rice husk, bagasse, saw dust, groundnut shells and jute wastes are produced annually. 80 per cent of these residues can be utilized in gasification for generation of 2700 MW of electricity; Wood produced from energy plantation on 1000 ha., can generate 3 MW of power. It is estimated that 62.6 million ha., of wasteland is available in the country for energy plantation, if wood wastes in 1 million ha., of energy plantation is utilized for power generation, this would mean an additional generating capacity of 3000 MW. In another study, exploitable potential for power from biomass is indicated as 17000MW.

Besides above, huge quantities of urban and industrial wastes (liquid wastes including effluents, fruit and vegetable wastes, distillery wastes, slaughter house wastes, municipal solid wastes, willow dust from textile mills, sugarcane press mud, hospital wastes, etc.) could also be recycled through biomethanation route and converted into energy and power.

A study carried out in 1986 has estimated that the potential for financially attractive industrial co-generation in states of Gujarat and Maharashtra alone could exceed 2000 MW during the 1986-96 period. About 1/3rd of the potential lies in existing plants, *i.e.* retrofit or replacement of existing steam generation equipment and almost all the potential comes from topping cycle where steam is used first for power generation and then for industrial processes. Over 2/3rd of the potential lies in Government owned industries. The estimates assume that natural gas is available only to fertilizer and petro-chemical industries. However, if natural gas could be made available to all industries, the cogeneration potential in the two states could be over 3000 MW. The potential for bottoming congeneration is estimated at only 50 MW mainly in refineries, fertilizer and petro-chemical plants. Bagasse and cane residue based congeneration was identified as the least expensive of all options with an estimated potential of about 350 MW in Maharashtra and 175 MW in Gujarat. The estimates could be lower or higher depending upon the utility purchase price for cogenerated power, regulations governing interconnection with the utility and the cost of back up power, among other things. Extrapolating above estimates to other states on a consistent basis, it is reasonable to assume a potential of 10,000 MW for financially attractive industrial congeneration in country. As per a study exploitable power generation potential from Municipal Solid Waste is estimated at 66 MW.

As many countries throughout the would including the industries in India produce large amounts of wastewater thereby creating environmental problems (if not properly treated). Besides health related problems for the populations near the sites where waste is dumped, the further degradation of waste in the environment will lead to uncontrolled release of methane - a potent green house gas. By the use of anaerobic reactors, the wastewater can be treated and the methane produced can be recovered. Besides reducing the amount of green house gases by controlled use of methane from waste, the substitution of oil and coal with bio-energy will result in saving the global environment by reduction of the use of fossil fuel.

The industries producing large amounts of wastewater with a high content of organic matter are specially suitable for installation of anaerobic reactors. Some obvious types of industries which are amenable for anaerobic treatment are as follows :

- Distilleries
- Tanneries
- Municipal wastewater
- Pulp and paper
- Breweries
- Fruit and food processing units
- Sugar effluents
- Dairies

- Combined effluent from small / medium scale industries
- Textile in general and cotton based pulp and paper in particular
- Leachate from large solid waste dumps
- Slaughter house
- Vegetable oil industries
- Chemical industries, etc.

Many of these industries are often located in clusters where a centralised treatment of wastewater could be possible. Production of electricity directly from gas is highly desirable as compared to a simple burning of the gas (heat). Therefore, a high priority should be given for industrial wastewater having a sufficient methane potential high enough to support co-generation of heat and electricity.

2.1 Potential of Biogas Recovery

The potential for methane production from 212 distilleries will be approximately 1.5×10^6 m^3 - CH_4 per day which potentially by co-generation can produce an effect of 4.35×10^6 kWh electricity and 7.4×10^6 kWh heat. The daily production of wastewater in 300 pulp and paper will be around 1.6×10^6 m^3 of wastewater per day having a potential of 0.21×10^6 m^3 CH_4 per day accounting for a production of 0.61×10^6 kWh electricity and 1.0 kWh heat. The daily production of wastewater from 2000 tanneries is approximately 52,500m^3– wastewater per day having a potential for production of 59,850 m^3 CH_4 per day accounting for a production of 152.250 kWh electricity and 293.265 kWh of heat.

The total estimated amount of methane generation from other sources except those illustrated above will be at least 50-80 times the production of methane from distilleries. Therefore, it is promising to tap the biogas form all sources particularly those industries listed in section 2.0.

Municipal wastewater generally have low content of organic matter compared to industrial effluent and methane potential is low. However, if the size of the plant is about 50 MLD, it is profitable to tap methane for energy recovery. In addition, some waste streams especially in parts of major cities will have high BOD. During anaerobic degradation, the nutrients will generally be preserved which means that nutrient-removing secondary treatment may be necessary for production of an acceptable effluent. Also by co-treatment of municipal wastewater together with other types of wastes (particularly chemical, drugs pharmaceutical etc.) has a high potential for methane production.

2.2 State of Art of Technology

Anerobic wastewater treatment technology has been implemented or are on its way about half of India's 212 distilleries. This large scale implementation of anaerobic technology has proceeded especially during the last five years assisted by the Ministry of Non-conventional Energy Sources (MNES). In the initial phase, 5 pilot plants were commissioned in different distilleries on a 50 per cent cost sharing basis. Public financial institutions such as IREDA have afterwards been active in financing the implementation of the technology in the different distilleries. The interest rate for this kind of activities varies between 9 and 15 per cent. The development of the anaerobic technology in the distilleries has taken place without any major co-ordination and National Environmental Engineering Research Institute, Nagpur has only been involved in design and research in a few of the distilleries. About 15 years ago, the All India Distillers Association had demanded a design package for the whole area and NEERI was involved in design. However, the time for a large scale implementation of anaerobic technology was not matured and due to former legislation for the area or the enforcement of the legislation for the area, the association was not pressed to come up with solutions to their waste problems. The results from the initial experiments were spread to other distilleries through the Indian Organization of Distilleries.

Different kinds of technology have been used for anaerobic treatment of distillery wastewater in different industries such as anaerobic fixed film reactor, CSTR, UASB and a hybrid type reactor where the UASB reactor is equipped with an anaerobic filter in the top part of the reactor. The average biogas production for the anaerobic reactors installed in the distilleries range between 20 to 30 $\times$ 10^3 Cum per day and the normal efficiencies for the systems are generally not more than 80 to 90 per cent.

The first pilot plant in the distillery was established in the Daurala Sugar Works in 1986. After intensive research and development, a full scale plant for treatment of 1500-1800m^3 of effluent per day has been made functional. Another anaerobic plant has just been commissioned in a distillery close to the Daurala Sugar work in Meerut. In this plant the methane-producing reactors are of fixed-film type using activated carbon as supporting material. NEERI was involved in the design of this plant and the process on this specific waste was studied intensively by NEERI before the plant was dimensioned.

The same extensive implementation of anaerobic reactors is presently not seen in other industrial branches outside the distilleries. However, India has a large potential for implementation of anaerobic technology in other industrial sectors. Especially, about 450 pulp and paper mills alongwith 3000 tanneries are under considerable pressure due to heavy pollution load discharged by these types of industries into the environment. Although all these industries are required to treat the effluent, this practice is seldom followed. In this few pulp and paper mills and tanneries where the effluents are not directly dumped into rivers, lakes or fields, the waste streams are often led into aerated lagoons or anaerobic lagoons. However, in many instances they get filled up and overflow into nearby fields. Serious pollutions of the groundwater is often observed and there are several reports concerning the damage caused by these, untreated effluents, particularly related to health and agricultural produce. The environmental problems are specially acute for the tanneries.

Many large scale pulp and paper mills have aerobic treatment of their effluent. Currently only one full scale plant of the CSTR type exist for the treatment of pulp and paper effluent in Padamjee Pulp and Paper Mills, Pune. The plant which is designed by a Swiss Company (Sulzer) treats 250 Cum effluent per day with a BOD removal of approximately 65 per cent. The low efficiency is mainly due to the high content of high molecular lignin-containing fibres in the effluent. With support from the Ministry of Environment and Forests under a programme for Cleaner Technology this plant will be reconstructed and transformed into the anaerobic fixed film process. NEERI has conducted intensive research and design package based on research has been given for modification of the existing facilities.

The effluent from a tannery depend on the type of process used in the industry such as chrome tanning, vegetable tanning or mixed type of tanning. Technologies are available in India for removal of chromium or oils and all fractions are suitable for anaerobic treatment. Large amounts of tanin will be present in the waste demanding a specialised microflora before it can be degraded. A demonstration scale plant for treatment of 65 Cum effluent per day has been designed by NEERI and will be implemented shortly in Ambur, Madras. This plant will be operated with a retention time of 4-5 days but a retention time as low as 1.5 day was found to be sufficient in laboratory experiments even with high COD removal efficiency upto 90 per cent.

Reports exist on a pilot plant for treatment of tannery waste water by the UASB process. The 40 Cum pilot plant was established in Kanpur in 1989 in the Pioneer Tannery situated close to a large cluster of tanneries. The tannery effluent were pretreated before the anaerobic process by recovering of the chromium salts from the collected chromium spent liquors by precipitation. Before applying the effluent to the UASB reactor, the tannery effluent was diluted with domestic sewage in an equalisation tank and the reactor was operated with a retention time of 8 to 10 hours. The full scale reactor has been constructed by a Dutch firm, consisting of a 25000 Cum reactor for treatment of the combined effluent from the Jajamau cluster of tanneries. An aerobic post-treatment of the digested effluent is further planned.

The dairy industries of India also produce large amount of wastewater suitable for anaerobic treatment and controlled methane production. A total number of 45 anaerobic reactors (UASB) have been installed throughout the country through National Dairy Development Board.

Anaerobic reactors for treatment of industrial wastewater are found throughout the world. The official number of anaerobic reactors only account for registered projects which sum up to around 400 where the vast majority are UASB reactors. However, a large number of especially anaerobic fixed film reactors exists in different industries throughout the world which never have been registrated as they have been constructed directly by the industries themselves.

Anaerobic treatment of liquid effluent generally use systems where the active biomass performing the anaerobic digestion process is retained in the reactor through immobilisation either on a supporting material as in the fixed film reactor or by attachment of the bacteria to each other forming heavy clumps (granula) which stay in the reactor even when a high linear flow is applied to the system as in the UASB reactor. By retaining the biomass within the reactor, the retention time of the effluent in the system often can be decreased to a few hours which has to be compared with the traditional continuous stirred tank reactor (CSTR) which the retention time traditionally will be from 4 to 20 days depending on the nature of the material and the temperature applied. The use of immobilised reactors are, however, only suitable for dissolved material free of particulate materials. Therefore segregation of the waste by screening, settling and may be, grit removal is necessary before the waste cam be handled by the system. Furthermore, pretreatment such as pre-hydrolysis are often necessary to assure proper performance. Also it is important to consider that nearly all industries producing a high strength wastewater will also produce a large quantum of organic solid waste with a large pollution load and a large methane potential.

The separation of gas and sludge is generally a problem in the anaerobic immobilised reactor systems leading to high concentrations of suspended solid in the effluent. This problem can be solved by post-treatment for instance by the use of membrane techniques such as ultra filtration. This kind of technique is also suitable when treating effluent with inhibitory elements such as effluent from distilleries, tanneries and pulp and paper mills where loss of the active biomass can be crucial for the process.

Secondary treatment will be necessary for removal of nutrients from the treated waste water or where the effluent should be re-cycled back in the industrial process after treatment which could be met by a additional aerobic step or by physico chemical methods such as reverse osmosis, ion exchange, etc.

Many international firms are active in commissioning of anaerobic immobilised system. Specialized firms within the UASB area exist in Holland, France, Belgium and USA.

Research on anaerobic immobilised systems are being done throughout the world. Holland with the group of Prof. Lettinga is famous for research in the UASB system, a system which actually was invented and implemented for the first time in South Africa. Basic research on granulation - the process whereby the catalysts of the UASB reactor are being formed - occur in countries like USA, Belgium, India and Denmark. A large EEC program has recently been started in cooperation between UK, Greece and Denmark for construction of a robust and flexible anaerobic system for anaerobic treatment of industrial effluent which varies throughout the season.

Table 2.1 highlights the R & D sponsored by MNES and covers quite a large range of substrates. Table 2.2 indicates the available technology packages for various substrates developed by different researchers. It is difficult to know how many wastewater treatment plants (anaerobic type) have commissioned by private companies or individual entrepreneurs. Therefore, this information has not been included. There are some, very critical polluted areas in country. Table 2.3 depicts these areas. In order to minimise pollution loads in these critical areas, it is necessary to assess the situation whether biomethanation route will be suitable. Therefore, extensives field work coupled with rigorous experimentation is required. If possible, pilot plants should also be commissioned if laboratory studies are successful.

Table 2.1 : MNES Supported Research and Development Programme (Completed/Ongoing)*

Category of Waste	*Sr. No.*	*Title of the project*	*PI/ORGN.*	*Objectives/Results*
(A) Completed Projects Urbanwastes				
Sewage	1	Pilot scale experiment with anafil digester plant for treating sewage	Sh. R.N. Joshi Civictrg. Institute Municipal Copn. of Creater Bombay	Aimed at verifying encouraging results obtained in Lab Scale model of anafil digester, process can treat raw dig-ritted sewage effective at low retention time of 9 hrs. The pilot plant commissioned in May 86 with flow rate of 500 CU.M/Day at 9.6 hrs R.T. Gas Generation is in order of 240-270/1/kg COD removed. Effluent has suspended solids value of 100 mg/L & BOD of about 20-40 mg/L Process developed can be tried at suitable sites for full scale plant (Small & Medium town) with designed Engg. Design.
Sewage	2	Sewage treatment cum biogas generation and its utilisation at ONGC township, Dehradun By NEDA, Lucknow	Director, NEDA Lucknow	Aimed at Installation of sewage biogas generation & utilisation system from 5000 contributing population for 0.625 MLD sewage discharged. The plant has been commissioned in April,

Contd...

Category of Waste	*Sr. No.*	*Title of the project*	*PI/ORGN.*	*Objectives/Results*
				90 & 16 cooking connections were provided to nearby houses of ONGC colony. All reports are being made to optimise the functioning of plants.
Municipal Solid Waste	3	New Energy Sources for biogas production	Prof. A.K. Verma JNU, New Delhi	Aimed at modifying the existing strains and developing new forms of bacteria responsible for depolymerization of cellulosic material into glucose before they are Metabolised to Biogas with ref. to municipal solid waste a large number of cellulolytic bacteria from live termite mould solids were screened for cellulose degradibility some of them have reported to have exhibited better activity at higher pH in digestion of cellulosic waste.
Municipal Solid Waste	4	Recovery of Biogas from Landfill	Sh. A.D. Bhide	Aimed at estimating amount of biogas from HSW in India & Recovering biogas from shallow uncontrolled landfill. Studies established that from HSW, 150-200 m^3/Ton biogas can be obtained. Biogas could be recovered from 5-7 years old uncontrolled landfill site for a period of 5-10 years. Methane content of recovered from experimental wells about 6-9 m^3/HR. for an extended period of time.
Slaughter	5	Studies for biogas on slaughter house refuse at thin cities of Hyderabad & Secunderabad	MEDCAP, Hyderabad	Aimed at preparation of Pre-Feasibility/C.U.M. DPR on establishment of biogas plant based on slaughter house refuse of twin cities of Hyderabad and Secunderabad with objective of safe disposal of slaughter house refuse and

Category of Waste	*Sr. No.*	*Title of the project*	*PI/ORGN.*	*Objectives/Results*
				energy recovery & utilisation. The pre-feasibility report was submitted in time, TRZ report includes data on lab. studies, elaborate account of waste management options and process alternatives including anaerobic digestion.
Kitchen Waste	6	Installation, monitoring and demonstration of biogas plants based on alternative feedstocks for filed evaluation.	Dr. C.S. Rao SPRERI V.V. Nagar	Aimed at installation, monitoring and Demonstration of biogas plant based on alternative feed stock like kitchen wastes, wh. etc., led to installation of 2 units of 10cu. M. Biogas plant evaluation based on kitchen wastes at Vallabh-vidya Nagar & 2 units of 5 CU.M pilot plant on wh. one at and municipality & the other at instt premises.
Agricultural waste Agrowaste	7	Lab. studies for assessing the feasibility of using Bi-phasic process for biomethanation of agrl. wastes	Prof.K.M. Dholakia SPRERI, V.V. Nagar	Refer ongoing project
Fruit & Veg. Waste	8	Microbiology studies on the conversion of food processing and other agro-industrial wastes for biogas production	Dr. Krishna Nand CFTRI Mysore	Refer Ongoing Project
Apple	9	Bioconversion of Apple waste into hydrogen and methane	Dr. A.P. Joshi Centre for Bio-chemical, Delhi	The project aims at Development of pilot plant based on apple pomace into hydrogen & methane and process optimisation. The three phase system for conversion of apple pomace into biofuels. In the lab scale each kg dry apple pomace yields 70 L of H_2 & CH_4 mixture (60:40) and about 200 L. of CH_4 : CO_2 mixture (70:30) at pH7.

Contd...

Category of Waste	*Sr. No.*	*Title of the project*	*PI/ORGN.*	*Objectives/Results*
				Direct Bio-methanation of apple pomace in two phases revealed 275 L biogas (57% CH_4) in lab and 178 L biogas (65% CH_4) in pilot scale.
Aquatic biomass water Hyacinth	10	Study on WH for metal pollution removal and subsequent utilization for energy in a biological system	Dr. Kaiser Jamil RRL., Hyderabad	Aims at designing waste water lagoons using WHS ability to absorb translocated, metabolise concentrate organic and inorganic ions, investigation reveals that biogas yield vary with different metal treatments three plants species studied showed variation with response to singal metal treatment. The gas yield in Pista & ceratophyllum was less than that of WH. These two plants showed higher lignin content than WH.
Industrial waste Distillery effluent	11	Biomethanation of distillery effluent a pilot plant study	Dr. M.C. Bardiya Daurala Sugar Works, Meerut	Refer ongoing project
Distillery effluent	12	Development of whole cell immobilisation technology for cost effective methane recovery from distillery spent wash	Dr. P. Khanna IIT, Bombay	The project aims at selection characterisation of suitable supporting material for max. Absorption of micro-organisms, designing of immobolised reactor for process stability. Study kinetics optimisation of immobolised cell reactor design, the project team left IIT Bombay & Joined NEERI. The project was short closed. The work carried out at IIT includes selection & characterisation of supporting materials for Max. Absorption of Micro-organisms, design of lab. Scale immobolised reaction, start up operation for IMMB. Cell reactions major thrust had been on design and operation of anaerobic fixed film process for methane recovery from distillery spent wash.

Contd...

Category of Waste	*Sr. No.*	*Title of the project*	*PI/ORGN.*	*Objectives/Results*
Distillery effluent	13	Biogas augmentation by *Cirtribactor intermidus* in industrial effluents distillery and fruit processing	Dr. T.M. Vatsala Sh. AMS Murugappa Chettiar Res. Centre, Madras	Aimed at study of biogas augmentations by *Citrobacter intermedis* in industrial effluents; A strain of *Citrobacter intermedius* was isolated from sewage and tested for producing hydrogen with various carbon sources, distillery & grape wastes showed production of hydrogen after innoculation with this culture. Studies also revealed that innoculum of distillery effluent with cowdung could produce high quantity of methane. It has demonstrated the possibility of evolving a microbiological method for augmentation of biogas from anaerobic digestion of distillery wastes.
Paper mill waste	14	Testing and development of industrial (Pulp & Paper mill) Effluent/waste using Immobilised cells and diphasic system	Dr. Harish Dak TCR & DC, Patiala	The project aims at development of biogas plant for utilisation of pulp and paper mills waste waste using diphasic immobilized cell reactor technology. The P.I. have tried three type of digester. Single stage digester, two stage digester & immobolised cell reactor, it has shown that two stage batch system is more efficient than single stage. Total energy recovery was approx. 78% in two stage while 62% in single stage. It has observed that immob, cell reactor is not feasible for E.T. plant sludge through more efficient due to high S.S. in sludge as effluent sludge is highly insoluble and contains 50-60% of Ash.

Contd...

Category of Waste	*Sr. No.*	*Title of the project*	*PI/ORGN.*	*Objectives/Results*
Textile ind. wastes willow dust	15	Setting up of a pilot plant of 25.05 CU.M. Using willow dust for biogas generation at Udaipur cotton mills	Prof. A.N. Mathur CTAE., Udaipur	Under the project a pilot plant of Capacity 25 CU.M. gas generation/day using willow dust has been commissioned in Udaipur Cottage mills in Jan. 1989. The average biogas production was 350-360 L/kg. willom dust having 63% methane as compared to 300 LIT/PAY cattle dung. It is a Batch process with pretreatment of willow dust with 10 urea and 3 batch digesters with CAP. To hold 150 to 2t willom dust for digestion over a period of 40-50 days. The gas is used in the canteen. The digested slurry on wheat and veg. crops observed significant improvement in yields.
Textile Ind. waste willow dust	16	Operational Research on the production of biogas and biomanure from willow dust by dry fermentation process	Dr. V. Sundaram Cotton Tech. Res. Lab., Bombay	Aimed at studies on dry fermentation of willow dust for biogas and bio-manure; successfully commissioned a pilot plant at Apollo textile mill, Bombay for treating 12 tons of willow dust. Gas is being used in the canteen and Lab. of the mill. Digested slury is also serving as a good manure for agriculture.
(B) Ongoing Urban wastes Sludge Gas	1	Operation of 2 MW AI-20 Aero-engine with sludge Gas (Phase-1)	Dr. B.R. Pal NAL, Bangalore	Aims at demonstrating through tests, feasibility of utilising sludge gas from sewage treatment plants to operate the time expired AI-20 aero gas turbine engines. The fabrication of engine stand modification of the engine to the ground run and procurement of auxiliary units have been completed attempts made to start the AI-20 engine on liquid fuel, some problems were faced during commissioning of the engine test & the same are being rectified.

Contd...

Category of Waste	*Sr. No.*	*Title of the project*	*PI/ORGN.*	*Objectives/Results*
Sewage engine	2	Biogas fulled small utility is engine genset	Dr. H.B. Mathur IIT, Delhi	Aims at developing a proportionately designed biogas injection system as a retrofittor enable total biogas operation of small engine retrofit system.
Municipal solid waste	3	Two stage biomethanation of MSW to improve bioleachate production Production and biogas generation	Dr. B.C. Bhattacharaya IIT, Kharagpur	Aims at optimising bioleachate production from MSW, optimising biomethanation process/ acid recovery process, 2 reactions developed. First leachate generator where acidogens grow to produce leachate containing VGA which is pumped to methane generator where methangens grow efficiently to produce methane gas from garbage. Lab, work has been initiated in 25 Lit. capacity bioreactor on the basis of data of obtained, work on developing a portable prototype model to be carried out.
Municipal solid waste	4	Mathematical modelling of Land-fill Gas system	MR. A.D. Bhide NEERI, Nagpur	Aims at mathematical model development for land fill gas system landfill gas system incorporating various factors like characteristics of solid wastes and site conditions; detailed planning work for actual execution of experiment on lab model and for similar landfill for biogas generation, completed effects of factors such as temperature. moisture, particle size organic content etc. On generation of biogas in landfill site to be studied for modelling the gas system.

Contd...

Category of Waste	*Sr. No.*	*Title of the project*	*PI/ORGN.*	*Objectives/Results*
Hospital sewage	5	Demonstration of FFR Technology for Bio-methanation of hospital waste water	Dr. B.K. Handa NEERI, Nagpur	Aims at installation of 2 pilot plants of 25 CU. M/ Day capacity from hospital waste water at Medical College, Nagpur on fixed film technology & collection of performance data and study of process kinetics involved. Lab scale studies completed. Biogas yield was between 0.08-0.24 CU.M/KG of COD/DAY, COD reduction was between 60-91 per cent; BOD reduction 80-94; & HRT 8 hrs. The 2 units of 25 CU.M/DAY capacity with brick granules & PVC corrugated rings as support media has been commissioned by NEERI at Govt. Medical College, Nagpur, plant performance is being monitored.
Horse Dung	6	Installation of horse dung based biogas plant at NDA premises at Kharakarsla, Pune	Mr. G.G. Sohani BAIF, Pune	Aims at installation of 2 units of 25 CU.M. Biogas plant based on horse dung and its performance evaluation; successfully designed and commissioned at NDA, Pune, gas is being used by 19 families of class IV quarters at NDA premises.
Kitchen waste	7	Development of biogas Plant for utilisation of vegetable waste from Agrl. produce market committee, Gultekadi market yard, Pune.	MGSN, Pune	Aims at installation of biogas plant of 2 units of 100 CU.M. capacity each on vegetable wastes from APMC, Pune, work is under advanced stage of construction. Digesters of 100 CU.M. capacity are ready; pulveriser machine. Small engine rubber gas ballon has been purchased. Simultaneously research work on lab. scale unit in progress for optimising. Process parameter to be used for operating full scale plant.

Contd...

Category of Waste	*Sr. No.*	*Title of the project*	*PI/ORGN.*	*Objectives/Results*
Agricultural wastes	8	Prototype development of a biogas plant for agrl. residues using biphasic process	Dr. C.S. Rao SPRERI, V.V. Nagar	Biphasic biogas reactor system for Agrl., residues. Aimed at lab, studies of assessing feasibility of using biphasic process for biomethanation of agri waste, results (Lab scale experiments) indicated that process was competitive in terms of methane yield; biphasic process could accept a range of solid agrl. wastes without major operating problems.Reactor based on agrl. residues.
Agro waste	9	Retropitting of existing 10 CU.M. cattle dung based biogas plant by agro waste based biogas plant	Dr. C.S. Rao SRRERI. V.V. Nagar	Aims of modifying the existing 10 CU. M. cattle-dung plant to work on agro waste and to study performance of the system. The system fully commissioned and being monitored for performance evaluation to obtain more meaningful data for a period of 6 months.
Food and agro waste	10	Microbiological, Biochemical and fermentation studies on Consortia Developed for Bio-gas production from food and Agro-industrial waste	Dr. Krishna Nand CFTRI, Mysore	Aims at evaluation of microbiological pre-treatment for fruits and veg. Processing wastes for developing simple rapid economical processes on solid study fermentation principle.
Fruit and Veg. waste	11	Pilot scale studies on methane generation from fruits and vegetable processing wastes	Dr. Krishna Nand CFTRI, Mysore	Aims at setting up of 2 Nos. of 25 CU.M. each pilot plant at CFTRI, Mysore based on food processing wastes, performance evaluation & optimization studies both units commissioned at CFTRI, Mysore, based on mango peals, loading rate is 85 and hydraulic retention time is 25 days. Gas being used in departmental canteen of

Contd...

Category of Waste	*Sr. No.*	*Title of the project*	*PI/ORGN.*	*Objectives/Results*
				CFTRI, Mysore. Aimed at studying the microbiological aspect of anaerobic digestion of food processing wastes objectives achieved studied & developed a consortia for processing food industries wastes. Led to commissioning of 2 units of 25 CU.M. capacity each based on mango peals at CFTRI.
				Developed several consortia, evaluated for production of biogas using stabilised bioreactor A no. of pure cultures also isolated, characterised and examined for cellulolised activities enzyme activities of different culture and consortia were determined and some were fourd to be promising for production of cellulases, efforts also made to collect samples from different habitants for isolation of XYLAN. Degrading and anaerobic bacteria, morphological and biochemical studies in detail were made to identify 14 methano trophic cultures.
Fruits and vegetable	12	Operation and process control for biogas production from fruit and vegetable processing wastes in 25 CU.M. Kvic digesters	Dr. Krishna Nand CFTRI, Mysore	Aims at performance evaluation, optimization studies, process economics, techno ecomomic feasibility report to 25 CU.M. digester from mixed fruits and vegetable processing wastes and utilisation of digester sludge for animal feed and fertilizer purposes.

Contd...

Category of Waste	*Sr. No.*	*Title of the project*	*PI/ORGN.*	*Objectives/Results*
Plant biomass	13	Development of flexi biogas production system suitable for Anaerobic bio-gasification of a variety of wastes and plant biomasses available in Ne India	Dr. H.D. Singh RPL, Jorhat	Aims at development of flexi biogas production system for anaerobic digestion of different feed materials like plant Biomass, Agro-industrial waste, animal wastes etc. In northern-eastern sector Lab. Scale studies for batch digestion in 10 CU.M. to evaluate biogas production from different mixes on feed stocks like paddy husk, banana stem, saw dust, tea waste sugarcane pressmud, aquatic weeds etc. At various loading rates and retention time. Digestion studies of mixed feed on pH and presumed also carried out, various plant biomass in horizontal fed-batch digester and vertical fed-batch digester being studied.
Industrial wastes Distillery	14	Fixed film reactor technology for distillery effluent pilot plant studies	Prof. P. Khanna NEERI, Nagpur	Aims at construction and operation of a pilot plan of 10 CU.M./Day treatment capacity for distillery effluent using fixed reactor technology and collection of performance data to develop appropriate control strategies for reactor operation. Unit designed and commissioned at WMDC/Distillery Maharashtra; commissioning data revealed COD and TVA reduction ranging between 60-86 per cent of 70-94 per cent respectively at OLR between 9.6-1.5 kg. of COD/CU.M/kg. of COD destroyed. Further studies planed to reduce HRT & increase OLR.

Contd....

Category of Waste	*Sr. No.*	*Title of the project*	*PI/ORGN.*	*Objectives/Results*
Distillery	15	Diphasic anaerobic digestion of distillery effluent studies on scaled up pilot plant	Dr. M.C. Bardiya Dauralla Sugar Works, Meerut	Aims at installation of $100M^3$/day capacity plot plant and to optimise process parameters for the same. Unit commissioned in Nov. 89 and has achieved organic loading rate more than 110 per cent of designed capacity with gas production of 35 CU.M/CU.M. of spent wash. The hydraulic retention time in acid phase digest is 36 Hrs and 10 days for methane phase digester methane content in biogas is between 62-68 per cent and hydrogen sulphide content is well below 0.55 V/V & BOD of the effluents is 8000 ppm. Spent wash loading is 110 CU.M/Day a pilot plant of 10 CU.M/Day effluent treatment capacity has been commissioned in Dec. 86 by Daurala sugar mill. The pilot plant has been operated at the designed capacity with organic loading rate of 6.5-9.0 kg of COD/CU.M per day-HRT 35 hrs. In Acid phase & 10 days in methane phase. The biogas production was 28 CU.M/CU.M. of spent wash methane content in biogas 638. Hydrogen sulphide content below 0.5%. Based on encouraging results of 10 CU.M. A pilot plant, an intermediate pilot plant of 100 CU.M/Day effluent treatment capacity was sanctioned by - to collect inquisite data for commercial plants.

Contd...

Category of Waste	Sr. No.	Title of the project	PI/ORGN.	Objectives/Results
Pressmud	16	Pilot study on Utilisation of pressmud as a supplementary feed stock for biogas generation	Prof. A.R. Lakshmanan, Annamalai Univ.	Aims at studies on utilisation of presumed as a supplementary feed stock for biogas generation and installation of a pilot plant of 60 CU.M. capacity. A. 60 CU.M. and 4 CU.M. capacity IBPS exclusively fed with pressmud, on KVIC model, at an experimental farm, annamalai University. Successfully commissioned; potential of sugarcane filter cake as an alternative feed stock has been proved; A method of strong sugarcane pressmud to be studied during off-season also developed, based on the design and encouraging results obtained. A 60 CU.M. capacity CBP at village Pinhahrur has also been commissioned and successfully operated on sugarcane pressmud. 29 families are using biogas for cooking purpose. Methane content of Biogas IS60-62 per cent.
Pressmud	17	Pilot plant of 340 CU.M. capacity for generation OD biogas from sugarcane pressmud at Padmashri Vithalrao Vikhe Patil Sahkar Karkhana Ltd. Pravara Nagar Distt. Ahmednagar	Sh. L.R. Subramanian	Aims at installation, monitoring and pilot demonstration plant of 340 CU.M. capacity for generation of biogas from sugarcane presumed
Pressmud	18	Pilot plant of 180 CU.M. capacity biomethanation of sugarcane pressmud Vasantdada Shetkari Coop. Sugar Mill Ltd. Sangali	Sh. V.R. Joglekar	Aims at installation, monitoring and pilot demonstration plant of 180 CU.M. capacity for generation of biogas from pressmud at Vasantdada Coop Sugar Mills Ltd. Sangali.

Contd...

Category of Waste	*Sr. No.*	*Title of the project*	*PI/ORGN.*	*Objectives/Results*
Pressmud	19	Pilot plant of 340 CU.M. capacity Bio-methanation of sugarcane pressmud at sugar works Ltd.	Sh. S.S. Shirgaokar	Aims at installation, monitoring and pilot demonstation plant of 340 CU.M. capacity for generation of biogas from sugarcane pressmud at Sugar Works Ltd. Sangli.
Tannery waste	20	Energy recovery from low/medium strength waste water through various fixed films reactors	Dr. S.N. Kaul NEERI, Nagpur	Aims at setting of a joint (5. CU.M./Day treatment capacity plant at Ampura (A Talco Unit). From tannery waste through various fixed film reactor technology for this benchscale studies completed by NEERI for identification of media, determination of performance limits for various operation parameters like organic loading, pH temp. retention time and develop engineering parameters and associated kinetic constants. Chromium removal/ recovery studies, formulation of scale up equation also completed with design details for the 65 CU.M. capacity pilot plant. However, construction of plant could not be started pending clearance from talco.
Tannery waste	21	Generation of biogas from industrial wastes tannery waste	Dr. C.A. Sastry IIT, Madras	Aims at studying the variables involved in generation of biogas from vegetable tannery wastes using anaerobic contact process and to set up a pilot plant of 2000 lit. On vegetable tannery wastes in one of the tanneries in TN; vegetable tannery waste water characterised, Lab scale anaerobic contact filter and the reactor developed being stabilised with cowdung slurry.

Contd....

Category of Waste	*Sr. No.*	*Title of the project*	*PI/ORGN.*	*Objectives/Results*
Paper mill waste	22	Use of immobilized cell technology for methane generation from prohydrolysate	Dr. S.S. Marwaha Punjab Univ. Patiala	Aims at degradation of Pre-Hydrolysate by cellulolytic acidogenic and methano genic bacteria from different sources, leading to production of biogas by Immobilization technique, various components involved in the degradation leading to methane generation and designing of immobilized cell reactor for methane gas lab scale studies for analysis of pre hydrolysate form the paper mill under way.
Willow dust	23	Setting of 25 CU.M. capacity willow dust based wages plant at Kharar Textile Mill Kharar Dist. Ropar	NTC, New Delhi	Aims at installation of 25 CU.M. and 90 CU.M. willow dust based biogas plant at Khahar (Punjab) Textile Mills and Kalyan Mal Mill, Indore, both units of NTC under technique guidance of CATE, Bilaspur NTC has shown interest to set up similar plants in other textile mills; 2 such plants of 25 CU.M. and 90 CU.M. capacity each are under advanced stage of construction in Kharar and Kalyanmal mills.
Willow dust	24	Setting up of 90 CU.M. capacity will dust based biogas plant at Kalyanal Mill	CMD., NTC., Indore	

* Source - MNES. N. Delhi, January, 1993.

Table 2.2 : List of substrates, Technology Developers TDPI & Addresses*

Sr.No.	*Substrate*	*TD/PI*	*Address*
1.	Pressmud	(a) Dr. H.D Singh	Pisum Chingamathak, Imphal 795001
		(b) Prof. A.R. Lakshmanan	9, Adhinarayan Road Annamali Nagar, Tamil Nadu- 608 002
2.	Willowdust	(a) Dr. V. Sundaram	Central Instt. for Research on Cotton Technology. Adenwala Road Matunga. Bombay - 400 019
		(b) Prof. A.N. Mathur	College of Technology and Agrl. Engg. Rajasthan Agri. Univ Udaipur - 313 001
		(c) Sh. Jagdish Raj Kalia	National Textile Crop. 9th Floor, Vandhana Bldg., 11, Tolstoy Marg, N. Delhi - 110 001.
3.	Tannery	Dr. S.N. Kaul	National Environmental Engg. Research Instt. J.L. Nehru Marg Nagpur 440 020
4.	Hospital Sewage	Dr. B.K. Handa	National Environmental Egg. Research Instt. J.L. Nehru Marg Nagpur-440 020
5.	Banana Stem	Dr. C.S. Rao	Sardar Patel Renewable Energy Res. Instt. Vallabh Vidyanagar Gujarat - 388 120
6.	Kitchenwaste	Dr. C.S. Rao	Sardar Patel Renewable Energy Res. Instt. Vallabh Vidyanagar Gujarat - 388 120
7.	Non-edible Seed Meals	Sh. G.G. Sohani	Bhartiya Agro Industries Foundation, "Kamdhenu" S.B. Marg, Pune - 411 016
8.	Horse Dung Seed Meals	Sh. G.G. Sohani	Bhartiya Agro Industries Foundation, "Kamdhenu" S.B. Marg Pune - 411 016
9.	Fruit & Vegetable Processing Waste	Dr. Krishnanand	Central Food & Tech Res. Instt. Mysore - 570 013

* Source MNES, N. Delhi

Table 2.3 : Critically Polluted Areas

Sr. No.	*Area*	*State*
1.	Mandi Gobindgarh	Punjab
2.	Pali	Rajasthan
3.	Najafgarh Drain Basin	UT of Delhi
4.	Vapi	Gujarat
5.	Singhrauli	U.P.
6.	Dhanbad	Bihar
7.	Durgapur	West Bengal
8.	Howrah	West Bengal
9.	Talcher	Odisha
10.	North Arcot	Tamil Nadu
11.	Manali	Tamil Nadu
12.	Korba	M.P.
13.	Digboi	Assam
14.	Chembur	Maharashtra
15.	Visakhapatnam	A.P.
16.	Bhadravati	Karnataka
17.	Greater Cochin Area	Kerala
18.	Kala Amb	Himachal Pradesh
19.	Parwanoo	Himachal Pradesh

*Source CPCB (MEF), N. Delhi, Jan. 1993

CHAPTER 3

BIODEGRADATION OF INDUSTRIAL ORGANIC COMPOUNDS BY ANAEROBIC PROCESSES

3.0 Introduction

Recent advances in chemical technology have led to the production of many and new potentially dangerous compounds. Some through normal usage eventually end up as a constituent in wastewater. These new substances, representing a wide array of compounds ranging from phenol to pesticides, each presents its own individual problem when dealing with its ultimate disposal. This has led to a great many problems and challenges for the people who must deal with these wastes, namely wastewater treatment industry. They have responded by introducing many new and innovative techniques for the treatment of these wastes, many of which employ anaerobic processes. As in the case of aerobic processes, not all organic compounds will satisfactorily under go breakdown through anaerobic conditions.

3.1 Microbiology and Biochemistry

For many millenia mankind has used methane as a source of energy and worshipped mud fed fires as divine signs. Natural gas, the most important reserve of methane in nature contains about 95 per cent of this flammable gas[1]. The origin of natural gas is closely related to the diagenetic and thermal alteration of organic matter. Microbial processes produce methane during the early stages of diagenesis whereas increasing thermal maturation of organic matter produces both methane, and C-2 and C-4 hydrocarbons[2]. Thus, natural gas is both biogenic and thermogenic origins. The differentiation is possible based on hydrogen and carbon isotopic composition.

After Volta discovered methane, 'the inflammable air', it took about a century before methane genesis was found to be connected to microbial activity. BeChamp[4] who advocated the theory of spontaneous generations, named 'organisms' responsible for methane formation from ethanol (Microzymacretae). This organism was apparently a mixed population since BeChamp was able to show that depending on substrate different fermentation products were formed. Popff[5], in 1875 was the first to systematically investigate the formation of marsh gas using many different substrate. For cellulose he found that the end products of anaerobic degradation were methane, some hydrogen and carbon dioxide. However, in 1876, Herter a collaborator of Hoppe-Seyler observed that acetate in sewage sludge was stoichiometrically equal to amounts of methane and CO_2[6]. Hoppe-Seyler followed this work[7] using mud but could not detect any organisms in the liquid above the precipitate. Based on Calorimetric considerations. Hoppe-Seyler concluded that small amount of energy which an organism gains from splitting of acetate is scarcely enough to allow it to grow. Omelianski[8] tried to separate from cellulose enrichments the hydroxygen-forming organism from those which form methane as an end product. Sohngen[9] was able to enrich two distinct acetophallic methanogenic bacteria, a sarcina and a rod, which could form long filaments often associated in bundles. He also found that formate, hydrogen and CO_2 could act as a methane precursors.

Omelianski[10] carried out further investigations on methane forming enrichments and confirmed Bechamp's observation that ethanol could be metabolized to methane and CO_2. Maze described a pseudo-sarcina always present in methane forming populations which had been cultured on dead chestnut leaves[11]. Similar organisms were later identified by him in acetone enrichment[12]. Micro-coccus was identified by Groenewege[13]. This organisms was claimed to be able to ferment methyl and butyl alcohols and acetone in presence of ethyl alcohol to methane. Coolhaas[14] worked with crude thermophilic enrichments with an optimum temperature for methane formation at 63°C; the substrates used were formic, acetic, isobutyric and other acids but not propionic and normal butyric acid.

Barkers studies upon methane-producing bacteria[15] opened up in 1936 a new era in this field. His highly enriched methanogenic cultures allowed him to perform some basic biochemical studies which he persued with his colleagues. Part of his work has been reviewed by himself[16] and Stadtman[17]. In 1947 Schnellen[18] was first to isolate two pure cultures of methane bacteria, viz; Methano-sarcina barkeri and Methano bacterium formicicum and eleven years later Smith *et al.*[19] Successfully applied the Huntgate-technique for isolation of another methanogen, Methanobacterium rumenantium. Huntgate technique proved to be ideal method for the work with these fastidious anaerobic bacteria and is still today with some modification, the method of choice. With the discovery of Byrant *et al.*[20] that conversion of ethanol to methane by Methane-bacterium-omelianskii was only feasible because this organism was a mixed synthrophic culture of a methanogenic and non-methanogenic bacterium, the spectrum of substrate for methanogen has been drastically reduced, leaving only acetate, H_2, CO_2, formate and methanol. Many reviews on methane bacteria have been written during last decade[21-26].

3.2 Flow of Substrate and Energy in Anaerobic Systems

Methane bacteria can only use a limited range of substrates for growth and energy production. The substances converted to methane are summarised in Table 3.1. As can be seen from these equations methanogens are not able to convert complex organic matter directly into methane. Therefore, combined action of physiologically distinct microorganisms is required to breakdown biopolymers or recalcitrant substrate to CH_4 and CO_2. In an environment where CO_2 and proton and the only inorganic electron acceptors available, methane is generated according to Fig. 3.1(a). At least three entirely different groups of organisms have to be active to convert complex organic materials into CH_4 and CO_2. In the first step polymers are hydrolysed and fermented often by the same organisms, and some acetate and H_2 are formed. During second step the various fermentation products are converted to acetate, H_2 and CO_2 by some poorly understood bacterial specialists, the obligate proton reducers[27]. These bacteria can only function if the partial pressure of H_2 is kept low by hydrogen consuming organisms. The last step is methanogenesis. The thermodynamic principles of anaerobic digestion, especially those which are important for mutualistic conversions of fatty acids and alcohol[28] also predict the same.

Table 3.1 : Energy Yielding Reactions Catalysed by Methanogens

Sr.No	*Reaction*		*Δ*G°*
1.	$4H_2 + CO_2$	$= CH_4 + 2H_2O$	– 139.2
2.	$4COO^- + 2H^+$	$= CH_4 + CO_2 + HCO^-_3$	– 126.8
3.	$HCOO^- + 3H_2 + H^+$	$= CH_4 + 2H_2O$	– 134.3
4.	$4CO + 2H_2O$	$= CH_4 + 3CO_2$	– 185.1
5.	$4CH_3OH$	$= 3CH_4 + CO_2 + 2H_2O$	– 102.5
6.	$CH_3OH + H_2$	$= CH_4 + H_2O$	– 121.5
7.	$4CH_3NH_2 + 2H_2O + 4H^+$	$= 3CH_4 + CO_2 + 4NH_4^+$	– 101.6
8.	$2(CH_3)_2NH + 2H_2O + 2H^+$	$= 3CH_4 + CO_2 + 2NH_4^+$	– 86.3
9.	$4(CH_3)_2N + 6H_2O + 4H^+$	$= 9CH_4 = 3CO_2 = 4NH_4^+$	– 80.2
10.	$2CH_3CH_2 - N(CH_3)_2 + 2H_2O$	$= 3CH_4 + CO_2 + 2CH_3CH_2NH_2$	– 70.0
11.	$CH_3COO^- + H_2O$	$= CH_4 + HCO^-_3$	– 28.2

*Free energy change at pH-7, RJ/CH_4-mole.

In the presence of electron acceptors such as metal oxides (FeOOH), MnO_2, etc.), nitrogen oxides (NO_3^-, NO_2^-) and/or oxidized sulphur compounds (SO^{2-}_4, SO_3^-, etc., including elemental sulphur), the fermentation pattern described above may be substantially altered. The products of first step are oxidized almost entirely to carbon dioxide and the electrons are transferred to one of these inorganic acceptors. For each acceptors there exists one or more specific bacterial species catabolysing the corresponding redox reaction. Methanogenesis usually occurs only after all these alternative electron acceptors are depleted. It is not known, whether obligate proton reducers are quantitatively important in anaerobic degradation of organic matter in the anaerobic reactions in absence of methanogensis.

In the digestive tract of living organisms, the percentage of ingested organic matter converted to CO_2 and CH_4 is rarely high since the soluble intermediate (*i.e.* fatty acids) are absorbed by animal to provide it with energy and biosynthetic intermediates. Therefore, the production of more soluble organic intermediates and less methane is essential to the host. It is probably achieved by keeping the mean HRT in digestive tract low enough (less than one day). As a result, the slow growing acetophilic methanogens and fatty acid oxidizing obligate proton reducers will washout. The few acidophilic methane bacteria[30] and obligate proton reducers[31] which are nevertheless found in these environments might just be transient or using substrate which allow then to grow rapidly, *e.g.* *Methanosarcina barkeri* can use hydrogen and CO_2 and CH_4 formation. The best studied digestive organ is the rumen[30], Fig 3.1(b) presents a scheme for the energy flow in a digestive tract, especially of ruminants. Methane, which is produced to a limited extent in these animals and released by belching, originates primarily from hydrogen oxidation. Hydrogen of geo-chemical origin, is under conditions for methogens, Fig 3.1(c). A generalised schematic conversion of lipids, proteins and carbohydrate is shown in Fig 3.2 and 3.3. The three stage methane fermentation and percentage flow of energy conversion content of complex organic materials through each stage into methane as represented by COD is shown in Fig. 3.4. The effect of hydrogen partial pressure on free energy conversion of ethanol, propionate, acetate and hydrogen during methane fermentation is shown in Fig. 3.5.

3.3 Taxonomy of Methanogens

Morphologically methanogens are a diverse group, however, physically they are quite similar since all are strict anaerobes and share the common metabolic capacity to produce methane. The uncertainty regarding their systematic relationship was reflected in early taxonomic schemes which initially dispersed the methane bacteria among better characterized bacterial groups according to their morphologies. Baker[16] emphasized the unique physiology of the group and thus gathered the methanogens into a single family, the Methanobacteriaceae. Although Baker's regrouping has been taken up in the most recent edition of Bergey's Manual of Determinative Bacteriology[32], the scheme still provided limited insight into the relationships among various methane bacteria.

The main reason was our limited knowledge concerning the metabolic, biochemical and molecular properties of these organisms. The dilemma posed by seemingly contradictory characteristic-similar physiology versus diverse morphology was not solved until Fox, *et al.*[33] applied comparative cataloguing of 16S rRNA to a variety of methanogens. The methanogenic bacteria were shown to be a an unique coherent group but phylogenetically as distinct from typical bacteria as from eukaryotes[34,35], Fig. 3.6. Subsequent studies on DNA structure, intermediate metabolism and lipid composition have confirmed and amplified these results and has led to a revision of the classification and nomenclature of methanogens[25], Fig. 3.7.

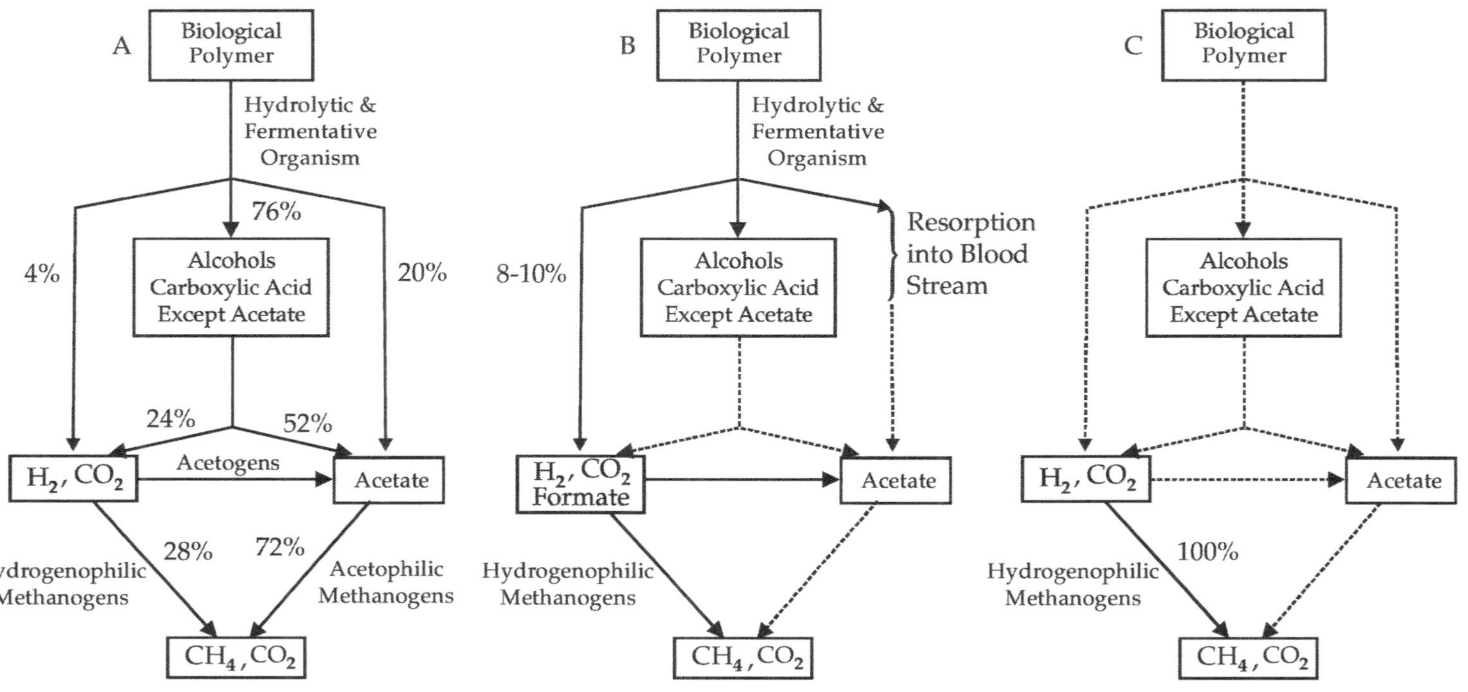

Fig. 3.1 : Flow of Substrate in Anaerobic System

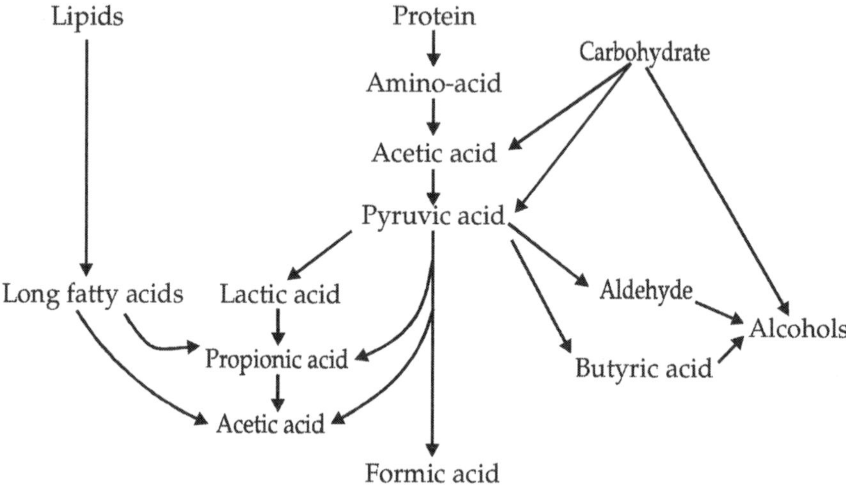

Fig. 3.2 : Major Reactions Performed by Acid Producing Bacteria

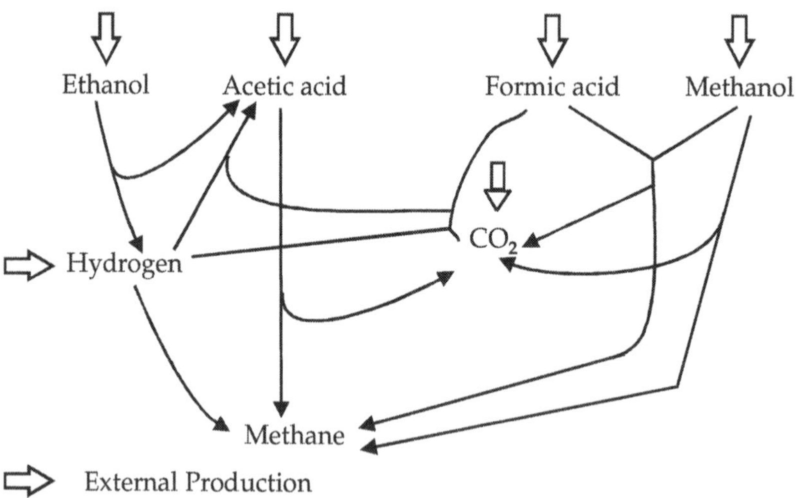

Fig. 3.3 : Major Reactions Performed by Methanogenic Bacteria

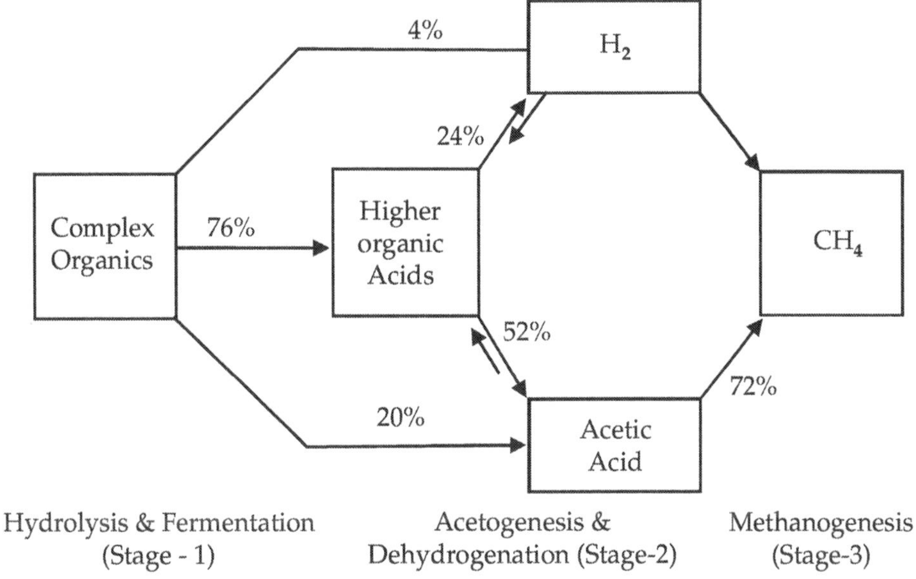

Fig. 3.4 : The three Stages of Methane Fermentations and Percentage Flow of Energy Content Represented by COD

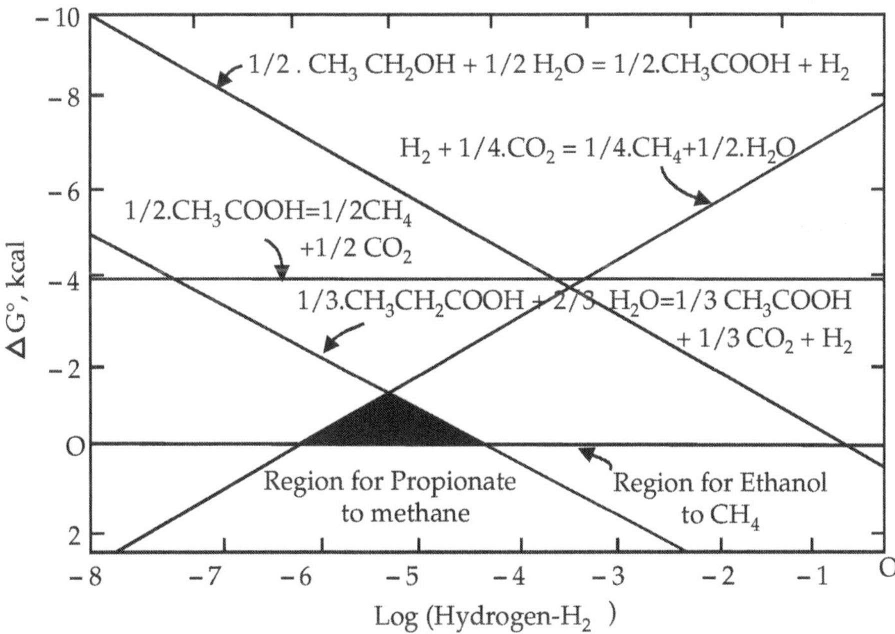

Fig. 3.5: Effect of Hydrogen Partial Pressure on the Free Energy of Conversion of Ethanol, Propionate, Acetate and Hydrogen during Anaerobic Reactions.

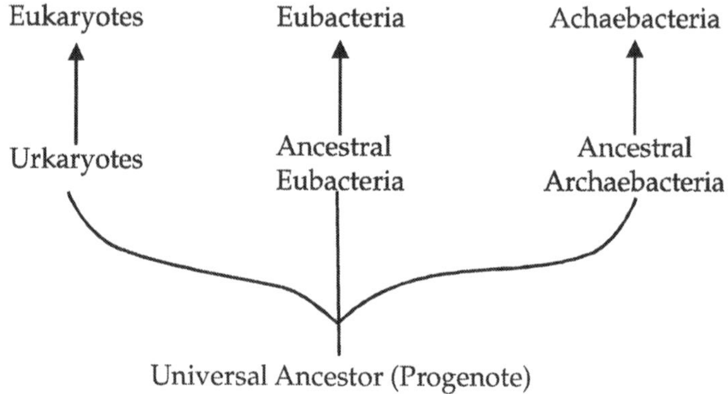

Fig. 3.6 : Three Primary Kingdom for Bacteria

Eukaryotes, eubacteria and archaebacteria are the categories of organisms. The methanogens belong to the archaebacteria. Both eubacteria and archaebacteria are alike in being prokaryotic cells (they are simple cells and lack nucleus and are very different in their structural properties from eukaryotic cells, which have a nucleus and several other subcellular organelles). The archaebacteria, the eubactria and an eukaryote-the original eukaryotic cell-stemmed from a common ancestor, the progenate, a very simple cell. The substrate spectrum of various methane bacteria is shown in Table 3.2.

Table 3.2 : Spectrum of Substrate for Various Methane Bacteria

S.N.	*Substrate*	*Micro-organisms*
1.	H_2	*Methanobacterium bryantii; M. formicicum; M. thermoautotrophicum*
		Methanobrevibacter arboriphilus; N. ruminantium; M. simthii Methanococcus mazei; M, vannielii, V voltae
		Methanomicrobium mobile; Methanogenium Cariaci; N. marisnigri, Methanospirillum hungatei; Methanosarcina barkeri
2.	HCOOH	*M. formicicum; M. ruminantium; N. smithii; M. vannielii; M. voltae M. mobile; M. cariaci; M. marisnigri; M. hungatei*
3.	CO	*M. barkeri*
4.	CH_3OH	*M. mazei; M. barkeri*
5.	CH_3NH_2,	*M. mazei; M. barkeri*
	$(CH_3)_2$ $NH_2(CH_3)_2N$	
6.	$(CH_3)_3$ N-N $(CH_3)_2$	*M. barkeri*
7.	CH_3COOH	*M. mazei; M. barkeri; Methanothrix sohungenii*

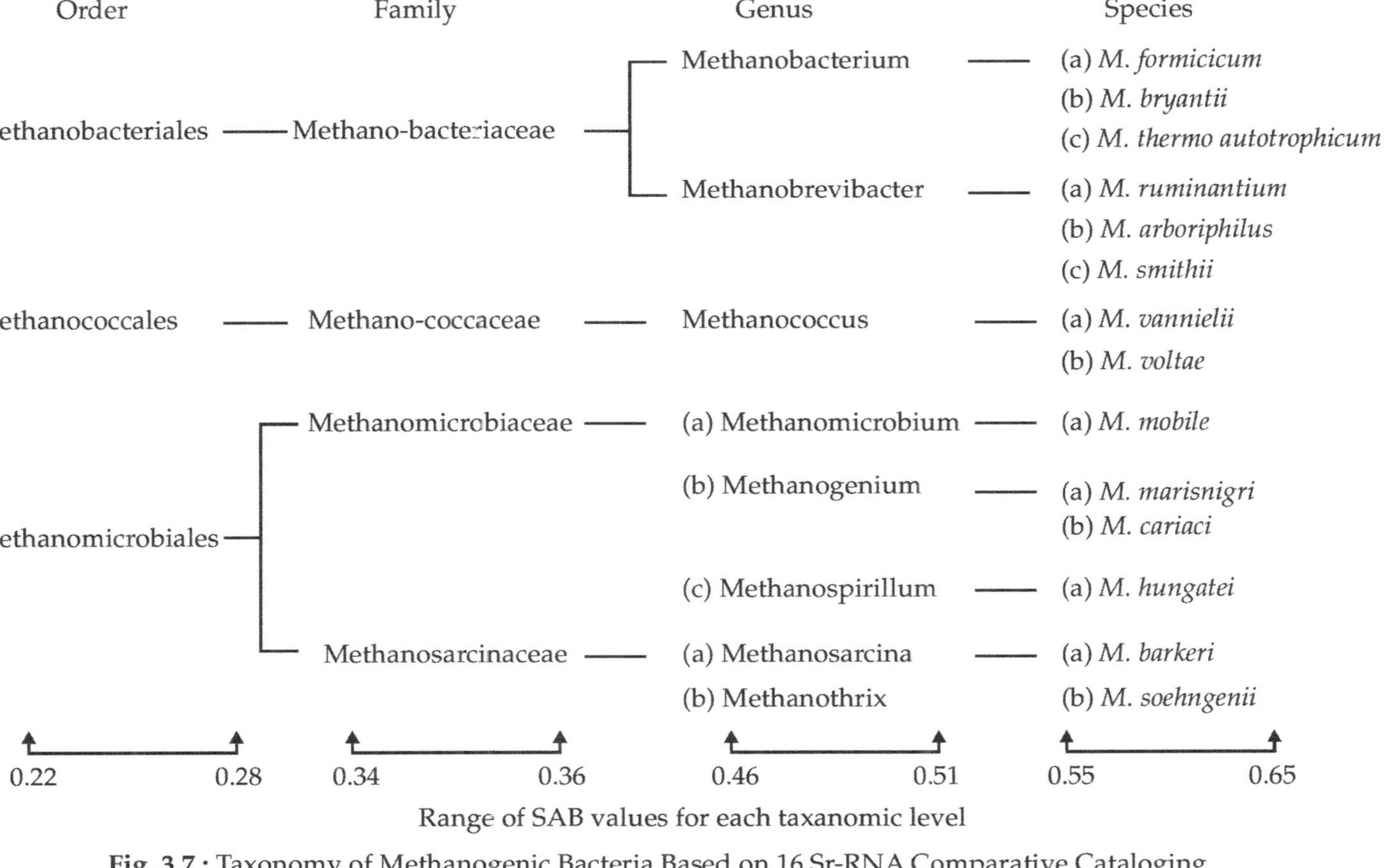

Fig. 3.7 : Taxonomy of Methanogenic Bacteria Based on 16 Sr-RNA Comparative Cataloging

3.4 Mechanisms of Anaerobic Energy Coupling

In absence of oxygen, as the terminal electron acceptor, are the identities of the alternative acceptors and the magnitude of the free energy potentially available from their use. With the exception of materials such as hydrocarbon (initial steps of whose bio-degradation require molecular oxygen), the same extensive range of organic nutrients is available to both aerobic and anaerobic Chemo-heterotrophic organisms as electron donors, and as source of energy as well as of carbon.

On the basis of their electron acceptors, therefore, anaerobic chemoheterotrophs can be divided into two broad classes using as their modes of energy metabolism either fermentation or anaerobic respirations. In fermentation, an organic substrate is oxidized in a series of anaerobic reactions in which the electron acceptor is a product of the metabolism of the substrate with the difference in redox potential of the substrate and the electron acceptor derived from it providing the energy for ATP synthesis. In fermentation, all component of the reaction schemes are normally soluble (fumarate reduction is an exception) and ATP is generated by the mechanism of substrate level phosphorylations (SLP). Table 3.3. The sole reason for accumulating the reduced fermentation product is to achieve redox balance. Common examples of electron acceptors in fermentation are pyruvate (reduced to lactate), acetyle CoA (reduced to ethanol, fumarate (reduced to succinate) and H^+ (reduced to H_2). By definition, therefore, fermentation products are both reduced and only partially degraded, and are consequently potential sources of energy, reducing power and (with exception of H_2) carbon for other organisms.

Substrate level phosphorylation (SLP) is the phosphorylation of ADP to ATP by a process of phosphate transfer from an 'energy rich' phosphorylated substrate produced during, degradation of the fermentation substrate. The initial incorporation of inorganic phosphate is as seen below, coupled to exergonic oxidation or lyase reactions.

The enzymes involved in these reactions are as follows : (*a*) pyruvate dehydrogenase; (*b*) phosphotransacetylase; (*c*) acetate kinase; (*d*) 2-oxoglutarate dehydrogenase; (*e*) succinate thiokinase; (*f*) acetaldehyde dehydrogenase; (*g*) glyceraldehyde phosphate dehydrogenase; (*h*) phosphoglycerate kinase; (*i*) pyruvate kinase; (*j*) pyruvate-formate lyase; (*k*) ketothiolase; (*l*) phosphoketolase; (*m*) orinithine transcarbamoylase (ornithine carbamoyl transferase); (*n*) carbamate kinase; (*o*) formyltetrahydrofolate synthetase.

Anaerobic respiration is the name given to the membrane-associated processes of electron transport (using the same or similar redox carriers but with a terminal acceptor other than oxygen) directly coupled to phosphorylation of ADP to ATP. Electron acceptors include nitrate (reduced to nitrite), nitrite (reduced to N_2); trimethylamine oxide (TMAO) (reduced to TMA); fumarate (reduced to succinate) : sulphate (reduced to sulphide) and CO_2 (reduced to

CH_4). As opposed to its role in fermentation, where it is an important product. H_2 is a common energy substrate in anaerobic respiration and this forms an important practical difference between two anaerobic classes of organisms, one of considerable significance with regard to the interspecies coupling in which hydrogen plays a key role.

Table 3.3 : Substrate Level Phosphorylation Reaction Occurring During Fermentation

(A) Oxidation Reactions Coupled to Phosphorylation Reactions

1. 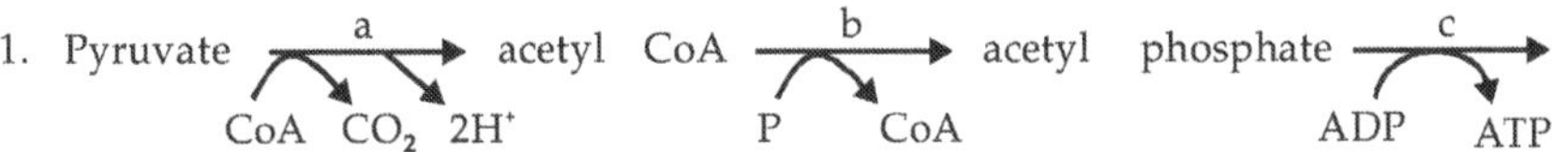

2. 2-Oxogutarate $\xrightarrow{d}$ Succinyl CoA $\xrightarrow{e}$ Succinate
   (d: Co-A in; CO_2, $2H^+$ out) (e: ADP → ATP; Co-A out)

3. Acetaldehyde $\xrightarrow{f}$ Acetyl Co-A $\xrightarrow{b}$ Acetylphosphate $\longrightarrow$ Acetate
   (f: Co-A in; $2H^+$ out) (b: P in; Co-A out) (ADP → ATP)

4. Glyceraldehyde –P $\xrightarrow{g}$ 1, 3 Diphosphoglycerate $\xrightarrow{h}$ 3-phosphoglycerate
   (g: P, NAD → NADH) (h: ADP → ATP)

5. Phosphoenol pyruvate $\xrightarrow{i}$ pyruvate
   (i: ADP → ATP)

(B) Lyase Reactions coupled to Phosphorylation Reaction

1. Pyruvate $\xrightarrow{j}$ Acetyl Co-A $\xrightarrow{b}$ Acetylphosphate $\xrightarrow{c}$ Acetate
   (j: Co-A in; Formate out) (b: P in; Co-A out) (c: ADP → ATP)

2. Aceto-acetyl-Co-A $\xrightarrow{k}$ Acetyl Co-A $\xrightarrow{b}$ Acetylphosphate $\xrightarrow{c}$ Acetate
   (k: Co-A in; Acetyl Co-A out) (b: P in; Co-A out) (c: ADP → ATP)

3. Xylulose-P $\xrightarrow{l}$ Acetylphosphate $\xrightarrow{c}$ Acetate
   (l: P in; Glyceraldehyde-P out) (c: ADP → ATP)

4. 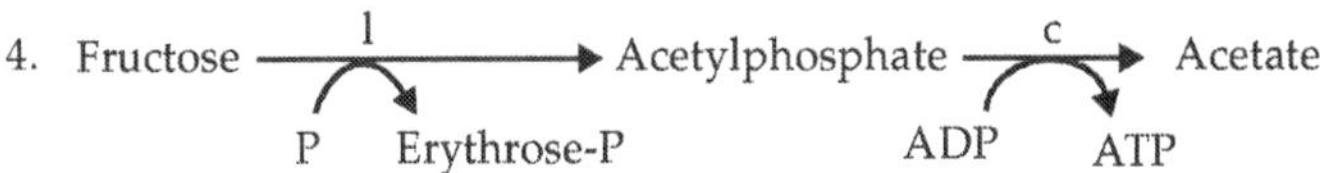

5. Citrulline $\xrightarrow{m}$ Carbamoyl-P $\xrightarrow{n}$ Carbonate $\longrightarrow$ $CO_2 + CH_4$
   (P → Ornithine; ADP → ATP)

6. Formultrahydroflolate $\xrightarrow{o}$ Formate + Tetrahydroflolate
   (ADP; P → ATP)

(C) Energy Released

1. ATP $\longrightarrow$ ADP; ΔA° = − 8000 Cal(pH-7)
2. ADP $\longrightarrow$ AMP; ΔG° = − 6500 Cal(pH-7)
3. AMP $\longrightarrow$ Adenosine + Phosphoric Acid; ΔG° = 2200 Cal (pH-7)

As with fermentation, the reduced product of anaerobic respiration are potential sources of energy and reducing power of other organisms. In most cases, however, these are obligatory aerobic chemolithotrophs and the methane, etc. must first diffuse across the aerobic/anaerobic (O_2/AnO_2) interface. As the electron acceptors such as CO_2 or substrate are themselves generally not products of anaerobe's own metabolism and are in fact more associated with aerobic environment, one can see the marked fluxes necessary across the O_2/AnO_2 interface, and so identify a major driving force for the carbon, nitrogen and sulphur cycles in nature.

3.5 Anaerobic Reactions*

Fermentation processes are capable of degrading almost the complete range of nutrients available for chemo-heterotrophic growth of micro-organisms including, sugars; fatty and amino acids; purine and pyrimidine bases; heterocyclic compounds; and polysaccaharide, protein and lipid polymers. Lignin and saturated hydrocarbons are generally considered to the recalcitrant in anaerobic environments due to the requirement for molecular oxygen in the initial catabolic reactions. There have been recent reports, however, of the slow breakdown of lignin (upto 15 per cent in 300 days), and of saturated hydrocarbons by mixed enrichment microbial populations.

Those organisms capable of fermentation also span a huge range. Although they include certain protozoa and fungi with the exception of yeast, however, here only anaerobic bacterial reactions will be considered. Anaerobic pathways for ethanol, lactate, butyrate and butanol/acetone have been shown in Fig 3.8-3.15. Fermentation reaction sequences are found amongst facultative (*e.g.* Enterobacteriaceae) and obligate (*e.g. Clostridia*) anaerobes. Whereas the *clostridia* are obligatarly fermentative, the sulphate reducers are facultative because in presence of sulphate as terminal electron acceptor, there mode of energy generating metabolism is anaerobic respiration, whereas in the absence of sulphate, they can ferment a somewhat restricted range of substrate. The

* Adapted from Hamilton, W.A. Enregy Transduction Anaerobic Bacteria, Society of Microbiologist, Cambridge; p. 84-146 (1991).

methanogens generally lack the normal respiratory cofactors. Despite this lack and its implications for mechanisms of energy transduction in these organisms, it is most convincing to deal with them under the general heading of anaerobic respiration with CO_2 fulfilling the role of alternative electron acceptor. In contrast to the extensive array of potential substrate for fermentation, only limited members of metabolic pathways is required for transformation into what is also a relatively small number of fermentation products. Classification is usually based on these products, rather than on substrate or organisms.

Some of the conventional pathways are summarised as follows :

Glycolytic pathway for conversion of pyruvate, with redox balance maintained by formation of ethanol as the reduced fermentation product formed by way of pyruvate decarboxylase. *i.e.,*

$$\text{Glucose} \longrightarrow 2\ CO_2 + 2\ \text{ethanol}\ (+\ 2\ \text{ATP})$$

Entner-Doudoroff pathway for fermentation of sugars is described by decarboxylation to acetaldehyde which is used as the Oxidant for NADH, *i.e.,*

$$\text{Glucose} \longrightarrow 2\ CO_2 + 2\ \text{ethanol}\ (+\ 1\ \text{ATP})$$

The heterolatic fermentation pathway is summarised the reaction sequences as follows :

$$\text{Glucose} \longrightarrow 2\ CO_2 + \text{lactate} + \text{ethanol}\ (+\ 1\ \text{ATP})$$

The Bifidum pathway of fermentation to lactate can be grossly represented as

$$2\ \text{Glucose} \longrightarrow 2\ \text{lactate} + 3\ \text{acetate}\ (+\ 5\ \text{ATP})$$

The fermentation of citrate to diacetal and/or cetoin *e.g.*, in *Streptococcus lactis* var. diacetylacetic or *Leuco – nostoc cremoria* is a special sequence of reactions where [TPP] - Enz is enzyme bound thiamine pyrophosphate, *i.c.*

$$3\ \text{Citrate} \longrightarrow 5\ CO_2 + \text{lactate} + 3\ \text{acetate} + \text{diacetyl}$$

Or, $$2\ \text{Citrate} \longrightarrow 4\ CO_2 + 2\ \text{acetate} + \text{acetion}$$

Fermentation of sugars to butyrate by way of glycolysis and pyruvate : ferredoxin oxidoreductase, *e.g.*, in *Clostridium butyricum* and *Cl. pasteurianum* is summarised as follows

$$\text{Glucose} \longrightarrow 2\ CO_2 + \text{butyrate} + 2H_2\ (+\ 1\ \text{ATP})$$

Butanol and acetone fermentation using *Cl. acetobutylicum* can be represented by

$$\text{Glucose} \longrightarrow 2\ CO_2 + \text{butanol} + 2H_2\ (+\ 2\ \text{ATP})$$

$$\text{Glucose} \longrightarrow 3\ CO_2 + \text{acetone} + 4H_2\ (+\ 3\ \text{ATP})$$

With the hydrogen partial pressure at 1 atm, the observed products of glucose by *Cl. pasteurianum* are (moles per mole glucose fermented) :

$$\text{Glucose} \longrightarrow 0.6 \text{ acetate} + 0.7 \text{ butyrate} + 2.0\, CO_2 + 2.6\, H_2 + 3.3 \text{ ATP}$$

$$\text{Glucose} \longrightarrow 2.0 \text{ acetate} + 2\, CO_2 + 4\, H_2 + 4 \text{ ATP}$$

(if H_2 pressure < 1 atm)

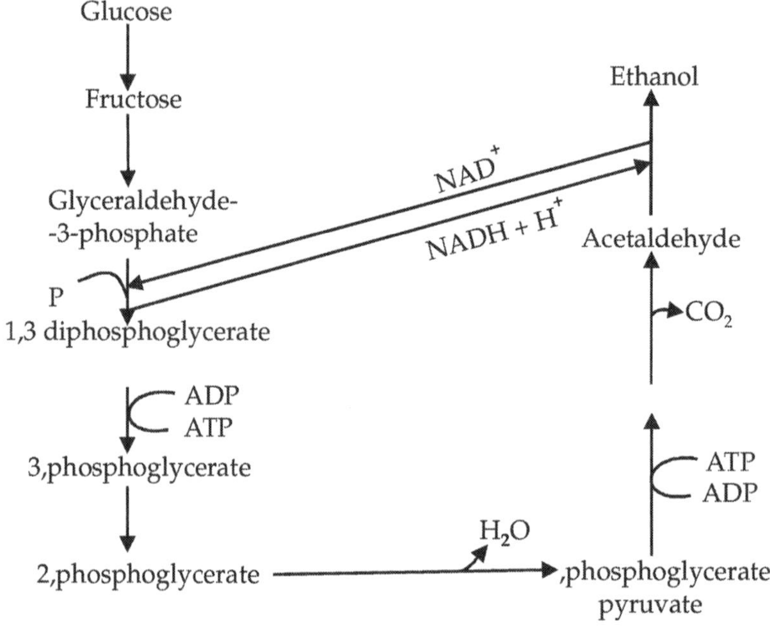

Fig. 3.8 : Glycolytic Pathway for Conversion of Glucose to Pruvate

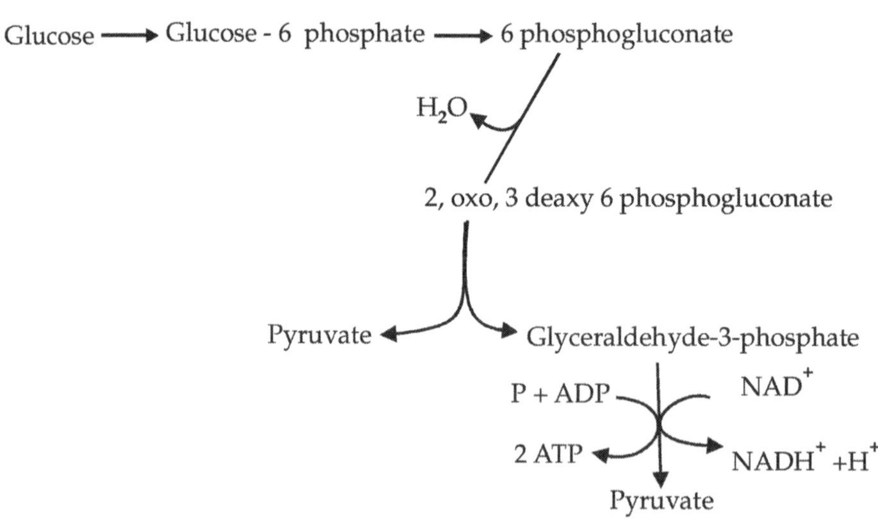

Fig 3.9 : Enter-Doudoroff Pathway for Fermentation of Sugar

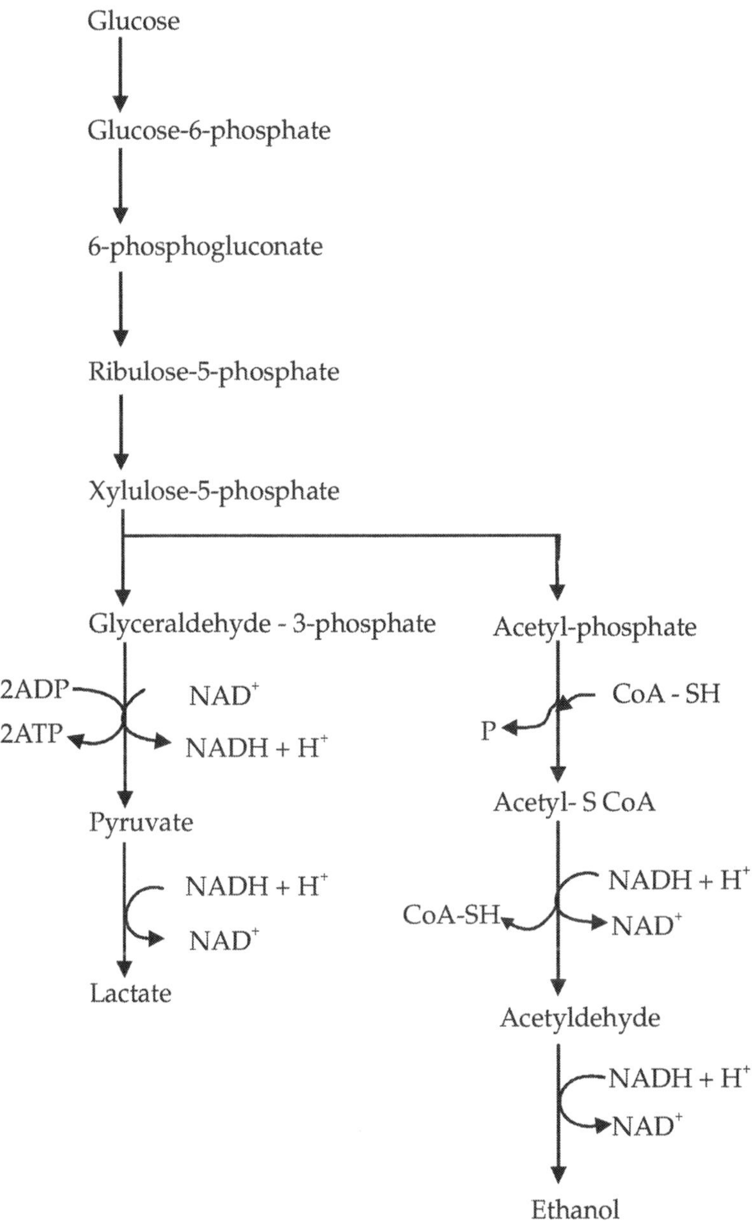

Balance : Glucose $\longrightarrow$ CO_2 + Lactate + Ethanol (+ 1 ATP)

Fig. 3.10 : The Heterolactic Fermentation Pathway

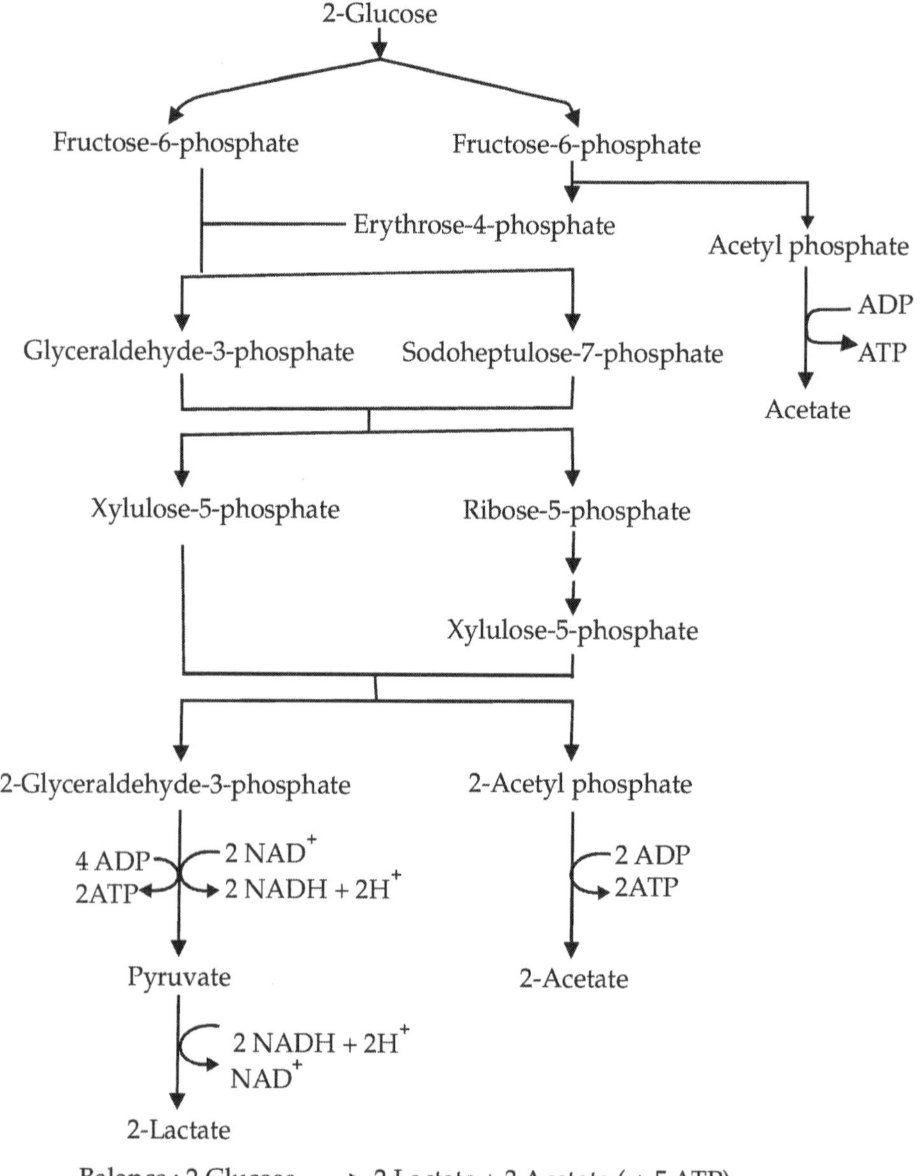

Fig. 3.11 : The Bifidum Pathway of Glucose Fermentation to Lactate.

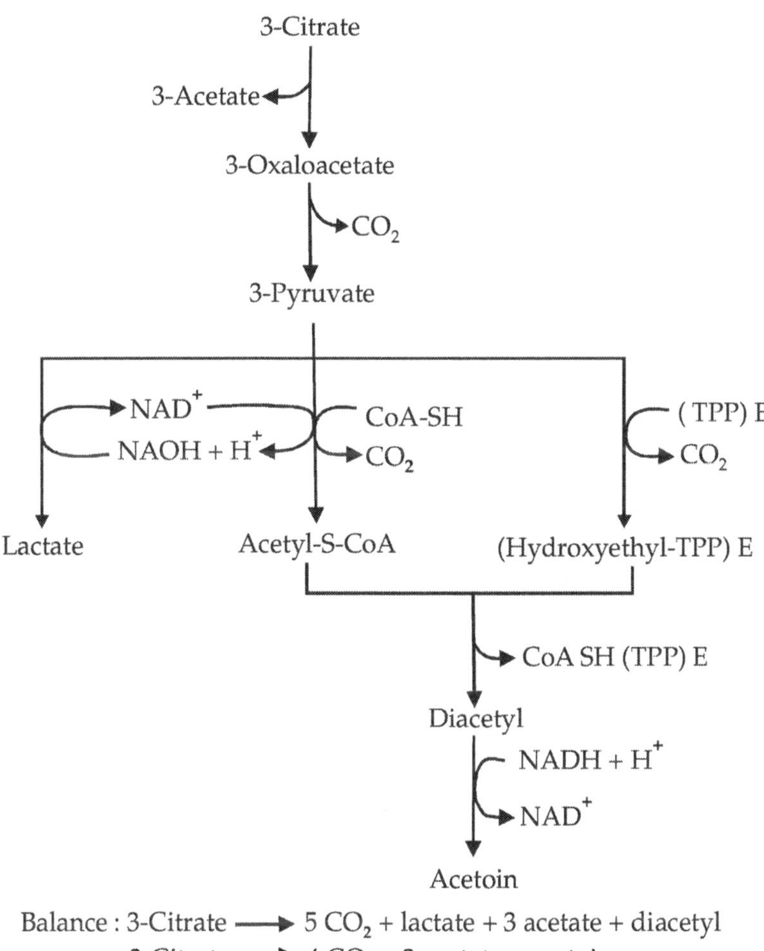

Fig. 3.12 : Fermentation of Citrate to Diacetyl and/or Acetoin. *e.g., Streptococcus lactis* var. *Diacetyl lactis or Leuconostoc cremoria* (TPP)- Enz. is Enzyme-Bound Thiamine Pyrophosphate.

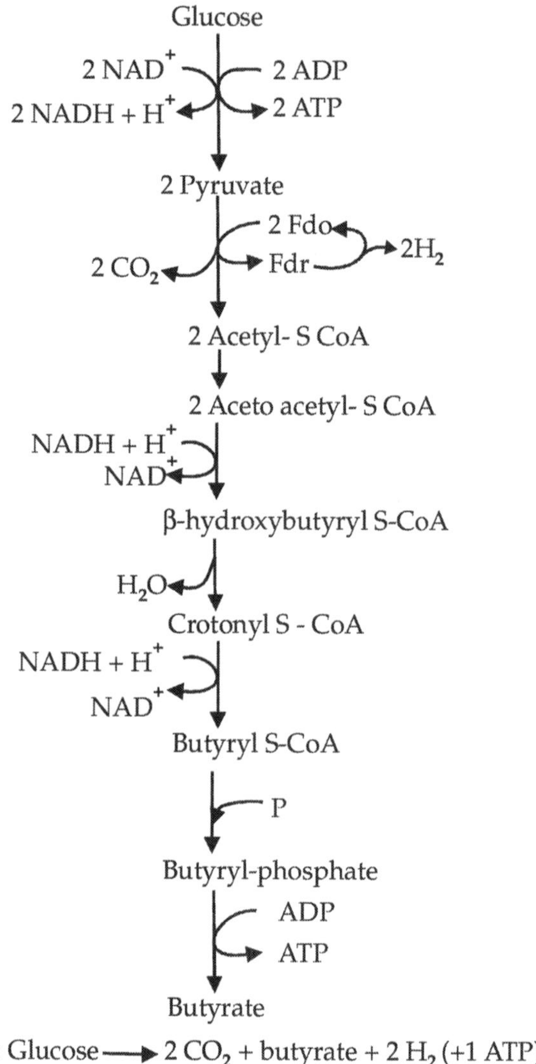

Fig 3.13 : Fermentation of Glucose by Way of Glycolysis and Pyruvate

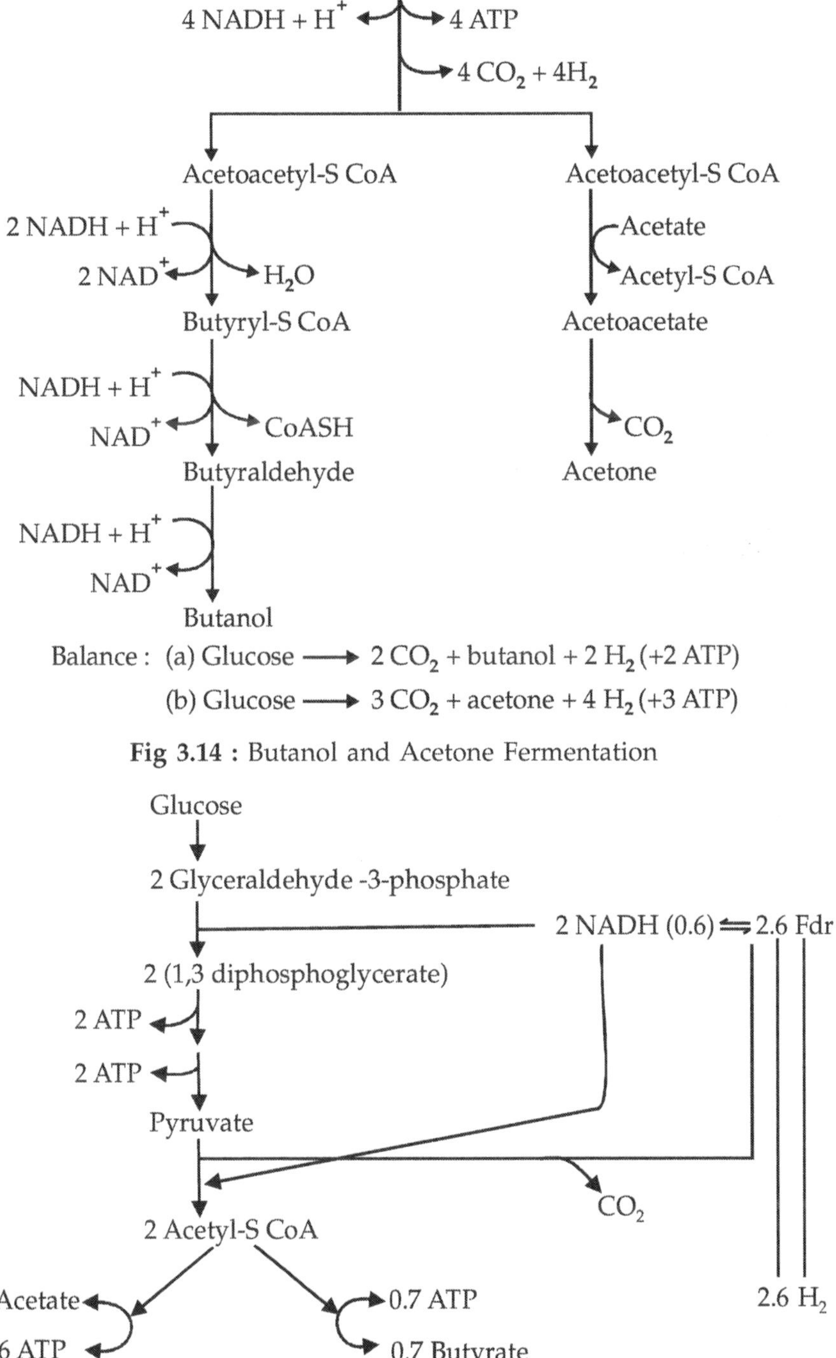

Fig 3.14 : Butanol and Acetone Fermentation

Fig 3.15: Glucose Fermentation by *Clostridium pasteruianum* at a H_2 pressure of 1 atm

3.4.1 Mixed-Acid Anaerobic Reactions

Like the *Clostridia*, the enterobacteria display as complex fermentation pattern involving an array of pathways and end-products. Like the *Clostridia*, they also employ the EMP glycolytic pathways for hexose breakdown but although acetyl CoA, CO_2 and H_2 are again key products of pyruvate catabolism, the mechanism of their formation is quite different in enterobacteria.

Unlike *Clostridia*, enterobacteria are facultative anaerobes and under aerobic conditions acetyl-CoA is formed from pyruvate by pyruvate dehydrogenase multi enzyme complex, with NAD^+ as cofactor. This enzyme is represented under anaerobre conditions while any pre-existing activity is inhibited by NADH. Instead, pyruvate-formate lyase is synthesized, giving rise to acetyl CoA and formate.

$$\text{Pyruvate} + \text{Enzyme} \longrightarrow \text{acetyl} - \text{enz} + \text{formate}$$

$$\text{acetyl} - \text{Enz} + \text{CoA-SH} \longrightarrow \text{Enz} + \text{acetyl CoA}$$

$$\text{Formate} \longrightarrow CO_2 + 2H^+ + 2e$$

$$2H^+ + 2e \longrightarrow H_2$$

Lactate is another significant product of the mixed fermentation and it is formed directly from pyruvate by the NADH-dependent lactate dehydrogenase.

$$\text{Pyruvate} + \text{Enz} - \text{TPP} \longrightarrow \text{Enz} - \text{Tpp} - \text{hydroxylethyl} + CO_2$$

$$\text{Enz} - \text{Tpp} - \text{hydroxylethyl} + \text{pyruvate} \longrightarrow \alpha \text{ acetolactage} + \text{Enz} \leftarrow \text{TPP}$$

Pathway for succinate and propionate fermentation is shown in Fig. 3.16.

3.4.2 Formation of Proton Motive Force in Fermenters

In addition to fumarates reductase there are two other mechanisms known to give rise to a proton motive force during fermentations and thus increase the overall energy yield.

3.4.2.1. Anaerobic Citrate Degradation

This is found among the lactic acid bacteria, some *Clostridia* and certain enterobacteria. Citrate is first cleaved to acetate and oxaloacetate which may then be decarboxylated to pyruvate and CO_2. Whereas the enzyme is generally soluble in *Enterobacter aerogenes*, Oxaloacetate decarboxylase is membrane-bound and contains biotin as a cofactor. Also, and most significantly, the enzyme is sodium-requiring and it has been shown that decarboxylase activity results in electrogenic translocations of sodium ions across the membrane with the establishment of an electro-chemical sodium gradient.

Other examples of sodium-translocating decarboxylases are methyl-malonyl CoA decarboxylase of *Vaillaonella alcalescens* and *Propiogenium modestum* (these propioni bacteria donot possess the methyl malonyl-CoA-Pyruvate transcarboxylase) and the glutaconyl CoA decarboxylase involved in glutamate fermentation by *Cl. symbosium*. The operation of an electroneutral Na^+/H^+

antiport would allow the transformation of electrochemical sodium gradient into a proton motive force and assuming a H^+ translocating stoichiometry of 3 for ATPase, it can be estimated on thermodynamic basis that the energy yield would be 1/3 ATP per substrate molecule decarboxylated. In case of *Pr. modestum* growing on succinate (the product being propionate + CO_2), this is the only energy-yielding reactions available to the cells.

3.4.2.2 Acid Fermentation Products

The formation and excretion of acidic fermentation can affect membrane energization, either detrimentally or beneficially. The undissociated forms of acetic and butyric acids (pKa values around 4.8) freely permeate the bacterial membrane and allow the accumulation of these fermentation products in the medium, whose relative amounts of the associated and undissociated forms will reflect the prevailing pH. As the increasing amounts are excreted by the cells, pH of the medium will be lowered until at values around 4, the great excess of the acids will be in the undissociated membrane-permeable form which will thus reenter the cell where they will tend to dissociate at higher intracellular pH values around 6. The effect of this process will be to translocate protons into the cell and so partially under-energise the membrane by collapsing the pH gradient (This is an electro-neutral phenomenon which does not affect the membrane potential and thus should not be referred to as 'uncoupling').

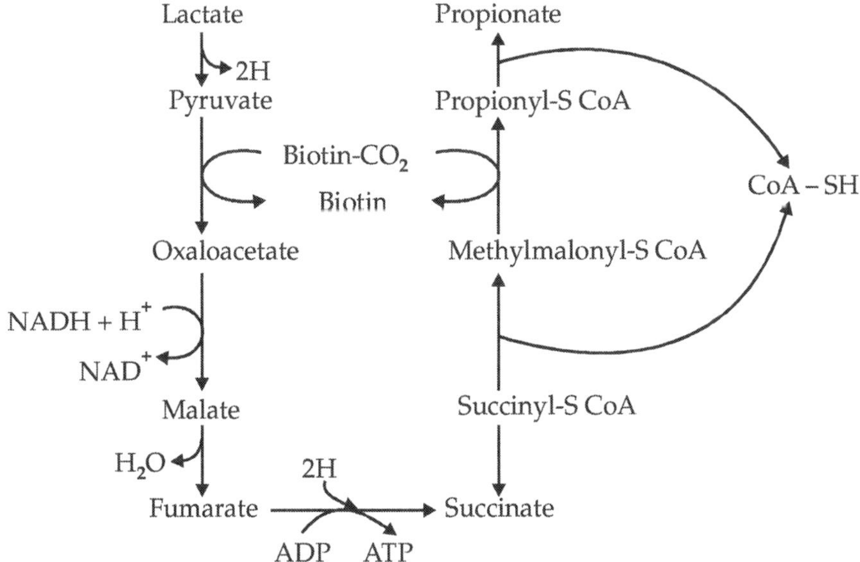

Balance : 3 lactate ⟶ 2 propionate + acetate + CO_2 (+ 3ATP)

Fig 3.16 : Pathway for Succinate and Propionate

Lactic acid, has its pKa value at 3.86 and its translocation across the membrane is dependent upon a carrier mechanism, which appears to be a proton support with proton/lactate of 2/1. This means, in effect, that the membrane translocation of lactate is electrogenic, with its intracellular accumulation being driven but the full proton motive force, while its efflux from the cell can at least potentially generate such a proton motive force. Evidence that this is indeed the case has been obtained from studies with lactic acid bacterium, *Streptococcus cremoris* and *E.coli*. Particularly significant is the observation that growth yield of *Strep*tococcus *cremoris* is increased when it is grown in co-culture with lactate utilizing organism.

In one respect, this situation parallels that referred to already where, in the co-culture of hydrogen producing and hydrogen consuming species, not only does the second organism gain from the provision of as energy source that is also a direct energy gain to the first organism resulting from the removal of its fermentation products. The molecular mechanisms of the two effects are, however, quite different.

3.4.3 Acetate Fermentations

It has been seen that acetate is a common product of several fermentation patterns in a range of organisms, and of particular importance in as much as its production is associated with an increased energy yield by SLP. When considering the sulphate-reducing and methanogenic bacteria, acetate will again be seen to play a key role, although this time predominantly as a substrate. In fact, the production and consumption of acetate and of hydrogen, by different species of microorganisms, forms the focal point in the complex series of interacting metabolic activities that characterize the anaerobic biodegradative microbial food chain. This chain leads from hydrocarbon leads from hydrocarbon, carbohydrate, protein and lipid polymers to the final products of mineralization-CO_2 and either sulphide or methane, depending on environmental conditions.

These processes are greatly facilitated by the so-called acetogenic bacteria, knowledge of whose existence and physiological activities has increased greatly in recent years. Operationally they can be considered as *two* subgroups : the homoacetogens which ferment hexoses, H_2 + CO_2, and occasionally carbon monoxide or methanol + CO_2, to acetate as sole fermentation product; and the hydrogen-producing acetogens which ferment alcohols and organic acids to acetate plus H_2. In concert, therefore with the other fermentative bacterial genera already described, the acetogens ensure that the full range of nutrient materials available to anaerobic microbial ecosystems can ultimately be converted to acetate and hydrogen.

With regard to the acetogens themselves, two metabolic processes are of particular interest : the synthesis of acetate from CO_2; and the wider implications of the production of H_2 in the fermentation of short-chain fatty acids such as propionate and butyrate.

3.4.3.1 The Homoacetogens : The Acetyl-CoA Pathway

The route whereby acetogenic bacteria such as *Cl. aceticum, C. thermoaceticum* and *Aceobacterium woodii* ferment hexoses to three molecules of acetate, involves glycolysis and the formation of acetate from both molecules of pyruvate by the action of pyruvate-ferredoxin oxidoreductase; the third acetate comes from the reduction of the two molecules of CO_2 arising from this last reaction. These organisms can also be grown on CO_2 as sole carbon source according to the equation :

$$2CO_2 + 4H_2 \longrightarrow \text{acetate} + 2H_2O$$

These mechanisms of heterotrophic and autotrophic CO_2 fixation are, in fact, identical and are referred to as the acetyl-CoA pathway, Fig. 3.17.

However, the acetyl-CoA pathway is confined neither to the acetogens nor to CO_2 fixation. It appears that in both eubacterial sulphate-reducing and archaebacterial methanogenic bacteria the route of carbon fixation from CO_2 is via the acetyl-CoA pathway (with the likely exception of the genera *Desulfobacter* and *Desulfuromonas*). At the same time, the acetyl-CoA pathway, or a slight variant of it, has also been found to be the basis for the reduction of acetate to methane, and for the oxidation of acetate to CO_2 again in a range of sulphate-reducing and methanogenic bacteria. Clearly, these reactions are central to all considerations of carbon and energy metabolism in these anaerobic chemotrophs and it would be seriously to miss the point, and to underestimate its crucial importance, to refer to the acetyl-CoA pathway solely as route of CO_2 fixation. Further, when it is appreciated that, even in the CO_2 fixation mode the pathway is associated with energy generation and ATP synthesis, it would clearly be wise to consider the reaction sequence rather as a mechanism for the oxidation of hydrogen with CO_2 as electron acceptor. In this respect, therefore, the acetogens directly parallel the methanogens.

Most information on the details of the acetyl-CoA pathway come from studies with *Cl. thermceticum,* and to a lesser extent with *Cl. formicoaceticum* and *A. woodii.* While the data are as yet incomplete and subject to some minor variations of interpretation, the general picture now appears to be clearly established. The complete reaction sequence can perhaps best be considered in terms of two partial reactions : the formation of methyl-tetrahydropteridine; and its carboxylation to acetyl-CoA through the mediation of corrinoid methyl-transferring protein and the enzyme carbon monoxide dehydrogenase.

The general scheme proposed for the acetyl-CoA pathway in *Cl. thermo-aciticum* is given in Fig. 3.17. The pteridinc derivative has been identified as tetrahydrofolate (THF); in methanogens it is tetrahydromethanopterin while it remains unidentified in those sulphate-reducers which employ the pathway. Formate is a free intermediate and the formate dehydrogenase catalysing its synthesis from CO_2 is an NADPH-dependent flavoprotein containing tungsten

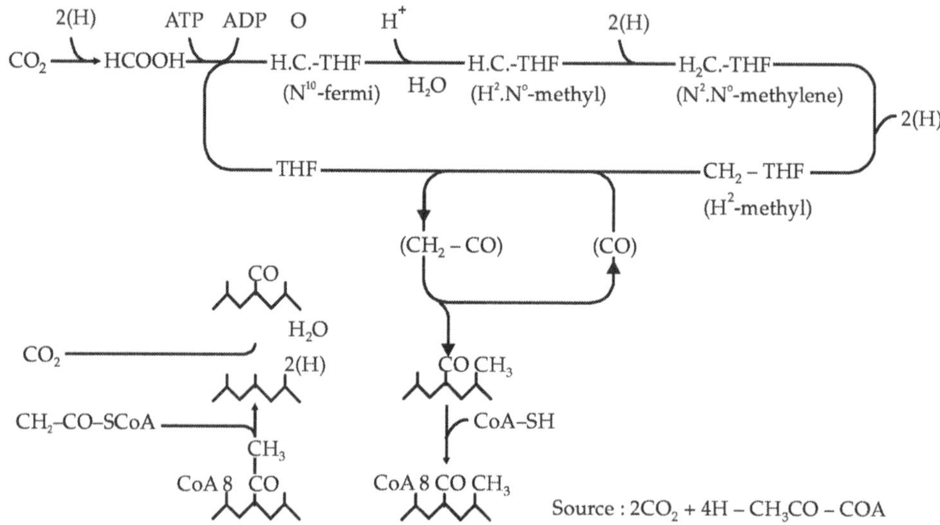

Fig 3.17 : The Proposed Acetyl-CoA Pathway in *Clostridium thermoaceticum*

selenium and iron-sulphur centres, with reduced ferredoxin being the likely physiological electron donor. The formation of N^{10} formyl-THF requires ATP activation. The transformation through N^5-methenyl-THF to N_5, N_{10}-methylene-THF is carried out by an NADPH dependent bifunctional enzyme; *Cl. formicoaceticum* there are two mono functional enzymes, the reduction being NADH-dependent. The N^5, N^{10} methylene-THF reductase catalysing the reduction to N^5-methyl-THF is flavoprotein containing zinc and iron-sulphur centres,' it is NAD(P)H independent with reduced ferredoxin again being the likely physiological electron donor.

The next stage of the reaction sequence involves a methyltransferase and a nonenzymic corrinoid protein. The methyl group is transferred to the reduced cobalt within the protein-bound corrinoid nucleus; Factor 111_m in *Cl. thermoaceticum*. But vitamin B_{12} in *A. woodii*, with cofactor identity not yet established for the sulphate-reducing bacteria and methanogens.

The heart of the acetyl-CoA pathway lies with the enzyme carbon monoxide dehydrogenase. This is a tetrameric protein, with nickel, zinc and iron-sulphur centres, which catalyses the addition of a carboxyl group to methyl-THF with the formation of acetyl-CoA, where the carboxyl donor may be CO_2 (plus H_2, hydrogenase and ferredoxin), pyruvate (plus pyruvate-ferredoxin oxidoreductase and ferredoxin) or carbon monoxide. Carbon monoxide dehydrogenase has three sites which can bind a C_1 group from carbon monoxide (possibly a nickel carbonyl), a methyl group and coenzyme A. Intramolecular transfer of these groups leads to the synthesis and release of acetyl-CoA.

Cells capable of the reactions of acetyl-CoA pathway possess very large amounts of the pteridine and corrinoid cofactors. *Cl. thermoaceticum* additionally has the standard respiratory hydrogen and electron carriers ferredoxin, flavodoxin (under iron-limitation, NAD^+, rubredoxin, menayl quinone and cytochrome b. *A. woodii*, however, lacks the quinone and cytochrome. The exact nature of the electron transport chains leading from H_2, reduced feredoxin or pyrimidine nucleotides to the reactions of accetyl CoA synthesis has still to be established, Equally, the site and mechanism of any electron transport-linked phosphorylation remains a matter for speculation; the yield of ATP by SLP from the formation of acetate from acetyl CoA is offset by the ATP used in the activation of formate. The largest drop in free energy is associated with the reduction of methylene-THF to methyl THF (E_{m7}-11mV., being the most positive in the sequence) and this reaction is therefore the one most likely to be coupled to the development of a protomotive force and the synthesis of ATP. The electron and energy donors for the reaction are either ferredoxin (from pyruvate-ferredoxin oxidoreductase) or NADH (from glyceraldehyde phosphate dehydrogenase) and there is a sufficient ΔE for the synthesis of at least 1 ATP although details of the mechanism remain unknown.

With reference to the methanogens. It has been suggested that membrane bound cobamides might have a role to play in electron transport and energy generation. If substantiated, such a mechanism might also be applicable to the acetogens.

The formation of acetyl-CoA from $H_2 + CO_2$ + methyl-THF + coenzyme A arises from a reaction sequence close to equilibrium, but the initial formation of enzyme-bound CO is likely to require the input of significant activation energy. This could account for an ATP yield of less than 1 for the pathway as a whole, and it has even been suggested that the mechanism of activation is by energy-driven reversed electron transport with a direct coupling between two proton-translocating enzymes, methylene-THF reductase and carbon monoxide dehydrogenase.

A second route of acetate synthesis is found in organisms characterized by the fermentation of purines and amino acids. It is known as the glycine synthase pathway and is present in, for example, *Cl. acidi-urici* and *Petococcus glycinophilus.* The reactions are identical to those of acetyl-CoA pathway leading to the formation of methylene-THF, which is then transformed to glycine synthase in a reaction involving the addition of a second CO_2, reducing equivalents and ammonium ions, and the release of free THF. The glycine reductase then forms acetate in a reaction sequence consuming two further reducing equivalents and giving rise to ATP by SLP form an as-yet unidentified phosphoryl intermediate. The glycine rise to ATP by SLP form an as-yet unidentified phosphoryl intermediate. The glycine synthase and reductase system is complex of proteins, at least one of which is a selenium metallo-enzyme.

3.4.3.2 The Hydrogen-Producing Acetogens

The most celebrated example of an H_2-producing acetogen is to be found in *Methanobacillus omelianskii.* This 'organism', first isolated in 1940, grows on ethanol with the formation of methane, but was shown in 1967 to be in fact an obligately syntrophic association of two bacteria; the so-called 'S' organism, an acetogen converting ethanol to acetate plus H_2; and the methogen *methanobacterium bryantii*, which oxidizes H_2 with the production of methane:

$$2CH_3CH_2OH + 2H_2O \longrightarrow 2CH_3COOH + 4H_2;\ \Delta G^\circ + 9.6 kJ/reaction$$

$$4H_2 + CO_2 \longrightarrow CH_4 + 2H_2O;\ \Delta G^\circ - 136 kJ/reaction$$

The ability of the S organism to obtain energy from the oxidation of ethanol is dependent upon the removal of H_2 by the methanogen, with the consequent alteration of the reaction from being endergonic under standard conditions to being exergonic under the conditions of the two-organism consortium. The term 'interspecies hydrogen transfer' is used to describe such an association, which has a structural as well as a metabolic manifestation in that the bacterial partners are found to be in close physical contact. In the cases cited earlier where co-culture with a H_2-oxidizing species was shown to confer an energetic advantage on the branched fermentation pattern found with *Cl. pasteurianum and Ruminococcus albus*, the association was seen to be optional. With the S organism and methanogen in *Methanobacillus omelianskii,* however, the association is obligate, and the S organism is unable to grow on ethanol in pure culture.

Other examples of H_2-producing acetogenic bacteria that can only exist in nature as components of syntrophic associations demonstrating interspecies hydrogen transfer are *Syntrophobacter wolinii* (propionate) and *Syntrophomonas wolfei* (butyrate).

$$CH_3CH_2COOH + 3H_2O \longrightarrow CH_3COOH + 3H_2 + H_2CO_3 \quad \Delta G^\circ + 76.1\ kJ/reaction$$

$$CH_3CH_2CH_2COOH + 2H_2O \longrightarrow 2CH_3COOH + 2H_2 \quad \Delta G^\circ + 48.1 kJ/reaction$$

Both in natural ecosystems and in the laboratory either a sulphate-reducing bacterium or a methanogen may serve as the H_2-oxidizing partner and by this means reduce the H_2 concentration to about 0.1 μM. A similar high affinity for acetate might also result in its concentration being maintained at around 10μM. Even under such ideal conditions, however, the fermentation of, for example, butyrate by *Syntrophomonas wolfei* would still only give a low net yield ($\Delta G'_{0'}$ = – 15 kJ/reaction).

In a thought-provoking article Thauer and Morris introduced the concept of ATP yield less than 1 per substrate molecule catabolized, and discussed the part that might be played by energy-driven reversed electron transport in integrating redox balance and net energy yield in anaerobic heterotrophic metabolism. These authors identified-44kJ as being the minimum free energy requirement for the synthesis of 1 ATP under quasi reversible conditions and suggested that, where–ΔG was numerically less than this figure, reversed electron transport reactions might feature in pathways leading to fractional ATP yields.

The catabolic route from butyrate most probably proceeds through butyryl-CoA, crotonyl-CoA, B-hydroxybutyryl-CoA and acetoacetyl CoA to two acetyl-CoA, with a net yield of 1 ATP by SLP from acetyl-CoA (the second acetyl-CoA being required for butyrate activation by a Co-ASH transferase). However, H_2 production from the butyryl-CoA to crotonyl-CoA step (E_{m7} - 15*m* V) is energy-requiring as, even at 0.1 μM H_2 the H^+/H_2 couple has an En of -200*m* V. Assuming a H^+–translocation to ATP stotichiometry from the ATPase of 3, the hypothesis is that 2/3 ATP is hydrolysed with the translocation of 2 H^+ which then re-enter the cell accepting two electrons from the butyryl-CoA to crotonyl-CoA reaction. By this means, the conversion of butyrate to two acetates is facilitated with an ATP yield of 1/3 which agrees with the ΔG for the overall reaction of -15kJ, Fig. 3.18.

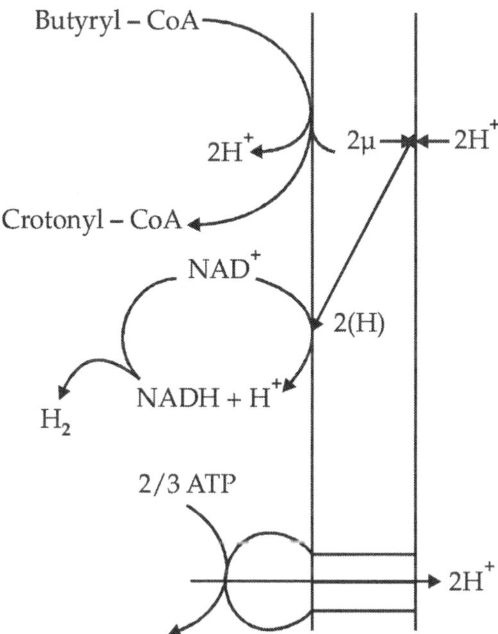

Fig 3.18 : Proposed Mechanism of Energy-driven Reversed Electron Transport during Production of Hydrogen by Oxidation of Butyryl-CoA to Crotonyl-CoA in *Syntrophomonas wolfei.*

3.4.3.3 Fermentation : General Comments

In this summary of microbial fermentation reaction sequences, the energetic consequences of the particular routes of carbon flux and mechanisms of redox balance have been stressed. An especially rich overall presentation of fermentation processes, including those concerned with amino acids, purines and pyrimidines, can be found in the second edition of Gottschalk's book, 'Bacterial Metabolism' (Gottschakl, 1986).

One particularly striking feature of fermentation as a source of energy for anaerobic microbial growth stems from the relative inefficiency of SLP and the limited occurrence of pmf-linked ATP synthesis (in terms of energy yield per mole of substrate used). For a given amount of cellular growth, large amounts of substrate are required with the consequent formation of similar quantities of a variety of reduced fermentation products. Where, in environmental or biotechnological processes, either the maximum turnover of substrate or the maximum formation of product along with the minimum wasteful production of microbial biomass is being sought, then clearly fermentative schemes have much to recommend them.

In terms of the thesis being developed here, however, the most significant feature is the excretion by organism A of reduced fermentation products which are potential sources of carbon, energy and reducing equivalents for organisms B, C and D, even within the same anaerobic environments. Ethanol, lactate, propionate, butyrate, for example, are all capable of acting as substrates for other fermentative species. Along with such food-chain considerations, there are the optional or obligate energetic benefits to be gained from lactate efflux and interspecies hydrogen transfer. It can be concluded that fermentative microorganisms are not likely to function in nature as single species in metabolic isolation, but that the natural habitat of anaerobic heterotrophic bacteria is within a closely functionally-integrated microbial consortium. In such a scenario, the acetogens take on a particularly important role in that they determine the potential for the anaerobic biodegradation of almost all organic primary substrates to the products acetate CO_2 and H_2.

These last tansformations as already seen, are often only possible in the presence of H_2-oxidizing bacteria, which therefore ultimately control the fermentative and biodegradative activities of the consortium as a whole. These key H_2-oxidizing bacteria, which incidentally further stimulate the overall process by their utilization of acetate, are the methanogenic and sulphate-reducing bacteria.

3.5 Anaerobic Respiration : Methanogenesis and Sulphate Reduction

3.5.1 Methanogenic Bacteria

The methanogenic bacteria are usually considered to belong to the group of bacteria that grow anaerobically using the process of anaerobic respiration for energy generation. The reducing substrate is usually hydrogen gas, with CO_2 acting as terminal electron acceptor. As noted later, in the absence of oxygen, a fairly wide range of facultative genera can use nitrate, nitrite, trimethylamine oxide or fumarate as alternative terminal electron acceptors in a redox mechanism that requires a standard, if slightly modified, electron transport chain coupled to ATP syntheses through a proton motive force. This also holds true for the sulphate reducing bacteria, save only that they are obligate rather than facultative anaerobes. In their reduction of CO_2 to methane, the methanogenic bacteria appear to be employing the same basic energetic mechanism and, certainly in ecological terms the sulphate-reducers, and methanogens can be seen to function, and therefore must be considered, in a parallel fashion. Both groups oxidise H_2 and metabolize acetate, with the methanogens predominating in more reducing (E_h = – 300mV) and sulphate-free environments.

Unlike the more conventional anaerobic respiratory bacteria, however, the methanogens are largely without cytochromes and other usual respiratory carriers. Both in respect of their oxidation of H_2 with CO_2 as electron acceptor and their use of the acetyl-CoA pathway, the methanogens bear a close physiological relationship to the fermentative acetogens : yet they clearly do not accord with our normal perception of fermentation since all their ATP arises from a chemiosmotic mechanism with no evidence for substrate level phosphorylation.

The most striking feature of the methanogens, however, lies not in their points of similarity but rather in the dramatic dissimilarity they show to other bacteria, whether aerobic or anaerobic. The methanogens are Arechaebacteria, a group of micro-organisms probably of early evolutionary origin, distinct from both eukaryotes and eubacteria and yet sharing certain characteristics with each group. The Archaebacteria were recognized as a separate Kingdom on the basis of studies of their 16S ribosomal RNA, and features of their uniqueness have now extended to various other ribosomal, transfer and messenger RNA properties, along with histones, ribosome architecture and the mechanisms of transcription and translation. From the point of view of cell structure and metabolism a characteristic feature identifying the Archaebacteria is the chemistry of their cell walls and membranes. Here, there is an absence of peptidoglycan in the wall, and the membrane comprises glycerolipids in an absence of peptidoglycan in the wall, and the membrane comprises glycerolipids in which 1-glycerol is ether-linked to long-chain phytyl groups. Along with the methanogens, the Archaebacterial Kingdom contains the extreme halophiles and the sulphur-dependent themoacidophiles.

Although methanogenesis is only found amongst the Archaebacteria, the methanogen group in fact displays a wide range of bacterial morphologies. There are four distinct types : rods, cocci, sarcinae and spirilla; demonstrating both positive and negative Gram reactions. The rods and spirillum type cells may grow as individual cells or as long filaments.

3.5.2 Chemolithotrophic and Methylotrophic Metabolism

With respect to their metabolism, the methanogens can be subdivided into two groups. The obligate chemolithotrophs obtain their energy and carbon from the reduction of CO_2 with H_2 as electron and energy donor. Most species can also grow on formate. On the other hand, the methylotrophic methanogens can grow on H_2 plus CO_2, but they are also able to grow on methanol, on mono-, di- or tri-methylamine, or on acetate as energy and carbon source. *Methanobrevibacter runinantium* is unusual in that it can grow on $H_2 + CO_2$, but additionally requires acetate as a source of carbon. Whereas there is no evidence for the presence of quinones or cytochromes among the chemolithotrophic methanogens, they are found in those methylotrophic methanogens able to oxidize methyl group to CO_2. For example, *Methanosarcina barkeri* has two b-type and one c-type cytochromes.

Recently, two intriguing lithotrophic methanogens have come to light. A freshwater *Methanospirillium* sp. and a marine *Methanogenium* sp. have been isolated in pure culture using propan-2-ol as hydrogen donor for methanogenic growth on CO_2 according to the reactions :

$$4CH_3.CHOH.CH_3 \longrightarrow 4CH_3.CO.CH_3 + 8[H]$$

$$CO_2 + 8[H] \longrightarrow CH_4 + 2H_2O$$

H_2 formate and butan-2-ol may also serve as hydrogen donor for both species, and ethanol and *n*-propanol additionally for the marine strain. Propan-2-ol has also been noted, however, to be a substrate for the growth of the syntrophic consortium designated *Methanobacterium omelianski,* and this raises a most interesting point regarding metabolic mechanisms and energy coupling at the cellular level. Where 'M. onmlianski' grows on ethanol or propan-2-ol it does so by the combined fermentative reactions of the acetogenic S organisms and the H_2-oxidizing capacity of its partner *Methanobacterium bryantii,* with energy coupling at the cellular level being achieved by interspecies H_2 transfer. However, in the newly isolated *Methanogenium* sp. the same two substrates can support growth by reactions that can be represented mechanistically and energetically by the same equations, but within a single organism and with no evidence for intermediary H_2 production.

An exactly analogous comparison will be drawn later between the growth on propionate by *Syntrophobacter wolinii* along with an H_2-oxidizing methanogen or sulphate-reducer, and by *Desulfobulus propionicus* in pure culture.

Detailed knowledge of metabolism and energetics of the methanogenic bacteria, is largely dependent on extensive studies with two particular species. *Methanobacterium themoautotrophicum* is an obligate chemolithotroph which grows at temperatures in excess of 40°, with an optimum between 65°C and 70°C. *M. barkeri* on the other hand is the most versatile methanogen known, although it is worth pointing out that, whereas growth on H_2 plus CO_2, methanol or methylamines may take from three to seven days, growth on acetate may take as three weeks. Despite this *M. barkeri* is generally isolated from such environments as anaerobic digesters which are high in acetate, and is not found in the rumen where acetate is at a very low concentration. Epithelial cells rapidly absorb volatile fatty acids, and the dilution rate in the rumen exceeds the maximum growth rate (μ_{max}) of the chemolithotrophic methanogens.

3.5.3 Carbon Assimilation in Methanogens

No methanogen uses the usual methylotrophic pathways for the assimilation of reduced C-1 compounds (ribulose monophosphate pathway, serine pathway, dihydroxyacetone pathway); none uses the Calvin (ribulose bisphosphate) pathway or reductive TCA cycles of CO_2 fixation. The route for assimilation of CO_2 in methanogens is the acetyl-CoA pathway, in which a major intermediate is methyl tetrahydromethanopterin ($H_4MPT-CH_3$), Fig. 3.19. As with acetogens (Fig. 3.17), the methyl group is transferred (by a methyltransferase) to a corrinoid protein. It is then carbonylated to acetyl-CoA by the nickel-containing CO dehydrogenase which produces the bound CO by reduction of carbon dioxide. An F_{420} dependent reductive carboxylation then gives pyruvate. The overall reaction is thus :

$$CO_2 \xrightarrow{3H_2} H_4MPT-CH_3 \xrightarrow{H_2 \;\; CO_2} CH_3CO-COA \xrightarrow[H_2F_{420} \;\; F_{420}]{CO_2} CH_3COCOOH$$

The pyruvate is then further carboxylated to oxaloacetate by the action of PEP synthetase and PEP carboxylase. In *M. barkeri,* 2-oxoglutarate is then produced by the condensation of oxaloacetate and acetyl-CoA to give citrate followed by the oxidative reactions of a conventional TCA cycle. These reactions are absent from *M. thermoautotrophicum,* however, and 2-oxo-glutarate is synthesized from oxaloacetae by the reductive reactions of the TCA cycle, with F_{420} again being the electron donor in the reductive carboxylation of succinyl-CoA to 2-oxoglutarate. In neither organism is there a complete cycle operating in an oxidative energy-generating mode.

Methanogenesis from acetate in *M. barkeri* proceeds by the reversal of the pathway described above : carbon monoxide dehydrogenase corrinoid protein, methyltetrahydromethanopterin, methyl-CoM, methane. Growth of *M. barkeri* on acetate induces a five-fold increase in the levels of carbon monoxide dehydrogenase as compared with H_2O/CO_2-grown cells.

3.5.4 Energy Coupling in Methanogens

3.5.4.1 General

The whole purpose of methanogenesis is to make ATP. All the evidence so far is consistent with the conclusion that methanogens obtain ATP by way of a protonmotive force (pmf) which is generated by electron transport, with hydrogen as the electron donor and CO_2 usually providing the electron acceptor. The overall reaction is :

$$CO_2 + 4H_2 \longrightarrow CH_2 + 2H_2O \; ; \Delta G° = -131 \text{ kJmol}^{-1}$$

In non-standard conditions, particularly with the usual low concentrations of hydrogen, the free energy change will be more positive and it is unlikely that more than one ATP per mole of methane could be produced. In 1956 Barker proposed a scheme for reduction of CO_2 to CH_4 which, in essence, remains a valid summary of this process :

$$CO_2 \xrightarrow[XH]{} XCOOH \xrightarrow[H_2]{+3.5} X\text{–}CHO \xrightarrow[H_2]{+23.4} X-CH_2OH \xrightarrow[H_2]{-44.8}$$

$$XCH_3 \xrightarrow[H_2 \quad XH]{-112.5} CH_4$$

The carrier of C_1 units (X) is not the same for every level of reduction. The numbers over each step are the standard free energy changes ($\Delta G°$, in kJ mol^{-1}) for each reaction (on the basis of the free intermediates). The hydrogen in methane and its precursors does not come directly from the H_2 energy source but from the protons of water. The process of electron transport has an extra degree of complexity in methanogens because there are, in effect, four different terminal electron acceptors which must be reduced in sequence; and, hence, four electron transport chains having the same donor but different acceptors. Not all of these chains can be coupled to ATP synthesis. As shown above, the first two reductions are endergonic and they must be driven by the second pair of reductions which are exergonic. This may be by way of a pmf established during one or both of the exergonic reactions.

It is remarkable that, in general, the methanogens lack quinones and cytochromes and that, where present, they do not appear to play a direct part in the generation of a pmf for ATP synthesis. None the less, the mechanism for driving the phosphorylation of ADP in methanogens is chemiosmotic, as has been shown by

- the measurement of a pmf;
- ATP synthesis driven by a valinomycin-induced K^+ efflux potential or an artificially generated proton potential; and
- sensitivity to protonophore uncouplers and the ATPase inhibitor DCCD.

3.5.4.2 Unusual Coenzymes in Methanogens

Some of the special coenzymes of methanogens appear to be functionally analogous to those involved in aerobic electron transport systems : they differ mainly in their low redox potentials. Other have the function of carriers of the reduced carbon precursors of methane; and yet others have a function primarily in biosynthesis. Some of the coenzymes function in more than one role. Their structures are given in Fig. 3.19 and their involvement in the reduction of CO_2 to methane is shown in Fig. 3.20.

Factor F_{420} is a 5-deazaflavin analogue of FMN. It functions as a soluble hydrogen carrier with a low redox potential (E_{m7} = – 360m V), is the hydrogen donor in the reduction of methyl-coenzyme M, and may also be the hydrogen donor in the earlier steps in reduction of CO_2. It can be reduced by hydrogen (hydrogenase), by formate (Formate dehydrogenase) and by NADPH ($NADP^+$ reductase). It may thus mediate between oxidation of hydrogen or formate and reduction and $NADP^+$. As mentioned above, coenzyme F_{420} is also the hydrogen donor in the reductive carboxylation of pyruvate (and oxoglutarate) during assimilation of cell carbon.

Factor F_{430} is a nickel tetrapyrrole which is also probably involved in the reduction of methyl-CoM to methane.

Methanofuran, previously called the CO_2 reduction factor (CDR), is the C_1-carrier at the level of oxidation of formate : it is thus the first bound intermediate in the reduction of CO_2.

(*a*) F_{420}

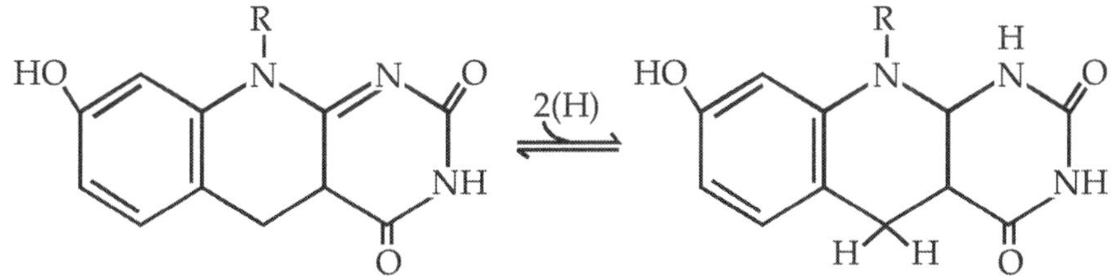
R
HO
N
N
O
2(H)
NH
O
R
H
HO
N
N
O
NH
H
H
O

Where

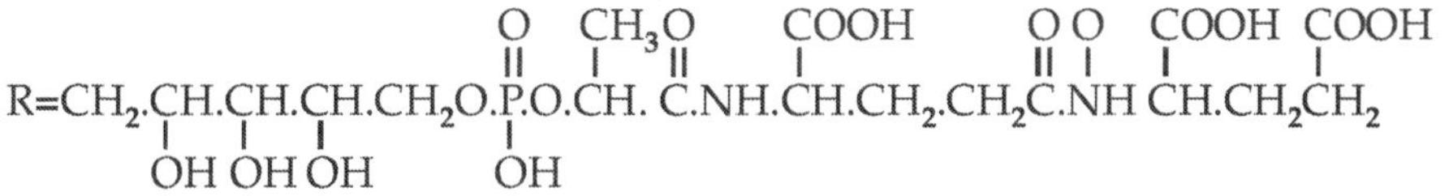
O CH3O COOH O O COOH COOH
R=CH2.CH.CH.CH.CH2O.P.O.CH. C.NH.CH.CH2.CH2C.NH CH.CH2CH2
OH OH OH OH

(*b*) F_{430}

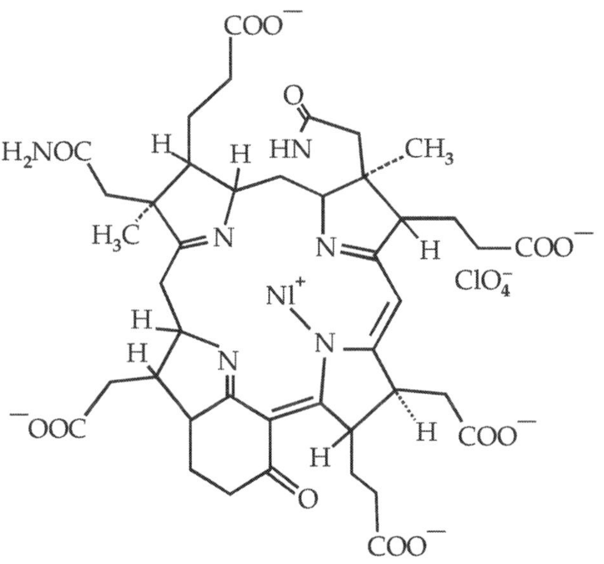
COO
O
H2NOC
H
H
HN
CH3
H3C
N
N
H
COO
ClO4
Ni+
H
N
H
N
OOC
H
COO
H
O
COO

(*c*) Methanofuran

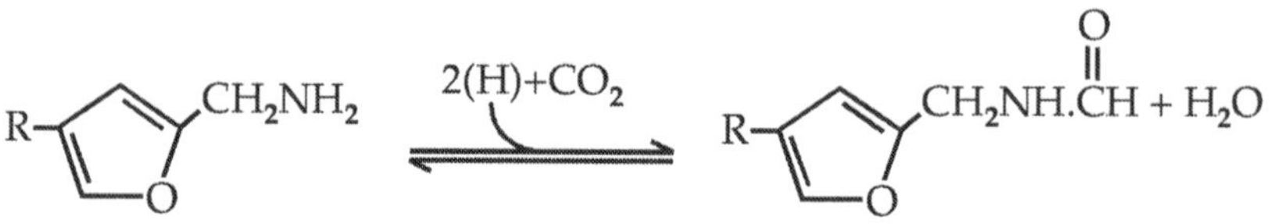

$$\text{R-(furan)-CH}_2\text{NH}_2 \xrightleftharpoons{2(H)+CO_2} \text{R-(furan)-CH}_2\text{NH.}\overset{O}{\overset{\|}{C}}\text{H} + H_2O$$

Where R =

$$HOOC.CH_2.CH_2.\overset{COOH}{\overset{|}{C}H}.\underset{COOH}{\underset{|}{C}H}.CH_2.CH_2.\overset{O}{\overset{\|}{C}}.NH.\overset{COOH}{\overset{|}{C}H}.CH_2.CH_2.\overset{O}{\overset{\|}{C}}.NH.\overset{COOH}{\overset{|}{C}H}.$$

$$CH_2.CH_2.\overset{O}{\overset{\|}{C}}\,NH.CH_2.CH_2.\text{(benzene ring)}\,OCH_2^-$$

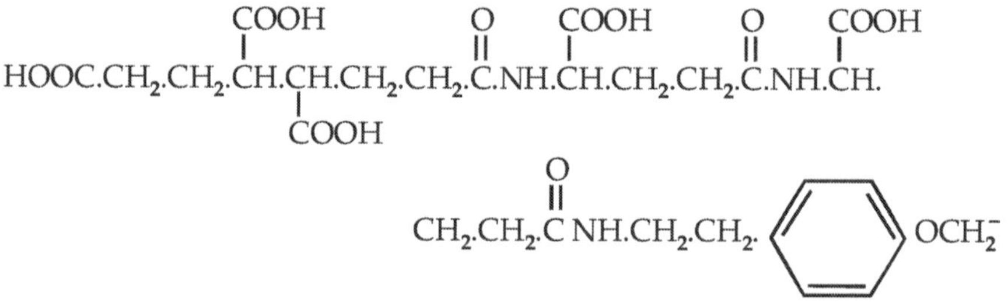

(*d*) Tetrahydro-methanopterin

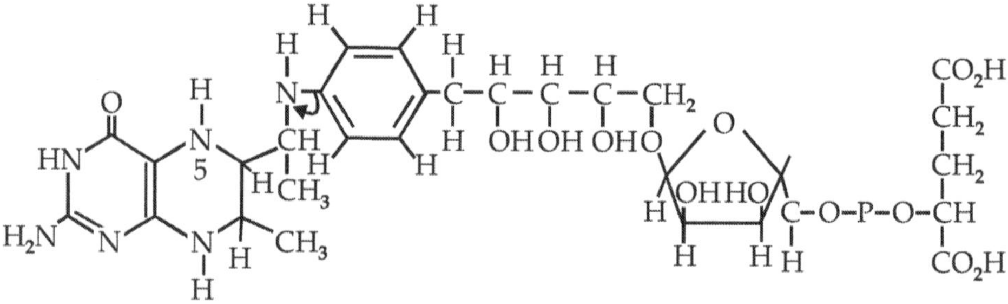

Fig. 3.19 : Structure of Cofactors Found in Methanogens

Tetrahydro methanopterin (H_4MPT) is the next carbon carrier, the formyl group being transferred from formyl-methanofuran to yield formyl-H_4MPT which is dehydrated to methenyl-H_4MPT before reduction to methylene H_4MPT and thence to methyl-H_4MPT.

The best-known of the novel coenzyme, M, is the smallest of all known coenzymes. Coenzyme M is 2-mercaptoethane sulphonate ($HSCH_2CH_2SO_3^-$). This acts as the methyl carrier (CH_3-$SCH_2CH_2SO_3$) in the last stage of methanogenesis catalysed by the methylCoM reductase complex.

Component B of the methyl-CoM reductase complex is also probably a coenzyme : it has recently been shown to be 7-mercaptoheptanoylthreonine phosphate (HTP).

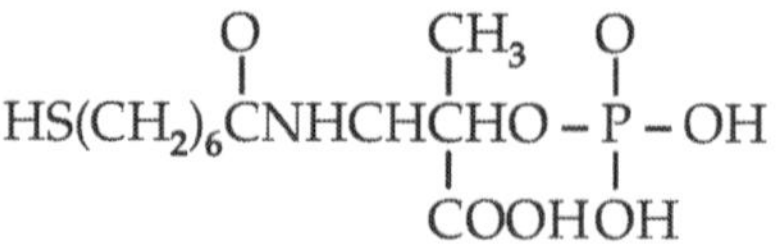

$$HS(CH_2)_6\overset{O}{\overset{|}{C}}NHCH\underset{COOH}{\underset{|}{C}}\overset{CH_3}{\overset{|}{}}HO-\underset{OH}{\underset{|}{\overset{O}{\overset{|}{P}}}}-OH$$

Its role is unknown, but it may act as a methyl carrier, the terminal thiol having the same role as that coenzyme M.

3.5.4.3. Energy transduction and methanogenesis from carbon dioxide Figure 3.20 summarizes the route for methanogenesis from CO_2; most of the steps of which have been discussed in the previous section.

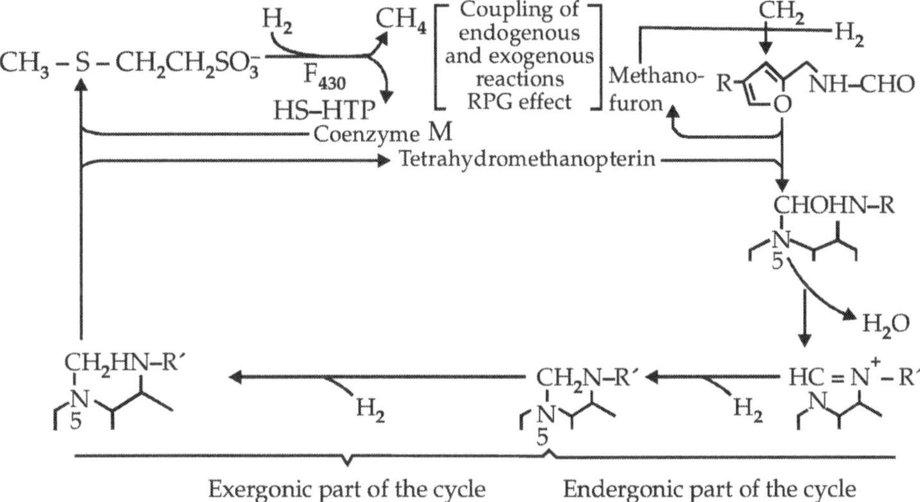

Fig 3.20 : Methanogenesis from CO_2. The 'RPG' Effect is the driving of Methanogenesis from CO_2 by addition of (for example) Methyl-CoM. It is probably an expression of the necessary coupling of the endergonic steps with the Exergonic Steps of Methanogenesis. This may well be achieved by way of a pmf. The Hydrogenases involved in each reduction step are not necessarily indentical. If they are arranged across the membrane so that only electrons pass across to a Hydrogen carrier on the inside then they will contribute to a pmf. MFR, Methanofuran; H_4MPT, Tetrahydromethanopterin; CoM, Coenzyme M.

The final step, which is the most complex is the formation of methane by the methyl-CoM reductase complex which is located in the cytoplasmic membrane. This complex consists of many proteins : component A-1 is a crude fraction containing hydrogenase; component A-2 is a single protein of unknown function; component A-3 consists of several proteins. Component C is the site of reduction of methyl-CoM to methane; it is a large protein associated with two molecules of the nickel tetrapyrrole F_{430} and two of coenzyme M. Component B is not a protein but another coenzyme of uncertain function (see above). Although a great deal is known about this part of methanogenesis, Ralph Wolfe (Who has been responsible for much of what is known) has pointed out that much is yet to be learned about this multienzyme system before any reaction mechanisms can be taken seriously.

In Fig. 3.20 the whole process of methanogenesis from CO_2 is drawn as a cycle. This emphasizes the essential fact that the later exergonic part of the process must be coupled somehow to the earlier endergonic part. The coupling of the first and last reactions has long been appreciated, and is illustrated by the RPG effect (named after its discoverer R.P. Gunsalus): the stimulation by methyl-CoM (and some other compounds) of CO_2 reduction to methane.

This coupling is almost certainly not via a direct chemical linkage of the reactions. The obvious way of coupling the last reaction of methanogenesis to the first reaction is by way of the pmf. If the methyl-CoA reductase complex is arranged suitably in the membrane then reduction of methyl-CoM could establish a pmf; this could then drive the endergonic reactions, which would also have to be arranged across the bacterial membrane.

3.5.4.4 *Energy Coupling During Methanogenesis from Methanol, Formaldehyde and Acetate*

Methanol can also act as methanogenic substrate for *M. barkeri* according to the reactions:

$$CH_3OH + H_2O \longrightarrow CO_2 + 6[H]$$
$$3CH_3OH + 6[H] \longrightarrow 3CH_4 + 3H_2O$$
$$\overline{4CH_3OH \longrightarrow 3CH_4 + CO_2 + 2H_2O}$$

One molecule of methanol is oxidized to supply the reducing equivalents required for the reduction of a further three molecules of methanol to methane, Fig. 3.21. The overall process is found to be Na^+ requiring and uncoupler-sensitive.

If, as seems likely, the electron carrier between the oxidation and reduction reactions is F_{420}, then the transfer of reducing equivalents from the redox level of methanol (E_{m7} = – 182mV) to F_{420} (E_{m7} = – 360mV) will be energy-requiring. Interestingly, however, *M. barkeri* can grow on methanol in the presence of H_2 according to the equation :

$$CH_3OH + H_2 \longrightarrow CH_4 + H_2O$$

Clearly this reaction, mediated by hydrogenase through F_{420} and coupled to the reduction of methanol to methane is exergonic and potentially capable of developing a pmf and consequent ATP synthesis. Under these conditions methanogenesis is neither Na^+-requiring nor uncoupler-sensitive. Treatment with the ATPase inhibitor DCCD, however, causes a rapid exhaustion of the ATP pool without the pmf being reduced from a measured value of -130mV. Methane production is inhibited under these conditions, and this inhibition can be relieved by the addition of an uncoupler. In other words, methanogenesis from methanol plus H_2 is under a from of respiratory control.

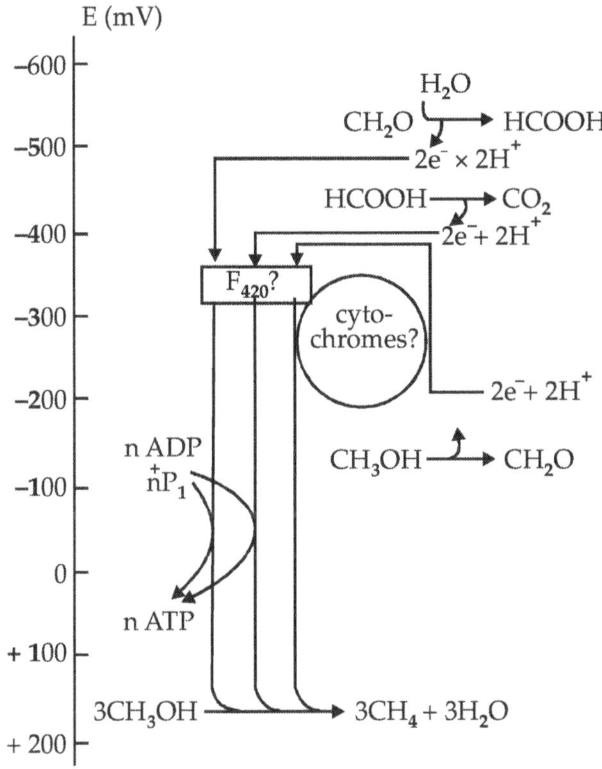

Fig 3.21 : Proposed Reaction Scheme for Methanogenesis in M. barkeri from methanol. The Redox Scale Indicates the Standard Mid-point Potentials of the Formal Redox Couples; the Intermediates are Most Likely Carrier-bound.

The reduction of methanol to methane involves only two enzymes : a methyltransferase catalysing formation of methyl-CoM from methanol and the methyl-CoM methyl reductase. This second reaction is strongly exergonic and it is therefore by way of the membrane-bound methyl-CoM reductase complex that redox energy must be transduced into a transmembrane proton gradient; although the molecular mechanism is not presently indentifiable.

It has been suggested that the cytochromes present in *M. barkeri* are required for an energy-driven reversed electron transport in the transfer of reducing equivalents from methanol to F_{420}, and that this is also the Na^+–requiring and uncoupler-sensitive step under growth on methanol alone. *Methanosphaera stadtmaniae* provides circumstantial support for this idea in that it can only grow on methanol under an H_2 gas phase and is found to lack any cytochrome. Fig. 3.21 presents a scheme for the growth of *M. barkeri* on methanol.

Methanogenesis from formaldehyde plus hydrogen mirrors the characteristics noted above for methanol plus hydrogen, *i.e.* respiratory control resistance to uncouplers, no requirement for Na^+, and coupled ATP synthesis by a chemiosmotic mechanism, presumably again at the methyl-CoM reductase. However, reduction of CO_2 to the level of formate is an endergonic reaction. This energy-requiring reductive C_1-transfer (to methanofuran) is thus a site of the uncoupler sensitivity of the overall process of methanogenesis from CO_2 plus H_2.

The considerations above have led to propose that Na^+ may have a role in the coupling of the endergonic and exergonic reactions of methanogenesis.

3.5.4.5 *The Role of Na^+ in the Reactions of Methanogenesis*

There role of Na^+ remains at least partially obscure in this instance. A Na^+ requirement has been identified for methane formation from H_2 plus CO_2, methanol, and acetate in a number of methanogenic species, and for the synthesis of ATP *M. themoautotrophicum* driven by a K^+ diffusion potential. However, the fact that methanogenesis and ATP formation from formaldehyde, and from methanol plus H_2 is Na^+-independent, suggests that Na^+ does not play a direct role in either the methyl reductase or the ATPase. The characterization of an Na^+/H^+ antiport in *M. thermoautotrophicum* has led to the proposal that most probably the regulation of intracellular pH may be the focus of the Na^+ requirement, which, from these data, appears to be closely linked with the various energy-dependent steps in methanogenesis. Inhibition of the antiport also inhibits methanogenesis from H_2 plus CO_2. Interestingly, the antiport is sensitive to amiloride, harmaline and NH_4^+, which are the characteristic inhibitors of the Na^+/H^+ antiport in eukaryotic cells.

The methanogenic pathway from acetate also poses some intriguing energetic questions. The need to expend 1 ATP in the formation of acetyl-CoA determines that methanogenesis from acetyl-CoA via carbon monoxide dehydrogenase must be associated with an ATP yield greater than 1. It is therefore particularly interesting that, where methanogenesis in *M. barkeri* has been inhibited with bromoethanesulphonate, the oxidation of carbon monoxide, gives

$$CO + H_2O \longrightarrow CO_2 + H_2$$

rise to an increase in the pmf from -100m V to -150mV. Addition of an uncoupler leads to rapid decreases in both the pmf and the ATP level, along with an increase in the rate of oxidation. This is directly paralleled by the demonstration that the oxidation of carbon monoxide to CO_2 in *Acetobacterium woodii* can be energetically coupled to histidine uptake. The reverse has also been demonstrated: the reduction of CO_2 with H_2 to carbon monoxide plus water is dependent upon methanogenesis and energy generation. Studies with uncoupler-insensitive methane formation from methanol plus H_2 in *M. barkeri* have shown that CO_2 reduction by carbon monoxide undehydrogense is uncoupler-sensitive and DCCD-insensitive, and thus appears to be driven directly by protonmotive force rather than by ATP. Unlike other energy-requiring reactions in methanogenesis, however, CO_2 reduction to CO is not Na^+-dependent.

The enzyme carbon monoxide dehydrogenase, like the methyl reductase, might therefore be directly responsible for energy transduction and the generation of a pmf, although once again the mechanism involved cannot yet be identified.

3.5.5 Sulphate Reducing Bacteria

3.5.5.1 General

Like the methanogens, the sulphate-reducing bacteria constitute a phylogenetically diverse group of organisms which none the less share a common physiological character which, in turn, dictates a common ecology; they from a so-called physiological-ecological group. They play a key role in the natural sulphur cycle, catalysing the process of sulphurication.

The sulphate-reducing bacteria are strict anaerobes requiring Eh values of –100mV or less for growth; they obtain their energy by an anaerobic respiratory process, using sulphate as terminal electron acceptor, *i.e.* dissimilatory sulphate reduction. The product of this reduction is sulphide and in parallel with the acetogens and methanogens there is now widespread adoption of the nomenclature 'sulphidogens' rather than 'sulphate-reducing bacteria.' Perhaps a more pressing reason for basing identification on the product of respiration is the recent appreciation that there are a number of sulphur-reducing bacteria (again giving rise to sulphide) which clearly belong within the same physiological-ecological group. The vast majority of the organisms within the group, however, use sulphate (and often other oxidized derivatives such as thiosulphate) as terminal electron acceptor, and are completely unreactive to sulphur itself. The term 'sulphate-reducing' bacteria retains by far the wider general usage. For these reasons, but while recognizing the degree of inexactitude involved, the author shall continue to refer to organisms within this group as the sulphate-reducing bacteria. (They are often in fact referred to by the abbreviation SRB, but this does not represent correct usage in the present context.)

As is well documented in the second edition (1984) of Postgate's excellent monograph 'The Sulphate-Reducing Bacteria,' before the late 1970s two genera only were recognized : Desulfotovibrio and the spore-forming Desulfotomaculum. The 12 or so species identified were characterized by a restricted metabolic capability, both with respect to the spectrum of nutrients that could serve as carbon and energy sources for heterotrophic growth (typically lactate), and the incomplete oxidation due to the absence of a functional TCA cycle with the consequent production of acetate as a metabolic end-product. The identification of two new species by Pfennig's laboratory in 1976 and 1977 heralded what has turned out to be something of a revolution in the knowledge and understanding of the sulphate-reducing bacteria : *Desulfuromonas acetoxidans* and *Desulfotomaculum acetoxidans* are both capable of oxidizing acetate and using it as a source of carbon and energy to support growth, while the former organism shows the other (at that time new) property of reducing elemental sulphur as terminal

Table 3.4 : Characteristics of Representative Sulphate-Reducing Bacteria

	Cell form	*Approximate Optimum Temperature for Growth* (°C)	H_2	*Compounds Oxidized* Acetate	Fatty Acids	Lactate	Others
Incomplete Oxidation							
Desulfovibrio							
desulfuricans	Curved	30	+	–	–	+	Ethanol
vulgaris	Curved	30	+	–	–	+	Ethanol
gigas	Curved	30	+	–	–	+	Ethanol
salexigens	Curved	30	+	–	–	+	Ethanol
sapovorans	Curved	30	–	–	C_4through C_{16}	+	
thermophilus	Rod-shaped	70	+	–	–	+	
Desulfotomaculum orient	Rod-shaped	30 to 35	+	–	–	+	Methanol
ruminis	Rod-shaped	37	+	–	–	+	
nigrificans	Rod-shaped	55	+	–	–	+	
Desulfobulbus							
propionicus	Oval	30 to 38	+	–	C_3	+	Ethanol
Complete oxidation							
Desulfobacter							
Postagater	Oval	30	–	+	–	–	
Desulfovibro baarsu	Curved	30 to 38	–	(+)	C_2 through C_{18}	–	
Desulfotomaculum							
acetoxidans	Rod-shaped	35	–	+	C_4, C_5	–	Ethanol
Desulfococcus							
multivorans	Spherical	35	–	(+)	C_2 through C_{14}	+	Ethanol, benzoate
niacini	Spherical	30	+	(+)	C_2 through C_{14}	–	Ethanol, nicotinate, glutarate
Desulfosarcina variabitis	Cell Packets	30	+	(+)	C_3 through C_{14}	+	Ethanol, benzoate
Desulfobacterium							
phenolicum	Oval	30	–	(+)	C_4	–	Phenol p-cresal, benzoate, glutarate
Desulfonema limicola	Filamentous	30	+	(+)	C_3 through C_{12}	+	Succinate

Symbols : + utilized; (+) = slowly utilized; – = not utilized.

electron acceptor. In the intervening ten years there has been a massive increase in both the umber and the activity of the research groups working with the sulphate reducers. For example, Pfennig and his co-workers, notably Widdel, have isolated and characterized many more new genera and species Table 3.4, cell physiology has been studied by many groups, in particular those of Thauer and Fuchs (H_2 oxidation and acetate metabolism) and of Peck and Le Gall (hydrogenase and electron transport carriers); Jorgensen and his colleagues have examined their ecological impact; Stetter's laboratory has extended the range of sulphur- and sulphate-reducers to include certain of the Archaecobacteria; and the whole has been leavened by the active interest of the oil-gas- and chemical-processing industries as a result of their involvement in problems such as corrosion and reservoir souring.

Recent studies of phylogenetic relationships among the sulphate- and sulphur-reducing bacteria, based on comparative oligonucleotide cataloguing of their 16 S ribosomal RNA, suggest that the Gram-positive sporeforming *Desulfotomaculum* and *Clostridium* genera are, in fact, closely related. The other genera tested represent a single distinctive, but not very coherent, cluster showing some relatedness to the aerobic myxobacteria and *Bdellovibrio; i.e. Desulfuromonas* (sulphur-reducing) and *Desulfovibrio, Desulfosarcina, Desulfonema, Desulfococcus, Desulfobulbur* (all sulphate-reducing). Whereas the majority of sulphate-and sulphur-reducers are mesophilic, there are now examples of thermophilic species being isolated from hot springs, hydrothermal vents and the formation waters associated with petroleum reservoirs. Both eubacterial and archaebacterial species have been recognized with examples of both sulphur and sulphate reduction. Amongst the extreme thermophilic subdivision of the Archaebacteria, *Thermoproteus, Thermodiscus* and *Pyrodictium* are capable of autotrophic growth on H_2, CO_2 and sulphur. Some species of Thermoproteus are facultative, and *Desulphurocoecus, Therofilum and Thermococcus* are other examples of heterotrophic sulphur-reducing themophilic Archaebacteria.

The other main subdivision of the Archaebacteria comprises the methanogens and the extreme halophiles, and here again, within the orders Methanobacteriales, Methanococcales and Methanomicrobiales, there are species demonstrating sulphur reduction, and indeed doing so in preference to methanogenesis in the presence of elemental sulphur. Sulphate reduction has now also been found in an extreme thermophilic Archebacterium, with the capability of growing heterotrophically on a wide range of relatively complex organic nutrients including glucose, peptone and bacterial cell homogenates. This latter property differentiates this organism markedly from the eubacterial sulphate reducers.

It is thus likely that the many and varied bacterial genera showing sulphur or sulphate reduction are examples of convergent evolution. For this reason one must guard against treating them as a uniform group and too freely extrapolating metabolic and energetic findings from one species to another. This warning is particularly apposite to the study of this bacterial group at the present time as, not only is information still fragmentary, but there is already clear evidence of divergent mechanisms of energy, generation and carbon flux in species that might otherwise be thought of as closely related.

However, it has been shown that the competition between sulphate reducing bacteria and methanogenic bacteria is very high. Sulphate reducing bacteria have a higher substrate affinity for H_2 than the methanogenic bacteria.[36]

3.5.5.2 Components of Redox Systems

The sulphate-(and sulphur-) reducing bacteria, unlike perhaps the methanogens, are true anaerobic respires in that substrate oxidation is coupled to the reduction of a terminal electron acceptor through the alternating reduction and oxidation of a chain of redox carriers, at least some of which of are membrane-associated. Although the molecular structure of many of these redox carriers is now known in some detail, knowledge of their structural organization and integrated functioning particularly with regard to the generation of a pmf, remains circumstantial and inconclusive.

The genus *Desulfotomaculum* has membrane-bound cytochrome *b* but no cytochrome c, whereas *Desulfovibrio* and the other Gram-negative sulphate reducers and the sulphur-reducing *Desulfuromonas* have various c-type cytochromes, and possibly also cytochrome *b*, depending on species and growth mode. Menaquinone would seen to be the principal quinone found in all species. Also, when present, the enzymes fumarate reductase nitrate reductase and nitrite reductase are membrane-bound : *Desulfovibrio gigas* and *Desulfuromonas acetoxidans* can grow with fumarate as terminal electron acceptor, while *Desulfobulbus propioniew* and *Desulfovibrio desulfuricans* can reduce nitrate and *D. gigas* reduces nitrite. Although the data are somewhat ambiguous, it is possible also that lactate dehydrogenase may be associated with the inner face of the cytoplasmic membrane.

The redox carriers ferredoxin and, where present, flovodoxin rubredoxin and an octa-haem cytochrome c_3 are all cytoplasmic as are the enzymes APS reductase and the bisulphite reductase.

Uniquely amongst terminal electron acceptors sulphate requires to be activated. This occurs by reaction with ATP and the formation of adenosine phosphosulphate (APS). This reaction, catalysed by ATP sulphurylase, is itself unfavourable thermodynamically and, therefore, has to be coupled to the hydrolysis of pyrophosphate which is catalysed by pyrophosphatase.

$$ATP + SO_4^{2-} \longrightarrow APS + PP_i \xrightarrow[H_2O]{} 2P_1$$

The need for this activation is primarily thermodynamic. The standard redox potential for the sulphate-bisulphite couple is -516 mV and so sulphate itself is quite unable to act as electron acceptor even from H_2 ($2H^+/H_2$, $E_{m.7}$ = –414mV) in an energy-generating reaction. However the redox potential of the APS bisulphite + AMP couple is -60mV. APS is thus the true terminal electron acceptor and its reduction to bisulphite plus AMP is the first step in the 8 electron reduction of sulphate to sulphide :

$$APS + 2[H] \longrightarrow HSO_3 + AMP$$

The net cost of two ATP (the product of APS reductase is AMP, not ADP) means that the 6 electron reduction of bisulphite to sulphide must yield in excess of 2 ATP, presumably by a chemiosmotic mechanism.

It has been suggested that *Desulfotomaculum* species do not carry out electron transport-linked phosphorylation (but see later). It is proposed that sulphate merely stimulates metabolism and facilitates redox balance by acting as an external electron acceptor with all ATP being generated by substrate level phosphorylation. Further, these cells lack high levels of inorganic pyrophosphatase but are said to possess instead pyrophosphate; acetate phosphotransferase and acetate kinase which trap the potential energy of pyrophosphate with the production of 1 ATP.

$$PP_1 + \text{acetate} \longrightarrow P_1 + \text{acetyl } P_1 \xrightarrow[\text{ADP}]{} \text{Acetate} + \text{ATP}$$

Such a reaction sequence would have the effect of reducing the net cost of sulphate activation from 2 to 1 ATP.

There are four bisulphite (sulphite) reductases recognized : desulphoviridin, the principal reductase in the genus *Desulfovibrio,* and also present in *Desulfococcus multivoraus* and *Desulfonema limicola;* desulforubidin, found in some *Desulfovibrio including D. desulfuricans.* Norway 4; desulfofuscidin from *Thermodesulfobacterium commune* and *D. thermophilus;* and P582 from *Desulfotomaculum nigrificans* and *Desulphonema magnom.* All forms of the enzyme have sirohaem (an iron tetrahydroporphyrin) as prosthetic group. It remains unresolved whether the reduction to sulphide is direct or via the trithionite pathway with trithionite and thiosulphate as either free or enzyme-bound intermediates.

Direct evidence for a chemiosmotic mechanism employing a pmf to drive ATP synthesis, reversed electron transport and the uptake of ions and nutrients in the sulphate-reducing bacteria is almost totally lacking. Although such data as are available are entirely consistent with a such a view :

- many species can grow on H_2 as a source of energy or on acetate as a source of energy and carbon, where substrate level phosphorylation is not possible with either substrate;
- H_2 oxidation with nitrite as electron acceptor in *D. desulfuricans* and *D. gigas* or with sulphite in *D. vulgaris* and *D. desulfuricans* has been shown to be accompanied by an uncoupler-sensitive transmembrane proton translocation, in each case with an approximate stoichiometry $\longrightarrow H^+/2e$ of 2;

- whereas the uptake of sulphate, sulphite and thiosulphate in *D. desulfuricans* appears to be by electroneutral proton support, sodium is pumped out of *D. vulgaris* by an electrogenic proton antiport ($H^+/Na > 1$) such that the addition of sodium acetate to resting cells creates an uncoupler-sensitive pmf of 60-90mV (inside negative);
- as discussed above, the presence of dehydrogenase and reductase enzymes and standard redox carriers, and their transmembrane organization is fully consistent with the operation of chemiosmotic energy transduction mechanisms.

3.5.5.3 Hydrogen Metabolism

A major part of our understanding of bioenergetics in the sulphate-reducing bacteria has come from studies of hydrogen metabolism. The ability to oxidize H_2 as a source of energy is widespread amongst all eubacterial and archaebacterial genera of sulphate and sulphur-reducing bacteria, growing either autotrophically or using accetate as carbon source, Table 3.4. At the same time, H_2 is also produced by many species either during fermentative growth in the absence of sulphate or even in the presence of sulphate as terminal electron acceptor this latter is particularly marked during growth on lactate.

Growth on H_2 is taken as incontrovertible evidence for the operation of electron transport linked phosphorylation and thus negates earlier suggestions that all ATP generation in *Desulfotomaculum* is by substrate level phosphorylation. (This is in accord with the presence of cytochrome *b* in this genus and the capacity of *Dt. acetoxidans* to grow on acetate as source of energy and carbon; it does not necessarily rule out the proposed acetyl phosphate pathway, although high levels of the pyrophosphate : acetate phosphotransferase have yet to be unequivocally demonstrated). From growth yield studies of *D. vulgaris* with H_2 as energy source and neither sulphate or thiosulphate (formally and energetically equivalent to sulphite) as electron acceptor, Thauer and his colleagues have deduced that the ATP yield from reduction of sulphate to sulphide is between 1.0 and 1.3, while for sulphite reduction the figures are between 3.0 and 3.5; the difference is due to the ATP cost in the required activation of sulphate to APS. Similar studies with *Desulfotomaculum orientis,* although of a more preliminary nature, also show a higher yield from thiosulphate or sulphite reduction than from that of sulphate but it is not yet possible to draw firm conclusions on either the ATP yields from reduction of electron acceptor, or the ATP cost for sulphate activation in this genus.

Reference has already been made to the ability of sulphate reducers to produce H_2 by fermentation in the absence of an external electron acceptor. Species (and substrates) capable of this mode include : *D. vulgaris* and *D. desulfuricans* (lactate and ethanol); *Desulfotomaculum orientis* and *Dt. nigrificans* (lactate and ethanol); *Thermodesulfobacterium commune* (pyruvate); *Desulfobulbus propionicus* (lactate, pyruvate and ethanol); *Desulfococcus multivorans* (lactate and pyruvate); and *Desulfosarcina variabilis* (lactate, pyruvate and fumarate). A number of studies have shown that, when one of these *Desulfovibrio or Desulfotomaculum* species are grown in co-culture with a H_2-oxidizing methanogen, the fermentation of lactate becomes characterized by the production of methane and acetate. This effect arises from interspecies H_2 transfer with a shift in the fermentation to acetate and H_2 production (exactly as with *Cl. pasteurianium* or *Ruminococcus albus* under reduced H_2 partial pressure), with the latter serving as the substrate for methanogenesis.

Intriguingly, a co-culture involving interspecies H_2 transfer has also been reported where the methanogen is the H_2 donor and the sulphate-reducer is the H_2-oxidizing partner. *D. vulgaris* cannot grow on either methanol or acetate whereas growth of *Methanosarcina barkeri* on these substrates is associated with a build up of H_2. CO-culture of these two organisms in sulphate medium gives substrate utilization with the production of sulphide, along with less methane and more CO_2.

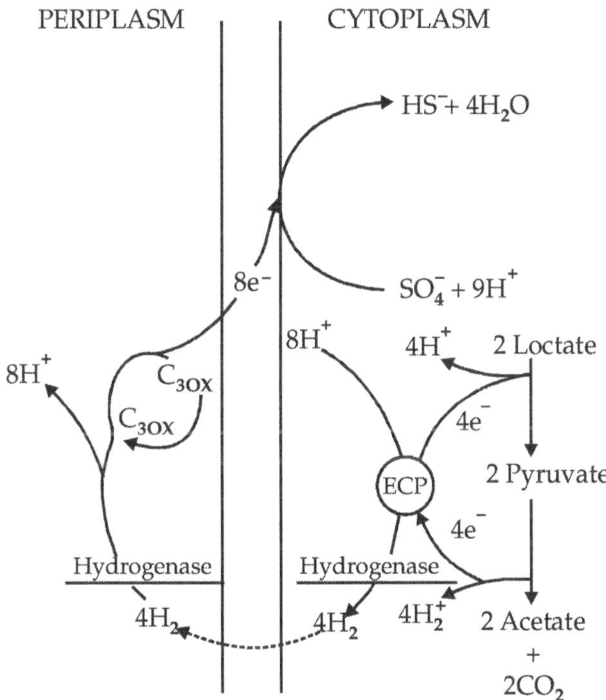

Fig. 3.22 : A schematic representation of enzyme localization, hydrogen cycling and vectorial electron transfer by Desulfovibrio with lactate and sulphate as substrates. ECP, Electron carrier proteins; C_3 cytochrome C_3

These phenomena, and more particularly the observation that the growth of *D. vulgaris* in sulphate medium with lactate (and to a lesser extent with pyruvate) is associated with significant H_2 production, led Odom and Peck to propose the hydrogen cycling model for energy coupling in the sulphate-reducing bacteria, Fig. 3.22. The essential features of the model are that the reducing equivalents from the two substrate dehydrogenation steps should reduce protons to produce H_2 in a reaction catalysed by a cytoplasmic hydrogenase. This H_2 would then diffuse access the membrane and be oxidized by second periplasmic hydrogenase. The specific electron acceptor for this hydrogenase is the low potential tetrahaem cytochrome C_3, which is also found in the periplasm. Transmembrane electron transport would then give both sulphate reduction in the cytoplasm and a transmembrane electrochemical gradient of protons : the protonmotive force. It is envisaged, therefore, that only periplasmic (or extracellular) H_2 oxidation is directly coupled to energy transduction and that organic substrate catabolism is fermentative in character, with ATP synthesis being mediated through an intra species H_2 transfer.

The principal points of evidence in favour of the H_2 cycling model are : the widespread occurrence of periplasmic hydrogenases and cytochrome C_3 in *Desulfovibrio;* the loss of lactate oxidation using sulphate in spheroplasts of *D. gigas* which have lost periplasmic hydrogenase and cytochrome C_3 : some degree of restoration of lactate oxidation by addition of partially purified hydrogenase and cytochrome C_3; and complete reduction of cytochrome C_3 added to *D. gigas* spheroplasts along with lactate and dehydrogenase with reduction being only 60% is sulphate is also present.

However, significant amounts of data are not consistent with the H_2 cycling model. A number of species are capable of growth on lactate and related organic substrates, but do not oxidize H, for example : *D. sapovorans, Desulfococcus multivorans* and a recently described hydrogen-inhibited mutant of *D. desulfuricans.* Further, the model cannot apply to the Grampositive *Desulfotomaculum* since the cell wall structure of this genus precludes a periplasmic space. It has been demonstrated that the H_2- oxidizing hydrogenase of *Dt. orientis* has an intracellular location. While these discrepancies from the model predictions might be considered as only species variations, the alterations proposed for their mechanism of energy coupling during the catabolism of organic nutrients are so fundamental that they could not be expected to be present in one species but absent from a closely related organism, that might even be from the same genus.

Other more serious problems arise from a consideration of the thermodynamics of the reaction sequence constituting the H_2 cycling model. The standard mid-point redox potentials of the pyruvate/lactate (–197mV) and $2H^+/H_2$ (–420V) couples are such that the production of H_2 from lactate dehydrogenation is a highly endergonic reaction. Even with a lactate : pyruvate ratio of 100 : 1 the reaction would only be energy-generating at H_2 partial pressures of less than 10^{-5} atm. Under these circumstances, therefore it would be predicted from the model that lactate oxidation, which might proceed in sulphate medium under a N_2 gas phase, would be inhibited by the addition of H_2. A number of studies have now shown this not to be the case. It is perhaps more reasonable, therefore, to suggest that H_2 production form lactate, during either fermentative or sulphate respiration modes, could arise by energy-driven, reversed electron transport between lactate dehydrogenase and hydrogenase, both of these enzymes being associated with the inner face of the cytoplasmic membrane.

The H_2 cycling model suggests some interesting speculation regarding the growth of *Desulfobulbus propionicus* on propionate (or equally the growth of *D. sapavorans* and other species on butyrate). Reference has already been made to the obligately syntrophic acetogens *Syntrophobacter wolinii* and *Syntrophomonas wolfei,* capable of growth on propionate and butyrate, respectively, only when in co-culture with a H_2 oxidizing sulphate reducer (or methanogen). The question therefore arises whether Dp. propionicus adopts what is formally the same mechanism but within a single organism, by virtue of H_2 cycling or intraspecies H_2 transfer. Since the pathway of propionate catabolism to acetate appears to involve a methylmalonyl CoA : pyruvate transcarboxylase and that part of the TCA cycle from succinyl-CoA to oxaloacetate (Fig. 3.23), the alternative mechanism would involve direct transfer of reducing equivalents to sulphate reduction, with energy-driven reversed electron transport from succinate (fum/succ $E_{m.7}$ = – 33mV). Even if molecular H_2 were to be produced from this dehydrogenation (and from malate to oxaloacetate). The thermodynamic make even more pressing the need to invoke energy-driven reversed electron transport.

($2H^+$ H_2, E_m = – 420mV; HSO_3^- /APS, $E_{m.7}$ = –60m V; HSO_3^-, HS, $E_{m.7}$ = – 116m V).

Whatever the final verdict on the H_2 cycling model, it has undoubtedly stimulated a very significant amount of research into the identification and characterization of hydrogenase enzymes in the sulphate-reducing bacteria. There appear to be three quite distinct periplasmic hydrogenases with high specific activity in the H_2 oxidation assay. *D. vulgaris* has a non-haem iron enzyme with no nickel and a two subunit structure (45.8 KD and 13.5 KD). The *D. gigas* enzyme also has two sub-units (62 KD and 26 KD) and non-haem iron but additionally it has 1 g atom nickel mol^{-1}. On the other hand, the hydrogenase from *D. desulphricans* Norway and the closely related *D. baculatus* is a nickel selenium non-haem iron enzyme.

In line with the predictions of the H_2 cycling mode, the search is on for unequivocal evidence for a cytoplasmic hydrogenase operating in the proton reduction mode. At least two studies have identified membrane enzymes from *D. vulgaris* antigenically different from the periplasmic hydrogenase. In one study three such enzymes were noted, none of which reacted with antibodies raised to the periplasmic enzyme. Two, however showed a degree of cross-reaction with antibodies to the nickel enzyme from *D. gigas* while the third reacted with antibodies to the *D. desulphuricans* nickel selenium hydrogenase. It remains to be demonstrated what are the true physiological functions of these enzymes and how, if at all, they relate to the H_2 cycline model.

The non-heam iron periplasmic hydrogenase of *D. vulgaris* has recently been the subject of some intriguing molecular biological analysis. The gene for the large submit has been cloned, although its expression in *E.coli* gives an enzymically inactive protein. Sequence studies show a considerable degree of homology between the N-terminal end of the molecule and (8Fe-8S) ferredoxin. The gene for the smaller subunit is downstream from that for the larger subunit and it is suggested the two might constitute an operon. Whereas there is no evidence of a signal peptide for the 45.8 KD submit, the mature 13.5 KD protein subunit lacks the hydrophobic N-terminal amino acid sequence encoded by its gene, which has sequence with the general characteristics of a signal peptide. Interestingly there is a hydrophobic region located between residues 99 and 132 in the large submit which has the general properties of an amino-terminal signal peptide; a situation similar to that with ovalbumin, a protein which is also post-translationally translocated across a membrane.

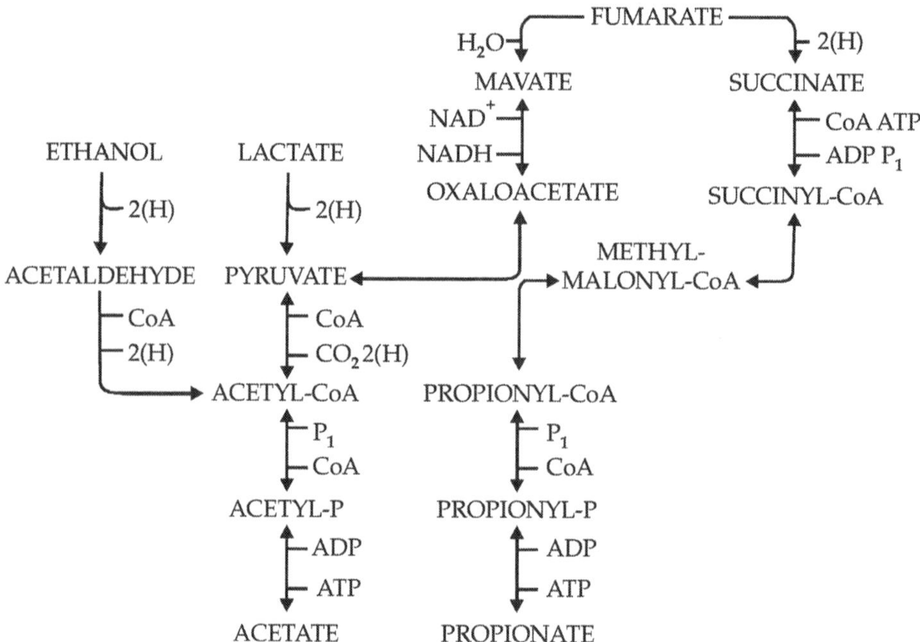

Fig. 3.23 : Proposed Pathway for the Formation and Degradation of Propionate in Dh. Propiosticular.

3.5.5.4 Nutrition and Carbon Flux

Table 3.4 clearly indicates how far our appreciation of the catabolic potential of the sulphate- and sulphur-reducing bacteria has advanced in the last ten or so years. The list of growth-supporting substrates continues to expand, and recent additions include : sugars (sulphate- and sulphur–reducing Archaebacterial thermophiles and *Dt. nigrificans*); methanol (*Dt. orientis* and a recent Gram-negative non-sporing isolate); catechol, resorcinol and hydroquinone (*Desulfobacterium catecholicum*); indole, anthranilate and hydroxy-substituted benzoate and phenylacetate (*Desulfobacterium phenolicum*).

The sulphate- and sulphur-reducing bacteria can be divided into two groups; those species capable of only incomplete substrate oxidation with acetate as a metabolic product, and those capable of complete oxidation to CO_2 with some species able to grow on acetate as the sole source of energy and carbon. It was long considered impossible, on thermodynamic grounds, that the TCA cycle could operate in an energy-generating mode with sulphate as terminal electron acceptor. The problem focused on the dehydrogenation of succinate to fumarate ($E_{m.7}$ = + 33mV). Clearly, the situation is equally adverse with sulphur reduction (S^o/HS-,$E_{m.7}$ = 270mV). Under these circumstances it should not be possible to oxidize acetate as a source of energy for growth. However, the discovery of *Desulfuromonus acetoxidans* and *Desulfotomaculum acetoxidams*, and the identification of acetate as a major respiratory substrate in sulphidogenic ecosystems has forced a radically altered view to prevail.

Enzymic and ^{14}C-labelling experiments have now clearly shown that an oxidative TCA cycle does operate in *Desulfobacter postgatei* and *Desulfuromonas acetoxidans*, with some relatively minor variations.

- Ferredoxin is the cofactor for 2-oxoglutarate dehydrogenase.
- Acetate activation and conversion of succinyl CoA to succinate are catalysed by a single enzyme : succinyl CoA; acetate CoA transferase.
- In *D. postgatei* the succinate and malate dehydrogenases are membrane-bound and menaquinone-dependent, whereas in *D. acetoxiodans* the malate dehydrogenase is NAD-dependent and located in the cytoplasm.
- The glyoxylate cycle is absent from both bacteria and the route of acetate assimilation is by reductive carboxylation to oxaloacetate, with ferredoxin being the electron donor for reductive carboxylation of acetyl-CoA to pyruvate (shown in *D. postagatei*).
- It was deduced for *D. postgatei* that the succinate to fumarate reaction is most likely to require energy-driven reversed electron transport, with the menaquinoe ($E_{m.7}$ = – 74mV) coupling of malate dehydrogenation rendering that reaction virtually irreversible and thus having the effect of 'pulling' the succinate dehydrogenation. ATP-dependent, and uncoupler and DCCD-sensitive dehydrogenation of succinate to fumarate has now been directly demonstrated in membrane preparations of *D. acetoxidans* with either sulphur or NAD as electron acceptor.

It turns out, however, that these are not the most common routes of acetate oxidation and assimilation amongst the sulphate-reducing bacteria. Other species, including *D. baarsii, Dt. acetoxidans, Desulfosarcina variabils, Desulfobacterium autotrophicum, Desulfococcus niacini* and *Desulfococcus multivorans,* lack 2-oxoglutarate dehydrogenase and a complete TCA cycle although they do use the reactions to 2-oxoglutarate for glutamate and amino acid synthesis. What they do possess is the enzyme carbon monoxide dehydrogenase and the acetyl-CoA pathway appears to be used for acetate oxidation and, where operative, CO_2 fixation. It is significant that the standard mid-point redox potential of the most positive step in the acetyl-CoA pathway (methyl-THF to methylene THF) is-117 mV which could thus readily be coupled to APS and bisulphite reduction in an exergonic reaction sequence.

Although *D. vulgaris* has also been shown to have an active carbon monoxide dehydrogenase, its physiological role remains unclear as *D. vulgaris* lacks the acetyl-CoA pathway and cannot oxidize acetate as a source of energy.

3.5.5.5 Sulphite Dismutation

Recently, a completely novel mechanism of energy generation has been reported in a newly identified isolate; *D. sulfodismutans.* The reaction sequence has been called an inorganic or chemolithotrophic fermentation since it involves the coupled oxidation and reduction of either sulphite or thiosulphate, according to the equations :

$$4SO_3^{2-} + H^+ \longrightarrow 3SO_4^{2-} + HS^-; \Delta G° = -58.9 \text{ kJmol}^{-1} \text{ sulphite}$$

$$S_2O_3^{2-} + H_2O \longrightarrow SO_4^{2-} + HS^- + H^+ \; \Delta G° = -21.9 \text{ kJ mol}^{-1} \text{ thiosulphate}$$

The mechanism of energy conservation remains unclear at present.

3.6 Anaerobic Microbial Consortia

A constantly recurring theme in this presentation of the bioenergetics of anaerobic microorganisms has been the relatively restricted metabolic potential of individual species, with the consequent interdependencies between species. Our consideration of the acetogens and methanogens has noted in particular the key roles taken by H_2 and acetate in these interdependencies. It can now be seen that, depending on species and conditions, the sulphate-reducing bacteria can act both as H_2–producing acetogens and as terminal oxidizers paralleling the methanogens. Any analysis of anaerobic microbial consortia, therefore, must be constantly aware of the interactions between individual components of the community, and expect the sulphate reducers to be of central importance in sulphate-containing environments. These considerations are not minor or confined to a few laboratory models. They are, in fact, the dominant and controlling features of a host of naturally occurring and economically important microbial ecosystems such as the rumen (methanogenic), anaerobic digesters (methanogenic), polluted estuarine or marine sediments (sulphidogenic) and microbial biofilms associated with metal corrosion (sulphidogenic.) A striking feature of such consortia is that, while they can be characterised in terms of the principal terminal oxidant species (methanogen or sulphate reducer), the primary nutrient for the consortium is invariably a substrate or substrates to which the methanogenic or sulphate-reducing bacteria are themselves totally unreactive. They have an absolute requirement for the heterotrophic and fermentative organisms to perform the initial partial decomposition, and often also to create anaerobic conditions within the consortium through reacting with any available oxygen. At the same time, the methanogens, or sulphate-reducers stimulate the overall conversion of substrate by oxidative removal of the products of fermentation.

In view of the central metabolic importance of H_2 and acetate, the acetogenic bacteria often act as intermediaries between the initial fermentative reactions and the terminal oxidative steps : As well as stimulating the degradative capacity of microbial ecosystems in this manner, significantly a major proportion of the energy input to the systems remains in the form of the reduced terminal electron acceptor. In a very real sense, methane and H_2S are extracellular energy currencies that are available to other micro-organisms, perhaps after diffusion to an aerobic micro-environment. Primary nutrient sources that are biodegraded in this manner-include complex biopolymers such as cellulose, hydrocarbons, and more or less recalcitrant molecules such as aromatics, phenols and sugars. Although individual sulphate-reducing species have recently been isolated that are capable of reacting with some of these substrates directly, their quantitative importance in most natural ecosystems remains to be established.

It is generally accepted that in sulphate-containing environments the terminal oxidant species will be sulphidogenic rather than methanogenic. This was previously thought to be due to a sulphide toxicity effect, but this is no longer considered a likely explanation. In fact, the production of sulphide is likely to be beneficial to both groups of organisms because of its effect of making the environment more reducing. It has now been shown that the sulphate reducers have higher affinities than the methanogens for both the major substrates for which they compete. The K values for H_2 for the sulphate reducers and methanogens are, respectively, around 2µM and between 6 and 20µM. For acetate the figures are approximately 200 µM and 3 µM respectively.

In a number of environments evidence has been obtained that methanogenesis and sulphidogenesis proceed simultaneously. Methanogenesis in these habitats is due to the methylotrophic methanogens (oxidizing methanol and the methylamines) that do not compete with the sulphate reducers for H_2. There have been recent reports of methanogenesis and sulphidogenesis in the non-traditional province for them of the open ocean; presumably this involves anoxic microparticulates. As understanding of these intriguing groups of anaerobic bacteria increases, along with improved techniques for enrichment, isolation and growth, one might reasonably expect such reports to be substantiated and extended.

3.7 Anaerobic Respiration : Reduction of Fumarate and Nitrate

Where fumarate or nitrate operate as terminal electron acceptors rather than CO_2 or sulphate, the differences in cell physiology are qualitatively major and not simply restricted to the quantitative effects resulting from the more positive mid-point potentials of the fumarate-succinate and nitrate-nitrite redox couples:

- The organisms capable of these activities are facultative anaerobes and will always use oxygen preferentially as terminal electron acceptor, fumarate and nitrate are therefore genuine alternatives to oxygen.
- A wide spectrum of bacterial genera have the capacity for fumarate and or nitrate respiration.
- At the cellular level, the mechanism of these energy conserving processes requires only minimal modification of the normal oxygen linked respiratory pathways. Consequently, the major thrust of research has been directed at elucidating the details of such modifications, and in the process much seminal information has been gained on the structural and functional bases of membrane-dependent energy conservation mechanisms in general.
- Cell-cell interactions and implications for microbial ecology are much less to the fore as, in terms of energy yield and consequent carbon flux, each organism is capable of being self-sufficient.

3.8 Anaerobic Biodegradation of Some Specific Type of Organic Compounds

This section places major emphasis on treatment of EPA priority pollutants, Table 3.5. Simple wastes like sewage, dairy, sago, etc. have not been covered in this section. Ratio of BOD/COD (R) defines the biodegrability of the wastewater, *viz.*

R = > 0.3; Easily biodegradable

0.20 – 0.25; Some acclimation is necessary

< 0.2 non-biodegradable

The range of compounds covered is divided into the following categories :

- aromatic compounds,
- pharmaceutical wastes,
- pesticides,
- other priority compounds,
- other selected organic compounds,
- compounds nonamenable to biodegradation and
- soaps and detergents.

Table 3.5 : Priority Pollutants Established by the EPA (1980)

Chlorinated Alkanes	*Chlorinated Ethers*	*Phenols*	*Pesticides*
Methyl Chloride	bis(Chloromethyl)Ether	Phenol	Aldrin
Methylene Chloride	2-Chloroethyl Vinyl Ether	2-Chlorophenol	Diedrin
Methyl Bromide	4-Bromophenyl Phenyl Ether	2, 4-Dichlorophenol	Chlordane
Chloroform	bis(2-Chloroethox)methane	Pentachlorophenol	4,4-DDT
Bromoform	bis(2-Chloroethyl)Ether	2-Nitrophenol	4,4-DDE
Carbon Tetrachloride	4-Chlorophenyl Phenyl Ether	2,4-Dimethylphenol	4,4-DDD
Dichlorobromomethane	bis(2-Chlorophenyl)Ether	4-Nitrophenol	a-Endosulfan-alpha
Trichlorofluoromethane		2,4-Dinitrophenol	b-Endosulfan-beta
Dichlorodifluoromethane	**Aromatics**	4,6-Dinitro-o-cresol	Endosulfan Sulfate
Chlorodibromomethane	Benzene	2, 4, 6-Trichlorophenol	Endrin
Chloroesthane	Toluene	Para-Chlorometa-Cresol	Endrin Aldehyde
1,1-Dichloroethane	Ethylbenzene		Heptachlor
1,2,-Dichloroethane	Napthalene	**Substituted Aromatics**	Heptachlor Epoxide
1,1,1-Trichloroethane	Fluoranthene	Nitrobenzene	α-BHC-Alpha
1,1,2-Trichloroethane	Acenaphthene	2, 4-Dinitrotoluene	β-BHC-Beta
1,1,2,2,-Tetrachloroethane	Benzo(a) Anthracene	2,6-Dinitrotoluene	ρ-BHC (Lindane)-Gamma
Hexachloroethane	Benzo(a)Pyrene	2,3,7,8-Tetrachlorodi	δ-BHC-Delta
1,1-Dichloroethylene	Chrysene	benzo-p-Dioxin	Toxaphene
1,2-trans-Dichloroethylene	Indeno(1,2,3-c,d)pyrene	Benzidine	
1,2,-Dichloropropane	3,4, Benzofluoranthene	3,3-Dichlorobenzidene	
1,2-Dichloropropylene	Benzo(K)fluoranthene	1,2-Diphenyl Hydrazine	
Trichloroethylene	Acenapthylene		

Contd...

Chlorinated Alkanes	*Chlorinated Ethers*	*Phenols*	*Pesticides*
Tetrachloroethylene	Benzo(gh1)perylene	**Polychlorinated Biphenyls**	**Metals**
Vinyl Chloride	Fluorene	PCB-1242 PCB-1221 PCB-1260	Antimony
Hexacholorobutadiene	Phenanthrene	PCB-1254 PCB-1232 PCB-1232	Arsenic
Hexachlorocyclopentadiene	Dibenzo(a, h anthracene Pyrene	Miscellaneous Acrolein	Beryllium Cadmium Chromium
Chlorinated Aromatics			
1,2,4-Tricholorobenzene	**Phthalate Esters**	Acrylonitrile	Copper
Chlorobenzene	bis(2-ethylhexy)phthalate	Asbestos	Lead
Hexachlorobenzene	butyl benzyl Phthalate	Cyanide	Mercury
2-Chloronaphthalene	Di-n-butyl Phthalate	Isophorone	Nickel
1,2-Dichlorobenzene	Di-n-octyl Phthalate	N-Nitrosodimethylamine	Selenium
1,3-Dichlorobenzene	Diethyl Phthalate	N-Nitroiodiphenylamine	Thallium
1,4-Dichlorobenzene	Dimethyl Phthalate amine	N-Nitrosodi-n-propyl	Zinc

3.8.1 Aromatic Compounds

The aromatic compound compose a wide variety of substances characterized by the presence of at least on benzene ring. Members of this group include phenols, chlorinated aromatics, and substituted aromatics. 41 out of 129 priority compounds listed in Table 3.5 are from this group. Many aromatic substances are toxic in small amounts, and many are carcinogenic. The use of aromatic compounds is wide spread in industry and they appear as constituents in wastewater originating from a number of industrial sources including coal gasification plants, oil refining, plastics, petrochemicals, pharmaceutical and paper mills. Many of these industries emit waste streams which are comprised of a mixture of aromatics, some of which have been found to be inhibitory to aerobic biological reactions[37].

This aspect makes anaerobic processes an attractive area to explore. It is known that anaerobic breakdown of the benzene nucleus can occur by two different pathways; they are :

- photometabolism
- methanogenic fermentation

3.8.1.1 Anaerobic Photometabolism of Aromatic Compounds

Procter and Scher[38], first to report first to report the isolation of a group of photosynthetic bacteria of the genus Rhodo-pseudomonas which were capable of metabolising benzoate anaerobically. Using both resting cell suspensions and cell extracts, they analysed for metabolic intermediates and from then data proposed the following pathway, with the implication that molecular oxygen could be used to replace the photochemically generated oxident :

$$\text{Benzoate} \longrightarrow \text{Protocatechuate} \longrightarrow \text{Catechol} \longrightarrow \alpha - \text{oxoacid}$$

Dutton, and Evans[39] challenged some conclusions of Procter, *et al.* based on research conducted with cultures of *Rhodopseudomonas plaustris*. It was shown that no significant oxygen uptake occurred when *R. patustris* was grown on benzoate and attempts to *R. patustris* under aerobic conditions with benzoate as a substrate resulted in failure. In addition, it was found that the direct addition of air into an actively photometabolising benzoate culture in the cessation of benzoate utilization.

Further studies by Dutton, *et al.*[39] resolved the question of how the anaerobic metabolism of benzoate occurred. Since the reaction was anaerobic, the use of an oxygenated ring cleavage was not plausible. The only other feasible biochemical option available for ring fissure would be hydrogenation or hydration. Proceeding under the assumption that hydrogenation was responsible, the expected intermediates were incubated with cells actively photometabolizing labelled ($O^{14}C$) benzoate. A pathway was proposed, which consisted of a reduction of benzoate (and derivatives) to cyclohexane-carboxylate derivatives followed by a coenzyme–A mediated beta oxidation, Fig. 3.24, this, by hydration results in ring cleavage forming the compound pimelate. It was posed that the reaction is catalyzed by reductases coupled to the redox potential component of the light induced electron transport system and it was further suggested that the compound ferredoxin served this function.

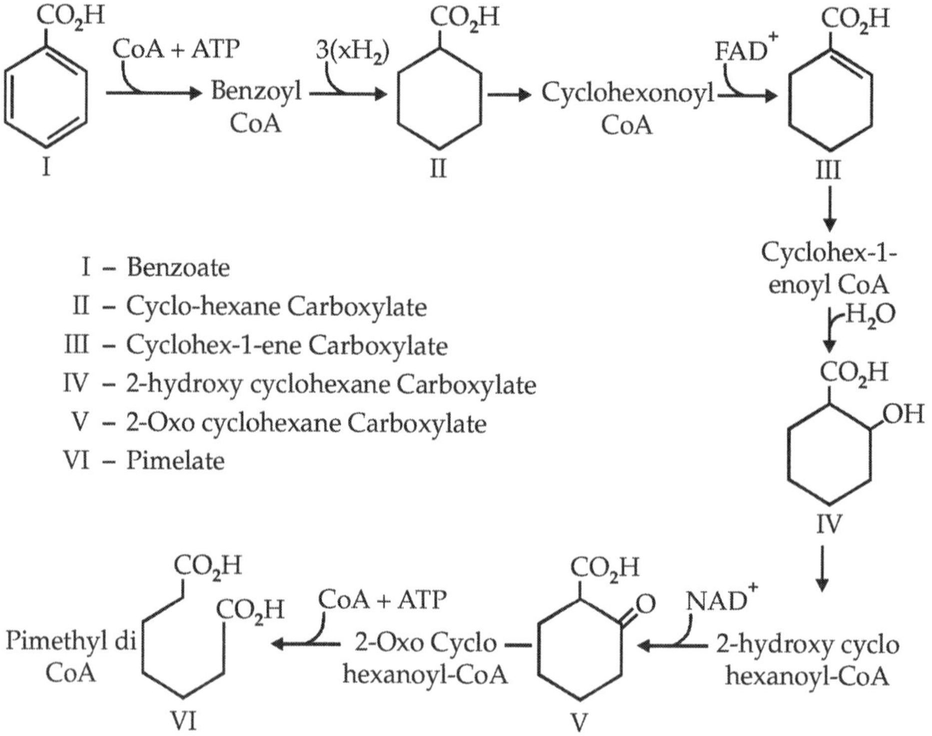

CO_2H I → CoA + ATP → Benzoyl CoA → $3(xH_2)$ → CO_2H II → Cyclohexonoyl CoA → FAD^+ → CO_2H III → Cyclohex-1-enoyl CoA → H_2O → CO_2H OH IV → 2-hydroxy cyclo hexanoyl-CoA → NAD^+ → CO_2H O V → 2-Oxo Cyclo hexanoyl-CoA → CoA + ATP → CO_2H CO_2H VI Pimethyl di CoA

I – Benzoate
II – Cyclo-hexane Carboxylate
III – Cyclohex-1-ene Carboxylate
IV – 2-hydroxy cyclohexane Carboxylate
V – 2-Oxo cyclohexane Carboxylate
VI – Pimelate

Fig 3.24 : Proposed Pathway for Reduction of Benzoate

3.8.1.2 Methanogenic Fermentation of Aromatic Compounds

Early evidence that biological destruction of benzoid structures in strictly anaerobic conditions was possible was provided by Travin and Bushwell[40]. Unit this time no evidence existed that the ring structure could be broken by anything other than oxygenating ring cleavage. Using a sewage sludge inoulum, it was shown that benzoate, phenylacetate, phenylpropionate, and cinnamate were completely broken down to CO_2 and CH_4. Lower fatty acids were detected as reaction intermediates. The amount of gas production was as high as 90 per cent of the theoretical yield. Extremely long acclimation times, some as long as 5 weeks, were required before gas production was initiated. This time period was found to be shortened by adapting the bacteria to an acetic acid substrate before adapting them to benzonic acids. Gas analysis showed that hydrogen gas was not present in appreciable amounts, indicating that the methane gas is formed by some mechanism other than CO_2 reduction. In an attempt to determine if either catechol or protocatechric acid functioned as an intermediate, both were fed to the stabilized benzoate utilising cultures. Cessation of gas production suggested that the pathway for anaerobic degradation of benzoic acid is different than the aerobic pathway proposed by standier[41–43].

Conclusive proof of production of CH_4 can CO_2 was provided by Fina *et al.*[44] Using radio isotopes they formed that the composition of the gases evolved under anaerobic conditions agreed with the following equation.

$$4C^{14}{}_6\,H_6CO_2 + 18H_2 \longrightarrow 15C^{14}H_4 + 9C^{14}O_2 + 4CO_2$$

Chemielowskii, *et al.*[45-47] conducted experiments on the decomposition of p-cresol, phenol and resorcinol. Using batch reactors with 100 mg/L of respective phenolic compounds, mixed populations of bacteria yielded complete conversion to CH_4 can CO_2. Analysis of accumulated intermediates revealed them to be saturated compounds and not Volatile acids as one would expect. It was, therefore, postulated that degradation of phenols proceeded first by hydrogenation of the aromatic structure to form alicyclic intermediates. These intermediates underwent ring fission, followed by the methanogenesis of the degradation products.

Keith[48] using isotopic trapping experiments identified cyclohexane carboxylic acid, 1-Cyclohexane-1-Carboxylic acid, heptanoate, valerate, butyrate, propionate, and acetate as reaction intermediates in the degradation of benzoic acid. Based on these results, he proposed the pathway shown in Fig. 3.25.

Studies by Ferry, *et al.*[49] enhanced the understanding on the degradation of benzoate. From actively degrading benzoate cultures they isolated three types of bacteria; *Methanobacterium formicum, Methanospirillum bungati* and an unidentified acetate degrader. *M. formicum* and *M. bungatii* were tested to determine, if either could actively degrade benzoate in pure culture. The negative results confirm Chemielowski's findings and suggest that the function of the methanogenic bacteria is not in the cleavage of the aromatic ring. Instead, they serve as terminal organisms of the food chain and produce methane from the intermediates formed by the breakdown of benzoic acid by some unknown bacteria.

Further evidence of this conclusion is provided by the fact that addition of Orchlorobenzoate inhibited benzoate degradation, while having no effect on methane production. The unidentified acetate degrader was found to increase its population significantly when acetic acid was added. Using the data based on rate of appearance and disappearance of the intermediates and the formation of products using labelled C^{14}, Ferry, *et al.* proposed this reaction scheme for the overall conversion of benzoate to methane :

- Degradation of benzoate to acetate, formate and H_2
- $4C_7\,H_6\,O_2 + 24H_2O \Leftrightarrow 12C_2H_4O_2 + 4CH_2O_2 + 8H_2$
- Conversion of acetate to methane and CO_2

$12C_2H_4O_2 \Leftrightarrow 12CH_4 + 12CO_2$

- Conversion of Formate to CO_2 and H_2

  $4CH_2O_2 \Leftrightarrow 4CO_2 + 4H_2$

- Reduction of CO_2 with H_2

  $12H_2 + 3CO_2 \leftrightarrow 3CH_4 + 6H_2O$

- Overall reaction

  $4C_7H_6O_2 + 18H_2O \leftrightarrow 15CH_4 + 13CO_2$

Evans[50] in a review article an biochemistry of aromatic compounds in anaerobic environments proposed the pathway shown in Fig 3.26 for the degradation of benzoate to methane and CO_2. Evans believed that the occurrence of heptanoic acid can be explained by something other than the reaction which has no bio-chemical precedent. The proposed intermediate for the precursor of heptanoic acid is I, methylcyclohexanone, which is converted to heptanoate by the addition of water.

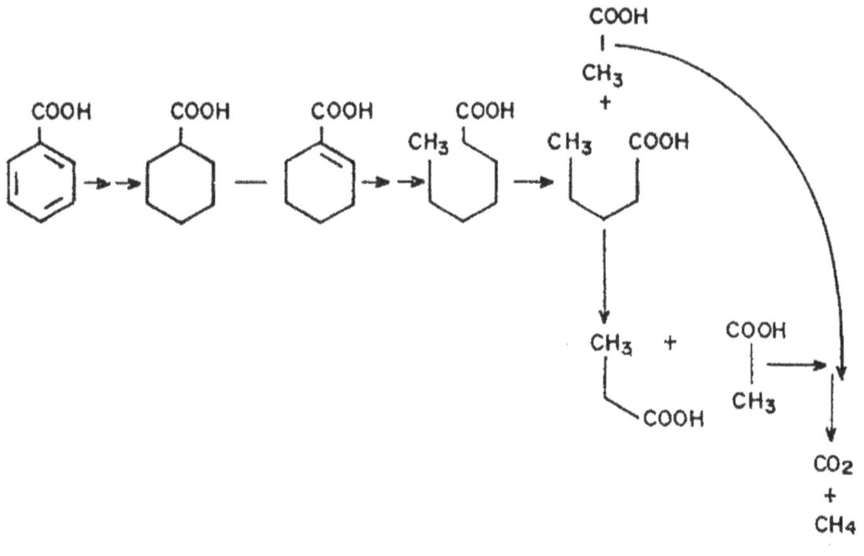

Fig 3.25 : Proposed Pathway for Degradation of Benzoate Acid to Methane and Carbon Dioxide

Neufeld[51] based on the work of Chemielowski and Williams, et al.[52], proposed a pathway for the degradation of phenols and benzoic acid, Fig 3.27. Voets, et al.[53] studied the breakdown of antimicrobial agents, widely used in hospitals and industry. Employing batch reactors operated on either mineral or organic medias, the fours antimicrobial phenolic compounds O-phenyl phenol, p-chloro-m-cresol 5, 5′ dichloro 2, 2′-dihydroxy-diphenyl methane and 5-chloro-2(2, 4 dichlorophenoxy) phenol, were tested for susceptibility to bacterial degradation. Of these four compounds, *O*-phenyl phenol at concentrations of 10 and 40 mg/L was the only one to undergo 100 per cent degradation in both the mediums. The compounds *p*-chloro *m*-cresol and 5, 5′ dichloro-2 and 2′-dihydroxy diphenyl-methane were not degraded at all, in either media. The compound 5-Chloro-2 phenol was degraded by 50 per cent in organic medium, but was not reduced to any detectable extent in mineral media.

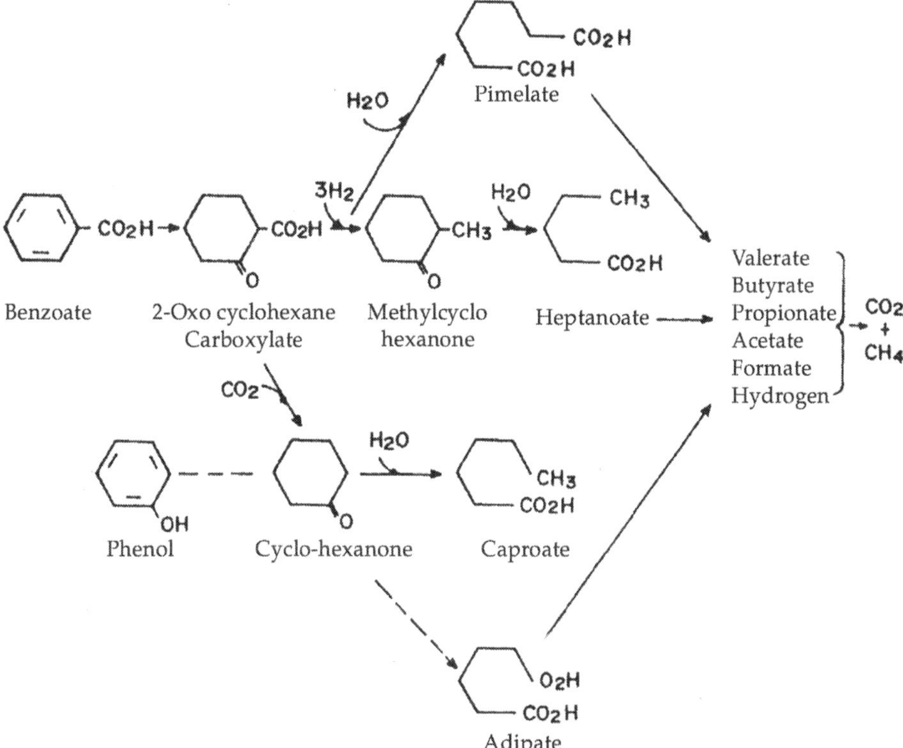

Fig 3.26 : Proposed Pathway for Degradation of Benzoate and Phenol.

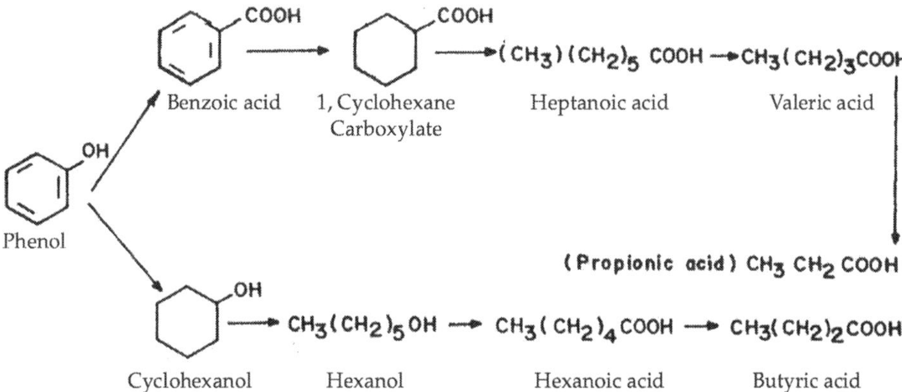

Fig 3.27 : Proposed Pathway for Phenol and Benzoic Acid Degradation

Healy and Young[54] presented evidence that the aromatics catechol and phenol could be degraded by a population of anaerobic bacteria to produce methane and CO_2. The reactions can be stoichiometrically described as :

Phenol : $C_6H_6O_2 + 4H_2O \longleftrightarrow 2.5\ CO_2 + 3.5\ CH_4$

Catechol : $C_6H_6O_2 + 3.5\ H_2O \longleftrightarrow 2.75CO_2 + 3.25\ CH_4$

The amount of gas produced in these experiments was approximately 79 per cent and 85 per cent of the theoretical gas production for catechol and phenol respectively.

Healy and Young[55] showed that eleven aromatics expected to be released in the disposal of a natural refractory compound ligand by heat treatment could be degraded by anaerobic processes. More than 80 per cent conversion of substrate carbon into gas was possible. Suidan, *et al.*[56-57] utilized an anaerobic granular activated carbon filter to simulate treatment of wastewater bearing catechol, *O*-cresol, and a mixture of *O*-cresol and glucose. In all cases the bacteria were able to degrade the above compounds. Kirsch, *et al.*[58] demonstrated the use of an anaerobic semi-continuous stirred tank reactor in the treatment of pentachlorophenol (PCP) at low concentration (less than 10 mg/L). The PCP was degraded anaerobically.

3.8.2 Pharmaceutical Wastes

Wastes from pharmaceutical manufacturing operations were ideal for treatment by anaerobic processes due to high COD. Jennett, *et al.*[59] employed unflow anaerobic filters in the treatment of a pharmaceutical waste. The influent waste had a high COD value, attributed to the presence of methanol in the wastes. Removal efficiencies were reported to be between 93 and 98 per cent, at a loading rate of 0.25 to 7 kg COD/m^3. d. Although, it was admitted that methanol is a readily degradable substance, some important results were attained. Attempts to disrupt filter performance by addition of sharp increases in the organic loading rate resulted in failure. Probably the most significant result of this experiment was the demonstration that solids were kept in the system at a manageable level, with little loss to the effluent.

Such's *et al.*[60] used an anaerobic filter in the treatment of wastes from a synthesized organic manufacturing plant. The filters were started on a methanol feed solution containing 2000 mg/L COD. After steady state conditions were reached, the feed was switched to the plant waste diluted down to 2000mg/L COD, at an organic loading rate of 0.5 kg COD/m^3d. At steady state conditions the COD removals ranged from 70 to 80 per cent. A three fold increase in the influent COD concentration resulted in disruption of filter performance, evident by a 18 per cent reduction in COD removal. When operated at the 200mg/L COD level the filters provided 33 per cent better removal of kg COD/m^3.d and gave less coloured effluent than the existing treatment system.

3.8.3 Pesticides

The fact that pesticide residue persist in the environment is well-established. Numerous cases exist where residues have been detected more than a decade after application. These residues are toxic to many organisms and have a potential for bioaccumulation. Many scientists believe that if a compound is resistant to aerobic condition, it will not degrade at all under anaerobic condition. The work on date on anaerobic breakdown of pesticides would lead us to think otherwise. It is well documented that anaerobic conditions existing in flooded rice fields facilitates degradation of pesticides better than aerobic conditions. Studies dealing with specific pesticides are described below :

The organochlorine insecticide 1, 2, 3, 4, 5, 6 hexachlorocyclohexane (γ-HCH) better known as lindane, gacutin, gammexane, or γ-benzene hexachloride (γ-BHC) and its related isomers α-BHC, β-BHC, δ-BHC, are on the EPA priority pollutants list, of the above only the γ-BHC isomer is effective as an insecticide. The other isomers, α-BHC, β-BHC, and δ-BHC, are formed in the manufacturing of γ-BHC and have no value in pest control. They are very resistant to biological breakdown, and yet they have been applied to agricultural areas in quantity in un-refined Kindane. Many countries including Japan, have banned the use of lindane altogether.

Hill *et al.*[61] in their study on anaerobic degradation of chlorinated pesticides, found that under anaerobic conditions at 35°C lindane degraded much more rapidly than under aerobic conditions. However, at 20°C, this difference was less pronounced. Prior to this, it was thought that none of the chlorinated hydrocarbons could be used as a carbon source for microbial metabolism. It was observed in the course of these experiments that degradation best occurred under conditions of high biological activity. Reaction rates appeared to be a function of the concentration of sludge and lindane. At high concentrations of both sludge and lindane the reaction rate followed first order kinetics, but when diluted solutions of sludge were used at low concentrations of lindane a reaction rate of zero-order was observed.

MacRae, *et al.*[62] studied the degradation of all four isomers of benzene hexachloride in submerged soils columns. Since any available O_2 present in the soil was soon utilized, the submerged soil quickly became anaerobic. Applications of the four isomers to the tubes were carried out at three times the expected dosage of the commercial applications. It was found that no evidence of prolonged existence of any isomer existed, when the columns were analysed for benzene hexachloride. All the four isomers were degraded at approximately the same rate, and degradation was faster in unsterilized soil than in sterilized soil, again suggesting the involvement of biological activity. To confirm these result C^{14} labelled γ-BHC was introduced into the soil and the soil surface measurement for release of labelled CO_2. The results obtained demonstrated that biological activity was indeed taking place, resulting in release of labelled CO_2.

MacRae, *et al.*[63] in further experiments isolated a bacterium identified as *Clostridium* sp. That was capable of degrading lindane in phosphate buffer using the lindane as the sole carbon source. It was shown that an initial concentration of 3.67 mg/L of lindane could be degraded to 0.02 mg/L in 27m hours. Ability of the organisms to degrade lindane was also provided by the release of the covalently linked chlorine of lindane as free chlorine ion. For the some of lindane degraded, 75 per cent of the theoretical amount of chloride expected to be released was measurement in the culture.

Sethunathan *et al.*[64] also worked with a species of *Clostridium*, in an attempt to identify degradation intermediates. The compound γ-penta chlorocyclohexene, a probable intermediate of the direct dehydrochlorination of γ-BHC, was not detected. This indicated that another mechanism was operating. Based on the similarity of the factors affecting lindane and DDT degradation and the fact that the *Clostridium* would convert DDT to DDD by reductive chlorination, it was suggested that γ-BHC is degraded by a similar pathway. However, scientific data to support this hyphothesis is unavailable.

Newland, *et al.*[65] obtained evidence indicating that degradation of lindane occurs faster under anaerobic conditions than under aerobic conditions. In simulated lake impoundments, anaerobic degradation of lindane was more complete (about 90 per cent Vs 15 per cent) and began earlier (264 hr vs 840 hr) than aerobic) degradation. In attempts to define degradation products, gas-liquid and thin layer chromatography were performed as samples from the impoundments. Both methods detected the presence of the α and δ isomers of BHC as the products of a biological meditated isomeric conversion. While this result was unexpected, it is acceptable on the grounds of thermodynamics since the γ isomer is the least thermodynamically stable of the four isomers (γ-BHC< α-BHC < δ-BHC < β-BHC). Since, in this reaction, no carbon of BHC is utilized, the reaction appears to be of no metabolic value. Instead this conversion may serve as a response by the organisms to reduce the toxicity in its environment.

Jagnow, *et al.*[66] performed screening studies on selected species of bacteria and recorded their capability to degrade γ-BHC anaerobically. The results are presented in Table 3.6. It is evident that Clostridia were highly active with some members of Bacillaceae and Enterobacteriacea also performing quite well. Nearly complete dechlorination in 4-7 days at a concentration of 2mg γ - BHC per 200mt of medium was observed for the three species, *Clostridium butyricum, Clostridium pasteurianum* and *Citrobacter freundii*. In addition tri and tetrachloro-benzenes were decided but not in sufficient enough quantities as major metabolites in the destruction of γ-BHC.

DDT (dichlorodiphenyl-trichloroethane) has the chemical formula 1, 1, 1, trichloro 2-2-bio-(4-chlorophenyl) ethane. Because of its highly intrinsic insecticidal value it has found use in the control of a variety, of pests. DDT is known to persists for many years in soil. Past evidence has decisively shown that DDT is degraded faster under anaerobic conditions than aerobic condition, and the degree of degradation is 67-69 inversely proportional to the amount of oxygen in the system. The extent of degradation is slight, with most of the DDT being reductively dechlorinated to DDD which is somewhat less toxic, but more bio-resistant than DDT, Fig. 3.28. Castra, *et al.*[67–69] have found DDD will degrade in flooded fields but at a much slower rate.

Table 3.6 : Results of Screening Studies of Selected Bacteria for their Capability to Degrade γ-BHC under Anaerobic Conditions

Species	*Strength**	*Species*	*Strength**
Clostridium butyrium	+++	*Citrobacter freundii*	+++
C. pasleruriumum	+++	*E. coli*	++
Bacillus polymyxa	++	*Enterobacter aerogens*	++
B. macerans	++	*E. cloacae*	+
B. laterosporous	+	*Serratia marcescens*	+
		Proteus mirabilis	+
Bacillus alvei	±	*Proteus vulgaris*	±
B. circulans	±	–	
B. brevis	–	*B. lentus*	–
B. cereus	–	*B. licheniformis*	–
B. coagulans	–	*Leuconostoc dextranicus*	–
Lactobacillus casei	–	*L. mesenterides*	–
L. plantarum	–	*Propisnibacterium*	–
L. brevis	–	*Shermanii*	–
Paracoccus denificans	–		

* +++, ++, - = Strong to moderate degradation
± = Ambigious degradation
- = No degradation

Guenti, *et al.*[70] first provided unequivocal evidence than DDT is degraded faster under anaerobic conditions than aerobic conditions. In an anaerobic soil system, the conversion of DDT to DDD was achieved, with ultimate minute traces of six other degradation products. Addition of organic material in the form of alfalfa increased the conversion rate. With alfalfa present, less than one percent of the original amount of DDT applied to soil was recovered after 15 weeks of incubation. The aerobic soil system on the other hand, showed 75 percent of the added DDT after a 6 months incubation period. Later research by Cuenzi, *et al.*[71] showed that other organic substrate as rice, straw, and cellulose can also enhance the conversion o DDT to DDD in anaerobic soils. Organic material added to non-flooded fields under aerobic conditions has no effect on this conversion.

The ability to degrade DDT to DDD has been demonstrated by many organisms. Previous research has shown that *E. coli, Aerobacter acrogenes*[72], *Serratia marcescens*[75] and *Proteus vulgaris*[74] all are capable of carrying out this conversion. Enzymes involved in this reaction have not been identified as yet, but the site of this reaction in *E. coli* was found to be in the membranous fraction of cells[75].

Besides lindane and DDT, most other organo-Chlorine pesticides are thought to by fairly stable. In some instances, however, they have been shown to undergo degradation by anaerobic processes. Hill, *et al.*[61] tested the compounds aldrin, dieldrin, heptachlor endrin and heptachlor epoxieldrin in biologically active anaerobic sludge. It was found that heptachlor and aldrin were degraded at a temperature of 35°C to unidentified products. Castro. *et al.*[68] showed that methoxychlor and helptachlor degraded faster in anaerobic conditions than aerobic conditions. It was shown that this degradation was the result of biological activity and not volatilization. The pesticides aldrin, endrin, dieldrin and chlordane were shown to be relatively resistant under both aerobic and anaerobic conditions.

3.8.4 Other Priority Pollutants

Bouwer *et al.*[76] studied the anaerobic breakdown of two concentrations of the priority pollutants chloroform, trichloroethylene, tetrachloroethylene, and bromodichloromethane. Another dibromochloromethane was also studied. It was found that chloroform degradation occurred completely at starting concentration of 16 and 34 μg/L. An initial concentration of 157 μg/L resulted in less pronounced degradation and occurred very slowly. Degradation of chloroform was only observed in cultures seeded with methanogenic bacteria. The authors proposed a possible degradation route consisting of both biological and chemical mechanisms. Chloroform combines with a suphoydryl group from the sodium sulphide (added to the media to maintain reducing conditions), forming a thiol. This resulting thiol is then attached by micro-organisms and degraded further.

Both trichloroethylene and tetrachloroethylene showed no conclusive evidence that degradation took place. This trend was observed at all starting concentrations. While it appeared that a small amount of decomposition was occurring, the results were not conclusive. Degradation of bromodichloromethane at initial concentrations of 16, 35 and 161 μg/L, was observed to take place quickly. Similarly, dibromochloromethane underwent rapid degradation at initial concentrations of 19, 45 and 205 μg/L. The removal of both bromodichloromethane and dibromochloromethane occurred in both the control and seeded cultures, which suggests a chemical degradation route. Biological activity may also play an important role in this degradation since the degradation rates were faster in the seeded cultured than in the control. A mechanism similar to that of chloroform degradation was suggested, where substitution of a sulphydral group for the bromine atom(s) produces a thiol which is then degraded by some biological mechanism which may increase the reaction rate.

3.8.5 Other Selected Organic Compounds

It has been established that some of the higher molecular weight hydrocarbons that makeup oil are decomposable by anaerobic bacteria. Novelli, *et al.*[77] reported that a species of *Desulfovibrio* could degrade eicosane, docosane, paraffin oil, and paraffin wax in the absence of other types of organic matter. Of the hydrocarbons used, part was found to be utilized in cell synthesis, part was oxidized into unknown compounds, while remainder was broken down into lower weight hydrocarbons. Rosenfeld[75] provided further evidence that members of the genus *Desulfavibrio* catalyse the anaerobic oxidation of hydrocarbons. It was found that the longer chain aliphatic constituents were most easily attacked. Insoluble fatty acids were detected as reaction intermediates but were transitory in nature and underwent further degradation. It was suggested that a dehydrogenase system is the mechanism by which *Desulfovibrio* is able to utilize hydrocarbons.

3.8.7 Compounds Nonamenable to Biodegradation

Most organic compounds are biodegradable to an extent. While they may not be degraded to the extent one would take, nevertheless, their exist bacteria that are capable of carrying out slight changes of these molecules. There, do exist, however, classes of extremely refractory organic compounds that because of their toxic, complex, or inert characteristics are resistant to microbial attack. This condition may manifest itself in the individual molecules or occur only above certain threshold concentrations. The resistance encountered can be the result of many factors, which include solubility, molecular size, amount of tertiary branching and the number, nature and position of the substituents. There exist general rules to determine the relative biodegradability/non-biodegradability of a compound. A few of them are as follows :

- materials that can pass through cell membrane move readily available to the microbes and are degraded faster;
- microbes prefer non-aromatic or cyclic aromatics over aromatics. The presence of a substituent in the benzene ring usually increases biodegradability;
- soluble compounds are more easily degraded than insoluble;
- a high degree of branching imparts greater resistance to degradation;
- dispersed compounds provide move surface area for attack and therefore, degrade better;
- compounds with unsaturated bonds are degraded more rapidly than saturated compounds;
- such materials are alcohols, aldehydes, acids, esters, amides, and amino acids are preferred over the corresponding alkanes, alkenes, ketones, dicarboxylic acids, nitriles, amines and chloroalkanes.

3.8 Soaps and Detergents

Much of the public's concern over biodegradability system from the subject of soaps and detergents and their resulting effects on receiving water. While must work has been done an aerobic breakdown of soaps and detergents, little had been done on anaerobic processes. Vath[79] conducted some preliminary work with several liner anionic and non-ionic ethyxylated surfactants. Using batch reactors operated between 22-28°C with initial surfactant concentrations of 20 and 100 mg/L, Vath[79-80] found that ethoxamers derived from both linear secondary alcohols and α-olefin alkylphenols, underwent degradation as observed by a loss of surfactant properties. The degradation proceeded by a similar mechanism to that of aerobic degradation, differing only in the hydrogen acceptor employed, Fig. 3.29.

Saldick[81] studied the breakdown of Cyanuric acid (S-triazine 2,4,6-triol), which is the end product of partial breakdown of sodium or potassium dichlorisocyanurate and trichloro-isocyanuric acid, both of which are used in the manufacture of household cleaners and dishwater detergents. Using anaerobic reactions, it was found that in about 72 hours, there was complete disappearance of cyanuric acid. In anaerobic media the overall degradation occurs by a biologically catalysed hydrolysis of the type :

$$C_3H_3N_3O_3 + 3H_2P \longrightarrow 3CO_2 + 3NH_3$$

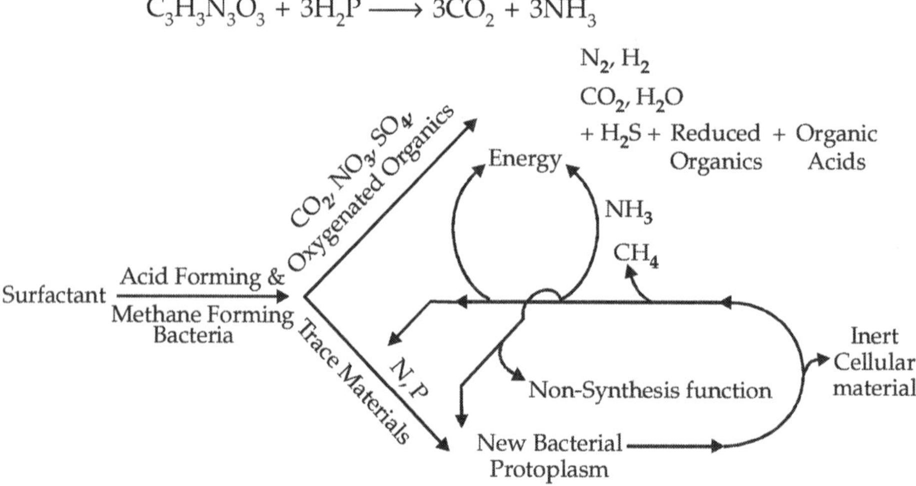

Fig. 3.29 : Pathway for Anaerobic Biodegradation of Surfactants

Barth *et al.*[82] studied the breakdown of sodium nitrilotriacetic acid (NTA), a possible substitute for phosphates in detergents. They resolved to settle conflicting reports on whether NTA was amenable to anaerobic breakdown. It was found that 20 mg/L of NTA could be totally breakdown by anaerobic reactions.

Trisodium carboxymethyloxysuccinate (CMOS) is a new detergent builder that contains neither phosphorous nor nitrogen, as the most commonly used builders do. Klein[83] showed that CMOS could be brokendown anaerobically. Using septic tank effluent as feed, 10, 40, and 100 mg/L concentrations of CMOS were completely removed after a 2 week acclimation period. Vicarro[84] found that a concentration of 20 mg/L CMOS would undergo degradation with CO_2 can CH_4 as virtually the only end products.

3.9 Definition of Biodegradation

The process of biodegradation is not an inherent characteristics of a specific organic compound. Rather, it is the result of highly complex chain of events revolving around interactions between the specific compound, micro-organisms, and environmental conditions where the two are combined[85]. Due to the highly variable nature of the process and broad scope and boundaries that must be included under the term biodegradation, a precise definition is difficult to formulate. Fundamentally, one might consider biodegradation as a change in the parent compound's physical or chemical properties, brought about by chemical processes.[56] This change usually involves conversion of a specific chemical compound or group of compounds and occurs by a series of discrete steps. It is accompanied by a corresponding decrease in the energy of the parent compond.[86] The degree of change can be divided into the following three operational categories :

- primary degradation-the minimal amount of change necessary in the chemical composition of the parent compound to make it no longer detectable by analytical measurement technique used in the detection of the parent compound;
- ultimate biodegradation-the most desirable end point in any biodegradation reaction, involving the total conversion of the parent compound by anaerobic processes into CO_2, H_2O and inorganic substance (if present in the parent compound). It broken down anaerobically, the end products would include; CH_4, and H_2O in addition to CO_2 and inorganic residuals;
- acceptable-biodegradation-in this category are placed biodegradation reactions where parent compound is changed to the minimum extent so that some desirable property such, as toxicity is no longer present.

The degree of stabilization strived for depends on the compound and its concentration. In some cases, primary degradation, may be perfectly acceptable for receiving waters while in other cases anything short of ultimate degradation would be undesirable.

3.9.1 Tests used to Predict Biodegradation

All present existing tests can be divided into two categories :

- those to assess biodegrability potential;
- those that simulate conditions in wastewater treatment processes.

Biodegradability potential tests (or presumptive tests) are designed to be used as quick screening procedures to assess the susceptibility of the test compound to biological breakdown. These tests are simple to run and equipment needs are minimal. Simulation tests on the other hand are move difficult to perform but provide more realistic conditions. It results of a potential test are inconclusive, a simulation test should be performed.

Regardless of the test types, it must be emphasized that the nature of test conditions will influence the test outcome to a degree[87] This necessitates proper control over environmental and nutritional facets of the test. Besides ensuring the promulgation of healthy population of micro-organisms, other factors such as pH, and temperature should be carefully controlled.

In testing for biodegradability one should keep in mind two additional processes, *viz.*,

- acclimation
- co-metabolism

Both processes can play a major role in determining the test outcome. Acclimation, the process by which bacteria adapt their metabolic activities to degrade foreign substrates, occurs by mutation and natural selection or by induction of specific enzymes. The degree of acclimation is highly dependent on the number and types of micro-organisms brought into contact with test compound. Co-metabolism is equally important. Some substances will not be degraded unless some additional factor is present. Such is the case if the ABS branched chain tetrapropylene benzene sulphonate and an un-identified species of *Pseudomonas*. In a pure culture the *Pseudomonas* will not actively degrade the surfactant. However, by the addition of glucose the surfactant will be degraded to isoproponal and catechol[88]. Testing systems involving co-metabolism are usually characteristic of environmental conditions encountered in nature and, therefore, provide more realistic estimates of biodegradation.

3.9.1.1 The River Die Away Test

The details of the test are gives elsewhere by Vath[79-80] It is a widely used test to determine biodegradability potential. The advantages and disadvantages of this test are as follows.

Advantages

- gives the relative rate as well as the completeness of the disappearance of the test compound;
- requires little equipment and is simple to run;
- large numbers of compounds can be screened in a relatively short time.

Disadvantages

- every river will not necessarily yield the same results. Factors such as pH, nutrient content, and the number and types of micro-organisms present in the water at the time of testing will significantly alter results;
- it may be necessary to analyse for specific degradation products, since the analytical procedures employed may not detect slight modifications of the parent compound to products that are resistant to further degradation;
- seasonal effects will substantially alter the results from the same river.

3.9.1.2 *Worburg Respirometer*

Kugelman, *et al.*[89] and Gossett *et al.*[90] have described the method quite extensively. The advantages and disadvantages are as follows :

Advantages

- the apparatus has capability to run many samples simultaneously;
- changes in rate and lag period can be measured.

Disadvantages

- the Warburg unit is costly and requires trained personnel to operate;
- small sample size make it difficult to obtain representative samples of the wastewater, which makes subsequent analysis difficult;
- it is difficult to collect samples from the gas or liquid phases during test.

3.9.1.3 *Flask Test Methods*

In this test sample biodegradation is determined by measuring the cumulative methane production from an incubated sample[91]. The results can be the related to sample volume (m^3CH_4/m^3) and sample mass (m^3CH_4/kg COD. The percentage organic matter converted to methane can be determined by using an empirical factor of 0.35 m^3CH_4 (at STP) produced per kg of COD[92]. The advantage and disadvantages are given as follows :

Advantages

- the test is simple to run and requires little equipment;
- reproducibility of the results are fairly good;
- its simplicity makes it deal as a presumptive test.

Disadvantages

- micro-organisms must have had prior exposure to test compound of an acclimation period must be incorporated into the test;

- a similar reference standard needs to be run as a control;
- specific analytical technique must be available to measure for suspected partial degradation products.

REFERENCES

1. Paolini, C. Storia del metano (1976) Milano
2. Tissot, B.P. and Welte, D.H. Petroleum Formation and Occurrence 1978) Springer-Verland, Berlin, Heidelberg, New York.
3. Schoell, M. Geochin. Cosmochim. Acta, 44, p 694 (1980).
4. Bechamp, A. Ann. Chim. Phy. 13, pl03 (1868).
5. Poff, L. Pfluger's Arch. Z. Ges. Physiology, 10, p113 (1975).
6. Hoppe-Seyler, F. Pfluger's Arch. Z. ges. Physiology, 12, p. 1 (1876).
7. Hoppe-Seyler, F. Hoppe-Seyler's Z. Physiol. Chem. 11, p. 561 (1887).
8. Omelianski, W. Centralbl. F. Babt. Abt. II, 11, p. 369 (1904).
9. Sohngen, N.L. Het Outstannwen Verdwijnenvan Waterstofen methanunonder deninolved Van het Organische leven Dissertation (1906), Tech, Univ. Delft.
10. Omelianski, W. Ann. Inst. Past. 30, p 56 (1916).
11. Mae, P. Comptes rendus Acad. Sc. Paris, 137; p887 (1903).
12. Maze, P. Comptes rendus Soc. Biol. 78, p 398 (1915)
13. Groenewege, J. Medo Burg. Geneesk. Dienst, Deel, 1 p 66 (1920).
14. Coolhaas, C. Centralbl. F. Bakt. A bt. II, 75, p 161 (1928).
15. Barker, H.A. Arch. Mikrobiol. 7, p 420 (1936).
16. Barker, H.A. Bacterial Fermentation (1956), J. Wiley, New York.
17. Stadtman, T.C. Ann. Rev. Microbiol. 21, p121 (1967).
18. Schnellen, ch. G.T.P. Onderzoekingen over de methaangisting Dissertation (1947), Tech. Univ., Delft.
19. Smith, P.H. and Hungate, R.E. H. Bacteriol. 75, p. 713 (1958).
20. Byrant, M.P and Wolfe, R.S. Arch. Mikrobiol. 59, p. 20 (1967).
21. Wolfe, R.S. Advan. Microbiol. Physiol 6, p. 107 (1971).
22. Zeikers, J.G. Bacteriol. Rev. 41, p 514 (1977).
23. Mah, R.A. Ann. Rev. Microbiol. 31, p. 3.09 (1977).
24. Wolf, R.S. and Higgins, I.J. Intervant. Rev. Biochem. 21, p 267 (1979).
25. Balch, W.E. Microbiol. Rev. 43, p 260 (1979).
26. Smith, M.R. Process Biochem. 34 (May 1980).
27. Mclnerney, M.J Appl. Environ Microbiol. 41, p. 1029 (1981).
28. Bryant, M.P.J Anim. Sci, 48 p 143 (1979).
29. Zehnder, A.J.B. Water Poll. Microbiol. Ed. Mitchell, R., Z., p. 349-376 (1978). J. Wiley and Sons, New York.
30. Hungate, R.E. The Rumen and its Microbes, (1966), Academic Press, Inc., London.

31. Mclnerney, M.J. Applied Environ. 41, p 826 (1981).
32. Byrant, M.P. (Ed), Bergey's Manual of Determinative Bacteriology, 8th Ed ; p 472-477 (1974). The Williams and Wilkinson Co. Inc; Baltimore.
33. Fox. G.E. Proc. Nall. Acad. Sci (U.S.A.) 74, p 4537 (1974).
34. Fox. G.E., Science. 209, p. 457 (1980).
35. Woese, C.R., Sci Amer. 244(6), p.64 (1981).
36. Kristyanson J.K. and Schonheit, P. Why do Sulphate Reducing Bacteria Out complete Methanogenic Bacteria For Substrate. Decolonia 60 p 264-266 (1983).
37. Newfeld, R.D. and Mack, J.D. Anaerobic Phenol Biokinetics, J. Wat. Poll. Fed. 52, p 9 (1980).
38. Procter, M.H. and Scher, S. Decomposition of Benzoate by a Photosynthetic Bacteria. J. of Bacteriol. 54, p 33 (1960).
39. Dutton, P.L. and Evans, W.C. Dissimilation of Aromatic Substrate by *Rhodopseudomonas palustris*. Proc. of the Biochem. Soc. 104, p. 30 (1967).
40. Travin, D and Buswell, A.M. The Methane Fermentation of Organic Acids and Carbohydrates. J. of Amer Chem. Coc. 56, p. 1751-1755 (1934).
41. Stanier R.Y. Jour. Bact. 59, p. 527 (1950).
42. Stanier, R.Y. Jour. Bact. 59, p. 137 (1950).
43. Stanier, RY. Jour. Bact. 59, p. 137 (1950).
44. Fina, L.R. and Fiskin. A.M. The Anaerobic Decomposition of Benzoic acid during Methane Fermentation II. Fate of Carbon one and seven, Arch. Biochem. 91, p. 163-165 (1960).
45. Chmielowski, J. Methane Fermentation of Some Phenolic Wastewater. Zesz. Nauk. Politech. Slaska, Inz. (Pol)., 8, p 95 (1965).
46. Chmielowski, J.A. Study of the Dynamics of Anaerobic Decomposition of Some Phenols in Methane Fermentation Zesz. Nauk. Politech. Slaska. Inz. (Pol,) 8, p95 (1965).
47. Chmielowski, J. Biochemical Degradation of Some Phenols During Methane Fermentation. Zesz. Nauk. Politech. Slaska. Inz. (Pol.) 8, p 97 (1965).
48. Keith, C.L. The Anaerobic Decomposition of Benzoic Acid during Methane Fermentation Arch. Microbiol. 118, p. 173-173 (1978).
49. Ferry J.G. Anaerobic Degradation of Benzoate to Methane by Microbial Consoritum. Arch Microbiol. /107, p. 33-40 (1976).
50. Evans, W.C. Biochemistry of Bacterial Catabolism of Aromatic Compounds in Anaerobic Environments, Nature, 270, (Nov. 3 1977).
51. Newfield, R.O. Anaerobic Phenol Biokinetics. J. Wat. Poll. Cont. Fed. 52, p9 (1980).
52. Williams, R.J. and Evano, W.C. Anaerobic Metabolism of Aromatic Substrates by Certain Micro-organisms. Biochem Soc. Trans., L, (1973).
53. Voets, J.P. and van Lancker, P. Degradation of Microbicides under Different Environmental Conditions. J. Appl. Bact. 40, p. 67-72 (1976).
54. Healy, J.B and Young, L.Y. Catechol and Phenol Degradation by a Methanogenic population of Bacteria Appl. Environ. Microbiol. 35, p216-218 (1978).
55. Healy, J.B. and Young, L.Y. Anaerobic Biodegradation of Eleven Aromatic Compounds to Methane Appl. and Envn. Microbiol 38(1), p 84-8 (1979).

56. Suidan, M.T and Calvert. H.W. Anaerobic Carbon Filter for Degradation of Phenols. J. Environ. Engr. div. Proc. am. Soc. Civil Engr. 107 (EE-3), (June 1981).
57. Suiden, M.T. and Calvert, J.W Anaerobic Activated carbon Filter for the Removal of Refractory and Toxic Compounds in Wastewater. Tech. Completion Report, ERC-08-79(1979).
58. Kirsch, E.J. and Grady, C.P.L. Protocol Development for the Prediction of the Fate of Organic Priority Pollutants in Biological Wastewater Treatment Systems. U.S Envn. Protect. Agency, (March 1981).
59. Jennett, J.C. and Dennis, N.D. Anaerobic Filter Treatment of Pharmaceutical Wastes. J. Wat. Poll. Con. Fed. 47(1), p104 (1975).
60. Sachs, E.F. and Jennett, J.C. Anaerobic Treatment of Synthesized Organic Chemical Pharmaceutical Wastes, 33rd Purdue Ind. Waste Conference, May 9-11, 1978, Purdue luid, West Lafayette, Indiana.
61. Hill. W.D. and Mc Carty P.L. Anaerobic Degradation of Selected Chlorinated Hydrocarbons Pesticides, J. Wat. Poll. Con. Fed. 39(8), p. 1259-1277 (1967).
62. MacRae, I.C. and Raghu, K. Anaerobic Degradation of Insecticide Lindane by *Clostridium* sp. Nature, 221, p859-860 (1969).
63. MacRae, I.C. and Raghu, K. Persistence and Biodegradation of Four Common Isomers of Benzene Hexachloride in Submerged Soils, J. Agri, Food Chem. 154(5), p 911-914 (1964).
64. Sethunathan N. and Yoshida, T. Degradation of Benzene Hexachloride by a Soil Bacteria. C. and J. Microbiol. 15(12), p. 1349-1354 (1969).
65. Newland, L.W. and Gerhard, B.L. Degradation of BHC in Simulated Lake Impoundments as affected by Aeration Wat. Poll. Con. Fed. 41(5) pt-2, R174-188 (1969).
66. Jagnow, G and Haider, K. Anaerobic Dechlorination and Degradation of Hexachloro-cyclohexane Isomers by Anaerobic and Facultative Bacteria. Arch. Microbiol. 115, p. 285-292 (1977).
67. Ko, W.H. and Lockwood, J.L. Conversion of DDT to DDD in Siol and Effect of these Compounds in Soil Micro-organisms. Cand. J. Microbiol, 14, p. 1069 (1968).
68. Hill, W.D. and McCarty P.L. Anaerobic Degradation of Selected Chlorinated Hydrocarbon Pesticides, J. Wat. Poll. Cand. Fed. 39(8), p 1258-12777 (1967).
69. Castro, T.F. and Yoshida, T. Degradation of Organochlorine Insecticides in Flooded Soils in Philippines J. Agr. Food Chem. 19, p 1168 (1971).
70. Cuenzi, W.D. and Beard, W.D. Anaerobic conversion of DDT to DDD and Aerobic Stability of DDT in Soil Siol. Sci. Soc. Proc. 35, p. 9110 (1971).
71. Cuenzi, W.D. and Beard, W.E. Influence of Solid Treatment an Persistence of Six Chlorinated Hydrocarbon Insecticides in the Field Soil Sci. Soc. Amer. Proc. 32; p522 (1968).
72. Mendel, J.L. Conversion of p, p-DDD by Intestinal Flora of Rat, Science 151, p 1527 (1916).
73. Stenerson, J.H.V. DDT-Metabolism in Resistant and Susceptible, Stable Flies and in Bacteria, Nature, 207 p 660 (1965).
74. Barker, P.S. and Whitaker, R.S. Conversion of DDT to DDD by *Proteus vulgaris* a Bacterium Isolated from the Intestinal Flora of Mouse, Nature 205, p621 (1965).

75. French, A.L. Dechlorination of DDT by Membranes Isolated from *E. coli.* J. Econ. Entomol. 63, p 756. (1970).
76. Bower, E.J. and Rittman, B.E. anaerobic Degradation of Halogenated I- and 2-Carbon Organic Compounds, Environ. Sci. and Tech. 15(5), p 596 (1981).
77. Noveli. G.D. and Zobell, C.E. Assimilation of Petroleum Hydrocarbons by Sulphate Reducing Bacteria. Jour. of Bact. 47, p447 (1944).
78. Rosenfield, W.D. Anaerobic Oxidation of Hydrocarbons by Sulphate-Reducing Bacteria Jour. of Bact. 54, p 664 (1947).
79. Vath, C.A. Biodegradation of Nonionics, pt-1, Soap and Chem. Spec. (Feb., 1964).
80. Vath, C.A. Biodegradation of Nonionics, pt-II, Soap and Chem. Spec. (March, 1964).
81. Saldick. J. Biodegradation of Cyanuric Acid, Appl. Microbiol, 28, (6), p. 1004-1008 (1974).
82. Barth, E.F. and Bunch, R.L. Biodegradation and treatability of Specific Pollutants, EPA-600/79-304 (1979).
83. Klein, S.A. and Jenkin, D.J. Wat. Cont. Fed. 46, p 2107 (1974).
84. Vicarro, J.P. Anaerobic Biodegradability of Carboxymethyl-Oxysuccinate, a Detergent Builder. Jour. Amer. Oil Chem. Soc. 54 (Jan., 1977).
85. Kirsch, E.J. and Grady, C.P.L. Protocol Development for the Prediction of the Fate of Organic Priority Pollutants in Biological Wastewater Treatment Systems. U.S. Environ. Protect. Agency, (March 1981).
86. Standard Methods Committee-Sub Committee on Bio-degradability, Required Characteristics and Measurements of Biodegradability J. Wat. Poll. Cont. Fed. 39(7), p. 1232-1235 (1967).
87. Gilbert, P.A. Biodegradation Test : Use and Value. Proc. of Workshop held at Mich. Bio. Station (Augm 1979), Amer. Soc. For Microbiology.
88. Pointer, H.A, Biodegradability, Proc. of Royal Soc. London. Ser. B, 185 (1079), p. 149-158 (1974).
89. Kulgelman, I.J. and McCarty, P.L. Cation Toxicity and Simulation in Anaerobic Waste Treatment, J. Wat. Poll. Cont. Fed. 37, p. 97 (1965).
90. Gossett, J.M. and McCarty P.L Heat Treatment of Refuse for Increasing Anaerobic Biodegradability Amer. Inst. of Chemical Engrs. series 72, (1976).
91. Owen, W.F. and McCarty P.L Bioassay for Monitoring Biochemical Methane Potential and Anaerobic Toxicity. Water Research, 13, p. 485-492 (1979).
92. McCarty, P.L. Anaerobic Waste Treatment Fundaments, Pt-I, Public Works p. 107 (Sept, 1964).

CHAPTER 4

TYPES OF ANAEROBIC REACTOR SYSTEMS

4.0 Introduction

The microbial culture consists of a number of cells asynchronously dividing and growing at a certain individual rate. Each cell metabolizes according to concentration gradients of substrate and metabolites in to close environment. The single microbial cell (biophase) as well as the culture as a whole can be considered as an open biological system. If this living system is compared with relatively single open chemical system, it can be characterized as a complex expanding system. The auto-reproductive or as it is occasionally called autosynthetic, capability of a living system can be compared to an autocatalytic reaction proceeding at a certain rate. The idea of the application of continuous processes in various microbiological is not a recent one. This is proved by the very extensive literature on the subject which cannot be grapsed in its entirety. The examples may be classified into related groups in accordance with different points of view, *viz.*:

- The nature of the fermentation process, *i.e.*, the microbiological approach, based on the analysis of the chemical physiological, morphological and genetic regularities and their application.
- The type of operation, *i.e.*, the biological technology and engineering approach.
- These are two faces of the same coin, but when a given process is being solved, the microbiological point of view is of primary importance.

4.1 Classification According to Nature of the Fermentation Process

The character of a fermentation process is determined by its purpose, *i.e.*, the production of micro-organism only (formation of biomas) or on the other hand effecting of a desired chemical transformation. The fermentation processes can be classified into different types according to the way the carbon compounds of the substrate are utilized and on the basis of comparison of the specific reaction rate (amount of product per unit cell mass per unit time) with regard to the utilization of carbon source. In one instance the intermediary or final products are formed as direct residue from the breakdown of the original molecule of carbon substrate in the course of the reaction supplying energy to the cells. In other instances the main product is formed beyond the period of rapid utilization of the basic carbon source. The energy metabolism supplies the energy for the biosynthesis of complex molecules containing carbon, nitrogen and other elements furnished by the substrate. However, on effecting any process under continuous conditions it must be considered whether product formation is connected with cell growth or whether it succeeds the completion of multiplication. The analysis of the kinetics of the corresponding reaction is therefore necessary for the determination of the character of a particular process.

The substrates are transformed by simple reactions into products in a defined stoichiometrical ratio without accumulation of intermediary products, *e.g.*, the growth of biomass or enzymatic transformation of the substrate by a suspension of previously cultivated cells. In the case of parallel reactions, the substrate is transformed into several products in different ratios according to variable stoichiometric relations. The relative rate of formation of products changes according to substrate concentration. The accumulation of polysaccharides or lipids at variable concentration of nitrogen source can serve as example. In such cases the nitrogen source evidently becomes the limiting factor in the continuous culture.

In the case of successive reactions one or more intermediary products accumulate before formation of the final product begins, but these are very complex processes.

In process consisting of stepwise reactions, either the organisms selectively transform the substrate into products one by one according to preferential selection or product formation starts only after complete transformation of the substrate into an intermediary product. The stepwise reactions are due to enzymatic selectivity and adaptation.

The chemical transformation must not necessarily be carried out exclusively by growing culture, it can also be affected by a mass of previously cultivated cells or by the enzymes produced by these cells. However, in most cases it is necessary to employ a growing culture, the energy necessary for the chemical transformation being gained through the breakdown of the basic substrate.

Though the given classification does not contain all the possible microbiological processes, one is able to predict whether a single-stage or multistage continuous process is concerned. However, the type of the particular stages is not determined. Whether the system will be composed of perfectly stirred homogenous or heterogeneous cultivators is decided by a detailed analysis of rate of growth and product formation. In most microbial processes two consecutive stages occur. First, the microbial mass is formed and only then does biosynthesis of the product ensure. The accumulation of the cell mass during the growth period is an autocatalytic process in which the rate of augmentation or expansion of biomass (reaction rate) increases with its concentration. For such a reaction a homogenous completely mixed cultivator is more effective. The accumulation of the product in microbial culture on the other hand is an indirect or pseudo-autocatalytic reaction since it lags behind the formation of cell mass in respect of time. With this type of reaction, where the reaction rate decreases with increasing concentration, optimum output should be obtained in a heterogenous type of reactor with no spread of residence times.

Since such an ideal reactor is not feasible for an microbial process, certain modifications are employed. However, it is always necessary to comply with the rule, that the distribution of the properties of the individual cells or the distribution of the individual residence time must be minimal, since, otherwise the actual output of the cultivator could be reduced.

4.2 A Classification Based on Type of Operation

For effectuating different type of chemical transformations, different types of continuous reactor systems are required, these being different from the reactors employed for the production of biomass. It is, therefore, necessary to classify the reactor systems according to their operation principles.

In open continuous reactor systems, the cells are constantly washed out with the outflowing fluid at a rate corresponding to the formation of new cells in the system. Under such conditions, it is therefore possible to attain their steady concentration. In a closed continuous systems the cells are retained in the system in a certain way and their amount progressively increases. Under these conditions several limiting factors interchange, finally the greater portion of the cell dies, and such a system is unable to attain a realistic dynamic steady state.

The biosynthesis of a certain product in a continuous culture cannot be understood as a classical chemical reaction. In many cases there is no linear relation between growth rate and formation of the product. This discrepancy follows from the fact that in the classical conception of chemical reactors, the reacting mixture is considered as a single phase. The microbiological process, on the other hand, has to be considered as a single phase. The microbiological process, or the other hand, has to be considered as two interdependent but separated phases. The bio-synthetic reactions whose results are manifested in the medium in one phase are effected by the activity of the living cells (the strictly defined other phase with its specific laws, properties and reactivity). Attempts to predict the behaviour of the microbial culture as a whole under continuous conditions and the course of the biosynthesis of the product, with the aid of specialized chemical thermodynamics based on the data from batch product cultivation would lead to incorrect conclusions.

Though the culture attains the steady state and its composition is quantitatively constant in time, the individual cells donot remain in an unchanging state. Due to growth and multiplication, the internal composition of the cells unceasingly and repeatedly changes qualitatively as well as quantitatively, however, not all the cells change simultaneously. It is, therefore, necessary to work out new thermodynamic conceptions for different types of microbial continuous process.

The scheme of classification of continuous culture systems is shown in Table 4.1 and 4.2. Presently both types of reactor systems are being used in methane fermentation (heterogenous and homogenous types of reactors), *i.e.,*:

- Heterogeneous reactors biofilm which include fixed film-fixed bed, fixed film rotating bed, fixed film fluidized bed and anaerobic upflow sludge blanket reactor,
- Homogenous reactors which include CSTR systems of various configurations.

Table 4.1 : Classification of Continuous Fermentation Systems

I Open Continuous System

1. Homogenous-spatially constant composition of the culture
   - single stage
   - multiple stage (volumes of the stage), (*i*) identical (*ii*) different

     simple chain / complex chain — inflow — into the first stage / into the first and consecutive stages (*i*) same substrate (*ii*) different substrate
   - single and multistage with partial feed back
2. Heterogeneous-spatially variable composition of culture
   - plug flow

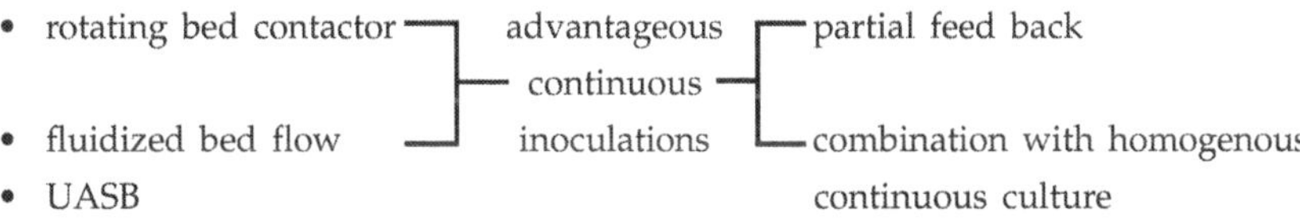

   - UASB

II. Closed Continuous System

- 100 per cent feed back of cell in open system (homogenous and heterogenous)
- growing cells mechanically separated from nutrient flow
- cells growing on interphase

  (*i*) liquid-gas (on surface (unstirred))

  (*ii*) liquid-solid (porous mass, soil, grates, wood shavings, etc.

III. Semi-continuous System

4.3 Structure and Function of Biofilm

Biofilms have been the subject of scientific observations since the end of the past century. However, biofilm received relatively little attention compared to suspended microbial systems. Only recently have researchers become conscious

Table 4.2 : Classification of Continuous Fermentation Systems

			Open		*Closed*	
			Single-Stage	*Multiple-Stage*	*Single unit*	*Multiple unit*
Homogenous		Without Recycle	One-CSTR	Series of CSTRs	Cellophane-Bags Culture CSTR with filter	Probably None
		With Recycle	CSTR with partial Recycle of cells	CSTR series with partial Recycle of cells	CSTR with 100 per cent Recycle of cells	CSTR series with 100 per cent Recycle of cells
Heterogenous	Single phase	Without Recycle	Plug flow tank	Partitioned	None	None
		With Recycle	Plug flow with partial Recycle of cells	Partitioned tank with partial Recycle of cells	Plug flow with 100 per cent Recycle of cells	Partitioned tank with 100 per cent Recycle of cells
	Multiphase		Fluidized bed Packed bed Rotating Contractor (*a*)	Series of (*a*)	Pellicle growth (*a*) with 100 per cent Recycle	Series of closed single units (No advantage)

of the almost universal association of micro-organisms with surfaces and with each other[1].

Systematic research on biofilms has been conducted in the last decade in various fields including microbiology, medicine, and engineering to gain a better understanding of the physiological and genetic processes involved in accumulation and activity. Engineers and medical researchers have focused on the nuisance role as well as the advantages of biofilms. The Dahlem Workshop on 'Microbial Adhesion and Aggregation' revealed that progress in biofilm process research was advancing significantly[2]. However, the workshop also revealed barriers which currently, limit further advances in biofilm science and technology. The various disciplines that focus on biofilms use different approaches, each having a rather narrow perspective of the problem. Exposure to other view would certainly be of great benefit and would help broaden our general understanding of biofilm systems.

4.3.1 Occurrence of Biofilms

Biofilms play an important role in nature and in technology. They can be found at almost any surface exposed to water irrespective of the prevailing trophic state. Whether biofilms are beneficial or detrimental, however, depends on the inclination of the observer.

Biofilms accumulating in a river bed or on suspended particles in rivers, lakes, and in marine environment are often considered advantageous since they contribute largely to the removal of contaminants from water. Biofilms growing in association with root cells enhance the availability of nutrients to plants. In sewage treatment systems, biofilms remove organic and inorganic pollutants. Biofilm reactors are used in the fermentation industry (quick vinegar process). The pharmaceutical industry increasingly is using biofilms that are artificially produced by immobilization of cells on specific carrier materials.

However, biofilm can also be problematic. Excessive biofilm accumulation on porous media such as sewage trickling filters or water well screens causes clogging which reduces their effectiveness. Biofilms on heat exchange surfaces are responsible for unfavourable increases of heat transfer resistance resulting in energy losses and thus reduced performance. Fluid frictional resistance increases when biofilms accumulate on pipe walls and ship hulls. Accelerated material deterioration (corrosion) may occur due to biofilm accumulation on water distribution system pipes, teeth (dental caries), semipermeable membranes, or on the walls of cultural monuments. In medical science, biofilms are responsible for a variety of health problems, such urinary tract injections, and infections related to implants. Biofilms may accumulate large amounts of hazardous substances such as heavy metals or chlorinated hydrocarbons, and in so doing become a hazardous material. In some instance, when biofilms grow on the surface of plastic materials (on the back side of the refrigerator door or on the seal of the shower stall), the occurrence of biofilms may be nothing but unsavory.

4.3.2 Systematics

A biofilm accumulates when micro-organisms colonize a surface and form a mono or multilayer of cells. In most cases, the organisms ultimately occupy the entire surface, but patchy growth is also often observed. The accumulation of biofilm is the net result of transport processes, and interfacial transfer processes (growth)[3]. The biofilm and its micro-environment are continuously exchanging specific components in this relatively complex way. To facilitate understanding and further discussions, it is proposed to subdivide the particular biofilm habitat of concern and introduce the concept of scale (Fig. 4.1) with the biofilm volume (micro-scale) representing the smallest unit of the biofilm system which in turn is considered to be the smallest unit of a biofilm reactor (macro-scale).

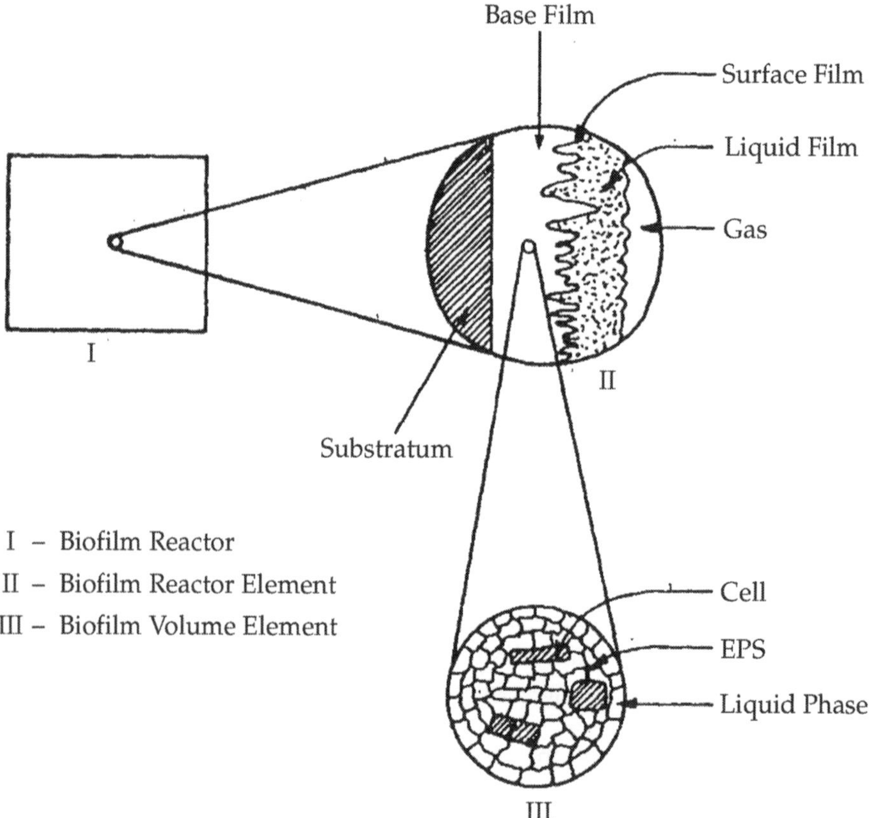

Fig. 4.1 : Schematic Representation of a Biofilm Reactor, the Biofilm Reactor Element and the Biofilm Volume Element.

The biofilm volume element contains the relevant microbial cells, water and in most cases the extracellular polymeric substances (EPS) which hold the cells together and to the surface. It may also contain exo-enzymes and inorganic inclusions such as clay particles and corrosion products.

The biofilm system consists of the biofilm, the substratum and the medium in which the biofilm is immersed–a liquid, a humid gaseous atmosphere, or a liquid film bounded by gas space.

The biofilm is the sum of the biofilm volume elements bounded by the surface of the substratum and the interface between biofilm and liquid film. The biofilm-liquid film interface is not always well defined. A transition area between biofilm and liquid film may exist, which is termed the surface film incontrast to the uniform and densely structured base film located underneath.

The surface with is colonzied is termed substratum. It may be an impermeable, non-porous material such as a sand grain, a tooth, or a piece of metal. In other cases, however the substratum may be porous and even gas permeable. It may be inert (glass), susceptible to biocorrosion (metal) or may in special cases serve as food source for the biofilm organism (cellulose).

In most cases, the biofilm organisms are supported with food and electron acceptors from the bulk liquid overlaying the biofilm. The deducts of cell metabolism (carbonaceous substances, nutrients, electron donors, electron acceptors) are collectively termed the substrate; the products are termed metabolities. Besides substrates, nutrients, and metabolites there are substances which may also be of importance.

The smallest unit of a biofilm reactor is termed the biofilm reactor element. It represents a biofilm system under the boundaries which are specific for the biofilm reactor in consideration. In special cases (fluidized bed), the reactor element may be represented by particles (sand grain) that are coated with a biofilm. Particles carrying a bio-film are termed biofilm support particles (BSP). Thus, a biofilm reactor appears to be a unique configuration of biofilm volume and biofilm systems. Distinctive criteria are the reactor geometry, and the case specific biofilm properties, transport and transformation processes, Fig. 4.2.

4.3.3 System Components

A biofilm volume element (BVE) consists of microbial cells which are in the most cases, embedded in an organic matrix of microbial origin. Furthermore, components of the BVE are soluble, colloidal and particulate substances, Fig. 4.3. The EPS are considered to be an open porous structure similar to a sponge, with a void space filled with water. The size of the pores may allow only substances to move freely in spaces which are small as compared to the pore size of EPS. In general, soluble substances belong to the category of substances that can move freely in space whereas colloidal and particulate substances remain entrapped.

Components which cannot move freely in space because they are attached to each other or entrapped in EPS structure are the solid phases of BVE. In this context, the term phase is used in a thermodynamic sense. Each category of solids (microbial, EPS, abiotic particles, etc.) constitute a different solid phase, but there is only one liquid phase (water).

Only the liquid phase contains substances which can move freely in space. In the bulk liquid and in certain areas of the surface film, particulate substances (solids) may be components of the liquid phase. In turn, liquids in the form of bound water may be a component of any solid phases.

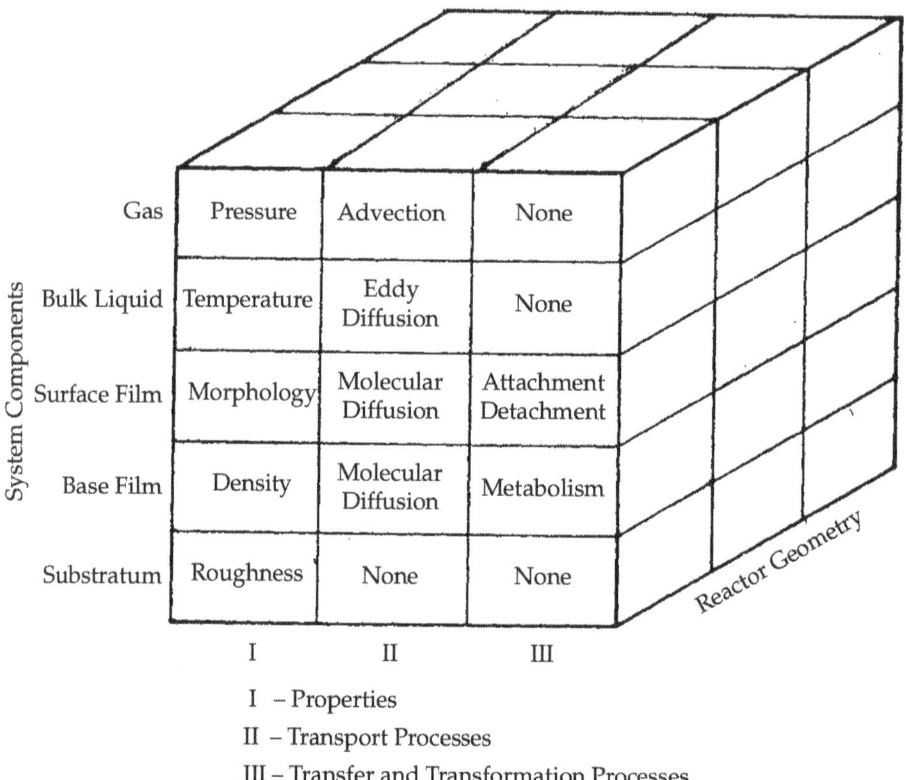

Fig. 4.2 : The System Components for Biofilm Reactors

Within the biofilm, particles of various solid phases are attached to each other. Thus, any change in volume of a particle (transformation processes such as synthesis of EPS, growth, or replication of cells) causes displacement of neighbouring and of adjacent BVE). As a result, the biofilm as a whole expands or contracts. Expansion and contraction of biofilms are classified as transport processes (advection). On the other hand, movement of solids from one phase to another (attachment) is an interfacial transfer process. Transfer processes between phases have some characteristics of a transformation process since educt and product do not belong to the same phase. However, the stoichiometric coefficients of these pseudo transformation processes are always equal to one.

4.3.4 *Biofilm System*

A biofilm system consists of various compartments, Fig. 4.3. Among these are the substratum on which the biofilm accumulates, the biofilm, the bulk liquid which overlays the biofilm and in special cases, the gas which overlays the bulk liquid. The biofilm compartment may be subdivided into two compartments : the base film and the surface film.

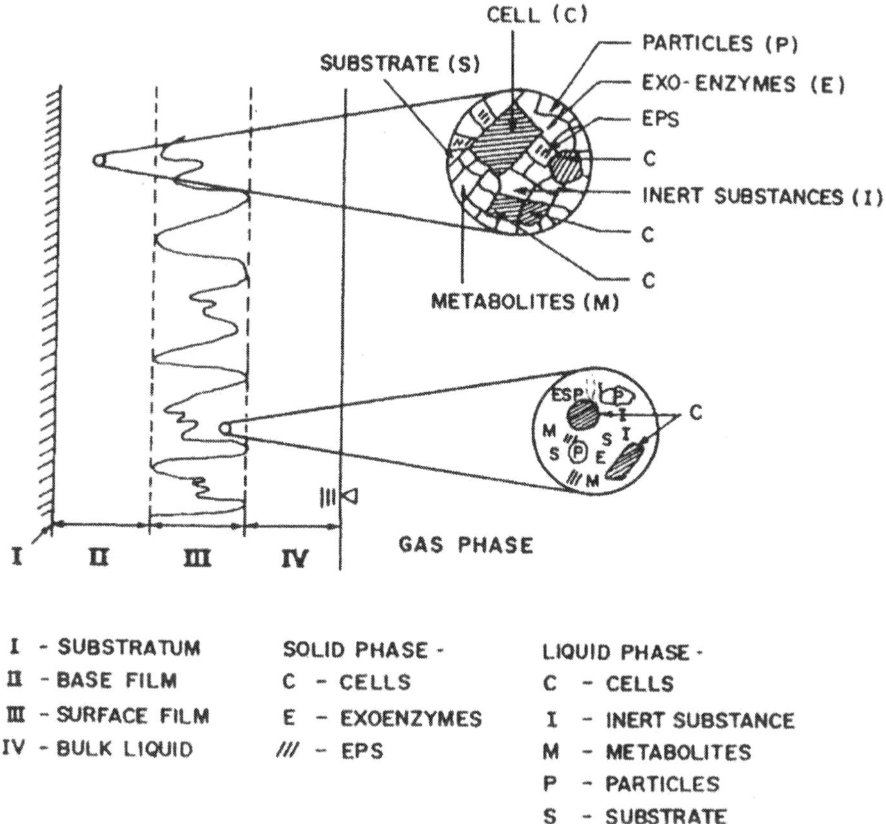

Fig. 4.3 : Schematic Representation of a Biofilm System

Each compartment is a characterized by atleast one phase : solid, liquid or gas. Thus, each compartment can be described in terms of its thermodynamics and transport properties as well as by the transport, interfacial transfer, and transformation processes which dominate within the respective compartment, Fig. 4.4.

In many cases, the substratum is an impermeable, non-porous solid material. Generally, a well defined interface exists between biofilm and substratum. In other cases, the substratum may contain pores, and thus a liquid phase or even a gas phase. If the pores are large enough, micro-organism may settle within the substratum and the biofilm-substratum interface becomes morphologically undefined. Semipermeable membranes allow feeding of the biofilm organism with the substrate from the substratum side of the biofilm.

The substratum may also be a liquid, as observed in oil water emulsions or as observed with biofilm accumulating at a gas-water interface. The characteristics of the substratum are decisive, especially during the early stages of biofilm accumulation, and may influence the rate of cell accumulation and the initial population distribution. Roughness of the surface and substratum's free energy are factors affecting adhesion of micro-organisms. Cavities provide protection against fluid shear stress, grazing micro-organisms, and abrasions.

The base film is relatively firm and structured. The surface film, on the other hand has a very irregular topography which provides a transition between the bulk liquid compartment and the base film. The transition region is characterized by the free movement of particulate substances in the macro-pores. Interfacial transfer dominate (particle deposition, entrapment, surface flocculation, erosion). The surface film may extend all the way to the substratum, especially when filamentous micro-organisms or stalked protozoa are dominant. In other cases, the surface film may not exist at all, as observed in many high velocity, oligotrophic environment as well as in monospecies biofilm systems.

The bulk liquid compartment processes affect the performance of biofilm systems primarily as a result of mixing and flow patterns characteristic of the specific biofilm reactor and its geometry. Mass-transfer from the bulk liquid compartment to the biofilm compartment is dependent and the fluid dynamic regime. Mass transfer under laminar flow conditions is much slower as compared to turbulent conditions. In some cases, the bulk liquid compartment is large to the thickness of biofilm (river top of biofilm accumulating on river bed); so that the mean distance over which substances are to be transported in the bulk liquid is large. In other cases, only relatively thin liquid films flows over the biofilm (sewage trickling filter) or the biofilm covered only by moisture (biofilm exposed to a humid gaseous atmosphere).

The gas compartment provides for aerobic anaerobic and or removal of gaseous metabolites such as CH_4, CO_2, etc.

4.3.5 *Biofilm Properties*

The physical, chemical and biological properties of a biofilm are dependent on the particular environment in which the biofilm accumulates. Abiotic and environmental factors are to be taken into account. Factors of importance are the organisms themselves, their presence and activity and interactions between the various populations forming the biofilm community. It is the concert of the environmental factors which determines the properties and function of the biofilm system. Methods to measure effectively the biofilm properties and relate them to biofilm performance are needed.

Thickness, density, and volume fraction are the major state properties of the system compartments, in particular of the biofilm compartment. Biofilm density is defined as the mass of dry solids per unit of the solid phase volume.

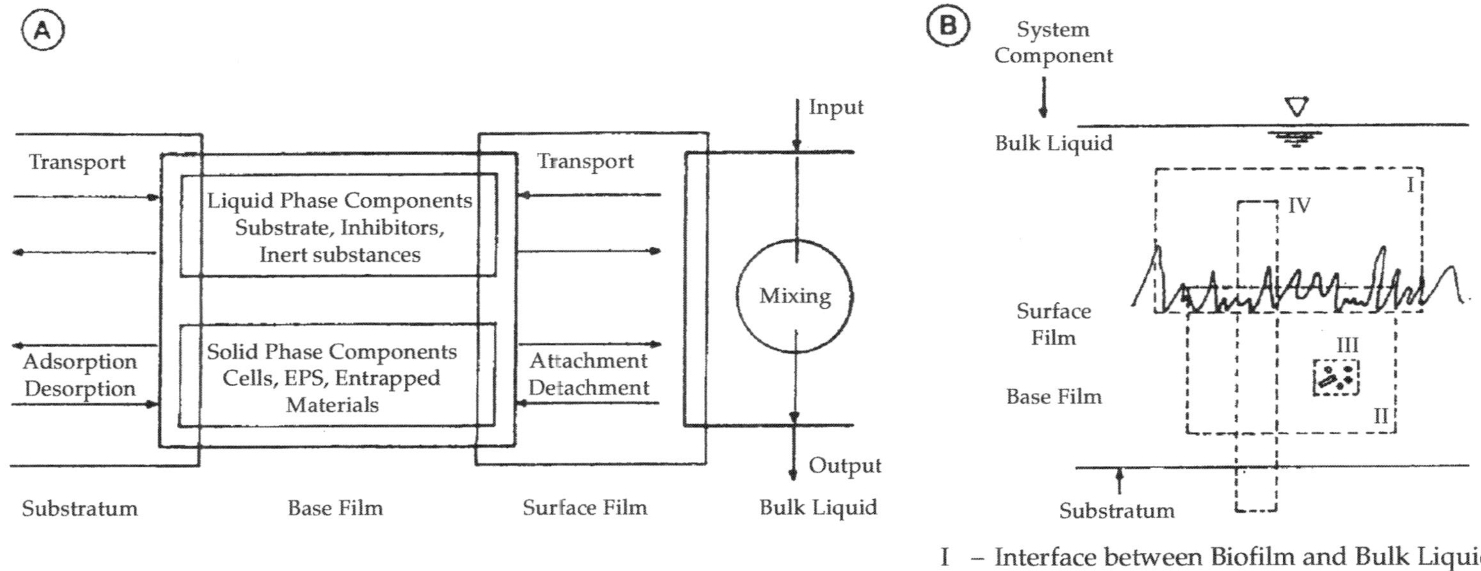

I – Interface between Biofilm and Bulk Liquid

II – Biofilm

III – Biofilm Volume Element

IV – Biofilm System

Fig. 4.4 : Components and Processes of a Biofilm System.

The volume fraction of a phase is defined as that portion of the BVE which is occupied by the particular phase of concern. The various volume fractions change significantly within the surface film compartment but are assumed uniform in the base film compartment.

Biofilm thickness, density, and volume fraction are difficult to measure *in situ*. Analytical methods are required to allow investigation of the biofilm properties in a non-destructive way.

The elemental composition of biofilm can be very different depending on the prevailing composition of the bulk liquid compartment. The inorganic composition of the biofilm may influence the physical properties of the biofilm. Tarakhia *et al.*[4] demonstrated the influence of calcium ion on the mechanical strength of biofilms by applying a calcium specific chelant which triggered rapid disruption of biofilms. Biofilm exposed to wastewater of industrial origin may contain large amounts of heavy metals. Accumulation of heavy metals in the biofilm is the result of flocculation at the biofilm-bulk liquid interface and of binding with extracellular polymers and in the cell envelope. The heavy metal accumulation may cause inhibition of cell metabolism and as a result transforms the biofilm into an environmentally hazardous material.

The biofilm can be visco-elastic because of its high fraction of extracellular polymer. The morphological structure of the surface film also contributes to the compliant nature of the biofilm and may cause an abnormally high hydraulic resistance. Consequently, the advective mass and heat transfer rate can be high as compared to other solid-liquid interfaces.

4.3.6 Interaction Between System Compartments

The various compartments of a biofilm system communicate with each other via transport and interfacial transfer processes (Fig. 4.4) which play a critical role in biofilm systems and in most cases, are rate limiting.

Transport processes are important in the bulk liquid compartment as well as in biofilm compartment. In the bulk liquid, transport delivers substrate to the biofilm compartment and caries away products of the metabolism. Interfacial transfer processes influence cell accumulation at the biofilm through attachment and detachment processes. Transport of soluble components within the biofilm compartment is primarily by molecular diffusion. Volumetric displacement as a result of growth is one mechanism for transport of cells within a biofilm. There may be other means of transport or interfacial transfer of particulate components within the biofilm and bulk liquid compartments, such as cell mobility and bioturbation by grazing micro-organisms.

4.4 Mechanism of Biofilm Formation

Insight into the mechanism of formation and composition of biofilms is essential for rational start up and design of fixed film reactors. Though present knowledge of biofilm properties is only rudimentary, some understanding of these aspects and their implications in fixed film reactor operation has been achieved.

A biofilm is a layerlike aggregation of micro-organisms attached to a solid surface[5]. In a review of effects of solid surfaces on microbial activity, Marshall[6] suggests that the initial attraction of microbes to solid surface is either due to accumulation of nutrients at the surface or due to physico-chemical forces. Initial colonization of a surface is prepared by adsorption of slimy substances such as extracellular proteins and polysaccharides. Non-slime producing organisms attach to the slime matrix produced by others and such a consortium builds up the biofilm. As the population density on the surface increases, there is enhanced opportunity for cross-feeding and co-metabolism and this further stimulates the growth of micro-colonies.

In a discussion on start up of anaerobic fixed film reactors, Salkinoga-Salonen *et al.*[7] reveal that aerobic reactors seldom have problems of biofilm formation, whereas anaerobic organisms attach to non-porous materials with difficulty. This is due to non-slime producing nature of anaerobic micro-organisms while aerobes commonly produce slime when carbonaceous waste is fed into a reactor. Clearly, the choice of support material seems more important for operation of anaerobic fixed film reactors. Porous materials such as granular activated carbon or porous polymers prove to be more successful compared to smooth materials such as plastic.

Several strategies have been used to solve the problem of poor attachment of seed biomass onto the carrier material in anaerobic fixed film reactors[7]. These include operating the reactor initially under aerobic conditions so as to allow a slimy layer to develop around carrier particles before anaerobic operation. The addition of slime-producing organisms and their substrates into the seed (*Leuconostoc* species plus sucrose) and the use of flocculating polymers such as arcylamide. Use of calcium salts rather than soda lime for wastewater neutralization is also reported to aid biofilm formation. The principal variables in design of fixed film reactors are as follows :

- media characteristics (affects head loss buildup)
  - (*a*) grain size
  - (*b*) grain size distribution
  - (*c*) grain size density and composition
  - (*d*) media charge
- reactor porosity—Determines the amount of biomass that can be stored
- reactor depth-affects head loss
- influent wastewater characteristics

(*a*) suspended solids

(*b*) floc or particle size and distribution

(*c*) floc strength

(*d*) floc or particle charge

(*e*) fluid properties

Possible Mechanisms operative in a biofilm reactor are give as follows:

- straining
  (*a*) mechanical-particles larger than the pore space of the media are strained out mechanically.
  (*b*) chance contact-particles smaller than pore space are trapped by chance contact
- sedimentation
- impaction
- interception
- adhesion
- chemical adsorption
  (*a*) bonding
  (*b*) chemical interaction
- physical adsorption
  (*a*) electrostatic forces
  (*b*) electrokinetic forces
  (*c*) van der-Waals forces
- flocculation
- biological growth

The composition of biofilms is solely dependent on the characteristics of the ecosystem in which they are formed. Moreover, within a particular ecosystem, there is a spatial and temporal variation in biofilm composition as the conditions change. With the knowledge of biofilm composition under varying operating conditions, it may be possible to increase the efficiency of fixed film reactors by manipulations of micro-flora to favour species that play crucial roles at various steps of substrate conversions.

A study of biofilm composition in anaerobic fixed film reactors has been provided for the first time by Robinson *et al.*[9]. The authors examined biofilms from 8 experimental fixed bed reactors packed with various support materials (plastic balls, rings, hollow spheres, blocks of pine and cypress wood). The reactors were operated on fresh swine waste at HRT of 2 days.

Light and UV microscopy, SEM and TEM studies revealed that the biofilms adhered more tightly to wood blocks than to plastic. Biofilm thickness was 1-3 mm through most of the reactor height, however, films did not form in the lower 5 cm and were very thin near the top.

The biofilms were composed of a heterogenous population of bacilli, cocci, spirilla and sarcina containing cysts. Two morphological types common to biofilms in all regions were the long filamentous rods observed at the film surface, and sarcina-containing cyst in the lower regions. Sheathed bacterial filaments have been reported in aerobic RBCs. It is possible at these filaments with the ability to intertwine are most conducive to the capture of organic solids and other bacterial forms. The structural arrangement may be beneficial when one cell type is physiologically dependent on the other. The diversity of methanogenic species may be of primary importance in utilization of different nutrients supplied from primary degradation by non-methanogens.

The lower layers of the biofilms were characterised by thick matrix material. Many crystals were embedded in the biofilms. The shapes of crystals varied and sizes ranged upto 100 μm. These were found to contain Ca, Mg, P or Ca, Mg, Mn were X-ray analysed.

An interesting observation is the random distribution of protrubenances, with openings of 100-500 μm diameter, in the thicker portions of the film. Cross-sections of the films revealed that the opening extended right through the biofilm. These open channels presumably facilitate gas and nutrient transport through the biofilm.

Similar observations regarding biofilm composition have also been made by Havey, *et al.*[9] The authors also report that with ageing, the mineral deposition in the film increases and this could be taken up as an indication of biofilm maturation during reactor start up.

The presence of calcium containing crystals within anaerobic biofilms supports the observation by Salkinoja-Salonen *et al.*[7] that calcium plays a vital role in biofilm formation.

The proper selection of support media is based on two factors–the concentration of particulates in the incoming wastes and the organic composition of wastewater[10]. Waste composition in important because certain types of wastes (such as carbohydrate containing wastes) produce larger biomass yields; also, high strength wastes produce more cell yield than low strength wastes. In general, the material selected should have a high surface to volume ratio so as to provide larger surface areas for biofilm attachment, while maintaining a sufficient void volume to prevent clogging of the reactor.

4.5 Types of Anaerobic Reactors and Processes

It is important to distinguish between reactor types and treatment processes (process lay-out). In the latter case one or several reactors are included in a total treatment plant scheme. In the following, the rather few different reactor types are discussed. These basic reactor types can be integrated in numerous treatment processes.

4.5.1 Basic Reactor Types in Anaerobic Processes

Various reactor types are used for anaerobic processes. Many of these have different names, which might result in the impression that there are numerous anaerobic reactor types. However, when looked upon in detail, the number of different reactor types reduces to rather few basic ones, which are discussed in the following, Table 4.3 lists a number of reactor types, which have been investigated or marketed during the last 5-10 years. The reactors have been divided into 8 basic types, which are believed to cover the range of reactors used at present. Some of the basic reactor types are known from aerobic processes (moving bed, fixed bed, recycled flocs cum activated sludge) while others (fluidised bed and expanded bed) primarily have been developed for anaerobic processes, including denitrification. The fluidized and expanded bed reactors can be used for aerobic processes as well, but in general this will not be advantageous due to oxygen transfer limitations. At present it is estimated, that approximately 100-150 full-scale plants using biofilm are in operation.

4.5.1.1. Fixed Bed

The micro-organisms are attached to an inert media, which can have any one of the media known from aerobic trickling filters. The wastewater passes the bed with vertical flow (up- or downflow). In all configurations, a substantial percentage of the biomass is present as suspended flocs entrapped in the voids of the inert media. In general, this reactor type is operated without recycle. This results in a plug-flow pattern in the reactor, although the gas production tends to stir-up the flow pattern by rising bubbles. Recycle may be used in order to control biofilm thickness to a certain degree, or to overcome toxicity and or pH problems. Excess sludge is removed from the reactor together with the treated wastewater. Means for backwash/excess sludge removal should be included in the design[12].

4.5.1.2. Moving Bed

The micro-organisms are attached to an inert (plastic) media, which is moved through the wastewater. An example of this type of reactor is the rotating disc reactor[13]. The media can be partially or fully submerged. The velocity between the media and the wastewater gives some sort of biofilm thickness control. Excess sludge leaves the reactor together with the treated wastewater.

4.5.1.3. Expanded Bed

The micro-organisms are attached to an inert support media, which might be sand, gravel, anthracite or plastic. The diameter is comparable to that used in fluidized beds but often slightly bigger. The biofilm media is expanded sufficiently high vertical velocity, obtained by high degree of recycle. The bed expansion is kept at a level where all particles still keep their place within the bed.

4.5.1.4. Fluidized Bed

The micro-organisms are attached to an inert media, which might be sand, activated carbon or garnet[14, 19]. The biofilm covered media is fluidized by a high vertical velocity, demanding a very high degree of recycle. The single particles do not have a fixed position in the bed, but are gently moved around. Still, each particle tends to stay located within a rather small bed volume.

The bed expansion is controlled by the vertical velocity and the recycle out take level. Biofilm thickness is controlled by the bed regeneration strategy and by the size and density of the inert media in combination with the vertical velocity. The gas production may create foaming and flotation at the top of the reactor. This must be controlled by hydraulic or by mechanical means in order to prevent particle escape together with the treated wastewater. Excess sludge can be removed from the bed regeneration stream taken out from the top of the fluidized bed where the biofilm tends to have maximum thickness.

4.5.1.5. Recycled Bed

At present only one investigation has been brought to the attention of the authors. In recycled bed reactors, the micro-organisms are attached to an inert media, which could be sand, antracite, iron particles, etc. A substantial part of the biomass is found in suspended flocs. The bed is kept suspended by mechanical stirring and/or gas stirring. Separation of the bed and the wastewater is performed in a separator (a sedimentation tank) from which the bed with the attached biofilm and suspended settled flocs are recycled to the reactor. Excess sludge can be drawn form the bed recycle pipe.

4.5.1.6. Recycle Flocs (Contact Reactor)

The reactor is basically identical to the recycled bed reactor except for the inert media. This means that the micro-organisms have to form flocs in order to stay in the reactor. Inert particles in the wastewater may act as carrier material and turn the reactor into a recycled bed reactor. The separation of wastewater and flocs is the critical part of the reactor. Again, gas stripping or cooling helps to control the separation.

4.5.1.7. Sludge Blanket Reactor (Clarigester Type Reactor)

The flocs are kept in suspension by the effect of gas bubbles. In order to avoid mechanical mixing, wastewater must be evenly distributed over the bottom of the reactor. An often used technique is to distribute the incoming water through bottom inlet pipes. These are by no means mandatory for the sludge blanket generation. The flocs have a biofilm structure and are formed as granules of 1-5 mm diameter. In the lower part of the reactor this dense layer of sludge granules is present. This results in a sludge concentration profiles with considerable concentration variations within the reactor[15]. Some sort of separation, either internal, in the reactor, or external, is needed. Full-scale sludge blanket reactors present delicate hydraulic problems in order to ensure an even flow distribution and avoid influent by-pass[16].

4.5.1.8. Digester

Ideally mix reactors with no specific means for biomass retention, necessitates a very high hydraulic retention time in order to keep the solids retention time high enough. This type is used for treatment of wastes dominated by suspended matter in high concentrations.

Table 4.3 : Basic Types of Reactors Anaerobic Processes

	Type of reactor	*Synonyms*	*Abbreviations*	*Commercial names (Company)*
Attached Biomass	Fixed bed	Fixed film/filter Submerged filter Stationary fixed film	SMAR/ANFIL (up-flow) AUF(up-flow) ADBR(down-flow)/ AF/DSFF	Anflow – –
	Moving bed	Rotating discs Rotating Biological Contractor	AnRBC RBC	(Autotrol) –
	Expanded bed	Anaerobic attached-film expanded-bed	AAFEB	–
	Fluidized bed	–	FBBR/IFCR	Anitron process (Door-Oliver) Hy-flo (Escolotrol) Enso-Fenox (Enso-Gutzeit)
	Recycled bed	Carrier-Assisted contact process	CASBER	
Non-Attached Biomass	Recycled flocs	Contact process		Bioenergy process (Biomechanics) Anamet process (Sorigor)
	Sludge blanket	Upflow sludge blanket	UASB	Biothan (Esmill) UASB (CSM)
	Digester	Clarigester type		Anodek(Sluzer)

4.5.2. Comparison Between the Various Reactors

The reactors described in the previous section have varying characteristics with respect to several factors of importance. Table 4.4 is an attempt to compare the reactors qualitatively with respect to factors such as biofilm control, recycle, solid/liquid separation, tolerance to hydraulic and organic overloading, etc. The Table 4.4 points out the differences between the various reactor types.

One problem when comparing the different reactors is that the borderline between the various reactor types is vague. The recycled bed and the recycled floc reactor resembles each other in many respects, *i.e.*, the, floc/biofilm structure of the biomass in the recycled bed reactor can be difficult to distinguish from the floc structure of the biomass in the recycled floc reactor. The sludge blanket reactor and the recycled floc reactor resembles each other to a high degree, the difference mainly being an internal vs. an external sludge separation device, etc.

Below, the various factors mentioned in Tables 4.4 to 4.6 are discussed briefly, well knowing that the differences between the reactors are not always as straightforward as the impression the tables might give.

4.5.2.1 Biofilm Structure

Expanded and fluidized beds and may be moving beds are believed to have distinct biofilm structure, whereas the other 'biofilm'-reactors do not seem to be dominated by homogeneous biofilms. Fixed beds have a loosely attached biofilm, responsible for a significant part of the bacterial activity.

The type of inert support material is believed to significantly influence the type of film formed. Porous surfaces promotes the development of a biofilm[17]. The development of granules or pellets without an inert carrier material has not at present been explained in details. Many environmental factors play a role, among these are the effect of liquid upflow velocity, either caused by the hydraulics[18] of the reactor or by rising gas bubbles[19].

4.5.2.2 Non-attached Biomass

Non-attached biomass plays an important role in fixed beds, recycled beds, and of course in recycled floc and sludge blanket reactors. Even in expanded bed reactors, non-attached biomass might be of some importance although only comprising a very little part of the total biomass, Jewell *et al.*[20] have observed many protozoa which are believed to be partly responsible for the low biomass yields observed. In fluidized bed reactors, the high vertical velocity is believed to make it difficult for free swimming organisms to exist. This could mean a draw-back with respect to effluent quality.

4.5.2.3 Biofilm Thickness Control

The fluidized beds have special devices for biofilm control. The material is taken out in a small recycle stream for biofilm/media separation. The anaerobic filter is frequently stated to need a special biofilm control device in the form of backwash. The moving bed has an inherent control in scour by speed of rotation. In all cases, sloughing off by a combination of bubble generation in the biofilm and hydraulic shear is believed to take place. Very little is known about this mechanism.

4.5.2.4 *Recycle*

Expanded and fluidized beds operate with very high and costly recycle rates, to whereas recycled floc and recycled bed reactors have moderate recycle rates comparable the aerobic activated-sludge reactors. Fig. 4.5 shows the effect of influent wastewater concentration, C_1 upon the degree of recycle required in expanded and fluidized bed reactors in order to achieve 80 per cent COD removal.

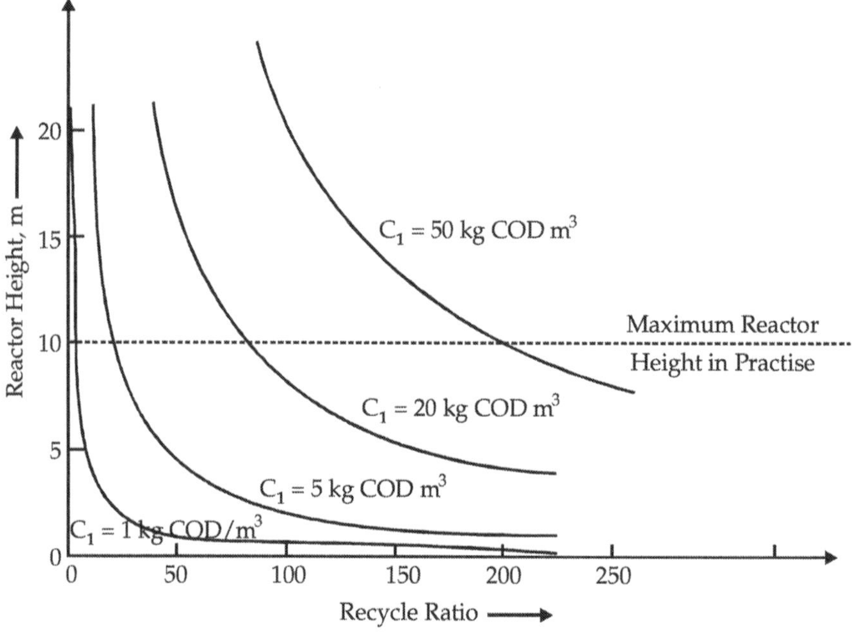

Fig. 4.5 : Recycle Rates for Expanded / Fluidized Beds Assumption.

r_x = 0.01 kg COD/kg VSS.h

X = 20 kg VSS/m^3

v = 10 m/h (vertical velocity)

E = 80 per cent (COD treatment efficiency)

P = 1.02 10^3kg/m^3 (total particle density)

It is seen that for concentrated wastes a very high degree of recycle is needed in order to keep the bed particles in suspension. For dilute wastes, like municipal wastewater, the recycle ratio is reduced to reasonably low values. The assumed upflow superficial velocity of 10m/h represents a conservative estimate based on an overall bed particle specific density identical for pure biological flocs. For very thin biofilms on high density media, the particle size will be of great importance in the considerations regarding the needed superficial upflow velocity[21].

Table 4.4 : Comparison Between the Various Types of Reactors

Sr.No.	*Parameters of Importance*	*Fixed bed*	*Moving bed*	*Expanded bed*	*Fluidized bed*	*Recycled bed*	*Recycled flocs*	*Sludge blanket*
1.	Biofilm structure of microorg., important	(+)	(+)	+	+	(+)	0	(+)
2.	Non-attached biomass important	(+)	(+)	0	0	(+)	+	+
3.	Biofilm thickness control in reactor	0	(+)	(+)	+	0	0	0
4.	Recycle necessary	0	0	+	+	+	+	0
5.	Mixing necessary[3]	0	0	0	0	+	+	(+)
6.	Separation equipment necessary	0	0	0	0	+	+	+
7.	Phasing[1] possible	(+)[2]	(+)[2]	+	+	+	+	(+)[2]
8.	Suitable for wastes with suspended organics	(+)	(+)	0	0	+	+	(+)
9.	Run-through of inerts in raw waste	0	0	(+)	+	0	0	+
10	Problems with foaming	0	0	+	+	(+)	(+)	+
11.	Problems with gas bubbles in reactor	(+)	0	(+)	(+)	0	0	(+)
12.	High microorganism wastewater contact	0	(+)	+	+	+	+	(+)
13.	Tolerates hydraulic overloading	+	+	+	+	(+)	(+)	(+)
14.	Tolerates organic overloading	+	+	(+)	(+)	+	+	(+)
15.	Suitable for high conc. of biodegr. toxics	+[2]	+[2]	(+)	(+)	(+)	+	(+)[2]
16.	Susceptibility to shock-dose toxicants	0	0	+	+	+	+	+
17.	Start-up problems	(+)	(+)	+	+	(+)	(+)	+
18.	Re-start-up easy	+	+	(+)	(+)	+	+	+

+ = yes 1. two reactors with separate acid and gas phases resp.

(+) = Partially 2. recycle or mixing mandatory

0 = no/none 3. in excess of the mixing caused by gas bubbles

Table 4.5 : Positive and Negative Factors* Determining the Active Biomass Retention of Various High Rate Anaerobic Wastewater Treatment Process Under Very High Loading Conditions+.

Sr.No.	*Factors*	*UASB*	*Upflow An-filter*	*Down Flow Stationary Fixed film*	*Expanded Bed*	*Fluidized Bed*
1.	Redistribution of the sludge due to high turbulence Flotation	– (at very high sludge loads)	–		–	–(?)
2.	Distribution of sludge aggregates	-(?) (possibly at very high lodas)				
3.	Detachment of biofilm		–(?)	–(?)	–	(- - -) (at incomplete acidogensis)
4.	Sludge bed expansion	±				
5.	Space occupied by packing/carrier material		–(---) (depending on the packing)	–(---) (depending on the packing)	(- - -) 70–80%	(- - -) 10–20%
6.	Surface area of the packing/carrier		+	++++ (1000 m^2/m^3)	+++ (100m^2/m^3)	++++ (2000-5000 m^2/m^3)
7.	Bed expansion				- - (10-20%	- - - - (approx50%)
8.	Film thickness		+	+++	++++	++
9.	Biomass conc., mg/L	>30,000	~10,000	~10,000	~35,000	>35,000

* It is assumed that all primary conditions for optimal application of various processes are met.

\+ Source Letting G, High Rate Anaerobic Wastewater Treatment using UASB Reactor Under a Wide Range of Temperature Conditions, Biotechnol. Genetic Rev., 2 (Oct), p253-284(1984).

Table 4.6 : Important Features of Various High Rate Anaerobic Wastewater Treatment Systems

Sr.No.	*Features*	*UASB*	*AF*	*Downflow Stationary Fixed Film Reactor*	*Expanded bed*	*Fluidized bed*
1.	Rate of start up					
	first start up	4-16 weeks	>3-4 weeks	>3-4 weeks	>3-4 weeks	About 3-4 weeks
	secondary start up	0-2 days	0-2 days	A few days(?)	A few days(?)	uncertain
2.	Performance w.r.t. removal and stabilization of SS	Satisfactory at low and moderate loading rates	Fairly good at low SS and when filter is not clogged	Very poor	Rather poor	Very poor
3.	Risk of channelling	Small, unless a poor feed inlet distribution system was installed	Great at high SS and in clogged filters	Small	Small	Almost non-existent
4.	Extent of effluent recycle required	Generally not required	Generally not required	Slight	Moderate	High recycle factor generally required
5.	Sophisticated feed inlet distribution system required	For low strength wastes and with dense sludge beds	Presumably beneficial	Not	Necessary	Essential
6.	Gas-solid separation device required	Yes, essential	Could be beneficial	Not required	Could be beneficial	Beneficial
7.	Carrier packing required	Can be beneficial specific cases	Essential	Essential	Essential	Essential
8.	Height-area ratio	Can be fairly high for high granular sludge beds	Moderate	Moderate	Moderate	Very high

4.5.2.5 Mixing

Recycled bed and recycled floc reactors need some sort of mixing in order to keep the bed and the flocs suspended. Mixing in order to obtain a higher degree of contact between the wastewater and the micro-organisms might be advantageous in sludge blanket reactors[22]. Mixing can be accomplished by mechanical stirring, gas mixing or by recycling of the wastewater, the latter as used in fixed beds by Chian and Dewalle[23] and Witt *et al.*[24]. Recycle may also be advantageous in relation to toxicity and pH-problems[25] and for general process improvement[22].

It should be noted that too intensive mechanical mixing may have an adverse effect on the process, by dispersing microstructures with organisms living in symbiotic relationships, like propionic/butyric acid bacteria and hydrogen utilizing methane formers.

4.5.2.6. Separation Equipment

For some reactors a solids-liquid separation equipment is an integrated part of the reactor. For all other types of reactors some sort of equipment might be necessary either in order to prevent occasional wash-out or to improve process performance. It is unlikely that anyone would build a fluidized bed reactor without some sort of separation equipment or other biomass security measures, as a process failure could result in total loss of biomass within 15 minutes. Expanded bed reactors have the same inherent risk, although smaller, whereas the other reactor type do not have this problem as they operate with much higher hydraulic passage time.

Ultrafiltration as a solid-liquid separation technique enhances process performance due to the removal of suspended solids from the effluent. For suspended floc processes ultrafiltration enables the maintenance of very high suspended solid concentrations in the reactor[26,27].

4.5.2.7 Phasing

In principle, phasing can be used for all reactor types. Fixed and moving beds must be equipped with recycle in the methane reactor in order to overcome pH problems.

4.5.2.8 Suspended Organics

Reactor types like expanded and fluidized beds and to a certain degree sludge blanket reactors might have their biomass structure damaged by significant amounts of suspended organics in the influent. Other reactor types may encounter problems due to clogging.

4.5.2.9 Inerts in Waste

Reactors able to selectively allow inert particles to pass through might be loaded higher than reactors where inerts are integrated in the biomass and thus influences solids retention time considerably. Fluidized beds–and partially expanded beds and sludge blanket reactors are believed to possess this property.

4.5.2.10 Foaming

Foaming problems are encountered in several reactor types. Fixed and moving beds are believed to have the smallest problems in this respect.

4.5.2.11 Gas Bubbles

Gas bubbles may cause different problems. They may adhere to flocs/bed particles and cause these to rise in the reactor, and may result in wash-out of biomass or deterioration of the effluent quality. Gas bubbles entrapped in fixed beds may result in clogging, and finally rising bubbles may cause channeling and short-circuiting in the reactor. Problems with gas bubbles are frequently met in fixed beds, expanded beds, fluidized beds and sludge blanket reactors.

4.5.2.12 Contact Wastewater/Biomass

The fixed bed reactors and possibly the moving bed and the sludge blanket reactors-have the lowest degree of contact between wastewater and biomass.

4.5.2.13 Hydraulic Overloading

Reactors with a separation equipment and no recycle are very susceptible to hydraulic overloading which will result in biomass washout. Other reactor types with separation equipment and recycle are less susceptible whereas expanded beds and fluidized beds are able to withstand even severe hydraulic overloading (of raw wastewater). The same holds for fixed and moving bed reactors, due to the biomass attachment to the fixed inert media.

4.5.2.14 Organic Overloading

Reactors where biofilm/floc characteristics are important for process performance (fluidized bed, expanded bed and sludge blanket reactors) are less suited for organic overloading, which tends to alter the characteristics of the film or the flocs.

Fixed film reactors are able to withstand high loading variations and to operate at a once per day or twice per day loading pattern[28,29].

4.5.2.15 Toxic Substances

For some environmental factors the influence or the various reactor types is considerably different. Fixed beds are superior to recycled floc reactors with regard to ability to withstand toxic substance[30]. Norrman[31] has observed that an expanded bed is broken down after exposure to toxic waste, whereas a fixed bed recovers without breakdown. Reactor types like fixed and moving beds are in general believed to withstand the effect of toxic compounds and other environmental factors better than expanded, fluidized and recycled beds plus sludge blanket reactors. The effectiveness of the last group of reactors depends to a large degree upon characteristics of bed or floc particles, characteristics which are easily changed by a changing environment.

Fixed and moving bed reactors tend to have a more or less hydraulic plug flow regime which may cause problems due to high concentrations of toxic substances in the influent or due to low pH. This can be overcome by recirculation as studied by Chian and DeWalle[31,32,33].

4.5.3 *Start-up*

All anaerobic processes have start-up problems. Fluidized bed and sludge blanket reactors seem to give more problems than the start-up of the other reactor types. Start-up after a longer close-down period is easy for most reactor types, but expanded and fluidized bed reactors are believed to be most troublesome in this respect.

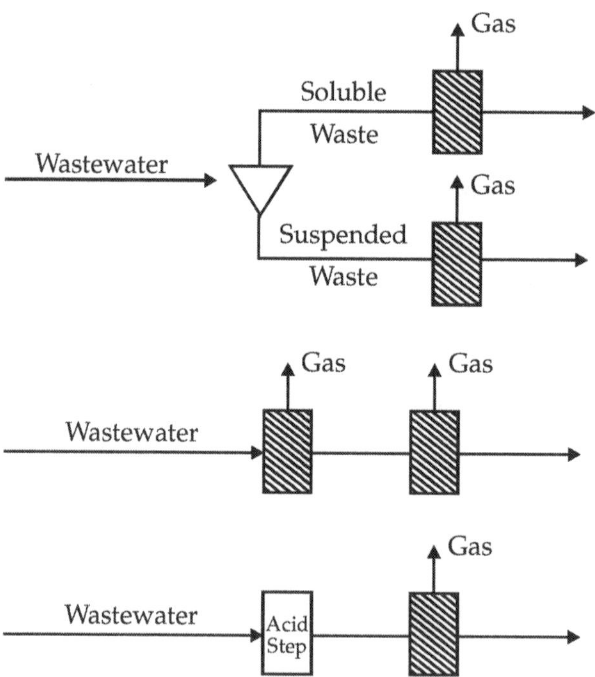

Fig. 4.6 : Staged and Phased Anaerobic Process

4.5.4 *Basic Process Types for Anaerobic Treatment of Wastewater*

Anaerobic reactors can be combined and equipped with various sludge/ water separators to a multitude of processes. In order to specify the differences in the processes, a few definitions are needed, *viz.* :

Parallel processes are two or more anaerobic processes, each single process comprising both the acid and the methane step, Fig. 4.6.

Staged processes means two (or more) anaerobic processes in series, each single process including both the acid and the methane step, Fig. 4.6.

Phased process means an anaerobic process where the acid and the methane step is separated, Fig. 4.6.

An important factor in process lay-out considerations is the solubility index of the wastewater. Wastewaters with a high solubility index (0.8–1.0) or low solubility index (0.0.-0.2) are basically treated in single pass processes, whereas wastes with a medium solubility index (0.2.–0.8) in many cases could be treated advantageously in parallel process.

Phasing of the anaerobic process[34-38] can be used in every case, and might result in reduced total reactor volume and better process control possibilities. Sulphate rich wastes might advantageously be phased or staged, in order to confine a possible inhibition caused by hydrogen sulphide, to the I step. In a combined process the inhibition will be effective on the slow growing methane bacteria, and thus results in a relatively bigger increase in necessary reactor volume. This comment also holds for other toxic materials, *e.g.*, those present in pulp and paper industry wastes[39].

Fig. 4.7 shows nine basic process lay outs. In order to increase process performance, solid-liquid separators could be used on all wastewater effluent streams, but in many cases this will result in a slight effluent quality improvement only. Reactor types like fixed and moving beds are in general believed to withstand the effect of toxic compounds and other environmental factors better than expanded, fluidized and recycled beds plus sludge blanket reactors. The effectiveness of the last group of reactors depends to a large degree upon characteristics of bed or floc particles, characteristics which are easily changed by a changing environment.

Fixed and moving bed reactors tend to have a more or less hydraulic plug flow regime which may cause problems due to high concentrations of toxic substances in the influent or due to low pH. This can be overcome by recirculation[31-33].

4.5.5. Design of Fixed Film Reactors

There are basically two different approaches to the design of wastewater treatment reactors, *viz.* :

- The traditional approach by which years of experience is synthesised into permissible loading figures on which the design in based and a resulting degree of purification is expected;
- The conceptual approach by which it is attempted to simulate the processes involved to such a degree that the purification result can be predicated.

Considering the complexity of the processes involved in anaerobic reactors it is in fact incredible how well the traditional approach has worked in practice for many years. However, numerous examples can be mentioned where the crude loading figures for design did not take into account relevant features of the processes. It can be said that the loading figures work well as long as they are applied to simple designs within the limits of the conventional experience. But it is dangerous to extrapolate to conditions outside the general range of experience. The conceptual approach strives to incorporate the main features of the processes in order to increase the generality of the applicability of the design approach. The difficulty inherent to this approach is the complexity that has given rise to a lot of scepticism from practitioners. This is very understandable because of the difficulty in understanding the content of the simulation, because of the number of unrealistic models proposed and because of a number of failures in practical application. It is time for bridge-building between the well proven practical approach and the fundamentally more correct approach. The least that can be demanded from the communication of the conceptual approach is an identification of the most fundamental features that have to be accounted for in design, whether based on the loading or the conceptual approach. The research results derived from the conceptual approach must be communicated in such terms and figures that they are applicable to practitioners when the results are ready to be so. Today, that can be done for a number of fixed film reactors and processes and is expected to influence design practice significantly.

The three potentially most relevant design parameters and appraisal parameters are related to either unit surface of the support media or to unit of sludge mass in the reactor or volume of reactor. It is an illustration of the failure of the conceptual approach in this case that today there is no indication as to which of these three parameters is the one of relevance under which circumstances.

If the biofilm of a fixed film anaerobic reactor is partially penetrated due to diffusional resistance, the removal has to be related to the surface area of the media in the reactor. If the wastewater is dominated by particulate matter that has to be adsorbed and hydrolysed, the removal has to be related to the surface area too.

However, if the biofilm is fully penetrated, either because the biofilm is thin, because the substrate concentrations are high or because of the methane production causes an increased diffusional exchange, the removal has to be related to unit of sludge mass. That again has to be related to other design parameters that influence the sludge mass accumulated per unit surface of the media or per unit volume of reactor.

No wonder that the practitioners look at extremely involved mathematical simulations with a lot of scepticism when not even such a fundamental question can be given an adequate answer.

PROCESS LAY-OUT	Type of process			Suitable for solubility Index		
	Parallel	Staged	Phased	0·0–0·2	0·2–0·8	0·8–0·1
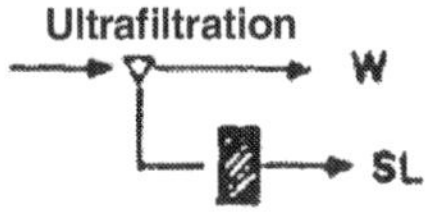 Ultrafiltration → W, SL				+	+	+
 → W				+	(+)	+
 → W			+	+	(+)	+
 → W		+		+	(+)	+
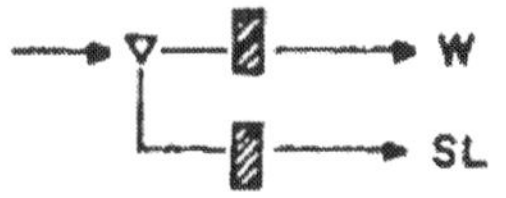 → W, SL	+				+	
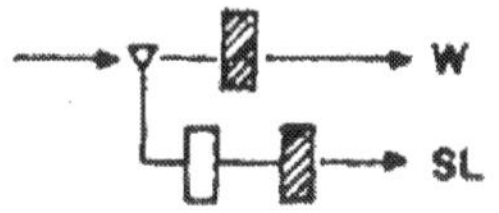 → W, SL	+		(+)		+	
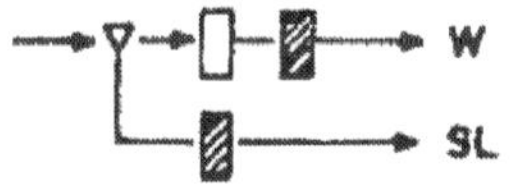 → W, SL	+		+			
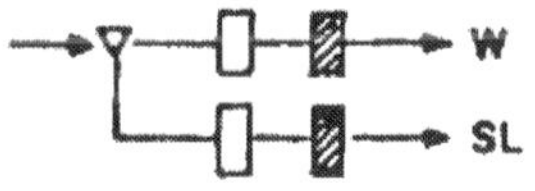 → W, SL	+		+		+	
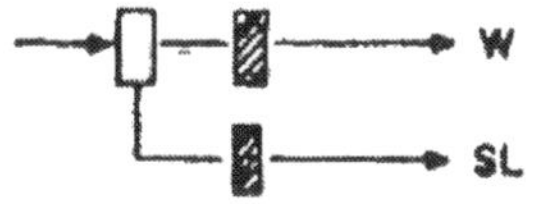 → W, SL		+		+	(+)	+

Methane producing reacter   Acid producing reacter   Solid/Liquid separator   W = Wastewater   SL = Sludge

Fig. 4.7 : Basic Process Lay-Outs, Anaerobic Processes

4.5.5.1 Biofilm Surface Area

The surface area present in the various fixed film reactors depends not only on the type of the reactor but also upon the mode of operation and the wastewater under treatment. The effect of the inert support material upon biofilm development have been studied by Murray and van den Berg[17] and Salkinoja *et al.*[40] The conclusion is that a porous inert media enhances biofilm development considerably as compared to a more smooth media. Investigations by van den Berg and Lentz and Young and Dahab[12] have demonstrated that in fixed bed reactors, a substantial part of the biomass is present in non-attached form, even pelletization granulation is observed[42]. Martensson and Frostell[43] have shown that in a recycled bed a biofilm structure is not found. The inert media is trapped within the biological flocs, resulting in higher density of the flocs. van den Berg and Lentz, have shown that in down-flow fixed bed reactors the major part of the biomass is attached to the inert surfaces. This difference between up- and down-flow reactors is not believed to be general for full-scale fixed beds but is probably a result of the experimental procedure and the specific inert media used.

The above mentioned investigations suggest that biofilm surface area might not be that important in fixed beds (and probably in moving beds as well). Young and Dahab[12], have demonstrated that the efficiency of the fixed bed reactors have no correlation to specific surface area. The implication of this is that the biofilm may be regarded as very homogeneous.

The anaerobic biofilm on fixed and moving beds tends to slough rather easily. This is either due to methane bubble generation in the inner part of the biofilm or to a general characteristic of anaerobic films. This may not always happen.

In sludge blanket reactors, where pelletization is more pronounced than in fixed bed reactors, the pellets are easy to break up by even slight mechanical stirring. This also indicates a generally loose structure of anaerobic films in accordance with the fact that anaerobic organisms are unskilled slime producers, Salkinoja Salonen *et al.*[44]

For expanded and fluidized beds, biofilm area might be of fundamental importance, but this still have to be demonstrated.

Even if the surface area of biofilms have no direct influence upon process performance, a biofilm structure might be important due to a selectivity towards suspended solids, which is discussed later on, and due to increased resistance to slug dose toxicants and finally due to a higher safety against biomass wash-out. Furthermore, a biofilm ecosystem might be well suited to degrade specific organic compounds, *i.e.*, pentachlorophenol.[40]

To avoid clogging of fixed bed reactors, it might be necessary to backwash at intervals or to drain off the suspended solids entrapped in-between the inert media[12]. From nitrifying fixed bed reactors it is known that backwash does affect process efficiency only to a minor degree and that the bed quickly recovers. An identical mechanism could be expected in anaerobic beds.

4.5.5.2 Biomass Concentration and Load

If the biofilm surface area has only minor influence upon reactor performance, then the objective of reactor constructors must be to maximize the biomass concentration within the reactor, eventually by the use of inert media to control hydraulics. Table 4.5 lists biomass concentrations found in various types of anaerobic reactors. The maximum substrate removal rate for a mixed anaerobic process was found to be approximately 1 kg COD/(kg VSS.d) at 35°C. This means that the upper loading limit for an anaerobic process with respect to biodegradable organics is 1-1.5 kg COD/(kg VSS.d). Loadings higher than this will result in significant fatty acid build-up, and consequently inhibition of the methane production. If a part of the organic substances are non-biodegradable or non-hydrolysible, then higher loadings can be permitted either if the inert material is soluble or if it (as suspended material) is not adsorbed. This is in accordance with general observations (van den Berg and Lentz[41]). The use of COD as the general parameter for organic substances in anaerobic processes has many disadvantages. The major disadvantage is that it does not reveal anything with respect to biodegradability. For this purpose BOD is still unsurpassed. This is true even for anaerobic processes.

Treatment of wastes containing high concentrations or inert material (organic or inorganic) will result in maximum organic loadings lower than 1 kg COD/(kg VSS.d) due to the influence of the inerts upon solids retention time. However, if fixed film processes are used instead of CSTR system, loading rates can be easily increased several folds.

4.5.5.3 Volumetric Load

The most widespread design parameter in anaerobic processes is volumetric load, B_v, expressed as kg COD/(m^3.d). This parameter is often used to demonstrate superiority of one type of reactor over another type. The maximum obtainable volumetric load is a function of several factors, including the organic matter in the wastewater and the observed yield coefficient. The influence of the type of reactor on the volumetric load is question of active biomass concentration present which includes the concept of solids retention time as a design parameter, the following equation can be established, assuming that effluent substrate concentration is small as compared to the influent concentration :

$$B_{v,\,COD} = \frac{(X_2/\theta_x) - B_{v,\,inert}}{Y}$$

Where

$B_{v,COD}$ is biodegradable volumetric load, kg COD/(m^3.d),

X_2 is reactor biomass conc., kg VSS / m^3

θ_x is solids retention time, d

$B_{v,inert}$ is load of non-biodegradable suspended organics, kg VSS (m^3.d) and

Y is yield coefficient, kg VSS/kg COD

Figure 4.8 depicts the influence of the various parameters, demonstrating the influence of yield and wastewater composition, and reactor biomass upon the maximum possible volumetric loading for a given solids retention time, $\theta_x = 15$ days. The volumetric loading is affected by the type of substrate through the yield coefficient. For wastes with a low yield coefficient, *e.g.*, acetic acid, the volumetric loading can be very high, whereas a waste with a high yield coefficient only allows for a small volumetric loading rates.

In the case of a two-phase process, the acid producing reactor can be loaded very high (4-6 times higher than a reactor for methane production) primarily due to the low solids retention time requirements of the acetogenic bacteria. It is seen that the load of non-biodegradable suspended organics plays only a minor role in these high-loaded processes. When the reactor biomass concentration is increased by a factor of 3 (from 10 to 30 kg VSS/m^3) then the maximum volumetric loading is increased almost by the same factor. This is true in Fig. 4.8 as well as in Fig. 4.9.

For wastes with high concentrations of inert suspended material, which leads to high inert loading, Bvinert, the result is, that the inert fraction of the suspended solids in the reactor is high. This the amount of active biomass is being limited and subsequently the maximum volumetric load is reduced as shown in Fig. 4.10 for a waste where 0, 5, 10 and 20 per cent of the influent COD is inert volatile suspended material. Rozzi and Verstraete[45] have theoretically discussed this for the recycled floc process (the contact process.) It is valid for the attached biofilm processes too, as long as the inert material is adsorbed, enmeshed or otherwise incorporated into the attached biomass.

As seen from the figures, the volumetric load can be used for design, and must in fact be a part of every process design. The problem is, that the parameter is only a secondary design parameter exemplified by the fact that reactors fed with volatile acid substrates can be loaded 4-5 times as high as reactors fed with more complex substrates carbohydrates, proteins, etc. This increased loading capability is caused primarily by the small yield coefficient for a volatile acid/ acetic acid substrate as compared to a more complex substrate and secondarily by an increased fraction of methane bacteria in the biomass. This effect should be taken into consideration, when comparing the results from different types of reactors.

4.5.5.4 Effect of Suspended Solids

The ability of anaerobic reactors to act selectively with regard to suspended solids in the influent is an important question. Suspended solids in the influent may be entrapped in the biofilm structure, but only loosely, which means that they are rather easy to wash-out either by the normal liquid flow and gas bubble disturbances or by some sort of backwashing. If this entrapment mechanism-which is well-known from nitrifying fixed beds-is functioning in anaerobic biofilms, then the possibility exist is that the biofilm has two layers, the inner layer the anaerobic active organisms, and an outer layer where suspended solids are loosely attached and by and then sloughed off. This double layer structure has the built in advantage of higher solids retention time for the inner layer than for the outer layer of the biofilm. Observation by Lettinga *et al.*[15] for sludge blanket reactors supports this theory. If it is correct, then this might be an important argument for having a biofilm structure in the reactor, whether it is supported by an inert media or not.

4.5.5.5 Design of Anaerobic Processes

As mentioned at the start of this chapter, it is still an open question how to design anaerobic processes in fixed film reactors. At present, there is no method for practical design using substrate loading per unit surface area, and it has not been demonstrated that this design parameter is the best one. The other choice for design is substrate loading per unit sludge mass. Recent knowledge of the fixed film processes indicates that this parameter can be used as primary design parameter. If the biomass is considered homogeneous, then solids retention time is just as good a design parameter. Together with the yield coefficient and the biomass concentration in the reactor, the basic design can be made. In case of suspended solids selectivity, as discussed in section 4.5–5.4, the solids retention time should not be used as a design parameter. The temperature effect should be taken into account. It is characteristic of stable anaerobic processes that the per cent removal of substrate is very high. Suspended solids in the effluent may deteriorate overall removal efficiency but this can be dealt with by solid liquid separation if needed. High-loaded processes are reported to give sludge with poor settling characteristics. Lower organic loading rates do not improve the substrate removal efficiency substantially, but might increase the methane production and reduce the yield coefficient through an improved stabilization of the solids. In recycled floc reactors, a lower organic loading rate (0.25 kg COD / kg VSS.d) improves the settleability of the sludge[46]. Lower organic loading may also be needed for wastes where the hydrolysis of complex organics is the rate limiting step. In contrast to this, Pol *et al.*[18], have found that a low loading rate deteriorates pelletization and that the loading rate-at least during start-up should exceed 0.6 kg COD/(kgVSS.d). A loading rate of this magnitude can not be used as initial start-up loading which means that very long start-up periods are needed before pelletization is accomplished.

The importance of the above mentioned findings in relation to biofilms attached to inert surfaces is impossible to assess at present, but in general, the significance of environmental factors are increased for reactors which depends strongly upon very specific biofilm characteristics (*e.g.*, fluidized beds, sludge blanket reactors and expanded beds).

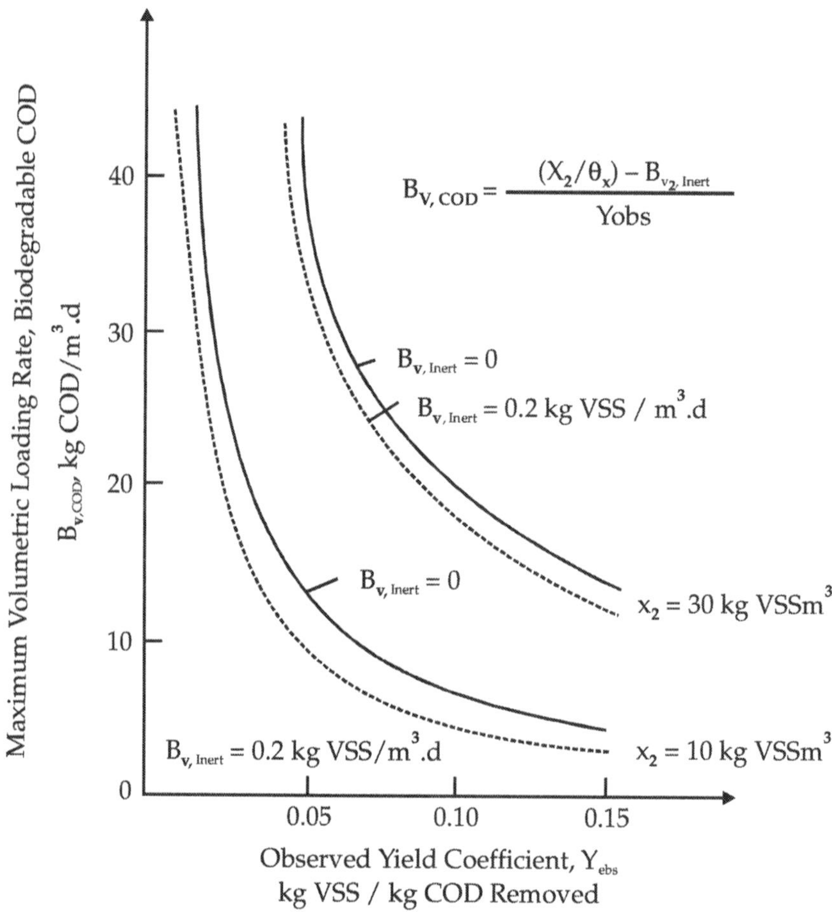

Fig. 4.8 : Volumetric Load (Biodegradable COD) for Anaerobic Methane Producing Processes, 35°C with a Solids Retention Time, θ_x, of 15 Days. ($B_{v, \text{ inert}}$ = Load of Non-biodegradable VSS, X_2 = Reactor Biomass Concentration)

The well functioning of most of the reactor types discussed here, depends upon certain biofilm characteristics, which means that a closer process control and a more sophisticated design is needed in order to have the processes working. A proper reactor operation is not only a question of the selection of one or another type of micro-organism (as in many aerobic processes) but a question of the selection of the highly specific micro-organism or-organisms that will make the reactor function. This means that the environmental factors in anaerobic reactors must be much closer supervised than in aerobic reactors.

4.5.5.6 Startup of Anaerobic Reactors

The start up of anaerobic processes has always been a problem due to the long startup periods involved, and is still considering a major area for future research. The main problem seems to be the generation of the exact microbial culture for the waste in question. If this culture is at hand, then start-up is a question of a few days, even if the culture have been stored for long periods (more than a year). In cases where the right microbial culture is not at hand, inoculation can be done with municipal digester sludge, manure, soil, etc. In such cases, it takes 1-2 month before the process are in operation, although not at maximum rate. Due to the low sludge yield of methanogens, it can take 4-8 months before a microbial steady state is arrived at. For thermophilic conditions, where the sludge yield is even smaller, longer periods-may be exceeding a year, may be needed in order to have the process working at a real steady state.

Increased start up velocity can be obtained, if biomass wash-out is minimized during start-up, either by some sort of temporary solid-liquid separation, like ultrafiltration or by process operation measures, like high recycle.

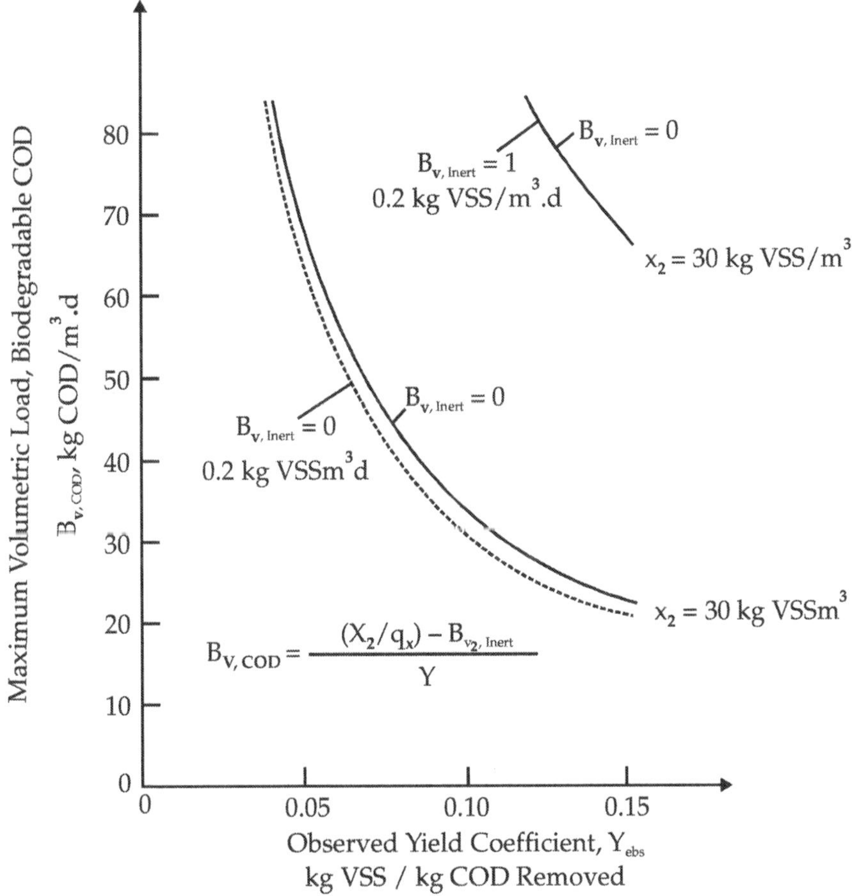

Fig 4.9 : Volumetric Load (Biodegradeable COD) for Anaerobic Production of Acetic Acid, 35°C (Solids Retention Time, θ_x = 3d). ($B_{v, inert}$ = Load of Non-biodegradable VSS) (X_2 = Reactor Biomass Concentration)

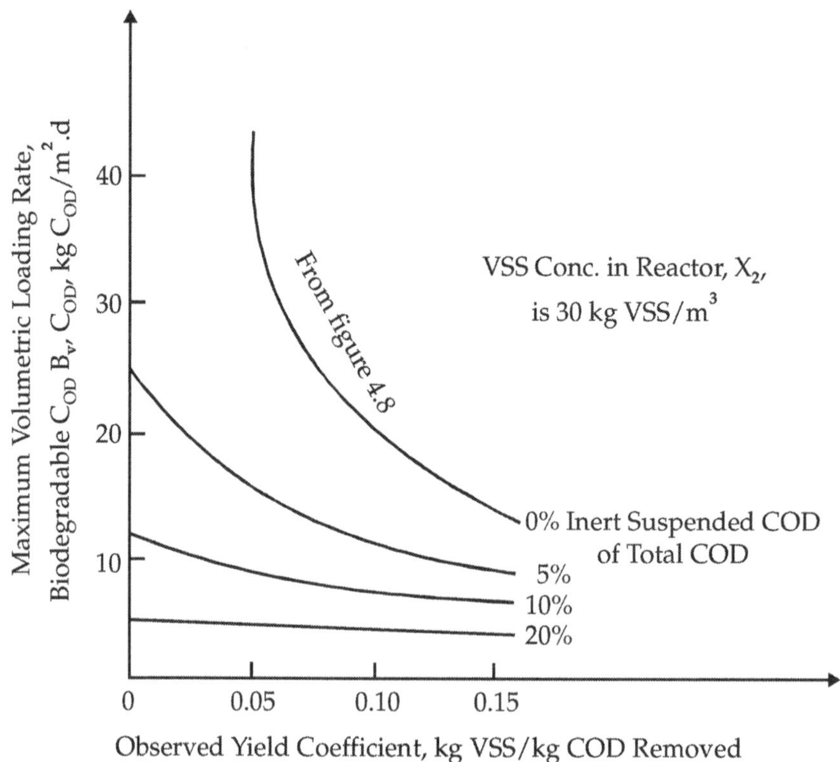

Fig 4.10 : Volumetric Load (Biodegradable COD) for Anaerobic Methanol Processes, 35°C with a Solids Retention Time, q_x of 15 days, and Various Percentages of Inert Suspended Material.

The initial loading during start-up must be low, e.g., approx. 0.1 kg COD/kg.d corresponding to 1-2 kg COD m^3.d for a reactor with a biomass concentration of 10-20 kg VSS/m^3. When gas production is increasing then the load can be increased step-wise (50% per week).

Volatile fatty acid should be monitored during start-up and if increased levels are observed (>1.0-1.5 kg HAc/m^3) then the loading should be reduced until the acid concentration is decreasing again.

The creation of a biofilm seems to be connected with special problems Salkinoja-Salonen *et al.*[7] A porous inert surface promotes biofilm growth. Slime producing organisms or substrates, may be under aerobic and high temperature conditions, can form the basic slime layer which can subsequently be used in an anaerobic process. The presence of Ca-ion seems to be advantageous and the same holds for high alkalinity.

During start-up, as during process operation in general, environmental factors must be kept within the limits for microbial comfort, that is pH, nutrient availability temperature, etc., must be taken into account.

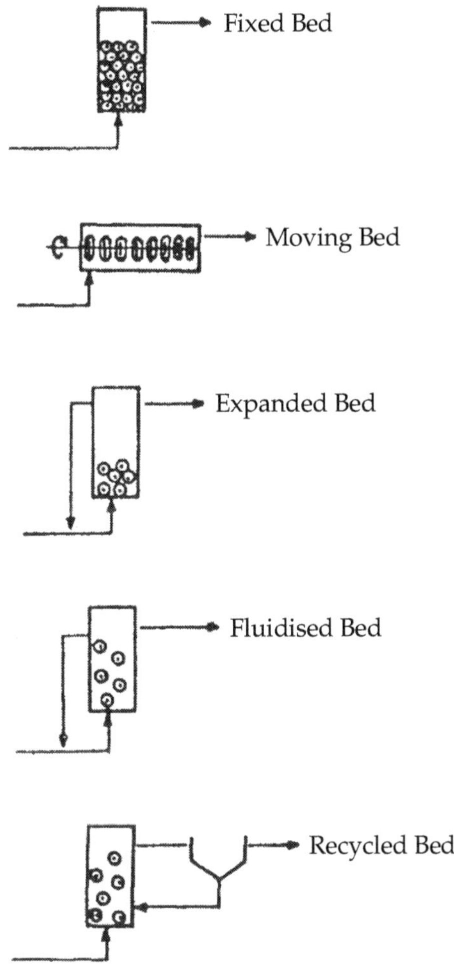

Fig. 4.11(A) : 5 Basic Reactor Types, Anaerobic Fixed Film

Basic design and operation parameters for anaerobic fixed film reactors are given in Table 4.7. Fig. 4.11 and 4.12 depicts the various types of anaerobic bioreactor systems.

The support material for biomass should possess following properties :

- durability,
- light in weight,
- no toxicity,
- good porosity to avoid clogging,
- availability of micro-pores,
- cost,

- media should be easily available,
- media should possess the property for easy attachment of biomass,
- large specific area, and
- media should be inert.

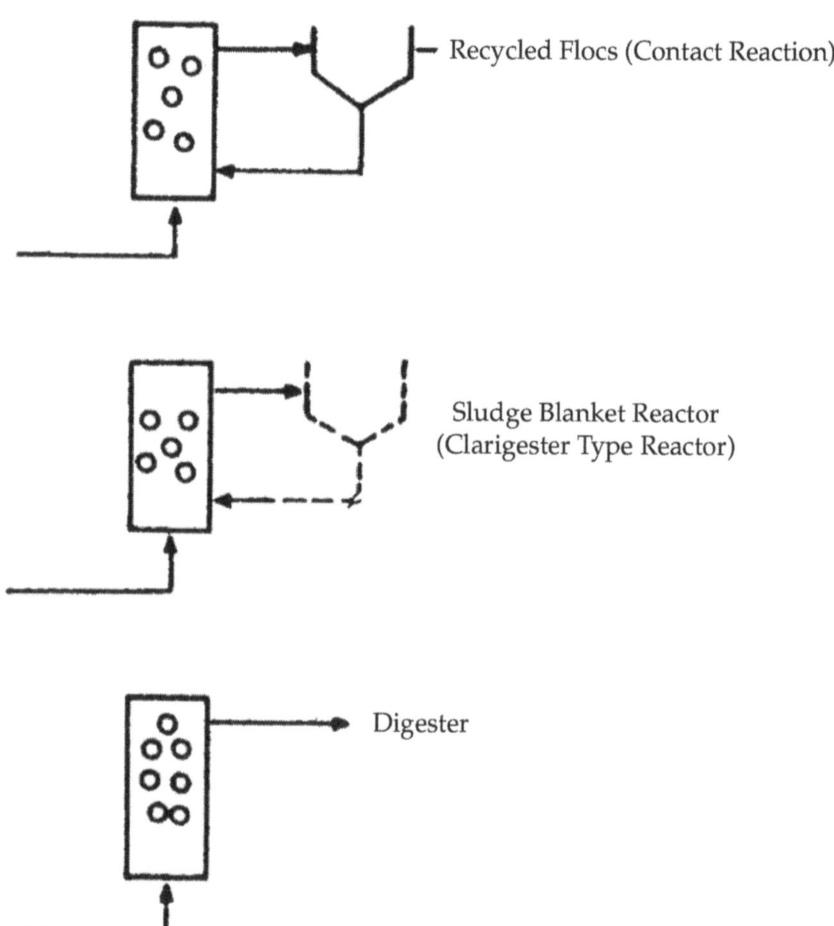

Fig. 4.11 (B) : 3 Basic Reactor Types, Anaerobic Flocs/Non-Attached Films

As compared to other fixed film reactors, fixed beds have been operated on a wide variety of support material types, shapes and sized. Material used include quartz, plastic, clay, oyster shells, stones, polymer foam, activated carbon, limestone, sand, needle punched polyester and PVC. The sized reported range from 1.5-100 mm.

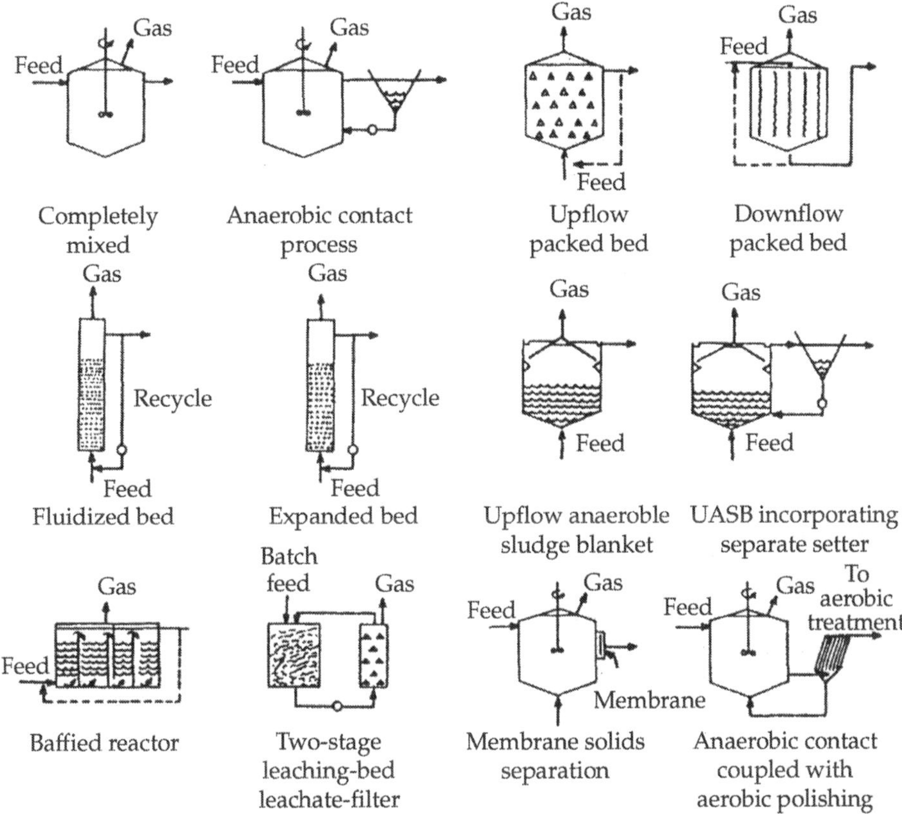

Fig. 4.12 : Different Anaerobic Reactor Configuration

The proper selection of support material is based on two factors the concentration of particulates in the incoming waste and the organic composition of wastewater. Wastewater composition is important because certain types of wastes (such as carbohydrate containing wastes) produce larger biomass yields; also, high strength waste produce more cell yield than low strength wastes. In general, the material selected should have a high surface to volume ratio so as to provide larger surface area for biofilm attachment, while maintaining a sufficient void volume to prevent clogging of the reactor.

The most important factors that must be considered are identified in Table 4.8. Before final decision is to be made, it is necessary that ranking of treatment options must be considered. Ranking of various treatment options must be carried out to arrive at the most appropriate alternative which minimizes adverse impacts and maximizes benefits through enhanced economic output. An integrated approach takes into account various criteria such as :

- environmental risks
- health risks

Table 4.7 : Basic Design and Operation Parameters for Anaerobic Fixed Film Reactors

Sr.No.	*Parameter*	*Unit*	*Fixed bed*	*Moving bed*	*Expanded bed*	*Fluidized bed*	*Recycled bed*	*Sludge blanket*
1.	Inert material type		Gravel/Plastic	Plastic	Garnet/Sand/ Carbon	Sand/Garnet/ Carbon	Sand	-
2.	Inert material dia.	mm	20-50	1000-3000	0.3-3	0.2-1	0.01-0.1	-
3.	Inert material, disc spacing	cm	-	10-20	-	-	-	-
4.	Inert material, rotation	rpm	-	2-5	-	-	-	-
5.	Inert material peripherical velocity	m/s	-	0.3	-	-	-	-
6.	Inert material, submergence	%	100	75-100	100	100	100	-
7.	Porosity, empty bed	%	40-98	-	-	-	-	-
8.	Porosity, operation	%	20-90	-	-	-	-	-
9.	Bed expansion	%	-	-	20-40	30-100	-	-
10.	Specific surface area	m^2/m^3	60-200	100-200	1000-3000	1000-2500	2000-5000	-
11.	Height of reactor	m	3-6	-	2-4	4-8	5-10	2-6
12.	Radius of reactor	m	5-20	1-3	2-3	2-3	5-20	5-20
13.	Vertical velocity, empty bed (incl. recycle)	m/h	0.01-0.10	-	2-10	6-20	-	0.05-0.30
14.	Recycle ratio	-	-	-	2-100	5-500	0.5-2	-
15.	Biomass concentration	$kgSS/m^3$	5-15	5-15	10-30	10-20	5-15	5-15
16.	Attached biomass	% of total	20-80* 50-90**	50-80	90-100	95-100	0	60-80
17.	Suspended biomass	% of total	20-80* 10-50**	20-50	0-10	0-5	100	20-40
18.	Suspended solids, effluent	gSS/m^3	20-300	20-300	20-100	20-100	20-100	20-100
19.	Gasflux, vertical	$Nm^3.m^2.d$	5-20	-	5-40	5-40	-	5-20
20.	Gasflux, vertical maximal	$Nm^3.m^2.d$	10-20	-	30-40	30-40	-	10-20
21.	Energy pumping	Wh/m^3	-	-	10-20	15-30	-	-
22.	Energy pumping, (incl. recycle)	Wh/m^3	20-40	5-10	20-1000	75-3000	10-30	15-30
23.	Energy, rotation	Wh/m^3 tank	-	20-80	-	-	-	-
24.	Energy, mixing	Wh/m^3 tank	-	-	-	-	5-15	10-30

- aesthetic risks
- annualised costs
- reuse potential
- institutional requirements
- land requirements
- process reliability

The exercise essentially involves ranking of the alternatives based on above mentioned criteria and comprises of the following steps, *viz.* :

- The various alternatives should be ranked on the basis of the defined environmental criteria in accordance with the laboratory findings, design data and cost estimates;
- a total score of 1000 is apportioned between the assessment criteria in each case based on their importance as well as subjective judgement;
- the treatment options to be evaluated against each criterion and assigned scores;
- total score for each alternative to be computed;
- the treatment options to be ranked by comparison of total scores.

Table 4.8 : Important Factors that must be Considered When Selecting and Evaluating Unit Operations and Unit Processes

Sr.No.	*Factor*	*Comment*
1.	Process applicability	The applicability of a process is evaluated on the basis of fact, experience, data from full scale plants, and pilot plant data from plant studies. If new or unusual conditions are encountered, pilot plant are necessary.
2.	Applicable flow range	The process should be matched to the expected flow range.
3.	Applicable flow variations	Most unit operations and processes work best with a constant flow rate, although some variation can be tolerated. If the flow variation is too great, flow equalization may be necessary.
4.	Influent-wastewater characteristics	The characteristics of the influent affect the type of processes to be used (chemical/biological) and the requirements for their proper operation.
5.	Inhibiting and unaffected constituents	What constituents are present that may be inhibitory and under what conditions? What constituents are not affected during treatment?
6.	Climatic constraints	Temperature affects the rate of reaction of most chemical and biological processes. Freezing conditions may affect the physical operation of the facilities.

Contd...

Sr.No.	Factor	Comment
7.	Reaction kinetics and reactor selection	Reactor sizing is based on the governing reaction kinetics. Data for kinetic expressions usually are derived from experience, the literature, and the result of pilot plant studies.
8.	Performance	Performance is most often measured in terms of effluent quality, which must be consistent with the given effluent discharge requirements.
9.	Treatment residuals	The type and amount of solid, liquid and gaseous residuals produced must be known or estimated. Often pilot plant studies are used to identify residual properly.
10.	Sludge handling constraints	In many cases, a treatment method should be selected after the sludge processing and handling have been explored.
11.	Environmental constraints	Nutrients requirements must be considered for biological treatment processes. Prevailing winds and wind directions may restrict the use of certain processes, especially where odours may be produced.
12.	Chemical requirements	What resources and what amounts must be committed for a long period of time for successful operation of unit operation or process?
13.	Energy requirements	The energy requirements as well as probable future energy costs, must be known if cost-effective treatment systems are to be designed.
14.	Other resource requirements	What, if any, additional resources must be committed to the successful implementation of proposes treatment.
15.	Reliability	What is the long-term record of the reliability of the treatment process? Is operation process easily upset? Can it stand periodic stock loadings? If so, how do such occurrence affect the quality of effluent?
16.	Complexity	How complex is the process under routine conditions and under emergency conditions such as shock loadings? What level of training must the operator need to have to operate the process?
17.	Ancillary process required	What support processes are required? How do they affect the effluent quality, especially when they become inoperative.
18.	Compatibility	Can the process be used successfully with existing facilities? Can plant expansion be accomplished easily? Can the type of reactor be modified?

4.6 Two Phase Anaerobic Fermentation

The objectives of this section are to present data to illustrate the potential in applying conventional high rate anaerobic digestion to high COD industrial wastes and to show how an advanced digestion process, two phase anaerobic digestion can provide superior performance in terms of waste stabilization efficiency and net energy recovery. The application of two phase process was demonstrated by pilot and full-scale operations at industrial sites.

4.6.1 Problems with Conventional Digestion

With high COD industrial wastes, unbalanced digestion and ultimately, process failure are encountered when attempts are made to operate conventional digesters, at high loadings and short hydraulic retention times (HRT)[47]. This is illustrated by data from conventional high-rate digestion of a real soft drink bottling waste, the properties of which are presented in Table 4.9. The high rate digestion run was conducted in a 7 litre completely mixed laboratory, digester having a culture volume of 5 litre. The digesting culture was developed from municipal refuse, sewage sludge, and biomass acclimated inocula, which provided a diverse group of digesting organisms and this assured the enrichment of an appropriate flora of anaerobic organisms on selected soft drink manufacturing waste.

The digester was operated in the so called high rate mode at increasing loading rates and decreasing HRTs, as indicated in Table 4.10. To obtain stable digestion, it was necessary to add an inorganic nutrient salt (nitrogen source) and an organic additive at the rate of 0.011g VS/g.feedVS to overcome nutritional deficiencies. Balanced digestion, as indicated by low concentration of residual volatile acids, high methane contents in the gas and high gas yield, was obtained upto a loading rate of 0.08lbVS ft^{-3} day^{-1} and an HRT of ten days, Table 4.11. Volatile acids accumulated and gas and methane productions dropped sharply at a 10 days HRT when the loading was increased to 0.125 lb VS ft^{-3} day^{-1} corresponding to feed VS and COD concentration of 20 g/L (2% weight) and 26,000 mg/L respectively. Since methane formers grow much more slowly than acid formers[48], a kinetic imbalance between acid production by acidogens and its conversion by methanogens ensured at a high COD concentration of 26,000 mg/L and an HRT of 10 days. The degree of this imbalance increased as the digester loading rate was increased from 0.125 to 0.21 to 0.42 lbVS/ft^3.d and the HRT decreased from 10 to 7.1 to 4.5 days as evidenced by accumulation of volatile acids and gaseous hydrogen upto 12,000 mg/L and 50 mol %; a virtual cessation of gas production occurred at a loading rate of 0.27 lb VS/ft^3.d and a HRT of 7.1 days, and a feed COD concentration of 38,000 mg/L. These observations indicated that stable and satisfactory high rate digestion of soft drink bottling waste is not feasible at a loading rate of 0.125 lbVS/ft^3.d or higher and an HRT of 10 days or lower.

4.6.2 Two Phase Anaerobic Digestion

The data presented above show that the conventional high rate digestion process, which harbour two physiologically different digesting organisms (the acid formers and methane formers) under the same environment, experiences retardation and eventually fails when the substrate loading rate is increased beyond an upper limit and the HRT is decreased below critical value. The reason for the observed failure is that higher loadings and shorter HRTs favour and the enrichment of acidogenic organisms and preclude the establishment and flourishing of the methane forming bacteria. Because higher loadings and feed COD concentrations encourage volatile acid production in an anaerobic digester, it is reasonable to promote the acidification reaction and develop a staged system in which this microbial phase is optimized in a first stage 'acid digester'. Because conditions promoting efficient substrate to acids conversions is not conducive to stable and efficient acid to methane fermentation, effluents from the first stage acid digester must be methanated in a separate second stage methane-phase digester to accomplish feed gasification and concomittant stabilization of waste COD. Thus, a multistage two phase, as described by Pohland *et al.*[49], and Ghosh, *et al.*[48] evolves naturally, when attempts are made to conduct the anaerobic digestion process at high loadings and short retention times to reduce digestion cost and enhance net energy recovery.[50,51]

In an attempt to search for operating conditions for optimum acidogenic conversion of the soft drink bottling waste, acid phase digestion runs were conducted at several loading rates and HRTs as indicated in Table 4.12. In the final phase of the two phase work, the acid-phase digester was operated in tandem with an upflow anaerobic filter to convert the volatile acids to methane. The two-phase system consisted of a completely mixed, continuously fed acid-phase digester (culture volume 2.5 litre), the effluent from which was continuously fed to the anaerobic filter having a gross volume of 5.5 litre and a porosity of about 90 per cent. Filter effluents were continuously recirculated to the influent and at a high volumetric recirculation ratio. The performance of two phase system is presented in Table 4.13. The data indicated almost ideal separation of acid and methane fermentation phases were achieved as indicated by the cassation of methane production in the first stage and conversion of acids to methane at high concentration, yield and production rate in second stage. The overall two phase was very stable, even at a feed COD concentration of 40,000 mg/L and afforded higher methane production rate and COD reduction at a much shorter HRT and a higher loading from the digestion process is estimated at 14 million Btu day^{-1} assuming the observed net energy production efficiency, of 65 per cent for two phase digestion, Table 4.15. Thus for this particular plant, energy recovered from the liquid waste could potentially displace 82% of the purchased fossil-fuel energy.

The advantage of two phase digestion are indicated in Table 4.14 which compares data from a conventional high rate and two phase digestion run.

Table 4.9 : Characteristics of Soft Drink Wastes

Sr.No	*Parameter*	*Value*
1.	pH	2.5
2.	Volatile Solids (VS), % dry solids by weight	99.2
3.	Total Suspended Solids (TTS), mg/L	<100
4.	Elemental analysis, % dry solids weight	
	Carbon	51.6
	Hydrogen	5.36
	Nitrogen	<0.50
	Sulphur	<0.07
	Potassium	0.03
	Calcium	0.10
	Magnesium	0.004
5.	Heating value, Btu/lb	8533

Table 4.10 : Characteristics of Feed Slurries for Conventional High Rate Digestion

Sr.No	*Parameter*	*Run (Value)*		
		I	*II*	*III*
1.	VS loading rate, kg $VS/ft^3 \cdot d$	0.04	0.08	0.128
2.	COD loading rate, $kg/m^3.d$	0.8	1.7	2.6
3.	Retention time, days	15	10	10
4.	VS concentration, g/L	9.6	12.8	20.0
5.	COD, g/L	12	17	26

Table 4.11 : Steady Stagte High Digestion of Soft Drink Waste

Sr.No	*Parameter*	*Run (Value)*		
		I	*II*	*III*
1.	VS loading, $lbVS/ft^3.d$	0.04	0.08	0.124
2.	COD loading, kg $COD/m^3.d$	0.8	1.7	2.6
	HRT, days	15	10	10
3.	Methane yield SCF/lbVS-added	6.28	6.99	0.95
	Std m^3/kgCOD-added	0.31	0.33	0.04
4.	Gas production rate (GPR) Std vol/vol.d	0.41	0.33	0.26
5.	Gas composition, mol %			
	Methane	61.1	61.7	45.1
	Hydrogen	0.0	0.0	0.6
6.	Volatile acids, mg/L (HOAC)	180	110	2480
7.	Effluent pH	6.8	6.4	5.0

Whereas both digestion processes had comparable gas and methane yield and COD stabilization efficiency, the methane production rate from the two phase run was more than seven times that from conventional digestion. Also, the methane content (and heat value) of the gases from the methane-phase digester of the two-phase system was considerably higher than that of the conventional digester. This implies that the gas clear up cost should also be lower for a two phase process. Gases from the acid phase digester had a heating value of 100 Btu/SCF and could be used for digester heating.

Table 4.12 : Operating Conditions for Acid Phase Digestion of Soft Drink Bottling Plant

Sr.No	*Parameter*	*Run (Value)*		
		I	*II*	*III*
1.	Retention time, days	7.1	4.5	2.2
2.	VS concentration, g/L	29.1	28.8	35.2
3.	COD, g/L	38	37	35.2
4.	VS loading rate lb/ft^3.d	0.26	0.40	1.00
5.	COD loading rate, kg/m^3.d	5.4	8.2	20.5

Table 4.13 : Gas Production and Effluent Quality from Mesophilic Two Phase Digestion of Soft Drink Waste

Sr.No	*Parameter*	*Run (Value)*		
		Acid Phase	*Methane Phase*	*System*
1.	HRT	2.2	5.2	7.4
2.	Volatile loading rate, lbVSS/ft^3.d	1.00	0.40*	0.30
3.	Feed VS concentration, g/L	35.2	-	35.2
4.	COD loading rate, kg COD/m^3.d	20.5	-	6.1
5.	Feed COD, g/L	45	-	45
6.	Gas production rate, Std Vol/Vol.d	1.03	3.68	2.90
7.	Gas composition, Mol%			
	Methane	0.2	70.5	63.1+
	Hydrogen	27.2	0.0	2.9+
8.	Methane Yield, SCF/lbVS added	0	6.66	1.16
	Std.m^3/kg COD,	0	0.42	0.32
9.	Total COD, mg/L	31,800	1900	1900
10.	Volatile acids, mg/L(HOAC)	7880	450	450
11.	Ethanol, mg/L	3540	33	33
12.	Effluent pH	4.7	7.5	7.5

* Based on packed volume; + Assuming that gases from both digesters are mixed

Table 4.14 : Advantages of Two Phase Mesophilic Digestion of Soft Drink Waste

Sr.No	Parameter	Conventional High Rate Digestion	Two Phase Digestion
1.	Loading, kg COD/m^3.d	0.8	6.1
2.	HRT, day	15	7.4
3.	GPR, Std Vol/vol.d	0.4	2.90
4.	Methane, mol %	61.1	70.5
5.	Hydrogen mol %	0	2.9
6.	Methane yield		
	SCF lb/VS added	6.3	6.2
	Std kg/COD added	0.50	0.48
7.	Effluent Volatile acids, mg/L	180	450
8.	Effluent pH	6.8	7.5
9.	Volatile reduction efficiency,%	72	64
10.	COD efficiency, %	84	96

4.6.2.1 Advantages of Two Phase Digestion

The most significant advantages of the two phase process are that it allows overall process operation at much higher feed concentrations and loading rates, and much shorter HRTs–operating condition under which the conventional digestion process often fails– and yet is provides the same or higher methane yields and COD stabilization efficiencies than those of the conventional high-rate digestion process. These advantages are reflected in the dramatic reduction in plant capital cost and enhancement of net energy production achieved by applying two-phase digestion.

Table 4.15 : Digester Volume Pre-requirement and Net Energy Production from Conventional and Two-phase Digestion of Soft Drink Bottling Waste

Sr.No	Parameter	Conventional Digestion	Two Phase Digestion
1.	Loading Rate, kg CODm^{-3}day^{-1}	2.1	6.1
2.	HRT, days	15	7.4
3.	Digester volume ft^3 × 10^3	198	66
4.	Methane energy production 10^6 Btu/day	125	125
5.	Operating energy requirement 10^6 Btu/day		
	Feed slurry heating	56.3	38.0
	Digester mixing	12.1	1.3
	Digester heat losses	6.7	3.9
	Pumping	0.5	0.5
6.	Total	75.6	43.7
7.	Net energy production (NEP) 10^6 Btu day^1	49.4	81.3
8.	NEP as % of digester methane energy	39.5	65.0

The soft drink bottling plant from which the waste for this study was collected used liquid propane and No.2 fuel oil as the major energy sources. The heating value of these fuels was 17 million Btu/day based on average consumption and a product output of 86.000 cases/day. Low- and high-Btu gases that could be produced by two-phase digestion of the plant waste have an energy value of about 21 million Btu/day. Net energy output.

4.6.2.1.1 Pilot and Full Scale Demonstration

Pilot Plants : The two-phase anaerobic digestion process has been successfully applied in pilot- and full-scale plants to several types of high-COD and low-SS (suspended solids) industrial liquid waste, Table 4.16. These two phase plants consist of a completely mixed acid-phase digester followed by an upflow anaerobic sludge blanket (UASB) methane digester Fig. 4.13. All plants are operated in the mesophilic temperature range.

Table 4.16 : Two Phase Pilot and Full Scale Plants

Sr. No.	*Industry*	*Location*	*Scale*	*Capacity (kg COD/day)*
1.	Distillery (enzyme and alcohol)	Belgium	Pilot	180
2.	Beet sugar	West Germany	Pilot	45
3.	Distillery (yeast and alcohol)	Belgium	Pilot	135
4.	Beet sugar	Belgium	Pilot	170
5.	Citric acid	West Germany	Pilot	120
6.	Beet sugar	West Germany	Pilot	45
7.	Flax retting	Belgium	Full-scale	350
8.	Starch-to-glucose	West Germany	Full-scale	20×10^3

The first pilot plant has a capacity of 180 kg COD day^{-1} and was built in 1977 to stabilize enzyme and alcohol wastes from a distillery in Belgium. Five other pilot plants varying in capacity from 45 to 170 kg COD/day were built and operated since then. These two-phase pilot plants were operated with feeds having CODs ranging from 10,000 to 32,000 mg/L the overall plant loading rates and HRTs ranged between 4 and 12 kg COD/m^3.d and 0.5 and 2.8 days, respectively. Gas production rates up to 8 standard (60°F and 32 inch Hg pressure) volumes per culture volume per day, and COD and BOD reductions up to 90 per cent were achieved Table 4.17. It is noteworthy that the observed pilot plant performances were comparable to and probably exceeded those that may be projected from bench-scale experience Table 4.17.

In contrast to the completely mixed and packed bed reactor designs originally proposed for methane fermentation during bench scale development of the two-phase process[52-53], the pilot plant described here utilized the UASB reactor developed by Lettinga *et al.*[54-55] This reactor design require no mechanical mixing or packing and performs satisfactorily as the methane fermenter of the overall two-phase process.

Typically, the UASB methane reactor contains a methanogenic sludge bed at the bottom inlet end. SS concentration in this bed is about 30 kg/m^3. Concentration above this zone ranges between 5 and 7 kg/m^3. Methanogenic activity of this sludge is 10 times higher than that of conventional anaerobic digesters. The incoming acidified feed flows up through the UASB sludge bed, and the effluent is decanted through an internal settler at or the reactor before discharge. The sludge bed contain non-methanogenic and methanogenic bacteria reside together in rapidly settling flocs. A detailed description of the UASB methane-digester operation and performance can be found in a separate publication[54].

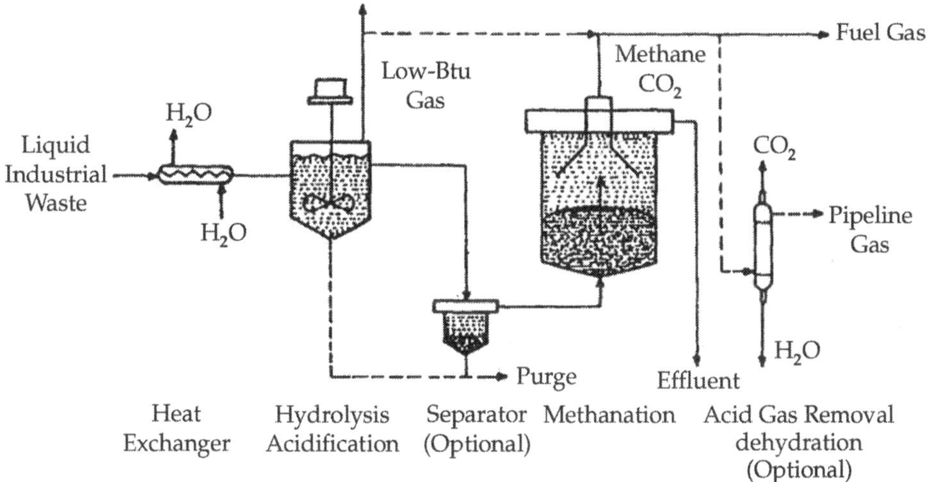

Fig. 4.13 : The Anthane/Anodek Process

Full Scale Plants : The first full-scale two phase plant was installed in Belgium in 1980 with support from the Ministry of Wildlife Conservation and the Association of Flax Manufactures to gasify and stabilize liquid wastes generated biological treatments (retting) of the fibre-containing flax plants at 30°C. The retting process facilitates separation of the fibre from the woody material by degradation of the natural binder. Volatile fatty acids are produced during the retting process and during storage of the waste in an equalization tank from which it is charged to the bottom of the UASB methane digester.

The flax operation is seasonal, so the plant operates only for about 6 months in a year. The digester has been operated for two seasons, and no difficulty was experienced in plant start-up at the beginning of each season. Plant capacity and average performance during the two seasons of operation are indicated in Table 4.17. A gas utilization system is being installed so that the biogas can be used for the flax retting operation.

Table 4.17 : Two-phase Pilot and Full Scale Plant Performance Data

Sr. No.	Substrate	*Plot Scale*			*Full Scale*	*Bench Scale*
		Enzyme and Alcohol	*Beet Sugar*	*Yeast and Alcohol*	*Flax Retting*	*Soft Drink*
1.	COD, mg/L	10,000	6000	32,000	7000	45,000
2.	Loading kg COD/m^3.d	4.1	10.8	11.6	10.0	6.1
3.	HRT, L	58	13.3	66	17	178
4.	GPR.std vol/vol.day	1.68	7.16	8.43	4.22	2.90
5.	Gas yield, std m^3/kg COD	0.41	0.66	0.73	0.42	0.48
6.	COD, removal %	79	90	72	87	96
7.	BOD, removal %	88	93	88	93	-

A much larger full-scale two phase plant having a capacity to process 20,000 kg COD/day is in the design stage at this time. This plant will gasify and stabilize wastewaters from a starch-to-glucose factory in West Germany.

The diphase anaerobic process is most appropriate for high strength spentwash, because of its widely noted advantages, e.g. the possibility of maintaining optimal environmental conditions for acid and methane forming organisms, attenuation of imbalances between organic acid production and consumption, stable performance and high methane content in the biogas produced in this process. The intensive bench scale studies have been performed on the diphasic anaerobic contact process preceded by lime neutralization and heat recovery from distillery spent wash[56-59]. The optimum hydraulic retention time (HRT) of 1.2 days in the acid phase with an organic loading rate[55] of 58kg COD/m^3.d as well as an HRT and MCRT of 17 and 40 days respectively, in the methane phase organic loading rate of 3.2 kg COD/m^3.d resulted in 70 per cent COD reduction with biogas production of 0.27 m^3/kg CODr (60 per cent CH_4) during the bench scale studies[56-57].

The toxic effect of K^+ ion was attenuated by acclimation of micro-organisms to K^+ ion and the antagonistic effect caused by Ca^{2+} and Na^{+}[60]. Further sulphide toxicity was directly proportional to H_2S concentration in the gas . H_2S in the gas can be controlled by enhancing pH of the methane phase to 7.7 or above, the pH control can help in process control strategy of for suppression of sulphide inhibition[61].

The pilot plant[59(a)] consist of three distinct steps, viz nutralization of spentwash, an acid phase and a methane phase. The acid phase comprises of two open horizontal plug flow reactors, each of 30m^3 capacity. The methane phase consists of two closed vertical continuously stirred tank reactor (CSTRs), each of 160m^3 capacity. The plant incorporates process control aids such as heat recovery from raw spentwash, on line pH metres, thermometers, and flow indicators. Necessary gas collection and flare facilities have also been installed.

Distillery spentwash, following neutralization to pH 6.0 with lime and heat recovery, enters the acid phase with a temperature of 45°C. The organic rich acid spentwash from the acid phase is charged to methane phase from a secondary settler, where solids are separated and then recycled into the methane phase. The influent spentwash flow rates have been regulated to provide desired HRTs in various units. The results presented in this study demonstrate the feasibility of diphasic anaerobic process for the biomethanation of distillery spent wash at a pilot plant scale. The methane phase without recycling could operate on a sustainable basis at an organic loading rate (OLR) of 3.25 kg COD/m^3.d with a COD reduction efficiency of 65 per cent and gas production of 0.3m^3/kg CODr at CH_4 concentration of 60%. The overall COD reduction efficiency of 77% was achieved during this mode of operation.

The contact phase enabled further enhancement of OLR to 5 kg COD/m^3.d without adversely affecting process performance. The COD reduction efficiency over the methane phase at an MCRT of 27 days is 70 per cent, with an overall COD reduction of 80%. The biogas contained CH_4, concentration of about 70 per cent. The contact mode of operation facilitates manipulation of MCRT, thereby enhancing process stability through a proper resistance of methane formers to sulphide inhibition. These studies also indicate the success of a pH control strategy to minimize sulphide inhibition to the methane phase[61].

The MLSS concentration in the methane phase have been observed to be low despite high rates of substrate utilization and gas productivity at high SRTs. This is in contrast to the concept of growth associated methanogenesis[62-64].

Karhadkar[65] has highlighted the comparison of anaerobic process for treatment of distillery spentwash and is presented in Table 4.18. Saraswat and Khanna[65] used the concept of phase separation, anaerobic activated sludge process and alkali pre-treatment have been incorporated in methane recovery from water hyacinth with the objective of developing rational and cost-effective designs of diphasic anaerobic activated sludge system. Evaluation of process kinetics and optimization analysis of laboratory data reveal that a diphasic with alkali treatment could be designed with an alkali pre-treatment step (3.6 per cent Na_2CO_3 + 2.5 per cent $Ca(OH)_2$(W/W) of water hyacinth (WH), 24 duration) followed by an open acid phase (2.1 days HRT) and closed methane reactor with sludge recycle (5.7 days HRT, 7.7 days MCRT) for gas yield of 50L/kgWH/d at 35-37°C. Likewise, a diphasic system without alkali treatment (AT) could be designed with an open acid phase (2 days HRT) followed by closed methane reactor with sludge recycle (3.2 days HRT, 6 days MCRT) for a gas yield of 32.5L/kgWH/d at 35-37°C, Table 4.19.

Economic analysis[66] for plant capacities between 2 and 71 m^3 biogas/d reveals that the unit cost of biogas production ranges from Rs. 4.6/m^3 to Rs. 5.2/m^3 and Rs. 0.6/m^3 to Rs. 1.09/m^3 respectively, for diphasic system with and without AT. Also the investment pay back periods range from 8.7 to 15.4 months and 4.5 to 8.5 months respectively. Notwithstanding lower capital cost for diphasic system with AT, in small capacity plants (2-4 m^3/d), the unit costs are still observed to be higher due to the high costs of chemicals. The costs for diphasic system without alkali treatment is on a lower side compared to diphasic system with alkali treatment. Cost estimates for existing single stage WH plant based on the KVIC design[67] operating at a hydraulic retention time of 30 days with a gas yield of 20 L/kgWH/d at 37°C bring out the unit cost of biogas production in this system as Rs. 5.30/m^3 with an investment pay back period of 25.2 months for a capacity of 2m^3 biogas/d. The design based on the present investigation are thus more cost effective than available in the literature.

The comparison of anaerobic and aerobic biotechnology per metric tonne COD destroyed is given below :

	Anaerobic	*Aerobic*
• Electricity	–	1100 KWh
• Methane	1.1×10^7 Btu	–
• Net cell production	20-50 kg	400-600 kg

Table 4.18 : Comparison of Anaerobic Process for Treatment of Distillery Spentwash

									Gas Composition		
Temp (°C)	*Volume (L)*	*HRT (day)*	*OLR (kd COD/ m^3.d)*	*COD (g/L)*	*SO_4 (g/L)*	*COD/ SO_4*	*%CODr*	*Biogass (m^3/m^3.d)*	*CH_4*	*CO_2*	*H_2S*
35	4	10	3.2	32	2.9	11.0	96	2	59	38	–
–	10	10	6.8	65	–	–	91	2.4	49	–	–
55	4	80	0.4	30	6.8	4.4	62	1.8	62	37	–
33	4	50	0.25	12	5.8	2.1	43	1.5	54	45	–
35	–	–	–	62	7.5	8.3	–	0.3	–	–	–
–	640	16	1.7	28	0.1	280	68	2.5	52	43	1
–	–	6.2	4.0	25	2-5	5-12	95	–	58	40	–
33	–	3.8	9.0	13	–	–	70	–	–	–	–
35	30	6.5	8.5	55	1.3	42.3	83	5.2	56	–	–
–	20	35	1.8	70-100	3-5	20-30	71	0.8	56	–	–
–	1.890	25	3.2	70-100	3-5	20-30	70	1.1	62	37	–
37	5	16	3.31	53	–	–	42	0.9	–	–	–
-	365	30	3.3	99	–	–	64	–	–	–	–
35	7.54	10.5	4.9	50	–	–	81	1.9	62	–	–
45	3×10^4	2.6	23	83	1.7	48.8	14]–79	–	–	–	–
33	1.6×10^4	20	3.5	72	1.4	51.4	66	0.7	61	36	4.5
45	3×10^4	2.6	36	95	1.7	55	16]–80	–	–	–	–
33	1.6×10^4	16	5.0	80	1.4	57	70	1.1	70	27	5

A well mixed suspended growth reactor system (CSTR) is normally not suitable for treating wastewater. There is always the problem of sludge settlement. It can be achieved either by applying vacuum in the settling unit or by freeze method. In practice, both of these methods cannot be applied. Therefore, fixed film reactors are always recommended for biogas recovery and treatment of wastewater.

Table 4.19 : Selected Physical and Chemical Properties of Methane

Sr. No.	*Parameter*		*Value*
1.	Formula		CH_4
2.	Molecular weight		16.042
3.	Boiling Point at 760mm		–161.49 °C
4.	Freezing Point at 760mm		–182.48 °C
5.	Critical pressure		47.363 kg/cm^2
6.	Critical temperature		–82.5 °C
7.	Specific gravity		
	– Liquid at (-164°C)		0.415
	– Gas at 25°C and 760mm		0.000658
8.	Specific volume at 15.5°C and 760mm		1.47 L/gm
9.	Calorific value at 15.5°C and 760mm		38,130.71 kJ/m^3(1,012Btu/ft^3)
10.	Air required for combustion(m^3/m^3)		9.53
11.	Flammability limits, % by volume		5 to 15
12.	Octane rating		130
13.	Ignition temperature, °C		650
14.	Combustion equation		$CH_4 + O_2 = CO_2 + H_2O$
	– O_2/CH_2 (complete), by weight		3.98
	– O_2/CH_4 (complete), by volume		2.0
	– CO_2/CH_4 (complete), by weight	50% reaction	2.74
	– CO_2/CH_4 (complete), by volume		1.00

Pure methane is a colourless and odourless gas and it constitutes between 50 and 70 per cent of gas produced by anaerobic digestion. The other 30-50 per cent is primarily CO_2 with a small amount of hydrogen sulphide. Table 4.19 lists some of the important properties of methane. The use of biogas in internal combination engines requires compression ratio of 8:1 or greater and special equipment and processes which include :

- Reducing H_2S content to less than 0.25 per cent to prevent corrosive damage to metal surfaces, particularly bearings and other working parts.
- A scrubbing system to remove CO_2; while CO_2 exerts no harmful effects on internal combustion engine, the presence of this non-combustible gas reduces heat content per unit volume and this lowers the operating efficiency of the engine.

- A compression capable of compressing gas to pressures between 2000-3000 psi (140-210 kg/Cm^2) when biogas is to be compressed, it is imperative that CO_2 be removed to prevent mechanical damage to compressor caused by liquefaction of CO_2 (Table 4.20 for critical constants of biogas components).
- High pressure cylinders, of the type used for oxygen, that can safely store gas at pressures upto 2400 psi (170 kg/cm$^{2)}$, to be installed on the vehicle.
- A set of similar high pressure storage cylinders equipped with a pressure control panel for filling vehicle cylinders.

The gas is most easily used in burning appliances that can use it directly from low pressure collection units. The biogas burns with a blue flame and has a heat value ranging from about 500-600 Btu/ft^3 (22,000–26,000 kJ/m^3) when its methane content ranges from 60 to 70 per cent. It can be used directly in gas burning appliances for heating, cooking, lighting and refrigeration.

Table 4.20 : Critical Temperature and Pressures of Biogas Components

Sr. No	*Gas Component*	*Critical Temperature (°C)*	*Critical Pressure (kg/cm^2)*
1.	CH_4	– 82.1	47.3
2.	CO_2	31.0	75.3
3.	H_2S	100.4	91.9
4.	NH_3	132.5	116.3

4.7 Management of Hydrogen During Anaerobic Biological Wastewater Treatment

4.7.1 General Perspective

The anaerobic biological conversion of organic wastes to methane is a complex process involving a number of microbial populations linked by their individual substrate removed to ensure the production or acetic acid.

Fig. 4.14 serves as a convenient basis for reviewing literature pertinent to the fate and effects of hydrogen throughout the overall anaerobic waste stabilization process. Accordingly, it was considered appropriate to begin with a review of the role of hydrogen in regulating acidogenesis [steps 2), 3), 4),] followed by a review of competition for methanogenic substrates by NRB and SRB [steps 5), 6), 7),] and the roles of hydrogen and acetic acid in regulating methanogenesis [steps 8 and 9] and to conclude with a review of the phase separation (or two-phase) process.

4.7.2 The Role of Hydrogen in Regulating Acidogenesis

4.7.2.1 Biochemical Interactions

The phasic nature of the anaerobic stabilization process illustrated in Fig. 4.14 as well as the sensitivity of the otherwise "rate-limiting" methanogenic step, has been demonstrated by recorded observations of organic acid and alcohol accumulations during periods of over loading or other stress.[68-71] That the accumulation of these intermediate organics is due to hydrogen sensitivities moderated by syntrophic associations between hydrogen-producing bacteria (acidogens) and hydrogen-consuming bacteria (methanogens, SRB, NRB) has only recently been recognized[72-77]. However, the full process engineering consequences of these relationships remain essentially unresolved, and some researchers have maintained that the physical separation of methanogens and acidogens (as presumably practiced in two-phase systems) cannot enhance anaerobic bioconversion rates since the necessary interspecies hydrogen transfer function is disturbed.

As illustrated in Fig 4.15 for the catabolism[78] of glucose, hydrogen in relatively low concentrations appears to regulate the overall conversion process by throttling the acidogenic reactions at several points in the glycolytic pathway. In order for catabolism to proceed continuously given a constant pool of NAD, the NADH produced during substrate-level phosphorylation of glyceraldehyde-3-phosphate and the oxidative decarboxylation of pyruvic acid to acetyl-CoA must be regenerated. This function is accomplished by the reduction of protons to form hydrogen gas, which is subsequently removed ("interspecies hydrogen transfer") by the hydrogenotrophs (methanogens, SRB, NRB). Accumulations of hydrogen beyond the collective assimilative capacity of these hydrogenotrophs necessitates an alternate method of electron disposal for NADH regeneration. This need is fulfilled by the fermentation of pyruvate to propionate, lactate, and ethanol and/or by the fermentation of acetyl-CoA to butyric acid. Since the methanogens cannot use these endproducts as substrates directly, their accumulation ensues, which often leads to a problematic depression in the pH of anaerobic treatment processes.

The thermodynamic considerations of the sequential reactions involved in the conversion of glucose to methane suggest that, if hydrogen is maintained at sufficiently low levels ($<10^{-4}$ atm), the production of propionic acid need never occur in practice, however, the nature of wastewater substrates is rarely as well defined or as simple as glucose. Moreover, an anaerobic reactor treating organic wastewater will likely receive a variable input. These inevitabilities serve to ensure the appearance of propionic and butyric acids in some proportion during the course of wastewater treatment. The subsequent degradation of these acids to acetic acid is also hydrogen dependent[79] and is mediated by obligate hydrogen-producing acetogens (OHPA) inextricably linked to the various hydrogen oxidizers. In addition to their hydrogen sensitivities, the number of OHPA and associated HOM bacteria existing in an anaerobic process may vary, depending upon the feeding schedule of a given reactor and its associated stability history. For example, following a shock loading or some other process upset, the proportions of OHPA-HOM will need to increase to accommodate accumulated hydrogen and propionic acid. However, if a long period of stable operation ensues, the relative proportions of substrate channeled through propionic acid may decrease, and the number of OHPA and HOM will decrease accordingly. This dynamic situation is considered central to the overall stability and efficiency of many anaerobic treatment systems and is, therefore, regarded as the key to stabilizing and improving anaerobic treatment. Moreover, the behaviour of the hydrogen system is believed to be reflective of the state of reactor stability, therefore, its measurement and quantitative interpretation within an operational perspective are considered important and are explored further in the following section.

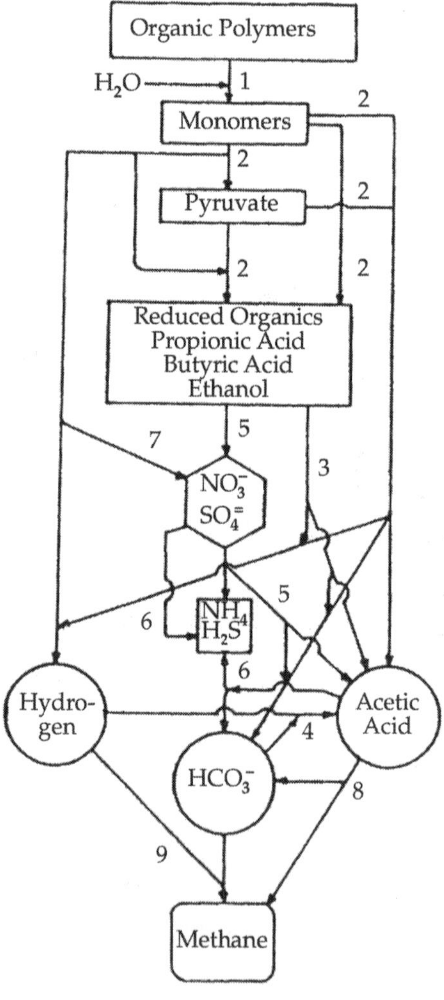

Fig. 4.14

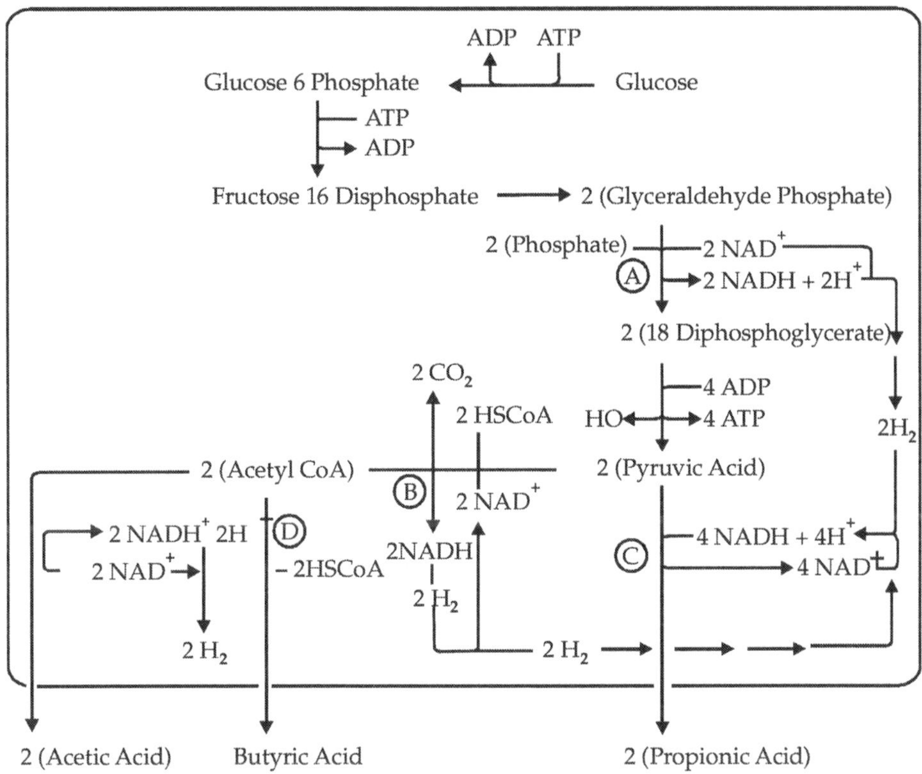

Fig. 4.15 : Hydrogen-Regulated Catabolic Pathways Possible for the Conversion of Glucose in Anaerobic Wastewater Treatment Systems.

4.7.2.2 Experimental Quantification of Hydrogen Effects

As could be expected from these rather obvious and probable population shifts encountered during the various studies on wastewater treatment, as well as from the differences in substrate composition and reactor configurations utilized, variable hydrogen partial pressures have been proposed at which the shunts and inhibitions leading to the accumulation of propionic acids and other acids occur. As shown in Table 4.21 Bryant *et al.*[72] observed inhibition of acetate production from ethanol at a hydrogen pressure of 0.5 atm. Smith[81] demonstrated that propionic acid degradation in digestion sludge was inhibited by 0.18 atm, but not by 0.09 atm H_2. Some investigators[80-82] reported an increase in volatile acids upon increasing the hydrogen partial pressure above digesting sludge from 10^{-4} to 10^{-3} atm; increases from 10^{-4} to 0.015 atm resulted in a linear increase in the propionic acid concentration despite stimulation of both methane production and acetate turnover.

Sykes[83] studied the production of hydrogen in laboratory-scale sewage sludge digesters subjected to pulse loadings of glucose and other carbohydrates. As indicated in Table 4.21, hydrogen accumulations were observed in varying concentrations in digesters subjected to different pulse loadings of glucose (4.4, 8.8, and 13.3 g/L). In all cases, hydrogen accumulated between 9 and 12h following the pulse; the time required for hydrogen assimilation and the subsequent degradation of propionic/butyric acids varied according to the strength of the glucose pulse applied. At the lower pulse loading (4.4 g/L glucose), hydrogen accumulated to 0.001 atm (0.11% by volume) at 12h but had disappeared by 16h following the pulse. Propionic and butyric acids accumulated shortly following the appearance of hydrogen, but did not disappear for several days following hydrogen disappearance. Accordingly, the pH decreased to 6.6 at 12h and returned to 7.1 following acid conversion at 3 days. At the higher loadings, pulsed digesters experienced similar, yet more exaggerated trends. The 8.8 g/L pulse caused a hydrogen accumulation to 2.6 per cent by volume; hydrogen did not disappear until the second day following the pulse. Total volatile acids accumulated to 2.5 g/L, the pH decreased to 6.11, and greater than 5 days were required for acid conversion and pH moderation. The highest load (13.3g/L) caused a hydrogen accumulation to 50 per cent; butyric acid and propionic acids accumulated, the pH decreased to 5.0, and the digester failed. Interestingly, and in contrast with the lower loadings, butyric acid was produced in greater amounts than propionic at the highest loading.

More recently, Barnes *et al.*[84], observed an increase in propionic and other volatile acids immediately following a hydrogen partial increase from 2×10^{-4} to 1.5×10^{-3} atm, as induced by a shock loading of 38.5 g/L BOD_5 applied to a fluidized anaerobic reactor treating a molasses wastewater. Hydrogen accumulations were moderated within several hours after shock load application, although the oxidation of propionic and other volatile acids required an additional 10-12h beyond the initial return of hydrogen to its normal operating level. Heyes and Hall[85] reported hydrogen partial pressures ranging from 4.3 to 6.5 ($\times 10^{-3}$) atm during normal propionic acid oxidation in digesting sludge at different retention times, and also postulated the existence of two subgroups of OHPA, one having a high growth rate and the other a scavenger (lowV_{max} lowK_s).

These results exemplify both the high sensitivity of anaerobic processes to hydrogen regulation as well as the necessary degree of undersaturation in the hydrogen removal system present. From their experiments, Kaspar and Wuhrmann[81-88] estimated that the hydrogen removal rate in well digesting sludge was at only 1 per cent of its maximum value. Similarly, the results of Shea *et al.*[90], indicated that the rate of hydrogen removal was only 3 per cent of its maximum value. However, a number of researchers have argued that hydrogen mass transfer may be rate-limiting in anaerobic treatment systems. Differences in reactor design and operation, therefore, serve to alienate hydrogen effects beyond variations expected due to population selections above.

4.7.3 Thermodynamic Quantification of Hydrogen Effects

In view of the reported differences in hydrogen effects, and the traditional difficulties in adequately defining the redox nature of the wastewater substrate and measuring hydrogen at levels of 10^{-6} to 10^{-4} atm, the regulatory effects of hydrogen on OHPA and HOM have been conveniently illustrated using thermodynamic models and equilibrium assumptions by several authors[91-95]. The associated redox half-reactions mediated during anaerobic stabilization of organic wastes to methane, based on biochemical standard free energy levels (pH 7, 1 atm, 1 kg/mol activity) obtained from Thauer[96] *et al.*; are presented in Table 4.16. Thermodynamic calculations associated with these reactions, and illustrated in Fig. 4.16 indicate that propionic acid oxidation to acetate (line1) becomes favourable only at hydrogen partial pressures below 10^{-4} atm, while butyric acid oxidation (line 2) becomes favourable at 10^{-3} atm H_2 or below. Similarly, ethanol and lactate oxidations (lines 3 and 4) are inhibited by hydrogen partial pressures approaching 1 atm.

In addition, inspection of the possible inorganic respirative couples provides an appreciation of the redox compatibilities of bicarbonate and sulphate reduction (lines 6 and 7), and also indicates the favourability of bicarbonate respiration over acetate cleavage (line 9) at hydrogen partial pressures above 10^{-4} atm, which is important since Methanosarcina can use both substrates, as discussed later.

Fig. 4.16 also provides possible insight into the product formation patterns to be expected in a two-phase process configuration. Without the application of hydrogen control in the first-phase reactor, the accumulation of hydrogen above 10^{-4} atm enhances the accumulation of propionic and butyric acids, but will also enhance the production of acetate from the respiration of bicarbonate (line 5). Moreover, if hydrogen is passed forward to a second-phase methanogenic reactor, the thermodynamic favourability of bicarbonate respiration may have adverse effects on acetate cleavage by methanogens which utilize both H_2/CO_2 as well as acetic acid. (This important issue is addressed in more details in the discussion of the Roles of Hydrogen and Acetate in Regulating Methanogenesis).

Also evident from the diagram in Fig. 4.16 is the favourability of sulphate reduction over bicarbonate respiration (line 7 vs. line 6) at all hydrogen partial pressures, and the favourability of acetate cleavage by SRB (line 10) over methanogens (line 9). Sulphate reduction by hydrogen (line 7) is favoured over the SRB acetate cleavage (line 10) at hydrogen partial pressures above 10^{-4} atm with the concentrations of other reactants indicated. Although the reactions mediated by NRB are not shown in Fig. 4.16, the standard free energy levels associated with the hydrogen or acetic acid mediated reductions of nitrates to ammonia are substantially higher than for sulphate reductions, Table 4.22. The exclusion of nitrogen species from the diagram was necessitated in order to keep the diagram legible, but their importance is not diminished. The standard free energies of both reactions (reactions 11 and 12), Table 4.22 are an order of magnitude higher the methanogenic reactions, which may have several important implications to engineering process applications including the redox

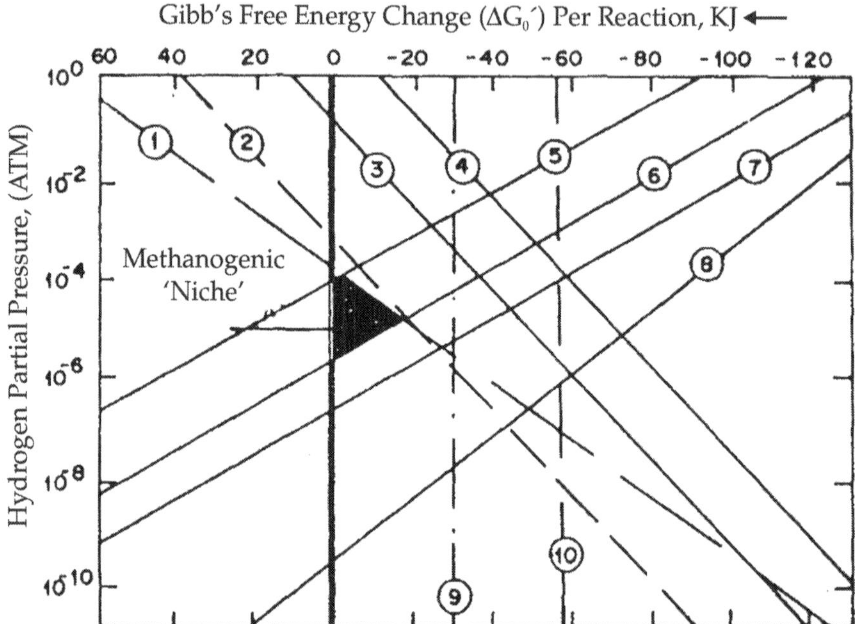

Fig. 4.16 : Graphical Representation of the Hydrogen-Dependent Thermodynamic Favourability of Acetogenic Oxidations and Inorganic Respirations Associated with the Anaerobic Degradation of Waste Organics.

incompatibility of methanogenesis and dissimilatory nitrate reduction, and the potentially superior kinetics of substrate degradations associated with the latter reaction. While numerous investigators have demonstrated the exclusivity of nitrate reduction over sulphate reduction and methanogenesis (as detailed in the following section), Rivard and Smith[97] have recently reported that methanogenesis may continue (36% inhibition) despite continuous additions (1.6 mmol/h of nitrate) to a semicontinuous, thermophilic[99] sewage sludge digester having a sludge volume of 4L and a 20 days retention time.

4.7.4 The Regulatory Roles of Inorganic Hydrogen Acceptors

Although the addition of high levels of nitrates and/or sulphates may have detrimental influences on methanogenesis, in lower concentrations, these ionic species may possibly be applied to moderate the detrimental influences of excess hydrogen on a stressed anaerobic reactor. Particularly, in a two-phase process, where the OHPA-HOM and acetate cleaving methanogenic population have been enriched in the second reactor, these species could be carefully added as chemicals to the first-phase reactor and serve as a hydrogen sink without adversely affecting the methanogens in the second-phase reactor. Obviously, these practices hold numerous practical challenges, including the control of the potentially inhibitory and corrosive hydrogen sulphide gas, and the addition of excess nutrients discharged in the form of ammonia[98]. However, from an experimental process-oriented perspective, such procedure may yield valuable information on the potential for manipulating hydrogen to control acidogenesis and methanogenesis. Moreover, a number of industrial wastewaters contain significant levels of sulphate for nitrates, in addition to high levels of waste organic materials, as emphasized in Table 4.24. The successful anaerobic treatment of these wastewaters requires an understanding of the competition for methanogenic precursors by SRB (and NRB), as well as the associated microbiology and substrate conversion kinetics.

The attainment of the highly reduced environment necessary for suphate reduction and methanogenesis in natural environments such as soils and sediment layers is mediated by a succession of redox couples involving organic electron donors and inorganic electron acceptors[121]. Practical examples of these succession, wherein the oxidation of organics is coupled first to oxygen reduction, followed by nitrogen, sulphur, and carbon reductions, include the results of landfill leachate treatment studies by Pohland[122] *et al.*, and Rees[123]. Other observations which are in accord with such thermodynamic predictions include the exclusion of nitrate reduction in the presence of oxygen,[124, 125] the preference for nitrate over sulphate as a final electron acceptor in the oxidation of acetate[126] and the addition of nitrates to putrid oxidation ponds.[127-128] Nitrogenous compounds have been demonstrated to inhibit methanogenesis in salt marsh sediments,[128,129] in soils[130,131] and in sewage sludge digesters[97]. Sulphate reducers have demonstrated a higher affinity for hydrogen than methanogens in marine sediments,[133,138] in freshwater sediments,[139,142] and in pure or mixed cultures of SRB and methanogens.[143]

In addition to competing for hydrogen, NRB and SRB also compete for organic electron donors such as formate acetate, propionate, butyrate, lactate, and other organic acids[140-150]. Sorensen *et al.*[137], predicted that this competition was particularly significant at the acetate and hydrogen concentrations of marine sediments, with acetate providing over 60 per cent of the electrons required for sulphate reduction, as compared with only 5-10 per cent provided by H_2.

Desulfovibrio,[146] Desulfotomaculum[151] and Desulfobacter[147] are all capable of terminal acetate oxidation, which places them in a commensal relationship with methanogens. Therefore, complete waste organic conversion may be possible despite total metanogenic inhibition. However, from a process engineering perspective, such an approach has decided disadvantages, including the loss of energy available from methane and the production of hydrogen sulphide or ammonia. Since sulphides and ammonia are much more soluble than methane, their dissolved components can contribute significantly to effluent ultimate COD and BOD levels. Nevertheless, a comparison of the kinetics of hydrogen and acetate uptake by methanogens (Table 4.25 and SRB Table 4.26) suggests that higher organic waste conversion rates may be available through sulphate

Table 4.21 : Relationships Between Hydrogen Concentrations (Partial Pressures) and the End Products of Anaerobic Catabolism.

Culture Condition	*Substrate*	*Hydrogen*	*Significance of Hydrogen Present (Ref)*
Rumen : dialysis sac	Rumen fluid, H_2	10^{-6}M	Normal methanogenesis (86)
Methanobacterium omelianskii	Ethanol media	0.5 atm	Ethanol degradation inhibited (72)
Sewage digester sludge	Glucose pulses 4.4 g/L	0.001 atm	Propionic acid accumulation to 0.3g/L., pH drop to 6.6,3 days to acid conversion and pH recovery (83)
	8.8 g/L	0.02 atm	Propionic acid accumulation to 1.2 g/L., pH drop to 6.1,5 days to pH recovery (83)
	13.3 g/L	0.50 atm	Butyric acid accumulation to 1.8g/L, pH drop to 5.0, no recovery (83)
Sewage digester	Acetic and propionic acids	not reported	Propionate to acetate degradation inhibited under hydrogen atmosphere (79)
Clostridium cellobioparum pure culture	Glucose rumen fluid broth	0.3 atm	H_2 production reduced 50% accumulation of butyrate, ethanol, lactate (74)
Sewage digester sludge	Sludge acetic acid	10^{-4}–<0.015 atm	Linear increase in propionic acid (81)
Sewage digester sludge	Sludge propionic acid ethanol	0.005 atm 0.07 atm	Propionate degradation inhibited, ethanol degradation inhibited (82)
Sewage digester sludge	Wastewater sludge	0.18 atm	Propionic acid degradation inhibited (81)
Thermoanaerobium brockii	Glucose	0.5 atm	Accumulation of lactate, ethanol, H_2 production inhibited (88)
Sewage digester sludge	Sludge	1.2μM	Normal operation (87)
Sewage digester sludge	Molasses and yeast	2×10^{-4} 1.5×10^{-3}	Organic shock load caused H_2 increase followed by propionic acid accumulation (84)
Cattle manure, batch digesters	Cattle manure	additions of 70:30 H_2:CO_2	Propionic acid accumulation (89)
Propionic acid enrichment	Propionic acid	$6.5,4.3\times10^{-5}$	"Normal operations at 8.2 days HRT (85) "Normal operations at 14.5 days HRT (85)

Table 4.23 : Some Redox Half-Reactions Responsible for Degradations of Selected Organics Curing Anaerobic Treatment of Industrial, Municipal, and Agricultural Wastes

Oxidations (Electron Donating Reactions)					ΔG°,kJ
1.	Propionate	– Acetate:	$CH_3CH_2COO^-+3H_2O$	– $C_3COO^-+H^+HCO_3$	+76.1
2.	Butyrate	– Acetate:	$CH_3CH_2CH_2COO+2H_2O$	– $2CH_3COO^-+H^+$	+48.1
3.	Ethanol	– Acetate:	$CH_3CH_2OH+H_2O$	– $CH_3COO^-+H^++2H_2$	+9.6
4.	Lactate	– Acetate:	$CH_3CHOHCOO^-+2H_3O$	– $CH_3COO^-+CHO_3^-+H^++2H_2$	-4.25
9.	Acetate	– Methane:	$CH_3COO^-+H_2O$	– $HCO_2^-+CH_4$	-31.0
Respirative (Electron Accepting Reactions)					
9.	HCO^-_3	– Acetate:	$2CHO_3^-+4H_2+H^+$	– $CH_3COO^-+4H_2O$	-104.6
6.	HOC^-_3	– Methane:	$HCO^-_3+4H_2+H^+$	– CH_4+3H_2O	-135.6
7.	Sulphate	– Sulphide:	$SO_4^-+4H_2+H^+$	– $HS+4H_2O$	-151.9
10.			$CH_3COO^-+SO_4^-+H^+$	– $2HCO_3+H_2S$	-59.9
11.	Nitrate	– Ammonia:	$NO_3^-+4H_2+2H^+$	– $NH^+_4+3H_2O$	-599.6
12.			$CH_3COO^-+NO_3^+ +H^++H_2O$	– $2HCO^-_3+H_4^+$	-511.4
13.	Nitrate	– Nitrogen gas:	$2NO_3^-+5H_2+2H^+$	– N_2+6H_2O	-1120.5

reduction than through methanogenesis. Moreover, sulphate and nitrate reducers are not limited to one and two carbon substrates, as are methanogens. Therefore, this approach may hold possibilities for reducting propionic acid and hydrogen, as well as acetic acid in a stressed reactor, in order to more rapidly the re-establish equilibrium with the existing hydrogen removal system. This potential advancement in process control, however, comes at the expense of potential methane energy recovery.

4.7.5 The Roles of Hydrogen and Acetate in Regulating Methanogenesis

4.7.5.1 Methanogenic Substrate Specificities

The majority of methanogens isolated from anaerobic wastewater treatment systems and natural anaerobic environments utilize hydrogen and single-carbon compounds such as carbon monoxide, carbon dioxide, and formate as substrates for methane production. In addition, there are two known genera of methanogens which can utilize the two-carbon compound, acetic acid. As indicated in Table 4.27 these include species of *Methanosarcina* and *Methanothrix*. The *Methanothrix* species are unable to use hydrogen in combination with CO_2 and, as such, are non-hydrogen-oxidizing acetotrophs (NHOA). In contrast, *Methanosarcina* can utilize H_2/CO_2 as well as acetate, carbon monoxide, methanol, and methylamines as growth substrates. Due to their ability to use both H_2/CO_2 and acetate, these bacteria are classified here as hydrogen-oxidizing acetotrophs (HOA). Hydrogen-oxidizing methanogens (HOM) which do not cleave acetate, but use H_2/CO_2 and formate as substrates, are also included in Table 4.25.

The HOA are unique in their capability to utilize multiple (one- and/or two-carbon) substrates. On the basis of energetics, this ability affords a higher potential for survival when competing with SRB and NRB for hydrogen acetate. However, their superior acetate utilization kinetics compared to NHOA and the regulation of substrate use by H_2/CO_2 are more important from an engineering perspective. These considerations are central to the integration of the specific capabilities of OHPA, HOA, NHOA, and HOM into an optimum second-phase anaerobic reactor as developed in more detail in the following sections.

4.7.5.2 Hydrogen Regulation of Acetate Cleavage

Species of *Methanosarcina* which utilize both acetate and H_2/CO_2 as substrates may be subject to catabolic repression of acetate cleave by low levels of hydrogen. A similar catabolic repression may be present for the metabolism of methanol and methylamines versus acetate, although these characteristics are not universally shared by all strains of *Methanosarcina*. Therefore, the behaviour of anaerobic reactor systems may be dependent on the cultures selected by seeding substrates, and operational procedures.

Table 4.24 : Review of Laboratory, Pilot, and Full-Scale Anaerobic Digestion of Sulphate-Containing Wastewater

Reactor type (Ref)	*Temp. (°C)*	*Volume (L)*	*H.R.T. (days)*	*O.L.R. (kgCOD)/ m^3.d)*	*COD (g/L)*	*SO_4 (g/L)*	*COD SO_4*	*COD Removal %*	*Gas Production (m^2/m^2.d)*	*Gas Composition %*		
										CH_4	*CO_2*	*H_2S*
Distillery Wastewaters												
Contact (98)	35	nr	nr	nr	62	7.5	8.3		0.3	nr	nr	nr
Semi-CSTR(100)	35	4	10.0	3.2	32	2.9	11.0	96	2.0	59	38	
Filter (101)	35	20	5.5	5.4	32	0.7	18.8	86	3.0	70	nr	
Filter (102)	40	500	1.2	40.0	45-55	0.3	158.3	48	14.2	61	nr	0.4
Filter (103)	40	500	1.0	50.0	45-50	nr	-		20.5	61	nr	
	(104)	18-29	120	5.0	10.0	45-56	2.3-3.5	17.4	68	4.8	20	70
2-7												
Fluidized Bed (105)	30	7	0.3	22.2	9.1	1.2	7.6	43	6.0	56	nr	
Baffled (106)	35	4	5.6	11.8	65	nr	-	72	2.3	nr	nr	nr
Contact (107)		640	16.0	1.7	28	0.1	280	68	2.5	52	43	1
Contact (108)			6.2	4.0	25	2-5	5-12	95		58	40	
Contact (109)	33		3.8	9.0	12.5		-	70				
Two-Phase (110)	39	5000	0.6	13.8	8	nr	-	79	5-7	75	23	
UASB												
Floc-Sludge Bed (111)		9		5.3		1.5	-	70				
Contact UASB (112)	35	30	6.5	8.5	55	1.3	42.3	83	5.2	56		
Series CSTR (113)		10	10.0	6.8	65			91	2.4	49		
Contact(114)		20	35.0	1.8	70-100	3-5	20-30	71	0.8	58		
Contact (114)		1890	25.0	3.2	70-100	3-5	20-30	70	1.1	62	37	

Contd....

Reactor type (Ref)	Temp. (°C)	Volume (L)	H.R.T. (days)	O.L.R. (kgCOD)/ m^3.d)	COD (g/L)	SO_4 (g/L)	COD SO_4	COD Removal %	Gas Production (m^2/m^2.d)	Gas Composition %		
										CH_4	CO_2	H_2S
CSTR (115)	55	4	80.0	0.4	30	6.8	4.4	62	1.8	54	45	
CSTR (116)	33	4	50.0	0.25	12	5.8	2.1	(43)	1.5	54	45	0.5
Filter (116)		132.2×10^6	8.0	13.6	80-100	4-10	10	71	2.8	nr		
Paper Mill Wastewaters												
Filter (117)		12	0.5	28.0	14-20	0.16	10	90	27.6	54		
Contact (118)	37	12	28	4.7	10-20	0.6-1.3	12	85	4.8	55	nr	nr
Contact (119)	52	58,000	4.0	3.5	12.5	0.6	20.8	88	2.2	60		
Winery Wastewaters												
CSTR (115)	33	640	7.2	3.3	22	nr		98	0.85	nr	nr	nr
CSTR (116)	55	4	80	0.16	13.0	0.5	26.0	97	1.1	58	42	nr
Batch (120)	35	1.6	20		26.5	25.4	1.0	50	1.2	nr	nr	nr

nr – not reported.

Methanosarcina sp. strain 277 may utilize H_2/CO_2, methanol, methylamines, or acetate as methanogenic substrates, although Smith and Mah[80,157] have demonstrated that methanol is preferentially metabolized in mixtures of these substrates. Similarly, Blaut and Gottschalk[173] demonstrated a preference for trimethyl-amine over acetate in *Methanosarcina barkeri* strain Fusaro. Contrary to these results, Weimer and Zeikus[158], Hutten *et al.*[104] and Krzycki *et al.*[172] shown that *Methanosarcina barkeri* are capable of simultaneous methanol and acetate metabolism in mixtures of these substrates.

In a similar fashion, as indicated in Table 4.28 H_2/CO_2 is preferentially utilized over acetate in mixtures of these substrates by *Methanosarcina* sp. to the extent that acetate cleavage is inhibited until H_2 is exhausted.[94,169,176] Hutten *et al.*[171], demonstrated that H_2 by itself inhibited acetate cleavage, although when H_2 was replaced by H_2/CO_2, acetate cleavage occurred simultaneously with bicarbonate reduction. Other investigators have reported a complete and rapid inhibition of acetate cleavage in the presence of H_2/CO_2, particularly in HOA such as *Methanosarcina barkeri*, Table 4.28. Hydrogen has also been demonstrated to inhibit acetate cleavage by reluctant hydrogen oxidizers such as *Methanosarcina mazeii*[169] and *Methanosarcina* strain TM-1[99]. These methanogens utilize hydrogen very slowly or not at all and, therefore, depend upon the participation of other methanogens to remove hydrogen and ensure the production of acetic acid. Similarly, *Methanothrix sp.* (which are NHOA) depend on other hydrogenotrophs to remove hydrogen. These bacteria differ from *M. mazeii* and *Methanosarcina* strain TM-1 in that they are unaffected by H_2/CO_2 and may continue to cleave acetate to methane and bicarbonate even under a hydrogen pressure of 1.3 atm[93].

The preferential substrate utilization patterns exhibited by HOA, *i.e.*, the preference for H_2/CO_2 followed by methanol, methylamine, and finally acetate metabolism are in accord with thermodynamic predictions. Standard free energy levels associated with the methanation of these substrates, listed in Table 4.29 indicated that the favourability of acetate cleavage is an order of magnitude lower than H_2/CO_2 and methanol conversions. The predicted energy inavailable from methylamine conversions is nearly triple that available from acetate cleavage at the standard state (*i.e.*, corresponding to 1 atm of H_2 in the gaseous phase). However, as indicated in the previous discussions, this high level of hydrogen is uncommon in well-operating anaerobic wastewater treatment systems. At more realistic hydrogen concentrations, the aceticlastic split completes more favourably with the methanogenic respiration of bicarbonate. As shown in Fig. 4.16 a hydrogen concentration in the gas phase below 10^{-4} atm will favour the aceticlastic reaction (line 9) over the respirative reaction (line 6).

4.7.5.3 Acetate Regulation of Acetate Cleavage

In addition to the aforementioned substrate specificities, aceticlastic methanogens are subject to kinetic competition for acetate, in comparison, growth kinetic parameters for the HOA and NHOA listed in Table 4.26 suggest that

Table 4.25 : Kinetics of Acetate Cleavage to Methane and CO_2 in Pure and Mixed Cultures

Substrate (Ref)	*Culture/Reactor*	*Temp. (°C)*	μ_{max} *(day^{-1})*	*Ks (mg (COD/L)*	k_d *(day^{-1})*	*Y (gVSS/) (gCOD)*	V_3 *(mgCOD) ------- (mg VSS day)*
Acetate (152)	Acetate enrichment	37	0.39	180	0.02	0.04	95
Acetate (153)	Acetate enrichment	35	0.40	180	0.04		
Acidified glucose(154)	Methane phase CSTR	37	3.4	642		0.04	10.0
Acetate (155)	Methane phase CSTR	36	0.49	4200			
Acidified glucose (156)	Methane phase CSTR		0.43	395			
Acetate (142)	*Methanobacterium* Sp.	30	0.26	11		0.01	26.0
Acetate (157)	*Methanosarcina barkeri* strain 277	36	0.60	320		0.04	15.0
Acetate (158)	*Methanosarcina barkeri*	37				0.03	
Acetate (80)	*Methanosarcina barkeri*	36	0.44			0.05	8.8
Acetate (99)	*Methanosarcina*	55	1.4	290		0.03	46.7
Acetate (93)	*Methanothrix soehngenii*	33	0.15	30		0.03	3.7
Acetate (159)	-do-	37	0.16	45		0.02	8.0

NHOA are acetate "scavengers" (low μ_{max}, Low K_5), while the HOA are capable of much higher acetate utilization rates. As illustrated in Fig. 4.17 the NHOA "scavengers" appear limited to approximately one-fourth of the maximum growth rate of HOA, although these bacteria have a much higher affinity (lower Ks) for acetate than the HOA. These Growth characteristics suggest that the NHOA

Table 4.26 : Kinetic Characteristics of Sulphate Reducers and Methanogens Growing on H_2 and Acetate

Culture (Ref)	*Substrate*	μ_{max} *(day^{-1})*	*Ks (mgCOD/L)*	*Y (gVSS/gCOD)*
Rumen bacteria, 37°C (86)	H_2/CO_2	5.4	$2x10^{-4}$	
Digesting sludge (87)	H_2/CO_2		$2.2x10^{-4}$	
Digesting sludge (81)	H_2/CO_2		0.105 atm	
Digesting sludge (90)	H_2/CO_3	1.1	0.75 atm	
Desulfovibrio vulgaris (160)	H_2/SO^-_2	3.6		0.09
Desulfovibrio vulgaris (160)	$H_2/S_2O^-_3$	5.0	—	0.15
Lake sediment (161)	H_2/SO^-_4		0.001 atm	—
Digester sludge (162)	Acetate/SO_4	8.3		0.05
Desulfovibrio vulgaris (163)	Lactate/SO_4			0.12
Desulfobacter postgatei (164)	Acetate/SO_4		13	
Desulfotomaculum orientis (163)	Lactate/SO_4			0.04
Desulfobacter postgatei (165)	Acetate/SO_4		4	

Table 4.27 : Substrates Used for Growth and Methane Production by Methanogenic Bacteria

Methanogen Type	*Methanogenic species*	*Substrates utilized*	*(Ref)*
(NHOA)	*Methanothrix soehngenii*	Acetate only	(93,159,166,167)
	Methanosarcina mazeii	All below, but H_2/CO_2 only reluctantly	(168,169)
	Methanosarcina Strain TM-1	All below except H_2/CO_2	(99)
(HOA)	*Methanosarcina bakeri*	Acetate	(170,158,157,171,173,169)
		Methanol	(174,175,157,171,172,169)
		Methylamines	(174,175,177,173,178)
		Carbon monoxide	(179,180,181,182)
		H_2/CO_2	(174,175,157,171,169)
(HOM)	*Methanobacterium*	H_2/CO_2	(182,182,184,185,186)
	Methanobrevibacter	H_2/CO_2, formate	(185)
	Methanococcus	H_2/CO_2, formate	(187)
	Methanomicrobium	H_2/CO_2, formate	(191)
	Methanogenium	H_2/CO_2, formate	(189)
	Methanospirillum	H_2/CO_2, formate	(190,191,192)

may outcompete HOA at acetate concentrations below 50mg/L, while above 250mg/L acetate, the HOA are more competitive. As a result of these comparative kinetics, Methanothrix (NHOA) have been found more predominant than methanosarcina in digesters having long retention times and in long-term acetate enrichments[95,194,197,198]. Methanosarcina are more prevalent in low retention-time reactors, such as in the lower reaches of plug-flow anaerobic filters[199-202] and probably in two-phase reactor systems as well, due to the higher acetate concentrations resulting from these digestion configurations. The high substrate affinity of NHOA may help them compete with SRB and NRB for acetate, although HOA may have a higher survivability due to their multiple substrate utilization capabilities. A similar situation may exist for OHPA on a variety of substrates, as suggested by Heyes and Hall[85], however, the kinetics and substrate specificities of the OHPA remain to be closely defined.

Table 4.28 : Effects of Hydrogen on Methanogenesis from Acetate in Pure and Mixed Culture.

Culture	*H_2(atm)*	*Effects noted due to hydrogen addition (Ref)*
	Examples of Inhibition due to Hydrogen	
Methanosarcina barkeri	1.0	Complete, rapid inhibition of acetate split (193)
Methanosarcina barkeri Strain 277	0.22	Hydrogen and methanol inhibition acetate split (57)
Methanosarcina Strain TM-1	0.1	Acetate degradation completely inhibited (99)
Acetate enrichment	0.2	Acetate degradation inhibited (194)
Methanosarcina barkeri Strain MS	0.02	Acetate cleavage inhibited until 0.22 atm H_2 reached (94)
M. Barkeri plus Rumen fungus	0.01-0.04	Acetate temporarily accumulated upon H_2 increase (95)
M.barkeri sp.277	0.8	Lowering H_2(aq) below 1mM starts acetate split (169)
Methanosarcina mazeii	1.75	Complete acetate inhibition, yet H_2 not oxidized (169)
	Examples of no Response to Hydrogen	
Acetate enrichment	0.5-0.8	No effect on acetate degradation (194)
Digester sludge	0.014	No effect on acetate degradation (81)
Methanothrix soehngenii	1.3	No effect on acetate degradation (93)
Digester sludge	0.7	No effect on acetate degradation (196)
M.barkeri, Rumen fungus *Methanobrevibacter*	0.01-0.04	No H_2 or acetate accumulation (195)

Table 4.29 : Standard Free Energies Associated with Methanogenic Conversion of Hydrogen, Methanol, Methylamine, and Acetic Acid

Substrate		*Reaction*	*ΔG°.kJ/mol CH_4*
Hydrogen	$4H_2+HCO_3H^+$	—> CH_4+3H_2O	– 135.6
Methanol	CH_3OH+H_2	—> CH_4+H_2O	– 121.1
Methanol	$4CH_3OH$	—> $3CH_4+HCO_3+H_2O+H^+$	– 102.5
Methylamine	$4CH_3NH_2+2H_2O+4H^+$	—> $3CH_4+CO_2 + NH_4^+$	– 101.6
Acetic acid	$CH_3COO^-+H_2O$	—> CH_4+HCO_3	– 31.0

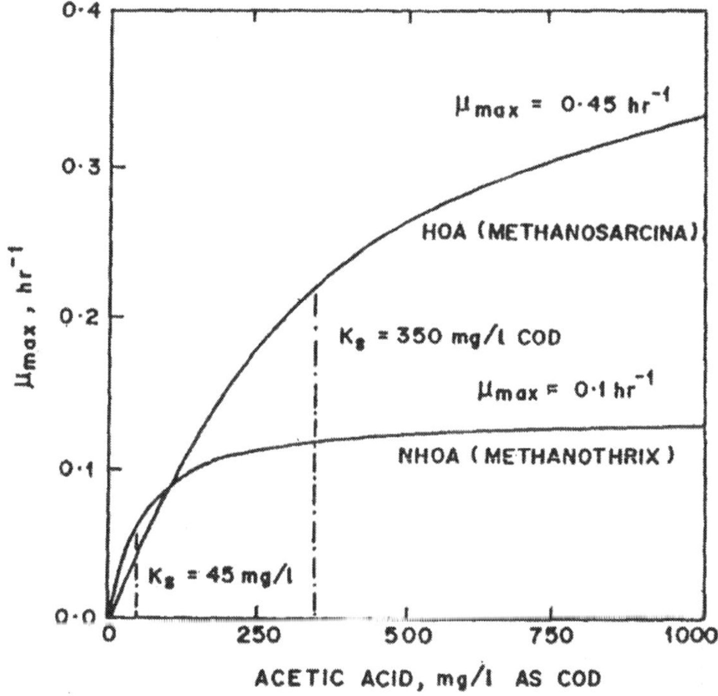

Fig. 4.17 : Comparative Kinetics of growth on Acetic Acid by HOA and NHOA.

4.7.6 Phase Separation

From the foregoing discussion, the importance of hydrogen and the approximate magnitudes at which it serves to control acidogenic and methanogenic processes can be appreciated. The primary tools available for controlling hydrogen, and thereby maximizing the rates of substrate utilization and/or methane production, arise from the kinetic capabilities and substrate specificities of distinct groups of micro-organisms. In single-phase processes, the proportions of OHPA and methanogenic bacteria are prey to the relatively uncontrolled production of acidogenic products, which may alternately reach inhibitory and substrate-limiting proportions.

The two-phase system was originally proposed to address these issues[154,203]. However, the role of hydrogen was not emphasized as the primary control factor in earlier investigations, and the methanogenic enrichments obtained were generally not characterized. Rather, these efforts at phase separation focused on methods of separating methanogenic from acidogenic species, and on achieving and characterizing the resulting environment in each phase. Hammer *et al.*[204] used a dialysis fermentor to evaluate optimum environmental characteristics, while Pohland and Ghosh[154] essentially washed the methanogens out of the first phase reactor using kinetics controls.

A number of researchers have also demonstrated that the environmental requirements of acidogenic and methanogenic (acetogens plus methanogens) enrichments are quite different from each other as reflected in terms of pH and ORP in Table 4.30. Moreover, many of these efforts have demonstrated that phase separation leads to improvements in both process kinetics and operational stability. Indeed, superior performance as well as improved quality of product gas from the second-phase reactor have been demonstrated for two-phase treatment of a variety of substrates. Although the two-phase process seems best suited for the treatment of soluble-type wastewaters, which present a high potential for volatile acids accumulation[110,210,215,216] superior performance has also been demonstrated for particulate-type substrates such as sewage sludge and agricultural wastes.[212,213,217,220] However, these studies failed to show definitive reasons for process improvements.

Several researchers have performed laboratory investigations on two-phase process applications using synthetically prepared, well-defined sugar wastes in efforts to characterize and/or quantitatively describe the reactions and/or enrichments responsible for two phase process improvements[153,221,224]. As before, the improvement in process kinetics and operational stability was clear in all cases; however, the reasons for these improvements were again not fully eludicated. In several cases, the experimental protocol included the addition of sulphates as nutrient salts,[101,221,224] therefore process improvement could have been due to factors associated with these additions (either by the removal of sulphate in the first phase, thereby protecting the methanogens from competitive inhibition, or by the superior acetate conversion kinetics of SRB versus methanogens, or both). Similarly, Ghosh *et al.*[208] noted significant denitrification in the acidogenic reactor of a two-phase system treating sewage sludge. Such improvement in process performance may have been due in part to analogous mechanisms related to NRB conversion (superior kinetics and protection against inhibition).

When the preceding result and observations are evaluated collectively, it is doubtful that sulphate and nitrate considerations were fully responsible for the demonstrated improvements in process efficiency. Instead, factors such as population selection (particularly for the OHPA and methanogens) were likely more important, as addressed by Cohen[224]. In particular, the constant production of propionic and butyric acids in the first-phase reactor established a healthy population of obligate hydrogen-producing acetogens (OHPA) and associated hydrogen-oxidizing methanogens, which ensured the rapid assimilation of these acids in the second phase, However, the methanogenic enrichments obtained (likely-rate HOM and HOA) were not fully characterized and credited with the improvement in overall substrate conversion demonstrated. Cohen[224] also recommended against the transfer of hydrogen gases produced in the first phase to the second phase, reasoning that this practice would inhibited the subsequent degradation of propionic and butyric acids. However, this approach does not effectively utilize all electrons available for methane production since hydrogen is wasted from the first phase. In one of the alternatives being suggested herein, these electrons are utilized to produce methane, and a higher degree of operational flexibility may also be afforded. Most importantly, the differences in OHPA and methanogenic populations enriched in the one and two phase systems need to be studied further in an attempt to better resolve the causes of the two-phase process improvements reported in the literature.

4.7.4 Summary of Hydrogen Management for Process Optimization

The preceding review and discussion provide the following summarized issues, which are central to the further development of state-of-the-art operational strategies for anaerobic treatment process applications:

1. Variations in wastewater loading rates applied to anaerobic processes may cause the accumulation of hydrogen, particularly for soluble-type, high-strength wastewaters.
2. Hydrogen at partial pressures $>10^{-4}$ atm leads to accumulations of propionic and butyric acids, and inhibits their oxidation by obligate hydrogen-producing acetogens (OHPA).
3. A process treating carbohydrates, which has reached a stable operating condition, may greatly reduce the number of electrons channeled through propionic and butyric acids. Therefore, the growth rates and relative populations of OHPA and hydrogen oxidizers (HOM/HOA) decrease also. The resulting condition represents an unstable system, *i.e.*, one susceptible to hydrogen and acid accumulations resulting from influent substrate variations.
4. Only two types of methanogens have been shown to be capable of acetate cleavage, *i.e.*, HOA and NHOA.
5. HOA are apparently capable of three times the acetate utilization rate of NHOA at acetate concentrations above several hundred mg/L.
6. AT hydrogen partial pressures 10^{-4} atm, HOA use H_2/CO_2 in favour of acetate, whereas acetate cleavage by NHOA is unaffected by hydrogen.

Table 4.30 : Optimum pH and ORP Values Determined for Methanogenic Populations

Acidogenic Reactor-Phase				*Methanogenic Reactor-Phase*			
Substrate (Ref)	*Reactor type*	*pH*	*ORP (mV,Ee)*	*Substrate*	*Reactor type*	*pH*	*ORP (mV,Ee)*
– (205)	—	—	—	Sewage sludge	CSTR	7.0	–530
– (206)	—	—	—	Meat peptone broth	CSTR	7.3	–220a
– (206)	—	—	—	Molasses residue	CSTR	8.3	–300a
– (206)	—	—	—	Fatty acids	CSTR	8.1	–270a
–	—	—	–	Strach	CSTR	7.2	–230a
Sewage sludge (204)	CSTR	6.8	–510	Sewage sludge	CSTR	7.0	–513
– (207)	—	—	—	Volatile acids	CSTR	6.8-7.2	–280 to -
520							
Sewage sludge (208)	CSTR	5.8	–240	Acidified sewage sludge	CSTR	7.1	–400
– (209)	—	—	—	Landfill lysimeter	Filter	7.2	–370a
Soft drink	CSTR	4.7	—	Acidified soft drink	CSTR	7.05	—
Wastewater (208)				wastewater			
Sewage sludge (211)	CSTR	6.7	—	—	—	—	—
Sewage sludge (212)	CSTR	6.8	–289	Acidified sewage sludge	Filter	7.2-7.4	–354
Dog food waste (213	CSTR	6.0	—	Acidified dog food	CSTR	7.5	—
Glucose (214)	UASB	6.0	400-500a	—	—	—	—

a : mV, E_H = mV, E_c + 240.

7. Sulphate- and nitrate-reducing bacteria may compete for methanogenic substrates (hydrogen and acetate), and are capable of higher growth rates than methanogens.

In view of these considerations, several treatment options for the management of hydrogen and volatile acids to avoid process instability can be conceived as illustrated in Fig. 4.18. These options, presently undergoing research investigations, may be described as follows :

Option 1 : The first system represents an anaerobic upflow filter in which a natural succession of acidogenic and methanogenic reactions may be expected to occur along the height of the column. Under a stable operating condition, hydrogen may be expected to be utilized rapidly by HOM and HOA residing in symbiotic association with OHPA, particularly in the lower half of the reactor. Under shock loading conditions, excess hydrogen will travel up the column and accumulate in the gaseous phase until the OHPA and hydrogen-oxidizing populations expand to accommodate the accumulated substrates.

Option 2 : The second system represents a classical two-phase process, incorporating a CSTR or an anaerobic upflow filter as the first phase and an anaerobic upflow filter as the second phase. In this system, hydrogen and CO_2 will accumulate in the first phase and stimulate the continuous production of propionic and butyric acids for the second phase. The introduction of these acids into the second phase will ensure a constant pool of OHPA and hydrogen oxidizers, even during stable operating periods. This enrichment strategy should maintain the propionic acid oxidation and hydrogen utilization capacity needed to moderate the effects of wastewater loading variations. The stability of this system is expected to be high, although the maintenance of OHPA in the second phase diminishes the space available for rate-limiting acetate cleaving methanogens. If this strategy proves optimum, variations in potential process applications which also decrease further testing are side-stream acidification and the simple "souring" of wastes in holding on equalization ponds prior to second-phase methane fermentations.

Option 3 : The third system employs the two phase configuration, but includes the capacity for the transferring the H_2/CO_2 produced in the first phase forward to several alternative injection points in the second phase methanogenic reactor. With this option, the transfer of H_2/CO_2 forward to the methanogenic reactor is expected to establish an enrichment of hydrogen oxidizers in the second phase near the gas injection point. Since the removal of hydrogen in this manner may lead to lower hydrogen levels in the first phase and therefore, lower propionic acid levels, a lower quantity of OHPA may be required in the second phase. It may be reasoned that this will leave more capacity in the second reactor for methanogens, thereby decreasing reactor size requirements in addition to providing additional methane from the H_2/CO_2 wasted from the first phase under Option 2. The stability of this system will depend upon the success of hydrogen moderation by gas-phase manipulation and the amount of

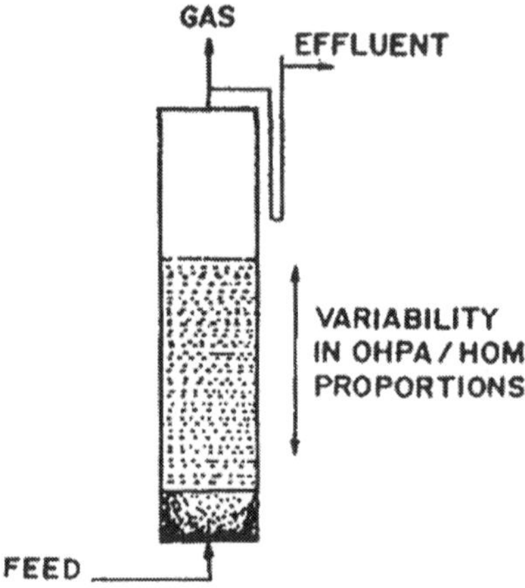

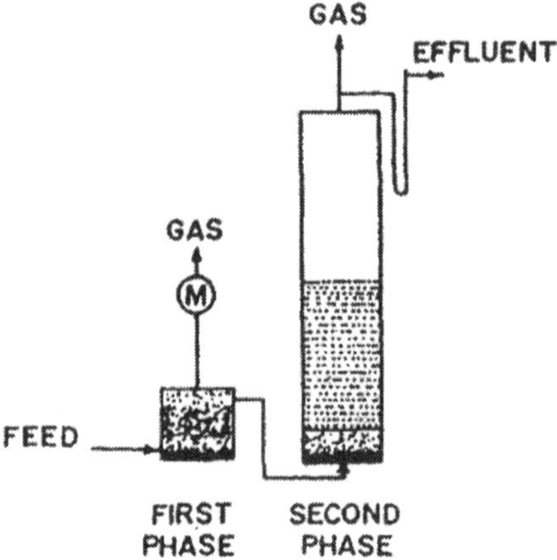

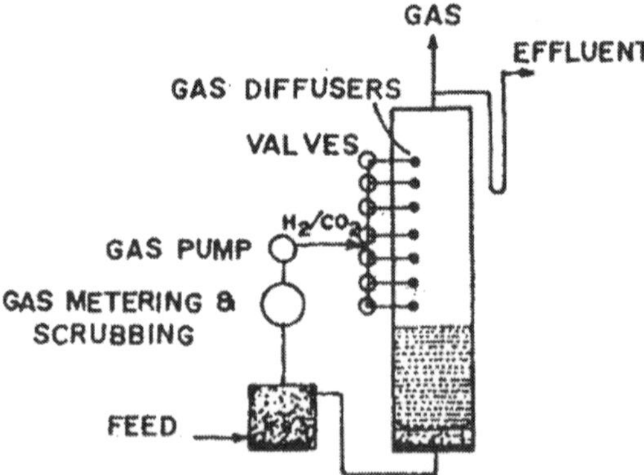

Fig. 4.18 : State of the Art Options in Anaerobic Industrial Wastewater Treatment.

hydrogen oxidizers present in the second phase. The point of hydrogen introduction then becomes an interesting engineering challenge. If hydrogen is introduced at the bottom of the second-phase filter, the OHPA residing there may be inhibited. HOA are expected to enrich in the midrage of the second-phase filter, due to the maintenance of high acetate levels established in a plug flow reactor. The introduction of H_2/CO_2 in this portion of the reactor may cause the inhibition of acetate cleavage; therefore hydrogen should be injected above these enrichments, and multiple injection ports at the upper reaches of the second phase reactor should be available. In the upper part of the reactor, acetate level should be low; therefore, NHOA should be enriched. These bacteria will be unaffected by hydrogen and will not be disturbed by its introduction from the first phase.

Each of these alternatives is currently under laboratory comparison to determine which affords the best practical mix of substrate turnover and process stability[225]. A close analysis of series reactors with a defined synthetic substrate (glucose) and operation under static and dynamic loading conditions is being performed together with studies of an actual industrial wastewater in larger reactors; results from these studies will be reported as they emerge.

4.8 Anaerobic Treatment of Various Wastewaters

During the last two decades, many investigations have been conducted with a large variety of organic industrial wastes. The tendency has been to try to treat increasingly more dilute wastes and in the recent years some effort has been laid upon the treatment of municipal and domestic sewage. This, of course, is bringing anaerobic treatment processes to the very heart of wastewater treatment. Concerning the economy in anaerobic treatment of dilute wastewater the general opinion is that the process today is economically attractive for wastewater with a COD concentration above 1-2 kg COD/m^3. Table 4.31 indicates wastewater amenable for anerobic biotechnology.

Table 4.32 depicts the characteristics of fixed film reactors presently being used. Literature pertaining to startup operation of anerobic processes are give in Table 4.33. Table 4.34 lists some investigation as municipal/domestic sewage. Table 4.35 highlights the types of industrial wastewaters investigated using anaerobic reactors. Table 4.36 shows the sludge mass and sludge load used for anaerobic reactors. Table 4.37 shows the performance of anaerobic filters using sanitary and dilute wastewaters. Table 4.38 highlights the performance of anaerobic filters treating septic tank effluent. Application of anaerobic filters has been shown in Table 4.39 and 4.40. Performance of full scale anaerobic filter installation is shown Table 4.41. Literature concerning anaerobic fixed bed is presented in Table 4.42. Table 4.43 shows the performance of anaerobic fixed film reactor of two different sizes. Performance of treating chemical industrial wastewater at two different temperatures is shown in Table 4.44. Table 4.45 shows the performance data for treating various substrates using anaerobic fixed bed reactor. Performance data for down flow stationary fixed film reactor treating various types of wastewaters is given in Table 4.46. The composition of waste treated in down flow stationary fixed film reactors is shown in Table 4.47.

Table 4.31 : Organics Amenable to Anaerobic Biotechnology+

Acetaldehyde	Formic acid	Isopropyl alcohol	Giant kelp
Acetic anhydride	Fumaric acid	Propionate	Animal wastes
Acetone	Glutamic acid	Propylene glycol	Cheese whey
Acrylic acid	Glutaric acid	Protocatechuic acid	Pear wastes
Adipic acid	Glycerol	Resorcinol	Pectin wastes
Aniline	Hexanoic acid	Sec-butanol	Meat packing
1-amino-2-propanol	Hydroquinone	Sec-butylamine	Corn milling
4-amino butyric acid	Isobutyric acid	Sorbic acid	Dairy
Benzoic acid	Isopropanol	Syringaldehyde	Brewery
Butanol	Lactic acid	Syringic acid	Rum distillery wastes
Butyraldehyde	Maleic acid	Succinic acid	Wine distillery wastes
Butylene glycerol	Methanol	Tert-butanol	Guar gum wastes
Catechol	Methyl acetate	Vanillic acid	Waster-soluble polymers
Cresol	Methyl acrylate	Vinyl acetate	Bean blanching
Crotonaldehyde	Methyl ethyl ketone	Corn	Pulp Mill evaporate
Crotonic acid	Methyl formate	Potato	Coking mill
Diacetone gulusonic acid	Nitrobenzene	Sugar cane	H_2-CO pyrolysis
Dimethoxy benzonic acid	Pentaerythritol	Bagasse	Wool scouring
Ethanol	Pentanol	Peat	Tannery wastes
Ethyl acetate	Phenol	Wood	Yeast
Ethyl acrylate	Phthalic acid	Corn stover	Heat-treated activated sludge
Ferulic acid	Propanal	Straw	
Formaldehyde	Propanol	Water hyacinths	

+ Personal communication : R.E. Speece-Advances in Anaerobic Biotechnology for Industrial Wastewater Treatment (1989).

The effect of overloading for 2 hours on the performance of stationary fixed film reactor is presented in Table 4.48. Rough guidelines for number of feed inlet nozzles required in a UASB reactor is presented in Table 4.49. Tentative rough design capacities for UASB reactors treating mainly liquid wastes in relation to temperature is shown in Table 4.50. Basic design features and performance of $30m^3$ and $800m^3$ plants treating liquid sugar and beet sugar using UASB reactor in shown in Table 4.51.

Table 4.32 : Characteristics of Fixed Film Reactors

Reactors Bed Type	*Inert Material*	*Characteristics*	*Specific Surface Area (empty bed) m^2/m^3*	*Empty Bed Porosity*	*Literature*
Fixed	Quartz	40mm diam		0.42	Arora *et al.* (1975)
Fixed	Quartz stone	20mm diam		0.42	Arora and Routh (1980)
Fixed	Plastic			>0.90	Benjamin *et al.* (1981)
Fixed	Rachig rings	10.16mm	45-49	0.76-0.78	Carrondo *et al.* (1982)
Fixed	Gravel	6-13mm diam		0.4	Coulter *et al.* (1957)
Fixed	Gravel	25-38mm diam			El-Shafie and Bloodgood (1973)
				0.96	Frostell (1981a)
Fixed				0.43	Frostell (1980)
Expanded	Sand	0.1-0.25 mm diam			Hall (1982)
Fixed	Plastic, actifill	9 cm			Hall (1982)
Fixed	Red clay blocs				Hall (1982)
Fluidized	Sand	0.3-0.4mm diam			Hall (1982)
Fluidized	Sand	0.7-0.8mm diam			Hall (1982)
Fixed	Oyster shells	60-100m diam		0.82	Hudson *et al.* (1978)
Fixed	Granite stone	25-38mm daim		0.53	Hudson *et al.* (1978)
Expanded	PVC-resin	1mm diam			Jewell *et al.* (1981)
Expanded	Al-oxide	0.5mm diam			Jewell *et al.* (1981)
Fluidized	Sand	0.5-0.7mm diam			Li and Sutton (1981)
Fixed	Polymer foam			0.95	Lindgren (1981)
Expanded	Sand	0.2mm diam		(0.35)	Lindgren (1981)
Fixed	Active carbon	1.5mm diam		(0.60)	Lindgren (1981)

Contd...

1	2	3	4	5	6
Fixed	Limestone	25-40mm diam			Lovan and Foree (1971)
Fixed	Stone/sand	3-20 mm diam/ 16-30 mesh			Pretorius (1971)
Fixed	Leca	6mm diam		0.40	Matsche and Ruider (1981)
Fixed	Plastic, Berl saddles			0.90	Mosey (1978)
Fixed	Polyurethane			0.96	Norrman (1982)
Expanded	Sand	0.8mm diam		(0.38)	Norrman (1982)
Fixed	Rachig rings/ Berl saddles			0.65-0.70	Plummer *et al.* (1968)
Fixed	Red clay blocs		53-397	(1.0)	van den Berg and Lentz (1979)
Fixed	Red clay blocs		120-140	(1.0)	van den Berg *et al.* (1980)
Fixed	Grey Potters clay	28mm ins. diam	141	(1.0)	van den Berg and Kennedy (1981)
Fixed	Red draintile clay	(25mm ins. diam)	149	(1.0)	van den Berg and Kennedy (1981)
Fixed	PVC	23mm ins. diam	174	(1.0)	van den Berg and Kennedy (1981)
Fixed	Needle punched Polyester		86	(1.0)	van den Berg and Kennedy (1981)
Fixed	Stone	40-50mm diam			Wilkie and Newell (1981)
Fixed	PVC	25mm diam			Walkie and Newell (1981)
Fixed	Quartzite stone	40mm diam		0.42	Young and McCarty (1967)
Fixed	Corrugated blocs		98	>0.95	Young and Dahab (1982)
Fixed	Corrugated blocs		132	0.95	Young and Dahab (1982)
Fixed	Pall rings	90.90mm	102	0.95	Young and Dahab (1982)
Fixed	Polypropylene spheres	90mm diam	89	>0.95	Young and Dahab (1982)

Contd...

1	2	3	4	5	6
Fluidized	Sand		2500-4000		Shieh *et al.* (1981)
Fixed	Limestone	25-38mm		0.49	Johanseen (1975)
Recycled		0.005-0.025 diam	5000-10,000		Martensson and Frostell (1982)
Fixed	Flocor R		230	0.975	Bories (1981)
Fixed	Surpac		206	0.94	Chian and DeWalle (1977)
Fixed	Granite rock	25-75mm diam		0.40	Richter and Mackie (1971)
Fixed	Intalox saddles (polypropylene)	25mm		0.83	Wilson and Timpany (1973)
Fixed	Stone	25-38mm diam		0.47	Dennis and Jennett (1974)
Fixed	Pall-rings	15mm	-	0.85	Mueller and Mancini(1975)
Fixed	Stone	25-38mm diam		0.43	Seeler and Jennett (1978)
Fixed	Stone	25-40mm diam		0.42	Brown *et al.* (1981)
Fluidized	Sand	0.7-0.8mm diam	3000	-	Hall (1981) and (1981a)
Fixed	Red clay tubes	25mm diam	70	(1.0)	Hall *et al.* (1981a)
Fixed	Norton Plastic rings	9cm	114	0.95	Hall *et al.* (1981a)
Fixed	Corrugated Blocs	75.50 pore size	100	>0.95	Dahab and Young (1981)
Fixed	Granite chips	12-25mm	-	0.40	Colleran *et al.* (1982)
	Mussel Shells	-	-	0.77	Colleran *et al.* (1982)
	Corals	-	-	0.72	Colleran *et al.* (1982)
	Reeds	-	-	0.66	Colleran *et al.* (1982)
	Limestone	-	-	-	Colleran *et al.* (1982)
Fixed	Red clay blocs size	28.28mm pore size	157	-0.70	Kennedy and vanden Berg (1982a)
Expanded	Ion exchange resin/PVC	0.5mm/ 1.5mm	5600	0.46	Jewell *et al.* (1978)

Table 4.33 : Literature on Start-up of Anaerobic Processes

Sr.No.	Method	Observations/ Recommendations during up-start	Initial Loading	Start-up Period Months	Type of Reactor	Literature
1.	VFA feed Municipal digester sludge as inoculum	*Granules (0.5-1mm diam) after 30-50 days *3stages during start-up 1. Adaption 2. Increased activity 3. Granules *Ca^{++}-conc. sufficiently high	0.05-0.1 kgCOD (kgVSS).d	2-(4)	Sludge blanket	de Zeeuw and Lettinga (1980)
2.	Methanol added to feed to promote (1980) growth the methanogens			1	Moving bed	Tait and Friedman
3.	VFA (C2,C3 and C4) from soured waste Sludge from fluidized floc reactor superior for start-up	*Long storage of culture possible (1-2 years)	0.2-0.4 kg COD/ (kgVSS.d)	2-3	Sludge blanket	Lettinga *et al.* (1979)
4.	Sewage sludge digester inoculum	*3 months storage gives decreased activity *For high loadings : Continuous inoculation advantageous	1-1.15kg VSS/m^3.d		Recycle flocs	van den Berg and Lentz (1977a)
5.	Floating filter to minimize biomass washout during start-up				Sludge blanket	Effers (1981)

Contd...

1	2	3	4	5	6	7
6.	Inoculumn : mixture of soil, wastewater and manure. 1 month without influent				Fixed bed	Arora *et al.* (1975)
		*Loading increased 25-100% per week	1kgCOD/ $(m^3.d)$	2	Fixed bed	Benjamin *et al.* (1981)
		*Start-up period depends upon type of inert media		1-3	Fixed bed	van den Berg and Kennedy (1981)
				8-13	Fixed bed	van den Berg and Lentz (1979)
		*4 times faster start-up with sludge from identical process than with municipal digester sludge		0.3-1.3	Fixed bed	Young and McCarty (1967)
7.	Seeding with sludge from another anaerobic reactor	*After 1 month increased loading possible	1kgCOD/ $m^3.d$	1-2	Recycled bed	Martenssowal Frostell (1982)
8.	Seeded with sludge digester biomass accumulation for 1 month	*Without seeding : no gas production and		1.5	Fixed bed	Chian and Dewalle (1977)
9.	Originally Seeded with digester sludge. If granules are added, this has only minor effect	*After initial wash out loading should be increased to >0.6kg COD/(kgVSS.d) in order to produce granules. *High Ca^{++} in waste-water increases biomass wash-out and start-up period	0.05-0.1 COD/kg VSS.d)	2-3	Sludge blanket	Pol *et al.* (1982)

Contd...

1	*2*	*3*	*4*	*5*	*6*	*7*
		* Approc. 2 months for granulation to take place *Granules does not act as precursers for new granules *High NH_3-N ($1kgN/m^3$) inhibits granulation *Initial sludge conc.: 10-20 kg VSS/m^3 *Initial start-up easier for fixed bed than for fluidized flocs			Fixed beds Sludge blanket	Lettinga *et al.* (1982)
10.	Digester sludge	*Acetic and >$1kgm^3$ *Polymers for better flocculation		4	Recycled flocs	Schlegal and Kalbskopf (1981)
		*Temperature >35°C (kgVSS.d)	>0.1kgCOD/			Salkinoja-Salonen *et al.* (1982a)
		*Porous surface of carrier material				
		*Carbohydrate for slime production *Aerobic operation ahead of start-up *Addition of slime producing organisms *VFA<1.5-2.5kgHAc/m^3 *HRT>1d *High alkalinity *No fluctuations in environmental conditions *$Ca(OH)_2$ for neutralization *Calcium may promote attachment	>2kg $COD/m^3.d$			

Main results obtained in a $6m^3$ UASB plant with raw domestic sewage influent are shown in Table 4.52. Results obtained with granular sludge UASB reactor and wastewater containing a high fraction of suspended matter is shown in Table 4.53. Results of slaughter house wastewater using UASB reactor are shown in Table 4.54.

Data concerning various types of substrates using laboratory scale UASB reactors are shown in Table 4.55. Results obtained in a $6m^3$ pilot plant and $200m^3$ scale plant are shown in Table 4.56 Table 4.57 shows the results obtained with USAB reactor using granular sludge for alcoholic and potato processing wastewater. Results obtained with three types of complex wastes using flocculant sludge UASB reactors are shown in Table 4.58 while Table 4.59 shows the results obtained using granular sludge. Suspended solids washout as found in $6m^3$ pilot plant and full scale plant are shown in Table 4.60. Performance of full scale installations of UASB in various parts of the world is shown in Table 4.61. Literature on sludge blanket reactor is highlight in Table 4.62.

Performance data for fluidized bed reactor treating chemical industry wastewater is presented in Table 4.63. Fig 4.19 shows the COD removal efficiency versus organic load for various types of wastewater. The effect of temperature on COD removal efficiency is shown in Fig. 4.20.

Table 4.34 : Studies of Anaerobic Treatment of Municipal/Domestic Wastewater

Sr.No.	*Type of Reactor*	*Reference*
1.	Fixed bed	Pretorius (1971)
2.	Fixed bed	Matsche and Ruider (1981)
3.	Fixed bed	Coulter *et al.* (1957)
4.	Fixed bed	Raman and Chakladar (1972 and 1972a)
5.	Fixed bed	Raman and Khan (1978)
6.	Expanded bed	Jewell *et al.* (1981)
7.	Fixed bed	Koon *et al.* (1980)
8.	Sludge blanket	Vanden Heuvel *et al.* (1981)
9.	Fixed bed	De Walle *et al.* (1981)
10.	Fixed bed	Genung *et al.* (1981)
11.	Fixed bed	Lindgren (1981)
12.	Sludge blanket	Lettinga and Vinken (1980)
13.	Recycled flocs	Schroepfer and Ziemke (1959)
14.	Sludge blanket	Roersma *et al.* (1981)
15.	Fixed bed	Wheatley and Cassell (1982)

The comparison of laboratory and pilot plant fluidized bed reactor treating sanitary wastewater is shown in Table 4.64. The performance data for treating food processing wastewater using fluidized bed are shows in Table 4.65. The performance data for anoxic fluidized and expanded bed reactors are shown in

Table 4.66. Industrial applications of expanded/fluidized bed reactors are shown in Table 467. Performance data for anoxic fluidized and expanded bed reactors for various types of wastewater are shown in Table 4.68. Data concerning full scale anaerobic fluidized bed installations are shown in Table 4.69. Summary of methane produced by using fluidized bed reactor is shown in Table 4.70 for various types of wastewater. Literature on anaerobic expanded bed is presented in Table 4.71 and similarly for fluidized bed in shown in Table 4.72.

Performance data for moving bed, recycled bed and recycled flocs reactors are shown in Tables 4.73 to 4.76.

Table 4.35 : Types of Industrial Wastes Investigated Using Anaerobic Reactors

Sr.No.	*Type of Waste*	*Influent Concentration, kg COD/m³*	*Type of Reactor*	*Reference*
1.	Dairy	1.5	Sludge blanket	Lettinga (1978)
		4.4	Recycled flocs	Anderson *et al.* (1980a)
		24	Fixed bed	Witt *et al.* (1980)
		4.4	Fixed bed	van den Berg *et al.* (1981a)
		4.4	Fixed bed	Kennedy and van den Berg (1981a)
		—	Fluidized bed	Jenkins *et al.* (1981)
2.	Milk waste	4	Fixed bed	van den Berg and Kennedy (1982)
		2.5	Fixed bed	Colleran *et al.* (1982)
3.	Distillery	13/22	Sludge blanket	Stander (1967)
		78-85	Recycled flocs	Bories (1980)
		26/34	Recycled flocs	Rose (1980)
		50	Recycled flocs	Anderson *et al.* (1980a)
		89	Recycled flocs	Frostell (1981)
4.	Rum stillage	50-70	Fixed bed	van den Berg *et al.* (1981a)
		27-85	Fixed bed	Bories (1981)
		53	Recycled flocs	van den Berg and Lentz (1977a)
		50-70	Fixed bed	Kennedy and van den Berg (1981a)
		~55	Fixed bed	Schafer (1982)
		53	Recycled flocs	van den Berg and Lentz (1980a)
5.	Alcoholic	12	Sludge blanket	Lettinga (1978)
6.	Cider	1.5	Recycled flocs	Anderson *et al.* (1980a)

Contd...

Sr.No.	Type of Waste	Influent Concentration, kg COD/m^3	Type of Reactor	Reference
7.	Brewery	—	Fixed bed	Tadman (1973)
		1-3	Sludge blanket	Eggers (1981)
		6-18	Fixed bed	Hall (1982)
		6-18	Sludge blanket	Hall (1982)
		6-18	Fluidized bed	Hall (1982)
8.	Brewery press liquor	6-27	Fixed bed	Lovan and Foree (1971)
9.	Soft drink bottling	6	Fluidized bed	Hickey and Owens (1981)
		5	Fluidized bed	Jeris (1982)
10.	Bettling	6.9	Fluidized bed	Barbara and Gasser (1979)
11.	Malting	4.2-12.2	Fixed bed	Wheatley and Cassell (1982)
12.	Food processing	7.2-9.4	Fluidized bed	Hickey and Owens (1981)
		8	Fluidized bed	Jeris (1982)
13.	Soy processing	9	Fluidized bed	Sutton and Li (1981)
14.	Sauerkraut	10-20	Sludge blanket	Lettinga *et al.* (1980a)
15.	Shellfish	0.4	Fixed bed	Hudson *et al.* (1978)
16.	Beet-sugar	3/17	Sludge blanket	Pette *et al.* (1980)
		4-9	Sludge blanket	Lettinga *et al.* (1980a)
		4-7	Recycled bed	Martensson and Frostell (1982)
		0.5-1.4	Sludge blanket	Heertjes *et al.* (1982)
		1-18	Sludge blanket	Lettinga (1978)
17.	Pectine	12	Recycled flocs	Sefried (1981)
		1-5	Recycled flocs	Werencke and Mudrack (1981)
18.	Whey	66	Fixed bed	van den Berg and Kennedy (1982)
		52-54	Fluidized bed	Jeris (1982)
		10	Expanded bed	Switzenbaum and Danskin (1982)
		10	Expanded bed	Switzenbaum (1981)
		8.1	Fixed bed	Hakansson *et al.* (1977)
		50-56	Fluidized bed	Hickey and Owens (1981)
19.	Whey permeate	5	Fixed bed	Hakansson *et al.* (1977)
		7/27	Fluidized bed	Sutton and Li (1981)
		7/27	Fluidized bed	Li and Sutton (1981)
		10/30	Fluidized bed	Li *et al.* (1982)

Contd...

Sr.No.	*Type of Waste*	*Influent Concentration, kg COD/m*^{3}	*Type of Reactor*	*Reference*
20.	Tomato peeling	—	Fixed bed	van den Berg and Lentz (1980d)
		10/20	Fixed bed	Stevens and van den Berg (1981)
21.	Pear peeling	110-140	Fixed bed	Kennedy and van den Berg (1981a)
		110-140	Fixed bed	van den Berg *et al.* (1981a)
		9-26	Fixed bed	van den Berg and Kennedy (1982a)
		50	Recycled flocs	van den Berg and Lentz (1977a)
		38	Recycled flocs	van den Berg and Lentz (1980a)
22.	Potato peeling	—	Recycled flocs	van den Berg and Lentz (1980a)
		51	Recycled flocs	van den Berg and Lentz (1980a)
23.	Potato blanching	11	Recycled flocs	Anderson *et al.* (1980a)
24.	Potato chips	-85	Recycled flocs	Viitasalo *et al.* (1981)
25.	Potato	38-70	Recycling flocs	van den Berg and Lentz (1977a)
		3.6-10.8	Digester + Fixed bed	Landine *et al.* (1982)
		—	Recycled flocs	van den Berg and Lentz (1980d)
26.	Potato	3	Fixed bed	Pailtrop *et al.* (1977)
		2-5	Sluge blanket	Lettinga *et al.* (1980a)
		2-12	Sludge blanket	Lettinga *et al.* (1979)
		-(3-8)	Digester + Fixed bed	Brown *et al.* (1981)
27.	Bean blanching	10.5	Fixed bed	van den Berg and Lentz (1980c)
		-10	Fixed bed	van den Berg and Kennedy (1981) and (1982a)
		9.5	Fixed bed	van den Berg *et al.* (19800)
		10/40	Fixed bd	van den Berg *et al.* (1981a)
		5.2	Sludge blanket	Lettinga *et al.* (1979)
		22	Recycled flocs	van den Berg and Lentz (1980a)

Contd...

Sr.No.	Type of Waste	Influent Concentration, kg COD/m^3	Type of Reactor	Reference
		20	Recycled flocs	van den Berg and Lentz (1979a)
		10	Fixed bed	Kennedy and van den Berg (1981)
		9-11	Fixed bed	Kennedy and van den Berg (1981a)
		—	Fixed bed	van den Berg and Lentz (1979a)
		—	Fixed bed	Kennedy and van den Berg (1979a)
		—	Fixed bed	van den Berg and Lentz (1980d)
		—	Sludge blanket	van den Berg and Lentz (1980d)
		—	Recycled flocs	van den Berg and Lentz (1980d)
		10/20	Fixed bed	Stevens and van den Berg (1981)
		10	Fixed bed	Kennedy and van den Berg (1982), (1982b) and (1982c)
28.	Vegetable/fruit	0.7-2.5	Sludge blanket	Goedvriend (1981)
29.	Bakery	9	Fluidized bed	Jeris (1982)
30.	Maize starch	10	Recycled flocs	Ross (1980)
31.	Wheat starch	6-13	Fixed bed	Richter and Mackie (1971)
		—	Recycled flocs	Anderson *et al.* (1980a)
		10 20	Fixed bed	Mosey (1978)
		—	Fixed bed	Taylor and Burm (1973)
		7-16	Fluidized bed	Pilot plant test (1980)
32.	Palm-oil	30-60	Recycled flocs	Chin (1981)
33.	Olive oil	(40)	Recylced flocs	Anderson *et al.* (1980)
34.	Slaughter house	1.5-2.0	Sludge blanket	Sayed *et al.* (1981)
		0.3-20.8	Fixed bed	Arora and Routh (1980)
		-2.0	Recycled flocs	Steffen and Bedker (1961)
		—	(Recycled flocs)	Lindauer and Sixt (1981)
		~1.5	Recycled flocs	Schroepfer and Ziemke (1959)
		.23-3.9	Fixed bed	Wheatley and Cassell (1982)
35.	Piggery waste	27-51	Fixed bed	van den Berg *et al.* (1981a)
		40-50	Fixed bed	Kennedy and van den Berg (1981a) and (1982d)
		59	Fixed bed	Colleran *et al.* (1981)

Contd...

Sr.No.	*Type of Waste*	*Influent Concentration, kg COD/m³*	*Type of Reactor*	*Reference*
		58	Fixed bed	Colleran *et al.* (1982)
36.	Calf manure	10	Sludge blanket	van Velsen *et al.* (1979)
37.	Cow manure	11.5/23	Expanded bed	Jewell *et al.* (1978)
38.	Silage	24	Fixed bed	Colleran *et al.* (1982)
39.	Molasses (Citric acid)	59 45-56	Recycled flocs Fixed bed	Anderson *et al.* (1980a) Carrondo *et al.* (1982)
		—	Fluidized bed	Binot *et al.* (1981)
40.	Molasses (Yeast)	9.1	Expanded bed	Frostell (1980)
		2-5	Recycled flocs	Anderson *et al.* (1980a)
41.	Molasses	9	Recycled bed	Martenson and Frostell (1982)
42.	Molasses distillery	—	Fixed bed/Sludge blanket/Recycled flocs	Braun (1981)
43.	Molasses distillery	70-100	Fixed bed/sludge blanket	Riera *et al.* (1981)
44.	Vegetable tanning	0.3-3.6	Fixed bed	Arora *et al.* (1975)
45.	Sulfide evaporator			
46.	Condensate	5	Fixed bed	Benjamin *et al.* (1981)
		—	Fixed bed	Tadman (1973)
		—	Fixed bed	Anaerobic (1973)
		0.5-3.5	Fixed bed	Norrman (1981)
		0.5-2	Fluidized bed	Norrman (1981)
46.	Kraft mill, black	1.1-1.25	Fluidized bed	Hakulinen *et al.* (1981)
	liquor condensate	2-7	Fluidized bed	Hakulinen and Salkinoja Salonen (1982a)
		1.4	Fixed bed	Norrman (1982)
		1.4	Expanded bed	Norrman (1982)
47.	Sulphite liquor	(2-8)	Fixed bed	Wilson and Timpany (1973)
48.	Sulphite pulping	5.6	Fluidized be	Sutton and Li (1981)
49.	Paper board	5.6	Recycled flocs	Jung (1949)
50.	Insulating board	~10	Sludge blanket	De Bekker (1981)
51.	Pulp (hard board (mill)	6.5	Fluidized bed	Sutton and Li (1981)
52.	Wood fibre	5.9	Recycled flocs	Schroepfer and Ziemke (1959)
53.	Chemical	24	Sludge blanket	Jans (1981)
		—	Fixed bed	Kennedy and van den Berg (1981a)
		14	Fixed bed	Kennedy and van den Berg (1982)

Contd...

Sr.No.	*Type of Waste*	*Influent Concentration, kg COD/m³*	*Type of Reactor*	*Reference*
		11	Fluidized bed	Jeris (1982)
		—	Fixed bed	Witt *et al.* (1980)
		14	Fixed bed	Kennedy and van den Berg (1981)
		—	Fixed bed	Kennedy (1980)
		12	Fluidized bed	Hickey and Owens (1981)
		14	Fixed bed	van den Berg *et al.* (1981a)
		14	Sludge blanket	van den Berg *et al.* (1981a)
		3	Fluidized bed	Sutton and Li (1981)
54.	Phenolic	0.4-1.0	Fixed bed	Cross (1981)
55.	Petrochemical	2-8	Fixed bed	Hovious *et al.* (1972)
56.	Pharmaceutical	1-6	Fixed bed	Seeler and Jennett (1978)
		~(3)	Fixed bed	Jennett and Rand (1980)
		1-16	Fixed bed	Dennis and Jennett (1974)
57.	Sludge heat treat liquor	~(4-10)	Fluidized bed	Pilot test (1979)
		10	Fluidized bed	Hickey and Owens (1981)
		10.5	Fixed bed	van den Berg *et al.* (1981a)
		8	Fluidized bed	Jeris (1982)
		4-10	Fluidized bed	Bell (1981)
		10	Fixed bed	Donovan (1980)
		10.5	Fixed bed	Kennedy and van den berg (1982)
		11	Fixed bed	Hall (1982)
		11	Sludge blanket	Hall (1982)
		11	Fluidized bed	Hall (1982)
		20	Fixed bed	Schwartz and De Baere (1981)
		10.5	Fixed bed	Kennedy and van den Berg (1981)
		10.5	Fluidized bed	Hall *et al.* (1981a)
		10.5	Fixed bed	Hall *et al.* (1981a)
		10.5	Sludge blanket	Hall *et al.* (1981a)
		15	Recycled flocs	Schlegel and Kalbskiof (1981)
		15	Recycled flocs	Schelgel and Kalbskopf (1982)
58.	Leachate	-	Fixed bed	Rydin and Haglund (1981)
		25-32	Sludge blanket	De Becker and Kaspers (1981)
		2.9-4.4	Fixed bed	Johansen (1975)
		6-62	Fixed bed	China and De Walle (1977)
59.	Textile	0.6-2.0	Fixed bed	Wheatley and Cassell (1982)
60.	Tannery	3.0	Fixed bed	Friedman *et al.* (1981)

Table 4.36 : Sludge Mass and Sludge Load for Anaerobic Reactors

Type of Waste/ Substrate	*Solubility Index* %	*Type of Reactor*	*Temperature* °C	*Reactor Volume* m^3	Biomass Conc. *kg SS/m^3*	Biomass Conc. *kg VSS/m^3*	*Sludge Mass kg COD/kg VSS.d*	*Literature*
Alcoholic	100	Sludge blanket	30	3.10^{-3}		36	0.5	Lettinga *et.al* (1979)
Sugar-beet, unsoured	95	Sludge blanket	30	61.10^{-3}		7	0.5-0.8	Lettinga *et al.* (1980a)
Sugar-beet, soured	60-80	Sludge blanket	30	18.10^{-3}		-	0.8-1.0	Lettinga *et al.* (1980a)
Bean blanching	90	Sludge blanket	30	3.10^{-3}		~13	0.6-0.8	Letting *et al.* (1980a)
Sauerkraut	97	Sludge blanket	30	3.10^{-3}		~9	0.8-1.2	Lettinga *et al.* (1980a)
Dairy	>75	Sludge blanket	30	18.10^{-3}		~15	0.4-06	Lettinga *et al.* (1980a)
Whey		Expanded bed				40		Switzenbaum and Danskin (1982)
Molasses, diluted	100	Recycled bed	37	40.10^{-3}		8.6-8.7	0.3-0.5	Martensson and Frostell (1982)
Sugar-beet	~90	Recycled bed	37	40.10^{-3}		9.3-18.8	0.6-1.3	Martensson and Frostell (1982)
Heat treat liquor/ Brewery	99	Fluidized bed	35	29.10^{-3}	5.7	4.6		Hall (1982)
Heat treat liquor/Brewery	99	Fixed bed, upflow	35	0.9	42.3	14.5		Hall (1982)
Heat treat liquor/Brewery	99	Fixed bed, downflow	35	620.10^{-3}	35.1	12.9		Hall (1982)
Heat treat liquor/Brewery	99	Sludge blanket	35	1.1	13.5	7.3		Hall (1982)

Table 4.37 : Anaerobic Treatment of Sanitary Wastewater and Dilute Wastewater⁺

Sr. No.	Nature of Wastewater	Characteristics	Process	Type of Study	Loading Rate	HRT(h)	%Removal	Temperature(°C)
1.	Settled Wastewater	COD-226 mg/L BOD-90 mg/L	Anaerobic Filter	Lab. Study	0.1	25	60 (COD)	- -
2.	Settled Wastewater	COD-124	Anaerobic Filter	Lab. Study	0.18	19	75 (COD) 90 (SS)	- -
3.	Diluted Raw Wastewater	COD-500 mg/L	Contact digester	Lab-and Pilot plant	0.48	24	90	-
4.	Dilute Synthetic Wastewater	COD-900 mg/L	Anaerobic Filter	Lab-Study	0.8 1.2 1.6 2.2	12.8 9.1 6.4 4.8	96 95 91 90	24-33 24-33 24-33 24-33
5.	Sanitary Wastewater	BOD-210mg/L SS-272 mg/L	Upflow filter	Full Scale	0.01	6-8	79.5 88.5	30
6.	Sanitary Wastewater	COD-288 mg/L	Upflow	Lab. Study	0.32	24	73	25-35
7.	Diluted City Wastewater	BOD-300 to 400 mg/L COD-1100 to 1300mg/L	Contact Digester	Pilot	0.6**	-	80(BOD)	-
8.	Synthetic Wastewater	COD-780mg/L	2 stage Anaerobic filter	Lab-study	0.5 1.0 2.0	15 8 4	83 79 71	21-26 21-26 21-26
9.	Diluted Synthetic Wastewater	COD-50 to 600 mg/L	Fixed film expanded bed	Lab. Study	upto 8	6-0.33	80	10-30
10.	Raw Wastewater	BOD-220 mg/L	Anaerobic filter	Pilot scale	0.24**	10	55	10-25
11.	Raw Wastewater	BOD-175 mg/L	do	Pilot scale	0.26*	5	73	28
12.	Raw Wastewater	BOD-120 mg/L SS-140 mg/L	do	Bench			61 (BOD) 64 (SS)	

+ Viraraghvan, T. and Kent, R, : Septic Tank Effluent Using an Anaerobic Filter, University of Regina, Regina, Saskatatchewan, Canada (1983).

* -kg $BOD/m^3.d$; **-kg $COD/m^3.d$.

Table 4.38 : Performance Summary for Anaerobic Filters Treating Septic Tank Effluent[+]

Sr. No.	*Location*	*Characteristics*	*Process*	*Type of Study*	*Loading Rate**	*HRT (h)*	*% Removal*	*Tempe-rature (°C)*
1.	India	BOD-240 mg/L	Upflow	Full Scale	0.02	6 days	71	-
2.	India	BOD-210 mg/L	Downflow and Upflow	Full Scale	low loading	High	75	-
3.	Washington State	BOD-241 mg/L COD-486 mg/L	Upflow	Full Scale			26 28	12.4
4.	India	BOD-290 mg/L	Upflow	Laboratory	0.34	-	76	23-33
5.	Thailand	COD-310 mg/L SS-170 mg/L	Upflow	Full Scale	0.5**	24	60 70	- -
6.	Washington State	BOD-103 mg/L SS-67 mg/L COD-305 mg/L	Upflow	Full Scale	0.04	38.4	28 42 23	-
7.	Washington State	BOD-217 mg/L SS-33 mg/L COD-542 mg/L	Upflow (trench)	Full Scale	0.027	-	39 33 21	7 and 14
8.	Washington State	BOD-120 mg/L	Upflow	Full Scale	0.19**	19	28 (COD)	11.8
9.	Washington State	-	Downflow filter	-	-	-	60 (COD) 28 (SS)	-

* - kg BOD.m^3.d ; ** - kg (COD/m^3.d'

\+ Viraraghvan, T and Kent, R; Septic Tank, Effluent Treatment using an Anaerobic filter; University of Regina, Regina, Saskatchewan, Canada (1983).

Table 4.39 : Anaerobic Filter Studies For Industrial Waste[+]

Sr. No.	*Waste*	*Influent (mg/L)*	*Effluent (mg/L)*	*% Removal*	*Loading Rate ($kg/m^3.d$)*	*HRT (h)*	*Tempe-rature (°C)*	*Remarks*
1.	Glucose	COD-1500	112-950	36.7-92.1	0.42-3.4	4.5-36	25	-
	nutrient broth	COD-3000	204-1105	63.0-93.4	0.42-3.4	9.72		
2.	Volatile acids	COD-1500	24-476	86.4-99.4	0.42-3.4	4.5-36	25	-
		COD-3000	42-240	92-98.6	0.42-0.85	36-72		
		COD-6000	139-794	86.9-97.7	1.7-3.4	18-36		
3.	Metrecal	COD-10,000	-	75-90	-	18	-	6 beds in series
4.	Pharmaceutical Waste	BOD-2000	-	94 (BOD) 70-80 (COD)	-	36	35	-
5.	Protein Carbohydrate	-	-	50-90 (COD)	3.2-27.2	3-24	35	
6.	Whey	COD-8100	-	98	1.9	-	22-25	
7.	Whey permeate	COD-5000	-	95	0.7	-	22-25	

+ *Switzerbaum. M. S.* : A Comparison of the Anaerobic Filter and Anaerobic Expanded Fluidized Bed Process; Proceeding of a Specialized Seminar of IAWPR, held in Copenhegan, Denmark, (16-18 June 1982), Water Sci. Technol., 15 (8/9), p345-358 (1982).

Table 4.40 : Anaerobic Filter Studies for Industrial Waste+

Sr. No.	*Waste*	*Influent (mg/L)*	*Effluent (mg/L)*	*% Removal*	*Loading Rate ($kg/m^3 \cdot d$)*	*HRT (h)*	*Tempe-rature (°C)*	*Remarks*
1.	Food processing	COD-8475	546-5000	41-93.6	-	0.54-3.45 (days)	35	-
	Carbohydrate waste	COD-3890	975-3090	25.2-81.3		0054-3.45 (days)		
2.	Waste sulphite	BOD-1300 to	-	27-58	-	3.7-4.0 (days)	-	
	liquor	BOD-5300						
3.	Potato processing	COD-3000	-	41-79	-	0.4-2.5	19-22	
4.	Fish processing	COD-466	90	80.7	-	3.1 (days)	26	
		COD-407	102	74.9		1.6 (days)	23.5	
		BOD-310	58	-	-	-	-	
5.	Sludge Heat Treatment liquor	COD-10,000	-	75-85 (BOD) 55-66 (COD)	6.5	1.5 (days)	-	Design values from pilot study
6.	Chemical plant High strength Industrial waste	-	-	65 (COD)	16.33	-	-	Design value for the celanese system
7.	Domestic wastewater	-	-	-	0.95	1.5 (days)	-	Design values, the Anflow system

+ Switzerbaum. M.S. : A Comparison of the Anaerobic Filter and Anaerobic Expanded/Fluidized Bed Process; Water Sci. Technol., 15 (8/9), p345-358 (1985).

Table 4.41 : Full Scale Upflow Anaerobic Filter Installation[+]

Sr. No.	*Location*	*Type of Waste*	*COD (mg/L)*	*Design flow (m^3/d)*	*HRT (day)*	*HRT (h)*	*Loading Rate- ($kg/m^3.d$)*	*%Removal*
1.	Spokane, Wa	Starch gluten	8800	490	0.9	Graded rock	3.8	64
2.	Vernon, Tx	Guar gum	9140	823	1	9cm pall rings	16	60
3.	Sar Juan, PR	Rum distillery	95000	1325	7-8	Synthetic (vinyl core)	8.9	75
4.	Bishop, Tx	Chemical	12,000	3785	1.5	9cm pall rings	9.6	80
5.	Pampa, Tx	Chemical	14,400	3785	1.5	9cm pall rings	10.4	90

\+ Switzerbaum. M.S. : Anaerobic Fixed Film Waste Treatment; Enzyme and Microbial Technol. 5(4) : p 242-250 (1983).

Table 4.42 : Literature of Fixed Beds Anaerobic Treatment

Sr.No.	*Type of Waste/ Substrate*	*Reactor Empty Bed Volume, m^3*	*Literature*
Up-flow			
1.	Municipal sewage	2.10^{-3}	Coulter *et al.* (1957)
2.	Glucose+Nutrient broth	29.10^{-3}	Young and McCarty (1967)
3.	Glucose+Nutrient broth	29.10^{-3}	Young (1968)
4.	Volatile acids		
5.	Domestic sewage	$8.10^{-3}/2$	Pretorius (1971)
6.	Synthetic	—	Plummer *et al.* (1968)
7.	Brewery press liquor	33.10^{-3}	Lovan and Foree (1971)
8.	Acetic acid	7.10^{-3}	Clark and Speece (1971)
9.	Wheat starch waste	600	Richter and Mackie (1971)
10.	Septic tank effluent	400.10^{-3}/ 600.10^{-3}	Raman and Chakladar (1972) and (1972a)
11.	Metrecal	3.10^{-3}	El-Shafie and Bloodgood (1973)
12.	Waste sulphite liquor	840.10^{-3}	Wilson and Timpany (1973)
13.	Pharmaceutical	130.10^{-3}	Dennis and Jennett (1974)
14.	Protein/carbohydrate	160.10^{-3}	Mueller and Mancini (1975)
15.	Leachate	9.10^{-3}	Johansen (1975)
16.	Vegetable tanning	—	Arora *et al.* (1975)
17.	Leachate	55.10^{-3}	Chian and De Walle (1977)
18.	Whey/Whey permeate	—	Hakansson *et al.* (1977)
19.	Domestic Sewage	0.01/0.6/3	Raman and Khan (1978)
20.	Pharmaceutical	8.10^{-3}	Seeler and Jennett (1978)
21.	Shellfish	23.10^{-3}	Hudson *et al.* (1978)
22.	Starch production	5.10^{-3}	Mosey (1978)
23.	Bean blanching	$(0.1–5).10^{-3}$	van den Berg and Lentz (1979a)
24.	Bean blanching	$(0.1–5).10^{-3}$	van den Berg and Lentz (1979)
25.	Chemical waste	1070	Witt *et al.* (1980)
26.	Pharmaceutical	—	Jennett and Rand (1980)
27.	Bean blanching	$(0.1-5).10^{-3}$	van den Berg and Lentz (1980c)
28.	Sludge heat treat liquor	—	Donovan (1980)
29.	Chemical	—	Ragan (1980)
30.	Domestic wastewater	—	Koon *et al.* (1980)
31.	Slaughter-house	26.10^{-3}	Arora and Routh (1980)
32.	Dog food/municipal	17.10^{-3}	Lindgren (1981)
33.	Synthetic Stillage	374.10^{-3}	Dahab and Young (1981)
34.	Synthetic	$7.5.10^{-3}$	Chappell (1981)
35.	Sulphite evaporator condensate	19.10^{-3}	Benjamin *et al.* (1981)

Contd...

Sr.No.	Type of Waste/ Substrate	Reactor Empty Bed Volume, m^3	Literature
36.	Heat treat liquor	18.10^{-3}	Schwartz and De Baere (1981)
37.	Domestic wastewater	$10\text{–}1400.10^{-3}$	Matsche and Ruider (1981)
38.	Heat treat liquor	1	Hall (1981) and (1981a)
39.	Molasses distillery	$0.25.10^{-3}$	Riera *et al.* (1981)
40.	Various	20	Wilkie and Newell (1981)
41.	Domestic Sewage	5.6	Genung *et al.* (1981)
42.	Tannery	18.10^{-3}	Friedman *et al.* (1981)
43.	Synthetic	7.10^{-3}	Frostell (1981a)
44.	Molasses distillery slops	$(16\text{-}500).10^{-3}$	Braun (1981)
45.	Pulp and paper	—	Norman (1981)
46.	Wine distillery	22.10^{-3}	Bories (1981)
	Up-flow		
47.	Phenolic waste	1.10^{-3}	Cross (1981)
48.	Potato waste	16.10^{-3}	Brown *et al.* (1981)
49.	Synthetic Stillage	354.10^{-3}	Young and Dahab (1982)
50.	Heat treat liquor/Brewery	900.10^{-3}	Hall (1982)
51.	Synthetic (dog food)	17.10^{-3}	Lindgren (1982)
52.	Black liquor condensate	12.10^{-3}	Norrman (1982)
53.	Molasses fermentation	120.10^{-3}	Carrondo *et al.* (1982)
54.	Pig slurry supernatant		
55.	Milk wastes	$9.10^{-3}/9$	Colleran *et al.* (1981)
56.	Pig slurry supernatant		
57.	Silage/Milk waste	9	Colleran *et al.* (1982)
	Down-flow		
58.	Paper mill	—	Anaerobic (1973)
59.	Bean blanching	$(0.1\text{–}5).10^{-3}$	van den Berg and Lentz (1979)
60.	Bean blanching	$(0.1\text{–}5).10^{-3}$	van den Berg and Lentz (1979a)
61.	Bean blanching/tomato		
62.	Peeling	$(2\text{–}120).10^{-3}$	van den Berg and Lentz (1982d)
63.	Bean blanching	$(0.1\text{–}5).10^{-3}$	van den Berg and Lentz (1980c)
64.	Bean blanching/synthetic		
65.	Sewage sludge	$(0.8\text{–}35).10^{-3}$	van den Berg *et al.* (1980)
66.	Bean blanching/Dairy		
67.	Piggery/Pear Peeling		
68.	Rum stillage/Synt. sew.		
69.	Sludge	35.10^{-3}	Kennedy and van den Berg (1981a)
70.	Bean blanching/Chem.ind.		
71.	Heat treat liquor	$0.7.10^{-3}$	Kennedy and van den Berg (1981)

Contd...

Sr.No.	Type of Waste/ Substrate	Reactor Empty Bed Volume, m^3	Literature
72.	Heat treat liquor	1	Hall (1981) and (1981a)
73.	Bark leachate	—	Rydin and Haglund (1981)
74.	Bean blanching	$(0.5–1.3).10^{-3}$	van den Berg and Kennedy (1981)
75.	Bean blanching/Chem.ind.		
76.	Dairy/Pear peeling/Rum		
77.	Stillage/Piggery/Synth.		
78.	Sewage sludge/Sludge heat		
79.	Whey	$(0.7–1.2).10^{-3}$	van den Berg and Kennedy (1982)
80.	Heat treat liquor/ treat liquor	$(0.8–35).10^{-3}$	van den Berg *et al.* (1981a)
81.	Brewery	620.10^{-3}	Hall (1982)
82.	Potato	8.10^{-3}	Landine *et al.* (1982)
83.	Rum distillery	11,300	Schafer (1982)
84.	Piggery waste	35.10^{-3}	Kennedy and van den Berg (1982a) and (1982d)
85.	Tomato peeling		
86.	Bean blanching	100.10^{-3}	Stevens and van den Berg (1981)
87.	Bean blanching	$17\text{-}58.10^{-3}$	Kennedy and van den Berg (1982b)
88.	Bean blanching	1.10^{-3}	Kennedy and van den Berg (1982c)
89.	Bean blanching		
90.	Pear peeling	35.10^{-3}	van den Berg and Kennedy (1982a)

Table 4.43 : Performance Data for Anaerobic Fixed Film Reactor of Two Different Sizes, Fed with Bean Blanching Waste+

Sr. No.	Parameter	Fixed Film Reactor Size 0.7 litre[a]	35 litre[b]
1.	Minimum HRT, days	0.5	1.0
2.	Max. COD loading Rate, $kg/m^3.d$	20[c]	10
3.	COD removal Efficiency,%	86	86
4.	Suspended COD of Effluent		
	1%	0.09	0.09
	2% of total effluent COD	65	65

(*a*) Surface of volume ratio–$140m^2/m^3$

(*b*) Surface to volume ratio–$120m^2/m^3$

(*c*) Independent of waste strength (0.5-2.0% COD)

\+ van den Berg and Armstrong, D.W. Anaerobic Waste Treatment Efficiency comparisons between Fixed Film Reactors, Contact Digesters and Fully Mixed Continuously Fed Digester, Proc. 35th Industrial Waste Conference, Purdue University, Ann Arbor Science, Ann Arbor, Michigan (1980).

Table 4.44 : Performance of Fixed Film Reactors Digesting Chemical Industry Waste at Maximum Loading Rates+

Sr.No.	Parameter	Temperature (°C) 25	35
1.	Loading rate (kg COD/m^3.d)	14	17.9
2.	Film surface loading rate (kg COD/m^3.d)	0.1	0.128
3.	HRT (days)	1.0	0.78
4.	COD removal (%)	81	84
5.	Methane gas (m^3-CH_4/m^3.d) at STP	3.7	4.9
6.	Methane gas yield (m^3-CH_4/m^3.d) at STP	0.026	0.036
7.	Volatile acids (mg/L)		
	(*a*) Acetic acid	280 ± 20	180 ± 20
	(*b*) Propionic acid	170 ± 20	180 ± 20
8.	Methane Content (%)	54	55

\+ Kennedy, K.J. and van den Berg, Effect of Temperature on Overloading on the Performance of Anaerobic Fixed Film Reactors, Proc. 36th Purdue Industrial Waste Conference, Purdue University, Ann Arbor, Michigan (1981).

Table 4.45 : Performance Data for Anaerobic Fixed Beds with Various Substrates

Sr. No.	Waste Type	Influent COD (g/L)	Loading (kgCOD/m^3.d)	HRT (day)	%CODr	Methane Yield (m^3/m^3.d)	Support Material	Reference
1.	Bean blanching Waste	10	10-18.5	1.0-0.54	87-88	2.5-5.4	Potter's clay	van den Berg *et al.* (1981)
2.	Molasses fermentation Waste	10-50	2.12	2.5-5.0	57-79	1.4-4.8	Raschig ring	Corrando *et al.* (1983)
3.	Kraft mill condensate	1.4	2	0.75	80	0.33*	Plastic	Norrman (1983)
4.	Thermal sludge conditioning liquor	12.4	20.6	0.59	67	0.37*	Plastic	Hall and Jovanavic (1982)
5.	Chemical industry waste	14	17.5	0.8	81	4.9	Potter's clay	van den Berg, *et al.* (1981)
6.	Pear peeling waste	110-140	18.9	5.8	54	3.4	-	van den Berg *et al.* (1981)
7.	Piggery	27-5	39.2	1.0	27	3.8	-	-do-
8.	Whey	66	5-20	3.3-13.2	87-98	3-6	NPP	van den Berg and Kennedy (1983)
9.	Water Hyacinth	9-11	15-20	0.5	62.65	0.29*	Insulation brick	Annachatre and Khanna (1987)

* m^3/kgCODr (Specific biogas yield).

Table 4.46 : Performance Data for Anaerobic Downflow Fixed Bed Fixed Film Reactor with Various Substrates[+]

Sr.No.	Type of Waste	Concentration (g/L)	Suspended Solid (g/L)	Reactor Size (L)	Temperature °C	Loading Rate (kg/m³.d)	% Removal
1.	Bean blanching	5.5-22 (TVS)	<0.1	110	35	9.4	75
		10 (COD)	1-3	0.7	10	4.2	88
					35	18.4	88
2.	Chemical industry	14 (COD)	0	0.7	25	14.0	81
3.	Cheese plant	1-4	<0.5	1.2	35	5-15	88-83
		1-3	<0.5	4×10^5	30	5	60
4.	Fish Processing	6-20 (COD)	3-10	1.2	35	2.5-13	70-92
5.	Liquor from heat	10.5 (COD)	<1	0.8-1.2	35	29.2	70
	Treated Sewage digester Sludge	11 (COD)	1	950	35	0-25	60
6.	Mansonite processing	9	<0.1	1.2	35	9	72
7.	Piggary waste	27-51 (COD)	16-33	35	35	6.1	70
8.	Pear peeling	110-140 (COD)	43-35	35	35	6.4	58
						18.9	54
9.	Rum stillage	50-70 (COD)	4.5-6.5	35	35	13.3	57
		70-105 (COD)	–	13×10^6	37-40	8-10	65-70
10.	Synthetic sugar	0.5	0	22.5	35	11.5	56
		1.0	0	22.5	35	9.0	60
		2.0 (COD)	0	22.5	35	4.6	79
11.	Synthetic sewage sludge	55 (COD)	47	35	35	7.4	77
12.	Tomato peeling waste	11-22 (TVS)	3.8-7.6	110	35	4.5-12.3	50-61
13.	Whey	66 (COD)	<3	1.2	35	5-20	87-98

+ van den Berg, L. Kennedy K.L. And Samson, R.; Anaerobic Downflow Stationary Fixed Film Reactor; Water Sci. Tech., 17 (1), p89-102 (1985).

Table 4.47 : Composition of Waste Treated in Downflow Stationary Fixed Film Reactors[+]

Sr. No.	Type of Waste	[a]COD (g/L)	Suspended COD (% of total)	Sodium (g/L)	Total Nitrogen (g/L)	Total Phosphate (g/L)	Ratio's	
							COD/N	COD/P
1.	Barley stillage	53	25	-	1.1	-	48	-
2.	Bean blanching	10 (4-40)	10-30	-	0.4	0.1[b]	25	100
3.	Chemical industry	14	0	-	2.5	0.3	5.6	47
4.	Citric acid	3.6	-	-	0.04	0.025	90	140
5.	Cheese plant	1-4	<15	0.15-0.58	0.05-0.58	0.01-0.06	20	67
7.	Fish processing	6-20	50	-	0.8-2.5	0.03-0.11	7.5	180
8.	Heat treated sewage sludge	10.5	<10	-	0.8	0.1	13	100
9.	Pear peeling waste	130 (110-140)	35-50	0.4	2.3[c]	0.47[d]	55	275
10.	Piggery waste	39 (27-51)	60-70	-	2.9	0.8	13	49
11.	Rum stillage waste	60 (50-70)	<10	0.7	1.1[c]	0.21[d]	55	285
12.	Skim milk waste	4	0	0.23[b]	0.2	0.04	20	100
13.	Synthetic sugar waste	10 (5-15)	0	-	0.29	0.09	35	108
14.	Synthetic sewage sludge	55	85	-	2.7	0.32	20	170
15.	Tomato peeling waste	15 (15-30)	<20	2.2	0.28	0.2	54	75
16.	Whey	66	<5	0.33	0.65	0.65	100	100

(*a*) - Average or most common Concentration used, range used in brackets

(*b*) - Sodium hydroxide added to increase alkalinity

(*c*) - NH_4 HCO_3 added

(*d*) - Sodium or potassium phosphate added

\+ van den Berg, L., Kennedy, K.J. and Samson, R.; Anaerobic Downflow Stationary Fixed Film Reactors; Water Sci. Tech., 17 (1), p 89-102 (1985).

Table 4.48 : Effect of 2 Hour Overloading on the Performance of Stationary Fixed Film Reactor[+]

Sr. No.	*Type of Waste*	*Type of Experiment*	*Tempera-ture (°C)*	*Loading Rate (kg COD/m³.d)*	*Film Loading Rate (kg COD/m³.d)*	*HRT (day)*	*%CODr*[(b)]	*%CH_4*	*Methane Yield ($m^3CH_4/m^3.d$)*
1.	Chemical	Control	35	10.9	0.078	1.19	83	53	3.0
	industry	Test	35	89.0	0.635	0.15	40	40	14.3
2.	HTL[(a)]	Control[(c)]	35	18.2	0.130	0.57	66	73	3.9
		Test[(d)]	35	94.2	0.673	0.11	30	69	9.2
3.	Chemical	Control	25	10.9	0.078	1.19	80	51	2.9
	industry	Test	25	61.4	0.438	0.23	44	42	9.1
4.	Bean blanching	Control	25	8.7	0.062	1.09	88	59	2.5
		Test	25	36.4	0.260	0.26	42	42	5.1

(*a*) - Liquor from heat treated sewage digester sludge

(*b*) - COD removal determined from methane rate, 0.33m³ CH_4/kgCODr

(*c*) - Control normal loading

(*d*) - Test-over loading

\+ Effects of Temperature and overloading on the performance of Anaerobic Fixed Film Reactors, Proc. 36th Purdue Industrial Waste Conference, Purdue University, Ann. Arbor, Michigan (1981).

Table 4.49 : Rough Guidelines for Number of Feed Inlet Nozzles Required in a UASB Reactor

Sr. No.	*Type of Sludge*	*Area (m^2) per nozzle*
1.	Dense flocculant sludge (exceeding 40 kg DS/m^3)	One at load less than 1-2 kg COD/m^3.d
2.	This flocculant sludge (less than 40kgDS/m^3)	Five at loads exceeding approximately 3 kgCOD/m^3.d
3.	Thick granular sludge COD/m^3.d	One at loads of approximately, 1-2 kg. COD/m^3.d

Table 4.50 : Tentative Rough Design Capacities for UASB Reactors Treating Mainly Liquid Wastes in Relation to Temperature+

Sr.No.	*Temperature (°C)*	*Design-Capacity (kg COD/m^3.d)*
1.	40	15–25
2.	30	10–15
3.	20	5–10
4.	15	2–5
5.	10	1–3

+ Lettinga, G, Design, Operation and Economy of Anaerobic Treatment, Water Sci. Technol. 15 (8/9), p177-195 (1983).

Table 4.51 : Basic Design Features of the Anaerobic Treatment Plant in Operation+

Sr.No.	*Parameter*	*Liquid Sugar ($30m^3$)*	*Beet Sugar ($800m^3$)*
(A) Composition of Wastewater			
1.	COD, mg/L	800-56,000	1,000-2,600
2.	BOD, mg/L	550-1450	—
3.	Sucrose (%COD)	—	50
4.	Acetate (%COD)	40	20
5.	Propionate (%COD)	45	5
6.	Butyrate (%COD)	4	5
7.	Suspended Solids (g/L)	0.1-1.0	0
8.	Kjeldahl-N (mg/L)	20-55	10-40
9.	pH	6.8-7.8	3.2-40
10.	Alkalinity (meg HCO^-_3/L)	10-40	—
(B) Basic Design Features			
1.	Tank configuration	Cyclindrical	Rectangular
2.	Building material	Steel	Concrete
3.	Height (m)	6	4.5
4.	Bottom surface (m^2)	5	178
5.	Depth of digesting zone (m)	4.9	3.3
6.	Depth of settling zone (m)	1.1	1.2

Contd...

Sr.No.	*Parameter*	*Liquid Sugar (30m³)*	*Beet Sugar (800m³)*
7.	Type of wastewater	Liquid Sugar	Beet Sugar
8.	Organic loading (kg COD/d), (kg COD/m^3.d)	400 13.3	13,000 16.25
9.	Influent concentration (mg COD/L)	17,000	3,000
10.	Purification efficiency (%)	94	86
11.	Hydraulic flow (m^3/h)	1.0	180
12.	Peak flow (m^3/h)	1.5	240
13.	Hydraulic settled surface load (m^3/m^2.d)	0.5	1.5
14.	Gas production (m^3/m^2.h)	1.7	1.2

+ Pette K.C., Full Scale Anaerobic Treatment Beet Sugar Wastewater Proc. 35th International Waste Conference, Purdue Univ. (1981), Ann. Arbor, Michigan.

Table 4.52 : Main Results Obtained in a $6m^3$ UASB Plant with Raw Domestic Sewage as Influent+

Sr.No.	*Parameter*	*Week Number*		
		1-11	*16-22*	*24-26*
1.	Temperature, °C	19-15	11-12	9.5-10
2.	Effluent COD (filtered), mg/L	100-200	150-175	175-250
3.	COD reduction, %(filtered)	65-80	55-70	55
4.	COD reduction, %(raw)	40-55	30-50	30
5.	CH_4 production, $m^3/kgCOD_a$	0.130	0.09	0.050
6.	Excess sludge production, $kgDS/kgCOD_a$	0.195	0.172	0.271

+ Lettinga, G., High Rate Anaerobic Wastewater Treatment using the UASB Reactor Under Wide Range of Temperature Conditions. Biotechnol. Genetic Engg. Rev., 2, p. 253-284 (1984).

Table 4.53 : Results Obtained with Granular Sludge UASB-Reactor and Wastewaters Containing a High Fraction of Suspended Matter+

Sr. No.	*Origin*	*Wastewater*		*Reactor*		*Amount of Seed Sludge (g)*	*Load (kg/m³.d)*	*Temp- (°C)*	*% COD*
		COD (g/L)	*Fraction of Dissol-ved (%)*	*Volume (L)*	*Height (m)*				
1.	Animal	3-7	75	60	2	1500	30	30	92
	Carrion destruction	3.5	75	60	2	1500	60	30	78
2.	Slaughter	1.5-2.2	50	30	1.3	1000	10	30	87
	house	1.5-2.2				1000	6	20	91
3.	Raw sewage	0.2-0.9	5-35	120	2	3300	0.7-2.7	8-20	50-85

+ Lettinga, G. : Design, Operation and Economy of Anaerobic Treatment Water Sci. Technol., 15 (8/9), p177-195 (1983).

Table 4.54 : Treatment of Slaughter House Wastewater Using UASB Reactor+

A : Characteristics of Wastewater

Sr.No.	*Parameters*	*Design-Capacity*
1.	Total COD, mg/L	1500-2000
2.	Total BOD, mg/L	490-650
3.	Kjeldahl Nitrogen (N), mg/L	120-180
4.	Total phosphate (P), mg/L	12-20
5.	pH	6.8-7.1
6.	Suspended solids as a percentage of total COD	40-50
7.	Acidification (percentage COD-VFA of total COD	12-15
8.	Grease, mg/L	50-100
9.	Temperature, °C	20

B : Organic Loading Rates, Liquid Detention Time and Process Temperature Over the Entire Experimental Period

Phase	*Number of Days*	*Loading Rates kg COD/m³.d*	*HRT (h)*	*Temperature (°C)*
I	0-142	0.5-1.5	40-20	30
II	159-296	1.5-2.5	18-12	20
III	297-324	2.5-3	10	30

\+ Sayed, S and Lettinga, G. : Anaerobic Treatment of Slaughter House Waste using Flocculant Sludge UASB Reactor, Agricultural Wastes, 11, p. 197-226 (1984).

4.9 Some Important Considerations

Site specific conditions control the process configuration of choice. Type of wastewater, temperature, concentration of organic pollutant, toxic or refractory components, required degree of treatment, shock loading characteristics of hydraulic, organic or toxic nature, space available, segregation of wastewater flows and many other factors dictate the most appropriate process configuration. Eventhough high rate anaerobic configuration results in the lowest required volume, it may be more economical to choose a low rate process that can be economically constructed in a membrane covered earthen basin. Low rate processes have inherent greater shock absorbing potential, requiring less supervision, control and often regate the need of flow or organic load equalization. In addition, there is a basic philosophical decision to be made on flow equalization. It may be more economical to incorporate the desired flow equalization volume in the biological reactor than to provide a comparable volume in a separate reactor. In some cases of organic acidic wastewater, a separate equalization tank of (stainless steel tank construction), whereas ordinary carbon steel or even earth may be used for the biological reactor which incorporates requisite volume.

Table 4.55 : Results of Some Laboratory UASB Experiments with various Type of wastes[(a)+]

Sr.No.	Origin	Water Soluble COD			Maximum COD Load					UASB Reactor		
		Total (mg/L)	*Disolved[(b)] (5)*	*VFA (mg/L)*	*Space Load kg/ m^2.d*	*Sludge Load kg/kg.d*	*HRT (h)*	*% COD*	*Sludge Yield kg/ kgCOD)*	*Temp. (°C)*	*Volume (L)*	*Length (m)*
1.	Sugar-beet Sap, Unsoured	5000-6000	95	Nil	4-5 (D)	0.5-0.8	48-24	95	0.15	30	61	1.05
2.	Sugar-beet Sap Soured (closed) circuit)[(c)]	6000-9500	80-60	4000-5000	8-10 (D) 10-14 (T)	0.7-1.1	12-24	84-95[(D)] 65-75 (T)	0.09-0.07[(D)] 0.1 (T)	30	18 30	0.7 1.00
3.	Sugar-beet Sap Soured (two stage)	6000-9000	95	5400-8000	8-9	0.8-1.0 (2nd stage)	24 (2nd stage)	90-97	0.04-0.03	30	18	0.7
4.	Bean-blanching	5200	90	500-1500	8-10	0.6-0.8	13-15	90-95	-	30	2.7	0.30
5.	Sauerkraut	10,000-20,000	97	400-1500	8-9	0.8-1.2	24	88-93	0.05-0.07	30	2.7	0.70
6.	Dairy (Skimmed milk)	1500	75	Nil	7-8	0.4-0.6	5	90	0.18[(d)]	30	18	0.70

(*a*) - Values mentioned for the COD load concern the maximum values that could be applied in the specific experiment

(*b*) - COD remaining after filtration over a filter SS-520b

(*c*) - Experiments conducted in closed simulated wastewater circuit

(*d*) - Higher values are obtained in case of substrate precipitation (pH fall or incase of overloading)

(D) - Dissolved

(T) - Total

\+ Lettinga, G, Use of Upflow Sludge Blanket-Reactor Concept for Biological Wastewater Treatment Biotechnol, Bioeng., (22 (4), p. 699-734 (1980).

Table 4.56 : Results Obtained in a 6-m^3 Pilot Plant and 200-m^3 Full Scale Plant Experiments[+]

Sr.No.	Waste Type	UASB Reactor (m^3)	Influent Characteristics		Maximum Loading Rate		Treatment Efficiency			
			CID[(a)] (mg/L)	Soured %	Organic (kg COD m^3.d)	Hydraulic (m^3/m^2.d)	Temp. (°C)	$E_{COD_{cent}}^{(b)}$	$E_{COD_{Diss}}^{(c)}$	$E_{COD_{cent}}$
1.	Liquid Sugar	6	4000-6000	15-25	20-25	4	28-32	92-95	93-95	-
2.	Campaign Waste	6	3500-4000	75	30-32	4-6	28-32	95-80	95-98	–
3.	Campaign Waste	200	4000-5200	70-90	14-16	3-4	30-34	97-95	-	90-95
4.	Potato Processing	6	2000-5000	25	3-5	1.2	19	95 (92)[(d)]	-	94
	(Lime used	6	2000-5000	30	10-15	3	26	95 (89)	-	98
	as neutrali-	6	2000-5000	12	15-18	4	30	95 (89)	-	97
	sing agent)	6	4000-16500	8	25-45	6-7	35	93 (89)	-	96

(*a*) - COD value based on centrifuged samples

(*d*) - E - based on centrifuged effluent samples and raw influent-COD values

(*c*) - $E_{COD_{cent}}$ based on centifuged influent and effluent COD values determined after flocculation of sample with 200 mg Fe^{+3}/L

(*d*) - Values in parentheses refer to effluent samples that have been allowed to settle for 30 minutes

\+ Lettinga, G. : Use of Upflow Sludge Blanket UASB-Reactor Concept for Biological Wastewater Treatment, Biotechnol Bio-eng., (22 (4), p699-734 (1980).

Table 4.57 : Results Obtained with UASB Reactors Using Granular or Mainly Granular Seed Sludge and UFA Solution, Alcoholic Wastewater and Potato Processing Waste as Feed+

	Substrate Characteristics				Experimental Conditions					COD reduction	
Sr.No.	*Type*	*COD (mg/L)*	*Socured (%)*	*Medium used for Growth of Seed Sludge inoculum*	*Reactor Volume*	*Volume of Sludge bed*	*COD Sludge Load kg/kg VSS.d*	*Temp. (°C)*	*COD Load (kg/m³.d)*	*Total (%)*	*Filtered %*
1.	VFA	1000 C_2	100	VFA	30L	15-20L	2-3	30	62	-	80-90
		1000 C_3									
2.	Alcoholic*	-	0	Sugar beet	2.7L	1.5L	0.6-0.7	30	22	-	>95
				Waste	28L	10L	0.7	30	14	-	85-90
3.	Potato	2.5-4.2	-	Digested	$6m^3$	$<2m^3$	0.27	19	3-5	88	95
	Processing	3.3-5	6-16	Sewage	$6m^3$	$<2m^3$	0.65	26	10-15	86	95
		3.5-4.5	-	Sludge	$6m^3$	$<2m^3$	0.97	30	15-18	83	95
		3.5-7.1	-	-	$6m^3$	$4m^3$	1.45	35	25-45	84	93

* - Methanol (51%) : Ethanol (27%); Propanol (12%); Butanol (10%)

+ Lettinga, G. High Rate Anaerobic Wastewater Treatment using the UASB Reactor under or Wide Range of Temperature Conditions; Biotechnol. Genetic Engg., Rev. 2 (Oct), p253-284 (1984).

Table 4.58 : Results Obtained with Three Types of Complex Wastes using Flocculant-Sludge UASB Reactor+

Sr.No.	Waste	Influent Total	COD* Soluble	Temperature (°C)	COD Load (Kg/m³.d)	%COD** Filtered	Unfiltered	Reactor Volume
1.	Domestic	322-950	(235-460)	15-20	2.0 (max)	30-80	-	6m³
	Sewage			9-12	2.0 (max)	-	-	
2.	Calf-Fattening	9500	(3800)	30	4	93	90	25L
				25	2	90	85	25L
3.	Slaughter house	1500-2200	(600-1100)	30	2.5-3.0	75-85	65-80	30m³
				20	1.5-2.5	78-75	55-75	

* - COD values for domestic sewage comprise average values of 5-15 composite daily samples

** -Filtered : COD reduction based on filtered effluent and raw influent

Unfiltered : COD reduction based on raw effluent and influent samples

+ Lettinga, G. : High Rate Anaerobic Wastewater Treatment using UASB Reactor Under a Wide Range of Temperature Condition; Biotechnol. Genetic Engg. Rev., 2 (Oct.). p253-284 (1984).

Table 4.59 : Results Obtained with Three Types of Complex Wastes using Granular Sludge UASB Reactors+**

Sr.No.	Type	Wastewater		Reactor		Amount of Seed Sludge (g)	COD Load (kg/m³.d)	Temp. (°C)	% CODr*	
		COD (g/L)	Fraction Dissolved (%)	Volume (L)	Height (m)				Filtered	Unfiltered
1.	Rendering waste	5500	70	60	2	1500	27	30	94	64
		3000	85	60	2	1500	63	30	80	63
2.	Slaughter house	1.5-2.2	50	30	2.3	1000	10	30	87	-
		1.5-2.2				1000	6	20	91	
3.	Raw sewage	0.2-0.9	5-35	120	2	3300	0.7-2.7	8-20	60-89	54-72

* - Based on filtered effluent samples and unfiltered influent samples

** - Average value measured over a periods of 5-12 days

+ Lettinga, G., : High Rate Anaerobic Wastewater Treatment using UASB Reactor Under a Wide Range of Temperature Condition, Biotechnol. Genetic Rev., 2 (Oct.), p. 253-284 (1984).

Table 4.60 : Suspended Solids Washout as Found in $6m^3$ Pilot Plant and Full Scale Experiments with UASB Process+

Period of Experiment (day)	COD-Space Load ($kg/m^2.d$)	HRT (h)	SS Washout		Total Amount of Sludge in Reactor	
			Total (mgSS-COD/ gCODinf)[(a)]	After 30min settling	(kg TS)	(kg TVS)
(A) Potato Waste ($6m^3$-Pilot Plant)[(b)]						
408-413	10-13	6.3	70-120	40-80	110-120	89-97
423-427	15-17	6.0	70-90	—	125-130	110-114
429-434	22-30	3.9	60-90	30-50	145-155	127-136
436-440	33-43	3.5	40-110	30-50	200-210	174-183
(B) Sugar-beet waste ($200m^3$-Full Scale plant)[(c)]						
54	11.2	6.7	100	—	6500	2280
61	13.9	7.6	60	—	6650	2530
68[(d)]	12.9	7.5	200	—	6650	2530
75	14.0	8	74	—	6500	—

(*a*) - SS content in potato waste varied between 100-150 mg $COD/gCOD_{inf}$ and in the sugar beet waste between 70-150 mg $COD/g\text{-}COD_{inf}$ during experimental periods considered

(*b*) - Sludge growth; 0.18-0.20 g sludge-COD/gCO_{inf} (30°C); 0.26-0.30 g sludge $COD/gCOD_{inf}$ (20°C)

(*c*) - Sludge growth; 0.08-0.16g sludge-$COD/gCOD_{inf}$

(*d*) - Only 20% of COD was present as VFA-COD whereas this value was 80-85 per cent during other days

\+ Lettinga, G. : Use of the Upflow Sludge Blanket (UASB)-Reactor Concept for Biological Wastewater Biotechnol, Bioeng., 22(4), p699-734 (1980).

Table 4.61 : Full Scale UASB Installations

Sr. No.	Location	Waste Characteristic			Design		
		Type	COD (mg/L)	Flow (m^3/d)	HRT (h)	COD Load ($kg/m^3.d$)	%CODr
1.	**USA (Partial List)**						
	(*a*) Locrosse, WI	Brewery	2,500	23,000	4.9	14.1	86
	(*b*) Caribou, ME	Starch	22,000	910	47	11	85
	(*c*) Plover, WI	Potato	43,000	3,000	17.5	6.0	80
2.	**Europe (Partial List)**						
	(*a*) Netherland	Sugar	17,000	3,000	24	1.3	94
	(*b*) Netherland	Sugar-beet	3,000	6,960	4.9	14.5	85
	(*c*) Netherland	Starch	7,700	1,750	2.4	8	85
	(*d*) Switzerland	Potato	2,200	2,210	6.5	8.3	80
	(*e*) Germany	Sugar-beet	7,500	2,400	15	12	86
	(*f*) Netherland	Alcohol	5,330	2,090	8	16	90

Table 4.62 : Literature on Sludge Blanket Reactors, Anaerobic Treatment

Type of Waste/Substrate	*Reactor Empty Bed Volume, m^3*	*Literature*
Wine distillery	650	Stander (1967)
Calf manure	$(22\text{-}25).10^{-3}$	van Velsen *et al.* (1979)
Alcoholic waste/Sugar beet/Bean blanching	$(3\text{-}6000).10^{-3}$	Lettinga *et al.* (1979)
Sugar-beet	0.06/30/200	van der Meer (1979)
Sauerkraut/Potato/Synthetic	$<120.10^{-3}$	Lettinga and Vinken (1980)
Municipal Sewage		
Sugar-beet	587	Pette *et al.* (1980)
Bean blanching	—	van den Berg and Lentz (1980d)
Liquified solid wastes	—	Rijkens (1981)
Brewery waste	4400	Richards (1981)
Brewery	4400	Ruppel *et al.* (1982)
Sugar-beet/Bean blanching	$(3\text{-}61).10^{-3}/200$	Lettinga *et al.* (1980a)
Sauerkraut/Dairy		
Vegetable and Fruit	6	Goedvriend (1981)
Bean blanching/Chemistry Industry	$(3\text{-}4).10^{-3}$	van den Berg *et al.* (1981a)
Municipal wastewater	—	Roersma *et al.* (1981)
Chemical industry	6	Jans (1981)
Synthetic	7.10^{-3}	Frostell (1981a)
Slaughterhouse	$10.10^{-3}/30$	Sayed *et al.* (1981)
Molasses distillery		
Slops	—	Braun (1981)
Brewery	100.10^{-3}	Eggers (1981)
Molasses distillery	50.10^{-3}	Riera *et al.* (1981)
Insulating board	6	de Bekker (1981)
Leachate	-6	de Bekker and Kaspers (1981)
Heat treat liquor	1	Hall (1981) and (1981a)
Heat treat liquor/Brewery	1.1	Hall (1982)
Sugar-beet	30/200	Heertjes *et al.* (1982)

Table 4.63 : Performance Data for Fluidized Bed Reactor for Chemical Industry Wastewater$^+$

(A) Operating Conditions for Single Phase System	**Value**
• Mean Sand size, (mm)	0.5
• Hydraulic loading, $m^3/m^2.h$	25-33
• Controlled bed expansion	90-110
• pH range	6.7-7.2
• Temperature, °C	30-35

(B) Operating Conditions for Two Phase System

	Phase-I	Phase-II
• Mean Sand size, mm	0.5	0.5
• Hydraulic loading rate, $m^3/m^2.h$	25-33	25-33
• Controlled bed expansion, %	40-60	90-110
• pH range	5.7-6.2	6.7-7.2
• Temperature, °C	30-35	30-35

(C) Design Parameter for Single Phase and Two Phase Fluidized Bed Reactors

			Loading Rate	
Sr. No.	*Operation*	*Biomass Concentration (gVSS/L)*	*(kgCOD/kgVSS.d)*	*(kgCOD/m^3.d)*
1.	Single Phase Organics to methane	15-25	0.4-1.0	10-20
2.	Two Phase organics of acetic acid	15-25	2.0-4.0	30-40
3.	Acetic acid to methane	8-15	2.0-4.0	25-35

(D) Single and Two Phase Fluidized Bed Design Values

			Two Phase Design Value	
Sr. No.	*Design Characteristic*	*Single Phase Reactor design Value*	*First Stage*	*Second Stage*
1.	Volumetric loading Rate*, kgCOD/m^3.d	15	40	30
2.	Controlled fluidized bed height, m	10.5	10.5	10.5
3.	Hydraulic loading rate, m/h	24	24	17
4.	Reactor area, m^2	127	48	51
5.	Fluidized bed volume, m^3	1333	508	536

* - A-20 per cent COD reduction assumed in the first stage of two phase system

\+ Sutton, P.M. and Li, A. : Single and Two phase Anaerobic Stabilization in fluidized Bed Reactor, Water Sc. Technol. (15 (8/9), p333-344 (1982).

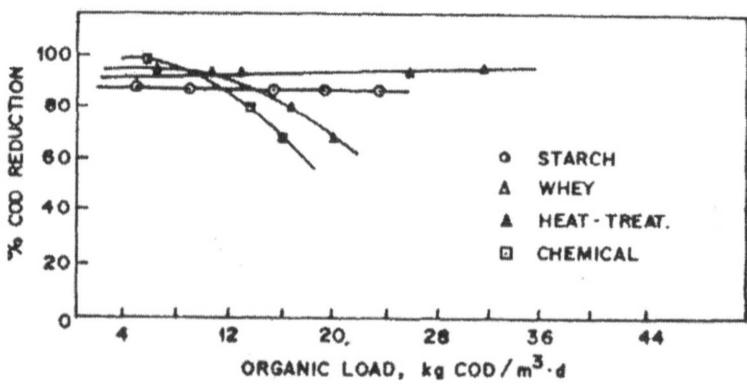

Fig. 4.19 : Anaerobic Fluidized Bed Reactor Treatment of High Strength Waste

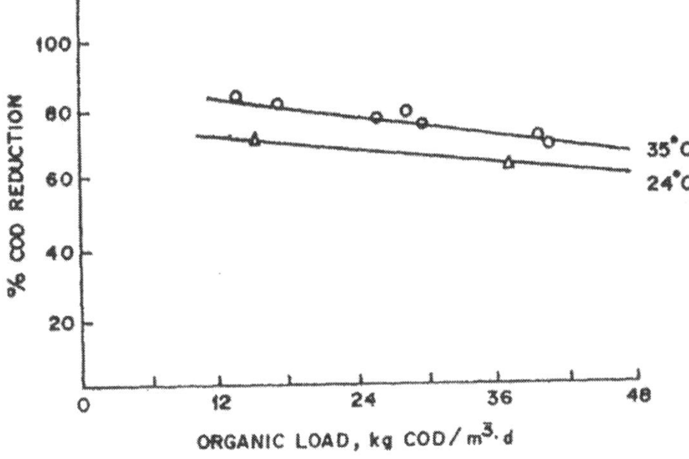

Fig. 4.20 : Acid Whey–% COD Removal Vs Organic Loading

Table 4.64 : Comparison of Laboratory and Pilot Fluidized Bed Scale Unit for Treatment of Municipal Wastewater

		Influent (mg/L)		*Effluent (mg/L)*	
Sr.No.	*Scale of Operation*	*Mean*	*Range*	*Mean*	*Range*
1.	Pilot Scale*				
	- Total COD	171.4	(70-1106)	101.7	(43-225)
	- Soluble COD	130.4	(55-225)	76.7	(11-178)
	- Suspended solids	35.5	(7-116)	30.5	(7-144)
	- BOD_5	74.2	(28-199)	47.2	(13-74)
2.	Lab. Scale**				
	- total COD	186	(80-306)	49.2	(22-126)
	- Suspended solids	86	(40-186)	16.5	(3-90)

* - Switzenbaum, M.S. and Hickey, R.F. : Anaerobic Treatment of Primary Effluent Environ. Technol. Letters, 5, p.189-200 (1984).

** - Jewell, W.J., Switzenbaum, M.S. and Morris, J.W., Municipal Wastewater Treatment with Anaerobic Attached Microbial Film Expanded Process, J. Wat. Poll. Cont., 53 (4), p.482-490 (1981).

Table 4.65 : Anaerobic Fluidized Bed Reactor Treatment-Food Processing Waste+

Sr. No.	*Loading Rate (kgCOD/m³.d)*	*HRT (h)*	*COD mg/L) Inf*	*Eff*	*%CODr*	*BOD(mg/L) Inf*	*Eff*	*%BODr*
1.	3.5	49.4	7210	1040	86	4370	320	93
	8.3	21.4	7390	1430	81	4700	440	91
	16.8	13.5	9450	1910	80	5900	630	89
	24.1	7.5	7530	1900	75	4775	690	87

\+ Jeries, J.S. : Industrial Wastewater Treatment Using Anaerobic Fluidized Bed Reactor, Water Sc. Technol. 15 (8/9), p. 169-176 (1982).

Table 4.66 : Performance Data for Anoxic Fluidized and Expanded Bed System+

Sr. No.	Medium	Particle Size (mm)	Upflow Velocity (m/h)	Influent Type	Oxidized-N Conc. in Influent (mg/L)	Super-ficial Retention (°C) Period (min)	Temp.	Biomass Conc. (g/L)	Vomuteric rate of Denitri-fication (b) (kgN/m³.d	Mass rate of Denitri-fication (mg NO_3-N/ gBTS.h)	Carbon Source	Comments
1.	Activated Carbon	0.65	29.5	$NaNO_3$	NR	6.4	24	NR	6.8	NR	CH_3OH	Fluidized bed
2.	-do-	-do-	28.8	-do-	25-40	7.0	22	-do-	4.7	NR	CH_3OH	-do-
3.	-do-	-do-	28.8	-do-	NR	7.1	20.5	-do-	4.9	NR	CH_3OH	-do-
4.	-do-	-do-	30.3	-do-	NR	5.4	16.2	-do-	4.1	NR	CH_3OH	-do-
5.	-do-	-do-	50.7	-do-	34-50	3.5	22.0	-do-	6.2	NR	CH_3OH	-do-
6.	-do-	-do-	19.5	-do-	NR	10.5	26.0	-do-	4.7	NR	CH_3OH	-do-
7.	Sand	0.6	37.2	Secondary Effluent	21.5	6.5	20.0	30-40	5.5	NR	CH_3OH	-do-
8.	-do-	-do-	58.8	-do-	18.7	4.0	19.0	-do-	7.2	NR	CH_3OH	-do-
9.	-do-	-do-	36.8	do+$NaNO_3$	55.0	6.5	25.0	-do-	13.5	NR	CH_3OH	-do
10.	-do-	1.0	4.6	Sec.Effluent	36.0	20	13-14	NR	2.7	NR	-do-	Expanded bed (lab.)
11.	-do-	1-2	11	-do-	NR	9.5	16-17	NR	4.0	NR	-do-	Expanded bed full scale
12.	-do-	-do-	15	-do-	19-32	7.0	12	NR	4.0	NR	-do-	-do-
13.	-do-	0.25	25	River water	10-15	8.3	6.16	15-20	2.2-3.0	NR	-do	-do-
14.	-do-	-do-	-do-	-do-	11-17	8.3	2-7	16-21	1.5-2.9	NR	-do-	-do-
15.	-do-	-do-	-do-	-do-	15-21.5	8.3	7-14.5	NR	3.4-4.2	NR	-do-	-do-
16.	-do-	0.2-0.5	20	-do-	8	NR	10	10-14	3.8	13	-do-	-do-
17.	-do-	0.2-0.5	20	-do-	8	NR	20	10-14	7.4	26	-do-	-do-

+ Cooper, P.F. and Wheeldon, D.H.V. : Fluidized and Expanded Bed Reactor For Wastewater Treatment, J. Wat. Poll. Control, 79 (2), p286-301 (1980).
(*a*)-Fluidized State; (*b*)-Based on volume of Fluidized bed; and NR-Not reported.

Table 4.67 : Expanded/Fluidized Bed Studies for Industrial Application+

Sr. No.	Waste Type	Influent (g/L)	Effluent (g/L)	%Removal	Loading Rate (kg/m³.d)	HRT	Temperature (°C)	Remarks
1.	Acid whey	50.3-56.1 (COD)	8.3-14.6 (COD)	72.0-83.6	13.4-37.6	1.4-4.9 (*a*)day	35	One Stage Fluidized bed
2.	Acid whey	52.2-55.4 (COD)	15.1-19.2 (COD)	65.2-71.4	15.0-36.6	1.5-3.5 (*a*)day	24	One Stage Fluidized bed
3.	Acid whey	52.2 (COD)	3.2 (COD)	94	10.5	5.0 days(*a*)	—	Two stage in series
4.	Food processing	7.2-9.4 (COD)	1.0-1.9 (COD)	75-86	3.5-24.1	7.5-49.4a(h)	35	Fluidized bed
5.	Chemical waste	12 (COD) 8.4 (BOD) (*b*)	—	79-93 (COD) 81-98 (BOD)	4.1-27.3	—	35	Fluidized bed
6.	Soft drink bottling	6 (COD) (*b*) 3.9 (BOD) (*b*)	—	66-89 (COD) 61-95 (BOD)	4-18.5	—	35	Fluidized bed
7.	Heat treatment liquor	10 COD (*b*) 5 (BOD)	—	52-75 (COD) 66-95.5 (BOD)	4.3-21.4	—	35	Fluidized bed
8.	Whey permeate	6.8 (COD)	2.2 (COD)	68 (COD)	8.6-10.4	—	30-35	Fluidized bed
9.	Whey permeate	27.3 (COD)	5 (COD)	82 (COD)	5.3-7.4	—	30-35	Fluidized bed
10.	Sweet whey	10 (COD)	1374-6071 mg/L (S-COD)	36.9-93.1	8.6-60.0	4.1-27.1 (*h*) (*c*)	25	Expanded bed
		5-20 (S-COD)	417-7958 mg/L (S-COD)	58.9-92.3	8.2-29.1	15-35(h)(c)	35	Expanded bed
11.	Black liquor condensate	1.4 (COD)	0.16 (COD)	89	10	3.5(h)	22	Expanded bed
12.	Black liquor	1.4 (COD)	0.28 (COD)	80	13	0.8h	22	Fluidized bed

(*a*) - Empty bed retention time; (*b*) Average valves; (*c*) Based on expanded bed volume

\+ Kaul, S.N. : Fluidized Bed Reactor for Wastewater Treatment, J. Chem. Engg. World, 23, p 230-238 (1992).

Table 4.68 : Performance Data for Anoxic Fluidized and Expanded Bed System[+]

Sr. No.	*Medium*	*Partical Size (mm)*	*Upflow Velocity (m/h)*	*Influent type*	*Oxidized-N in influent (mg-N/L)*	*Superficial Retention Period (min)*	*Temp. (°C)*	*Biomass(a) Conc. (g/L)*	*Loading rate (kgN/m³.d)*	*Carbon Source*	*Comments*
1.	Anthracite	2-3	0.1-8.0	Industrial effluent	226-2260	120	22-34	NR	3.9-27.5	CH_3OH	Tapered Fluidized bed
2.	Sand	1.0	78.6	-do-	1450	NR	20	NR	10.6	Molasses	Fluidized 10:1 recycle
3.	Sand	1.0	78.6	-do-	1450	NR	38	NR	38.4	Molasses	-do-
4.	Activated sludge	1-2	0.39	-do-	7550	1143	NR	25	6	Petro-chemical	Fluidized bed (stirred) oxygenous & endogenous
5.	Activated Sludge	NR	0.12	Secondary Effluent	11	72	NR	29.5	NR	Sewage	Fluidized bed, Endogenous
6.	Sand	0.45	42.2	Secondary Effluent	3-10	2.8	13	5.5-11.0	4.6	CH_3OH	Fluidized bed

(a) - In fluidized state,

(b) - based on volume of fluidized bed

NR - Not Reported

\+ Cooper, P.F. and Wheeldon, D.H.V. : Fluidized and Expanded Bed Reactor for Wastewater Treatment, J. Wat. Poll. Control, 79(2), p. 286-301(1980).

Table 4.69 : Full Scale Anaerobic Fluidized Bed Installation+

Sr. No.	*Location*	*Type*	*COD (mg/L)*	*Flow (m^3/d)*	*HRT (h)*	*Loading Rate ($kg/m^3.d$)*	*Media*	*%COD*
1.	Birmingham AL	Soft-drink bottling waste	6900	380	6	9.5	0.6mm sand	77
2.	Mid-west	Soyprocess waste	9000	—	Less than 24	13	0.4mm sand	—

+ Switzenbaum, M.S. Anaerobic Fixed Film Waste Treatment, Review Enzyme and Microbial Technol., 5(4) p242-250 (1983).

Table 4.70 : Summary of Methane Produced by Using Fluidized Bed Reactor+

Sr. No.	*Waste*	*COD (g/L)*	*m^3CH_4/kg CODr*	*%CH_4*	*Loading Rate ($kgCOD/m^3.d$)*
1.	Food process	7-10	0.4	70	3.5-24.1
2.	Chemical	12	0.41	82	3.5-5.7
3.	Soft drink	4-18	0.41	60	3.5-5.7
4.	Zimpro supernatant	7.8	—	72	3.4-16.7

+ Jeris, J.S. : Industrial Waste Treatment using Anaerobic Fluidized Bed Reactor. Wat. Sci. Technol., 15 (8/9), p169-176 (1982).

Table 4.71 : Literature on Expanded Beds (Anaerobic Treatment)

Type of Waste/Substrate	*Reactor Empty Bed Volume (m^3)*	*Literature*
Cowmanure	$3.8.10^{-3}$	Jewell *et al.* (1978)
Synthetic	1.10^{-3}	Switzenbaum and Jewell (1980)
Molasses fermentation	$(7\text{-}13).10^{-3}$	Frostell (1980)
Domestic wastewater	1.10^{-3}	Jewell *et al.* (1981)
Domestic wastewater/synthetic	1.10^{-3}	Jewell (1980)
Synthetic	1.10^{-3}	Lindgren (1981)
Algae	—	Jewell and Schraa (1981)
Whey	—	Switzenbaum and Danskin (1982)
Black liquor condensate	1.10^{-3}	Norrman (1982)
Synthetic	1.10^{-3}	Morris and Jewell (1982)

4.9.1 Cell Immobilization

The success of biological treatment is directly related to the efficiency of cell immobilization and/or retention. The cells must be retained in the system and higher cell concentrations result in lower reactor volume requirements. The original anaerobic processes utilized completely mixed reactors with no cell separation and recycle. Consequently, the hydraulic retention times were equal to the solids retention times and the reactor volumes were necessarily very large. The development of the anaerobic contact process for meat packing wastewaters was a milestone because it incorporated cell separation and recycle, and uncoupled solids retention time from hydraulic retention time and reactor volume.

Table 4.72 : Literature on Fluidized Beds (Anaerobic Treatment)

Type of Waste/Substrate	*Reactor Empty Bed Volume* (m^3)	*Literature*
Heat treat liquor	—	Pilot plant test (1979)
Bottling	135	Barbara and Gasser (1979)
Wheat starch	~380.10^{-3}	Pilot plant test (1980)
Heat Treat Liquor	~50.10^{-3}	Bell *et al.* (1981)
Kraft bleach liquor	1	Hakulinen *et al.* (1981)
Kraft bleach liquor	70.10^{-3}	Hakulinen and Salkinoja Salonen (1981a)
Kraft bleach liquor	~700	Hakulinen and Salkinoja Salonen (1982a)
Whey permeate	63.10^{-3}	Sutton and Li (1981)
Whey permeate	63.10^{-3}	Li and Sutton (1981)
Whey permeate	63.10^{-3}	Li *et al.* (1982)
Citric acid fermentation	0.5.10^{-3}	Binot *et al.* (1981)
Dairy waste	60.10^{-3}	Jenkins *et al.* (1981)
Acid whey/Food Proc/Chem.	—	Hockey and Owens (1981)
Waste/Soft drink bottling/ Heat treatment liquor Whey/Food Industry/Chem.	~50.10^{-3}	Jeris (1982)
Soft drink/Bakery/Heat treat liquor/Heat treatment liquor	60.10^{-3}	Hall (1981) and (1981a)
Heat treatment liquor/Brewery	29.10^{-3}	Hall (1982)
Soybean	1300	Sutton *et al.* (1982)

Table 4.73 : Performance Data for Anaerobic Moving/Recycle Bed Reactors

Sr. No.	*Reactor*	*Waste*	*Influent COD*	*Loading Rate*	*HRT (h)*	*%CODr*	*Methane Yield* ($m^3/m^2.d$)	*Reference*
1.	Recycled bed	(a) Molasses Waste	9.3	2.3-3.9	4	87-90	0.7-1.2	Martensson and Frostell (1983)
		(b) Beet Sugar Waste	4-7	25	4.6	86-89	7-8	-do-
2.	An RBC	Synthetic Waste	1.105*	1.5	17.5	96	—	Tait and Friedman (1980)
			2.39*	6.4	8.7	79	—	-do-

*–TOC/L

Table 4.74 : Literature on Moving Beds (Anaerobic Treatment)

Type of Waste/Substrate	*Reactor Empty Bed Volume, (m^3)*	*Literature*
Synthetic	1.10^{-3}	Tait and Friedman (1980)
Synthetic	1.10^{-3}	Friedman and Tait (1980)
Synthetic	1.10^{-3}	Friedman *et al.* (1981)

Table 4.75 : Literature on Recycled Beds (Anaerobic Treatment)

Type of Waste/Substrate	*Reactor Empty Bed Volume, (m^3)*	*Literature*
Molasses/Sugarbeet	40.10^{-3}	Martensson and Frostell (1982)

Table 4.76 : Literature on Recycled Flocs (Anaerobic Treatment)

Type of Waste-Substrate	*Reactor Empty Bed Volume (m^3)*	*Literature*
Packinghouse/Wood fibre/Municip/Sewage	$(1\text{-}2).10^{-3}$	Schropfer and Ziemke (1959)
Packinghouse Bean Blanching/Pear	~2600	Steffen and Bedker (1961)
Peeling/Potato/ Rum/Stillage	30.10^{-3}	van den Berg and Lentz (1977a)
Bean Blanching	30.10^{-3}	van den Berg *et al.* (1980)
Bean Blanching/ Potato Peeling	—	van Den Berg and Lentz (1980d)
Synthetic	—	Anderson and Duarte (1980)
Yeast/Molasses/ Potato/Dairy/ Cider/Wine dist./ Starch gluten	~1-100	Anderson *et al.* (1980a)
Wine distillery	$(22\text{-}26).10^{-3}$	Borries (1980)
Starch	1000	Ross (1980)
Wine distillery	1200	Ross (1980)
Wine distillery	2200	Ross (1980)
Distillery	24	Frostell (1981)
Sugar/potato/ Starch/Bear	—	Pette and Versprille (1981)
Pulp & Paper	—	Norrman (1981)
Molasses Distillery Slops	—	Braun (1981)
Pectin	3600	Seyfried (1981)
Heat Treat Liquor	18000	Schlegal and Kalbskopf (1981) & (1982)
Potato Chips	106	Viitasalo *et al.* (1981)

Young and McCarty's development of the attached growth upflow filter was another milestone, while Jeris' development of the fluidized bed biological system was still another milestone. Lettinga's development of the UASB significantly advanced the alternative available for process configuration. The UASB was also an economically alterative to the relatively expensive packing requirement of upflow filters. As an aside, it appears to have been rather fortuitous that sugar-beet wastewater was the candidate with which the UASB process was developed, because it provided the required hydrogen generation and calcium which appear to be requisites for granule formation.[226]

Due to the inherently lower efficiencies and loading rates of the anaerobic filter and the occasional instability of the UASB a hybrid of these two systems is being proposed, an unpacked lower section and a packed upper section. Such a hybrid avoids the potential plugging and higher costs of a fully packed reactor and yet maintains the high loading rates of a UASB while incorporating stability to the blanket by insuring against high solids loss from a filamentous blanket and complete loss of a granular blanket which for some reason floats instead of settles.

There is a fundamental deficiency in providing such as an inefficient solids device as a crude gravity sedimentation chamber to such a low synthesis process as anaerobic treatment. A crude gravity sedimentation chamber in a high synthesis ratio aerobic process merely degrades the quality of the effluent. However, when synthesis rates are nominally less than 5 per cent, as can occur in anaerobic processes, instead of the nominal 50 per cent characteristic of aerobic processes, biomass inventory can be significantly depleted and start up and or recovery times can be greatly prolonged. A hybrid reactor greatly decreases the dependence on development of a granular sludge for process stability, since the packing aids in retention of either flocculent or filamentous biomass should it predominate.

The greatest need in anaerobic treatment is to reduce start up times and increase process stability. It is unusual that a vendor can market a process which may require six months and even longer in some cases to achieve design loading rates, even more so, that a customer would tolerate such a deficiency. Enhanced biomass inventory build up technique should warrant the highest priority in development of anaerobic treatment.

To illustrate the crucial role of biomass retention efficiency, consider the following example of two scenarios :

- Complete retention of the original seed and all subsequent synthesized biomass.
- 20% retention of original seed and 20% retention of all synthesized biomass.

Assume in both cases that the design loading rate is 10 kg/m^3-day, the bacterial synthesis rate is 5 per cent and both reactors are started with 100 per cent seed whose concentration of biomass is $3kg/m^3$ which is characteristic of digested domestic sludge.

In the first case, the time required to achieve design loading rate is :

$$3kg/m^3\ (1+0.05)^n = 10kg/m^3$$

$$n = 25 \text{ days}$$

In the second case, which is somewhat characteristic of seeding an upflow filter or unpacked UASB, required time to achieve the design loading rate is :

$$20\% \times 3kg/m^3\ (1 + .2 \times .05)^n = 10kg/m^3$$

$$n = 280 \text{ days}$$

If the biomass retention rate could be improved to only 50 per cent, the required start up time could be reduced to n = 75 days as shown in this tabulation :

% Biomass Retention	Start up Time-Days
10	700
20	280
30	160
40	107
50	75
75	40
100	25

Unit processes and techniques which have been used to provide improved solids removal are :

- Membrane filtration
- Granular media filtration

Dorr-Oliver offers a membrane solids separation process, but indicates it is only economical on high strength wastewaters. Crossflow microfiltration has been successfully used for solids separation in aerobic processes in Japan, but when applied to anaerobic processes, undue plugging is reported. Dynatech has used a pilot scale, pressurized reactor with membrane solids separation for anaerobic conversion of hydrogen, carbon monoxide and carbon dioxide to methane. Alcohol production schemes have employed hollow fibre techniques for cell immobilization.

Dissolved air flotation has reportedly been successfully used for over 20 years for solids separation from an anaerobic contact process treating meat packing plant wastewaters in Minnesota. Apparently the natural gas evolution from the mixed liquor is adequate to provide the gas bubbles for flotation.

Considerable advances have been made in granular media filtration of water. Moving bed filters which are continuously cleaned have been developed to successfully filter water with high suspended solids concentrations. There is the potential for adapting such a process to cell separation in anaerobic treatment. Upflow through granular media increases the capacity for solids removal and prevents plugging beyond a head loss comparable to the submerged weight of the media. It appears promising that adaptation of existing technology for solids separation can be practically adapted to anaerobic treatment, resulting in more efficient solids retention, higher biomass/loading rate characteristics, and improved process stability. It is considered to be the most crucial development need for exploitation of anaerobic treatment.

4.9.2 Enhanced Specific Activity

It is difficult to assay the specific activity of anaerobic processes being fed particulate slurries such as domestic sludge, because the actual biomass cannot be distinguished from the refractory particulars. However, with soluble feedstocks, essentially all of the suspended solids within the reactor is of biological synthesis origin. Commonly the specific activity of prototype anaerobic processes treating soluble industrial wastewaters is approximately 1kg COD utilized/kg biomass-day. For the biomass converting acetate to methane, there are two classes of Methanogens-Methanothrix and Methanosarcina. The first class has a low specific activity and a low half velocity constant which causes it to predominate in systems with low steady state acetate concentrations. This is rue of most operating systems because they are lowly loaded. The second class–Methanosarcina–have relatively high half velocity constants and specific activities. In addition, there is a growing body of evidence to indicate their nutritional requirements are more difficult to satisfy.

In highly loaded systems, it has been consistently demonstrated that a shift in predominance from the low rate Methanothrix to the high rate Methanosarcina can be effected by supplementation of trace nutrients. Even though the acetate concentration is high enough to favour the shift in predominance–if only kinetics were involved, no such shift is observed unless nutrients are supplemented. Thus, it appears that both nutrient requirements and kinetics control the shift in predominance.

The specific activity of the Methanosarcina has been observed to be three to five times as high as Methanothrix. Acetate to methane conversion rates in 20 days SRT/CSTR reactors were found to be as high as 90 kg/m^3-day when Methanosarcina predominated. In addition, Methanothrix systems fail by washout at SRT of 10 days, while Methanosarcina systems can convert 35kg/m^3-day of acetate to methane at as low as 5 days SRT.

Dramatic results have also been noted from short-term supplementation of trace nutrients. Wilkie, *et al.*[287] noted rapid declines in what were previously pernicious, high volatile acids concentrations in Napier grass digesters when nickel, cobalt, molybdenum selenium and sulphate were supplemented. Ghosh (1986) noted similar results in the digestion of Bermuda grass. A starch plant anaerobic process in Ireland showed a rapid and dramatic stimulation from the supplementation of trace metals. A municipal sludge digester with elevated volatile acids concentration showed a rapid decline after iron supplementation. A "stuck" digester in Baltimore had a 167 per cent increase in gas production the day after iron was supplemented even though he soluble iron content was very high at 12mg/L.

Iron, cobalt, nickel, molybdenum, selenium, calcium, magnesium and ppb levels of vitamin B_{12} have been proven to cause dramatic stimulation in methane production rates. Their relative costs have been minor, but their impact has been great.

4.9.3 Microbial Interactions

Anaerobic reactors present a unique ecosystem in which diverse groups of bacteria catalyze the conversion of complex organic compounds to methane and carbon dioxide in a highly controlled and coordinated fashion.[227] Methanogenesis was initially considered to be a two-phase process in which the volatile fatty acid (VFA) and other fermentation end-products of hydrolytic/fermentative bacteria were directly covered to CH_4 and CO_2 by methanogenic species. The multiphase nature of the process was subsequently revealed by the discovery of hydrogen-producing acetogenic bacteria and by a better appreciation of the limited substrate capabilities of methanogens.

Advances in our knowledge of the microbiology of methanogenesis have highlighted the role of interspecies H_2 transfer in anaerobic digestion. Obligate H_2 producing acetogens (OHPA species) oxidize VFA fermentation products, such as propionate, butyrate, etc., to acetate, CO_2 and H_2. Their importance in methanogensis is evidenced by the finding that propionate and butyrate are the primary precursors of acetate and H_2 in sewage sludge digesters.[228] Acetogenic dehydrogenation of VFA is thermodynamically unfavourable when the H_2 partial pressure (pH_2) exceeds 10^{-3} to 10^{-4} atm[229]. VFA conversion in digesters is rendered feasible, however, through simultaneous removal of H_2–utilizing species, such as the hydrogenotrophic methanogens.[230] Untill recently, OHPA isolates could only be maintained in the laboratory in syntrophic coculture with H_2–utilizing methanogens or sulphate-reducing *Desulphovibrio* spp. (which provide an alternative electron sink).

To date, only a limited number of OHPA species has been isolated. These include the mesophilic species, *Syntrophomonas wolfei*[231] and *Syntrophobacter wolinii*[232], which oxidize butyrate and propionate, respectively. The generation times of these organisms are extremely long, with respective doubling times of 84 and 161 hours for *S. wolfei* and *S. wolinii* in methanogenic syntrophic coculture. The recent finding of Beaty and Bclnerney[233], that *S. wolfie* can be maintained in pure culture on a crotonate medium, may be expected to accelerate the biochemical study of this key digester group. OHPA species are also known to be involved in the β-oxidation of longer-chain fatty acids (stearate, oleate, etc.) arising from lipid hydrolysis[234] and in the anaerobic degradation of aromatic compounds.[235] The doubling time of the benzoate oxidizer isolated by Mountfort *et al.*[235] is 166 hours in methanogenic coculture. A thermophilic OHPA species which oxidizes butyrate in coculture with a H_2–utilizing methanogen has been isolated by Henson and Smith.[236] Interspecies H_2 transfer is also involved in the oxidation of acetate to CH_4 and CO_2 by a thermophilic syntrophic coculture.[237]

Maintenance of low PH_2-values in digesters is primarily dependent upon the activity of the H_2 utilizing methanogens. Almost all known methanogens convert H_2/CO_2 to CH_4[238]. Mesophilic species exhibit doubling times of the order of 3 or more hours and have K_a values for H_2 in the range 2.5-13 μM[239]. They appear to be greatly undersaturated for H_2 in stably operating digesters[240]. Methanogenic species which can utilize, as direct methanogenic substrates, the C-2 compound, acetate; C-1 compounds, such as formate, methanol and carbon monoxide and methylated amines have also been isolated[238]. Of these, only acetate is regarded as an important precursor of CH_4 digesters, being generally recognized as the source of two-thirds or more of the CH_4 evolved, with most of the remainder coming from CH_2 reduction[241]. Aceticlastic methanogensis has been documented for only two methanogenic genera : Methanosarcina[241,242] and Methanothrix[243]. With acetate as substrate, doubling times for the mesophilic species, *Methanosarcina barkeri* and *Methanosarcina mazei*, are 24 hours and reported K_a values range from 3 to 5mM[241,242, 244]. The filamentous mesophilic accticlast, *Methanotrix soehengenii* displays a considerably lower K_a value for acetate (0.5-0.7 mM) but also exhibits a much slower growth rate, with a doubling time of 3.5-9.0 days[243, 245]. *M. soehngenii* is reported to be more abundant in digesters operated at long hydraulic retention times whereas Methanosarcina tend to predominate at shorter retention times[246]. A second mesophilic *Methanothrix* species, designated *Methanothrix concilii*, has been isolated from sewage[247]. Thermophilic *Methanothrix* spp. have been described by Zinder *et al.*[248] and by Nozhevnikova and Yagodina[249]. A thermophilic *Methanosarcina* species, *Methanosarcina thermophila*, has also been isolated and characterised[250,251].

4.9.4 Monitoring and Process Control

Parameters in current engineering use for routine monitoring of digester performance include/organic/hydraulic loading rate, biogas/CH_4 productivity, COD/BOD reduction, alkalinity, pH and VFA concentrations[252]. A fall in biogas production rate, an increase in VFA concentration or a change towards acidity are accepted indicators of potential digester failure. Appreciation of the central role of hydrogen in the overall process, however, suggests that these changes may be regarded as the result rather than the cause of digester imbalance. Monitoring of the partial pressure of H_2 is being considered as a potentially sensitive means of early detection of fermentation failure[229]. Studies are currently in progress to develop reliable methods for estimation of H_2 levels in digester gases and mixed liquors[253] and to correlate fluctuations in H_2 concentration with changes in feed composition, loading rate and other process variables, for both soluble and particulate feedstocks[254]. Propionic acid may also be used as an indicator of digester balance. The PH_2 range in which acetogenic dehydrogenation of propionate and hydrogenotrophic generation of CH_2 can both proceed exergonically is extremely narrow[229, 230]. Any digester imbalance which results in H_2 partial pressure outside that range should, therefore, cause an immediate increase in propionate. This could be used to signal corrective measures before the microbial community becomes more drastically disturbed by acid accumulation and the consequent fall in pH.

Recent studies on the digestion of complex organic wastes have suggested that other process intermediates may also be used as indicators of the efficiency of coupling between the hydrolytic/acidogenic and the methanogenic phases of the overall process. Phenylacetic acid (PAA) is an intermediate of the degradation of aromatic amino acids and lignocellulosic materials. The conversion of PAA to CH_4 is catalyzed by the syntropic activity of the OHPA species, *Syntrophus buswellii*, in association with hydrogenotrophic and aceticlastic methanogen[235]. Iannotti *et al.*[255] reported that effluent PAA levels in digesters treating swine manure appeared to provide a more sensitive indicator of digester stress than volatile acids. Although the data reported is preliminary in nature, it implies that key process intermediates (other than VFA) may be used to monitor digester balance in applications where the feed is rich in lipids or more refractory organic compounds, such as aromatic or halogenated aliphatic compounds, lignocellulose, etc.

4.9.5 *Effective Biomass Measurement*

The assay procedure most commonly used in engineering practice for seed sludge quantification and biomass measurement in operational digesters is the determination of volatile suspended solids (VSS). These analyses provide crude estimates of effective microbial biomass since they also measure non-microbial organic material contributed by wastewaters containing insoluble organics (e.g. animal manures, crop residues, starch wastewaters, etc.) Furthermore, these methods give no indication of the population density or functional activities of different trophic groups in seed sludges or in digester biomass. Direct enumeration techniques are not a viable alternative since, they are tedious and time consuming, require access to anaerobic culture facilities and may yield misleading data, as species of importance in digesters may not necessarily flourish under the diagnostic enrichment/isolation conditions used, and vice versa[256].

Since methanogenic bacteria exhibit a unique biochemistry, the possibility of quantifying the methanogenic biomass in digesters by measurement of unique methanogen components has been investigated by a number of researchers. Coenzyme F_{420} a deazaflavin analogue which plays a key metabolic role in methanogens as an electron carrier–has received most attention. It is present in all methanogens tested to date and does not occur in other bacteria, with the exception of *Streptomyces griseus* and *Anacystis nidulans*. Although simple, reliable procedures are now available for F_{420} measurement[257,258], wide variation in individual methanogen F_{420} levels (Table 4.77) complicates interpretation of measured sludge F_{420} concentrations[258,259]. Observations by can Bruggen *et al.*[260], that the cocoid form of *Methanosarcina mazei* exhibits a weak F_{420} fluorescence compared to the strong fluorescence of the sarcina aggregate, suggest that F_{420} levels may also vary with morphological form, thus further complicating interpretation of sludge F_{420} data.

Table 4.77 : Coenzyme F_{420} Content of Mesophilic Methanogens[a].

Organism	F_{420} *(nMole.gVSS^{-1})*
H_2 - Utilizing Methanogens	
Methanobacterium bryantii	1400-2400
Methanobrevibacter ruminantium	40
Methanobrevibacter arboriphilus	1800
Methanospirillum hungatei	1900
Formate-utilizing methanogens	
Methanobacterium formicium	3000-3600
Acetate-utilizing methanogens	
Methanosarcina acetivorans	27
Methanosarcina barkeri, 277	9[b]
Methanosarcina barkeri, MS	19[c]
Methanothrix soehngenii	330

[a] Data adapted from Dolfing and Mulder[261].

[b] Expressed as nMole.g protein^{-1} [262].

[c] Expressed as nMole.g (wet wt) cells^{-1} [263].

Correlation of F_{420}, concentrations with potential methanogenic activity for individual substrates is also extremely difficult, since the F_{420} levels maintained by acetate-utilizing methanogens are much lower than those of the H_2/CO_2 or formate-utilizing species (Table 4.77). A recent study by Dolfing and Mulder[261], using granular sludges cultivated on various direct and indirect methanogenic substrates (acetate, ethanol, propionate, etc.), revealed a good correlation only with formate a substrate. Measurement of sludge F_{420} levels may, however, be useful in determining the vertical distribution of methanogenic bacteria in upflow or downflow fixed-bed reactors or the distribution of methanogens between the biofilm and mixed liquior in fixed-bed and hybrid reactors[264].

The distinctive archaebacterial membrane lipids present in all methanogens contain unique phycanyl glycerol ethers which can be extracted, purified and quantified by HPLC procedures[215]. Consequently, measurement of the phytanyl glycerol ether content of anaerobic sludges has potential as a technique for estimating methanogenic biomass content. For non-methanogenic trophic groups, measurement of membrane phospholipid esters has been shown to provide a good estimate of microbial biomass and to correlate well with extractable ATP measurements[266]. In addition, analysis of individual phospholipid fatty acids may permit more detailed characterization of microbial community structure since some fatty acids are unique to individual bacteria, or to groups of bacteria, and may function as molecular signature for these species[267]. Measurement of both phytanyl glycerol ethers and membrane phospholipids and identification of individual phospholipid fatty acids, in anaerobic sludges, may prove useful in determining the relative contents of methanogenic and non-methanogenic biomass; in monitoring changes in the ratio of the two populations and in assessing individual species distribution in response to changes in loading rate, presence of toxicants, etc. Lipid analyses may also be used to distinguish between microbial and non-microbial VSS in digesters treating plant biomass or animal waste feedstocks[266].

4.9.6 Biomass Activity Measurements

Details knowledge of the activities of different trophic groups in seed sludges, mixed, liquor, biofilm and granule samples should assist in defining start-up conditions and assessing maximum loading capacities for operational reactors. Earlier methods, which were confined to estimating methanogenic activity levels with acetate as substrate, required large amounts of sludge and were both time-consuming and error-prone[258]. Routine test procedures have now been developed, based on gas chromatography and pressure-lock syrings sampling[268] or pressure-transducer monitoring[258] of the headspace in test vials. These methods are sensitive, rapid and reliable, and can be utilized to determine the potential methanogenic activities of digester sludges against the full range of direct and indirect methanogenic substrates. Activity test procedures may also be used to assess the inhibitory effect of toxicants on individual methanogenic and OHPA population in crude sludges; to study the effect of changes in process parameters on the distribution and activity of these populations and to characterize the microbiological composition of biofilms and granules developed in reactors under different loading rates and with varying feed compositions.

4.9.7 Anaerobic Biofilm Formation

Start-up and long-term operation of fixed-bed reactors, particularly if operated in downflow mode with modular or channelled support materials, is dependent upon the initial development of a competent biofilm on the support surface and upon its continued maintenance in a form which ensures efficient contact and reaction of the influent substrate with the retained flora. It is important, therefore, to determine the influence of the support material on the attachment of different trophic groups and of individual species within these groups. Such studies have already been initiated[269, 270] and may be expected to provide a more valid basis for choice of support material type and packing arrangement. For example, efficient biofilm retention of slow-growing organisms, such as the OHPA bacteria or the aceticlast, *Methanothrix soehngenil,* could accelerate start-up and ensure more stable operation and more efficient COD removal at possibly higher loading rates. Ongoing studies have shown that *Methanosarcina mazei* MC3 does not attach to a wide range of plastic and clay supports at the initial adhesion stage of biofilm formation[271]. *Methanosarcina* sp. have, however, been reported to play a dominant role in anaerobic biofilms during the start-up phase[258, 272]. The possibility that the presence of *Methanosarcina* sp. results from passive entrapment, rather than active involvement in biofilm formation, merits further study. Studies of this kind should assist in the definition of start-up regimes and could be economically beneficial if they result in significantly shorter start-up times for fixed-film reactors.

4.9.8 Sludge Granulation

Studies on mesophilic granule formation have shown that varied granular forms may be cultivated on different wastewaters and under different start-up conditions [273-274]. Granules developed on mainly VFA feeds tend to be of the "filamentous" type, up 5 mm in size and mechanically fragile. They are reported to contain inert carrier material and to be dominated by a highly filamentous form of *Methanothrix,* presumed to be *M. soehngenii.* More robust, "rod"- type granules develop on sugar-beet or potato processing wastewaters. These are generally smaller (upto 3mm in size), contain no detectable inert carrier and are again dominated by *M. soehngenii* like species, but in a much shorter chain-length form[270,271]. Granules containing *Methanosarcina* sp. as the dominant aceticlast can also be cultivated under high VFA concentrations and low pH[270,272]. However, Lettinga *et al.*[275] consider their development in reactors to be undesirable since they exhibit poor activity at low acetate concentrations and are prone to washout because of their small size (<0.5 mm). A fourth type of granule has been reported from a full-scale plant treating maize starch production wastewater which contained an exceptionally high Ca^{++} content (>700 $mg.L^{-1}$). These "spiky" granules (c. 1 mm in length) were dominated by filamentous organisms and contained up to 60 per cent by weight of $CaCO_3$[270, 271].

At concentrations below 200-300 ppm, Ca^{++} has been shown to exert a positive effect on granule formation[276]. All references shown in brackets should be shown as superscript on the word preceeding them but without a bracket. This has been attributed to chelating effect of Ca^{++} bridges and to the increased density arising from $CaCO_3$ precipitation[277]. Physico-chemical studies by Mahoney *et al.*[276] also provide further evidence of the importance of divalent cation bridging in the initial stages of granulation. Higher concentrations of Ca^{++} adversely affect granule formation due to excessive precipitation of $CaCO_3$ and $(Ca)_3\ (PO_4)_2$[274]. This causes scaling of the outer surfaces of pre-formed granules, increasing granule density and decreasing sludge activity. The high sludge concentrations which result at the base of the reactor make efficient contacting of the wastewater with the sludge impossible[276].

Formation of thermophilic granules has also been investigated. Wiegant *et al.*[279] initially failed to obtain granulation during a 300 days study with pure VFA mixtures. Granulation occurred rapidly, however, with glucose or sucrose feedstocks. The granules obtained differed from the common mesophilic forms, being smaller in size (c. 1.3 mm) and consisting mainly of thermophilic *Methanosarcina* sp.[279] Successful formation of very small thermophilic granules (c. 0.2 mm), on a mixture of acetic, propionic and lactic acids, has recently been reported by Endo and Watanabe[280]. Much larger aggregates (c. 3.0 mm) has been obtained by Bochem *et al.*[281] in chemostant studies of acetate enrichments. These granules consisted of densely-packed *Methanosarcina* clusters surrounding a more loosely-packed central area which contained at least two non-methanogenic species.

Adaptation of granular sludge cultivated initially on one type of industrial wastewater, to a different waste stream has not always been successful at laboratory or at full-scale. Early studies by Lettinga *et al.*[282] showed that granules with high specific activity and excellent settling properties form readily on sugar-beet processing wastewater. Transfer of granules, developed on sugar beet or corn starch wastewaters, to a sucrose feedstock resulted in the development of voluminous layers of bacterial biomass and gelatinous exopolymers on the granule surfaces causing a severe decrease in sludge settleability and ultimate washout from the reactor[283]. These authors concluded that a sudden change in feed composition cannot always be accommodated by granular sludges developed initially on mainly acidified effluents. Granules with excellent settling properties could, however, be formed *ab initio* with sucrose-containing feedstocks[283].

Similarly, initiation of granulation with more recalcitrant wastewaters has not always been successful, as shown by pilot-scale on sludge conditioning liquor in Canada[284]. Granulation was not obtained during 360 days of operation, despite repeated re-seeding, modification of the gas-liquid-solids separator and eventual operation in anaerobic contact mode with effluent solids settlement and recycle. Change to a brewery wastewater, however, resulted in granulation within six weeks[284].

Although granulation occurs with a variety of wastewaters, further studies are required in order to ascertain the roles of individual bacterial groups or species and of insoluble inorganic materials (e.g. calcium) in effective granule formation. Identification and study of the Methanothrix species which dominate mesophilic "filamentous" or "rod"-type granules is of particular interest, in view of reported differences in doubling times and K_a values for acetate between *M. soehngenii* and *M. concilii*[243,247]. The possible involvement of carbohydrate decomposing bacteria, as suggested by Dolfing and Bloemen[285], also merits investigation.

4.9.9 Micronutrient Effects

An increasing number of digester studies indicates considerable stimulation of methanogenesis by supplementation with trace elements[286-288]. Bioavailability of minerals to digester micro-organisms is difficult to measure and what appears to be sufficient concentration of individual micro nutrients in a feedstock may not equate with adequate bioavailability[287].

Recent studies by Speece *et al.*[285], on methane production from acetate in mixed reactors receiving unlimited substrate and operated at 10 or 20 days HRT/ SRT, highlighted interesting effects of nutrient supplementation. The enrichment culture, which had been maintained in the laboratory for eight years on a synthetic medium containing a variety of inorganic nutrients, appeared to be dominated by *Methanothrix* like rods. Additional supplementation with 13 mg. L^{-1} Fe^{++} above the basal level was shown to change the microbiological dominance of the enrichment within 10 days to a *Methanosarcina mazei*-like organism and to increase the maximum acetate utilization rate from 3-4 to 30-40g acetate L^{-1} d^{-1}. This change was inhibited if either phosphate or sulphide anions were added with the extra Fe^{++}, indicating that bioavailability of the iron was essential for the population change[288]. The higher acetate utilization rates were substainable for only a short period. (c. 7 days), during which the *Methanosarcina*-like cells were shown to become associated in large aggregate form. The acetate utilization rate then dropped abruptly to c. 5g.$L.^{-1}d^{-1}$. This was shown to correlate with a break-up of the aggregates to individual coccoid cells. Maintenance of Ca^{++} and Mg^{++} levels above 200 mg. L^{-1} prevented de-aggregation and maintained high acetate utilization rates for prolonged periods[288].

Pure culture studies by Scherer *et al.*[289] have yielded conflicting results. Omission of Ca^{++} from the growth medium resulted in flocculant growth of Methanosarcina barkeri strain Fusaro (flocs of 1 mm size), whereas addition of 2mm $CaCl_2$ caused an immediate change to a dispersed growth from. By contrast, *M. barkeri* strain Julich grew in large aggregate from (c. 3mm) in the presence of C^{++}. Ability to aggregate appears to be a strain-specific property of strain Julich and to be associated with a significantly higher polysaccharide content, rich in glucuronic acid residues[289].

These, and other studies, suggest that the bioavailability of variety of trace elements, vitamins and other nutrients may have dramatic effects on them microbial composition of digester biomass. A better understanding of the microbial requirements for, and the bioavailability of, trace elements and other nutrients in aerobic reactors may permit greater control of both the composition and the activity of the microbial populations involved.

4.9.10 Biomass Characterization

The immunological procedures developed by Conway de Macario and Macario[290, 291] appear to offer most promise for the rapid identification of methanogenic species active in anaerobic digesters. These authors have assembled a comprehensive bank of antisera against 30 reference methanogens available in pure culture[292] and have developed a series of calibrated polyclonal antibody probes from the antisera bank. Testing of individual methanogens against all of the available probes has provided a sequence of data, distinctive for each methanogen and referred to as its antigenic fingerprint. Arrangement of the fingerprint data in a matrix formed by the phylogenetic organization of methanogens clearly showed that antigenic relatedness coincides with phylogenetic kinship[290]. Arrangement of the data in this way permitted selection of representative methanogens for development of panels of monoclonal antibodies using mouse hybridoma techniques[291]. Many of the monoclonal antibodies studies to date are monospecific, reacting exclusively with a single antigenic determinant on a particular methanogen immunotype[292].

The availability of polyclonal and monoclonal antibody probes will permit precise identification and quantification of individual methanognes - not only at the species or strain level but at the immunotype level - in microlitre sample of digester mixed liquor, biofilm and granule preparations. In particular, this will permit identification of immunotypes active under start-up conditions; assessment of immunotype changes in response to duration of operation, alteration of HRT or increase in loading rate; predominant immunotypes in granules and in biofilms developed on different support materials, etc.[239]. Ultimately, such information should prove invaluable for definition of seeding, start-up and operational regimes and for optimization of reactor design with respect to individual wastewater types.

4.9.11 General Consideration

Developments in rector design and operation have established anaerobic digestion as an accepted process for industrial wastewater treatment. Parallel advances in our understanding of the complex microbiology of the process are providing new insights into microbial interactions and into factors governing the dominance, activity and maintenance of individual species in digester mixed liquors, biofilm and granule. This knowledge will contribute significantly to improve design, start-up, process control and operation of anaerobic digesters, with corresponding effects on the efficiency, stability and cost effectiveness of full-scale anaerobic waste treatment applications.

Use of monoclonal antibody probes may be expected, in the long term, to identify the methanogen immunotypes most promising for genetic engineering manipulation[293]. The discovery of methanogen plasmid[294] the identification of a phage specific for a *Methanobrevibacter species*[295] and the more recent isolation and characterization of a *Methanothrix phage*[296] have increased the possibility not developing genetic exchange systems for methanogens. In addition, cloning studies have been successful in obtaining expression of functional methanogen gene products in *Escherichia coli* and *Bacillus* subtilis[297-298]. These development suggest that future microbiological research will continue to complement process and engineering studies, expanding the application of anaerobic digestion technology in the areas of wastewater treatment and renewable energy production.

Anaerobic treatment of industrial wastes in fixed film reactors is already widespread. The process is cost effective for warm wastes with a high concentration of biodegradable organic matter (1-2 kg COD/in^3). For colder and more dilute wastes, like domestic sewage, the process is not cost effective at present, but the trend in process development indicates that this might be changed in the future.

The composition of the substrate is important in relation to design of anaerobic processes. The removal rate, measured as kg COD/kg VSS.d, is 5-6 times higher when acetic acid is the substrate as compared to a process where glucose-or more complex organics is the substrate. This illustrates the importance of making biomass balance to account for biomass fractions.

Many biofilm reactors operates with significant amounts of suspended biomass which obscures interpretation of experimental data and renders design of such reactors difficult. It is an open question whether biofilm surface area or total biomass is the most important reactor property.

The question of diffusional resistance within the biofilm, is crucial for the understanding of the kinetics of anaerobic fixed films. Provisional theoretical calculations indicate that it does not play a significant role, but there is no experimental evidence to support this. In the literature the same set of data have been used for verification of models both with or without diffusional resistance.

The existence of a biofilm structure in a reactor is beneficial with respect to resistance against toxic substances and inert material in the waste and acts as a safeguard against biomass wash-out. The mechanisms governing the development of a biofilm both on inert support material and in pellets/granules is not well understood.

The start-up problems often encountered in anaerobic fixed film reactors emphasise the lack of basic knowledge of the microbiology and the biochemistry of the processes involved. Time consumption during start-up might be reduced considerably through increased knowledge of biofilm build-up, micronutrients, pH and temperature effects.

Under Indian context the choice of the reactor system may be given as follows : Fixed bed (up flow or down flow) >UASB> Rotating Bed > Fluidized ro Expanded bed reactors. This sequence is based on the requirement of skill for operation of anaerobic reactor system.

REFERENCES

1. Marshall, K.C., Interfaces in Microbial Ecology, Cambridge, M.A., Harvard University Press (1976).
2. Marshall, K.C.; Ed. Microbial Adhesion and Aggregation, Dahlem Kowfernen Berlin; Spiringer (1984).
3. Charaklis, Biofilm Development : A Process Analysis In Microbial Adhesion and Aggregation, Ed. Marshall, Dahlem Konferenzen, Berlin : Springer p. 137-157 (1984).

4. Tarakhia, M.H. and Characklis, W.G. Influence of Calcium Specific Chelant on Biofilm Removal, Appl. Envrn. Microbial. 46, p. 1236-1238 (1983).
5. Rittmann, B.E. and Mc, Carty, P.L., Model of Steady State Biofilm Kinetics; Biotechnol. Bioengg., Vol. XXII, p. 2343-2357 (1980).
6. Marshall, K.C. The Effect of Surfaces on Microbial Activity, Water Pollution Microbiology, R. Mitchell (Ed.), John Wiley and Sons, New York, p. 51-70 (1987).
7. Salkenoja-Salonen, M.S. and van den Berg. L., Starting up of an Anaerobic Fixed Film Reactor; Wat. Sci. and Technol., 15 (18/9), p. 305–308 (1983).
8. Robinson, W.E. and Aldrich, H.C., Light and Electron Microscopic Examination of Methane Producing Biofilm in Anaerobic Fixed Bed Reactor, Appl. Even. Microbial. 48, p. 127-136 (1984).
9. Harvey M and Ogilvie, J.R., Methanogenic Activity and Structural Characteristics of Microbial Biofilm on a Needle Punched Polyester support, Appl. Envn. Microbol. 48(3), p. 633–638 (1984).
10. Switzenbaum, M.S., Anaerobic Fixed Film Wastewater Treatment, Enzyme and Microl. Technol. 5(4), p. 241–320 (1983).
11. Jeris, J.S. and Owens, R.W., Pilot Scale, High Rate Biological Denitrifications; JWPCF, 47, p. 2043-2057 (1975).
12. Young, J.C. and Dahab, M.F., Effect of Media Design on the Performance of Fixed Bed Anaerobic Reactors, Presented at IAWPR-Specialized Seminar on Anaerobic Treatment, June, Copenhagen, Denmark (1982).
13. Antonie, R.L., Fixed Film Surfaces-Wastewater Treatment CRC Press Inc. Cleveland Ohio, (1976).
14. Shieh, W.K. and Mulcahy, L.T., FBBR Kinetics-A Rational Design and Optimization Approach Presented at IAWPR Specialized Seminar on Anaerobic Treatment, June, Copenhagen, Denmark (1982).
15. Lettinga, G. and Vinken, J.N., Feasibility of Upflow Anaerobic sludge Blanket (UASB) Process for the Treatment of Low Strength Wastes, Proc. 35th Industrial Waste Conference, May 13/15 Pardu Univ., Lafayette, Indiana. Ann. Arbor Science, Ann Arbor, Mich., p 625-634 (1981).
16. van den Meer, R.R., Anaerobic Treatment of Wastewater containing Fatty Acids in Upflow Reactors, Delft Univ. Press, the Netherlands (1970).
17. Murray, W.D. and Uan den Berg, L., Effect of Support Material on the Development of Microbial Fixed Films Converting Acetic acid to Methane, J. Appl. Bacteriology, 51, p 257-265 (1981).
18. Pol, L.W., Granulation in UASB Reactors. Presented at IAWPR-Specialized Seminar on Anaerobic Treatment, Copenhhagen, Denmark, (June 1982).
19. Li, A. and Sutton, P.M., Door-Oliver's Anitron Syatem : Fluidized Bed Technology For Methane Production From Dairy Wastes Presented at Whey Products Institute Annual Meeting, Chicago, Illinois (April 1981).
20. Jewell, W.J. and Scharaa. G., Micro-algae Separation, Concentration and Conversion to Fuel with an Anaerobic Expanded Bed Rector., U.S. Department of Energy (RE. XB-9-8263-1-3. (1981).
21. Jewell, W.J., Anaerobic Attached Film Expanded Bed Fundamentals, Presented at First Int. Conf. on Fixed Film Biological Processes, Kings Island, Ohio,. (April, 1982).

22. Letting G. Design, Operation and Economy of an Anaerobic Treatment Presented at IAWPR-Specialized Seminar an Anaerobic Treatment, Copenhagen, Denmark, (June 1982).

23. Chian, E.S.K. and De-Walle, F.B., Treatment of High Strength Acidic Wastewater with a Completely Mixed Anaerobic Flow, Wat. Res., 11, p295-304 (1977).

24. Witt., E.R., and Humphrey, W.J., Full Scale Anaerobic Filter Treats High Strength Wastes, Proc. 34th Industrial waste conference (May 8/10 1979), Purdue Unvi., Ann Arbor Mich., p 229-234 (1980).

25. Ferguson, J.F. and Bengamen, M.M., Neutralization in Anaerobic Treatment of an Acidic Waste, Presented at IAWPR-Specialized Seminar on Anaerobic Treatment Copenhagen, Denmark, (June 1982).

26. ven den Heuvel, J.C. Boelhouver, C., Purification of Municipal Wastewater by Subsequent Reverse Osmosis and Anaerobic Digestion Biotechnol. Bioengg., 23, p 2001-2008 (1981).

27. Sutton, P.M. and Li, R.R., Dorr-oliver's Fixed Flow and Suspended Growth Anaerobic System For Industrial Wastewater Treatment and Energy Recovery. Presented at 37th Annual Purdue Industrial Waste Conference, Purdue Univ., Lafayette, Indiana (1982).

28. Kennedy, K.J. and van den Berg, L., Use of Fixed Film Reactors For Production of Methane from Waste, Presented at 3rd Bioenergy R and D Seminar, March, Ottawa, Canada (1982).

29. van den Berg, L. and Kennedy, K.J., Comparison of Intermittent and Continuous Loading of Stationary Fixed Film Reactors for Methane from Waste, J. Chem. Tech. Biotechnol., 32, p 427-432 (1982).

30. Parkin, G.F. and Speece, R. E., Attached Vs Suspended Growth Anaerobic Reactors: Copenhagen, Denmark, (June, 1982).

31. Norrman, J., Treatment of a Black Liquor Condensate from a Pulp Mill in an Anaerobic Filter and an Expanded Bed. Presented at IAWPR Specialized Seminar an Anaerobic treatment, Copenhagen, Denmark, (June, 1982).

32. De-Walle. F.B. and Chian, E.S.K. Kinetics of Substrate Removal in a completely Mixed Anaerobic Filter Biotech. Bioengg., 18, p. 1275-1297 (1976).

33. Speece, R. E. and Kem, J.A. The Effect of Short Term Temperature Variations on Methane Production, J. WPCF, 42, p 1996-1997 (1970).

34. Bories, A., Fermentation Methanique Avec Separation des phases Acidosene et Methanogene Appliquee Au Traitement des Effuents a Forte Charge Polluante (Distilleries). Annales de Technologie Agricole, 29, p 50-528 (1981).

35. Asinare, C.M., Production of Methane From fresh Water Macro-Algae by An Anaerobic Two Step Digestion System. In Energy from Biomass, W. Palz. P. Chartier, and D.O. Hall (Eds.), Applied Science Publishers, London, (1981).

36. Li, a and Sutton, P.M., Energy Recovery From Pre-treatment of Industrial Wastes in the Anaerobic Fluidized Bed Process, Presented at 1st conference on Fixed Film Biological Processes, Kings Island, Ohio., (April, 1982).

37. Verstrate, W., Phase Separation in Anaerobic Digestion : Motives and Methods., Trib Cebedeau, 34, p. 453-454 and p. 367-325 (1981).

38. Vilta Salo, I., Full Scale Biogas Production and Sewage Treatment for Potato Chips Industry Presented at the Second Int. Symposium an Anaerobic Digestion, Travemude, Germany, (Sept. 1981).

39. Bengamin, M. M., Toxicity and Biodegradibility of Pulp Mill Waste Constituents to Anaerobic Bacteria, Presented at IAWPR Specialized Seminar on Anaerobic Treatment, Copenhagen, Denmark, (June, 1982).
40. Salkinoja-Salogen, M., Biodegradability of Recalcitrant Organochlorine Compounds in Fixed Film Reactors, Presented at a LAWPR Specialized Seminar an Anaerobic Treatment (June, 1982), Copenhagen, Denmark, (June, 1982).
41. van den Berg, L. and Lentz, C.P., Comparison Between up and Down flow Anaerobic Fixed Film Reactors of Varying Surface to Volume Ratio for the Treatment of Bean Blanching Waste, 34th Industrial Waste Conference (May 8/10, 1979), Purdue Univ., Lafayette, Indiana. Ann Arbor Science, Ann Arbor, Mich, p. 319-325 (1980).
42. Young, J.C., The Anaerobic Filter for Waste Treatment, Ph. D. Thesis, (1968) Dept. of Civil Engg, Stanford Univ. U.S.A., (1968).
43. Martensson, L. and Frostell, B., Anaerobic Wastewater Treatment in a Carrier Assisted Sludge Bed Reactor, Presented at a IAWPR Specialized Seminar an Anaerobic Treatment, (June, 1982), Copenhagen, Denmark (June, 1982).
44. Salkinoja-Salonen, M.S., Starting Up of an Anaerobic Fixed Film Reactor, Proc. Presented at a LAWPR Specialized Seminar on Anaerobic Treatment, Copenhagen, Denmark, (June, 1982).
45. Rozzi, A. and Verstraete, W., Calculation of Active Biomass and Sludge Production Vs Waste Composition in Anaerobic Contact Processes, Trib. Cebedeau. 34. (455), P 421-427 (1981).
46. Donnelly, T., Industrial Effluent. Treatment with the Bioenergy Process, Process Biochemistry, 13(6), p 14-16 (1978).
47. Ghosh S. and Henry M.P., Stabilization and Gasification of Soft-drink Manufacturing Waste by Conventional and Two-phase Anaerobic Digestion *Proc. 36th Ind. Waste Conf. Purdue Univ.* (1981).
48. Ghosh S. Conrad J.R. and Klass D.L., Anaerobic Acidogenesis of Sewage Sludge. J. *Wat. Pollut. Control. Fed.* 47. 30, (1975).
49. Pohland F.G. and Ghosh S., Developments in Anaerobic Treatment Processes. *Biological Waste Treatment* (Edited by Canale R.P.), Interscience New York, p. 85-106(1971).
50. Ghosh S., Alleviation of Environmental Problems of Waste Disposal with Production of Energy and Carbondioxide, Paper presented at the 28th Annual Meeting Society of Soft Drink Technologists. Colorado Springs. CO., (1981a).
51. Chosh S., Energy Production Efficiencies in Anaerobic Gasification Processes. *Proc. Biochem.*, June/July, (1981b).
52. Cohein A., Anaerobic Digestion of Glucose with Separated Acid Production and Methane Formation; Wat. Res. 13(7), p. 571-580 (1979).
53. Massey, M.L. and Polland, F.G., JWPCF 50(9), p. 2204-2222 (1978).
54. Lettinga G. *et al.*, Anaerobic Gistuig Voor. de Zuivering van Afvalwater, *Extern IV* 6, 379, (1977b).
55. Pipyn P., Verstraete W. and Ombrejt J.P., Aplostscale Anaerobic Upflow Reactor treating Distillery Wastewaters, *Biotechnol. Lett.* 1, 495, (1979).
56. Agharkar, R.S., Optimal Design of Diphasic Anaerobic Digestion of Wastewater, M. Tech. Thesis, I.I.T. Bombay (1983).
57. Karhadkar, P.P., Optimal Design of Diphasic anaerobic Digestion of Distillery Spentwash. M. Tech. Thesis, I.I.T. Bombay (1983).

58. Oatnaik, P.C., Wastewater Management of Molasses based Distillery Spentwash. M. Tech. Thesis I.I.T. Bombay (1984).

59. Raman, N.S. Wastewater Management for Molasses Based Distillery Spentwash M. Tech. Thesis, I.I.T. Bombay (1984).

59(a). Karhadkar, P.P., Handa, B.K. and Khanna, P. Pilot Plant Distillery Spentwash Biomethanation, J. Envirn. Engg. (ASCE), 116(6), p 1029-1045 (1990).

60. Kulgelman, I.J. and Chin., K.K, Toxicity, Synergism and Antagonism in Anaerobic Wastewater Treatment Chemistry, Advances in Chemistry Series No. 105 (1971), F.G. Pohland, Ed. Association of Chemical Society, Washington, D.C. (USA).

61. Kahadkar, P.P Sulfide and Sulphate Inhibition of Methanogenesis, Water Res., 21(9), p 1061-1066 (1987).

62. Bhatia, D., Vieth, W.R. and Venkatasubramaniam, K., Steady State and Treatment Behaviour in Microbial Methanification II, Biotechnol. Bioengg., 27(8) p. 1109-1207 (1985).

63. Bhatia, D., Vieth, W.R. and Venkatasubramaniam, K., Steady State and Treatment Behaviour in Microbial Methanification II, Biotechnol. Bioengg., 27(8) p 1109-1207 (1985).

64. Karhadkar, P.P. Methane Recovery from Distillery Spentwash, Ph. D. Thesis, I.I.T. Bombay (1988).

65. Sarswat, N. and Khanna, P., Methane Recovery from Water Hyacinth through Anaerobic Activated Sludge Process, Biotechnol. Vol XXVIII, p 240-246 (1986).

66. Khanna, P. Annachhtre, A.P. Naik, N.S. and Saraswat, N., Ministry of Energy and Syndicate Bank Competition on Low Cost and Efficient Model of Biogas, Environ. Science Engg. Group, I.I.T. Bombay, (1984).

67. KVIC Report, Gobar Gas Scheme, Research and Development Work under Biogas Programme Retrospect and Prospect-Khadi and Village Industries Commission, Bombay, India, (July) 18, 1983.

68. Kaplovsky, A.J., Volatile Acids Production During Digestion of Seeded, Unseeded and Limed Fresh Solids, Sewage Instr. Wastes, 23, p 713 (1951).

69. Pohland, F.G. and Bloodwood, D.D., Laboratory Studies an Mesophilic and Thermophilic Anaerobic Sludge Digestion, JWPCE, 35, p 11(1963).

70. McCarty, P.L., individual volatile Acids in Anaerobic Treatment, JWPCF, 35, p. 1501 (1963).

71. Andrews, J.F. and Pearson, E.A., Kinetic and Characteristics of Volatile Acid Production in Anaerobic Fermentation Processes, Int. J. Air Wat. Pollut., 9, p438 (1965).

72. Bryant, M.P. *Methanobacillus omelianskii*. A., Symbiotic Association of Two Species of Bacteria, Arch. Microbiol, 59, p 20 (1967).

73. Iannotti, E. L., Glucose Fermentation Products of *Ruminococcus albus* Grown in Continuous culture with *Vibrio succinogens*; Changes caused by interspecies Transfer of H_2, J. Bact., 114, p 1231 (1973).

74. Chung, K.T. Inhibitory Effects of H_2 on Growth of *Clostridium cellobioparum*. Appl. Environ. Microbiol., 31(3), p 342 (1976).

75. Scheifinger, C. C., H_2 production by *Selenomas ruminatium* in Absence and Presence of Methanogenic Bacteria, Appl. Microbiol., 29, p 480 (1975).

76. Byrant, M.P., Growth of Desulfovibrio in Lactate or Ethanol Media Low in Sulfate in Association with H_2-utilizing Methanogenic Bacteria, Appl. Environ. Microbiol., 33, p 1162 (1977).
77. McInerney, M.J., Anaerobic Bacterium that Degrades Fatty Acids in Synthrophic Association with Methanogens, Arch. Microbiol, 122, p 129 (1979).
78. Mosey, F.E., Mathematical Modelling of Anaerobic Digestion Process : Regulatory Mechanism for formation of Short-chain Volatile Acids from Glucose, Wat. Sci. Tech., 15, p. 209 (1983).
79. Smith P.H. and Shuba, P.J., Terminal Anaerobic Dissemination of Organic Molecules Proc. Bioenergy Res. Conference, Amherst, M.A. (June 1983).
80. Smith, M.R. and Mah, R.A., Acetate as Sole Carbon Energy Source for Growth of Methanosarcina strain 227, Appl. Environ Microbiol., 39, p 993 (1980).
81. Kaspar, H.F. and Wuhramann, K, Kinetic Parameters and Relative Turnovers of Some Important Catabolic Reactions in Digesting Sludge., Environ. Microbiol., 36. p 1(1978).
82. Kaspar, H.F. and Wuhramann, K. Product Inhibition in Sludge Digestion, Microbiol Ecol., 4, p 241 (1978).
83. Skyes, R. M., Hydrogen Production in the Anaerobic Digestion of Sewage Sludge. Ph.D Thesis, Purdue Univ., Lafayette. (IN. (1970).
84. Barnes D., *et al.*, Influence of Organic Shock Loads on the Performance of an Anaerobic Fluidized Bed System, Paper presented at the 37th Purdue Waste Conference, Purdue In (April 1983).
85. R.H. Heyes and R.J. Hall., Kinetics of Two Subgroups of Propionate-Using Organisms in Anaerobic Digestion, "*Appl. Environ. Microbiol.,* 46(3), p 710 (1983).
86. Hungate. R.E., Hydrogen an in Intermediate in the Rumen Fermentation, *Arch. Microbiol.,* 59, p. 20 (1967).
87. J.A. Robinson and J.M. Tiedje, "Kinetics of Hydrogen Consumption by Rumen Fluid. Anaerobic Digester Sludge, and Sediment, *Appl. Environ. Microbiol.,* 44(6), 1374 (1982).
88. A. Ben-Bassat *et al.*, Melabolic Control for Microbial Fuel Production During Thermophilic Fermentation of Biomass, 4th Symp. on Energy from Biomass and Waste, Lake Vista.
89. D.R., Boone, Terminal Reactions in the Anaerobic Digestion of Animal Waste, *Appl. Environ, Microbiol.,* 43, p57 (1982).
90. T.G. Shea *et al.*, Kinetics of Hydrogen Assimilation in the Methane Fermentations, *Water Res.,* 2, p833. (1968).
91. P.L. McCarty, *Energetics of Organic Matter Degradation,* Advances in Chemistry Series 105, (American Chemical Society, Washington, DC., p 91, (1971).
92. P.L. McCarty, One Hundred Years of Anaerobic Treatment, Paper presented at the 2nd International Symposium on Anaerobic Digestion. Travemunde, Germany. (Aug. 1981).
93. A.J.B. Zhender, *et al.*, Characterization of an Acetate-Decarboxylating Non-Hydrogen-Oxidizing Methane Bacterium, *Arch. Microbiol,* 124, 1 (1980).
94. M.J. Melnereney and M.P. Bryant, Review of Methane Fermentation Fundamentals, *Fuel Gas Productions from Biomass,* CRC Press, Inc. Boca Raton. FL. (1981).

95. W, Gujer and A.J.B Zhender, Conversion Processes in Anaerobic Digestion, *Progr. Wat Technol.*, 15(8/9), p. 127-167 (1983).
96. R.K. Thauer *et al.*, Energy Conversion in Chemotrophic Anaerobic Bacteria, *Bacteriol, Rev.*, 41(1), p. 100 (1977).
97. C.J. Rivard and P.H. Smith, Effects of Continuous Addition of Nitrate to a Thermophilic Anerobic Digestion System, Poster 0-6, Presented at the 84th Annual Meeting of the American Society for Microbiology, St. Louis, MO. (Mar. 1984).
98. G.K. Anderson, Identification and Control of Inhibition in the Anaerobic Treatment of Industrial Westewaters, *Process Biochem.* 28(4) (1982).
99. A.K. Basu and E. LeClere, Mesophilic Digestion of Beet Molasses Distillery Wastewater, Adv. *Water Pollut. Res.*, 6, p. 581 (1972).
100. A.K. Basu and E, LeClere, Mesophilic Digestion of Beet Molasses Distillery Wastewater, *Adv, Water Pollunt. Res.*, 6, p. 581 (1972).
101. A Bories *et al.*, Methanisation des Eaux Residuaries de Distilleries, *Trib. Cedebeau*, 34(456), 475 (1981).
102. R. Braun and S. Huss, Anaerobic Digestion of Distillery Effluents, *Process Biochem.*, 6, 25 (1982a).
103. R. Braun and S. Huss, Anaerobic Filter Treatment of Molasses Distillery Slops, *Water Res.*, 16, 1167 (1982b).
104. M.J.T. Carrondo *et al.*, Anaerobic Filter Treatment of Molasses Fermentation Wastewater, *Wat. Sci Technol.*, 15, 117 (1983).
105. B. Frostell, Anaerobic Fluidized Bed Experimentation with a Molasses Wastewater, *Process Biochem.*, 37(6), 17 (1982).
106. W.C. Hiatt *et al.*, Anaerobic Digestion of Rum Distillery Wastes, Proceedings of the 28th Purdue Industrial Waste Conference, p 966, (1973).
107. C.J. Jackson, Whiskey and Alcohol Distillery Wastes, *J. Proc. Inst. Sewage Purification*, 207 (1956).
108. H.A. Painter *et al.*, The Treatment of Malt Whiskey Distillery Wastes by Anaerobic Digestion, *Breewer's Guardina*, 9 (1970).
109. A.E. Pattet *et al.*, "The Treatment of Strong Organic Wastes by Anaerobic Digestion, J. Inst. Public Health Engg., 6, 170 (1959).
110. P. Pipyn *et al.*, A Pilot Scale Anaerobic upflow Reactor Treating Distillery Wastewaters, Biotechno. Lett., 1, 495 (1979).
111. I.G. Prince *et al.*, Power Ethanol Production Using Tower Fermentation, Ph. D. Thesis, Univ. Sydney, Australia (1982).
112. L.A. Roth and C.P. Lentz, Anaerobic Digestion of Rum Stillage, J. Can., Inst. Food Sci. Technol. 10(2), 105(1977).
113. B.P. Sen and T.R. Bhaskaran, Anaerobic Digestion of Liquid Molasses Distillery Wastewaters. J. Wat Pollut. Contr. Fed., 34, 1015 (1962).
114. T.G. Shea *et al.*, Investigation of Rum Distillery slops Treatment by Anaerobic Contact Process, Proceedings of the 5th National Symposium on Food Processing Wastes, Montery, CA, p 155 (1974).
115. G.J. Stander, Effluent from fermentation Industries : 5 A review of Methods of Disposal and Utilization; II, The Significance of solid phase and of volatile acid Development

in the Anaerobic Method of Treatment, III The Influence of High concentrations of sulphates on the Anaerobic Digestion Method of Treatment at Thermophilic (SSC) and Mesophillic (33°C) temperatures J. Inst. Sewages Purification, 3, 286 (1950).

116. L.M. Szendry, startup and operation of Balardi Anaerobic Filter, Paper presented at the 3rd International Symposium on Anaerobic Digestion, Boston, MA (Aug, 1993).
117. A. Aivasidis *et al.*, Anaerobic Treatment of Acetic Acid and sulfite Evaporator Condensate in a Fixed Bed Loop Reactor, Paper presented at the Poster Session, 3rd International Symposium on Anaerobic Digestion, Boston, MA (Aug. 1983).
118. G. Brune *et al.*, Anaerobic treatment of an Industrial Waste Containing Acetic Acid, Furferal and Sulfite, Process Biochem., 3, 20 (1982).
119. N.E. Takeshita and N. Mimito, Energy Recovery by Methane Fermentation of Pulp Mill Waste Water and Sludges Pulp and Paper Can. 82 82(5), 99 (1981).
120. B.J. Obarsky *et al.*, Sulfus Removal of Polysulfide Rubber Manufacturing waste waters by Anaerobic Treatment, Proce of the 33rd Purdue Industrial Wastes Conf. Lafayette, In P. 402 (1978).
121. W. Stumm and J.J Morgan Aquatic Chemistry Wiley New York (1981).
122. F.G. Pohland *et al.*, Leachate and Gas Quality During Landfill, Stabilisation of Municipal Refuse, Paper Presented at 3rd Intt. Symp. on Anaerobic Digestion, Boston, MA (Aug. 1983).
123. J.F. Ress, Biochemical Characterization of Landfill, Paper presented at the 3rd Intt. Symp. on Anaerobic Digestion Boston MA (Aug. 1983).
124. W.J. Payne, Reduction of Nitrogenous Oxides by Micro-organisms, Bactriol. Rev., 37, 409 (1973).
125. J.M. Krul, The Relationship Between Dissimilatory Nitrate Reducing and Oxygen Uptake by Cells of an Alcaligenes Strain in Flocs and Suspension and by Activated Sludge Flocs, Water Res., 10, 337 (1976).
126. R.E. McKinney and R.A. Conwey, Chemical Oxygen in Biological Waste Treatment, Sewage Industrial Wastes, 29 1097 (1957).
127. H. Hukelekian, Effects of the Addition of Sodium Nitrate on Sewage Hydrogen Sulfide Production and BOD Reduction, Sewage Works, J. 15(2), 255 (1943).
128. E. Schulze, Nitrate Treatment of Waste waters containing Hydrogen Sulfide, Gas-u-Wasserfaih, 95, 4 (1954).
129. W.L. Balderston and W.J. Payne, Inhibition of Methanogenesis in Salt Marsh Sediments and whole Cell suspensions of Methanogenic Bacteria by Nitrogen Oxides, Appl. Eviron. Microbiol., 32, 264 (1976).
130. J. Sorensen, Reduction of Ferric Iron in Anaerobic, marine sediment and Interactions with Reduction of Nitrate and Sulfate, Appl. Environ. Microbiol., 43(2), 319 (1982).
131. J.M. Bollag and S.T. Czlokowski, Inhibition of Methane Formation in Soil by Various Nitrogen-containing Compounds, *Soil Biol. Biochem.*, 5, 673 (1973).
132. R.J. Buresh and W.H. Patrick, Nitrate Reduction to Ammonium in Anaerobic Soil, *J.Am. Soil Sci.*, 42, 913 (1978).
133. R.S. Ormeland *et al.*, Sulfate Reduction and Methanogenesis in Marine Sediments, *Geochim. Cosmochim-Acta*, 42, 209 (1978).
135. J.W. Abram and D.B. Nedwell, Hydrogen as a Substrate for Methanogenesis and Sulphate Reduction in Anaerobic Saltmarsh Sediment, *Arch. Microbiol.*, 117, 93 (1978a).

136. J.W. Abram and D.B. Nedwell, Inhibition of Methanogenesis by Sulfate Reducing Bacteria Competing for Transferred Hydrogen. *Arch. Microbiol.*, 117, 89 (1978b).

137. J. Sorensen etal, Volatile Fatty Acids and hydrogen as Substrates for Sulfate-Reducing Bacteria in Anaerbic Marine Sediment, *Appl. Environ. Microbiol.*, 42, 5 (1981).

138. R.S. Ormeland *et al.*, Methane Production and Simultaneous Sulphate Reduction in Anoxic Saltmarsh Sediments, *Nature* (London), 296, 143 (1982).

139. M.R. Winfrey and J.G. Ziekus, Effect of Sulfate on Carbon and Electron Flow During Microbial Methanogenesis in Freshwater Sediments, *Appl. Environ. Microbiol.*, 33. 275 (1977).

140. R.F. Strayer and J.M. Tiedje, Kinetic Parameters of the Conversion of Methane Precursors to Methane in a Hypereutrophic Lake Sediment, *Appl. Environ. Microbiol.*, 36, 330 (1978).

141. T.E. Cappenberg, Interrelationships Between Sulfate-Reducing and Methane-Producing Bacteria in Bottom Deposits of a Fresh Water Lake. I Field Observations: II. Inhibition Experiments; III. Experiments with ^{14}C Labelled Substrates, *Antonie van Leeuwenhoek*, 40, 285, 297, 457 (1974).

142. T.E. Cappenberg, A Study of mixed Continuous Cultures of Sulphate-reducing and Methane-Producing Bacteria, *Microbial Ecol.*, 2, 61 (1975).

143. J.K. Kristjansson *et al.*, Different K_a, values for Hydrogen of Methanogenic and Sulphate Reducing Bacteria : An Explanation for the Apparent Inhibition of Methanogenesis by Sulphate, *Arch. Microbiol.*, 131, 278 (1982).

144. Y.I. Sorokin, Role of Carbon Dioxide and Acetate in Biosynthesis by Sulphate-Reducing Bacteria, *Nature*, (London), 210, 55 (1966a).

145. Y.I Sorokin, Sources of Energy and Carbon for Biosynthesis in Sulphate-Reducing Bacteria, *Microbiol.*, 35, 643 (1966b).

146. W. Baziong *et al.*, Acetate and Carbon Dioxide Assimilation by *Desulfovibrio vulgaris* growing on Hydrogen Plus Sulfate as the sole Energy Source, *Arch. Microbiol.*, 123, 301 (1979).

147. F. Widdel and N. Pfenning, Studies on Dissimilatory Sulfate Reducing Bacteria that Decompose Fatty Acids, I. Isolation of New Sulfate-Reducing Bacteria Enriched with Acetate from Saline Environments, *Arch. Microbio.*, 129. 395 (1981).

148. F. Widdel ad N. Pfenning, Studies on Dissimilatory Sulphate-Reducing Bacteria that Decompose Fatty Acids. II Incomplete Oxidation of Propionate by *Desulfobulbus propionicus* ge. nov. sp. nov., *Arch. Microbiol.*, 131, 360 (1982).

149. H.J. Laanbroek and N. Pfenning, Oxidation of Short-Chain Fatty Acids by Sulphate-Reducing Bacteria in Freshwater and Marine Sediments, *Arch. Microbiol.*, 128, 330, (1981).

150. R.L., Snyder and A.L. Mills, Freshwater Sediment Acetate Turnover Along a Sulphate Gradient, Poster N-21, presented at 84th Annual Meeting of the American Society for Microbiology, St. Louis, MO, Mar. 1984.

151. F. Widdel and N. Pfenning, A New Anaerobic Spore Forming. Acetate Oxidizing, Sulphate-Reducing Bacterium, *Arch, Microbiol.*, 119, 112 (1977).

152. A.W. Lawrence and P.L. McCarty, Kinetics of Methane Fermentation in Anaerobic Treatment, *J. Water Pollut, Contr. Fed.*, 42(2), p3 (1969).

153. I.J. Kugelman and K.K. Chin, Toxicity, Synergism and Antagonism in Anaerobic Waste Treatment Processes, Symposium Series (105, (American Chemical Society, Washington, DC), p55 (1971).
154. F.G. Pohland and S. Ghosh, Developments in Anaerobic Stabilization of Organic Wastes-The Two-Phase Concept, *Environ. Lett.,* 1(4), 255 (1971).
155. S. Ghosh and D.L. Klass, Two-Phase Anaerobic Digestion, *Process Biochem.,* 5, 2 (1978).
156. M.I. Massey and F.G. Pohland Phase Separation of Anaerobic Stabilization by Kinetic Controls, *J. Water Pollut. Contr. Fed.,* 50(2), 2204 (1978).
157. M.R. Smith and R.A. Mah, Growth and Methanogenesis by *Methanosarcina* Strain 227 on Acetate and Methanol, *Appl. Environ. Microbiol.,* 36(6), 870 (1978).
158. P.J. Weimer and J.G. Ziekus, Acetate Metabolism in *Methanosarcina barkeri, Arch. Microbiol.,* 119, 175 (1978b).
159. B.A. Hoser *et al., Methanothrix songheniigen,* Nov. sp. Now. A New Acetatrophic Non-Hydrogen-Oxidizing Methane Bacterium, *Arch. Microbiol.,* 132, i (1982).
160 W. Badziong and R.K. Thauer, Growth Yields and Growth Rates of *Desulfovibrio vulgaris* (Marburg) Growing on Hydrogen Plus Sulphate and Hydrogen Plus Thiosulfate as the Sole Energy Sources, *Arch. Microbiol.,* 117, 209 (1978).
161. D.R. Lovely *et al.,* Kinetic Analysis of Competition Between Sulfate Reducers and Methanogens for Hydrogen in Sediments, *Appl. Environ. Microbiol.,* 43. 1373 (1982).
162. A.C. Middleton and A.W. Lawrence, Kinetics of Microbial Sulphate Reduction, *J. Water Pollut. Contr. Fed.,* 49, 1959 (1977).
163. C.L. Liu and H.D. Peck, Competitive Bioenergetics of Sulphate Reduction in *Desulfovibrio and Desulfotomaculum* sp. *J. Bacteriol.,* 145, 2 (1981).
164. P. Schonheit *et al.,* Kientics Mechanism for the Ability of Sulphate Reducers of Out-Compete Methanogens for Acetate, *Arch Microbiol.,* 131, 285 (1982).
165. K. Ingvorsen *et al.,* Kinetics of Sulfate and Acetate Uptake by *Desulfobacter postgatei. Appl. Environ. Microbiol.,* 47 (2), 403 (1984).
166. G.B. Patel, Some Characteristics of *Methanothrix* Species Strain (GP6 Isolated from Sewage Sludge, Poster 1-7, Presented at the 84th Annual Meeting of the Ammerican Society for Microbiology. St. Louis, MO (Mar. 1984).
167. S.H. Zinder and T. Anguish, Isolation of a Thermophilic Gas Vacuolated *Methanothrix,* St. Louis, MO, (Mar. 1984).
168. R.A. Mah, Isolation and Characterization of *Methanococcus mazeii, Current Microbiol.,* 3, 321 (1980).
169. T.J. Ferguson and R.A. Mah, Effect on H_2-COD_2 on Methanogenesis from Acetate or Methanol in *Methanosarcina* sp., *Appl. Environ. Microbiol.,* 46(2), 348 (1983).
170. T.N. Zhilina, Biotypes of *Methanosarcina, Microbiol.,* 45(3), 414 (1976).
171. T.J. Hutten *et al.,* Acetate, Methanol and Carbon Dioxide as Substrates for Growth of *Methanosarcina barkeri, Antonie van Leeuwenhoek,* 46, 601 (1980).
172. J.A. Krzycki and J.G. Zeikus, Characterization and Purification of Carbon Monoxide Dehydrogenase from *Methanosarcina barkeri, J. Bacteriol.,* 158(1), 231 (1984).
173. B. Blaut and G. Gottschalk, Effects of Trimethylamine on Acetate Utilization of *Methanosarcina barkeri, Arch. Microbiol.,* 133, 230 (1982).

174. T.N. Zhilina and G.A. Zavarzin, Trophic Relationship Between *Methanosarcina* and Its Associates, *Microbiol.*, 42(2), 235 (1973).

175. P.J. Weimer and J.G. Ziekus, One-Carbon Metabolism in Methanogenic Bacteria, *Arch, Microbiol.*, 119, 49 (1978a).

176. R.A. Mah *et al.*, Studies on an Acetate-Fermenting Strain of *Methanosarcina, Appl. Environ. Microbiol.*, 35, 1174 (1978).

177. H. Hippe *et al.*, Utilization of Trimethylamine and other *N-Methyl* Compounds for Growth and Methane Formation by *Methanosarcina barkeri, Proc. Natl. ACod. Sci.*, 76(1), 494 (1979).

178. J.A. Patterson and R.B. Hespell, Trimethylamines and Methylamine as Growth Substrates for Rumen Bacteria and *Methanosarcina barkeri, Current Microbiol.*, 3, 79 (1979).

179. A.J. Kluvyver and C.G. Schnellen, On the Fermentation of Carbon Monoxide by Pure Cultures of Methane Bacteria, *Arc. Biochem.*, 14, 57 (1947).

180. L. Daniels *et al.*, Carbon Monoxide Oxidation by Methanogenic Bacteria, *J. Bacteriol.*, 132, 118 (1977).

181. J.M.O' (Brienetal, Association of Hydrogen Metabolism with Unitrophic or Mixotrophic Growth of *Methanosarcina barkeri* on Carbon Monoxide, J. Bacteriol., 158(1), 373 (1984).

182. J.A. Krzycki *et al.*, Comparison of Unitrophic and Mixotrophic Substrate Metabolism by an Acetate-Adapted Strain of *Methanosarcina barkeri, J. Bacteriol.*, 149(1); 247 (1982).

183. P.H. Smith and R.E. Hungate, Isolation and Characterization of *Methanobacterium ruminantium* n. sp., *J. Bacteriol.*, 74, 713 (1958).

184. J.G. Zeikus and R.S. Wolfe, *Methanobacterium thermoautotrophicus* sp. n., an Anaerobic, Autotrophic, Extreme Thermphile, *J. Bacteriol.*, 109, 707 (1972).

185. J.G. Zeikus and D.L. Henning, *Methanobacterium arbophilicum* sp. nov., An Obligate Anaerobe Isolated from Wetwood of Living Trees, *Antonie von Leeuwenhoek*, 41 543 (1975).

186. N.L., Schauer and J.G Ferry, Metabolism of Formate in *Methanobacterium formicium, J. Bacteriol.*, 142 800 (1980).

187. T.C. Stadtman and H.A. Barker, Studies on the Methane Fermentation IX. The Origin of Methane in the Acetate and Methanol Fermentations by *Methanosarcina, J. Bacteriol.*, 61, 269 (1951).

188. W.E. Balch *et al.*, Methanogens Reevaluation of a Unique Biological Group, *Microbiol. Rev.*, 43, 260 (1979).

189. J.A. Romesser *et al., Methanogenium,* A New Genus of Marine Methanogenic Bacteria, and Characterization of *Methanogenium cariaci,* sp. nov. and *Methanogenium marisnigri* sp. Nov., *Arch, Microbiol.*, 121, 147 (1979).

190. J.G. Ferry *et al., methanospirillium,* A New Genus of Methanogenic Bacteria, and Characterization of *Methanospirillum hungatii,* sp. *Nov, Int, j. Syst. Bacteriol.*, 24, 465 (1974).

191. G.B. Patel *et al.*, Characterization of a Strain on *Methanospirillum hungatii, Can. J. Microbiol.*, 22, 14004 (1976).

192. J.G. Ferry and R.S. Wolfe, Nutritional and Biochemical Characterization of *Methanospirillum lungatii, Appl. Environ. Microbiol.*, 34, 371 (1977).

193. L. Baresi *et al.*, Methanogenesis from Acetate : Enrichment Studies, *Appl. Environ. Microbiol.*, 36(1), 186 (1978).
194. L. van den Berg *et al.*, Factors Affecting the Rate of Methane Formation from Acetic Acid by Enriched Methanogenic Cultures, *Can. J. Microbiol.*, 22, 1312 (1976).
195. D.O. Mounfortetal, Fermentation of Cellulose to Methane and Carbon Dioxide by a Rumen Anaerobic Fungus in Triculture with *Methanobrevibacter* sp. strain RAI and *Methanosarcina barkeri, Appl. Environ. Microbiol.*, 44(1), 128 (1982).
196. P.H. Smith, Studies of Methanogenic Bacteria in Sludge, U.S. EPA 600/2-80-093, 1980.
197. R.E. Speece *et al.*, Nickel Stimulation of Anaerobic Digestion, *Water Res.*, 17(6), 677 (1983).
198. R.A. Mah, Interaction of Methanogens and Non-Methanogens in Microbial Ecology, Proceedings of the 3rd International Symposium on Anaerobic Digestion, Boston, M. A. (Aug. 1983).
199. M. Harvey *et al.*, Mechanism of Methanogen Retention in an Anaerobic Reactor with a Polyster Fixed Film, Poster 0-10, presented at the 84th Annual Meeting of the American Society for Microbiology, St. Louis, MO, (Mar. 1984).
200. R. Robinson *et al.*, Light and Electron Microscopic Examinations of Methane-Producing Biofilm from Sewage Digesters, Poster 1-82, Presented at the 84th Annual Meeting of the America Society for Microbiology, St. Louis, MO, (Mar. 1984).
201. M.G. Hitton *et al.*, Methanogenesis from Acetate by a Mesophilic Anaerobic Pilter, *Process Biochem.*, p.2, (Dec. 1983).
202. T. Noike *et al.*, Characteristics of Carbohydrate Degradation and the Rate-Limiting Step of Anaerobic Digestion, *Biotechnol. Bioeng.*, 27(10) 1482 (1985).
203. H.E. Babbitt and E.R. Baumann, *Sewage and Sewage Treatment* John Wiley, London, p. 588, (1958).
204. M.J. Hammer *et al.*, Dialysis Separation of Sewage sludge Digestion, *J, Am. Soc. Civil Eng.*, SA5, 907 (1968).
205. H.A. Dirasian *et al.*, Electrode Potentials Developed During Sludge Digestion, *J. Water Pollut. Contr. Fed.*, 35, 424 (1963).
206. G.A. Nikitin, Change in Oxidation-Reduction Potential During Fermentation in Different Media by Batch Cultures of Methane-Forming Bacteria, *Microbiol.*, 4(5), 510 (1958).
207. F.F. Blanc and A.H. Molof, Electrode Potential Monitoring and Electrolytic Control in Anaerobic Digestion, *J. Water Pollut. Contr.* 45, 655 (1973).
208. S. Ghosh *et al.*, Anaerobic Acidogenesis of Sewage Sludge *J. Water Pollut. Contr. Fed,* 47, 1 (1975).
209. E.S.K. Chian and F.B. DeWalle, Treatment of High Strength Acidic Wastewater with a Completely Mixed Anaerobic Filter, *Water Res.*, II, 295 (1976).
210. S. Ghosh and M.P. Henry, Stabilization and Gasification of Soft Drink Manufacturing Waste by Conventional and Two Phase Anaerobic Digestion, Proceedings of the 36th Purdue Industrial Wastes Conference, Lafayette, IN, p292. (1981).
211. J.A. Eastman and J.F. Ferguson, Solubilization of Particulate Organic Carbon During the Acid Phase of Anaerobic Digestion, *J. Water Pollut. Contr. Fed.*, 53, 523 (1981).

212. S. Ghosh *et al.*, Two-Stage Upflow Anaerobic Digestion of Concentrated Sludge, Proceedings of the 5th Symposium on Biotechnology for Fuels and Chemicals, Gatlinburg, TN, (May 1983b).

213. J.D. Keenan, Multiple Staged Methane Recovery from Solid Wastes, *J. Environ. Sci. Health* A11(8/9), 525 (1976).

214. R.J. Zoetemeyer, Acidogenesis of Soluble Carbohydrate Containing Wastewaters, Ph.D. Thesis Univ. Amsterdam, Amsterdam, The Netherlands, 1982.

215. A. Bories, Fermentation Methanique avec Separation des Phases Acidogene et Methanogene Applique and Treatment des Effluents a Forts Charge Follunate (Distilleries), *Ann. Technol. Agricole,* 29, 509 (1980).

216. S. Ghosh *et al.*, Methane Production from Industrial Waste by Two-Phase Anaerobic Digestion, in Proceedings of the 6th Symposium on Energy from Biomass and Wastes, Orlando, FL, (Jan. 1983a).

217. J. Normann and B. Frostell, Anaerobic Wastewater Treatment in a two-Stage p. 387, (1977).

218. B.A. Rijkens, A Novel Two-Step Process for the Anaerobic Digestion of Solid Wastes, Paper presented at the 5th Symposium on Energy from Biomass and Wastes, Orlando, FL, (Jan. 1981).

219. R.K Datta, Acigogenic Fermentation of Corn Stover, *Biotechnol. Bioengg.*, 23, 61 (1981).

220. E. Colleran *et al.*, One and Two-Stage Anaerobic Digestion of Agricultural Wastes, in Proceedings of the 3rd International Symposium on Anaerobic Digestion, Boston, MA, (Aug. 1983).

221. E.P. Willimon, Jr. and J.F. Andrews, Multi-stage Biological Processes for Waste Treatment, Proceedings of the 22nd Purdue Industrial Wastes Conference, West Lafayette, IN, p. 645, (May 1967).

222. P.M. Heertjes and R.R. Va der Meer, Comparison of Different Methods for Anaerobic Treatment of Dilute Wastewaters, Proceedings of the 34th Purdue Industrial Wastes Conference, p790, (1979).

223. A. Cohen *et al.*, Anaerobic Digestion of Glucose with Separated Acid Production and Methane Formation, *Water Res.*, 13, 571 (1979).

224. A. Cohen, Optimization of Anaerobic Digestion of Soluble Carbohydrate Containing Wastewaters by Phase Separation, Ph. D. Thesis, Univ. Amsterdam, The Netherlands, (1982).

225. S.R. Harpes and F.G. Pohland, Enhancement of Anaerobic Treatment Stability Through Process Modification, Presented at the 58th Annual Meeting of the Water Pollution Control Federation, Kansas City, MO, (October 1985).

226. Lowenthal, R., Dold, P. and Marais : An Hypothesis for Pellitization in Upflow UASB Reactor Processes, Proc. of Symposium - Anaerobic Digestion, Bioemfontein, South Africa, (Sept., 21/24, 1986).

227. Mclnerney, M.J. And Bryant, M.P., Fuel Gas Production from Biomass, Vol. I, p19-46 D.L. Wise (ed.) CRC Press, Inc., Boca Raton, Florida, (1981).

228. Smith, P.H. Report No. EPA-600/2-80-093. U.S. Environmental Protection Agency, Cincinnati, Ohio, (1980).

229. Wolin, M.J. And Miller, T.L., ASM News 48:561-565, (1982).

230. Boone, D.R., Biotechnological Advance in Processing Municipal Wastes for Fuels and Chemicals, p85-96. A.A. Antonpoulos (ed.), ANL/CNSV-TM-167, Argone National Laboratory, Argonne, Illinois, (1985).

231. McInerney, M.J., Bryant, M.P., Hespell, R.B. and Consterton, J.W. *Appl. Environ. Microbiol.* 41:1029-1039, (1981).

232. Boone, D.R. and McInerney, M.J., Abstr. Annu. Meet. Am. Soc. Microbiol., 40, p636-632 (1986).

233. Beaty, P.S. and McInerney, M.J., Abstr Annu. Meet. Am. Soc., Microbiol., I-132, p.187(1986).

234. Roy, F., Albagnac, G. and Samain, E., *Appl. Environ. Microbiol.* 49:702-705).

235. Mountfort, D.O., Brulla, W.J., Krumholz, L.R. And Bryant, M.P. (1984). *Int. J. Syst. Bacteriol.,* 34p 216-217, (1984).

236. Henson, J.M. and Smith, P.H. *Appl. Environ, Microbiol.,* 49, p1461-1466, (1985).

237. Zinder, S.H. and Koch, M., *Arch. Microbiol.,* 138, p. 263-272, (1984).

238. Balch, W.E. and Wolfe, R.S., *J. Bacteriol.,* 137, p. 264-273, (1979).

239. Robinson, J.A. and Tiedje, J.M., *Arch. Microbiol.,* 1,37 p. 26-32, (1984).

240. Kaspar, H.F. and Wuhrmann, K., *Appl. Environ, Microbiol.* 36 p. 1-7, (1978).

241. Smith, P.H. and Mah, R.A., *Apple Microbiol.,* 14p 368-371, (1966).

242. Smith M.R., Z. Zinder, S.H. and Mah. R.A., Process Biochen, 15 p. 39-39, (1980).

243. Mah, R.A. (1980) *Curr. Microbiol.,* 3p 321-326, (1966).

244. Huser, B.A., Wuhrmann, K. and Zehrinder A.J.B. Arch. Microbiol, 132 p. 1-9, (1982).

245. Zinder, S.H. *ASM News* 50 p 294-298, (1984).

246. Krzcki, J.A. and Zeikus, J.G. Biotechnological Arvances in Processing Municipal Wastes for Fuels and Chemicals, p15-34, (1985).

A.A. Antonopoulos (ed.), ANL/CNWS-TM-167, Argonne National Laboratory. Argonne, Illinois.

247. Mah, R.A., Third International Symposium on Anaerobic Digestion-Proceedings, p. 13, Massachusetts, (1983).

248. Patel, G.B., *Can. J. Microbiol.,* 30p 1383-1396, (1984).

249. Zinder, S.H., Cardwell, S.C., Anguish, T. Lee, M. and Koch, M., *Appl. Environ. Microbiol.,* 47p 796-807, (1984).

250. Nozhevnikova, A.N. and Yagodina, T.G., *Microbiology* 51p 534-541, (1982).

251. Zinder, S.H. and Mah, R.A., *Appl. Environ. Microbiol.,* 38p 96-1008, (1979).

252. Zinder, S.H., Sowers, K.R. and Ferry, J.G. *Int. J. Syst. Bacteriol.* 35p 522-523, (1985).

253. Archer, D.B., *Enzyme Microb. Technol.* 5p 162-170, (1983).

254. Scott, R.T., Williams, T.N., Whitmore, T.N. and Lloyd, D., *Eur. J. Appl. Microbiol. Biotechnol.* 18p 236-241, (1983).

255. Smith, P.H., Personal communication. University of Florida, Gainesville, Florida, (1986).

256. Iannotti, E.L., Mueller, R.E. Sievers, D.M., Georgacakis, D.H. And Gerhardt, K.O., *J. Ind. Microbiol.* 1p 57-61, (1986).

257. Dolfing, J., Griffioen, A., Van Neerven A.R.W. And Zevenhuizen, L.P.T.M., *Can. J. Microbiol* 31p 744-750, (1985).

258. Hutschemackers, J., Delafontanie, M., Naveau, H.P. And Nyns, E.-J., *Biomass* 2p 115-125, (1982).
259. Reynolds, P.J., Ph.D. Thesis. National University of Ireland, Dubin, Ireland, (1986).
260. Zabranska, J. Schneiderova, K. and Dohanyos, M., *Biotechnol. Lett.* 7p 547-522, (1985).
261. van Bruggen, J.J.A., Goosen, N.K. and Stumm, C., Antonie van *Leeuwenhoek* 50p 80-81, (1984).
262. Dolfing, J. and Mulder, J.W., *Appl. Environ. Microbiol.* 49p 1142-1145, (1985).
263. Baresi, L. and Wolfe, R.S., *Appl. Environ. Microbiol.* 41p 388-391, (1981).
264. Eirich, L.D., Vogels, G.D. And Wolfe, R.S., *J. Bacteriol.* 140p 20-27, (1979).
265. Reynolds, P.J. and Colleran, E., Anaerobic Treatment A Grown-Up Technology, p. 515-531, Industrial Presentations (Europe) B.V., Schiedam, The Netherlands, (1986).
266. Tornabene, T.G. and Langworthy, T.A., *Science* 203p 51-53, (1979).
267. Henson, J.M., Smith, P.H. and White, D.C., *Apple. Environ. Microbiol.* 50p 1428-1433, (1985).
268. White, D.C., Microbes in Their Natural Environments, p. 37-66. J.H. Slater, R.Whittenbury and J.W.T. Wimpenny (ed.), Cambridge University Press, Cambridge., England, (1983).
269. Dolfing, J. and Bloemen, W.G.B.M., *J Microbiol. Methods* 4p 1-12, (1982).
270. Verrier, D and Albagnac, G., Energy from Biomass, p. 537-541. W. Palz, J. Coombs and D.O. Hall (ed.), Elsevier Applied Science Publisher, London, (1985).
271. Switzenbaum, M.S., Report No. Env. E. 90-86-1. Department of Civil Engineering, University of Massachusetts, Amherst, Massachusetts, (1986).
272. Verrier, D., Personal communication. INRA, Villeneuve d'Asoq, France.
273. Robinson, R.W., Akin, D.E., Nordtedt, R.A., Thomas, M.V. and Aldrich, H.C., *Apple Environ. Microbiol.*, 48p 127-136, (1984).
274. Hulshoff Po, L.W., de Zeeuw, Q., Dolfing, J. and Lettinga, G., Anaerobic Wastewater Treatment p40-43. W.J. van den Brink (ed.), TNO, The Hague, The Netherlands, (1983).
275. Lettinga, G., Hulshoff Pol, L.W., Hobma, S.W., Grin, P., de Jong, P., Roersma, R. and IJspeert, P., Anaerobic Wastewater Treatment, p411-429. W.J. van den Brin (ed.), TNO, The Hague, The Netherlands, (1983).
276. Lettinga, G., de Zeeuw,, Hulshoff Pol, L., Wiegant, W. And Rinzema, A., Anaerobic Digestion, p279-301. The China State Biogas Association, Beijing, China, (1985).
277. Hulshoff Pol, L.W., de Zeeuw, W.J., Velzeboer, C.T.M. and Lettinga, G., *Wat. Sci. Tech.* 15(8/9)p 291-304, (1983).
278. Brunetti, A., Boari, G., Passino, R. and Rozzi, A., Anaerobic Wastewater Treatment, p317-334. W.J. van den Brink (ed.), TNO, The Hague, The Netherlands, (1983).

ADDITIONAL REFERENCES

1. Anaerobic contact filter for treatment of waste sulphite liquor. T.W. Beak consultants Ltd., Montreal, Quebec. (Report 103-1) (1973).
2. Anderson, G.K., And A.C. Duarte. Research and Application of anaerobic processes. Environmental Technology Letters, 1, 484-493. (1980).

3. Anderson, G.K., T., Donelly, and D.J. Letten, Anaerobic treatment of high-strength industrial wastewaters. In K. Curi (Ed.), (1980a).
4. Treatment and disposal of liquid and solid industrial wastes. Pergamon Press, Oxford, UK.
5. Andrews, J.K., and E.A. Pearson. Kinetics and characteristics of volatile acid production in anaerobic fermentation processes. Int. J. Air and Water Pollut., 9, 439-461., (1965).
6. Andrews, J.F. Dynamic model of the anaerobic digestion process. J. San, Engng, Div., Proc. ASCE, 95, 95-16., (1969).
7. Andrews, J.F., and S.P. Graef. Dynamic modeling and simulation of the anaerobic digestion process. In Anaerobic biological treatment process. Advances in Chemistry series 105. ACS, Washington, D.C. pp. 126-162., (1971).
8. Antonie, R.L. Evaluation of a rotating disk wastewater treatment plant. Water Pollution Control, 46, 498., (1974).
9. Antonie, R.L. Fixed biological surfaces - wastewater treatment. CRC Press Inc., Cleveland, Ohio., (1976).
10. Aris, R. (1975). The mathematical theory of diffusion and reaction in permeable catalyst, vol. 1. Clarendon Press, Oxford, UK., (1975).
11. Arora, H.C., S.N. Chattopadhya, and T. Routh. Treatment of vegetable tanning effluent by the anaerobic contact filter process. Water Pollution Control, 74, 584-596., (1975).
12. Arora, H.C., and T. Routh. Treatment of slaughterhouse effluents by anaerobic contact filter. IAWPC Tech. Annual IV and VII, 67-78, (1980).
13. Arvin, E., and G.H. Kristensen, Precipitation of calcium phosphate and pH-effects in denitrifying biofilms. Vatten, 35, 329-337, (1979).
14. Arvin, E., And G.H. Kristensen. Effect of denitrification on the pH in biofilms. Wat. Sci. Tech., 14, 833-848, (1982).
15. Asinari, C.M. di San Marzano, H.P. Naveau, and E.J. Nyns. Production of methane from freshwater macro-slage by an anaerobic two step digestion system. In Energy from biomass, W. Palz, p. Chartier, and D.O. Hall (eds.). Applied Science Publishers, London, (1981).
16. Atkinson, B. And others. Kinetics, mass transfer, organism growth in biological film reactor. Trans. Inst. Chem. engrs., 45, T257-264 (1967).
17. Balch, W.E. And others, Methanogens : Revaluations of a unique biological group. Microbiological Reviews, 43, 260-296 (1979).
18. Barbara, M.A., and R.F. Gasser. Hy-flo fluidized bed treatment of a bottling waste. Presented at Wastewater Equipment Manufacturers Association, Meeting, June, Philadelphia. (in Hall, 1980).
19. Bauchop, T., and S.R. Elsden. The growth of micro-organisms in relation to their energy supply. J. Gen. Microbiol., 23, 457-469 (1960).
20. Bell, B.A., H. Jeris, and J.M. Welday. Anaerobic fluidizied bed treatment of thermal sludge conditioning decant liquor. In Proceedings of the seminar/workshop : Anaerobic filters : An energy plus for wastewater treatment, p 171-178. Argonne National Laboratory, Argonne, Illinois. (Rep. ANL/CNSV-TM-50 (1981).
21. Benjamin, M.M., J.F. Ferguson, and M.E. Buggins. Treatment of sulfite evaporator condensate with an anaerobic reactor. In TAPPI environmental conference, April 29, New Orleans. p1-10 (1981).

22. Benjamin, M.M., S.L. Woods, and J.F. Ferguson, Toxicity and biodegradability of pulp mill waste constituents to anaerobic bacteria. Presented at IAWPR-specialised seminar on anaerobic treatment June, Copenhagen, Denmark (1982).
23. Binot, R.A., H.P. Naveau, and E.J. Nyns, Biomethanation of soluble organic residues by immobilised fluidised cells. Presented at Second international symposium on anaerobic digestion, 6-11 September, Travemunde, Germany (1981).
24. Boone, D.R. And M.P. Bryant. Propionate-degrading bacterium syntrophobacter wolinii sp. nov. gen. nov., from methanogenic eco systems. Applied and Environmental Microbiology, 40, 626-632 (1980).
25. Bories, A. Fermentation methanique avec separation des phases acidogene et methanogene appliquee au traitment des effluents a forte charge polluante (distilleries). Annales de Technologie Agricole, 29, 509-528 (1981).
26. Bories, A. Methanisation des eaux residuaires de distilleries. Presented at Theme journee internationales du CEBEDEAU "La methanisation des residue organique". 25-27 Mai, Liege, Belgique (1981).
27. Braum, R. Anaerobic filter treatment of molasses distillery slops. Presented at Second International Symposium on Anaerobic Digestion, 6-11 September, Travemunde, Germany. (1981).
28. Brown, G.J., and others. Sludge accumulation in an anaerobic lagoon-anaerobic filter system treating potato processing waste-water. In Proceeding of the 35th industrial waste conference, May 13, 14 and 15, 1980, Purdue University, Lafayette, Indiana. Ann Arbor Science Publishers Inc., Ann Arbor, Mich. pp. 610-616.
29. Buhr, H.O., and J.F. Andrews. Review paper : The thermophilic anaerobic digestion process. Water Res., 11, 129-143 (1977).
30. Cappenberg., T.E. A study of mixed continuous cultures of sulphate-reducing and methane producing bacterias, Microbiol Ecol., 2, 60-72 (in Zehnder, 1978) (1975).
31. Carrondo, M.J.T., and others. Anaerobic filter treatment of molasses fermentation wastewater. Presented at IAWPR-specialised seminar on anaerobic treatment, June, Copenhagen, Denmark (1982).
32. Cecil, D. Vakstinetik and anaerobic brydning. (Growth kinetics of anaerobic digestion) (in Danish). Aalborg University, Denmark, (1981).
33. Chappel, K. Anaerobic digestion of synthetic feeds. Presented at Second International Symposium on Anaerobic Digestion, 6-11 September, Travemunde, Germany (1981).
34. Chian, E.S.K., and F.B. de Walle. Treatment of high strength acidic wastewater with a completely mixed anaerobic filter. Water Res., 11, 295-304 (1977).
35. Chin, K.K. Anaerobic treatment kinetics of palm oil sludge. Water Res., 15, 199-202 (1981).
36. Chou, W.L., and others. The effect of petrochemical structure on methane fermentation toxicity. Prog. Wat. Tech., 10, (5/6), 545-558 (1978).
37. Chou, W.L., R.E. Speece, and R.H. Siddiqi. Acclimation and degradation of petrochemical wastewater components by methane fermentation. Biotechnology and Bioengineering Symposium No. 8, 1978. John Wiley and Sons, London, UK. pp. 391-414, (1979).
38. Clark, R.H., and R.E. Speece. The pH tolerance of anaerobic digestion In Advances in water pollution research. Proceedings of the fifth international conference hold in

San Francisco and Hawaii 1970. vol. 1. Pergamon Press, Oxford, UK. ppH-27/1-14 (1971).

39. Colleran, E.A. Booth, M. Barry, A. Wilkie, P. J : Newell, and L.K. Dunican. In Energy from biomass., W. Palz, P. Chartier, and D.O. Hall (eds.), p 416-422. Applied Science Publishers London. (1981).
40. Colleran, E.M. Barry, A. Wilkie, and P.J. Newell. Anaerobic Digestion of agricultural wastes using the upflow anaerobic filter design. Process Biochemistry, (March/April), 12-70 + 41. (1982).
41. Colvin, J.R., L.C. Sowden, and L. van den Berg. The ultrastructure of the major species of an enriched methanognic culture utilizing acetic acid. Canadian Journal of Microbiology. 25, 826-832 (1979).
42. Coulter, J.B., S. Soneda, and M.B. Ettinger. Anaerobic contact process from sewage disposal. Sewage and Ind. Wastes, 29, 468-477 (1957).
43. Cross, W. H. Anaerobic activated carbon filters for the treatment of wastewaters containing phenol. In Proceedings of the seminar/workshop : Anaerobic filters : An energy plus for wastewater treatment. Argonne National Laboratory, Argonne, Illinois. (Rep. ANL/CNSV-TM-50) (1981).
44. Dahab, M.F., and J.C. Young. Energy recovery from alcohol stillage using anaerobic filters. Presented at the 3rd Symposium on Biotechnology in energy production and conservation, May, Gatlinburg, Tennesse.
45. De Bekker, P.H.A.M.J. Anaerobic treatment of leachate from controlled tips of municipal solid waste. In EAS-ISWA Symposium 81. 22-26 June, Munich, Germany. Gesellschaft zur Forderung der Abwassertechnik e. V (GFA), St. Augustin, Germany. pp. 621-639 (1981).
46. De Bekker, P.H.A.M.J. Anaerobic treatment of waste water of a soft insulating board industry. Presented at the Second International Symposium on Anaerobic Digestion, 6-11 September, Travemunde, (1981a).
47. Demuynck, M.G. Anaerobic fermentation : Environmental biotechnology - furture prospects, Presented at FAST Environmental Biotechnology Workshop, Versailles.
48. Dennis, N.D., and J.C. Jannett. Pharmaceutical waste treatment with an anaerobic filter. In Proceedings of the 29th industrial waste conference, May 7, 8, and 9; 1974 Purdue University, Lafayette, Indiana. pp. 36-43 (1974).
49. De Walle, F.B., and E.S.K. Chian. Kinetics of substrate removal in a completely mixed anaerobic filter. Biotechnology and Bioengineering, 18, 1275-1295 (1976).
50. De Walle, F.B., J.C. Kennedy, T. Zeising, R. Seabloom. Treatment of domestic sewage with the anaerobic filter. In Proceedings of the seminar/workshop : Anaerobic filters: An energy plus for wastewater treatment, p. 143-157. Argonne-National Laboratory, Argonne, Illinois. (Rep. ANL/CNSV-TM-50) (1981).
51. De Zeeuw, W., and G. Lettinga. Acclimation of digested sewage sludge during start-up of an upflow anaerobic sludge blanker (UASB) reactor. In Proceedings of the 35th industrial waste conference, May 13, 14 and 15, 1980, Purdue University, Lafayette, Indiana. Ann Arbor Science, Ann Arbor, Mich. pp39-47. (1980).
52. Donnelly, T. Industrial effluent treatment with the bioenergy process. Process Biochemistry, 13, (6), 14-16 (1978).
53. Donovan, E. J. Jr. Treatment of high strength wastes by an anaerobic filter. In Anaerobic of the seminar/workshop, January-9-10, 1980. Howey-In-The-Hills, Florida, Argonne National Laboratory, Argonne, Ill., USA, 1981. pp. 170-198 (1980).

54. Dunn, I. J. Tech.-Chem. Lab. ETH, Schweiz. Personal communication (1981).

55. Eastman, J.A., and J.F. Ferguson. Solubilization of particulate organic carbon during the acid phase of anaerobic digestion. Journal WCPF, 53, 352-366 (1981).

56. Eggers, E. Improvement of the solid-liquid-gas separation efficiency of the anaerobic upflow column by a floating filter. Presented at Second International Symposium on Anaerobic Digestion, 6-11 September, Travemunde, Germany.

57. Elmaleh, S. Epuration et eneergie, Trib. Cebedeau, 34, 83-92 (1981).

58. El-Shafie, A. T., and D.E. Bloodgood, Anaerobic treatment in a multiple upflow filter system. Journal WPCF, 45, 2345-2357, (1973).

59. Ferguson, J.F., B.J. Eis, and M.M. Benjamin. Neutralization in anaerobic treatment of an acidic waste. Presented at IAWPR-specialised on anaerobic treatment, June, Copenhagen, Denmark, (1982).

60. Friedman, A.A., and S.J. Tait. Energy recovery from anaerobic rotating biological contractor (AnRBC) treating high strength carbonaceous wastewaters. In D. Smith and others (Eds.), Ist National Symposium/workshop on rotating biological contractor technology, Proceedings, June, Champion, Pennsylvanis, Vol. 1. pp. 759-789, (1980).

61. Friedman, A.A., K.S. Young D.G. Bailey, and S.J. Tait. New observations with anaerobic fixed film reactors. In Proceedings of the seminar/workshop : Anaerobic filters : An energy plus for wastewater treatment. Argonne National Laboratory, Argonne, Illinois, (Rep. ANL/CNSV-TM50) (1981).

62. Frostel, B. Wastewater treatment and energy recovery in an expanded bed system. Presented at 3rd International congress on industrial waste water and wastes, February, Stockholm, Sweden (1980).

63. Frostel, B. Pilot scale anaerobic-aerobic biological treatment of distillery waste. Chemistry and Industry, 465-469 (1981).

64. Frostel, B. Anaerobic treatment in a sludge and bed system compared with a filter system. Journal WPCF, 53, 216-222 (1981a).

65. Genug, R.K., W.W. Pitt, G.M. Davis, and J.H. Koon. Development of a wastewater treatment system based on a fixed-film, anaerobic bioreactor. In Proceedings of the seminar/workshop : Anaerobic filters : An energy plus for wastewater treatment, p.199-243. Argonne National Laboratory. Argonne Illinois. (Rep. ANL/CNSV-TM 50) (1981).

66. Ghosh, S., and F.G. Pohland. Kinetics of substrate assimilation and product formation in anaerobic digestion. Journal WPCF, 46, 748-758, (1979).

67. Ghosh, S. and others. Anaerobic acidogenesis of wastewater sludge, Journal WPCF, 47 30-45 (1975).

68. Ghosh, S., and D.L. Klass, Two-phase anaerobic digestion. Process Biochemistry. 13, 16 (1978).

69. Goedvriend, G.J. The application of anaerobic treatment of fruit and vegetable waste-water. Presented at Second international symposium on anaerobic digestion, 6-11 September, Travemunde, Germany (1981).

70. Gujet, W., and A.J.B. Zehnder. Conversion processes in anaerobic digestion. Presented at IAWPR-seminar on anaerobic treatment, June, Copenhagen, Denmark. (1982).

71. Gonenc, E. Nitrification in rotating disc reactors. Department of Sanitary Engineering. Technical University of Denmark, Lyngby. (1982).

72. Hakulinen, R., M.S. Salkinoja-Salonen, and M.L. Saxelin. Purification of kraft bleach effluent by an anaerobic fluidized bed reactor and aerobic trikling filter. In Proceedings of TAPPI Environmental Conference, New Orleans, USA, pp197-203 (1981).

73. Hakulinen, R., and M. Salkinoja-Salonen. Rening av Cellulosaindustrins blekeria vlopps vatten : Jamforelse mellan 5 biologiska reakorer. (Treatment of kraft bleach effluents : Investigations with 5 different biological reactors) (in Swedish). NORDFORSK, Helsingfors, Finland, (Publication 1981:1).

74. Hakulinen, R., and M. Salkinoja-Salonen. An anaerobic bed reactor for the treatment of industrial wastewater containing chlorophenols. In P.F. Cooper and B. Atkinson (Eds.), Biological fluidised bed treatment of water and wastewater. Ellis Horwood Ltd., Chichester, UK. (1981b).

75. Hakulinen, R., and Salkinoja-Salonen, Treatment of pulp and paper industry wastewaters in an anaerobic fluidised bed reactor. Process Biochemistry, (1982).

76. Hakulinen, R., and M. Salkinoja-Salonen. Treatment of kraft bleaching effluents : Comparison of results contained by enso-fenox and alternative methods. In Proceedings of the Technical Association of the Pulp and Paper Industry 1982 international pulp bleaching conference. Tappi Press, Atlants, USA. (1982a).

77. Hakulinen, R., and M. Salkinoja-Solonen, Treatment of pulp and paper industry wastewaters in an anaerobic fluidised bed reactor. Process Biochemistry, 17, (march/April), 18-22 (1982b).

78. Hall, E.R. Recent developments in high rate anaerobic processes. Presented at CIC-CSChE Symposium on new technology for pollution abatement, December, Toronto, Canada, (1980).

79. Hall, E.R. Evaluation of alternative anaerobic systems for the treatment of high strength wastewaters. Presented at Second international symposium on anaerobic digestion, 6-11 September, Travemunde, Germany (1981).

80. Hall, E.R., B.E. Jank, and M. Jovanovic. Energy production from high strength industrial wastewater. Presented at 3rd Bioenergy R and D seminar, March, Ottava, Canada. (1981a).

81. Hall, E.R. Biomass retention and mixing characteristics in fixed-film and suspended growth anaerobic reactors. Presented at IAWPR-specialised seminar on anaerobic treatment June, Copenhagen, Denmark (1982).

82. Hanaki, K., T. Matsuo, and M. Nagase, Mechanism of inhibition caused by long-chain fatty acids in anaerobic digestion process. Biotechnology and Bioengineering, 23, 1591-1610 (1981).

83. Harremoes, P. The significance of pore diffusion to filter denitrification Journal WPCF, 48, 377-388 (1976).

84. Harremoes, P. Biofilm kinetics. In R. Mitchell (Ed.), Water pollution microbiology, vol. 2. John Wiley and Sons, Chichester, UK. Chap. 4 pp. 71.109 (1978).

85. Harremoes, P., J. la Cour Jansen, and G.H. Kristensen. Practical problems related to nitrogen bubble formation in fixed film reactors. Prog. Wat. Techn., 12, 253-269 (1980).

86. Harremoes, P., I. Sekoulov, and L. Bonomo. Design of fixed film nitrification and denitrification units based on laboratory and pilot scale results. In EAS-ISWA Symposium 81. June, Munich, Germany Gesellschaft zur Forderung der Abwassertechnik e. V. (GFA), St. Augustin, Germany pp423-451 (1981).

87. Harremoes, P. Criteria for nitrification in fixed film reactor. Wat. Sci. Tech., 14, 167-187 (1982).

88. Hartmann, L.W. Anaerob spildevandsbehandling. (Anaerobic treatment of waste water) (in Danish). M.Sc. Thesis, Department of Sanitary Engineering, Technical University of Denmark, Lyngby (1981).

89. Heertjes, P. M., L.J. Kuijvenhoven, and R.R. van der Meer. Fluid flow pattern in upflow reactors for anaerobic treatment of beet sugar factory wastewater. Biotechnology and Bioengineering, 24, 443-459 (1982).

90. Henze, M. Gas production by anaerobic digestion of liquid farm manure and sewage sludges. In Elmia-Avfall 79. Conferences in connection with the international trade fair on recycling, refuse, collection and waste treatment "Elmia-Avfall 79," September 17-21, Jonkoping, Sweden. p324-334 (1979).

91. Henze, M. Sewage treatment by activated sludge - A model with emphasis on oxygen consumption and sewage composition. Prog. Wat. Tech., Suppl. 1, 41-60 (1979a).

92. Henze, M. Anaerobic fermentation of agricultural and industrial wastes. Presented at FAST Environmental Biotechnology Workshop Versailles. (1982).

93. Hickey, R.F., and R.W. Owens. Methane generation from high strength industrial wastes with the anaerobic biological fluidized bed,. Presented at 3rd Symposium on biotechnology in energy production and conservation, May, Gatlinburg, Tennessee. (in Switzenbaum, 1982 (1981).

94. Hoban, D.J. And L. van den Berg. Effect of iron on conversion of acetic acid to methane during methanogenic fermentations. Journal of Applied Bacteriology, 47, 153-159 (1979).

95. Hovious, J.C., J.A. Fisher, and R.A. Conway. Anaerobic treatment of synthetic organic wastes. EPA Report, Project 12020 DIS, January, (In Mueller, 1975) (1972).

96. Hudson, J.W., F.G. Pohland, and R.P. Pendergrass. Anaerobic packed bed treatment of shellfish processing wastewaters. Proceedings of the 33rd industrial waste conference, May 9, 10 and 1, 1978, Purdue University, Lafayette, Indiana, Ann Arbor Science, Ann Arbor, Mich. 1979. pp560-574 (1979).

97. Hakansson, H., B. Frostell, and J. Norrman. Anaerobic treatment of whey and whey permeate in submerged filters. Institute for vatten-ochLuftvards for skinning, Stockholm, Sweden, Publ. B, 406 (1977).

98. Jans, A.J.M. Detoxification and anaerobic treatment of formaldehyde containing wastes. Presented at Second international symposium on anaerobic digestion, 6-11 September, Travemunde, Germany (1981).

99. Jenkins, B.G.M., D.A. Stafford, and J. Quickenden. The fluidised bed digester (attached film) - for treatment of dairy waste. Presented at Second international symposium on anaerobic digestion, 6-11 September, Travemunde, Germany (1981).

100. Jennett, J.C., and M.C. Rand. A comparison of anaerobic Vs. aerobic treatment of pharmaceutical waste. In Anaerobic filters : An energy plus for waste-water treatment. Proceedings of the seminar/workshop, January 9-10, 1980. Howey-In-The-Hills, Florida. Argonne National Laboratory, Argonne, Ill. pp. 77-93 (1980).

101. Jeris, J.S., and P.L. McCarty. The biochemistry of methane fermentation using C^{14} tracers. Journal WPCF, 37, 178-192 (1965).

102. Jeris, J.S., and R.W. Owens. Pilot-scale, high-rate biological denitrification. Journal WPCF, 47, 2043-2057 (1975).

103. Jeris, J.S. Industrial wastewater treatment using anaerobic fluidized bed reactors. Presented at IAWPR-seminar on anaerobic treatment, June, Copenhagen, Denmark. (1982).

104. Jewell, W.J., H.R. Capener, S. Dell'orto, K.J. Fanfoni, T.D. Hayes, A.P. Leuschner, T.L. Miller, D.F. Sherman, P.J. van Soest, M.J. Wolin, and W.J. Wujcik, Anaerobic fermentation of agricultural residue : Potential for improvement and implementation. New York State College of Agriculture and Life Sciences, Cornell University, Ithaca, N.Y. (Rep. EY-76-S-02-2981-7) (1978).

105. Jewell, W.J. Development of the attached microbial film expanded bed process for aerobic and anaerobic waste treatment. Chapter 15 in Biological fluidized bed treatment England. (1980).

106. Jewell, W.J., and G. Schraa. Microalgae separation, concentration, and conversion to fuel with an anaerobic expanded bed reactor. U.S. Department of Energy. (Re. XB-9-82-63-1-3) (1981).

107. Jewell, W.J. Anaerobic attached film expanded bed fundamentals. Presented at "First international conference on fixed film biological processes." Kings Island, Ohio, April 20-23 (1982).

108. Jewell, W.J., and J.W. Morris. Influence of varying temperature, flow rate and substrate concentration on the anaerobic attached film expanded bed process. In Proceedings of the 36th industrial waste conference, May 1981, Purdue University, Lafayette, Indiana. Ann Arbor Science, Ann Arbor. Mich., 1982.

109. Johansen, O.J. Treatment of leachates from sanitary landfills. Ph. D. Report. University of Washington, Seattle (1975).

110. Jung, H. Schnellfaulung zur Behandlung stark organisch verschmutzter Abwasser. Vom Wasser, 17, 38-48 (1949).

111. Jorgensen, J.B. The anaerobic treatment of leachate. M.Sc. Thesis, University of Wiscosin, (1972).

112. Kandler, O., J. Winter, and U. Temper. Methane fermentation in the thermophilic range. In Energy from biomass, W. Plaz. P. Chartier, and D.O. Hall (eds.), p. 472-477. Applied Science, London (1981).

113. Kaspar, H. Ph. D. Thesis, Swiss Federal Institute of Technology, Nr. 5984. (in Zehnder, 1978), (1977).

114. Kennedy, K.J., and L. van den Berg. Effects of temperature and overloading on the performance of anaerobic fixed film reactors. Presented at the 36th industrial waste conference, Purdue University, Lafayette, Indiana (1981).

115. Kennedy, K.J., and L. van den. Berg. Use of fixed film reactors for production of methane from waste. Presented at the 3rd Bioenergy R and D seminar, March, Ottawa, Canada (1981a).

116. Kennedy, KJ., and L. van den Berg. Stability and performance of anaerobic fixed film reactors during hydraulic overloading at 10 to 35°C. Water Res., (1982).

117. Kennedy, K.J., and L. van den Berg. Anaerobic digestion of piggery waste using a stationary fixed film reactor. Agricultural Wastes, 4, 151-158 (1982a).

118. Kennedy, K.J., and L. van den Berg (1982b). Effect of height on there performance of anaerobic downflow stationary fixed film (DSFF) reactors treating hean bleaching waste. Presented at the 37th Purdue Industrial Waste Conference, Purdue University, Lafayette, Indiana (1982b).

119. Kennedy, K.J. and L. van den berg. Thermophilic downflow stationary fixed film reactors for methane production from bean blanching waste. Biotechnology Letters, 4, (3), 171-176 (1982c).

120. Kennedy, K.J., and L. van den Berg. Continuous vs. Slugloading of downflow stationary fixed film reactors digesting piggery waste. Biotechnology Letters, 4, (2), 137-142 (1982d).

121. Koon, J.H., and others. The feasibility of an anaerobic upflow fixed-film process for treating small sewage flows. In Proceedings of U.S. Department of Energy Optimization of water and wastewater for municipal and industrial applications conference, December, 1979, New Orleans, vol. 1, pp193. (in Switzenbaum, 1982a) (1980).

122. Kotze, J.P., P.G. Thiel, and W.H.J. Hattingh, Anaerobic digestion 2. The characterization and control of anaerobic digestion. Water Res., 3, 459-494 (1969).

123. La Motta, E.J. Internal diffusion and reaction in biological films Environmental Science and Technology, 10, 765-769 (1976).

124. Landine, R.C., and others. Anaerobic fermentation-filtration of potato processing wastewater. Journal WPCF, 54, 103-110 (1982).

125. Lawrence, A.L., and P.L. McCarty, Kinetics of methane fermentation in anaerobic treatment. Journal WPCF, 41, R1-R7 (1969).

126. Lettinga, G. Feasibility of anaerobic digestion for purification of industrial waters. In 4. Symposium EAS 1978, June, Munich, Germany, Gesellschaft zur Forderung der Abwassertechnik e. V. (GFA), St. Augustin, Germany, pp226-256 (1979).

127. Lettinga, G. and others. The application of anaerobic digestion to industrial pollution treatment. Presented at Ist International Symposium on Anaerobic Digestion, September, Cardiff, UK (1979).

128. Lettinga, G., and J.N. Vinken. Easibility of the upflow anaerobic blanket (UASB) process for the treatment of low strength wastes. In Proceedings of the 35th Industrial Waste Conference, May, 13, 14 and 15, 1980, Purdue University, Lafayette, Indiana, Ann Arbor Science, Ann Arbor, Mich., 1981. pp625-634 (1980).

129. Lettinga, G., and others. Use of the upflow sludge blanket reactor concept for biological wastewater treatment, especially for anaerobic treatment. Biotechnology and Bioengineering, 22, 699-734 (1980a).

130. Lettinga, G., W. De Zeeuw, and E. Ouborg (1981). Anaerobic treatment of wastes containing methanol and higher alcohols. Water Res., 15, 171-182 (1981).

131. Lettinga, G., and others. Design, operation and economy of anaerobic treatment. Presented at IAWPR-specialised seminar on anaerobic treatment, June, Copenhagen, Denmark (1982).

132. Li, A., and P.M. Sutton. Dorr-Oliver's anitron system : Fludized bed technology for methane production from dairy wastes. Presented at Whey Institute annual meeting, April, Chicago, Illinois (1981).

133. Li, A., P.M. Sutton, and J.J. Corrado. Energy recovery from pretreatment of industrial wastes in the anaerobic fludized bed process. Presented at First International Conference on Fixed Film Biological Processes, Kings Island, Ohio, April 20-23 (1982).

134. Lindauer, G.D., and H. Sixt. Application of recent developments in anaerobic waste water treatment to slaughterhouse wastes. In Energy from biomass, W. Plaz, P. Chartier, and D.O. Hall (eds) p440-447. Applied Science. London (1981).

135. Lindgren, M (1981). Anaerob avloppsvattenrening (Anaerobic treatment of wastewater) (in Swedish). Kungliga Tekniska, Hogskolan, Stockolm Sweden (1981).
136. Lindgren, M. (1982). Mathematical modeling of the anaerobic filter process. Presented at IAWPR-specialised seminar on anaerobic treatment, June, Copenhagen, Denmark.
137. Lovan, C.R., and E.G. Foree. The anaerobic filter for the treatment of brewery press liquor waste. In Proceedings of the 26th Industrial Waste Conference. May 4, 5, and 6, 1971. Purdue University Lafayette, Indiana, Part 2, pp1074-1086 (1971).
138. Matsche, N., and E. Ruider. Belebungsverfahren mit anaerobem Schwimmfilter als Vorstufe. O Oster, Wasserwirtschaft, 33, 408-20 (1981).
139. McCarty, P.L., and R.E. Mckinney. Volatile acid toxicity in anaerobic digestion. Journal WPCF, 33, 223-232 (1961).
140. McCarty, P.L., and R.E. McKinney. Salt Toxicity in anaerobic digestion. Journal WPCF, 33, 399-415 (1961a).
141. McCarty, P.L. Anaerobic waste treatment fundamentals, pt. 1-4. Public Works, 95, Sept. - Dec. (1964).
142. McCarty, P.L. Anaerobic treatment of soluble wastes. In W.W. Eckenfelder Jr. and Austin, Texas. (1968).
143. McCarty, P.L. Energetics of organic matter degradation. In R. Mitchell (Ed.), Water pollution microbiology. Wiley-Interscience, London, UK, 1972. pp91-118 (1971).
144. McCarty, P.L. Energetics and kinetics of anaerobic treatment. In Anaerobic biological treatment processes, Advances in chemistry series 105. ACS, Washington, D.C., pp.91-07 (1971a).
145. McCready, P.L., and others. Trace organics in groundwater. Environmental Science and Technology, 15, 40-51 (1981).
146. McCarty, D. Preliminary study of the hydrolyzation of insoluble organics and acid production during anaerobic respiration. Department of Sanitary Engineering, Technical University of Denmark, Lyngby (1978).
147. McInery, M.J., M.P. Bryant, and N. Pfenning. Anaerobic bacterium that degrades fatty acids in syntrophic associations with methanogens. Arch. Microbiol., 122, 129 (1979).
148. Morris, J.W., and W.J. Jewell. Organic particulate removal with the anaerobic attached film expanded bed process. In proceedings of the 36th Industrial Waste Conference, May 1981, Purdue, University, Lafayette, Indiana. Ann Arbor Science, Ann Arbor, Mich., 1982.
149. Mosey, F.E. Anaerobic filtration : A biological treatment process for warm industrial effluents. Wat Pollut. Control, 77, 370-378 (1978).
150. Mosey, F.E. Anaerobic biological treatment of food industry waste waters. Wat. Pollut. Control, 80, 273-291 (1981).
151. Mosey, F.E. Mathematical modelling of the anaerobic digestion process : Regulatory mechanisms for the formation of shortchain volatile acids from glucese. Presented at IAWPR-seminar on anaerobic treatment, June, Copenhagen, Denmark (1982).
152. Mosey, F.E. New developments in the anaerobic treatment of industrial wastes. Water Pollution Control, 81, 540-552 (1982a).
153. Mueller, J.A., and others Oxygen diffusion through zoogloel flocs. Biotechnology and Bioengineering, 10 331-358 (1968).

154. Mueller, J.A., and J.L. Mancini, Anaerobic filter kinetics and applications. In Proceedings of the 30th Industrial Conference May 6, 7, and 8, 1975; Purdue University, Lafayette, Indiana. Ann Arbor Science, Ann Arbor, Mich., 1977. pp423-447 (1975).

155. Murray, W.D., and L. van den Berg. Effect of support material on the development of microbial fixed films converting acetic acid to methane. Journal of Applied Bacteriology, 51, 257-265 (1977).

156. Murray, W.D., and L. van den Berg. Effects of nickel, cobalt, and molybdenum on performance of methanogenic fixed-film reactors. Applied and Environmental Microbiology, 42, 502-505 (1981a).

157. Martensson, L., and b. Frostell. Anaerobic wastewater treatment in a carrier assisted sludge bed reactor. Presented at IAWPR-Specialised Seminar on Anaerobic Treatment, June, Copenhagen, Denmark (1982).

158. Norrman, J. Anaerobic treatment of wastewaters from pulp and paper industries. Presented at Kemian Paivat, November 18-20, Helsinki, Finland (1981).

159. Norrma, J. Treatment of a black liquor condensate from a pulp mill in an anaerobic filter and an expanded bed. Presented at IAWPR-specialised seminar on anaerobic treatment, June, Copenhagen, Denmark (1982).

160. Novak, J.T., and D.A. Carlon. The kinetics of long chain fatty acid digestion. Journal WPCF, 42, 1932-1943 (1970).

161. O'Rourke, J.T. Kinetics of anaerobic waste treatment at reduced temperatures. Ph.D. Thesis, Stanford University, Stanford, USA. (in Zenhnder, Ingvorsen, Marti, 1981) (1968).

162. Pailthrop, R.E., G.A. Richter, and J.W. Filbert. Anaerobic secondary treatment of potato processing wastewater. Presented at 44th Annual Conference WPCF, San Francisco, California (1971).

163. Parkin, G.F., R.E. Speece, and C.H.J. Yang. A comparison of the response of methanogens to toxicants : Anaerobic filter vs. Suspended growth systems. In Proceedings of the seminar/workshop : Anaerobic filters : An energy plus for wastewater treatment, p37-57 : Argonne National Laboratory, Argonne. Illinois, (Rep. ANL/ CNSV-TM50) (1981).

164. Parkin, G.F., and R.E. Speece. Attached vs. suspended growth anaerobic reactors : Response to toxic substances. Presented at IAWPR-specialised seminar on anaerobic treatment, June, Copenhagen, Denmark (1982).

165. Parkin, G.F., and R.E. Speece. Modeling toxicity in methane fermentation systems. Journal of the Environmental Engineering Division, ASCE, 108, (EE3), 515-531 (1982a).

166. Petersen, J.H., and S. Fogh. Anaerob forgering at kulhydrater til metan (Anaerobic fermentation of carbohydrates to methane) (in Danish). Plantefysiologisk Institut, Kobenhavns Universitet, Copenhagen, Denmark (1980).

167. Pette, K.C., and others Full-scale anaerobic treatment of beetsugar wastewater. In Proceedings of the 35th industrial waste conference, May 13, 14 and 15; 1980, Purdue University, Lafayette, Indiana, Ann Arbor Science, An Arbor, Mich., 1981. pp635-642 (1980).

168. Pette, K.C., and A.I. Versprille. Application of the UASB-concept on wastewater treatment. Presented at Second international on anaerobic digestion, 6-11 September, Travemunde, Germany (1981).

169. Pilot Plant test of a heat liquor using a hy-flow fluidized bed treatment system. Summary report. Ecolotrol Environmental Systems, New York, N.Y., (1979).

170. Pilot plant test of a starch waste using a hy-flow fluidized bed treatment system. Summary Report. Ecolotrol Environmental Systems, New York, N.Y., (1980).

171. Plummer, A.H., J.F. Malina, and W.W. Eckenfelder. Stabilization of low solids carbohydrate, waste by an anaerobic submerged filter. In Proceedings of the 23rd Industrial Waste Conference, May 7, 8 and 9, 1968. Purdue University, Lafayette, Indiana, pp. 462-473 (1968).

172. Poels, J., A. Vermeiren, and W. Verstraete. Zuivering van voorgecentrifugeerde varkensmergmest in een anaerobe opstroomreactor. H_2O, 14, 337-339 (1981).

173. Pol, L.W. Hulshoff and others. Granulation in UASB-reactors, Presented at IAWPR-specialised seminar on anaerobic treatment, June, Copenhagen, Denmark (1982).

174. Pretorius, W.A. Anaerobic digestion. 3. Kinetics of anaerobic fermentation. Water Res., 3, 545-558 (1969).

175. Pretorius, W.A. Anaerobic digestion of raw sewage. Water Research. 5, 681-687 (1971).

176. Popel, F. Leistung, Berechung and Gestaltung von Tauchtropfkorperanlagen. Stuttgarter Berichte zur Siedlungswasserwirtschaft, 11 (1964).

177. Ragan, J.L. Celanese experience with anaerobic filters. In Anaerobic filters : An energy plus for wastewater treatment. Proceedings of the seminar/workshop, January 9-10, (1980). Howey-in-The Hills, Florida. Argonne National Laboratory, Argonne, Ill. pp. 129-141 (1980).

178. Raman, V., and N. Chakladar. Upflow filters for septic tank effluents. Journal WPCF, 44, 1552-1559 (1972).

179. Raman, V., and Chakladar, N. Low cost treatment of effluent from septic tanks by reverse flow (upflow) filters. In Proceedings of the low cost waste treatment symposium, NEERI, NAGPUR, India (1972a).

180. Raman, V., and A.N. Khan. Upflow anaerobic filter : A simple sewage treatment device. In International conference on water pollution control in developing countries, Bangkok, Thailand, 21-25 February (1978).

181. Richards, E. A. Bioenergy from anaerobically treated wastewater, Brewers Digest, 56, (9) (1981).

182. Richter, G.A., and J.A. Mackie. Low cost treatment for high strength waste CH_2M Hill Report, October. (in Meuller, (1971).

183. Riera, F.S., and others. Anaerobic treatment of stillage from sugar cane molasses. Presented at Second international symposium on anaerobic digestion, 6-11 September, Travemunde, Germany (1981).

184. Ritman, B.E., and P.L. McCarty. Design of fixed-film processes with steady-state-biofilm model. Prog. Wat. Tech., 12, 271-281 (1980).

185. Rijkens. B.A. A novel process for the anaerobic digestion of solid wastes leading to biogas and a compost-like material. In Energy from biomass, W. Palz, P. Chartier, and D.O. Hall, (eds.) p. 435-439 (1981).

186. Roels, J.A. Bioengineering report. Application of macroscopic principles to microbial metabolism. Biotechnology and Bioengineering, 22, 2457-2514 (1980).

187. Roersma, R., P. Grin, and G. Lettinga. Anaerobic treatment of sewage. Presented at Second international symposium on anaerobic digestion, 6-11 September, Travemunde, Germany (1981).

188. Ross, W.R. Treatment of concentrated organic industrial wastes by means of the anaerobic digestion process : A review of South African experience and current

research. Presented at 3rd international congress of industrial wastewater and wastes, February, Stockholm, Sweden (1980).

189. Rozzi, A., and W. Verstraete. Calculation of active biomass and sludge production vs. waste composition in anaerobic contact processes. Trib. Cebedeau, 34, (455), 421-427 (1981).
190. Ruppel, W., M. Biedron, B. Thornton, and R.J. Swientek. Upflow anaerobic sludge blanket reduces COD 75-85%, produces methane gas. Food Processing, (Sept), 66-68 (1982).
191. Rydin, S., and P. Haglund. Control of bark landfill leachates with an anaerobic treatment system. TAPPI, 64, (8), 59-62 (1981).
192. Salkinoja-Salonen, M.S., and De. E. Hughes. Analytical techniques in anaerobic digestion. Presented at Second international symposium on anaerobic digestions, 6-11 September, Travemunde, Germany (1981).
193. Salklinoja-Salonen, M. and Others. Biodegradability of recalcitrant organochlorine compounds in fixed film reactors. Presented at IAWPR-specialised seminar on anaerobic treatment, June, Copenhagen, Denmark (1982).
194. Salkinoja-Salonen, M. The behaviour of chloroorganic chemicals in the environment and their biological treatment. Presented at FAST environmental biotechnology workshop, Versailles (1982).
195. Salkinoja-Salonen, M.S., E.J. Nyns, P.M. Suton, L. van den Berg, and A. D. Wheatley, Starting-up of an anaerobic fixed-film reactor. In Proceedings of IAWPR-specialised seminar on anaerobic treatment, June, Copenhagen, Denmark (1982a).
196. Satterfield, C.N. The role of diffusion in catalysis. MIT Press (1963).
197. Satterfield, C.N. Mass transfer in heterogeneous catalysis. MIST Press (1970).
198. Sayed, S., and others. Anaerobic treatment of slaughterhouse waste. Presented at Second international symposium on anaerobic digestion, 6-11 September, Travemunde, Germany (198).
199. Schafer, P. Personal communication (1982).
200. Schlegel, S. and K.H. Kalbskopf. Treatment of liquors from heat treated sludges using the anaerobic contact process. Presented at Second international symposium on anaerobic digestion, 6-11 September, Travemunde, Germany (1981).
201. Schlegel, S., and K.H. Kalbskopf. Die Behandlung van Filtratwassern termisch koditionierter Schlamme nach dem anaeroben belemungsvberfahren. GWF-Wasser/ Abwasser, 123, 203-208 (1982).
202. Schroepfer, G.J., and N.R. Ziemke. Development of the anaerobic contact process. Sewage and Ind. Wastes, 31, 164-190 (1959).
203. Schwartz, L.J. and L.A. De Baere. Thermophilic anaerobic digestion of a high strength liquid waste stream. Presented at Second international symposium on anaerobic digestion, 6-11 September, Travemunde, Germany (1981).
204. Seeler, T.A., and J.C. Jennett. Treatment of a wastewater from a chemically synthesized pharmaceutical manufacturing process with the anaerobic filter. In Proceeding of the industrial waste conference, May 9, 10, and 11, 1978, Purdue University, Lafayette, Indiana. Ann Arbor Science, Ann Arbor, Mich., 1979, pp. 687-695 (1978).
205. Seyfriend, C.F. Behandlung organisch verschmuzter Abwasser mit anaeroben Verfahren zur Minimierung des Energieaufwands. Gewasserschutz, Wasser, Abwasser,

45, Inst. fur Siedlugswasserwirtschaft, Technische Hochs Aachen, Aachen, Germany (1981).

206. Shieh, W.K. Suggested kinetic model for the fluidized-bed biofilm reactor. Biotechnology and Bioengineering, 22, 667-676 (1980).

207. Schieh, W.K., and L.T. Mulcahy and E. J. LaMotta. Unified design basis for the fludized bed biofilm reactor. Presented at 2nd World Congress of the Inter-American Confederation of Chemical Engineers, October, Montreal, Canada (1981).

208. Shieh, W.K., and L.T. Mulcahy, FBBR kinetics - a rational design and optimization approach. Presented at IAWPR-specialised seminar on anaerobic treatment, June, Copenhagen, Denmark (1982).

209. Sixt, H. Anaerobe Reinigung organisch hochverschmutzer Abwasser mit dem anaeroben belebungsverfahren der Nahrungsmittleherstellung. Dissertation, Universitat Hannover, Germany (1979).

210. Sixt, H., S. Wernecke and K. Mudrack> Neuere Entwicklungen *Venue :* DDA Ground, Next to BD-Block, Shalimar Bagh, Club Road, Shalimar Bagh, Delhi-88 aufdem Gebiet der Anaeroben Abwasserbehandlung. Korrespondenz (Abwasser, 27, 22-27 (1981).

211. Smith, M.R., and R.A. Mah. Growth and methanogenesis by Methanosarcina strain 227 on acetate and methanol. Appl. Environ. Microbiol., 36, 870-879 (1978).

212. Smith, P.H. and R.A. Mah. Kinetics of acetate metabolism during sludge digestion. Appl. Microbiol., 14, 368-371 (1966).

213. Speece, R.E. And P.L. McCarty. Nutrient requirements and biological solids accumulation in anaerobic digestion. In Advances in water pollution research. Proceedings of the international conference, September, 1962, London, UK, vol. 2 Pergamon Press, Oxford, UK. pp. 305-322 (1964).

214. Speece, R.E. and K.A. Kem. The effect of short-term temperature variations on methane production. Journal WPCF, 42, 1990-1997 (1970).

215. Speece, R.E., G.F. Parkin, J. Yang, W. Kocher. Methane fermentation toxicity responce : Contact mode. In Proceedings of the seminar/workshop : Anaerobic filters : An energy plus for wastewater treatment, 11-35. Argonne National Laboratory, Argonne, Illinois, (Rep. ANL/CNSV-TM50) (1981).

216. Speece, R.E., G.F. Parkin, and D. Gallagher. Nickel stimulation of anaerobic digestion. Environmental Studies, Drexel University, Philadelphia (1982).

217. Sprott, G.D., K.F. Jarrell, and R. Knowler. Potent inhibitory action of acetylene on methanogenic bacteria. Presented at Second international symposium on anaerobic digestion, 6-11 September, Travemunde, Germany, (1981).

218. Stafford, D.A., D.L. Hawkes, and R. Horton. Methane production from waste organic matter. CRC Press, Boca Raton, Florida, USA (1980).

219. Stander, G.J. Treatment of wine distillery wastes by anaerobic digestion. In Proceedings of the 22nd industrial waste conference, May 2, 3, and 4, 1967. Purdue University, Lafayette, Indiana. pp. 305-322 (1967).

220. Steffen, A.J., and M. Bedker. Operation of full-scale anaerobic contact treatment plant for meat packing wastes. In Proceedings of the sixteenth industrial waste conference, May 2, 3, and 4, 1961. Purdue University, Lafayette, Indiana. pp. 423-437 (1961).

221. Stephenson, J.P., and K.L. Murphy. Biological fluidized bed wastewater denitrification. Prog. Wat. Tech., 12, 159-171 (1980).

222. Stevens, T.C., and L. van den Berg. Anaerobic treatment of food processing wastes using a fixed-film reactor. In Proceedings of the 36th industrial waste conference, May 1981, Purdue University, Lafayette, Indiana. Ann Arbor Science. Ann Arbor, Mich. (1982).

223. Stuckey, D.C. and Others. Anaerobic toxicity evaluation by batch and semi-continuous assays. Journal WPCF, 52 720-729 (1980).

224. Sutton, P.M. and A. Li. Anitron system and oxidron system : High-rate anaerobic and aerobic biological treatment systems for industry. Presented at 36th industrial waste conference, Purdue University, Lafayette, Indiana (1980).

225. Sutton, P.M., A Li, R.R. Evans, and S. Korchin. Door-Oliver's fixed film and suspended growth anaerobic systems for industrial wastewater treatment and energy recovery. Presented at 37th annual Purdue industrial waste conference, Purdue University, Lafayette, Indiana (1982).

226. Switzenbaum, M.S., and W.J. Jewell. Anaerobic attached-film expanded-bed reactor treatment. Journal WPCF, 52, 1953-1965 (1980).

227. Switzenbaum, M.S. Anaerobic expanded bed treatment of wastewater. In Proceedings of the seminar/workshop : Anaerobic filters : An energy plus for wastewater treatment, p. 115-127. Argonne National Laboratory, Argonne, Illinois. (Rep. ANL/ CNSV-TM50) (1981).

228. Switzenbaum, M.S. and S.C. Danskin. Anaerobic expanded bed treatment of whey. Presented at 36th industrial waste conference, 1981, Purdue University, Lafayette, Indiana (1982).

229. Switzenbaum, M.S. A comparison of the anaerobic filter and the anaerobic expanded/ fluidized bed processes. Presented at IAWPR-specialised seminar on anaerobic treatment, June, Copenhagen, Denmark (1982a).

230. Sykes, R.M. Theoretical heterotrophic yields. Journal WPCF, 47, 591-600 (1975a).

231. Tadman, J.M. Dow Chemical, personal communication. (in Muller, 1975) (1973).

232. Tait, S.J. and A.A. Friedman. Anaerobic rotating biological contractor for carbonaceous wastewaters. Journal WPCF, 52, 2257-2269 (1980).

223. Taylor, D.W. and J.R. Burm. Full-scale anaerobic filter treatment of wheat starch plant wastes. AIChE Symp. Set., 69, 30-37 (1973).

234. Termofil undradning of spildevandsslan og gylle. (Thermophilic digestion of wastewater sludge and farm (manure) (in Danish). COWI consult, Consulting engineers and Planners AS, Virum, Denmark, (1981).

235. Thauer, R.K. And others. Energy conservation in chemotrophic anaerobic bacteria. Bacteriological Reviews, 41, 100-180 (1977).

236. Thauer, R.K. Biochemistry and energetics. Presented at Second international symposium on anaerobic digestion, 6-11 September, Travemunde, Germany (1981).

237. Therkelsen, H.H., J. E. Sorensen, and A.M. Mielsen. Thermophilic anaerobic digestion of wastewater sludge and farm manures. Presented at Second international symposium on anaerobic digestion, 6-11 September, Travemunde, Germany (1981).

238. Toerien, D.F. and W.H.J. Hattingh. Anaerobic digestion. 1. The microbiology of anaerobic digestion. Water Res., 3, 385-416 (1969).

239. van den Berg, L., G.B. Patel, D.S. Clark, and C.P. Lenz. Factors affecting rate of methane formation from acetic acid by enriched methanogenic cultures. Canadian Journal of Microbiology, 22, 1312-1319 (1969).

240. van den Berg, L. Effect of temperature on growth and activity of a methanogenic culture utilizing acetate. Canadian Journal of Microbiology, 23, p 897-902 (1977).

241. van den Berg, L., and C.P. Lentz. Food processing waste treatment by anaerobic digestion. In Proceedings of the 32nd industrial waste conference, May 10, 11, and 12, 1977, Purdue University, Lafayette, Indiana. Ann Arbor Science, Ann Arbor, Mich., 1978. pp. 252-8 (1977a).

242. van den Berg, L. and C.P. Lentz. Comparison between up- and down-flow anaerobic fixed film reactors of varying surface-to-volume ratios for the treatment of bean blanching waste. In Proceedings of the 34th industrial waste conference, May 8, 9, and 10, 1979, Purdue University, Lafayette Indiana, Ann Artor Science, Anno Artro, Mich., 1980 pp. 305-322 (1979).

243. van den Berg, L. and C.P. Lentz. Performance of fixed film reactors in anaerobic digestion of wastes. Presented at Ist International Symposium on Anaerobic Digestion, September, Cardiff, UK. (1979a).

244. van den Berg, L. and C.P. Lentz. Developments in the design and operation of anaerobic fermenters. Presented at Ist Bioenergy R and D Seminar, March, Ottawa, Canada (1979b).

245. van den Berg, L. and C.P. Lentz. and D.W. Armstrong. Anaerobic waste treatment efficiency comparisons between fixed film reactors, contact digesters and fully mixed, continuously fed digesters. In Proceedings of the 35th industrial waste conference, May 13, 14, and 15, 1980, Purdue University, Lafayette, Indiana. Ann Arbor Science, Ann Arbor, Mich., 1981. pp788-793 (1980).

246. van den Berg, L. and C.P. Lentz. Performance and stability of the anaerobic contact process as affected by waste composition, inoculation and solids retention time. In proceedings of the 35th industrial waste conference, May 13, 14, and 15, 1980, Purdue University, Lafayette, Indiana. Ann Arbor Science, Ann Arbor, Mich., 1981. pp. 305-322 (1980a).

247. van den Berg, L. and others. Effects of sulphate, iron and hydrogen on the microbial conversion of acetic acid to the methane. Journal of Applied Bacteriology 48, 437-447 (1980b).

248. van den Berg, L. and C.P. Lentz. Effects of film area-to-volume ratio, film support, height and direction of flow on performance of methanogenic fixed film reactors. In Anaerobic filters : An energy plus for wastewater treatment. Proceedings of the seminar/workshop, January 9-10, 1980. Howey-In-The-Hills, Floriada. Argonne National Laboratory, Argonne, Ill., 1981. pp1-10 (1980c).

249. van den Berg, L. and C.P. Lentz. Effect of digester configuration, waste composition and inoculum on rates of production of methane from waste. Presented at 2nd bioenergy R and D seminar March, Ottawa, Canada (1980d).

250. van den Berg. L. and K.J. Kennedy. Support materials for stationary fixed film reactors for high-rate methanogenic fermentations. Biotechnology Letters, 3, 165-170 (1981).

251. van den Berg, L. and K.J. Kennedy., and M.F. Hamoda. Effect of type of waste on performance of anaerobic fixed film and upflow sludge bed reactors. Presented at the 36th industrial waste conference, Purdue University, Lafayette, Indiana (1981a).

252. van den Berg, L. and K.J. Kennedy. Dairy waste treatment with anaerobic stationary fixed film reactors. Presented at IAWPR-specialised seminar on anerobic treatment June, Copenhagen, Denmark (19682).
253. van den Berg, L. and K.J. Kennedy. Comparison of intermittent and continuous loading of stationary fixed-film reactors for methane production from wastes. J. Chem. Tech. Biotechnol., 32, 427-432 (1982a).
254. van den Heuvel, J.C., R.J. Zoetemyer, and C. Beolhouver. Purification of municipal wastewater by subsequent reverse osmosis and anaerobic digestion. Biotechnology and Bioengineering, 23, 2001-2008 (1981).
255. van den Meer, R.R. Anaerobic treatment of wastewater containing fatty acids in upflow reactors. Delft University Press, The Netherlands (1979).
256. van Velsen, A.F.M., E. van't Oever, and G. Lettinga. Toepassing van de opstro-omreaktor bij de anaerobe behanding van kalverdrijfmest. H_2O, 12, 59-63 (1979).
257. Verstraete, W., L. de Baere, and A. Rozzi. Phase separation in anaerobic digestion: Motives and Methods. Trib. Cebedeau, 34, (453-454), 267-375 (198).
258. Viitasalo, I., and others. Full scale biogas production and sewage treatment for potato chips industry. Presented at the Second International symposium on anaerobic digestion, 6-11 September, Travemunde, Germany (1981).
259. Wernecke, S., and K. Mudrack. Undersuchungen zum anaeroben abbau von Starke und Pektin in Abhargigkeit von den Fermenter-Belastung. GWF, Wasser - Abwasser, 122, 1-9 (1981).
260. Wheatley, A.D., and L. Cassell. Effluent treatment by anaerobic biofiltration. Pollution research unit, UMIST, Manchester (1982).
261. Wilkie, A., and P.J. We well. The operation and economic assessment of full-scale anaerobic fixed bed reactors. Presented at the Second international symposium on anaerobic digestion, 6-11 September, Travemunde, Germany (1981).
262. Williamson, K., and P.L. McCarty. A model of substrate utilization by bacterial films. Journal WPCF, 48 9-24 (1976).
263. Wilson, A.W., and P.L. Timpany. Anaerobic contact filter for the treatment of waste sulphite liquor. Presented at the 59th annual meeting, Canadian Pulp and Paper Association, Montreal, Canada (1973).
264. Witt, E.R., W.J. Humphrey and T.E. Roberts. Full-scale anaerobic filter treats high strength wastes. In Proceedings of the 34th industrial waste conference. May 8, 9, and 10, 1979, Purdue university, Lafayette, Indiana, Ann Arbor Science. Ann Arbor, Mich. pp. 305-322 (1980).
265. Yang, J., and others. The response of methane fermentation to cyanide and chloroform. Prog. Wat. Tech., 12, 977-987 (1980).
266. Young, J.C. and P.L. McCarty. The anerobic filter for waste treatment. In Proceedings of the 22nd industrial waste conference, May 2, 3, and 4, 1967. Purdue University, Lafayette, Indiana. pp. 305-322 (1967).
267. Young, J.C. The anaerobic filter for waste treatment Ph. D. Dissertation, Department of Civil Engineering, Stanford University, Standfor, USA (1968).
268. Young, J.C. Performance of anaerobic filters under loading and operating conditions. In Anaerobic filters : An energy plus for wastewater treatment. Proceedings of the seminar/workshop, January, 9-10, 1980. Howey-In-The-Hills, Florida, Argonne National Laboratory, Argonne, Ill. pp. 305-322 (1980).

269. Young, J.C. and M.F. Dahab Effect of media design on the performance of fixed-bed anaerbic reactors. Presented at IAWPR-specialised seminar on anaerobic treatment, June, Copenhagen, Denmark (1982).

270. Zehnder, A.J.B. Ecology of methane formation. In R. Mitchell (Ed.), Water pollution microbiology, vol. 2, John Wiley and Sons, new York N. Y. Chap. 13, pp. 305-322 (1978).

271. Zhender, A.J.B., K. Ingvorsen, and T. Marti. Microbiology of methane bacteria. Presented at the Second international symposium on anaerobic digestion, 6-11 September, Travemunde, Germany (1981).

272. Zeikus, K.G., and R.S. Wolfe. *Methanobacterium thermoutotrophicus* sp.n., an anaerobic, autotrophic, extreme thermophile. J. Bacteriol., 109, 707-713 (1972).

273. Zeikus, J.G. The biology of methanogenic bacteria. Bacteriological Reviews, 41, 514-541 (1977).

274. Zinder, S.H., and R.A. Mah. Isolation and characterization of a thermophilic strain of Methanosarcina unable to use H_2-CO_2 for methanogenesis. Appl. Environm. Microbiol., 38 996-1008 (1979).

275. Zehnd, A.J.B., Huser, B.A., Brock, T.D. and Wuhrmann, K. Characterization of an acetate-decarboxylating non-hydrogen-dioxidizing methane bacterium. Arch. Microbiol., 124, 1-11 (1980).

276. Zehnder, A.J.B., Ingovorsen, L. And Marti, T. In Huglies, D.E., and others (Eds.). Anaerobic Digestion 1981. Elsevier Biomedical Press, Amsterdam, pp. 45-68 (1982).

277. Zhilina, T. N. and Zavarzin, G.A. Cyst formation by Methanosarcina. Microbiology, 48, 451-456 (1979).

278. Zoetemeijer, R.J. Matthijsen, A.J.C.M., Van der Heuvel, J.C., Cohen, A and Beolhouwer, C. Acidogenesis of soluble carbohydrate-containing wastewater. Trib. Cebedeau, 34, 444-465 (1981).

CHAPTER 5

FACTORS AFFECTING ANAEROBIC PROCESSES

5.0 Introduction

Wide spread application of the anaerobic process has been hampered somewhat by its reputation as being easily upset and unreliable. Further development of anaerobic process technology is dependent on a better understanding of the factors associated with the stability of the biological processes involved. Process instability is usually indicated by a rapid increase in the concentration of volatile acids with a concurrent decrease in methane gas production, indicating that the fastidious methane bacteria are most susceptible to upset.

Optimum conditions and ranges for anaerobic digestion have been studied by many investigators who are not always in agreement. One reason for this may be that their studies are conducted using different feed materials as well as different methodologies. The nature and composition of the substrate material dictate the microbial regime present and it appears that a single set of parameters is not valid for all situations[1].

Many of the laboratory and pilot-scale studies reporting process failures have used acclimation periods of a month or less. Adaptation of the micro-organisms to a substrate has been reported to take more than five weeks[2] and sufficiently acclimated bacteria have shown unusual stability toward stress-inducing events such as hydraulic loads, fluctuations in temperature and in volatile acid and ammonia concentrations[3].

5.1 General Consideration

The physical and chemical environment influences the growth and metabolism of microbes. It should be emphasized that the effect of each factor is influenced by others. The term 'Optimal temperature', for example is a loose expression unless all the environmental conditions are takern into account.

Considerable interest has been devoted recently to physiology of micro-organisms in a biological reactor. The cells and especially the culture which consists of billions of cells, is a very complex system. The question arises whether such a complex system can be defined by the exact state in the biological sense. This system is also able to change in a very wide range. When the state of culture is considered with regard to generation times, it is apparent that a certain distribution of multiplication rate of particular cells exists in the continuous culture. If, for example, the generation time is 60 minutes, then culture contains micro-organisms with generation times of 30 minutes and 90 minutes. If appearance, suppression or another change of a certain property is considered the manifestation of physiological state, then two factors must be taken into consideration which cause such a change *viz.*

- the environment
- the age of the cells

The environment causes changes of the properties of culture by changes of the environmental conditions. These changes manifest themselves in the micro-organisms by :

- immediate action on the activity of enzymes,
- formation or disappearance of conditions for the formation of certain enzymes; this causes quantitative change of different enzymes present in the cells and qualitative changes as well as adaptive enzymes.

A permanent change of cell type can occur through selection of mutants. Age can manifest itself in several ways. Aging, however, is characterised primarily by a decrease of the synthetic activity of the cell which can lead to interruption of growth and multiplication.

The problem arises whether the culture itself exists in a single physiological state or whether different physiological states of individual organisms of very different generation times are described. With cultures of same organisms having an average generation time of 30 minutes, many cells will have a generation time of 30 minutes. This raises the question of whether the micro-organisms of 30 minutes age are in the same physiological state in the first as well as in second culture. It must further be taken into account that culture possesses a wide dispersion of generation times of individual cells in continuous cultivation at a very high dilution rate. Such a population is less uniform physiologically from this point of view that for example a batch culture at the start of the stationary phase. These relations of generation time are valid, although in continuous cultivation micro-organisms with unusually long generation times are discriminated.

It is also of interest to know whether all organisms consume a certain minimum of the limiting factor on limitation by certain nutrient and thus multiply more slowly or whether the more active organisms deprive the less active ones of this factor. Whatever the situation, the resulting generation time corresponds to the average and by itself would not give a true picture of the physiological state of the organisms present.

It follows that influence of age and environment overlap, and the manifestation of a cell of the same age from cultivation of different average times need not necessary be identical. Since the culture contains cells of different generation times which are in a different physiological state with regard to their age and various influences of changes of the surrounding environment, which we determine in a homogenous sample, such a state is called 'average physiological state'. This state in the stirred fermentations is the same at all points of the medium. Again, it is important that this average state of the continuous culture is independent of time where as in batch system it changes with time. Fixed film reactors are difficult to describe in the above context because of the complexity of the reactor and biomass attached on the media.

It follows that the average physiological state of continuous culture (suspended or fixed film) is a descriptive term which must be further corroborated by actual biochemical, biophysical and cyto-architatonic proof, characterizing a certain property or sum of properties and their changes, studied under the conditions of continuous cultivations at the normal manifestations of the healthy organisms.

There are some independent variables and also dependent variables which are given as follows :

	Variables	
Sr. No.	*Independent*	*Dependent*
1.	Time	Test of mathematical Theory
2.	Growth rate	Physiological states
3.	Nutrient concentration	Rate of metabolism and product formation
4.	Product concentration	Metabolic pathways and regulation of metabolic processes
5.	pH	Induction enzymes
6.	Inhibitor	Mutation rates
7.	Enzymatic inductors	Composition of Cells
8.	Mutagenic substance	Cell morphology
9.	Agitation (wherever essential)	Virulence
10.	Temperature	Development of micro-organisms
11.	-	Nutrition of
12.	-	micro-organisms

Gaden[4] classified the fermentation schemes in three categories, *viz.*

- ***Growth associated :*** Product formation is in parallel with growth, *i.e.*, the rate of production is in linear relationship with the growth rate over the course of fermentation (Alcohol production).
- ***Non-growth associated*** : Product formation is delayed so that the product begins to appear after the microbial growth ceases and the substrate is consumed (Antibiotics fermentation).

- ***Intermediate type :*** Combination of first two types mentioned above (Lactic and Glutamic acid fermentation).

Luedeking, *et al.*[5] obtained the following empirical relationship between the rate of product formation (dP/dt) and the rate of cell growth (dN/dt).

$$\frac{dP}{dt} = \alpha\frac{dN}{dt} + \beta N \qquad ...(5.1)$$

Where,

N bacterial concentration

α, β empirical constants, the values mostly depend on the pH value Equation 5.1 can be rearranged by assuming the value of μ as μ_{max} and is given as follows:

$$\frac{dP}{dt} = \left(\alpha + \beta\frac{N}{dN/dt}\right)\frac{dN}{dt}$$

$$= (\alpha + \beta / \frac{d\ln N}{dt})\frac{dN}{dt}$$

$$= (\alpha + \beta / \mu_{max})\frac{'dN}{dt}$$

$$= (\text{Cons}\tan t)\frac{dN}{dt}$$

It is interesting to note that equation 5.1 has both growth associated and non-growth associated part. The first term (αdN/dt) of right hand side of equation 5.1 represents the growth associated effect while the second (βN) represents the non growth effect. During logarithmic growth, where $\mu = \mu_{max}$ represents a growth associated fermentation. For this reason, confusion often occurs when one talks about the character of fermentation process Indeed, it can change drastically throughout its course.

In general, methane as a product produced during anaerobic digestion of organics follows a course quite similar to category-1. Two conditions must be considered, *viz.*

- Conditions for growth and production are different,
- In continuous system, microbes flow from the reactor which could still produce the required product and their production capacity is thus lost. This invariably happens with well mixed systems and not with fixed film reactors.

Some papers concerning batch cultivation prove that the microbes must undergo the whole development in order to maintain high production capacity. Degeneration is assumed to take place and microbes stop producing, particularly in the phase of acid formation. This aspect may not occur in methanogenesis.

The problem of toxicity of the product must be solved particularly about H_2S which is also a product which is generated during anaerobic process.

A completely continuous process cannot be effected in a single stage with regard to the phase character and the particular phases may be divided into several stages. This may not be necessary for anaerobic process for the production of biogas. However, if this concept is applied, it may be advantageous.

The important factors affecting anaerobic processes can be listed as follows :

- growth and nutrition of micro-organisms
- morphological changes during growth
- influence of dilution rate and limitation on composition of cells
- enzymatic changes and regulation of metabolic processes
- problems concerning culture development
- mutagensis

5.3 Description of Factors Affecting Anaerobic Process

5.3.1 Temperature

Digestion and gas production can occur over a wide range of temperatures (4-60°C) if the temperature is held constant. Once an effective temperature range is established, small fluctuations can result in process upset.

Although digesters are operated in mesophilic range (20-40°C), methanogenesis is known to occur at temperature as low as 4°C. The effect of increasing temperature in the range of 4-25°C range is profound, the rate gas production has been reported to change from 100-400 per cent for a 12 degree increase in temperature. Ultimate COD stabilization at 25°C was shown to be 95 per cent of that at 35°C and the substrate resisting decomposition at 25°C were those of a complex nature such as lipids and long chain carbohydrates. At 20 and 15°C an increasing proportion of these substances failed to degrade[6].

Although low temperature digestion (20-25°C) requires solids rentention time (SRT) approximately twice as long as for mesophilic digestion, gas production, and quality and other parameters indicative of process stability were favourable and digestion at these temperatures may provide a viable alternative in cold climates.

The possible merits of thermophilic (40-60°C) digestion have also been investigated. It has the advantages over mesophilic digestion of shorter SRTs, increased digestion efficiency, better sludge dewaterability characteristics and increased destruction of pathogenic organisms. Its disadvantages are that it has a higher volatile acid/alkalinity ratio which must be buffered and the thermophilic organisms are more sensitive than mesophilic organisms to temperature fluctuations. For example, a temperature variation of $\pm$2.8°C was allowable for mesophilic process at 38°C, but only much lower fluctuations of $\pm$0.8 and $\pm$0.3°C could be tolerated at the thermophilic operating temperatures of 49 and 52°C[7]. There is also a possibility of lysis of bacterial cell at higher temperatures if the process is not controlled rationally.

Thermophilic operation must also provide superior mixing to ensure better food and heat distribution and more uniform feeding throughout the day. The extra energy required to maintain thermophilic condition is an additional disadvantage. However, this does not apply to distellery spent wash.

Table 5.1 lists some investigations on temperature[8]. The most common range for process operation is 25-40°C. Investigations show that anaerobic process can operate in temperature interval of 10-45°C without major changes in microbial ecosystem. At temperatures of 40-45°C, the microbial activity is still significant, but due to a high decay rate the observed yield coefficient of methane bacteria approaches zero and thus renders continuous operation difficult.

Temperature changes in a process must be done slowly, which means by steps of 1°C per day. If this is observed, the micro-organisms will adapt without fail in the metabolic processes, although metabolic rates will change. It is important to distinguish between temperature dependency of various growth constants. Substrate removal rate (or methane production rate) is of primary interest, but decay rate coefficient, the yield coefficient, and specific growth rate of the organisms are also temperature dependent.

From the data of Kennedy, *et al.*[10], Lettinga, *et al.*[11], Stander, *et al.*[12]; Lawrence, *et al.*[13], van den Berg, *et al.*[14], and O' Rouke, *et al.*[15], the temperature constant from 10-30°C can be estimated as defined by the equation given as follows :

$$\frac{R(T_1)}{R(T_2)} = \exp\,[k\ (T_1\text{-}T_2)] \qquad ..(5.3)$$

Where,

$K = 0.10°C^{-1}$

Thermophilic process have a rather constant methane production rate, independent of temperatures in the range of 50-70°C. The rate is, depending upon the substrate, somewhat 25-50% higher than the messophilic rate at 35°C. The major problem with thermophilic processes is the low net yield (50 per cent of the yield at 35°C) which results in very slow startup and very slow accommodation to loading variations, substrate changes or toxic substances. The tendency to lyse quickly at high temperature makes it obligatory that the reactor system should work under exponential growth phase[16]. Another problem is the few bacterial species able to grow at high temperature. This might result in odd temperature dependencies, when *e.g.* only two species are present. van den Berg, *et al.*[17] report on thermophilic treatment of bean blanching waste without any operational problem within 2-3 months of the experiment, except slow startup. It is thus not possible to give any general conclusions regarding thermophilic temperature dependency.

The response to quick temperature changes will be a temporary stoppage of activity[18] which might result in a process breakdown if the process loadings are initially high and is not decreased totally stopped, simultaneously with the change in temperature. Jewell, *et al.*[19] investigated the affect of temperature shock on expanded bed reactors and demonstrated that very low loaded processes are not very sensitive to temperature shocks. Temperature shocks can be exploited in order to facilitate separation of solids and liquids by suspending gas production temporally[20].

The need for a constant temperature is explained by different kinetic behaviour of the acidogenic and methanogenic micro-organisms. Owing to their (acidogen) factor growth rates, the acidogens acclimatize more rapidly to changed conditions, causing accumulation of metabolic products. This results in imbalance.

Theoretically, a long SRT should allow for larger fluctuation in temperature without disturbance. This occurs since, provided the growth rate of the micro-organisms is well below the maximum, they may simply adjust their metabolic activity when environmental conditions are changed. Frostell[21] (unpublished) found that an accidental increase in temperature of 10°C from one day to another was readily tolerates in an anaerobic filter.

Table 5.1 : Temperature Dependency of Anaerobic Processes

Sr. No.	*Temperature Range, °C*	*Reactor Type*	*Type of Waste*
1.	10-35	Fixed bed	Bean blanching/Chemical
2.	10-45	Sludge blanket	Synthetic/Potato
3.	10-30	Expanded bed	Synthetic
4.	10-35	Expanded bed	Synthetic (glucose)
5.	53-62	Digester	Sewage Sludge
6.	10-35	Fixed bed	—
7.	5-5	Fixed bed	Bean blanching
8.	10-44	—	Acetate
9.	35/50-60	Digester	Sewage Sludge
10.	35-60	—	Acetate
11.	40-80	—	Acetate
12.	5-55	—	Acetate
13.	15-45	Recycled flocs	Wine distillery
14.	20-50	Digester	Acetate
15.	11-23	Digester	Leachate
16.	10-40	Digester	Acetate
17.	8-20	Digester + Fixed Bed in series	Potato
18.	25-35	Fluidized bed	Kraft bleach
19.	0-70	Digester	—

Collectively, these results indicate that the temperature dependence cannot be predicted unless inocula from several different sources are mixed to ensure that an optimal flora develops. An optimal flora should be able to take advantage of extra input of energy provided at elevated temperatures.

If the substrate concentration is 5000 mg/L or above, methane production could be sufficient to raise the waste temperature significantly. Wastes with organic concentration less than 2000-5000 mg/L must be warm to begin with for optimum treatment, or else must be treated at less than optimum temperatures[22].

Discussion on the influence of temperature on anaerobic treatment must also take into consideration the SRTs. The influence of SRT on gas production may be explained by a simple analogy : two working at half their maximum capacity to the same job as one person working maximally. Since SRT is a measure of biomass of the system in the reactor, an increased SRT can compensate for lower temperatures.

5.2.3 Nutrients

The stability of the process is largely dependent on wastewater composition. It is desirable that wastewater contains carbohydrates, proteins and lipids as well as various growth factor such as vitamins and trace elements.

High concentration of readily degradable carbohydrate may cause acidification and process failure[33] and large amounts of protein may bring about ammonia toxicity[24]. Many industrial wastewaters are not nutritionally balanced and there is often a lack of nitrogen, phosphorus and trace elements. These are needed for the synthesis of all mass and if not present in sufficient amounts, must be added before treatment. The word 'sufficient' is somewhat arbitrary since the need for external nutrition depends on how the anaerobic treatment is carried out. With a long SRT, nutrients will be recycled.

Inspite of low observed yield coefficients, it is seen that the anaerobic process might result in nutrient deficiency unless surplus nutrients are added. Speece, *et al.*[25] have shown that nitrogen and phosphorus content of the volatile suspended solids (VSS) during anaerobic digestion is about 10.5 per cent and 1.5 per cent respectively. With knowledge of yield coefficient, the necessary nutrient concentration in the field can be calculated. Often COD/N or COD/N/P/ ratio is used to described the nutrient requirements.

The theoretical minimum COD/N ratio is seen to be 350/7. A value around 400/7 must be regarded reasonable for high loaded anaerobic processes (0.8-1.2 kg COD/kg VSS.d)[26]. For low loaded processes (<0.5 kg COD/kg VSS.d), the COD/N ratio is seen to increase dramatically to values 1000/7 or more. This is consistent with the observations by many workers, that the often quoted necessary COD/N ratio of (200-300)/7 are too small for normal process operations[25–27].

A microbial cell contains a C:N:P:S ratio of approximately 100 : 10 : 1 : 1[28]. Therefore, in order for microbial growth activity to occur, these elements must be present and available and their absence or scarcity can, infact, be rate limiting. Appropriate C:N and C:P ratios of 25:1 and 20:1 respectively have been suggested[7]. Normally a C:N:P ratio for an anaerobic system is defined as 100:2.5:0.5.

It must be kept in mind that C:N and C:P ratios are of use only when referring to fermentable substances and available nitrogen and phosphorus. Lignin and cellulose as well as any components the lignin may protect from microbial attack must be considered unavailable. Table 5.2 summarises the requirements for nitrogen and phosphorus in anaerobic treatment.

Table 5.2 : Summary of Nitrogen and Phosphorus Requirements in Anaerobic Processes

Sr. No.	*Degree of Nutrition*	*Source*	*Type of Process*	*C:N:P*	*Ref.*
1.	Sufficient	Theory	General	500:5:1[a,b]	-
2.	Optimal	Laboratory	Conventional digester	60:5:1[b]	29
3.	Sufficient	Laboratory	Contact-process	110:5;1[c]	25
4.	Sufficient	Pilot-plant	Anmet Process	750:5:1[c,d]	30
5.	Sufficient	Full-scale	UASB-Process	850:5:1	31

Assumption

a - Sludge production = 0.1kg sludge/kg CODr
Empirical formula considered for sludge = $C_5H_9O_3N$

b - need for phosphorus is one-fifth that for nitrogen

c - (0.375) × COD = C (carbon)

d - (2) × BOD_5 = COD

The preceeding account about nitrogen and phosphorus requirements shows that the values differ widely and a judicious and sound judgement must also be applied alongwith actual experimental findings. However, from the available data it appears that the nutritional requirement of UASB is minimum.

At concentration of 9 milli-moles, all inorganic sulphur compounds, other than sulphate inhibit both cellulose degradation and associated methanogensis in order of :

thiosulphate > sulphite > sulphide > hydrogen sulphide

The sulphate reducing bacteria evidently complete with methane bacteria for hydrogen. The ability of hydrogen utilizing sulfate reducing organisms to inhibit the methanogens has been demonstrated, however, both sulphate reduction and methane production occurred in presence of excess hydrogen[32]. Sulphate and sulphide is found in many industrial wastewaters (distillery, pulp and paper, pectine, etc.). The toxicity of sulphide is related to free H_2S concentration. This means that at low pH (<6.5) increases toxicity whereas presence of iron reduces toxicity due to precipitation of iron salts.

Sulphate is reduced to sulphide at higher redox potential than that at which the fermentation process operates. This can be observed in upflow anaerobic fixed beds where sulphate bacteria predominates the eco system at the bottom of the bed[33]. This should make it possible to operate a 2-stage reactor with sulphide generation in the first one (including precipitation with added iron salts or gas stripping).

Another method of overcoming sulphide toxicity is to strip H_2S from the gas[20]. For general operation, sulphate-sulphur concentration in the influent below 0.3-0.6 kg/m^2 should be regarded as harmless, but as mentioned previously pH and precipitation can change the picture radically. Free H_2S will inhibit at concentration above 0.1 kg S/m^2. Alternatively, operate the anaerobic reactor system with pH values above 7.5 but less than 8.5 to overcome sulphide problems.

Nutrients other than nitrogen and phosphorus are essential for anaerobic processes. Table 6.3 lists some of these compounds. The need of sulphur seems to lie within the same magnitude as the need for phosphorus[27]. The importance of nickel for methanogens has been realised only recently and has been included in Table 5.3.

5.3.3 Cations

All cations are capable of producing toxic effects in any organism if the concentrations are high enough, but relative toxicity of cations varies. Generally, toxicity increases with valency and with atomic weight. The results on cation effects in anaerobic processes indicate that methane bacteria show the same basic responses to the presence of cations as do other organisms, however, they exhibit a greater sensitivity to toxic effect of cations than do the acid formers.

Three basic effects are observed : toxicity, antagonism and stimulation. The toxicity of a cation can vary with the presence of other cations. This ability of one cation to affect (increase or decrease) the toxicity is called antagonism. Low concentrations of cations have a stimulating effect on microbial metabolism. For example, although sodium, potassium, ammonium, calcium and magnesium ion can be toxic to micro-organisms (especially to methane producers) stimulation is achieved at concentration approximately one half toxic levels.

Cations play a nutritional role in metabolism of all micro-organism, serving as metabolic activators for wide variety of enzymes. Popular theories all postulate some sort of interaction between cations and enzymes. Such interaction is suggested to produce stimulation where the correct metallic activator unites with the enzymes, but toxicity results when the enzyme unites with an improper cations. Antagonism can be described as a sort of competition between the functional and non-functional cations for enzyme surface[34]. Cation effect can be summarised as follows :

- Cation-effects in anaerobic treatment are function of all types and concentration of cations present.
- Optimum ionic concentrations are : 0.01M for monovalent ions (Na^+, K^+, NH^+_4) and 0.005 M for divalent ions (Ca^{2+}, Mg^{2+}).
- Concentrations lower or higher than optimum results in less than maximum efficiency.
- Inhibition caused by excessive concentration of any one ions can be antagonised (minimized) by addition of optimum concentrations of another ion.
- Maximum antagonism of an inhibiting cation can be obtained by the addition of optimum concentration of several other cations, rather than only one.

Table 5.3 : Micro-nutrients for Anaerobic Processes[5]

Sr. No.	*Compound*	*Beneficial Concentration**	*Effect*
1.	Fe^{++}	0.2	Precipitation of sulphide, flocculation/ biofilm structure
		12-120 (Soluble)	do
2.	Na^{++}	0.01	Build up of F-430 Co-factor in methanogens
		0.006	Increase in activity
3.	Mg^{++}	0.01-0.02	Flocculation
4.	Ca^{++}	0.01-0.04	Flocculation
5.	Ba^{++}	0.01-0.10	Flocculation
6.	C^{++}	0.01 0.003	Vitamin - B_{12} Increase in activity
7.	SO^-_4	0.02	Increase in activity
8.	B	Not definite	Possible electron acceptor
9.	Sn	Not definite	Transmethylation
10.	Mo and Se	Not definite	Methane production from formate which involves formate dehydrogrease.

* All values in mg/L

The heavy metals are more toxic than light metal ions, although at very low concentrations even extremely toxic cations such as lead and mercury produce stimulating effects. The toxicity of a heavy metal in anaerobic digestion depends on its chemical form, *i.e.*, heavy metal in the form of precipitated sulphides are of little consequence to biological system and high concentrations of toxic heavy metal can be tolerated if sufficient sulphide can be provided to act as a precipitant[7].

When sulphates of copper, zinc, nickel and iron were fed to laboratory digesters, there was little evidence of effects on gas production[35]. The sulphate was apparently reduced to sulphide which combined with metals to form the insoluble heavy metal sulphide. When metal chloride salts were substituted for the sulphates, thereby, eliminating the sulphate precursor marked reduction of digester performance was observed.

Apparently the bacteria are able to concentrate the soluble heavy metal ions around the cell wall through complexation with proteins and toxic effects coincide with near maximum uptake or a bonding site saturation of metal by bacteria. The order of decreasing toxicity on a weight per weight or molar basis was found to be as follows[36]:

$$Ni > Cu >> Cr\ (VI) \sim Cr\ (III) > Pb > Zn$$

The method of feed had an affect on the toxicity levels, generally, pulse or shock feedings produced a lower toxic limit that the corresponding step-fed method. Table 5.4 shows the results of Hayes, *et al.*[36] Toxic limits were taken to be that concentration at which total gas production was reduced by 70 per cent from baseline values.

To control these effects, it is possible to add sulphide for precipitation and to operate at maximum allowable pH (because heavy metal hydroxides are only very slightly soluble). If subsequent leaching of heavy metals is anticipated problem due to ultimate land disposal of solids remaining after digestion, it should be kept in mind that there is some evidence that heavy metals associated with biomass are less available to environment than inorganic precipitate, particularly in acid environments.

5.3.4 *pH, Acidity and Alkalinity*

These parameters are treated together since they are closely inter-related. One of the most important environmental requirements is that for a proper pH[37]. Anaerobic treatment can proceed quite well with a pH varying from about 6.6 to 7.6 with an optimum range of about 7.0 to 7.2. Beyond these limits, digestion can proceed but with less efficiency. At pH values below 6.2, the efficiency drops off rapidly, and the acidic conditions produced can become quite toxic to the methane bacteria. For this reason, it is important that the pH not be allowed to drop below this value for a significant period of time as this parameter is very important[38].

The pH of liquor undergoing anaerobic treatment is related to several different acid-base chemical equilibria. However, at the near neutral pH of interest for anaerobic treatment (between 6 and 8) the major chemical system controlling pH is the CO_2-bicarbonate system which is related to pH or hydrogen ion concentration through the following equilibrium equation.

Table 5.4 : Heavy Metal Toxicity Limits for Anaerobic Digestion

Sr. No.	*Metal*	*Step Fed*		*Pulse Fed Toxic Limit**
		*Inhibiting Concentration**	*Toxic Limit**	
1.	Cr (III)	130	260	<200
2.	Cr (VI)	110	420	<180
3.	Cu	40	70	<50
4.	Ni	10	30	>30
5.	Cd	—	>20	>10
6.	Pb	340	>340	>250
7.	Zn	400	600	<1750

* All values in mg/L

$$[H^+] = K_1 \frac{[H_2CO_3]}{[HCO^-{}_3]} \qquad ...(5.4)$$

The carbonic acid concentration (H_2CO_3) is related to the percentage of carbon dioxide in the digester, K_1, is the ionization constant for carbonic acid and bicarbonate ion concentration ($HCO^-{}_3$) forms a part of the total alkalinity system[38].

The bicarbonate ion concentration or bicarbonate alkalinity is approximately equivalent to the total alkalinity for most wastes when the volatile acid concentration is very low. When the volatile acids begin to increase in concentration, they are neutralised by the bicarbonate alkalinity and in its place form volatile acid alkalinity[39]. Under these conditions, the total alkalinity is composed of both bicarbonate alkalinity and volatile acid alkalinity. Under these conditions, the bicarbonate alkalinity can be approximated as follows :

$$BA = \left[\frac{TA}{TVA}\right] - (0.85)\,(0.833) \qquad ...(5.5)$$

where,

BA – bicarbonate alkalinity (as $CaCO_3$), mg/L

TA – total alkalinity (as $CaCO_3$), mg/L

TVA – total volatile acid concentrations (as acetic acid), mg/L

This formula is similar to that used by *Pohland* and *Bloodgood*[39], but includes a factor (0.85) to account for the fact that only 85 per cent of the volatile acid alkalinity is measured by titration of total alkalinity to pH-4. The equation also assumes that there is no significant concentration of other materials such as phosphates, silicates, or other acids salts which will also produce a significant alkalinity[38].

When the bicarbonate alkalinity is about 1000 mg/L and the percentage of carbon dioxide is between 30 and 40 per cent, the pH will be about 6.7. If the alkalinity drops below this values, the pH will drop to undesirable levels. Such a low alkalinity does not give much safety factor for a small increase in volatile acids; will result in a significant decrease in bicarbonate alkalinity and digester pH[38].

On the other hand, a bicarbonate alkalinity in the more desirable range of 2,500 to 5000 mg/L provides much buffer capacity so that a much larger increase in volatile acids can be handled with a minimum drop in pH[40]. This gives a good factor of safety and allows time for control if an upset results[38].

If an increase in volatile acid concentration drops the bicarbonate concentration too low as calculated by equation 5.5 and a serious drop in pH threatens, then the bicarbonate alkalinity should be controlled. This may be done by reducing the feed rate to allow the volatile acids to be utilized and decrease in concentration or it may be done by addition of alkaline material such as lime or sodium bicarbonate.

The non-methanogenic organisms are not nearly as sensitive and are able to function in a range of pH from 5 to 8.5. Process pH results from interaction of carbon dioxide-bicarbonate buffering system present with volative acids and ammonia formed by the process. The acid-base state of a digester can be monitored by measuring pH and the partial pressure of CO_2. Anaerobic fixed film particularly anaerobic filters have a higher tolerance against low pH than suspended growth system[21].

A high level of alkalinity is desirable since it provides a good margin against sudden increases in the concentration of volatile acids. The major acid-base system encountered in anaerobic treatment are listed in Table 5.5[41]. Since a high buffer capacity in a given acid-base system is achieved only near its pK_a, value. Table 5.5 shows that the bicarbonates, the sulphide, the phosphate and the ammonia systems are most important at or near neutral pH values. The last two are usually present only in very low concentration compared with bicarbonate; normally, therefore, only the bicarbonate system need be considered[38].

Table 5.5 : Acid-base System Providing Buffer Capacity in Anaerobic Treatment

Sr. No.	*Acid-base System*	*pK_a at 25°C*
1.	H_2CO_3 - HCO^-_3	6.4
2.	NH_4^+ - NH^-_3	9.3
3.	H_2S - HS	6.9
4.	$H_2PO_4^{2-}$ - HPO_4^-	7.2
5.	Acetate - propionate-butyrate	4.5-4.9

The concentration of volatile acids and alkalinity during normal treatment depend on the concentration of waste water and on its concentration. In a dilute waste water, and volatile acids and alkalinity are removed at a relatively higher rate with the effluents. The volatile acids to alkalinity ratio is therefore a better criterion for the stability of anaerobic system than their absolute values. A ratio of total volatile acids (as acetic acid) to total alkalinity (as $CaCO_3$) of less than 0.1 is desirable[42].

5.4 Ammonia

Ammonia is usually formed in anaerobic treatment from the degradation of waste containing organic nitrogen, proteins or urea. Inhibitory concentrations may be approached in industrial wastes containing high concentrations of these materials or in highly concentrated sludges (municipal). Ammonia may be present during treatment either in the form of the NH_4^+ ion or as dissolved NH_3 gas. The equilibrium equation is as follows :

$$NH_4^+ \rightleftharpoons NH_3^+ + H^+$$

When the H^+ concentration is sufficiently high (pH of 7.2 or low), the equilibrium is shifted to the left so that the inhibition is related to NH_4^+. At higher pH values, the equilibrium shifts to the right and NH_3 gas concentration may become inhibitory. The NH_3 gas is inhibitory at a much lower concentration[38] than NH_4^+.

The ammonia-nitrogen analysis gives the total sum of NH_4^+ ion and NH_3 gas concentration. If the NH_3-N concentration is between 1500 and 3000 mg/L and the pH is greater than 7.4 and 7.6 the NH_3-N gas concentration can become inhibitory. This condition is characterised by an increase in volatile acid concentration which tends to decrease pH, temporarily relieving the inhibitory condition. The volatile acid concentration here will then remain quite high unless the pH is depressed by some other means such as by adding hydrochloric acid to maintain pH between 7.0 and 7.2. When the NH_3-N concentration exceeds 3000 mg/L, then the ammonium ion itself becomes quite toxic regardless of pH and the process can be expected to fail. The best solution is either dilution or removal of the source of ammonia-nitrogen from waste itself (or pretreating the waste).

Free undissociated NH_3 is the most toxic compound[3] (inhibition at 0.1-0.2 kgN/m^3). Total ammonia plus ammonium ion concentration of 5-8kgN/m^3 can be tolerated if the pH of the reactor is low enough.

There has been considerable debate concerning the toxicity of the ammonia and the volatile solids themselves, regardless of pH. Current thinking is that only the unionized volatile acids in the concentration range of 30-60 mg/L are toxic and the process inhibition by ammonia is a result of excessive concentration of free ammonia rather than NH_4^+ ion[5,43]. It is not clear which type of bacteria attacks ammonia. Bhattacharya[44] feels that acetoclastic bacteria is inhibited. Wiegant[45] has found that hydrogen utilizing methanogens are inhibited. The relationship between unionized ammonia concentration (UAN) and total ammonia concentration (TAN developed by Heinrichs, *et al.*[46] is as follows :

$$UAN = \frac{TAN}{[1+10^{(pKa-pH)}]} \qquad \text{... (5.6)}$$

5.5 Volatile Acids

The two major volatile acid intermediates formed in anaerobic treatment are acetic acid and propionic acid[47,48]. The importance of these two acids as precursors of methane which shows the pathways by which mixed complex organic materials are converted into CH_4 gas. The percentages normally shown are based on COD conversion. The percentages would be different for different wastes[49].

The complete methane fermentation of complex waste has been compared to a factory assembly line operation in that the processing of raw waste material to the final methane product requires the help of several workers. The raw material must be worked on by each group of micro-organisms to prepare it for handling by the next. For example, 30 per cent of complex waste becomes propionic acid through the action of the acidic bacteria and if propionic using bacteria are not functioning, this portion cannot be converted into methane gas. This is true even though the propionic acid bacteria themselves directly produce only 13 per cent of the methane. They convert the remaining 17 per cent to acetic acid.

The acetic acid fermenting methane bacteria are also very important, since if they fail, 72 per cent of the waste cannot be converted to methane gas. It is interesting to note that acetic acid is formed by several routes and through action of different bacteria. Only about 20 per cent of the waste is converted directly to acetic acid during acid formation stage. A much larger portion (52-per cent) is formed from the action of various methane producing bacteria which ferment propionic acid and other intermediate to acetic acid and methane.

For different industrial waste, the percentages may be different. However, the largest percentage will still result from acetic acid fermentation which is the most prevalent volatile acid produced by fermentation of organics. The other volatile acids, although significant, are of minor importance.

Pyruvic and acetic acids occupy key intermediate positions from which further biochemical reactions originate. They may serve as building blocks for synthesis of more complicated organic molecules, converted to the intracellular storage product polybetahydroxybutyric acid (PBH) or further degraded to ethanol by yeast or to methane and CO_2 by methane bacteria.

Thus, although many different organisms are required in anaerobic treatment two groups of methane bacteria which handle acetic acid and propionic acid, are the most important in methane fermentation. Unfortunately, they also appear to be among the slowest growing methane bacteria and most sensitive to environmental changes.

Some fermentation products that have been detected during anaerobic digestion have been listed in Table 5.6. Table could be extended but under normal conditions only of the few compounds are produced in significant amounts *viz.* H_2, CO_2, acetate, H_2S, propionate, butyrate, and ammonia. H_2S is usually precipitated by heavy metals present and this gives anaerobic sludge a characteristic black colour. Hydrogen gas is rapidly used in methane formation. When the process is disturbed, on the other hand, the gas may contain large amounts of H_2S gas. Acetic, propionic and butyric are the most common volatile acids found in anaerobic treatment[23]. Acetate is readily further metabolised, while butyrate and especially propionate, are converted more slowly into other products. Probably the last two must first be degraded into acetate they can be converted to CH_4 and CO_2[42]. Presence of higher molecular volatile fatty acids result in operational problems.

Table 5.6 : Fermentation Products from Anaerobic Digestion

Sr. No.	*Product*	*Formula**	*Significance***
1.	Hydrogen	H_2	+++
2.	Carbondioxide	CO_2	+++
3.	Formate	$HCOO^-$	+
4.	Bicarbonate	HCO_3^-	+++
5.	Acetate	CH_3COO^-	+++
6.	Ethanol	C_2H_5OH	+
7.	Propionate	$CH_3CH_2COO^-$	++
8.	Lactate	$CH_3CH(OH)COO^-$	+
9.	Butyrate	$CH_3(CH_2)_2COO^-$	++
10.	Valerate	$CH_3(CH_2)_3COO^-$	+
11.	Caproate	$CH_3(CH_2)_4COO^-$	+
12.	Ammonia	NH_3	++
13.	Hydrogen sulphide ion	HS^-	++

* Most common formation at digestion pH

** (+++) of primary importance

(++) always present

(+) may be present

5.6 Suitability of Different Chemicals for pH and Alkalinity Control

Lime is the most important widely used material for controlling pH in anaerobic treatment, mainly because it is readily available and fairly inexpensive. However, occasionally some problems have arisen from its use which are related as the relative solubility of some of the calcium salts which form in the digester. Because of this problem, close control over lime addition is required and a knowledge of the solubility problems with lime is helpful.

Control of pH is usually considered when it appears likely to drop below 6.5 to 6.6. If lime is added, it initially increases the bicarbonate alkalinity by combination with CO_2 present.

$$Ca\ (OH)_2 + 2\ CO_2 = Ca\ (HO_3)_2$$

However, $Ca(HCO_3)_2$ formed is not very soluble, and when the bicarbonate alkalinity reaches some point between 500-1000 mg/L, additional lime additions result in the formation of insoluble $CaCO_3$ as follows :

$$Ca\ (OH)_2 + CO_2 = CaCO_3 + H_2O$$

Lime additions beyond this point do not increase the soluble bicarbonate alkalinity and so have little direct effect on digester pH. The pH remains between 6.5-7.0 until CO_2 concentration has decreased to less than 10 per cent by reaction with lime. The pH then suddenly increases above 7.0 and approaches 8.0 as a result of decrease in CO_2 per cent. After a short period of time, biological reaction occurs, the percentage of CO_2 in the gas will begin to increase again. As soon as it exceeds 10 per cent, the pH will drop below 7.0. This may occur even without the formation of additional volatile acids. If lime is then added again, the cycle repeats itself.

Thus, nothing beneficial is obtained if additional lime is added to raise the pH above 5.7 to 5.8. After this point, the lime simply combines with CO_2 to form insoluble $CaCO_3$. This insoluble $CaCO_3$ is ineffective for neutralization of excessive volatile acids or for raising pH.

Thus, for effective use of lime, it should not be added until the pH drops below 6.5. A quantity should be added which is sufficient only to raise the pH to about 6.7 to 6.8. Good mixing of the lime is required in the digester and caution must be exercised to prevent the creation of vacuum from the removal of CO_2 from the gas by combination with lime.

Sodium bicarbonate, although seldom used, is one of the most effective materials for pH control in anaerobic treatment. This material has significant advantages over other materials. It is relatively inexpensive when purchased in large quantities. It does not react with CO_2 to create vacuum in the digester and there is little danger that it will raise pH to undesirable levels. It is quite soluble and can be dissolved prior to addition to the digester for more effective mixing. This material can be added to give alkalinity in the system of 5000 to 6000 mg/L without producing any adverse or toxic effects. However, it is more expensive than lime but less quantities are required. In addition, no precipitation occurs. The ease of control, addition and handling make it a very desirable material for pH control in the digesters.

For fixed film reactors where possibilities of clogging always exist, it is preferable to use $NaHCO_3$ than lime for reasons mentioned above.

5.7 Mutagenic and Antimutagenic Substances[49]

In most papers concerning studies of the mechanism of mutation, mutagenic substances or effects are employed possessing an strong toxic influence on the micro-organisms besides the mutagenic activity. In the chemostat it was possible to investigate successfully the influence of mutagenic substances whose presence in the substrate does not lethally affect the growing bacterial culture. The best results with regard to the production of mutants resistance to bacteriophage T-5 were obtained by purines and purine derivatives. These substances cause a much higher mutation rate than the spontaneous mutation with *E. coli* B/I in the chemostat, Table 5.7. These substances, however, show as much smaller effect on mutation to resistance to bacteriophage T-6. Theophylline for example increases mutation rate T-5 resistance from 1.3 to 11×10^{-8}/h/bact. The mutation rate to T-6 resistance on the other hand increases from 0.3 to 1×10^{-8}/h/bact. It is of interest that pyrimidines and their analogoues do not display this effect.

It was surprising that purine nucleosides suppressed the mutation rate in the chemostat, *i.e.*, their activity is antimutagenic. The most effective antimutagens were adenosine, guanosine, and inosine ribonucleosides. Pyrimidine ribonucleosides on the other hand did not possess any antimutagenic activity. The effect of different guanosine concentrations on mutation rate to T-5 and T-6 resistance caused by theophylline show that small concentration of about 2 mg/L, reduces mutation rate to one half. An analogous 50 per cent reduction of the mutation rate is caused by adenosine in an amount of 0.4 mg/L and by inosine at 2 mg/L.

These nucleosides have no effect on mutation caused by ultraviolet radiation. Purine and purine nucleosides are normal compounds of the cell. The question, therefore, arose whether the described effect of these substances on mutation is the manifestation of physiological processes occurring within the cell under the influence of the mentioned substances during normal metabolism. It may be assumed, therefore, that nucleosides and purines during spontaneous mutation also act as regulators of processes involved in mutation, Table 5.8.

Table 5.7 : Chemical Compounds Appearing Mutagenic in the Chemostat

Sr. No.	*Compound**	*Rate***	*Compound*	*Rate***
1.	Caffeine	19	8 Methoxycaffeine	5.2
2.	Theophylline	11	8 Chlor coffeine	4.2
3.	Paraxanthine	8.4	Azaguanine	3.4
4.	Theobromine	7.5	Benzimidazol	3.0
5.	Tetramethylureic Acid	7.0	Adenine	2.8

* All compounds were tested in a concentration of 150 mg/L

** Mutation rate investigated as transition of sensitivity to resistance to T-5 and expressed in units $\times 10^{-8}$/h/bact.

Table 5.8 : Influence of 150mg/L of Adenosine Mutation Rate of *E. coli*-B on Transition of Sensitive Form into T-5 and T-6 Resistant Form

Sr. No.	*Item*	*Resistance to T-5**	*Resistance to T-6**
1.	Spontaneous mutation	4.0	1.0
2.	Adenosine	1.2	0.4

* $\times 10^{-8}$/h/bact.

Adenosine, however, in this case does not appear to suppress spontaneous mutation completely, even at higher concentrations (500mg/L) in contrast to the effect of adenosine on mutation activity of theophylline or coffeine whose mutation effect is completely suppressed by adenosine. Further, proof for the correctness of the assumption that cells mutation is controlled by some nucleotides and nucleosides is given by the considerable decrease of spontaneous mutation in the chemostat under anaerobic conditions where a large amount of adenosine accumulates in the cells and further by the absence of the mutagenic effect of theophylline under these conditions, Table 5.9.

Further, studies are required in this area for several different types of compound, particularly for methanogenic bacteria.

5.8 Rate Limiting Process

A deficiency in any of the essential nutrients or co-factors will, of course, have a rate limiting effect on the reaction involved. Clausen[50], for example, found the microbical reaction kinetics to be first order reaction based on carbon concentration at low carbon levels.

The concern, however, of this section, is assuming sufficient nutrients, substrate and micro-organisms, what are the rate limiting processes in methanogenesis based on physical barrier and/or chemical equilibrium.

Table 5.9 : Influence of Anaerobiosis Under Different Conditions on Mutation Rate of *E. coli*-B for Obtaining Resistance to T-5 and T-6

		*Anaerobiosis**	
Sr. No.	*Item*	*T-5*	*T-6*
1.	Spontaneous mutation	0.6	0.4
2.	Adenosine	0.8	0.6
3.	Theophylline	1.1	1.3
4.	Theophylline + adenosine	—	—
5.	X-irradiation	7.0	5.4

* $x0^{-8}$/h/bact.

There are four potential rate limiting steps in the anaerobic conversion of substrate to methane, *viz.* :

- Insoluble substances to be converted to soluble substances by extra-cellular enzymes (hydrolysis).
- Formation of volatile acids by acid forming bacteria.
- Conversion of volatile acids to CO_2 and CH_4 by methane bacteria.
- Transfer of dissolved products from liquid to gas phase.

The third step has been considered to be the rate limiting since decreasing the average microbial residence time has increased the volatile acids concentration. It was theorized that these build ups occurred because of the slow metabolic and consequently slow reproduction rate of the methanogenic bacteria. These bacteria were thought to be largely displaced from the digester at SRTs of 2 days or less. This evidence, however, does not, distinguish between possibilities 3 and 4 or biological conversion and phase transfer, as the rate limiting step.

Finney *et al.*[51] postulate that transfer of gas phase is actually rate limiting, and, further, that normal metabolic activity of the methanogenic bacteria is inhibited by product gases. In support of this, their hypothesis, their experiments using vigorous agitation at elevated temperature and reduced pressure showed a substantially increased production rate. They suggest that the micro-bubbles in the vicinity of the cell wall decrease the bacterial surface area available for permeation process. We do know that due to its unique hydrogen binding ability, water is highly oriented or bound to surrounding micro-organisms. We also know that water can form dodecahydral lattice or cage around methane molecule[51].

Since little attraction exists between methane and water, the presence of methane molecule within the cage or clathrate affords additional strengths to the dodecahydral lattice. Such a clathrate crystal will not collapse at normal melting point of water; infact clathrates are known to crystallize at temperature as high as 20°C in methane gas lines[11,52]. In case of solid waste treatment, first step appears to be rate limiting[53].

5.9 Cause and Control of Treatment Unbalance in Anaerobic Process

In order to summarise the arguments in this chapter, it was felt that some salient features must be presented.

5.9.1 Optimum Conditions for Anaerobic Treatment

- Optimum Temperatures
  - Mesophillic range (20-40°C)
  - Thermophillic range (45-61°C)
- Strict Anaerobic Conditions
- Sufficient Biological Nutrients

    - Nitrogen
    - Phosphorus
    - Others
- Optimum pH (6.6 to 7.6)
- Absence of Toxic Material
    - Organic
    - Inorganic
    - Complex (inorganic-organic)

5.9.2 *Indicators of Unbalanced Treatment*

- Parameters Increasing
    - Volatile acids concentration
    - CO_2 per cent gas
- Parameters Decreasing
    - pH
    - Total gas production
    - Waste stabilization

5.9.3 *Steps to Follow in Controlling Unbalance*

- Maintain pH near neutrality
- Determine cause of unbalance
- Correct cause of unbalance
- Provide pH control until treatment unit returns to normal

5.9.4 *Factors Causing Unbalanced Treatment*

- Temporary Unbalance
    - Sudden change in temperature
    - Sudden change in organic loading rate
    - Sudden change in nature of substrate
- Prolonged Unbalance
    - Presence of toxic materials
    - Extreme drop in pH
    - Slow bacterial growth during start-up

5.9.5. *Important Waste-water Characteristics for Anaerobic Treatment Evaluation*

- Organic strength and composition
- Alkalinity
- Inorganic nutrient content
- Temperature
- Content of potentially toxic materials

In Table 5.10, important control parameters and frequency of control have been suggested and may be useful for all those persons connected with anaerobic treatment of wastewaters.

Table 5.10 : Important Control Parameters in Anaerobic Treatment

Sr. No.	*Parameter*	*Frequency**
1.	Influent	+++
	* Rate	
	* Concentration	
	* Composition	
	* Toxicity	
2.	Solids Retention Time (SRT)	+++
3.	Temperature	+++
4.	pH	+++
5.	Gas production and composition	++
6.	Volatile acids and alkalinity	+

a -+++ Should be continuously controlled
++ Daily control
+ Intermittent control

REFERENCES

1. Clark, C.S. Laboratory scale Composting Studies, J. of Envn. Engg. Div. (ASCE), 10 (EE-1) (1978).
2. Kotze J.P. A Biological and Chemical Study of Several Anaerobic Digesters. Wat. Res.; p. 195 (1968).
3. Krocker, E.J. Anaerobic Treatment Process Stability. J. Wat. Poll. Cont. Fed. 51(4) 1997.
4. Gaden, E.L. Fermentation Process Kinetics J. Biochem. Microbiol. Tech. and Engg. 1, p413 (1959).
5. Luedeking, R and Piret, E.L. A Kinetic Study of Latic Acid Fermentation Batch Process at Controlled pH, J. Biochem. Microbiol. Tech. And Eng. 1, p. 393 (1959).
6. Stevens, M.A. and Schalte, D.D. Low Temperature Anaerobic Digestion of Swine Wastewater, J. Envn. Engg. Div. (ASCE), 105 (EE-1) (1979).

7. Gosh, S. Anaerobic Process-Literaute Review, J. Wat. Poll. Cont. Fed. 50(10) (1978).
8. Heuze, M. and Harremoes, P. Anaerobic Treatment of Wastewater in Fixed Film Reactors A Literature Review, Water Science and Technology, IAWPRC (1983), Pergamon Press Ltd. U.K.
9. van den Berg, L. Effect of Temperature on Growth and Activity of a Methanogenic Culture Utilizing Acetate, Canadian J.L. Microbiology, 23, p898-902 (1977).
10. Kennedy, K.J. and van den Berg, L. Effect of Temperature and Overloading on the Performance of Anaerobic Fixed Film Reactors, 36th Purdue Univ. Waste Conference (1981), Purdue, Indiana.
11. Lettinga, P.G. Feasibility of Anaerobic Digestion for the Purification of Industrial Wastes. In 4 Symposium EAS (1978), June, Munich, Germany, Gesellschaftzur Forderung der Abwassertechnik e.v.(GFA), St. Augustin, p226-256, (1978), Germany.
12. Stander, G.J. Treatment of Wine Distillery Wastes by Anaerobic Digestion 22nd Industrial Waste Conference, May 2-3, p892-907 (1967), Purdue Univ., Purdue, Indiana.
13. Lawrence, A.L. and McCarty, P.L. Kinetics of Methane Fermentation In Anaerobic Treatment, J. Wat. Poll. Cont. Fed., 41, R_1-R_7 (1969).
14. O' Rourke, J.T. Kinetics of Anaerobic Waste Treatment at Reduced Temperature s, Ph.D. Thesis (1968), Stanord Univ., Stanford.
15. van den Berg, L. Factors Affecting Rate of Methane Formation from Acetic Acid by Enriched Methanogenic Cultures, Canadian J. Microbial 22, p1312-319 (1976).
16. Zindev, S.H. and Mah, R.A. Isolation and Characterization of a Theromophilic Strain of Methanosarcina Unable to Use H_2-CO_2 for Methanogenesis, Appl. Envn. Microbiol. 38, p996-1008 (1979).
17. van den Berg, L. and Lentz, C.P. Effects of Film Area to Volume Ratio, Film, Support, Height and Direction of Flow on the Performance of Fixed Film Reactors, Proc. Workshop, Jan 9/10/(1980), Howey-In-The Hills, Florida.
18. Speece, R.E. and Kim, J.A. Effect on Short Term Temperature Variations on Methane Production, J. Wat. Poll. Cont. Fed., 42, p1990-1997 (1970).
19. Jewell, W.J. and Morris, J.W. Influence of Varying Temperature, Flow Rate and Substrate Concentration on the Anaerobic Attached Film Expanded Bed, 36th Purdue Univ., Industrial Waste Conference, (May 1981), Purdue Univ., Indiana.
20. Anderson, G.K. and Durate, A.C. Research Application of Anaerobic Process Envn. Tech. Letters, 1, p. 484-493 (1980).
21. Forsetell, B., (Unpublished Data), 1980, Sewedish Water and Air Pollu. Inst., Sweden.
22. McCarty, P.L. Anaerobic Waste Treatment Fundamentals, Part. IV, Public Works, p95-99 (Dec. 1964).
23. Mahr. I., Untersuchungen iiber die Rolle der niederen Fettsauren beim anaeroben Fauprozess and Einblicke in servic Biozonose, Water Res., 3, p. 507-517 (1969).
24. McCarty, P.L. Anaerobic Waste Treatment Fundamentals, Part-I, Chemistry and Microbiology, Public Works, 94, p107-112 (1964).
25. van den Berg, L. and Lentz, C.P. Food Processing Wastewater Treatment by Anaerobic Digestion, 32nd Industrial Waste Conference, May 10-12, p252-258 (1977), Purdue, Indiana.
26. Benjamin, M.M. Treatment of Sulphite Evaporator Condensate with an Anaerobic Reactor, TAPPI Envn. Conf., p1-10 (April 1981), New Orleans.

27. Martensson, L. Anaerobic Wastewater Treatment in a Carrier Assisted Sludge bed Reactor, Presented at IAWPR Specialized Seminar on Anaerobic Treatment (June-1982), Copenhagen, Denmark.

28. Mitchell, R. Introduction to Environmental Microbiology, (1974), Englewood Cliff, N.J. Prentice Hall, Inc.

29. Sroczynski, A. Ocyszczanie Sciekow Krochmalni Zesz. Nauk. Polit. Lodzkiej, nr 169, Chem Spz., 23, p149-169 (1975).

30. Hus, L. Karakteristik pa avloppsvatten struktioner. Ant. fran STU Anaerobseminarium 1978-10-18, STU-119, p27-23 (1979).

31. Pette, K.C. Anaerobic Wastewater Treatment at CSM-Sugar Factory, Paper presented at 16th General Assembly CITS. p1-8 (1979), Amsterdam.

32. Chynoweth, D.P. Anaerobic process, J. Wat. Poll. Cont. Fed. (June 1979).

33. Carrondo, M.J.T. Anaerobic Filter Treatment of Molasses Fermentation Wastewater, Presented at IAWPR specialized seminar on Anaerobic Treatment (June 1982), Copenhagen, Denmark.

34. Kugelman, I.J. and McCarty, P.L. Cation Toxicity and Stimulation in Anaerobic Waste Treatment, J. Wat. Poll. Cont. Fed., 37 (1), (1965).

35. Lawerence. A.W. and McCarty, P.L. Role of Sulphide in Preventing Heavy Metal Toxicity in Anaerobic Treatment, J. Wat. Poll. Cont. Fed., 37(2), p. 393 (1965).

36. Hayes, T.P. and Theis, T.L. The Distribution of Heavy Metals in Anaerobic Digestion J. Wat. Poll. Cont. Fed., 50(1), (1978).

37. Barker, H.A. Biological Fermentation of Methane, Indust., and Engg. Chemistry, 48, p1438-1442 (1956).

38. McCarty, P.L. Anaerobic Waste Treatment Fundamentals, Part-II, Public Works, p123-126 (Oct. 1964).

39. Pohland, F.G. and Bloodgood, D.E. Laboratory Studies on Mesophillic and Thermophillic Anaerobic Sludge Digestion, J. Wat. Poll, Cont. Fed., 1, p. 11-42 (1963).

40. Sawyer, C.N. and Howard, F.S. Scientific Basis for Liming of Digesters, Sewage and Industrial Wastes, 26, p935-944 (1954).

41. Capri, M.G. and Marais, G, V.R. pH Adjustment in Anaerobic Digestion, Water Res., 9, p307-313 (1975).

42. Kasper, H.F. and Wurrmann, K. Product Inhibition in Sludge Digestion, Microbial Ecology, 4, p241-248 (1978).

43. Hobson, P.N. and Snow, B.G. Inhibition of Methane Production by *Methano bacterium Formicicum*, Water Res., 9, p649-652 (1976).

44. Bhattacharya, S.K. Toxic Substances in Methane Fermentation; Fate and Effect on process Kinetics; Ph.D. Thesis, Drexel University (1986), Philadelphia.

45. Wiegant, W.M. The Mechanism of Ammonia Inhibition in Thermophilic Digestion of Livestock Waste, Agric. Wastes, 16(4), p243-253 (1986).

46. Heinrichs, D.M. Effect of Ammonia on Anaerobic Digestion of Simple Organic Substrates. J. Env. Engg. (ASCE), 116(4), p698-709 (July/Aug. 1990).

47. Jeris, J.S. and McCarty P.L. The Biochemistry of Methane Fermentation using C-14 Tracers, 17th Industrial Waste Conference, Purdue Iniv. Engg. Ext. Series 112 (1963).

48. McCarty, P.L. and Jeris, J.S. Individual Volatile Acids in Anaerobic Treatment, J. Wat. Poll. Cont. Fed., 35, p. 393 (1963).

49. Novick, A. Mutagens and Antimutagens, Brookhaven Symp. Biol. No. 8, p201 (1956).
50. Clausen, E.C. Biological Production of Methane from Energy Crop, Biotech. Bio-engg. XXI (1979).
51. Dugan, P.R. Biochemical Ecology of Water Pollution, Plenum Press (1972), New York.
52. Buswell, A.W. and Rodebush, W.H., Water, Scientific Amer., 194 (1956).
53. Dewalle, F.B., and Chian, E.S.K. Kinetics of Substrate Removal in a Completely Mixed Anaerobic Filter, Biotech. Bio. engg., XVII (1956).

CHAPTER 6

TOXICITY EFFECTS IN ANAEROBIC PROCESSES

6.0 Introduction

Sufficient knowledge about toxicity of important toxins and inhibitors in any biological process is imperative for an application of process. It has been found that engineers designing treatment systems for removal of organics from wastewaters will often prefer conventional aerobic systems where as in many cases, anaerobic option would have been satisfactory. While considering many advantages of anaerobic treatment, then the question must be asked about the limited application of anaerobico processes in this country.

It has been the feeling that the use of anaerobic options cannot tolerate continuous or slag doses of toxic substances in wastewaters. There are several well documented cases in the literature where presence of toxicants may have resulted in inhibition of bio-oxidation processes which ultimately lead to the failure of the process, particularly in the case of anaerobic sludge digestion. It is worthwhile to mention that it is not necessary to presume that methanogens are more sensitive to toxicants than facultative organisms.

During the past decade, there have been many studies concerning the possible inhibitory effects on anaerobic digestion process due to several compounds found in the present industrial wastewaters. In many cases where municipal digester performance had decreased, the source of the upset could be related to presence of a chemical that caused inhibition to the micro-organisms. These compounds are released from different streams generated by different industrial setups, Table 6.1.

These wastewaters may not be subjected to pre-treatment at the industrial complex itself, and are released into the sewers from where they are transported to the local treatment plant for final treatment. At the plant, these chemicals eventually wind up in the sludge being digested. Due to large number of the compounds generated and the geographical specificity of the industries that generate them, the available literature contains numerous but isolated cases.

McCarty,[1] has been able to summarise the available literature on toxic materials and their control particularly in relation to the following :

Sodium, Potassium, Calcium, Magnesium, Ammonical nitrogen, Sulphide, Heavy metals and few organic compounds concentration levels.

During this period only conventional type of anaerobic digestion was being practised.

6.1 Inhibition, Toxicity, Antagonism and Synergism

Fig 6.1 is a schematic representation of rate of biological reaction rate versus increasing concentration of compounds depicted by McCarty[1] and is quite analogous to the graphical representation by Kugelman and Chin[2].

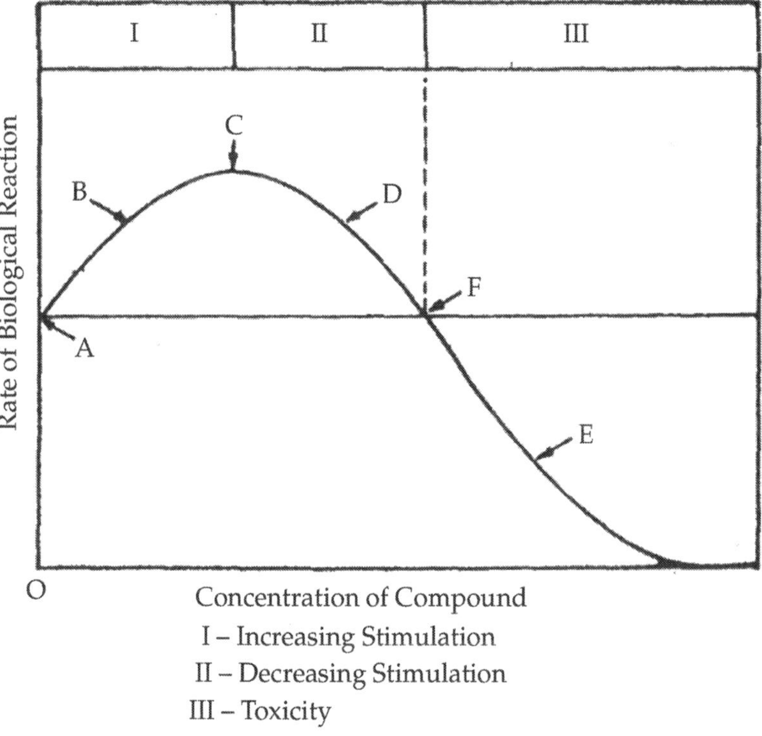

Fig. 6.1 : Schematic Representation of Effect of Compound on Biological Activity

Point-A represents the rate at which biological reactions occur prior to addition of identified compound. Addition of the chemical in limits that produce values below the optimum concentration usually result in a stimulation of biological activity as indicated by an increase in the reaction rate of the micro-organisms present. Point-C is considered the optimum concentration value since the micro-organisms are at their highest rate of biological activity. Any additional amount of compound added that results in concentration greater than optimum but less than the cross over concentration (threshold limit point-F), will produce a decrease in biological activity of the system. As was the case with region-B where micro-organisms were being stimulated, the same micro-organisms in region-D were now considered to be inhibited by the compounds added at those concentrations.

If the compound is presented in concentration greater than cross-over concentration (point-F), the biological activity of the system is now at a value less than when the compound was absent. At region-E, the compound (at those concentrations) is considered toxic to the micro-organisms present, and, if added to produce high enough concentration, the microbial activity will totally cease.

Table 6.1 : Types of Industries that Produce Toxic Compounds*

1. Steam electric power plant
2. Leather tanning and finishing
3. Iron and steel mills
4. Petroleum refining
5. Petro Chemical complexes
6. Inorganic Chemical industries
7. Textile mills
8. Organic chemical industries
9. Non-ferrous metal industries
10. Paving and roofing materials
11. Plant and ink formulation and printing
12. Soap and detergent industry
13. Auto and other laundries
14. Plastics and synthetic materials
15. Pulp/paperboard and converted paper products
16. Rubber processing
17. Adhesives, gum and wood chemicals
18. Pesticides industry
19. Pharmaceutical industry
20. Explosive manufacturing units
21. Mechanical products units, viz Aluminum forming, battery manufacturing, coil coating, coffer forming, foundaries, plastic processing, porcelain enamel, electrical and electronic products etc.
22. Electroplating
23. Ore mining and dressing
24. Coal mining

* Source Keith : L.H and Telliard, W.A., Priority Pollutants–A perspective view, Engg. Sci. Tech. 13 p. 416-423(1979).

To generalise, toxicity effects are highly dependent on concentration and type of material present in the system. When a compound is added to a biological system, the concentration level that is maintained dictates whether or not the biological activity within the process is stimulated, inhibited, or will cease altogether. In addition, toxicity may be dependent on the type of material added to the system, because some compounds are more toxic than others. Considering the graphical representation, Fig. 6.1, the compound considered more toxic will inhibit the micro-organisms present at lower concentrations than the less toxic materials.

A careful attention is necessary about environmental conditions under which toxic levels are observed. Several mechanisms are known that have the ability to decrease or increase the magnitude of the toxicity exemplified by a biological process due to the presence of certain materials. This is important for any researcher if he is not aware about these mechanisms at the time the observation is made and, therefore, erroneous conclusions about toxic effects on biological system from a specified compound are likely to be made. The most important mechanisms can be categorised as follows :

- antagonism and synergism
- complex formation
- microbial growth
- acclimation
- volatilization

6.1.1 Antagonism and Synergism

Antagonism is the condition where the toxicity effects caused by the original material is reduced to the presence of a second, while synergism is the condition where toxicity effects caused by the original material is increased by the second material in the system. Both antagonism and synergism are closely dependent upon concentration[2]. Fig. 6.2 presents a schematic diagram to illustrate these effects and was developed by McCarty, *et al.*[3] based on arbitrary dual system of compounds A and B. Compound A is introduced into a system at concentrations that produce toxic conditions, the biological reaction rates of the micro-organisms at this point are measured and used as a control value. Compound B is now introduced in increasing amounts in order to measure the pattern of the antagonistic or synergistic relationship between the two compounds.

When compound-B is added, if the effects are antagonistic, the biological reaction rate of the micro-organism involved will definitely increase. While on the other hand, if the addition of compound-B results in the reduction of biological reaction rate of the micro-organisms further, then the effects are synergistic.

The antagonistic response pattern can be determined as follows. As the concentration of compound-B increases (even at low concentration of B) and the concentration of compound A stays constant, the toxic effect produced by compound-A decreases as shown by the increases in the biological reaction rate of the micro-organisms involved in the process. After increasing the concentration of B still further, the antagonistic effects reach a peak value followed by a decrease in biological reaction rate. This decrease continues as the concentration of compound-C increases until the crossover concentration is observed. The cross over concentration is defined in terms of biological reaction rate in the system at which the biological reaction rate in the system containing both compounds-A and B equal the biological reaction in a system containing only compound-A. Further increase in compound-B will decrease the reaction rates even lower.

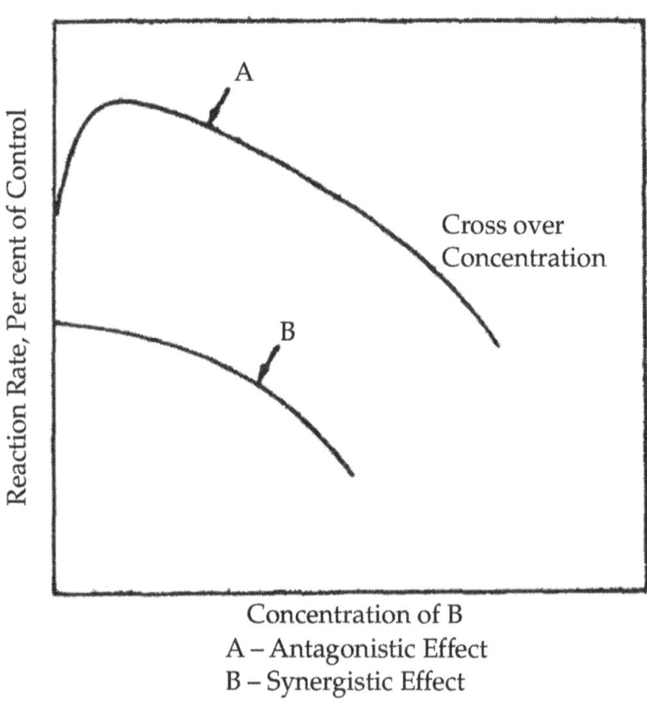

Fig. 6.2 : Schematic Representation of Antagonistic and Synergistic Relationship in Two Cation System with the Concentration of Cation–A Held at a Toxic Level.

For synergistic systems, low concentrations of compound-B have little effect on the biological reaction rates of the micro-organisms in the system. Further increase in the concentration of compound-B (continually), there is steady decline in the biological reaction rate. This was a typical case when two compounds are encountered. However, there are cases where combinations of compound interact to antagonise the toxic compound. In this case, the first compound is described as primary antagonist while the second is identified as secondary antagonistic. A primary antagonist is a compound which produces antagonism in a dual system while a secondary antagonistic is a material which antagonizes the toxic compound only when the primary antagonist is present[3]. Although antagonistic and synergistic behaviour have been observed for only few compounds, it should not be presumed that they do not exist for other compounds.

The information available about antagonism and synergism is centrted around light metal cations such as potassium, magnesium, calcium and sodium[2,3] online. It was concluded that only small concentration of antagonistic or synergistic compounds were required to invoke a change in the biological reaction rate of the micro-organism. In addition, combinations of materials were found to produce either antagonistic or synergistic effects, but not both.

The effects of some of studied-cations are summarised below :

Sr.No	*Cations**	*Stimulatory*	*Moderately inhibitory*	*Strongly inhibitory*
1.	Sodium	100-200	3500-5500	8,000
2.	Potassium	200-400	2500-4500	12,000
3.	Calcium	100-200	2500-4500	8,000
4.	Magnesium	75-150	1000-1500	3,000

*All values in mg/L.

Similar effect has been observed for ammonia-nitrogen in an anaerobic system and is presented below :

Sr.No.	*Ammonia-N**	*Effect*
1.	50-200	Beneficial
2.	200-1000	No adverse effect
3.	1500-3000	Inhibitory at high pH
4.	Above 3000	Toxic

* All values in mg/L.

The quantity of sulphide salts required for precipitation are given below :

Sr.No	*Sulphide salt added*	*Concentration of heavy metal precipitated*
1.	1mg/L Sulphide (S^{-2})	1.8-2.00mg/L
2.	1mg/L Sodium Sulphide (Na_2S)	0.75-0.84mg/L
3.	1mg/L Sodium Sulphide ($Na_2S.9H_2O$)	0.24-0.27mg/L

Soluble heavy metals + Sulphide ⟶ Insoluble heavy metal sulphides

Cu, Ni, Zn

[Toxic] Non-Toxic

Inhibitory, toxic, etc. effects have been studied by Hovious, *et al.*[4] for organic compounds commonly found in waste-waters of petro-chemical industry on the mechanism of the anaerobic digestion process. One particular aspect of the study was concerned with the effect a combination of these compounds had on anaerobic digestion.

There are three mechanisms which are said to be favouring the inhibition caused by sulphide. It is supposed that sulphate is acting as an alternative electron acceptor (rather than sulphide) and is primarily responsible for the inhibition of methanogens in a sulphate containing environment. Recent studies, however, indicate both processes (methanogenesis and sulphate reduction) can occur simultaneously and at relatively high (10mM) sulphate concentrations. Fig 6.3 shows the effect of H_2S in an anaerobic environment.

Initially, slowly increasing but equal quantities of four known inhibitors, *viz.*

- phenol
- acrylonitrile
- ethyl acrylete
- formaldehyde

were injected into a synthetic waste being fed to an anaerobic filter reactor. They were able to conclude that the compounds act synergistically causing the reactor to be inhibited at concentration level below those that caused inhibition when each of the compound had acted alone.

In order to investigate the true nature of toxic effects that compounds exerts on the above mechanisms identified for anaerobic systems, synergistic and antagonistic relationship between compounds other than the metal cation group must be examined. No information is available at this time regarding this aspect and needs extensive research work.

6.1.2 Complex Formation

This aspect like antagonistic relationship between compounds is a mechanism that reduces the magnitude of toxicity experienced by the system. To form a complex, the compound is not in the proper chemical form it gain access into the cell. This tends to keep the biological activity at its normal condition.

In some cases, complex formation can be easily predicted according to the stoichiometric relationships. While in other cases, it becomes difficult to predict complex formation because some of the compounds will not bind until they have been transformed into a suitable condition by microbial activity[2].

An example of complex formation in relation to anaerobic waste treatment is the complexation of heavy metals and sulphides. Heavy metals have been blamed for many anaerobic treatment failures[5–7]. Low but soluble concentrations of the following salts are toxic in their free ionic form, *viz.*,

- zinc
- copper
- chromium
- nickel

and it is these metals which have been associated with most of the problems of heavy metal toxicity in an anaerobic process. Table 6.2 highlights metal solubility product constant for some of the commonly occurring metal ions in the wastewater[8]. If the metal concentration is not unusually high, then there are usually sufficient sulphides present to precipitate them in non-toxic form. Sulphides are biologically produced in the anaerobic system by reduction of sulphates or other sulphur containing organics or released from degradation of sulphur containing organic compounds such as proteins[6].

On the other hand, if heavy metals concentrations are quite high and sufficient sulphides are not formed during anaerobic reactions then sodium sulphide or sulphate salt (which will be reduced to sulphide under anaerobic conditions) can be added.

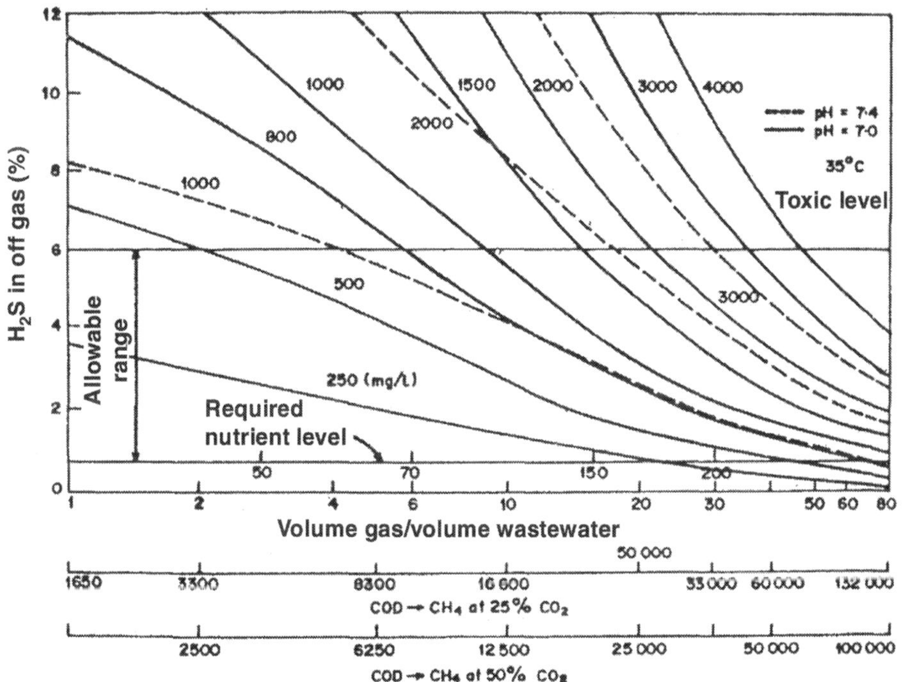

Fig. 6.3 : Effect of H_2S on the Anaerobic System.

Second example identifies a natural process by which heavy metals are removed and/or reduced by precipitation of metals as an insoluble carbonate salt. This is considered proper for certain metals only if the pH value of the anaerobic reactor is high enough (*i.e.* pH ≥ 6.2 for Cd and pH ≥ 6.7 for Zn). An ideal illustration of this protection is against iron toxicity and is good at all pH values greater than 6.4 which permits any sulphide initially present as FeS to be released to precipitate other metals without simultaneously releasing high concentration of ferrous ions into solution.

Table 6.2 : Solubility Products of Some of the Commonly Occurring Metals in Anaerobic System[8]

Sr.No.	*Heavy Metal*	*Sulphide Salt*	*Solubility Product*	*Solubility, mg/L*
1.	Cd	CdS	1.0×10^{-28}	1.0×10^{-9}
2.	Cu	CuS	8.0×10^{-37}	9.0×10^{-14}
3.	Co	CoS	5.0×10^{-22}	2.0×10^{-6}
4.	Fe	FeS	4.0×10^{-19}	6.0×10^{-5}
5.	Pb	PbS	7.0×10^{-29}	2.0×10^{-9}
6.	Ni	NiS	3.0×10^{-21}	5.0×10^{-6}
7.	Zn	ZnS	2.5×10^{-22}	2.0×10^{-6}

6.1.3 Microbial Growth

Micro-organisms involved in biological waste treatment processes can be rendered metabolically inactive as a results of reactions occurring between inhibitory compounds present in the system and some essential protein or molecular group of the organisms. Another inhibition mechanism involves the absorption of toxic compound on to the surface of the cell. A reaction of this type could possibly change the cell's permeability. Inhibition of enzymes is a result of the inhibitory substance and substrate competing for the same attachment sites on the enzymes, destruction of an essential functional groups of the enzyme or an attraction of the inhibitory compound for the part of the enzyme[4].

Hovious, *et al.*[4] found that the ability of a compound to inhibit a biological waste treatment process is also a function of the type of growth exhibited by the micro-organism involved. In a system containing fast growing organisms, there is greater possibility that inhibition can occur because the substrate uptake and conversion rate are greater than on a system of slow growers. In additions, it would be expected that inhibition effects would be greater in flocculated growth systems than slimes or attached growth system. It is because the flocculated growth systems are more exposed to inhibitory substances (more cells exposed) and in attached growth systems, only outside layer of the cells are exposed to the inhibitory compound. An extensive research programme is needed on this aspect.

6.1.4 Acclimation

Inactivation of the metabolic activity of the micro-organisms results when an inhibitory substance is added in high concentration into a biological system. Acclimation is defined as the ability of the micro-organism to overcome the adverse effects caused by the inhibitory substance. It is often observed that there the complete removal of toxic compound is not possible from a system. It implies that the mechanism of acclimation represents a re-alignment of the metabolic pathways of the micro-organisms instead of a development in a mutation or survival of fittest[2].

The rate of addition of inhibitor substance into a system can have an effect on the micro-organisms ability to acclimate to the toxic substance. This can be achieved either by gradual dose or slug input. It has been the experience worldover that inhibition caused by toxic substance is not as severe if the specific substance is added gradually to the system instead of in a slug dose.

McCarty, *et al.*[10] and Kelgelman, *et al.*[2] undertook studies to determine the nature of salt toxicity in the anaerobic digestion process. Conventional type digesters were used and injected with varying levels of cations, *viz.*, calcium, sodium, potassium and magnesium in order to assess the toxic concentration of each of the above metal ions. It was found that the cations injected on a slug basis were more toxic than when added gradually. It was possible to acclimate the micro-organisms to cation concentrations 2 to 3 times the slug dose concentrations that caused toxicity.

Parkin, *et al.*[11] Yang, *et al.*[12] tried to evaluate the response of methane fermentation systems to exposure of some 30 organic and metallic pollutants commonly found in industrial wastewaters. Toxic substances were added to the anaerobic reactors operating in different modes of hydraulic flow, *viz.* batch, semi-continuous and continuous. Batch and semi-continuous operation used similar procedure as adopted by Owen, *et al.*[14] *i.e.* using serum bottles. Upflow laboratory sized anaerobic filters were used for continuous studies.

The batch systems were fed only beginning of that test during which a given dose of toxic material was also added. Several units for each compound tested was required so that the effects over a wide range of concentration could be observed. In the semi-continuous studies serum bottles were fed once a day with the proper amounts of substrate, and test compounds. Also, at this time, a pre-determined amount of solids was removed to provide the system with desired solids retention time. The toxicant was added to the continuous flow units in two different fashions. In the first method, the toxicant was added continuously into to the units with substrate while with the second method the toxicant was added for only one day periods. For all the tests, the amount of gas produced was used as a measure of metabolic activity and was compared against a control unit to determine the degree of inhibition.

In the batch experiments, unacclimated methanogenic organisms were fed nickel concentrations ranging from 50 to 500 mg/L. A concentration of 200mg/L of nickel showed 86 per cent of the gas production of the control value while a concentration of 300mg/L of nickel showed only 43 per cent of the control value. In the semi-continuous system, gas production was inhibited by concentration of 80 to 100 mg/L of nickel. By the end of 10 days, however, all systems recovered and were producing gas values equal to control. Nickel was added continuously in the feed to an anaerobic filter over 30 days. The results indicated that once acclimated, 250mg/L of nickel could be handled by anaerobes. Using a different filter, one day slug addition were made to the filter. The results showed that slug doses of upto 500 mg/L could be handled with only a 25 per cent decrease in gas production. After a period of about 9 days the gas production had returned to the value before the compound was introduced into the system.

Parkin *et al.*[13] were able to conclude that methanogenic bacteria have considerable potential to acclimate to toxic compounds found in waste treatment systems. It was also found that with acclimation process, toxic compounds, 2.4 to 12 times those causing inhibition to unacclimatized bacteria could be tolerated with no decrease in gas production.

It is interesting to mention that concentration and exposure time of a toxic substance play a vital role in the micro-organism's ability to acclimate to the compound. Gradual introduction of a toxic substance in an anaerobic system allows time for the micro-organism to acclimate to the toxic substance. It may also be added that even though slug doses to biological treatment systems sometimes do occur, the general tendency shown by the inhibitory material is that it tries to build up gradually in the system instead of suddenly. This aspect does allow some acclimation to occur in actual systems. It may be mentioned that a special research programme is needed to study the entire gambit of microbial acclimation.

6.1.5 Volatilization

This refers to gas stream flows through the reactor, with some of the toxic substances transferred to the gas phase. This helps to remove the substance alongwith exiting gas stream. This process of volatilization can be useful in aiding an anaerobic system to recover from inhibitory, toxic effects caused by organic toxicant additions.

Kirsch, *et al.*[15] mentions that volatization mass transfer process phenomenon and the rate of removal of pollutant is equal to the product of mass transfer coefficient and a driving gradient. This expression can be represented as follows:

$$R_v = K_v (S_0 - S_0) \quad ......(6.1)$$

where,

R_v – rate of removal of material by volatilization, M/L.T

K_v – overall mass transfer coefficient for volatilization, T^{-1}

S_0 – effluent concentration in liquid if there exists an equilibrium between liquid and gas above M/L^3

S_0 – influent concentration to the system, M/L^3

This expression involves determination of overall mass coefficent (K_v). In an anaerobic system gas following through the system is primarily methane and carbon-dioxide generated by the anaerobic micro-organisms as a by-product of waste utilization.

In some of the cases where the anaerobic system was not working well but after a suitable interval of time, the system returned to normal condition. It seems that volatilization mechanism is the only possible answer to this behaviour. Pearson, *et al.*[16], while developing inhibition models, had determined that an anaerobic system severely inhibited due to the presence of formaldehyde was able to recover. This was related to volatilization mechanisms which reduced the formaldehyde concentration to non-inhibitory levels. Stucky, *et al.*[17] attributed the recovery of semi-continuous anaerobic reactor (initially inhibited by vinyl acetate) to the removal of the pollutant by volatilization. The literature on this aspect of removal mechanisms to reduce the concentration of toxicant to a level where it becomes non-inhibitory is limited. Therefore, exhaustive investigation is required particularly for organic compounds which are volatile but also toxic.

6.2 Anaerobic Tests Used for Measuring Toxicity

Failures of anaerobic systems are ascribed to the presence of toxic substances in the reactor system. Unless it is certain that failure has occurred due to certain compound in the waste, chemical analysis alone is useless. Therefore, there is a need to develop suitable tests to determine the toxicity of organic and metallic compounds. These toxicity tests can be categorised as follows:

- batch test,
- semi-continuous test,
- reversibility test, and
- continuous flow test.

With the exception of reversibility test, all other tests can be used to determine toxicity of a compound to the micro-organisms. To accomplish this, the performance (metabolic activity) of the units containing the material to be evaluated is compared against that of a control unit. The most important parameters of measuring biological activity are as follows :

- cell growth,
- organic reduction, and
- gas production.

It is worth while to mention that most of the studies to date have used gas production as a means to monitor the metabolic activity just the opposite of aerobic studies where oxygen uptake rates are considered. The other two parameters related to cell growth and organic reduction are considerably more time consuming and consistency of data obtained is also questionable.

The main purpose is to determine whether or not the test compound is toxic to the anaerobic organisms and at what concentration levels. Focus has to be diverted to methanogenic bacteria because formation of methane from volatile acids is considered to be the rate limiting step in the process[16]. It is well documented in the literature that most anaerobic process failures stem from inhibition of methanogenic bacteria.

6.2.1 Batch Toxicity Test

In bio-degradation tests the compound being tested is the sole carbon source for bacteria. In case the micro-organisms are able to metabolize the compound directly, then it is considered bio-degradable. The batch test on the other hand determines the effect a test compound has on micro-organisms directly metabolizing a carbon source other than the test compound. The basic procedure for all anaerobic batch toxicity tests are by and large similar.

It has been determined extensively that acetic acid is responsible for 70 per cent of methane gas produced in the second stage with 15 per cent from propionic acid[19,20]. It must be mentioned that, to date, only pure cultures of methanogens growing on formate or acetic acid have been isolated. This is another area of research.

Batch toxicity tests are primarily concerned with the response of those anaerobic bacteria which utilize acetic acid when the test material is present. A reactor vessel of volume 125mL to 1.5 litres is required. Under anaerobic conditions, everything required to sustain an anaerobic population, i.e., seed organisms, dilution water, carbon source and nutrients is added to the unit at the beginning of the experiment with no further additions made. The seed micro-organisms are developed from anaerobically digested sludge. The digested sludge is placed in a batch laboratory-sized digester and is fed with substrate to be used in the assay. The digester is operated on a semi-continuous basis with hydraulic and solids retention time ranging from 15 to 150 days. This unit is usually operated at a temperature of 35°C which is optimum for mesophilic phase.

Jeris, *et al.*[19] has found that anaerobic toxicity is determined as the adverse effect of a substance on the predominant methanogens which implies those micro-organisms that convert acetic acid to methane gas.

In order to estimate the effect that the added compound has on the methanogens present, the performance of the unit containing the test compound is compared with that of the control test. The activity is measured in term of methane or total gas generated. The amount of methane produced from acetic acid can be determined from the following chemical reaction. In place of sodium, potassium acetate can also be used.

$$CH_3\ COONa + H_2O = NaHCO_3 + CH_4 \qquad ...(6.2)$$

To fully test a compound, the substance is added at increasing concentrations to different units, to provide a range of effects increasing from non-inhibitory to severely toxic. Parkin, *et al.*[13] found that with increasing dose of nickel from 0 to 500mg/L, the total gas production over an 8 days period, the rate of methane gas was reduced including the total quantity of gas produced. The maximum rate ratio (MRR) as defined by Owen, *et al.*[14] is given as follows:

$$\text{MRR} = \frac{\text{Gas production with test compound over specific time perid}}{\text{Gas production of control}} \quad ....(6.3)$$

There are three ways by which the metabolic activity of the microbes can be ascertained in presence of a toxicant and are given as follows :

- No difference in gas production between the unit containing the test compound and the control unit.
- A lower gas output in unit containing the test material than in the control.
- A higher gas output in the unit containing the test material than in the control unit.

Case-1 refers to condition where the test compound present neither inhibits the methanogenic bacteria nor degrades them. Case-2 identifies the situation where the presence of test material at that particular concentration inhibits the methanogenic output. Case-3 highlights the condition where the test material is not inhibitory to the micro-organisms and at the same time either stimulates or is degraded by them.

Although the basic objective of toxicity tests is to determine toxic concentration levels, in some cases biodegradation can also be predicted. The batch toxicity test is the only toxicity test that can determine toxic levels and also whether or not the compound is being degraded. In batch toxicity tests the control unit and the test unit are fed equal amounts of acetic and/or propionic acid at the beginning of the experiment. Since no further substrate is added through the course of the experiment and if no inhibition occurs in the test unit, then equal total gas amounts within experimental error should be achieved in both units. If the gas produced is the test unit is greater than that in the control, this means that the test compound in being degraded since it is the only other available carbon source left in the system.

There are three methods that have been used to date and are given below :

- anaerobic Warburg respirometer,
- anaerobic toxicity assay, and
- batch toxicity assay.

6.2.1.1 Anaerobic Warburg Respirometer

To determine the amount of oxygen consumed in aerobic microbial degradation of organic compounds, the Warburg respirometer is a readily available and widely used device. It can also be adopted to measure gases evolved from the breakdown of organic compounds by anaerobic processes[3,4,20–22]. It is extremely useful when no specific analytical technique for the measurement of the test compound exists[23]. The test compound along with a suitable acclimated seed is placed in a test flask. At a constant temperature and volume, the amount of gas consumed or produced is represented by changes in pressure. By measuring pressure changes, and knowing the gas and fluid volumes of the flask, temperature, the gas being exchanged and density of the fluid in the manometer, the amount of gas evolved can be calculated[24]. This method allows to run many samples simultaneously. Rate of change and lag period can be accurately measured. However, this equipment is costly and requires trained personnel to operate. Sample size required in small which makes subsequent chemical analysis difficult. In addition, it is difficult to collect samples from the gas or liquid phase during the test.

6.2.1.2 Anaerobic Toxicity Assay (ATA)

ATA test was developed by Owen, *et al.*[14] This test utilizes the theory and procedures developed by Miller, *et al.*[25] The reactor unit is a stoppered 125 mL serum bottle and contains a combination of nutrients, test material, seed organisms and a carbon source for a total volume of 50mL. Owen, *et al.*[14] used acetate (750 mg/L) and propionate (250mg/L) as the principal carbon source. Stuckey, *et al.*[17] added 0.10mL of ethanol to the serum bottle volume of 50mL. Liquid and gas samples were extracted by syringe for analysis.

6.2.1.3 Batch Toxicity Assay (BTA)

BTA is a modified version of ATA method to study toxicity to unacclimated methanogenic bacteria[11-13]. Unlike ATA test which uses both propionate and acetate as the main carbon sources, the batch toxicity assay uses only acetate and at much higher overall concentration (3000mg/L added as CH_3COOK). The rest of the procedure is by and large the same as described in ATA.

6.2.2 Semi-Continuous Assays

There are two basic differences between batch toxicity test and semi-continuous assays test and are given as follows :

- a solid retention time is maintained in the system,
- substrate levels are renewed daily on batch basis.

Batch toxicity tests utilize small quantity of organisms to consume a finite amount of substrate in a short time. The semi-continuous assay that closely resembles the operation of a conventional digestion process and requires 24-hour cycle of feeding the unit, wasting solids and measuring the daily gas production. Each day fresh substrate is added to the unit to replace the substrate used by the organisms from the previous day. Excess solid removal is essential to maintain a specific solids retention time.

The test unit are inoculated with seed micro-organisms and necessary nutrients alongwith carbon source are added. The unit is maintained at constant temperature of 35°C. The test unit should be operated for a period of atleast 7 times the solids retention period before addition of test compound[17].

There are two procedures to determine the relative amount of inhibition caused by the presence of test compound. The effect on the methanogenic bacteria is determined by the amount of methane gas produced.

6.2.2.1 Method-I

This test does not use a control as outlined in the batch toxicity test section. This procedure requires a substrate to be used by the anaerobic micro-organisms which will produce methane gas in a well defined stoichiometric relationship. Excess solids are wasted on a daily basis to maintain proper solids retention time, and fresh substrate is added in an amount only to replace that used from the previous day and also to maintain a constant substrate concentration in the system.

The steady state unhibited gas (without any test compound present) is determined at the beginning of the experiment and is the control amount of gas production. The amount of added substrate that is needed can be determined by the amount of methane gas produced for that day, which is based on the proper stoichiometric relationships. When a test material is added to the system and the gas production is less than the previously determined steady-state value, the system is considered to be inhibited.

6.2.2.2 Method-II

This procedure requires conventional control unit, when the added substrate is of a complex nature and the stoichiometry is not obvious. Therefore, the only precise method to measure the relative amount of inhibition when using complex substrate is to compare the performance of units containing test materials with that of a control unit when both are receiving equal daily amount of the substrate.

6.2.3 Reversibility Assays

Parkin, *et al.*[5] conducted studies on reversing inhibition and toxicity effect anaerobic systems. Experiments are initiated by injecting the toxicant at different concentrations into serum bottles. After the desired exposure time to the toxicant (may be 1 hour or 1 day) the serum bottle was placed into a centrifuge to separate the active biomass from the toxicant. Immediately following, toxicant free acetate enrichment media was injected into the serum bottle to replace the withdrawn supernatant. Daily feeding and gas measurements were then continued as outlined in semi-continuous mode section for the purpose of monitoring the recovery of the system.

Parkin, *et al.*[13] found that a recovery from toxicant exposure was dependent on both toxicant concentration and exposure times. With nickel concentration of about 800 mg/L and an exposure time of less than one day resulted in only a slight decrease in gas production once the toxicant in the liquid phase is removed. In addition, they found that a concentration of 2,400 mg/L (1 hour exposure) and an exposure to 800 mg/L, for 4 days resulted in zero gas production for the duration of the experiment.

6.2.4 *Continuous Assay*

Parkin, *et al.*[13] also conducted continuous flow experiments using laboratory-sized anaerobic upflow filters containing void volumes of one litre. The filters were fed synthetic solution containing inorganic nutrients, buffers and acetate as the only carbon source. The loading rate to the filter was initially 5 kg/m^3.d (2000 mg/L influent acetate concentration) which was later increased to 10 kg/m^3.d (3000 mg/L influent acetate concentration). Nickel was continuously added to the filter for over 30 days. After that the system stabilized at a nickel concentration of about 250 mg/L. One day slug additions of nickel were made to another anaerobic filter. At slug doses of up to 500 mg/L, only a 25 per cent decrease in gas production was observed. After a 1,000 mg/L dose, however, gas production almost ceased with the system not recovering significantly within five days.

6.3 Evaluation of Toxicity Tests

In general to predict the response of the anaerobic process to toxicant exposure, the toxicity test should simulate the full scale process as closely as possible. For this reason, to accurately predict the response of toxicant exposure to continuous digestion processes, the continuous toxicity test much in the same manner as semi-continuous toxicity test, should utilize laboratory sized test reactors that closely resemble the full scale reactor. To date, the only documented continuous toxicity tests have been conducted at Drexel University[26]. As with the semi-continuous test (and the semi-continuous full-scale reactor) the solids retention time also plays as important role in the recovery of continuous anaerobic processes when exposed to toxicant materials. It is generally conceded that these new continuous flow anaerobic processes provide solids retention times that are several magnitude higher than those maintained in the semi-continuous processes. Young, *et al.*[27] reported solids retention times in excess of 100 days while they were developing the anaerobic upflow filter process. Gorgan, *et al.*[28] while studying temperature effects on start up of high SRT anaerobic processes estimated the solids retention time to be greater than 100 days in an upflow anaerobic sludge blanket reactor for loadings upto 13.5 kg COD/m^3.d at 25°C.

The effect of solids retention times maintained in the continuous digestion process is similar to that experienced in semi-continuous reactors but to a greater extent. Due to the greater solids retention times maintained, the continuous reactors systems have a greater capacity to recover from toxicant exposure. The longer the SRTs maintained in the system, the longer the system has to acclimate to the toxicant. This means that some concentrations that produce irreversible inhibition in the semi-continuous process (due to washout of all viable organisms) may be sustainable if applied to the continuous process since there are more organisms to be washout before complete system failure. The continuous system has a slight edge over semi-continuous system because the micro-organism have a longer period of time of acclimation, adaption, etc., to occur, due to higher SRTs maintained in the continuous reactors.

Unlike the semi-continuous and continuous toxicity tests, the value of batch toxicity tests is severely limited when used as a method for prediciting system behaviour when exposed to toxicant materials and determining toxic concentration levels, since batch tests (their mode of operation) are not representative of any continuous flow process.

The main purpose of reversibility test was to determine the maximum concentration level and exposure time of the added toxicant that would exhibit reversible toxic effects once the liquid supernatant was separated from the biomass and removed from the system. During the experimental conducted at Drexel University, most of the materials used in reversibility tests exhibited that inhibition was reversible to some degree.

The results produced by a test will not have significance if the procedure used in the test cannot be carried out in a full scale anaerobic process. In order for the results produced by the reversibility test to have any significance, a procedure must be developed so that the toxicity effects produced in a full scale systems by toxicant addition can be reversed.

6.5 Toxicity Models

Several attempts have been made to develop models which could predict toxic response. There are some salient features which are considered by the investigators for developing the toxicity models, *viz.*

- Model should account for observed data and provide insight on the process behaviour.
- Most of the models assume steady state conditions.
- Models should attempt to describe unsteady-state behaviour.

At present there are two types of toxicity models available, *viz* :

- Empirically based models.
- Kinetically based models.

Empirically derived models are developed by observing the response curve (gas production versues time) of a system after it has been exposed to a toxic material in concentration that produce inhibition. The model is then developed by mathematically fitting a best-fit line through the response curve.

Kinetically based models on the other hand, attempt to explain system response due to toxicant exposure by trying to model the effects that the toxicant has on the metabolic activity of the micro-organisms involved in the process.

6.5.1 Empirical Models

Parkin, *et al.*[29] tried to empirically fit an expression based on the experimental data. This equation is quite similar to the D.O. Sag Curve equation. The equation is given as follows :

$$M = A \exp(-k_1 t) + B \exp(-k_2 t) \quad \text{......(6.4)}$$

where

M = methane generated, mL/day

A,B = empirical constant

t = time after addition of toxicant, day

k_1 and k_2 = toxicity rate constant and acclimation rate constant respectively, day^1

The expression for two constants and two rate constants are as follows :

$K_1 = C(T)^c$

$K_2 = D(T)^d$

$B = E(T)^e$

A = control methane generation minus B

where,

T = concentration of toxicant in the serum bottle immediately following slug addition, particularly cyanide (1.0-10.5 mg/L)

C,D,E,c,d, and e = Constants determined from regression analysis.

This model as the ability to predict downtime (periods of zero gas production) as well as to predict the concentration that initiates inhibition (threshold concentration). However, it is difficult to generalise this model for all applications. Further research is required to prove its utility.

Pearson, *et al.*[16] based his analysis on the data from septic tank. The toxic chemicals are introduced into the septic tank when recreational vehicles discharge their holding tanks into the septic tanks. These chemicals (formaldehyde, zinc, phenol, and their different combinations) are present in holding tanks to act as preservatives in order to control obnoxious odours. The activity ratio (A) is defined by equation 6.5.

$$A = G/V \quad \text{...(6.5)}$$

where,

A = relative anaerobic gasification activity

G = total volume of gas produced from spiked sludge, mL

V = total volume of gas produced from unspiked sludge, mL

The best-fit line of the data can be summarised by the following equation 6.6

$$A = [1 + \sum_{i=1}^{m} (Ci/Ki)^a] \quad \text{....(6.6)}$$

where,

m	=	number of preservatives
Ci	=	concentration of the (i)th preservative, mg/L
Ki	=	half kill dose, mg/L, *i.e.*, value of Ci for which A = 0.5 when m = 1
a	=	sensitivity coefficient

The following empirical equation has been described by them

$$A = \left[1 + \left(\frac{\text{Formaldehyde}}{200} + \frac{\text{Zinc}}{400} + \frac{\text{Phenol}}{500}\right)^{1.314}\right]^{-1} \quad ....(6.7)$$

It may be mentioned that the equation assumes the effects of combinations of toxicants to be strictly additive. In addition, this expression has been developed for unacclaimated micro-organisms operating in batch mode of operations.

The anaerobic gas production rate in the semi-continuous test was computed as the hourly gas production per unit volume of sludge. The activity ratio (A) was also redefined to equal the rate of anaerobic gas production per unit volume of toxicant-spiked sludge, divided by the rate of anaerobic gas production per unit volume of unspiked sludge, at the same temperature.

The activity ratio (equations 6.8 to 6.11) at any temperature can be compared against the activity ratio of a control unit, *i.e.* :

$$A = \left[1 + \left(\frac{\text{Formaldehyde}}{100}\right)^{0.2}\right]^{-1} \quad ....(6.8)$$

$$A = \left[1 + \left(\frac{\text{Zinc}}{700}\right)^{0.5}\right]^{-1} \quad ....(6.9)$$

$$A = \left[1 + \left(\frac{\text{Phenol}}{700}\right)^{0.5}\right]^{-1} \quad ....(6.10)$$

$$A = \left[1 + \left(\frac{\text{Quaternary } NH_3 \text{ as Org,} -N}{50}\right)^{0.25}\right]^{-1} \quad ....(6.11)$$

The semi-continuous system were found to respond more gradually to increasing toxicant dose than did the batch system which was reflected in a higher value of the sensitivity exponent 'a' for the batch system.

6.5.2 *Kinetic Models*

Neufeld, *et al.*[30] was the first to develop a model based on kinetics, in order to find whether or not anaerobic degradation was a viable process. This exercise related to phenol removal.

At high substrate concentration, the specific bio-oxidation rate $\left(\frac{1}{x}\frac{ds}{dt}\right)$ approaches a maximum value. The biokinetic model which takes into account all the phases of microbial growth is given as follows :

$$V = \frac{V_o}{[1+K_s/S+(S/K_1)^N]} \quad ...(6.12)$$

Where,

V = substrate utilization rate

V_o = maximum specific substrate utilization rate

S = substrate concentration

K_s = half velocity saturation constant

K = inhibition constant

N = order of inhibition

A synthetic waste containing phenol as organic carbon source and carbonate as the only inorganic source of carbon was used by Neufeld, *et al.*[30] The values of various constants and rate constants are given as

V_o = 0.08 phenol/g VSS/day

N = 4

K_s = 966 mg/L

K = 700 mg/L

The best-fit equation for substrate inhibition model of anaerobic degradation of phenol was found to be :

$$V = \frac{0.08}{[1+700/S+(S/966)^4]} \quad ....(6.13)$$

At the conclusion of the experiment the following carbon balance was obtained.

1.0 mg phenol – C + 0.315 mg carbonate – C = 0.382 mg biomass – C + 0.261 mg volatile acid – c + 0.671 mg soluble TOC (intermediate – C) ...(6.14)

No methane gas was observed to be produced or any alkalinity utilized, indicating that reactions observed are for non-methanogenic anaerobic organisms only and the model represents first stage anaerobic degradation of phenol.

Parkin, *et al.*[26] used traditional Monod type equation as a starting point for developing a method of describing toxicity kinetics

$$dS/dt = (k\ S\ X_a/K_s + S) \quad .....(6.15)$$

where,

k = specific utilization rate

X_a = active micro-organism concentration

K_s = half velocity constant

It is an observed fact that acetic acid utilization is normally considered the rate limiting step in anaerobic process. Therefore, modelling based on acetate utilization should adequately describe the kinetics of methane fermentation of complex organics. Normal design procedure involves predicting minimum generation time (θ_c^{min}). The two values are defined as follows :

- minimum bacterial generation time (θ_c^{Lim}) = $(Y\,k - b)^{-1}$ ....(6.16)
- bacterial washout time (θ_c^{min}) $= \dfrac{(YkS_o - b)^{-1}}{K_s + S_o}$ .....(6.17)

where

Y = yield coefficient

b = decay coefficient

It is a general practice to apply safety factor to θ_c^{min} or θ_c^{Lim} for design purposes. Addition of toxicant material changes the kinetics of methane fermentation. This is reflected by decrease in k and/or increase in K_s values. A return to non-toxic values of K_s and V_{max} can be obtained after the system goes through a suitable period of acclimation, metabolism and toxic washout. Formaldehyde was used as a toxicant material.

This model points out the need for large SRTs when treating industrial wastewaters containing chromic and transient toxic doses. Not only does a proper SRT value insure process stability, it allows for prolonged operation under conditions of washout prior to ultimate failure. The anaerobic filter, expanded bed, anaerobic rotating biological contractor and UASB all have the ability to provide high SRTs and low HRTs. These reactors, therefore, have the built in capacity better than other types of anaerobic processes[26]. However, no description has been given to determine the toxicity kinetics for these types of reactors.

Inhibition coefficient Model developed by Parkin, *et al.*[26] utilizes the concepts developed by Lehninger[30] and is basically based on enzyme kinetics. There are 3-basic types of reversible inhibition as described by Lehninger, *i.e.*, competitive, non-competitive and un-competitive. The models developed by Parkin, *et al.*[26] combine inhibition coefficients with substrate utilization equation (dS/dt) and are given as follows for plug flow at steady state conditions or batch system.

- competitive

  $- dS/dt = (k\,S\,X_a) / [K_s(1 + T/K) + S]$ .....(6.18)

- non-competitive

  $- dS/dt = (k\ S\ X_a) / [K_s + S\ (1 + T/K)]$ .....(6.19)

- un-competitive

  $- dS/dt = (k\ S\ X_a) / (K_s + S)\ (1 + T/K)$ .....(6.20)

Where,

T = toxicant concentration

K = inhibition coefficient

Similarly, models were developed for continuous stirred tank reactors at steady state conditions, *viz.* :

- competitive

$$S = \frac{Ks(1+b\theta_c)(1+T/K)}{\theta_c(Yk-b)^{-1}} \quad ....(6.21)$$

- un-competitive

$$S = \frac{Ks\,(1+b\theta_c)}{\theta_c Y\,k - b\,(1+T/K)(1+T/K)} \quad ....(6.22)$$

- non-competitive

$$S = \frac{Ks\,(1+b\theta_c)(1+T/K)}{\theta_c Y\,k - b\,(1+T/K)(1+T/K)} \quad ....(6.23)$$

An increase in θ_c^{min} is caused by toxicant addition. If proper efficiency is not to be severely decreased after toxicant exposure, sufficiently long SRT values are required.

Volatile acids and higher fatty acids (HA) are potential inhibitors[10,31]. The mechanism must still be regarded as rather obscure, but the undissociated acids apparently play a role. The inhibition function proposed by Andrew[31] is associated with specific growth rate of the micro-organism and is represented as follows:

$$\mu = \frac{\mu_{max}}{[1+Ks/C_{HA}+C_{HA}/K]} \quad ....(6.24)$$

During the investigation of distillery spent wash treatment, it was found by Karhadkar, *et al.*[32] that sulphide particularly, inhibits the anaerobic digestion process (diphasic) and the model used by them is an improvement over the model developed by Bhatia, *et al.*[33]

$$SUR = \frac{RS}{(1+K_sS)(1+K[H_2S])} \quad .....(6.25)$$

Where

SUR = substrate utilization

R = maximum substrate utilization

$[H_2S]$ = hydrogen sulphide concentration in gas

K = inhibition coefficient.

Equation 6.25 can be simplified and is represented by substrate concentration (S)

$$S = \frac{S_o}{[1/Ic + R/K[H_2S]]} \quad ....(6.26)$$

$I_c = K(\theta/\theta_c)$ [contact phase]

Significant characteristics of many xenobiotic organic compounds is that they are inhibitory to their own bio-degradation at high concentration. Thus, it seems appropriate to use a kinetic expression that characterises such as the one proposed by Andrew[31].

$$SUR = \frac{\mu_{max} S}{Y[k_s + S + S^2/K]} \quad ....(6.27)$$

Equation 6.27 reduces to Monod's expression when K becomes infinity. The striking feature of an inhibitory substance is the existence of a substrate concentration above which the specific growth and substrate removal rates has important implication for reactor design and operation.

The main argument against the Andrews equation is its inability to predict the typical substrate concentration above which growth ceases. It seems important that models should accurately reflect the ability of a substrate to totally inhibit its own biodegradation. Luong[34] and Han *et al*[35] proposed model capable of depicting this phenomenon and is given as follows:

$$SUR = \frac{\mu_{max}}{Y}(1 - S/S^*)^n \frac{S}{[S + K_s(1 - S/S^*)^m]} \quad ....(6.28)$$

Where

S^* = critical substrate concentration above which growth ceases

m, n = exponents

Luong equation is somewhat simpler in form because m = 0. The model of Han, *et al.*[35] should be applicable to situation where xenobiotic compounds cause inhibition to bio-oxidation of biogenic compounds we are interested in, equation 6.29 can be used to describe the effect of such a situation.

$$SUR = \frac{\mu_{max}}{Y}(1 - S_I/S_I^*)^n \frac{S}{(S + K_s(1 - S_I/S_I^*))} \quad ....(6.29)$$

where,

S_I = concentration of xenobiotic compounds,

S = concentration of biogenic compound, and

S_I^* = critical concentration of xenobiotic compound above which degradation of biogenic substrate ceases.

Types of inhibition, can be described in terms of μ_{max} and K_s and is summarised as follows :

		Effect on	
Sr. No.	*Type*	μ_{max}	K_s
1.	Competitive	None	Increase
2.	Non-competitive	Decrease	None
3.	Un-competitive	Decrease	Decrease
4.	Mixed	Decrease	Increase

Grady[36] presents an excellent review on the present status of toxicity of compounds for bio-oxidation of organics. Simple single substrate can provide much useful information, but more research is needed for interactions among organic compounds in biological systems. Although it well never be possible to study all possible substrate combinations, basic study of a limited number of key compounds should do much to define fundamental principles that can be applied broadly. It must also be mentioned that unsteady state situation has not been considered mathematically. In addition, toxicity studies have been limited to well mixed systems and fixed film reactions have been ignored so far in terms of mathematical models. Multicomponent substrate situation also needs further research.

6.6 Compounds Known to be Toxic to Anaerobic Systems

A summary of toxic substrates which severally inhibit the various anaerobic treatment processes has been presented in this section. The description is presented in Table 6.3. The types of systems that have been considered are as follows :

- batch system;
- semi-continuous system;
- semi-continuous system with high SRT values;
- continuous systems.

Table 6.4 describes the limits of concentration of toxic compounds that can be tolerated in an anaerobic system. Inhibition/toxicity to methanogenic bacteria or mixed cultures is presented in Table 6.5.

Table 6.3 : Concentration Levels that Cause Toxicity to Different Anaerobic Treatment Processes

Sr. No.	*Toxicant*	*Batch System 50% Inhibition*	*Semi-Continuous System[1]*	*Semi-Continuous System With High SRT[2]*	*Continuous System[2]*
1.	Acrolein	11.5(37)*	—	—	—
2.	Acrylonitrile	100(4)*3	5(22)*	—	—
3.	Ammonium	7500-10000(11)*	—	12000(11)*	—
4.	Cadmium	–	1(22)*	200(11)*	—
5.	Calcium	8000(3)*3	—	15000-20000(11)*	—
6.	Carbon tetrachloride	—	10(22)*	—	—
7.	Chloroform	—	0.1(22)*	7.5(11)*	400(12)*
8.	Chromium-III	—	—	100(11)*	—
9.	Chromium-IV	—	—	40-60(11)*	—
10.	Copper	—	10(22)*	60(11)*	—
11.	Crotonaldehyde	50-100(4)*	—	—	600(4)*
12.	Cyanide	1(12)*	—	5(11)*	20(12)*
13.	1,2 Dichloroethane	—	1(22)*	—	—
14.	1,1, Dichloroethene	—	—	50-100(11)	—
15.	Ethyl-Benzene	340(37)*	—	1000(11)	—
16.	Ethylene-Dichloride	50(17)*	2-8(17)*4	—	—
17.	Formaldehyde	200(16)*	100(16)*4	150-200(11)	400(11)
18.	Hydrazine	—	—	150(11)	—
19.	Lead	—	50(22)	—	—
20.	Magnesium	50-70(3)	—	—	—
21.	Methylene Chloride	14(17)*	>10(17)*	—	—
22.	Nickel	300(11)*	40(22)*	>100(11)*	>250(11)
23.	Nitrobenzene	12.5(37)*	—	—	—
24.	Phenol	500(16)*	700(16)*4	—	—
25.	Potassium	12000(3)*	—	—	—
26.	Sodium	8000(3)*	—	—	—
27.	Sulphide	>400(11)*	200(6)*	—	—
28.	Tetrachloro Ethylene	80-160(38)*	—	—	—
29.	1,1,1 Tetrachloro	—	1(22)*4	—	—
30.	1,1,2 Trichloro 1,2,2, Trifloroethane	10-20(38)*	—	50-250(11)*	—
31.	Trichlorofluoroethane	—	0.7(22)*	—	—
32.	Vinylacetate	11.5(17)*	>1.2(17)*4	—	—
33.	Vinyl Chloride	40(17)*	65(17)*	—	—
34.	Zinc	—	10(22)*	—	—

( )* refers to reference number; All values in mg/L.

1. toxic concentration refers to concentration that causes process failure unless otherwise stated.
2. toxic concentration is defined as no methane gas production for a period greater than 40 days.
3. concentration that is contained in biomass being digested.
4. represents concentration that produces 50% inhibition.

Table 6.4 : Toxic Levels for Some Xeno-Biotic Compounds[36]

Sr. No.	*Toxicant*	*Concentration (COD)mg/L*
1.	Aniline	100
2.	Benzoic Acid	100
3.	Phenol	100
4.	2-Chlorophenol	100
5.	4-Chlorophenol	100
6.	m-Cresol	100
7.	4-Nitrophenol	100
8.	2,4 Dimethylphenol	100
9.	2,4 Dinitrophenol	100
10.	Benzene	100
11.	Chlorobenzene	100
12.	Nitro-benzene	100
13.	1,2-Dichlorobenzene	30
14.	1,3 Dichlorobenzene	30
15.	1,4 Dichlorobnenzene	30
16.	1,2,3 Trichlorobenzene	30
17.	Dimethylphthalate	100
18.	Diethylphthalate	75
19.	Di-n-butylphthalate	20
20.	Butylbenzephthalate	20

Table 6.5 : Inhibition/Toxicity for Anaerobic Process (Methanogenic Bacteria or Mixed Culture)[20]

Sr. No.	*Toxicant*	*Concentration*	*Effect*	*Accli-mated*	*Unaccli-mated*	*Suspended*	*Attached*
1.	Acetaldehyde						
2.	Acetate						
3.	Acetylene	8μM	Total Inhib		+	+	
4.	Acrolein	0.2mM	50% Inhib	+			
		$0.5kg/m^3$	Inhib begins		+		+
5.	Acrylic acid	12mM	50% Inhib	+			
6.	Acrylonitrite	4mM	50% Inhib	+			
7.	Ammonia	$8kg/m^3$	Inhib begins pH not known	+		+	
		$6kg/m^3$	Inhib begin pH not known	+			+
		$24kg/m^3$	Inhib begin pH not known		+	+	
		$4\text{-}6kg/m^3$	Partial Inhib	+		+	
		$0.4 kg/m^3$	Inhib	+		+	

(Contd...)

Sr. No.	Toxicant	Concentration	Effect	Accli-mated	Unaccli-mated	Suspended	Attached
8.	Free ammonia	0.05-1.1 kg/m^3	Acceptable			+	
9.	Aniline	26μM	50% Inhib				
10.	Carbon-tetrachloride						+
11.	Catechol	24μM	50% Inhib	+			
		0.2-0.4kg/M^3	No Inhib	+			
12.	Propane	1.9μM	50% Inhib	+			+
13.	Propene	0.1mM	50% Inhib	+			
14.	Propinic acid	8mM		+			
15.	Difurfurydisulphide	0.001kg/m^3	Inhib		+	+	
16.	2,4 Dinitrophenol	0.25kg/m^3	Recovery after 1 day		+		+
17.	Ethyl acetate	11mM	50% Inhib	+			
18.	Ethylbenzene	3.2mM	50% Inhib	+			
19.	Ethylene-dichloride	0.005kg/m^3	Inhib begins	(+)			
20.	Eugenol	0.1-0.5kg/m^3	Inhib	+		+	
		0.1kg/m^3	Total Inhib	+		+	
21.	Formaldehyde	0.5-1kg/m^3	Total kill				
22.	Furfural	0.5kg/m^3	Inhib		+	+	
		2kg/m^3			+	+	
		5kg/m^3		+		+	
23.	Hydrogen	200kPa	Inhib	+		+	
24.	Penta Chlorophenol	0.27kg/m^3	Inhib	+			
25.	Resorcinol	29mM	50% Inhib	+			

+ refers to yes

(+) refers to partially

Blank not known.

REFERENCES

1. McCarty, P.L. Anaerobic Waste Treatment Fundamentals, Parts I-IV; Pub. Works p. 95-197-112 (1964).
2. Kulgelman, I.J. and Chin, K.K.; Toxicity, Synergism and Antergonism in Anaerobic Waste treatment Processes; Advances in Chemistry Series No. 105 p. 55-90 (1971); American Chemical Society, Washington.
3. McCarty, P.L.; Kulgelmen, I.J. and Cation Toxicity and stimulation in Anaerobic waste Treatment; J. Wat. Poll. Con. Fed; 37, p. 97-116 (1965).
4. Hovious, J.C.; Waggy, G.T. and Conway, R.A.; Identification and control of Petrochemical Pollutants Inhibitory to Anaerobic Process. Environmental Protection Series, EPA-R2-73-194(1973).
5. Regan, T.M. and Peterx, M.M. Heavy Metal in Digesters : Failure and care; J. Wat. Poll. Cont. Fed. 42, p. 1832-1839 (1970).

6. Lawerence, A.W. and McCarty, P.L. The Role of Sulphide in Preventing Heavy Metal Toxicity in Anaerobic System; J. Wat. Poll. Control Fed. 37, p. 392-402 (1965).
7. Rudgual, H.T. Effect of Coffe-Bearing Wastes on Sludge Digestion; Sew. Works J. 18 p. 1130-1137 (1946).
8. Dickerson, R.E. Gray, H.B. and Haight, G.P.; Chemical Principles, W.A. Bengamin, Inc. Menlo Park, California (1974).
9. Mosey, F.E. Assessment of the Maximum Concentration of Heavy Metals in Crude Sewage Which will not Inhibit Anaerobic Digestion. J. Wat. Poll. Control, 75, p. 10-20 (1976).
10. McCarty, P.L. and Mckinney, R.E. Salt Toxicity in Anaerobic Digestion. J. Watt. Poll. Con. Fed. 33, p. 399-415 (1961).
11. Parkin, G.F. and Kocher, W.M. Microbial Methane Fermentation Kinetics for Toxicants Exposure. Prepared for the National Science Foundation Resources, Urban Environmental Engineering Programe, July 1 (1980), Philadelphia, Pennsylvania.
12. Young, J., Speece, R.E., Parkin, G.F. Gossett. J., and Kocher, W. The Response of Methane Fermentation to Cyanide and Chloroform Prog. Wat. Tech. 12, p. 977-989 (1980).
13. Parkin, G.F. and Kocher, W.M. Microbial Fermentation kinetics for Toxicant Exposure. Preferred for Air Force Office of Scientific Research, September 1, 1980. Philadelphia, Pennsylvania.
14. Owen, W.F. Stuckey, D.C. Healy, J.B., Young, L.Y. And McCarty, P.L. Dioassay for Monitoring Biochemical Methane Potential and Anaerobic Toxicity. Water Research 13, p. 485-492 (1979).
15. Krisch, E.J., Grady, C.P.L., Wukasch, R.F. and Tabak, H. Protocol Development for the Prediction of the Fate of Organic Priority Pollutants in Biological Wastewater Treatment Systems. Purdue Univ. Conference (1981), West Lafaye the Indiana.
16. Pearson, F., Shiun-Chung, C and Gautier, M. Toxic Inhibition of Anaerobic Biodegradation. J. Wat. Poll. Cont. Fed. 52, p. 472-482 (1980).
17. Stuckey, D.C., Owen, W.F. Parkin, G.F. and McCarty, P.L. Anaerobic Toxicity Evaluation and Semi-Continuous Assays. J. Wat. Poll. Cont. Fed. 52, 720-729 (1980).
18. McCarty, P.L., Kinetics of Waste Acclimation in Anaerobic Treatment. Development in Industrial Microbiology, I, 144. 1966 Inct. Biol. Sci.; Washington. D.C.
19. Jeris, J.S. and McCarty, P.L. The Biochemistry of Methane Fermentation using C-14 Tracer J. Wat. Poll. Cont. Fed. 37, p. 178-186 (1965).
20. McCarty, P.L., Jeris, J.S. and Murdoch, W. Individual Volatile Acids in Anaerobic Treatment J. Wat. Poll. Cont. Fed. 35, p. 1501-1515 (1963).
21. Gossett, J.M. and McCarty, P. L. Heat Treatment of Refuse for Increasing Anaerobic Biodegradation AICHE Series 72(158), p. 64-71 (1976).
22. Barth, E.F. and Bunch, R.L. Biodegradation and Treatability of Specific Pollutants, EPA 600/79-034 (1979).
23. van den Berg, L. Assessment of Methanogenic Activity in Anaerobic Digestion : Apparatus and Method, Biotech, Bio-eng., 16, p. 1459-1469 (1974).
24. Umbreit, W.W. Burris, R.H. and Stauffer, J.B. Manometric Techniques, Burgess Publishing Company, Minneapolis (1957).

25. Miller, T.L. and Wolin, M.J. A Serum Bottle Technique for Cultivating Obligate Anaerobes, App. Microbial, 27. p. 985-987 (1974).
26. Parkin, G.F. and Speece, R.E. Response of Methane Fermentation to Industrial Toxicants, Paper Presented at 53rd Annual Water Pollution Control Federation Conference, Sept. 1980, Las Vegas, Nevada.
27. Young, J.C. and McCarty P.L. The Anaerobic Filter for waste Treatment, Technical Report No 87 (1980) Dept of Civil Engg, Standfood University, California.
28. Gorgan, J. and Obayashi, A. Unpublished Data (1980) IIT, Chicago, Illinois.
29. Parkin, G.F. and Speece, R.E. Modelling Toxicity in Methane Fermentation Systems, Paper Precented at ASCE National Conference on Environmental Engineering (July 1980), New York.
30. Lehninger, A.L. Biochemistry p. 197-206 (1976), Worth Publishers, New York.
31. Andrews, J.F. Dynamic Model of Anaerobic Digestion in Anaerobic Fermentation Processes. Int. J. Air and Water Poll. 9, p. 439-461 (1969).
32. Karhadkar, P.P. Handa, B.K. and Khanna, P. Pilot Scale Distillery Spent Wash Biomethanation. J. Envn, Engg. (ASEC), 116 (6), p. 1029-1045 (1990).
33. Bhatia, D., Vieth, W.R. and Venkatasubramaniam, K. Steady State and Transient Behaviour in Microbial Methanification, Biotech-Bioengg., 27 (8), p. 1199-1207 (1985).
34. Luong, J.H.T. Generalization of Monod Kinetics for Analysis of Data with Substrate Inhibition, Biotech-Bioengg., 29(2), p. 242-248 (1987).
35. Han. K. and Levenspiel, O. Extended Monod Kinetics for Substrate Production and Cell Inhibition, Bio-tech Bio-engg. 32(4), p. 430-437 (1988).
36. Grady, C.P.L. Biodegradation of Toxic Organics : Status and Potential, J, Envn. Engg. (ASCE), 116(5) p. 805-827 (1990).
37. Lin Chou, A. W. and McCarty, P.L. The Role of Sulphide in Preventing Heavy Metal Toxicity in Anaerobic Treatment J. Wat. Poll. Cont. Fed. 37, p. 382-406 (1965).
38. Benson, E.N. Comparative Effects of Halogenated Hydrocarbon Solvents on Waste Disposal Process; Paper Presented at 31st Purdue University Waste Conference, (1976), Purdue Univ. Indiana.
39. Henze, M. and Harremos, P. Anaerobic Treatment of Wastewater in Fixed Film Reactor-A Literature Review, Water Science and Technology, IAWPRC (1983), Pergamon Press Ltd., U.K.

CHAPTER 7

PRODUCTION OF HYDROGEN AND SULPHUR RECOVERY

7.0 Production of Hydrogen*

In all organic matter hydrogen atoms are bound to carbon, nitrogen , sulphur and other elements gaseous hydrogen is produced as well as consumed by living organisms. While the conversion of hydrogen is widely distributed among micro organisms it is virtually absent from higher organisms. Though large amounts of hydrogen are produced in the biosphere only a small proportion of it reaches the atmosphere as such. It is rather continuously produced and decomposed in the troposphere[1–2].

One of the new areas of fuel production which put the least strain on our environment is biological conversion of solar energy to hydrogen. Biophotolysis, *i.e.* photosynthetic decomposition of water into hydrogen and oxygen, is one of the approaches to utilize this unlimited energy resource, the solar radiation. Hydrogen is considered to be an energy medium of the future, because it is easily converted into electricity through fuel cells and burns without pollutants. In this review, hydrogen production by cell-free systems and whole cell systems of algae, cyanobacteria, photosynthesis and non-photosynthetic bacteria has been discussed.

Hydrogen is considered as a possible energy source in the future are energy shortage is one of the serious problems the world faces today. It is an ideal fuel, not only as an alternative energy source but also as highly efficient energy carrier. The heating value of the gas per unit volume is less than other gaseous fuels ($1/3 \times CH_4$), however, the heating content per unit mass of liquid volume is about 2.75 times greater than that of hydrocarbon fuels. It has great potential for use as a primary or secondary energy source, for chemical synthesis or for electrical storage and generation with fuel cells[3].

Hydrogen can be produced from natural gas, petroleum product, and coal by hydrogasification. It can also be produced by electrolysis and thermochemical decomposition of water.

* Adapted from Dr. S.B. Rahalkar's Thesis on Biodegradation of Organic Pollutants by Photosynthesis, Nagpur University (1991) and T.M. Vatsala and C.V. Seshadari, Microbial Production of Hydrogen-A Review; Proc. Indian National Academy B51(2), p. 282-295 (1985).

The application of biological hydrogen evolution to the production of fuel would result in a new solar energy based technology. The advantages of this approach to hydrogen production, includes: the system could be operated at low physiological temperatures (*i.e.* 10-40°K); the only major input into the system would be solar energy and a hydrogen donor, possibly organic wastes; the production of hydrogen would not involve the evolution of pollutants, as in the case of fossil fuel refineries; the fuel produced, hydrogen gas, would be clean burning (*i.e.* yielding water); the would readily lend itself to a potentially beneficial and profitable programme of multiple utilization, including the production of food for human and animal consumption.

Biological production of hydrogen has been observed in a great number of microbial species [3–5]. Gaffron and Rubin[6] discovered the formation of hydrogen by the green algae *Scenedesmus*. This ability was observed subsequently with several other species of unicellular green algae[7]. Heterocystous, filamentous blue-green algae (cyanobacteria) seemed to be less sensitive to oxygen[8], but high rates of hydrogen evolution were observed in an anaerobic environment. Weissman and Benemann[9] described continous hydrogen evolution by *Anabena cylindrica* for 18 days and Jeffries *et al.*,[10] observed it for 30 days, both under limited light conditions. Photoproduction of hydrogen by some *Oscillatoria* sp. has been reported by Sharvan Kumar *et al.*[11] In this immobilized *Oscillatoria* sp. was tried for hydrogen production from sewage water. Algae and cyanobacteria are less advantageous in hydrogen production because (1) their hydrogen generation rate is less, (2) besides hydrogen, oxygen is also produced and its separation from hydrogen is an energy consuming process, and (3) hydrogen production decreased after a short period due to production of oxygen.

For the nonphotosynthesis organotrophic bacteria, the conversion efficiency of sugars to hydrogen is theoretically restricted to 33 per cent[12]. In practice, Segers, *et al.*,[13] found that only 6.5 per cent of the reducing equivalents entrapped in the sugar molecules were set free in the form of hydrogen gas.

The light dependent production of hydrogen by photosynthesis bacteria was first observed with cultures of *R. rubrum*[14]. Rhodospirillaceae produce hydrogen and carbon dioxide by a light-dependent decomposition of several organic compounds[15–16].

Hydrogen production by photosynthetic micro-organisms is rather a classical discovery but it was only recently that biological systems were investigated as potential solar energy converters[17–25].

However, applications of biological hydrogen production are still in an early stage of development and the following are some of the important factors that should be borne in mind *viz.*:

- Hydrogenase and nitrogenase with catalyse hydrogen production are unstable under aerobic conditions.

- Hydrogen producing photosynthetic organisms also usually exhibit high hydrogen uptake activity.
- Competition for electrons with other biological reactions may diminish hydrogen production.
- Electron flow between photosystems I and II may become the rate limiting step.[26]

Two basic directions could be taken up to approach these problems, namely living-cell systems and cell-free systems[27].

7.1 Artifically-Reconstituted Systems

The natural, intact-cell, biochemical systems operate under a network of checks and balances, which help to maintain the organisms within a state of homostasis. This precludes excessive build-up of end-products and it also sets up an upper limit for the rate of hydrogen production. One of the most efficient ways of removing these constraints would be to isolate the hydrogen producing systems from its cellular environments, thus creating a cell-free systems. The general format of cell-free approach is given in Fig. 7.1[27–29].

The first-cell-free systems studied was the chloroplast-ferredoxin hydrogenase reaction, wherein spinach chloroplasts were bacterial hydrogenase and ferredoxin to mediate electron-flow between the chloroplasts and the hydrogen-evolving enzyme[17]; hydrogen evolution was observed but it could not be sustained due to its affinity towards oxygen.

Hydrogen photoproduction by reconstituted cell components has been demonstrated by several worker[30–32], wherein chloroplasts, bacterial hydrogenase and methylviologen were used, the electron-donors being cysteine and 2,6-dichlorophenol indophenol (DPIP). Chloroplasts and methylviologen could be replaced with blue-green algal lamella (*Nostoc muscorum*) and ferredoxin respectively. Bacterial hydrogenase was obtained from species of *Chromatium, Desulfovibrio* and *Clostridium*. This system demonstrated that hydrogen photoproduction was directly coupled to PS-I and hydrogenase when an electron mediator, such as ferredoxin or methylviologen was present. In 1964, Whatley and Grant[33] suggested that a complete photosystem consisting of PS-I and PS-II could produce hydrogen using water as the electron-donor. Hydrogen photoproduction from a cell-free preparation of the green algae, *Chlamydomonas eugametos,* using reduced pyridine nucleotides as an electron-donor[34], was reported.

In the 1970s, production of hydrogen through biophotolysis was proposed as an alternative energy resources and thus earlier experiments were re-examined. For example, Krampitz[35] and Benemann *et al.*[17] demonstrated hydrogen photoproduction from water.

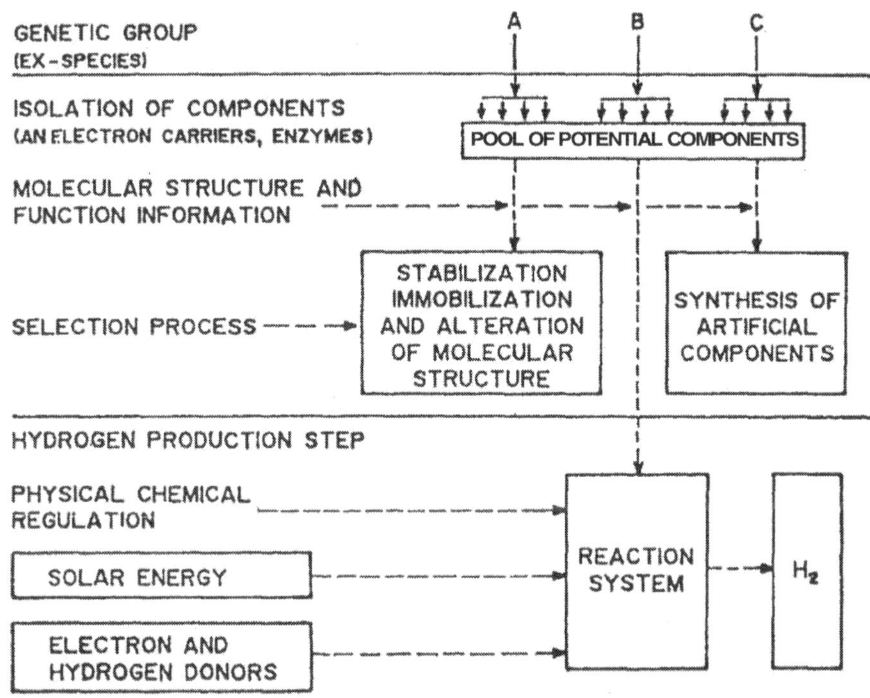

Fig. 7.1 : Biochemical Process of Hydrogen Production in a Cell-Free System

In using systems similar to those mentioned above, glucose oxidase and/ or ethanol-catalase were added as an oxygen trap to protect hydrogenase from oxygen inactivation; H_2 was therefore evolved through the biophotolysis of water. The cell-free systems studied included use of dithiothreitol as an electron donor[36–38], the stability of photosystems by micro encapsulation[39] or glutaraldehyde treatment[19] immobilization of chloroplast[40–41], ferredoxin[42], hydrogenase[43–44] and an examination of interactions between cell components from different origins. An enzymic electron cell techniques was applied to spatially separate the hydrogenase from the oxygen evolving site[45].

The key to successful construction of the biophotolytic hydrogen production system will be the stabilization of photosynthetic particles and hydrogenase against, O_2 inactivation. Stabilization of hydrogenase[45] and photosynthetic particles[46] and use of synthetic electron mediator[47] and/or inorganic or synthetic catalysts instead of hydrogenase[48], have been suggested as solutions. Table 7.1 summarises the work available on cell-free hydrogen production systems.

7.2 Hydrogen Production by Living Cell Systems

Hydrogen production can best be examined under three heads, *viz* :

- Hydrogen production by green algae.

Table 7.1 : Hydrogen Photoproduction by Cell-Free Systems

Cell-free systems	*Hydrogenase*	*Rate of H_2 production*	*Electron donor*	*Reference*
Chloroplast (spinach) + MV (Fd) + CMU Lamellae (blue-green alga Nostoc) + MV (Fd) + CMU	Hydrogenase *Chromatium*, *Desulfovibrio*, *Clostridium*	5.5 µmol/hr/2ml reaction mixture	(31,32,102) (31)	Cysteine-DPIP Ascorbate-DPIP
Chlamydomonas cell-free + Preparation + DCMU Chloroplast (spinach) + DCMU	*Chlamydomonas*	3.2 µmol/hr/2.2mL reaction mixture 4.3 µmol/hr/mg Chl.	(34) (37)	NADH OR NADPH Dithioetheritol
Chloroplast (*C. Limicola thiosulfatophilum*) + gramicidin + MV + tetramethyl-p-phenylene diamine glucose-glucose-oxidae-ethanol catalase.	*Clostridium*	450 µmol/hr.mg bChl.	(103)	Ascorbate-Dithioetheritol
Chalmydomonas cell-free Preparation + DCMU PS-I particle + MV micro-encapsulation of the system	*Chromatium*	7.4 µmol/hr/mg Chl.	(38) (39)	Dithioetheritol or NADH Ascobate-DPIP
Choloroplast (spinach) + MV	*Clostridium*	125 µmol/hr/mg Chl.	(104)	Ascorbate-TMPD
Chloroplast (spinach) + MV + glucose-glucose oxidase and ethanol - catalase as an oxygen trap	*Clostridium*	2.1 µmol/hr/mg Chl.	(35) (35)	H_2O
Chloroplast (spinach) + Fd glucose-glucose oxidase and catalase as an opxygen trap	*Clostridium*	14 µmol/hr/mg Chl.	(35)	H_2O
Chloroplast (spinach) + Fd (spinach) glucose - glucose oxidase and ethanol-catalase as an oxygen trap.	*Clostridium*	20 µmol/hr/mg Chl.	(46) (105)	H_2O
Chloroplast (spinach, lettuce, tobacco, chenopodium etc.) + Fd (spinach, *Spirulina* etc.) + glucose - glucose oxidase as an oxygen trap, stabilization of chloroplast by glutaraldehyde fixation	*Clostridium* *Escherichia coli* *Thiocapsa* etc.	94 µmol/hr/mg Chl.	(19) (106) (107)	H_2O

CMU, 3 – (4 – Chlorophenyl) -1, 1 - dimethyl urea; DCMU, 3 - (3,4 - dichlorophenyl - 1, 1 - dimethyl urea, DPIP, 2-6, dichlorophenol indophenol; Fd, Ferredoxin; NADH, nicotinamide adenine dinucleotide (reduced form); NADPH, nicotinamide adenine dinucleotide phosphate (reduced form); TMPD, N-tetramethyl - p - phenylene diamine; MV, Methylviologen.

- Hydrogen production by cyanobacteria; the term blue-green algae is not being used here as in the reviewed literature. This group has been treated as algae because of the presence of PS-I and PS-II.
- Hydrogen production by photosynthetic bacteria.

The capability of algae to produce hydrogen and the effect of low light intensity on this process was reported by Gaffron *et al.*[49] In photosynthetic bacteria, hydrogen production under dark and aerobic conditions was first reported by Roelofsen[50] and Nakamura[51]. The capacity of photosynthetic bacteria to carry out photoproduction of hydrogen was first reported by Gest and Kamen[52–53]

7.2.1 Hydrogen Production by Algae

Algae have two photosystems (PS-I and PS-II) and are capable of generating reducing potential from water. Bacteria, on the other hand, have only one photosystem and require exogenous donors for photosynthesis[(22)]. The major difference in hydrogen photoproduction in the two groups, is that algae are theoretically capable of producing hydrogen directly through the photolysis of water or photosynthetic products whereas photosynthetic bacteria requires external electron donors such as organic compounds or reduced compounds for this purpose.

7.2.1.1 By Green Algae

Gaffron and Rubin[49] suggested two possible mechanisms for the generation of electron donors for hydrogen production : (i) generation from endogenous stored compounds photochemically metabolized, and (ii) generation through photolysis of water. Hydrogen photoproduction coupled to endogenous fermentation has been demonstrated by many researchers[54–59]. The sources of reductants suggested are from Embden-Meyerhof pathway[54] anaerobic Kreb cycle pyruvate[56] hydrogenase reaction under dark[58]. There are others who have suggested that generation of reductants occurs by way of biophotolysis of water[60].

Using sodium dithionite as oxygen scavanger, the influence of different light intensities and periods of anaerobic preincubation in the dark mH_2 photo-production were studied with *Chlorella fusca*[61]. For high rates of H_2 production 30,000 lux were necessary. Although, H_2 was immediately produced after transition to anaerobic conditions, the optimum rate of H_2 production was reached after 5 hr dark adaption period only. It appears that the production of H_2 in *Chlorella* is an anaerobic photosynthetic process, which occurs in the absence of CO_2 and can be experimentally stabilized by exogenous oxygen scavengers.

7.2.1.2 By Cyanobacteria

Hydrogen production by cyanobacteria has been investigated in both heterocystous and non-heterocystous forms. Benemann and Weare[62] used the heterocystous *Anabaena Cylindrica* to demonstrate light-driven simultaneous production of H_2 and O_2 from water. Since then many successful experiments have been made with this alga[60,63–66]. Hydrogen production is possible for upto several weeks with a relatively high conversion of light energy to hydrogen (up to 3%) under laboratory conditions. The performance of outdoor converter receiving solar radiation was one order of magnitude lower than that of a laboratory converter.

Table 7.2 : Hydrogen Photoproduction Rates in Algae

		H_2 produced/hr (μmol) intensity			
Sr.	*Algae*	*dry wt^{-1}*	*mg Chl.$^{-1}$*	*Light (Lux)*	*Reference*
1.	Green Algae				
	Ankistrodesmus brauni		0.49	4400	(56-57)
			11.1		
	Chlamydomonas reinhardii	0.25	17.0	300	(94)
			3.2	4400	(56-57)
	Chlorella fusca		0.49	4400	(56-57)
			30.0	3000	(61-95)
	Chlorella protothecoides		16.4		
	Chlorella sp.	0.16	11.0	400	(94-96)
	Coelastrum proboscideum		24.6		
	Selenastrum sp.		11.8		
	Senedesmus obliquus	0.22	15	500	(94-96)
			20.2		
	Kischneriella lunaris		32.1		
2.	Cyanobacteria				
	Anabaena cylindrica	1.4	320	106000	(63)
		0.7	35000	(64)	
		0.17	11.0	7000	(97)
		0.58		8800	(71,98)
	Anabaena flos-aquae	0.17			
	Calothrix scopulorum	0.13	8.7	4000	(99)
	Nostoc muscorum	0.03		5300	(100)
	Oscillatoria brevis	0.17	11.0	4000	(99)
	Oscillatoria sp. Miami BG7	0.18	230	3500	(27,68)
		0.29	260		
	Synechpepccis sp. Miami BGO				
	BGO 43511	56.0		6000	(101)

Table 7.3 : Electron Donors Used for Hydrogen Production by Photosythetic Bacteria

Species/Electron Donor Used	*Reference*	*Species*	*Reference*
RHODOSPIRILLACEAE		*Rhodospeudomonas palustris*	
(Puple non-sulphur bacteria			
		Formate	108
Rhodospirillum rubrum	(118)	Acetate, lactate,	
Acetate	(119)	malate and succinate	(109)
Lactate	(86)	Pyruvate, α-Ketoglutarate,	74
		succinate, malate Oxal-acetate,	(110)
		glucose, thiosulphate	
Pyruvate	(52)	*Rhodopseudomonas sphaeroides*	(111)
	(84)	Glucose	
	(120)	Chromatiaceae (purple-	
	(121)	sulphur bacteria)	
	(88)		
	(123)		
	(90)		
Succinate	(96)	*Chromatium* D	(16,85,102)
	(52)	Thiosulfate	(1961)
	(84, 124)		(19615)
	(119)		
Fumarate	(52)		
	(84,124)	*Chromatium* sp.	
	(119)	Malate and CO_2	(112)
Malate	(96)		
	(125)		
	(53)	*Chromatium* sp. (marine)	
	(84,124)	Thiosulphate	(113)
	(117)		(114)
	(119)	Sulfide	(114)
	(116)		
	(89)	Succinate, acetate, furmarate	(114)
	(126)	and malate	
	(127)	*Thicocapsa roseopersicina*	(114)
Oxalacetate	(52)	Thiosulphate, oxalacetate,	(115)
	(84)	acetate, pyruvate.	
Rhodopseudomonas acidophila		Chlorobiaceae (Green-sulphur	
Lactate	(137)	bacteria)	
Rhodopseudomonas capsulata		*Chloropseudomonas ethylica*	
Propionate, pyruvate, succinate,	(135)	Lactate, pyruvate, citrate,	(116)
fumarate, butyrate, glucose,		α-ketoglutarate,	(117)
fructose and sucrose.		xylose, mannitol glucose	(116)
Lactate	(135)	*Chloropseudomonas* sp.	
	(129)	Formate	(116)
Malate	(130)		(117)
	(135)		

H_2 production with immobolized cells of *Mastigocladus laminosus* in Polyurethane has been studied. Similar studies have been made with a thermophilic organisms *Chlorogloeas fritschii* for the development of high temperature H_2 production [67]. Marine algae have also been shown to evolve H_2[68,69]. It has been suggested that electrons for hydrogen production are transported from vegetative cells to heterocysts[70]. Production of hydrogen is catalyzed by hydrogenase[72]. The oxidative Pentose Phosphate cycle[73] isocitric dehydrogenase reaction coupled to ferredoxin NADP oxidoreductase[74] and the pyruvate-ferredoxin oxidoreductase reaction[65,75] are believed to supply electrons. Ps-I in hetorocysts may function as a supplier of ATP[72] and as a reducer of ferredoxin from endogenous reductant[74]. Hydrogen production from sulphide *via* Ps-I hydrogenase pathway has been reported in blue-green algae[76]. Aerobic hydrogen production was demonstrated in *Anabaena* strains CA and IF when cells were growing under N_2-fixing conditions[77].

Light induced hydrogen evolution by *Nostoc muscorum* is stimulated by sulfide with a rate of about 20μmol H_2/mgChl and it is stable for about 1 day at low light intensities (1860 lux). Oxygen evolution does not impair sulphide-stimulated formation[78]. Table 7.2 given the performance of biophotoreactions with green algae and Cyanobacteria.

Table 7.4 : Hydrogen Production Rate in Photosynthetic Bacteria

Bacteria	*μmol*	*H_2 Produced hr^{-1} mg^{-1} ptn**	*Electron Donor*	*Light Intensity, Lux*	*Reference*
Rhodospirillum rubrum	10.7/mg N	1.7	Malate	10800	128
		1.68	Malate	35948	126
		0.96	Malate	12249	117
R. rubrum mutant-C		4.0	Formate	Dark	131
R. rubrum ATCC-1170	5.0/mg dry wt		Malate	6500	87
	10.3/mg dry wt		Malate-Propionate 1 : 2	6500	87
Rhodopseudomonas capsulata	5.8/mg dry wt	12.0	Lactate pyruvate	10800	135
R. palustris		0.78	glucose	7722	110
Chromatium D	0.09/mg cells	1.2	Thiosulphate	50000	102
Thiocapsa roseopersicina		1.84	pyruvate	5325	115

* Converted to protein (ptn) with following factors: 2 for dry wt to ptn. 13 for cell O ptn; 6.25 for N to ptn

7.2.1.3 Photosynthetic Bacteria

The photosynthetic bacteria comprise two essentially different groups: purple bacteria (sulphur and non-sulphur) and green sulphur bacteria. The photosynthetic metabolism occurs under anaerobic conditions.

The initial observation on light-dependent evolution of H_2 by PSB was made by Gest and Kamen[52–53]. Since then, a great deal of research has been done on the effect and mechanism[79–80], nutritional requirements[81] and the control of the hydrogen-evolving system in order to improve hydrogen production rates[82]. An understanding of the hydrogen metabolism has been facilitated mainly by the use of purple non-sulphur bacteria like *Rhodospirillum rubrum* and *Rhodopseudomonas capsulata.* H_2 photoproduction by PSB is mainly associated with nitrogenase rather than hydrogenase activity[83] and it is known to be inhibited by nitrogen gas or ammonium ion[84]. PSB contain photosystem-I and require organic substrates to form H_2 and CO_2 Table 7.3 gives a list of organic substrates used as electron-donors.

Relatively higher rates of hydrogen production with PSB have been observed (Table 7.4) and this does not require continous argon sparging. Pure H_2 could be available without difficulty in gas separation as CO_2 is the only contaminant. Thus, the use of PSB for H_2 production from organic compounds seems to be very promising.

Bennett and Weetall[85] made hydrogen from glucose and showed that the cost of H_2 production was at least one order of magnitude higher than the present cost of natural gas. Zurrer and Bachofen[86] demonstrated continous production of hydrogen by *R. rubrum* in batch cultures upto 80 days using wastes from whey and yoghurt containing lactic acid. H_2 evolution in continous culture of *R. rubrum* was observed for 20 days with significantly higher production rates than those obtained in batch cultures. However, utility of organic wastes rich in nitogenous compounds in doubtful, because nitrogenase is repressed by combined nitrogen[36]. A mixed culture concept might be more fruitful to overcome such problems.

We found that *R. rubrum* could utilize organic and nitrogen compounds present in the effluents of biogas digesters (using cowdung) to evolve, H_2 photosynthetically. Besides, the same species utilizes malic acid or a combination of it with propionic acid to photoproduce H_2 the amount of H_2 evolved in the latter case is however more[87]:

7.2.1.4 Hydrogen Production in Dark

Several strains of blue-green algae as also temperate and cold-water green and red marine algae reported to produce H_2 in the dark, of course in a relatively low extent[36–37]. Photosynthetic bacteria too produce H_2 in the dark[88–90].

A non-photosynthetic bacterium, Citrobacter intermedius produces 1.19 mole of H_2 per mole which can be increased by agitation with inorganic sulphur salts[81,91,93]

7.3 Thermodynamic Aspect

Thermodynamical aspects of hydrogen production and the theoretically possible ways in which glucose may be decomposed to hydrogen or methane and carbon dioxide are shown below[132]. (Combustible energy of glucose compared to the combustible energy of the formed methane and hydrogen).

$$\text{Glucose} + H_2O \longrightarrow 2 \text{ acetate} + 2H^+ + 2CO_2 + 4\ H_2$$

$$4\ H_2 + 2O_2 \longrightarrow 4\ H_2O;\ G = -\ 227 \text{ kcal/mole}$$

$$\text{Glucose} + 6\ O_2 \longrightarrow 6\ CO_2 + 6\ H_2O;\ G = -\ 686 \text{ kcal/mole}$$

$$\text{Glucose} \longrightarrow 3\ CO_2 + 3\ CH_4$$

$$3\ CH_4 + 6\ O_2 \longrightarrow 3CO_2 + 6\ H_2O;\ G = -\ 586 \text{ kcal/mole}$$

$$\text{Glucose} + 6\ H_2O \longrightarrow 6\ CO_2 + 12\ H_2$$

$$12\ H_2 + 6\ O_2 \longrightarrow 12H_2O;\ G = -\ 680 \text{ kcal/mole}$$

Hydrogen formation in the dark is not regarded as an efficient process to transfer energy. Only 33 per cent of the combustible energy of organic compounds is conserved in the fermentation products. In contrast to hydrogen, methane formation is more efficient. Apparently 85 per cent of the energy is conserved assuming that 3 moles of methane are formed from 1 mole of glucose. In photosynthetic bacteria complete dissimilation of carbohydrates through a light dependent anaerobic Krebs cycle is thought to operate[15]. The decomposition of 1 mole of glucose yields 12 moles of hydrogen. Almost 100 per cent of the combustible energy of glucose would be conserved. Thus photosynthetic bacteria offer a better group of microorganism for hydrogen production.

7.4 Advantages of Anaerobic Photosynthetic Production of Hydrogen

The anaerobic photoproduction of hydrogen by photosynthetic bacteria offers several advantages over other systems from the technical and economic points of views. Photosynthetic bacteria produce hydrogen at much higher rates than other types of micro organisms and the fundamental mechanisms by which hydrogen is produced are becoming well understand[133]; molecular hydrogen is produced by an irreversible reaction catalysed by nitrogenase and proceeds even under an atmosphere of 100 per cent hydrogen gas; the reaction is anoxygenic; there is no residual oxygen-evolving activity which would cause problems of inactivation of the catalyst or of the separation of oxygen from hydrogen. The only gaseous contaminant produced is carbon dioxide which can be scrubbed easily; photosynthetic bacteria trap light energy over a wide spectral range and can withstand high light intensities; the capacity of photosynthetic bacteria to use, as reductants, a wide range of carbon compounds as metabolizable substrates (sugars, short chain fatty acids, organic acids) make them able, in principle, to consume organic wastes. The use of these wastes may render the process economically feasible with hydrogen as a useful byproduct.

Photosynthetic bacteria possess a diverse and evolutionarily ancient metabolism, which is reflected in different ways in which they can metabolize hydrogen. Three enzymes have been implicated in hydrogen metabolism in

PSB[133]. These are : (1) nitrogenase, which catalyses unidirectional, ATP dependent hydrogen evolution and can function either in the light, or in the dark under anaerobic or micro-aerobic conditions; (2) hydrogenase, which is membrane bound and, although capable of both hydrogen evolution and uptake, functions physiologically in the direction of hydrogen oxidation and (3) "Classical" or reversible hydrogenase, which may be either soluble or membrane bound and functions mainly during dark anaerobic fermentation. Nitrogenase is not inhibited by carbon monoxide, an inhibiter of hydrogenase and is dependent on ATP. The classical hydrogenase, as mentioned above, catalyses the reversible reaction $H_2 \longrightarrow 1H^+ + 2e^-$. It seems that this enzyme catalyses mainly hydrogen uptake *in vivo*. Dixon[134] has suggested that a function of hydrogenase is to reutilize the hydrogen which is evolved as a by product of the nitrogenase reaction, retaining reducing equivalents for nitrogen or carbon dioxide reduction. They have further mentioned that it is not known whether hydrogen reutilization can occur throughout growth when excess of substrate is present. Hillmer and Gest[135] have demonstrated that the utilization of hydrogen for photoreduction process in *R. capsulate* is inhibited by the substrate lactate. Kelly *et al.*[136] observed that hydrogen uptake is only observed at low substance concentrations. This could be of physiological importance to the bacteria during growth in natural habitats. Rapid depletion of substrate could allow the expression of an autotrophic mode of growth through reutilization of the hydrogen produced by nitrogenase. Thus, the interplay between nitrogenase and hydrogenase could also be of importance for cell maintenance under conditions of substrate limitation and in the transition between photoheterotrophic and photoautotrophic growth.

A genetic or regulatory linkage between nitrogenase and hydrogenase has been proposed in a study with nif mutants of *R. acidophila*[137]. It has been reported that in *R. capsulate,* although nitrogenase may influence hydrogenase synthesis by supplying inducers (*e.g.* hydrogen), there is no strict correlation between hydrogenase and nitrogenase synthesis[138]. The exact mechanism of electron transfer in hydrogen metabolism and nitrogen fixation is not resolved so far. A light driven electron flow generates ATP. It is assumed that NAD and other substances of negative redox potential are reduced in a reversed electron flow utilizing ATP. When reducing power and energy for nitrogen fixation in the cell is produced in excess nitrogenase evolves hydrogen, which can be probably *via* the ferredox in-hydrogenase.

Sensitivity of nitrogenase towards oxygen can be decreased by immobilization. Immobilization techniques for enzyme and bacteria have been developed for the industrial application of these biocatalyst. The immobilization method offers considerable advantage and continous use of enzyme and bacteria. Stabilization of hydrogenase and nitrogenase has been successfully achieved by immobilizing nonphotosynthetic bacteria[139–140]. The hydrogenase and nitrogenase of immobilized bacteria were protected from the deleterious effect of oxygen because the diffusion of oxygen was limited by the gel matrix[141].

In addition to the stabilization effect, immobilized systems using photosynthetic bacteria have the following advantages over the free cell system. These are : (1) cells can grow to high biomass yield in immobilized gel, thus, all harvesting procedures (*e.g.* centrifugation) are eliminated and hydrogen production can be carried out in the same reactor; and (2) continuous hydrogen production can be carried out without the loss of cells. *Rhodopseudomonas* sp. Miami 2271 cells, immobilized with agar on an agarose – coated polyster film, produced hydrogen at a rate as high as 445 ml hr^{1} g dry $cells^{-1}$ from lactate and were markedly protected from inhibition by oxygen and nitrogen. Immobilized cells also exhibited salt tolerance than aqueous cell suspensions[142].

7.5 Application of Photosynthetic Bacteria in Waste Treatment

A number of organic wastes such as orange peel waste, sugar refinery waste, whey waste, other lactate containing waste, etc. have been tried for hydrogen production. Some of the important references for hydrogen production by Rhodospirillaceae are summarized in Table 7.5.

Table 7.5 : Application of Photosynthetic Bacteria in Waste Treatment

Sr.No.	*Type of Waste*	*Organisms Used*	*References*
1.	Agriculture waste	*Rhodopseudomonas gelatinosa*	Shipman R.H. *et al.* Biotech Bioeng 17, 1561 (1975)
2.	Industrial wastewater *e.g.* from Starch industry, Wool washing factory, Canned food Industry	Mixed	Kobayashi M and Chan Y.T. Water Research 7, 1219 (1973)
3.	Soyabean waste	*Rhodopseudomonas gelatinosa* (SCP Production)	Sasaki K, *et al.* JFT; 59, 471 (1981)
4.	Cassava starch	*Rhodopseudomonas gelatinosa*	Noparatnaraporn N *et al.* JFT 61, 515 (1983)
5.	Orange processing Wastewater	*Rhodopseudomonas* sp., *Chromatium* sp.	Mitsui A, *et al.* Dev. Ind. Microbiol. 26, 206-209 (1985)
6.	Pineapple peel waste	*Rhodopseudomonas shaeroides*	Noparatharapon, N *et al.* JFT, 64, 137 (1986)
7.	Lactate containing waste	*Rhodopseudomonas rubrum*	Zurrer H. and R. Bachofen Appl. Environ. Microbol. 37, 789(1979)
8.	Sugar refinery waste and straw paper mill effluent	*Rhodopseudomonas palustris and Rhodospirillium molischianum*	Vincenzini M. *et al.* Int. J. Hyd. Energy 7, 725-728, 1982
9.	Animal waste	*Mixed*	Ensign, J.C. Microbial Energy Conversion Ed. H.G, Schleggel and J. Barnea pp. 455
10.	Cow-dung	*Ropseudomonas capsulata*	Vrati S. and Verma J. JFT, 61, 157-162, 1993

(Contd...)

Sr.No.	Type of Waste	Organisms Used	References
11.	Photosynthetic bacteria in waste treatment	*Rhodopseudomonas capsulate*	Swada H. and P. L. Rogers, JFT, 55, 295-310 (1977)
12.	Photosynthetic bacteria in waste material	Role of *R. capsulata* with Agricultural/Industrial effluent	Swada H. and P. L. Rogers, JFT, 55, 326-336 (1977)
13.	Photosynthetic bacteria in waste treatment.	*Rhodopseudomonas capsulata* and *Klebsiella* Sp.	Swada H. and P. L. Rogers, JFT, 55, 311-315(1977)

JFT – Journal of Fermentation Technology

7.6 Microbial Sulphur Recovery

7.6.1 Introduction

Problems with sulfur disposal are universal. They are related primarily to energy production and, to a smaller extent, to the mining industry. As more emphasis is placed on the use of hydrocarbon such as oil and coal for energy production, sulphur control and disposal will become a more pressing problem.

One form of sulphur residue is through the generation of $CaSO_4$ sludges, which are a result of the scrubbing of flue gases from coal fired plants. In addition, H_2S waste streams are generated from the desulfurization of coal and oil. This problem is generally handled by using physical-chemical processes (*e.g.*, Claus and Wellman-Lord processes), which convert the H_2S to elemental sulphur. Other, more limited, areas in which high sulphates (acidic wastes) are a problem, include acid mine drainage (coal mining), acid wastes from mining operations, and wastes from ethanol distilleries

Currently, H_2S waste streams are handled by physical-chemical processes, with the major objective being the recovery of elemental sulphur. Presently, high sulphate wastewaters are neutralized and are either discharged or stored in lagoons.

The biological conversion of high sulphate waste or hydrogen sulphide waste streams to the more desirable form of elemental sulphur has not been developed beyond the laboratory stage. Studies by Cork[143] and Cork and Cusanovich[143, 144] on the kinetics of sulphate reduction and on the kinetics of hydrogen sulfide conversion to elemental sulphur are, to date; the only published material on the applied bio-process production of elemental sulphur. These studies will be discussed in more detail in a subsequent section of this chapter. In the next section, a brief summary of the microbiology of sulphur transformations will be presented. A more detailed summary has been conducted by Cork[143].

7.6.2 Microbiology of Sulphur Transformation

All organisms require sulphur, with the major need in the incorporation of sulphur in proteins. In addition, microorganisms can use sulphur in a similar manner that organisms use the various forms of nitrogen; that is as both an energy source and as an electron acceptor. The various oxidation states and the forms that sulphur exists in the environment are illustrated in Fig. 7.2, which represents the sulphur cycle. Of particular interest is the anaerobic portion of the cycle, the reduction of sulphate to hydrogen sulphide, and the oxidation of sulphides to elemental sulfur.

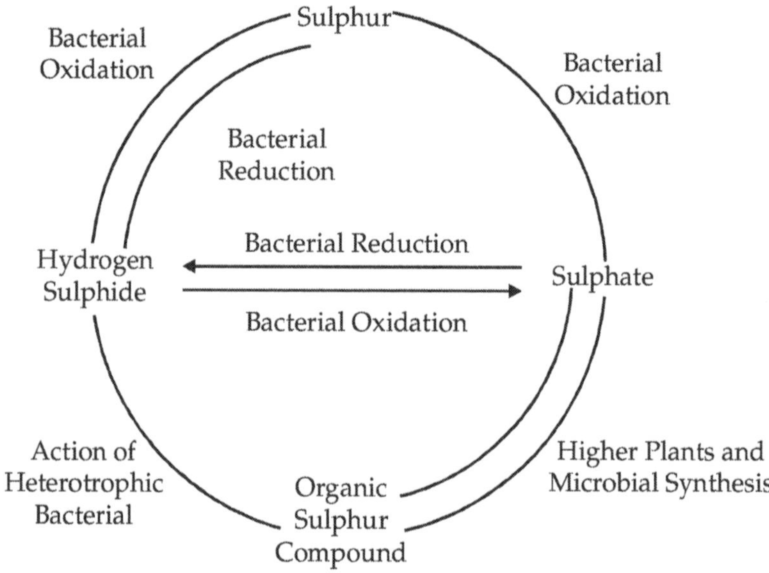

Fig. 7.2 : The Sulphur Cycle (Cork, 1978)

There are a wide variety of bacteria which can utilize sulphate as an electron acceptor in the oxidation of organic carbon. Of particular interest is the genus *Desulfovibrio* which is the group of bacteria most widely mentioned in studies of sulphate reducing bacteria in both natural waters and wastewater. Cork[143] in his studies on the kinetics of sulphate reduction used *Desulfovibrio desulfuricans.* This particular organisms can use a variety of organic compounds including lactate, pyruvate and ethanol. *D. desulfuricans* oxidizes lactate to CO_2 and acetate while reducing sulphate to hydrogen sulfide. This process is expressed as follows:

$$\underset{\text{(lactate)}}{2CH_3CHOCOO^-} + H_2SO_4 \longrightarrow \underset{\text{(acetate)}}{2CH_3COO^-} + 2HCO_3^- + H_2S$$

However in studies conducted by Middleton and Lawrence[145] in which a mixed culture of sulphate reducing bacteria was used, the substance was acetate. Furthermore, complete oxidation of acetate to CO_2 was achieved in contrast to the partical oxidation of lactate to CO_2 by a pure culture of *D. desulfuricans.* The implications of these differences in terms of treatment processes will be pointed out in a later section. In both studies the sulphate reducing bacteria were found to be relatively fast growing. At 30°C the *D. desulfuricans* species was found to have a maximum specific growth[143] rate of 1.0 day^{-1}, while the mixed culture had an estimated maximum specific growth[145] rate of 0.55 day^{-1}, or roughly 50 per cent of the pure culture. In addition, the cell yield for the mixed culture was found to be low, at 0.065 mg/mg acetate. Interestingly, the endogenous coefficient was calculated to be 0.0 day^{-1}.

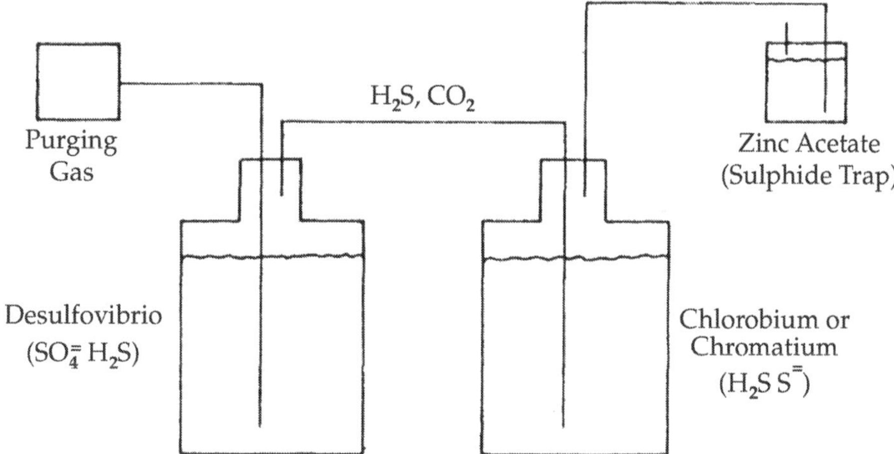

Fig. 7.3 : The Gas Purged Mutualistic System Utilizing Desulfovibrio and Either Chlorobium or Chromatium (Cork and Cusanovich, 1979)

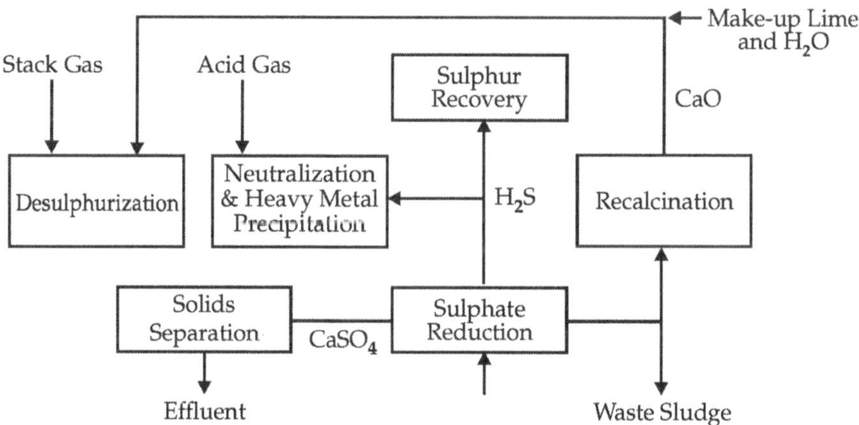

Fig. 7.4 : Possible Process Schemes using Microbial Sulphate Reduction for Treatment of Sulphuric Acid Wastes or Reclamation of Calcium Sulphate Sludges (Middleton and Lawrence, 1977).

7.6.4 Some Important Process Treatment Options

The following two process schemes developed in collaboration with Dr. Douglas J. Cork[(150)] Department of Biology, Illinios Institute of Technology. As shown in Fig. 7.5, the first schematic depicts the proposed process scheme for treatment of a high sulphate wastewater. The two basic unit processes are the sulphate reduction reactor and a photolithotrophic fermentor which is similar to the process used by Cork[143]. One major difference is the use of the photolithotrophic fermentor to provide the organic material required to drive the sulphate reduction step. The essential feature of the process is the recirculation of enough organic carbon to reduce the sulphate in the wastewater. Another major difference is the use of reactors in order to induce the conversion of synthesized and stored polyglucose to acetic acid. Physical processes would be required for the separation of the sulphur and acetate from the bacteria, and subsequent recycle of biomass back to the fermentor.

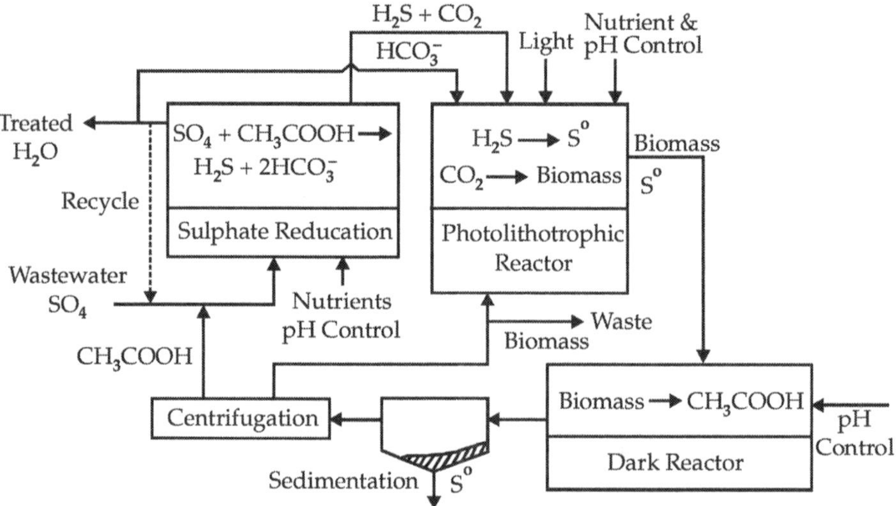

Fig. 7.5 : Integration of Anaerobic Processes for Sulphur Recovery from a High Sulphate Wastewater.

The second proposed system involves the biological (Fig. 7.6) treatment of H_2S waste gas streams which are normally found in the desulphurization of hydrocarbon and coal gasification[(147)]. In this case, the sulphate reduction step would not be required and the organic carbon (acetate) from the dark reactor could then be channeled into an anaerobic reactor for the recovery of methane, a potential energy source. Thus, light energy could be directly converted into a readily usable form of energy (methane), while oxidizing H_2S into the Claus process, in which methane is not recovered.

The process schemes cover the majority of pollution problems associated with sulphur, *i.e.* high sulphates and H_2S, and can potentially convert them to elemental sulphur, the most commercially desired form of sulphur. In additon, energy recovery is possible. Both processes use anaerobic microorganisms and thus have the advantages of no aeration equipment required, low sludge production, and the processes appear not to be energy intensive.

Because of the impact both processes would have on controlling sulphur in the environment, there appears to be justification for fundamental research in two major areas. In the first area, studies should be conducted on the *Chlorobium* species to determine such things as :

- maximum overall substrate utilization rate (optium organism concentration)
- minimum light requirements
- kinetics of acetate conversion in the dark reactor
- factors affecting the storate of polyglucose.

A second area of research would involve the sulphate reducing step and would include such things as :

- optimization of the sulphate reduction process.
- toxicity problems and control strategies

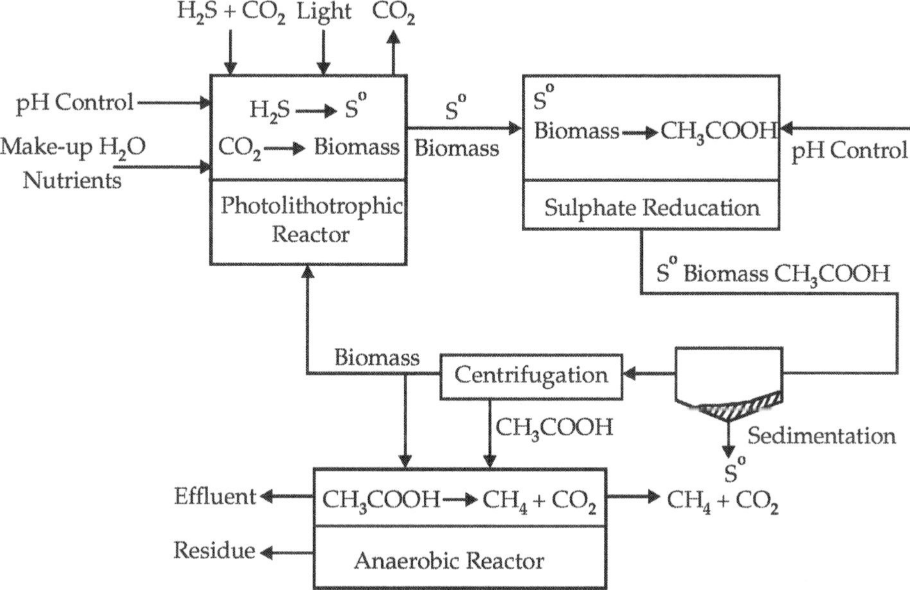

Fig. 7.6 : Integration of Anaerobic Processes for Sulphur and Energy Recovery from an Acid Gas Stream

The control of sulphur pollution is a widespread problem and mainly associated with hydrocarbon processing and power production. For example, the mining of coal results in acid mine drainage, the burning of coal results in $CaSO_4$ sludges (SO_2 scrubbing), and coal and oil desulphurization results in H_2S waste streams. Of the two forms of sulphur pollution SO_4^{-2} is a major problem with no reliable method of control.

A review of the literature indicates that microbial sulphur control is only in the formative stages. There is also an indication that potential exists for the use of both the sulphate reducing bacteria and a photolithotrophic organism *Chlorobium* which oxidises H_2S to S° using light. Two process schemes were proposed for both types of sulphur problems, *i.e.*, high sulphate wastewaters and H_2S gas streams. The process scheme for the treatment of high sulphate waters is particularly promising since the normal requirement for an outside source of organic carbon for sulphate reduction would not be required. However, before process development can begin, work needs to be accomplished on photolithotrophic bacteria, *Chlorobium* and the sulphate reducing bacteria.

REFERENCES

1. Schelegel, H.G. A Schneider, K. Microbial Metabolism of Hydrogen in Comprehensive Biotechnology, Vol. I, 439-457 (1985) Ed. Murry, Moo-young, Pergamon Press, New York.
2. Conrad, R. and Seiler, W. The Role of microbiological processes for the atmospheric hydrogen cycle. Forum Microbiologie. 4 : 219-225 (1980).
3. Zajie, J.E., Kosaric, N. and Brosseau, J.D., Microbial Production of Hydrogen. In Advances in Biochemical Engineering, Ed. T.K. Ghose, A. Fiechter and N. Blakebrough, Vol. 9, pp. 57-109, (1978) Springer-Verlag, Berlin.
4. Gray C.T. and Gest, H. Biological Formation of Molecular Hydrogen Science, 148, 186-192, (1965).
5. Kessler E. Hydrogenase, Photoreduction and Anaerobic Growth, In Algal Physiology and Biochemistry Ed., W.D.D. Stewart, pp. 456-473, (1974) Blackwell Scientific Publication, London
6. Geffron, H. and Rubin, J. Fermentative and Photochemical production of Hydrogen in Algae J. Gen. physiol., 26, 219-240, (1942).
7. Healey, F.P. Hydrogen Evolution by several Algae, Planta. 91, 220-226, (1970)
8. Benemann, J.R. and Weare, N.M. Hydrogen Evalution by Nitrogen Fixing *Anabena cylindrica* Cultures Science, 184, 174-175, (1974).
9. Weissman, J.C. and Benemann, J.R. Hydrogen Production by Nitrogen Starred Cultures of *Anabena cylindrica,* Appl. Environ. Microbial., 33, 123-131, (1977).
10. Jeffries, T.W., Timowrian, H. and Ward, R.L. Hydrogen Production by *Anabena cylindrica* Effects of Varying Ammonium and Ferric ions, pH and Light. Appl. and Environ., Microbial., 35, 704-710, (1978).
11. Shravan Kumar, Subhashini, T., Venkataramana, Renuka, B.R. and Vinayakumar Photoproduction of Hydrogen by some *Oscillatoria* sp. J. Microbial Biotechnol., 5, 25-31, (1990).
12. Thauer R.K., Jangerman K. and Decker K. Energy Conservation in Chemotrophic Anaerobic Bacteria, Bactriol. Rev. 41, 100-180, (1977).
13. Segars, L., Verstrynge L. and Verstraete, W., Production Patterns of Non-oxenic Fermentation as a Function of pH., Biotechnol. Ltt. 3, 635-640, (1981).
14. Gest H. and Kamen, M.D, Photoproduction of Molecular Hydrogen by *Rhodospirillum rubrum* Science 100, 558-559, (1949).

15. Gest H., Ormerod J.G. and Ormerod K.S. Photometabolism of *Rhodospirillum rubrum* Light Dependent Dissimilation of Organic Compounds to Carbon-di-oxide and Molecular Hydrogen by an Anaerobic Citric acid Cycle. Arch. Biochem. Biophys. 97, 21-33 (1962).

16. Ormerod, J.G., Ormerod, K.S. and Gest H. Light Dependent Utilization of Organic Compounds and Photoproduction of Molecular Hydrogen by Photosynthetic Bacteria Relationship with Nitrogen Metabolism., Arch. Biochem. Biophys, 94, 449-463(1961).

17. Benemann J.R., Berenson J.A., Kaplan N.O. and Karone M.D. Hydrogen Evolution by Chloropose Ferredoxin-hydrogenase System; Proc. Natl Acad. Sci. USA 70 2317-2320 (1973).

18(a). Ormerod, H.G. and Gest, H. Symposium on Metabolism of Inorganic Compounds IV-Hydrogen Photosynthesis and Alternative Metabolic Pathways in Photosynthetic Bacteria; *Bacteriol. Rev.* 26, 51-66 (1962).

18(b). Ormerod, J.G., Ormerod K.S. and Gest H. Light Dependent Utilization of Organic compounds and Photoproduction of Molecular Hydrogen by Photosynthetic Bacteria. Relationship with Nitrogen Metabolisms; *Arch. Biochem. Biophys.* 94, 449-463, (1961).

19. Rao, K.K., Rosa L. and Hall D.O., Prolonged Production of Hydrogen Gas by a Chloroplast Biocatalytic System; *Biochem. Biophys.*, Res. Commun. 68, 21-28, (1976).

20. Gray, C.T. and Gest, H. Biological Formation of Molecular Hydrogen; Science 186-192 (1965).

21. Pfennig, N. Photosynthetic Bacteria; Ann. Rev. Microbiol. 21, 285-324, (1967).

22. Gest, H. 1963 Metabolic Aspects of Bacterial Photosynthesis in *Bacterial photosynthesis* Eds H Gest, S. San Pietro and L.P. Vernon 129-150., (1963) Yellow Springs, Ohio; The Artioch Press.

23. Yoch, D.C. and Arnon D.I. Biological Nitrogen Fixation of Photosynthetic Bacteria; In The Biology of *Nitrogen Fixation* Ed. A. (1974) (New York).

24. Meyer,J., Kelley, B.C. and Vignais, P.M. Aerobic nitrogen fixation by *Rhodopseudomonas capsulata* FEBS Lett. 85, 224-228, (1978).

25. Seibert, M., Lien, S. and Weaver, P.S. Photobiological Production of Hydrogen. A Solar Energy Convention. SER.I/TR-33-122. Golden Colarodo Seifert and Pfenning N. 1978. Hydrogen Metabolism and Nitrogen Fixation in Wild Type and Nif Mutants of *Rhodopseudomonas acidophila; Biochimie.* 60, 261-265, (1980).

26. Mitsui A, Biological and Biochemical Hydrogen Producer; in Solar *Hydrogen Energy System* Ed. T. Data, Chapter 8, 171-191; (1979) (Oxford: Pergaman Press).

27. Mitsui, A and Kumazawa Hydrogen Production by Marine Photosynthetic Organisms as a Potential Energy Source in *Biological Solar Solar Energy Converson* Ed. A. Mitsui, A. Miyachi, and A. San Pietro 23-51, (1977) (New York: Academic Press).

28. Mitsui, A. Utilization of Solar Energy for Hydrogen Production by cell-free System of Photosynthetic Organisms in *Hydrogen Energy* Part I pp. 309-316, (1975) Ed. T.N. Veziroglu (New York Plenum Press).

29. ____, 1976 Photoproduction of Hydrogen *via* Microbial and Biochemical Processes in Proc. Symp Course *Hydrogen Energy Fundamentals* Ed. T.N. Veziroglu 77-99, (1975) (Miami: University of Miami Press).

30. Arnon, D.I., Losada, M., Nasaaki, M and Tagawa, K Photoproduction of hydrogen, photofixation of nitrogen and a unified concept of photosynthesis; *Nature,* 190, 601-606, (1961).

31. Mitsui, and Arnon D.I., Photoproduction of Hydrogen Gas by Isolated Chloroplasts in Relation to Cyclic and Non-cyclic Electron Flow; *Plant Physiol* 375 IV, (1962).

32. Lien, S. and San Pietro, A., In *An Inquiry into Biophotolysis of water to Produce Hydrogen* NSF-RANN Report 50 (1962).

33. Whatley, F.R. and Grant, B.R., Photoproduction of Methyl Viologen by Spinach Chloroplasts; Fed. Proc. 23, 227, (1964).

34. Abeles F.B. Cell-free Hydrogenase from *Chlamydomonas; Plant Physiol.* 39, 169-176, (1964).

35. Kramptitz L.D. Hydrogen Production and Photosynthesis and Hydrogenase Activity NSF RANN Report No. HAL, HA3, HA5 N-73-1013(1973) Biophotolysis of Water, NSF-RANN Report No. HA2 N-73-014, (1973).

36. Ben-Amotz, A., Erbes. D.L., Rioderer-Henderson M.A., Peavey, D.G. and Gibbs M., Hydrogen Metabolism in Photosynthetic Organisms. 1 Dark Hydrogen Evolution and Uptake by Algae and Mosses. *Plant Physiol.* 56, 72-77, (1975).

37. do and Gibbs, M. Hydrogen Metabolism in Photosynthetic Organisms. II Light dependent hydrogen evolution by preparations from *Chlamydomonas, Scenedesmus* and Spinach; *Biochem., Biophys. Res. Commun.* 64, 355-359(1975).

38. King, D., Erbes, D.L., Ben-Amotz, A. and Gibbs, M. The Mechanism of Hydrogen Photoevolution in Photosynthetic Organisms in *Biological Energy Conversion* Eds. A. Mitsui, A. San Pietro and S.Tamura 69-80, (1977) (New York: Academi Press).

39. Kitajima, M. and Butler W.L. Microencapsulation of Chloroplast Particles; *Plant Physiol.* 57, 746-750 (1976).

40. Ochiai, H, Shibata, H., Matsuo, T., Mashinokuchi, K. and Yakawa, M. Immobilization of Chloroplast Photostems; *Agri. Bio. Chem.* 41, 721-722 (1977).

41. Ochiai, H, Shibata, H., Matsuo, T., Mashinokuchi, K. and Yakawa, M. Immobilization of Chloroplast Photosystem within Polyacrylamide Gel formed by Redox Polymerization; *Agri; Bio. Chem.* 42, 683 (1978).

42. Shin, M. and Oshino, R.A. Ferredoxin-Sepharose 4Bas a tool for the purification of Ferridoxin-NADP-reductase; *J.Biochem.* (Tokyo) 83, 357-361 (1978).

43. Yagi, T. Separation of Hydrogenase Catalysed Hydrogen Evolution System from Electron Donating System by means of Enzymic Electric Cell, *Proc. Natl. Acad. Sci., U.S.A* 73, 2947-2954, (1976).

44. Use of an Enzymic Electrol Cell and Immobilized Hydrogenase in the Study of the Biophotocysts of Water to Produce Hydrogen; in *Biological Solar Energy Conversion* Eds. Mitsui, S. Miyachi. A San Pietro and S. Tamara pp. 686-687 (1977) (New York, Academic Press).

45. Lappi, D.A. Stolzenbach, F.I., Kaplan N.Q and Kamen M.D. Immobilization of Hydrogenase on Glass Beads; *Biochem. Res. Commynio* 69, 878-884, (1976).

46. Packer, L. Problems in the Stabilization of the *in vitro* Photochemical activity of Chloroplasts used for H_2 Production; *FEBS LETT.* 64, 17-19 (1976).

47. Adams, M.W.W., Reeves, S.G., Hall, D.O., Chrostou, C., Ridge B and Rydon, H.N., Biological Activity of Synthetic Tetranuclear Iron Sulfur Analogues of the Active Sites of Ferrodoxin; *Biochem. Biophys. Res. Commun* 79, 1184-1191 (1977).

48. Krasna, A.I., Catalytic and Structural Properties of the Enzyme Hydrogenase and its role in Photolyses of water in *Biological Solar Energy Conversion* Eds. A. Mitsui, S. Miyachi, A. San Pietro and S Tamara 13-22, (1977) New York: Academic Press).

49. Gaffron, H. and Rubin, J. Fermentative and Photochemical Production of Hydrogen in Algae; *J. Gen. Physiol.* 26, 219-240 (1942).
50. Roleofsen, P.A. Metabolism of Pure Sulfur Bacteria; *Proc. Acad. Sci. Amsterdam* 37, 660-669, (1934).
51. Nakumara, H., Uber das Vorkommen der Hydrogenlyase in *Rhodobacillus pollustris* Und Uber ihre role in mechanisms der bacterielen photosynthese; *Acts Phytochim.* 10, 211-218 (1937).
52. Gest and Kamen, M.D. Studies on the Metabolism of Photosynthetic Bacteri. IV. Photochemical Production of Molecular Hydrogen by Growing Cultures of Photosynthetic Bacteria; *J. Bacterial.* 58, 239-245 (1949).
53. Gest and Kamen, M.D. Photoproduction of molecular Hydrogen by *Rhodospirillum rubrum; Science* 109, 558-559, (1949).
54. Kaltwasser, H. Stuart T.S. and Gaffron H. Light Dependent Hydrogen Evolution by *Scenedesmus Planta* 89, 309-322, (1969).
55. Healey, F.P. The mechanism of Hydrogen Evolution by *Chlamydomonas moewusii Plant Physiol.* 45, 153-159 (1970).
56. Stuart, T.S. and Gaffron, H., The Kinetic of Hydrogen Photoproduction by Adapted Algae; *Planta* 100 228-243 (1971).
57. Stuart, T.S. and Gaffron, H., The mechanism of Hydrogen Photoproduction by Several Algae I. The Effect of Inhibitors of Photophosphorylation; *Plant* 106, 91-100 (1972).
58. Klein, U and Betz, A. Fermentative Metabolism of Hydrogen Evolving *Chlamydomonas moewusii; Plant Physiol.* 61, 953-956 (1978).
59. Senger, H. and Bishop N. Observations on the Photohydrogen Producing Activity during the Synchronous Cell cycle of *Scenedesmus obliques; Plants* 145, 53-62 (1979).
60. Jeffries, T.W. and Leach, K.L. Intermittent Illumination Increases Biophotolytic Hydrogen Yield by *Anabaena cylindrica; Appl. Environ. Microbiol* 35, 1228-1230 (1978).
61. Mahro, B. and Grimme, L.M. Hydrogen Photoproduction by Green Algae. The Significance of Anaerobic Pre-incubation Periods and of High Light Intensities for Hydrogen Photoproduction of *Chlorella fusca; Arch. Microbiol* 132, 82-86 (1982).
62. Benemanny and Weare, N.M. Hydrogen Evolution by Nitrogen Fixing *Anabaena cylindrica* culture, *Science* 184, 174-175 (1974).
63. Weissmann J.C. Benemann J.R. Hydrogen Production by Nitrogen Starved Cells of *Anabaena cylindrica; Appl. Environ. Microbiol.* 33, 123-131 (1972).
64. Bothe, Tennigkeit, J. and Eisbrenner, G. The utilization of Molecular Hydrogen by the Blue-green Algae *Anabaena cylindrica; Planta,* 133, 237-243 (1977).
65. Bothe and Loos, E., Effect of Far Red Light and Inhibhitors on Nitrogen Fixation and Photosynthesis in the Blue-Green Algae *Anabaena cylindrica; Arch; Microbiol,* 86, 241-254 (1972).
66. Telor, E., Luijk, L.W. and Packer An Inducible Hydrogenase in Cynobacteria enhances N_2 Fixation; *Arch. Biochem. Biophys* 185, 185-194 (1978).
67. Murallem Photoproduction of Hydrogen and $NADPH_2$ by Polyurethane Mobilized Cyanobacteria. Proc. Int. Conf. Commer; *Appl. Implication Biotechnol.* 1037-1050 (1983).
68. Kumazawa, S. and Mitsui A., Charaterization and Optimization of Hydrogen Photo Production by a Salt Water Blue-Green Algae *Oscillatoria* sp. Miami B.G. 7. I. Enhancement through Limiting Supply of Nitrogen Nutrients; *Int. J. Hydrogen Energy* 6, 339-348 (1981).

69. Phillips, E.J. and Mitsui A., Role of Light Intensity and Temperature in the Regulation of Hydrogen Photoproduction in the Marine Cyanbobacterium *Oscillatoria* sp. Stain Miami BG7; *Appl. Environ. Microbiol.* 45, 1213-1220 (1983).

70. Wolk C.P. Movement of Carbon from Vegetative Cells to Heterocysts in *Anabaena cylindrica: J. Bacterial* 96, 2138-2143 (1968).

71. Jones, L.W. and Bishop N.I. Simultaneous Measurement of Oxygen and Exchange from the Blue-Green Alga *Anabaena; Plant Physiol.* 57, 659-665 (1976).

72. Bothe H., Falkenberg B. and Nolteernsting V. Properties and Function of the Pyruvate: Ferrodozinoxide-reductase from the Blue-Green Algae *Anabaena cylindrica; Arch. Microbiol.* 96, 291-304 (1974).

73. Winkenbach, F. and Wolk, C.P. Activities of Enzymes of the Oxidative and the Reductive Pentose Phosphate Pathways in Heterocysts of a Blue-Green Algae: *Plant Physiol.* 480-483(1973).

74. Smith, R.V. Noy R.J. and Evans, M.C.W. Physiological Electron Donor Systems to the Nitrogenase of the Blue-Green Algae *Anabaena cylindrica: Biochem. Biophys. Acta.* 253, 104-109 (1971).

75. Leach, C.K. and Carr, N.G. Pyruvate: Ferrodoxin Oxidoreductase and its Activation by ATP in the Blue-Green *Ababaena variabilits; Biochem. Biophys Acta.* 245, 165-174 (1971).

76. Belkin. S. and Padan, E. Sulfide Dependents Hydrogen Evolution Cyanobacterium *Oscillatoria limmetica; FEBS Lett.* 94, 291-284 (1978).

77. Xiaukong, Z., Maskell, J.B. Tabita, F.R. and Van Bualen C. Aerobic System Hydrogen Production by the Hetrocystous Cyanobacteria, *Anabaena* sp. Strains C.A. and IF; *J. Bacterial* 156, 1118-1122 (1983).

78. Weisaar, H. and Boger, P., Sulfide Stimulation of Light Induced Hydrogen Evolution by the Cyano bacterium *Nostoc muscorum; Z. Waterforsh,* 38C, 237-242 (1983).

79. Gar, N.G. *Methods in Microbiology,* Eds., J.R. Norris and D.W. Ribbons pp. 53-57 (1969) (New York: London: Academic Press).

80. Gottschalk, G., *Metabolism* Ed, M.T. Starr in *Bacterial* 237-249 (1979) (New York: Heidelberg: Berlin: Spinger Verlag).

81. Hillmer, P. and Gest, H., Hydrogen Metabolism in the Photosynthetic Bacterium *Rhodopseudomonas capsulata,* Hydrogen Production by Growing Cultures; *J. Bacteriol.* 129. 724-731(1977).

82. Joanneau, Y., Kelley B.C., Berlies, Y., Lespinad, P.A. and Vignais, P.M. Continous Monitoring by Mass Spectrometry of H_2 Production and Recycling in *Rhodopseudomonas capsulata; J. Bacterial.* 143, 628-636 (1980).

83. Carithers, R.P. Yoch, D.C. and Arnon, D.I. Two Forms of Nitrogenase from the Photosynthetic Bacterium *Rhodospirillum rubrum; J. Bacterial* 137, 779-789 (1979).

84. Gest, H. and Bregoff H.M., Studies on the Metabolism of Photosynthetic Bacteria and Photoproduction of Hydrogen and Nitrogen Fixation of *Rhodospirillum rubrum; J. Biol. Chem.* 182, 153-170 (1950).

85. Bennett, M.A, and Weetall, H.H. Production of Nitrogen using Immobilized *R. rubrum; Solid Arch. Biochem.* 1, 137-142 (1976).

86. Zurrer, H. and Bachofen, R., Hydrogen Production by the Photosynthetic Bacterium *Rhodospirillum rubrum; Appl. Environ. Microbiol.* 37, 789-793 (1979).

87. Vatsala T.M. Influence of Nutritional factors on – Hydrogen Production by *Rhodospirillum* rubrum ATCC 11170 communicated to MIRCEN; *J. Applied Microbiol. Biotechnol* (1985).

88. Uffen, R.A., Growth Properties of *Rhodospirillum rubrum* Mutants and Fermentation of Pyruvate in Anaerobic Dark Conditions; *J. Bacteriol.* 116, 874-884 (1973).

89. Paschinger, M.A. Changed Nitrogenase Activity in *Rhodospirillum rubrum* after Substitution of Tungsten for Molybdenum; *Arch. Microbiol.* 101, 379-389 (1974).

90. Voelskow, H. and Schon, G. Pyruvate Fermentation in Light-grown Cells of *Rhodospirillum rubrum* during Adaptation to Anaerobic dark Condtitions *Arch. Microbiol.* 119, 129-133 (1978).

91. Brosseau J.D. and Zajic, J.E. Hydrogen Gas Production and Growth of *Citrobacter intermedius or Glucose, Development Industrial Microbiol.* 19, 553-567 (1978).

92. Brosseau J.D. and Zajic, J.E. Anaerobic co-oxidation of acetate and glucose by *Cirobacter intermedius* and a species of *Pseudomonas; Can. J. Microbiol.* 26, 1503-1505 (1980).

93. Brosseau J.D. and Zajic, J.E. 1982 Agitation Effects on Hydrogen Gas Production by *Citrobacter intermedius; Biotech. Bioengg.* 24, 1469-1472 (1982).

94. Healy Hydrogen Evolution by Several Algae; *Planta* 91, 220-226 (1970).

95. Bishop N.I., Frick M, and Jones K.W. Photohydrogen Production in Green Algae; Water Serves as the Primary Substrate for Hydrogen and Oxygen Production in *Biological Solar Energy Conversion* pp. 3-22 (1977) Eds. A Mitsui, S. Miyachi, A. Sen Pietro and S. Tamura (New York: Academic Press).

96. Bose, S.K. and Gest H. Hydrogenase and Light Stimulated Electron Transfer Reactions in photosyinthesis. *Nature.* 195, 1168 (1962).

97. Daday, A., Platz, R.A. and Smith, C.D. Anaerobic and Aerobic Hydrogen Gas Formation the Blue-Green Alga *Anabaena cylindrica; Appl. Environ. Microbiol.* 34, 478-483 (1977).

98. Timourian, H. and Ward H.L. Hydrogen Production by *Anabaena cylindrica.* Effects of Varying Ammonium and Ferric ions, pH and Light *Appl. Environ. Microbiol.* 35, 704-710 (1978).

99. Lambert, G.R. and Smith G.D. Hydrogen Formation by Marine Blue-Green Algae; *FEBS Lett.* 83-1591 (1977).

100. Spiller, H., Ernst. A., Kerfin, W., and Bogar. Increase and Stabilization of Photoproduction of Hydrogen in *Nostoc muscorum* by Photosynthetic Electron Transport Inhibitors; *Z. Naturforsh. Tiel. C.* 33, 541-457 (1978).

101. Mitsui Philips, E.J., Kumazawa. S., Reddy K.J., Ramachandran S., Matsunaga, T., Haynes, L. and Ikemoto, H. Progress in Research towards Outdoor Biological Hydrogen Production using Solar Energy. Sea Water and Marine Photosynthetic Organisms; *Annals New Acad. Sciences,* 514-530(1988).

102. Aronan, Mitsui, A., and Paneque, A Photoproduction of hydrogen gas coupled with photosynthetic photophosphorylation; *Science* 134, 1425-1427 (1961).

103. Bernstein, J.D. and Olson J.M. Photoproduction of Hydrogen by Membrance of green Photosynthetic Bacteria Photosynthesis; *Proc. Int. Cong. 5th.* 6, 475-650.

104. Hoffmann P., Thauer R. and Trebst A 1977 Photosynthesis Hydrogen evolution by spinach chloroplast coupled to a *Clostridium* hydrogenase; *Z. Naturforsch. Teil.* 32, 251-262 (1977).

105. Fry. I., Papgeorgiou G., Telor, E and Packer L.E. 1977 Reconstitution of a system for Hydrogen Evolution with Choloroplasts, Ferredoxin and hydrogenase; *Naturforsch. Tel. C.* 32, 110-117 (1977).

106. Gogotov, I.N. and Hall, D.O. Hydrogen Evolution by Chloroplast-Hydrogenase system; Improvement and Additional Observation; *Biochimie* 60, 291-295 (1978).

107. Reeves, S.G. Rao., K.K. Rose, L. and Hall D.O. Biocatalytic Production of Hydrogen in *Microbiol Energy Conversion* Eds. H.G. Schlegel and J. Barnea, 235-242 Forenare, 1977 (New York: Pregamon Press).

108. Quadri, S.M.H. and Hoare D.S. Formic Hydrogenlyase and Photoassimilation of Formate by a Strain of *Rhodospirillum palustris; J. Bacterial.* 95, 2344-2357 (1968).

109. Vicenzini, M., Matterasui, R., Tredici, M.R and Florenazare, G., Hydrogen Photoproduction by Immobilized Cells. Light Dependent Dissimilaration of Organic Substances by *Rhodopseudomonas Plause Int. J. Hydrogen Energy* 7, 231-236 (1982).

110. Mitsui T.V. and Glinshil V.P. Effect of Ammonium on Hydrogen Evolution and Nitrogen Fixation in *Rhodopseudomonas palustris; Mikrobiologiya* 43, 586-591 (1974).

111. Macler, B.A. Polroy R.A. and Bassham, J.A. Hydrogen Formation in Nearly Stoichiometins Amounts from Glucose by a *Rhodopseudomonas sphaeroides* mutant, *J. Bacterial.* 138, 446-452 (1979).

112. Newton J.W. and Wilson, P.W. Nitrogen Fixation and Photoproduction of Molecular Hydrogen by Thiordaceae; *J. Microbiol. Serol* 19, 71-77 (1953).

113. Mitusi, A., Survey of Hydrogen Producing. 1-68 (1975) NSF. Annual Report.

114. Ohta, Y., Frank J. and Mitsui A., Hydrogen Production by Marines Photosynthetic Bacteria. Effect of Environmental Factors and Substrate Specificity on the Growth of Hydrogen Producing Marine Photosynthetic Bacterium *Chromatium* sp. Miami PBS 1071; *Int. J. Hydrogen Energy* 6, 451-460 (1981).

115. Gogotov, I.N. Relationship in Hydrogen Metabolism between Hydrogenase and Nitrogenase in Phototrophic Bacteria; *Biochemic.* 60, 267-275 (1978).

116. Kondratieva, O.N. and Gogotov, I.N. Production of Hydrogen by Green Photosynthetic Bacteria (*Chloropseudomonas);* Nature 221, 83-84 (1969).

117. Gogotov and Zorin N.A., Hydrogen Metabolism and Hydrogenase Activity in *Rhodospirillum rubrum; Mikrobiologiya* 41 947-952 (1972).

118. Energy Conversion and Generation of Reducing Power in Bacterial Photosynthesis; *Adv. Microbiol. Physiol.* 7, 243-282 (1971).

119. Kohlmiller, E.F. and Gest, H., A Comparative Study of the Light and Dark Fermentation of Organic Acids by *Rhodospirillum rubrum; J: Bacterial.* 61, 261-282 (1956).

120. Gorrel, T.E. and Uffen R.L. Fermentative Metabolism of Pyruvate by *Rhodospirillum rubrum* after Anaerobic Growth in Darkness; *J.Bacterial.* 131, 533-543 (1977).

121. Schon, G. and Biedermann, M. Growth and Adaptive Hydrogen Production of *Rhodospirillum rubrum* (F_1) in Aerobic Dark; *Biochem. Biophys. Acta* 304 65-75 (1973).

122. van Nod, C.B., The Predent Status of the Comparative Study of Photosyntheses Ann. Rev. Plant Physiology 13, 1-2 (1962).

123(a)Uffen Symbesma C. and Wolfe R.S. Mutants of *Rhodospirillum rubrum* Obtained after Long Anaerobic Dark Growth; *J. Bacteriaol.* 108. 1348-1356 (1971).

123(b) Van Niel C.B. 1962 The Present status of the Comparative study of Photosynthesis; *Ann. Rev. Plant Physiology* 13, 1-26 (1971).

124. Gest H. Ormed J.G. and Ormerod, K.S. Photometabolism of *Rhodospirillum rubrum*. Light Dependent Dissimilation of Organic Compounds to CO_2 and Molecular H_2 by Anaerobic Citric Acid Cycle; *Arch. Biochem. Biophys.* 97, 21-33 (1962).

125. Bregoff, H.M. and Kamen M.D. Studies on the Metabolism of Photosynthetic Bacteria. XIV. Quantative Relations Between Malate Dissmilation Photoproduction of Hydrogen and Nitrogen Metabolism in *Rhodospirillum rubrum; Arch. Biochem. Biophys* 36, 202-220 (1952).

126. Schick, H.A. Interrelationship Between Nitrogen Fixation; Hydrogen Evolution and Photoreduction in *Rhodospirillum rubrum: Arch. Microbiol.* 75, 102-109 (1971).

127. Miyaki, J., Tomzuka N. and Kamobayashi, A.J. Prolonged Photo-hydrogen Production by *Rhodospirillum rubrum: J. Fermn. Tech.* 60, 199-203 (1982).

128. Stiffler, H.J., Gest H. Effects of Light Intensity and Nitrogen Growth Source on Hydrogen Metabolism in *R. rubrum. Science* 120, 1024-1026 (1954).

129. Kelley, B.C. Meyer, C.M. Gandy, C.G. and Vignais P.M. Hydrogen Recycling by *Rhodospirillum capsulata; FEBS. Lett.* 81, 281-285 (1977).

130. Weaver, F.F. Wall J. and Gest, H. Characterization of *Rhodopseudomonas capsulata: Arch. Microbiol.* 105, 207-216 (1975).

131. Gonel and Uffe Light Dependent and Light Independent Production of Hydrogen Gas by *Rhodospirillum rubrum* Mutant, C. Photochen; *Photobiol.* 27, 351-358 (1978).

132. Zurrer, H. and Bachofen, R. Hydrogen Production by the Photosynthetic Bacterium *Rhodospirillum rubrum* Appl. Environ. Microbiol. 37, 789-793 (1979).

133. Vignais, P.M., Colbear A., Willison J.C. and Jouanneau Y. Hydrogenase. Nitrogenase and Hydrogen Metabolism in the Photosynthetic Bacteria Adv. *Microbiol. Physiol.* 26, 155-234 (1985).

134. Dixon R.D.D Hydrogenase in Legume root Nodule bacteriols Occurence and Properties *Arch. Microbiol* 85, 193-201 (1972).

135. Hillmer P. and Gest H. H_2 Metabolism in the Photosynthetic Bacterium *Rodopseudomonas capsulata;* Production an Utilization of H_2 by Resting Cells *J. Bacteriol.* 129, 732-739 (1977).

136. Kelly, B.C., Meyer C.M. Gamdy C. and Vignoir, P.M. Hydrogen Recycling by *Rhodopseudomonas capsulata, FEBS Letters,* 1, 281-285 (1977).

137. Siefert, E. and Pfennig N. Hydrogen Metabolism and Nitrogen Fixation in Wild type and Nif-mutants of *Rhodopseudomonas acidophla Biochem* 60, 261-265 (1978).

138. Cobeau, A., Kelly, B.C. and Vignais P.M. Hydrogenase Activity in *Rhodopseudomonas capsulats;* Relationship with Nitrogenase Activity *J. Bacteriol.* 141, 141-148 (1980).

139. Karube, I., Aizawa, K., Ikeda, S. and Suzuki S., Continous Hydrogen Production by Immobilized Cells of *Clostridum butyricum Biochem. Biophys Acta,* 444, 338-343 (1976).

140. Metsunage, T., Karube, I and Suguki S., Some Observations on Immobilized Hydrogen Producing Bacteria. Behaviour of Hydrogen in Membrances Biotechnol. Bioengg. 22, 2607-2615(1980).

141. Matsunaga, T., and Mitsui A., Sea Water Based Hydrogen Production by immobilized Marine Photosynthetic Bacteria. Biotechnol. Bioengg. Symp. 12, 441-452 (1982).

142. Mitsui, A., Matsunaga T., Ikemoto H and Renuka B.R. Organic and Inorganic Waste Treatment and Simultaneous Photoproduction of Hydrogen by Immobilized Photosynthetic Bacteria Dev. Ind. Microbol. 26, 206-209 (1985).

143. Matsunaga, T. and Mitsui, A. Sea Water Based Hydrogen Production by Immobilized Marine Photosynthetic Bacteria. Biotechnol. Bioengg. Symp. 12, 441-450 (1982).

144. Cork, D.J. *Kinetics of Sulfate Conversion to Elemental Sulfur by a Bacterial Mutualism; A Hydrometallurgical Application.* Ph. D. Thesis, University of Arizona (1978).

145. Cork., D.J. and Cusanovich M.A. Continous Disposal of Sulfate by a Bacterial Mutualism. *Dev. Ind. Microbiol.* 20, 591-602 (1979).

146. Middleton, A.C and Lawrence A.W. Kinetics of Microbial Sulfate Reduction. *J. Wat. Poll. Con. Fed.* 49 1659-1670 (1977).

147. Sirevag, R. and Ormerod J.C. Synthesis, Storage and Degradation of Polyglucose in *Chlorobium thiosulfatophilum. Arch. Microbiol.* 111, 239-244 (1977).

148. Cork, D.J. and Cusanovich, M.A. Sulfate Decomposition : A Microbiological Process. In *Applications of Bacterial Leaching and Related Microbiological Phenomena,* Ed. L.E. Murr, pp. 207-222. (1978) Academic Press, Inc. New York.

149. Pugsley, E.B., Cheng, C.Y., Updegraff, D.M and Ross, L.W. Removal of Heavy Metals From Mine Drainage in Colorado by Precipitation *Chem. Eng. Ser.* 67 75-89 (1971).

150. Cork, J.D. Acid Waste Gas Bioconversion An Alternative to the Claus Process. Dev. Ind. Microbiol. 23 (1982).

151. Cork, J.D Design and Utilization of a Bench Scale Gaslift Fermentor Dev Ind. Microbiol. 22, 733-738 (1981).

CHAPTER 8

KINETICS OF ANAEROBIC TREATMENT PROCESSES

8.0 Introduction

The problems of non-ideal flows are intimately tied to those of scaleup because the question of whether to pilot plant or not rests in large part on whether we have knowledge of all the major variables involved in the process. If we do, there is no reason to pilot plant. Often the factor that is uncontrolled in the design of bioreactors is the magnitude of the nonideality of flow, and unfortunately this factor very often differs widely between large and small units. Therefore, ignoring this factor may lead to gross errors in design. In reality, the flow of fluid is never ideal. It is, therefore, necessary to consider the non-ideal flow to acquire an intuitive feel for the magnitude of this phenomenon in reactors of various types, to know whether or not it has to be treated and when it does have to be treated, to approach its solution in a rational manner consistent with the present by knowledge of the subject.

8.1 Residence Time Distribution of Fluid in Vessels

To predict the exact behaviour of a reactor vessel, it is necessary to know what is happening in it. There is only one way and that is to tag and follow each and every molecule as it passes through the vessel. This can be done essentially by having a complete velocity distribution picture of the fluid within the vessel. Though fine in principle, the attendant complexities make it impractical to use this approach. The information of ages of molecules in the exit stream or the distribution of ages of the molecules in the exit stream or the distribution of residence times of the molecules within the vessel can be found easily and directly by widely used experimental technique, the stimulus response technique. The information can them be used in one of the two ways to account for the non-ideal flow behaviour of fluid in a chemical flow reactor.

The basis of this approach is a set of typical (or standard) operators which depict elementary fluid mechanic models (the well mixed or complete mixing model, the plug flow model, the cellular or cascade model and their combinations) from which one can determine the completion time directly. With this approach, a mathematical description involves, in addition to the deterministic aspect, also a search for combination of elementary operators so matched that the resultant model is a sufficiently accurate replica of the actual unit operation or of a process.

- The plug flow model refers to a flow pattern in which the particles travel axially without intermixing, whereas the carrier fluid is displaced like a plug and its composition varies from position to position along the reactor. With this form of flow, the apparatus is a distributed-parameter unit.
- The dispersion model refers to a flow in which the particles move axially and intermix in part both axially and radially as they would in the case of diffusion. In analysis, one considers either only the axial mixing or both the axial and radial mixing (in which case we have a two parameter dispersion model). The parameters of the model are the turbulent dispersion coefficient or the longitudinal dispersion coefficient. D_L and/or the radical dispersion coefficient, D_R. In experimental determinations the two coefficients, D_L and D_R are usually represented as dimensionless groups or products, know as Peclet numbers,

$$P_e = \frac{VL}{D_L} \text{ or } \frac{VL}{D_R} \qquad \text{.... (8.1)}$$

Where, V is the linear velocity of the fluid and L it the length of the tubular reactor. An apparatus in which the flow has the above pattern is referred to as a distributed-parameter unit.

The well mixed model refers to a tank reactor in which the contents are mixed equally well throughout the tank. At the inlet the stock is mixed so intimately that the concentration is equalized all over the reactor at once, being of the same value equal to that at the outlet. This is a unit with lumped (or concentrated) parameters.

The cascade reactor model is a cascade of well mixed single tank reactors with no mixing between the tanks. The parameters of the model is the number 'n' of the well mixed tank reactors. With 'n' tending to infinity, this model reduces to a plug flow model; with n = 1, to a well mixed tank reactor model.

The combination models refer to cases which may combine more than one of the flow patterns described earlier along with stagnation zones, short circuiting, recycle and so on.

The parameters of a combination model are the volumes of the various zones (with mV_r-referring to volume of a well stirred zone; bV_r-to a plug flow zone; dV_r to a stagnation zone), and the fractions of the flows connecting the cascaded tanks (with λ referring to short circuiting fraction, R to the recycle fraction, and V_r to the reactor volume).

The unknown parameters of a model are usually found by experiment. To this end, an amount of tracer material is introduced at the inlet to the reactor so as to disturb the flow composition, and the transient response is plotted as a curve at the exit. Most often, the tracer is a salt or an acid solution, a dye or a radio-isotope. The standard singals are impulse (Dirac delta) input function, the unit step function and a random sinusoidal input function.

For combination model there is a specific relation between parameters. At D_L tending to zero (dispersion model) or 'n' tending to infinity (the cascaded reactor model), the flow pattern is that of the plug flow. At D_L tending to infinity or at n=1 we have a well mixed tank reactor. If the number of well-stirred tank reactors is in excess of from six to ten, the relation between n and D_L for a closed tank reactor takes the form, *i.e.* :

$$\frac{1}{n} = \frac{2D_L}{VL} \text{ or } n = \frac{P_e}{2} \qquad .....(8.2)$$

The parameters of a combination model are usually the geometric dimensions of various zones and the parameters of the constituent flows. The various forms of testing for a systems response is given in the following sections.

8.1.1 Inpulse Testing and C-Diagram

In this case, a mass of tracer is instantaneously injected at time t = 0 into the feed stream. The concentration of the tracer material is then observed at the system exit for a series of time intervals. These data are plotted to give a C-diagram which is equivalent to the distribution density of tracer particles in the flow as a function of their residence time in the reactor.

$$C = \int^{\infty} t\, C(t)\, dt \qquad .....(8.3)$$

Where C(t) is the residence time distribution describing the tracer's fraction residing in the reactor for a time shorter than t.

The impotant numerical characteristics of the distribution function are the average residence time and the variance which gives the spread of a variable about its mean. These numerical characteristics can readily be found from the C-diagram by statistical method.

Let C_i be the concentration of tracer observed at some section of the reactor at time t_i when the tracer is injected into the flow instantaneously. Then, the mean residence time will be :

$$\bar{t} = [\int_0^{\infty} t\, c\, dt] / \int_0^{\infty} c\, dt \approx \sum_{i=1}^{n} t_i c_i / \sum_{i=1}^{n} c_i \qquad ......(8.4)$$

The variance will be

$$\sigma_t^2 = \left(\int_0^{\infty} t^2 c\, dt\right) / \left(\int_0^{\infty} c\, dt\right) - \left(\int_0^{\infty} t\, c\, dt / \int_0^{\infty} C\, dt\right)^2 \qquad .....(8.5)$$

$$= \sum_{i=1}^{n} t^2 c_i / \sum_{i=1}^{n} c_i - \left(\sum_{i=1}^{n} t_i c_i / \sum_{i=1}^{n} c_i\right)^2 \qquad .....(8.6)$$

The dimensionless time is defined as

$$\theta = t / \bar{t} \qquad .....(8.7)$$

and the dimensionless variance as

$$\sigma^2 = \sigma_t^{\,2} / (\bar{t})^2 \qquad .....(8.8)$$

For the C-distribution in dimensionless time (θ), σ^2 will be as follows :

$$\sigma^2 = \left(\sum_{i-1}^{n} \theta^2 C_{i/} \sum_{i=1}^{n} C_i\right) - 1 = \sum \theta_i^{\,2}\, C_i\, \Delta\theta - 1 \qquad .....(8.9)$$

For accuracy of computation to be satisfactory, the upper limit of summation, n must be sufficiently large. Also, the interval along the time axis, $\Delta t = t_{i+1} - t_i$, should be chosen sufficiently small, and the concentration of the tracer should be observed over the entire sensitivity range of the instrument recording C-diagram.

By using analytical solution for the equation of a particular model (Table 8.1 and 8.2), one can deduce analytical expression which contain the desired model variables for the numerical characteristics listed above. Then, the unknown model variables are found by comparing experimental and numerical values (usually presented graphically).

8.1.2 Step Testing and the F-Diagram

A step change in the tracer amount produces what is known as the F-response or F-diagram. Denoting as F_i the value of the response function of the system to a step disturbance at time, t_i, the probabilistic characteristics of the distributing function can be found as follows :

The mean residence time will be

$$\bar{t} = \int_0^{\infty} [1 - F(t)]\, dt = \frac{\Delta t}{2} \sum_{i=1}^{n} \left[(1 - F_i) + (1 - F_{i+1})\right] \qquad ......(8.10)$$

and the variance will be

$$\sigma_t^2 = 2\int_{\bar{}}^{\infty} t[1-F(t)]dt$$

$$\approx \frac{\Delta t}{2}\sum_{i=1}^{n}[(1-F_i)+(1-F_{i+1})](t_i+t_{i+1})-\overline{t^2} \qquad .....(8.11)$$

In equations 8.11 and 8.12, it is assumed that the F-diagram is normalized, that is, its value at $t \to \infty$ is taken as unity. Further computation is the same as with C-diagram.

The two responses are related as follows :

$$F = \int_0^{\theta} . \, C d\theta \qquad ..... (8.12)$$

$$C = dF/d\theta \qquad ....(8.13)$$

8.1.3 Sinusoidal Disturbance

To find the variables involved it is necessary to observe the ratio of the amplitude of the concentration waves at the inlet to outlet from the reactor, $A(I)/A(\theta)$ and the phase angle Φ.

8.1.4 Random Disturbance

Let a random change in tracer be applied to the inlet of a system. Then variations in the tracer concentration at the system's exit will likewise be random. On denoting the mutual correlation function of the input and out signals as R_{yx} (t) and the auto-correlation function of the output as $R_x(t-\bar{t})$, one may then find the unknown distribution function C(t) by solving an integral equation of the form.

$$R_{yx}(t) = \int_0^{\infty} R_x(t-\bar{t})\, C(t)\, dt \qquad ... (8.14)$$

The unknown model parameters are found by comparing the distribution function found by experiments with that obtained by calculations.

Table 8.1 : Types of Flow Patterns

(A) The Plug Flow Model : It is applicable to tubular units in which length to diameter ratio is over 20:1. The model can be described by an equation of the form

$\delta c/\delta t + V\ \delta c/\delta x = 0$

The initial and boundary conditions are

$C\ (x, o) \quad = \quad C_o; \qquad (0 \leq x \leq L)$

$C\ (o, t) \quad = \quad C_o\ (t);\ (t > 0)$

Its solution is

$C\ (x, t) \quad = \quad C_o,\ t > L/v$

$\qquad C_o\ (t - L/v),\ t > L/v$

x=0 x=L

v v

C_o C

(B) The Well Mixed Tank Model : It is applicable to cylindrical units with dished bottom fitted with an agitator and baffles. It can be described by an equation of the form

$$V\frac{dc}{dt} = Q\ (C_o - C)$$

The initial and boundary conditions are

$C_o = C_{in}\ \delta\ (0)$

$C_o = C_o\ (t)$ (an arbitrary function)

The solution of the equation is

$C/C_o = \exp\ (-\ t/\bar{t})$

$$C(t) = \exp\ (-\ (t/\bar{t})[C_o(t) + \frac{1}{t}\int_0^t \exp\ (\xi/\bar{t})\ C_o(\xi)d\xi]$$

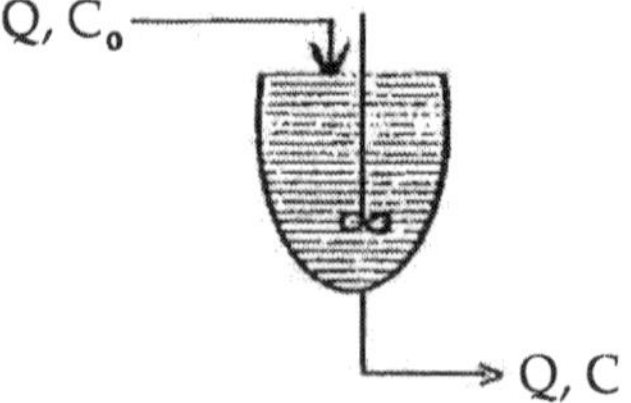

(C) The One Parameter Dispersion Model : This applies to tubular reactors, packed and unpacked towers or columns and involves axial dispersion of the feed. The equation describing the model has the form

$$D\ \delta^2\ C/\delta x^2 - v\ \delta c/\delta x = \frac{\delta c}{\delta t}$$

The initial and boundary conditions are

– total length $-\infty < x < +\infty$

– the length of the experimental portion $0 \leq x \leq L$

$C\ (x, o) = 0\ (-\infty < x < +\infty)$

$C\ (o, t) = C_{in}\ \delta\ (0)\ (t \geq 0)$

The solution of the equation is

$C\ (L, Q)/C_{in} = 1/2\ (P_e/\Pi\theta)^{1/2} \exp[-P_e(1-\theta)^2/4\theta]$

Where,

$P_e = vL/D$

$\theta = t/\bar{t}$

$t = L/v$

x = 0 x = L

– ∞ SD δc/δx QC + ∞

(D) The Two Parameters Dispersion Model : It applies to tubular reactors, towers with a small length to diameter ratio and an appreciable longitudinal and radial dispersion. Its equation has the form

$D_x\ \delta^2c/\delta x^2 - v_x\ \delta c/\delta x + (Dr/r)\ (\delta(r\delta c/\delta r)/\delta r) = \delta c/\delta t$

$C = (x, r, t)$

The initial and boundary conditions are as follows

– total length $0 \leq x \leq L$

– the radius of the reactor $0 \leq r \leq R$

$$C(x, r, o) = 0$$

$$C(o, o, t) = C_o\delta(0)$$

$$v\ C\ (o, r, t) - D_x[\delta c(o, r, t)/\delta x] = 0$$

$$\delta c(L, r, t)\delta x = 0$$

$$\delta c(x, r, t,)\delta r = 0$$

The solution of the equation is

$$C\ (Z, \rho, \theta) = \sum_{n=1}^{\infty} \frac{C_o(k_o - 1/2\ Dz)}{2k_o} \exp\ (-\lambda^2\theta) J_o(X_n\rho)\ X$$

$$[\exp\ (1/2\ Dz - k_o)^z + \frac{k_o + Dz/2}{k_o - Dz/2} X \exp\ (1/2\ Dz + k_o)^z]$$

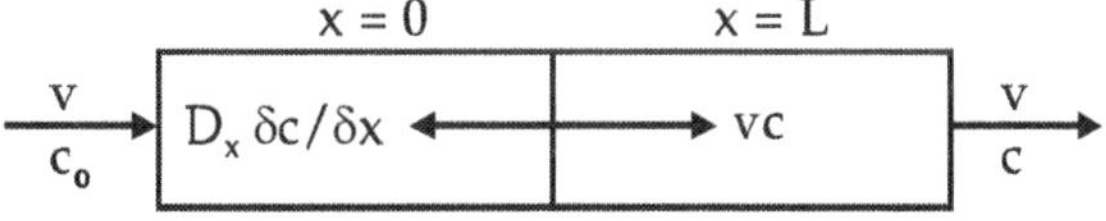

where $z = X/L;\ \rho = r/R;\ \theta = t/\bar{t};\ Dz = D_x\ \bar{t}/L$

J_o = Bessel's zero order function of the first kind

X_n = root of the Bessel first order function of the final kind and k_o is the root such that $e^k = (D_z/2 + k)/(D_z/2 - k)$

(E) The Cascade Model : This model applies to as series of equally sized well mixed vessels, plate type towers, fluidized bed reactors and packed towers. The equation of the model has the form

$$\frac{dC_i}{dt} = Q/V_i(C_{i-1} - C_i); (i = 1,2,.....n)$$

The initial and boundary conditions are

$C\ (0) = C_o\delta(0)$ at $V_1 = V_2 = V_n = V$

The solution of the equation is

$$\frac{C_n}{C_n} = \exp\ (n\theta)n^n\theta^{n-1}/(n-1)!$$

where, $\theta = t/\bar{t},\ \bar{t} = V/Q$

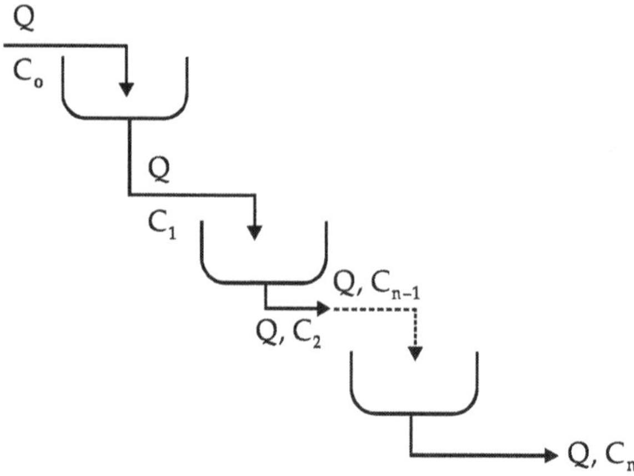

(F) The Dispersion Model with Stagnation : This applies to flows through a particulate medium and a packing. Its equation is

- at $k_1 = k_2 = k$
- $Q\ \delta C_1/\delta x + SH_1 D\delta^2 C_1/\delta x^2 - kS(C_1 - C_2) = SH_1\delta C_1/\delta t$
- at $k_1 = k_2$
- $Q\delta c_1/\delta x + SH_1\ D\dfrac{\delta^2 C}{\delta x^2} - S(k_1 C_1 - k_2 C_2) = SH_2\ \delta C_1/\delta t$
- $k_1\ C_1 - k_2 C_2 = H_2\ dc_2/dt$

The initial and boundary conditions vary according to the type of the experimental set up.

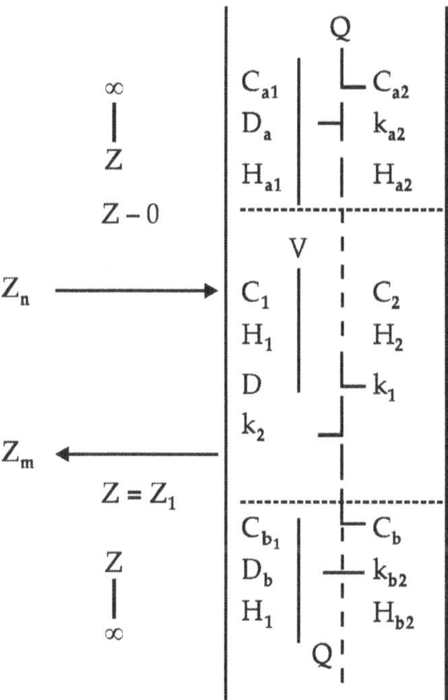

The solution for the first two moments of the residence time distribution involving various experimental setups is given by Kafarov, *et al.*

Source : Kafarvov, V.V. and Dorokhov, I.N. Systems Analysis of Operations in Chemical Engineering.

Book - 1 : Fundamentals of Strategy, Moscow, Nauka (1976).

Book - 2 : Topological Principle of Formalism, Moscow, Nauka (1979).

Book - 3 : Statistical Methods of Plant Identification in Chemical Engg., Moscow, Nauka (1981).

(G) The Cascaded Reactor Model with Stagnation : This model is true of flow in a particulate medium and packings and applies to any chemical reactor where relatively stationary 'dead spaces exist'. The equation of the model is

$V_{1i}\ dX_1/dt = Qx_{i-1} - Qx_1 + v_i\ k_2\ y_i - \ v_i\ k_2\ y_i - v_i\ k_1\ x_i$

$V_{2i}\ dY_i/dt = v_i\ k_1\ x_i - v_i\ k_2\ y_i$

where, $V_i = v_{1i} + v_{2i}$

The initial and boundary conditions are

– at t = 0

$X_1\ (0) = x_{in}\ \delta(0)$

$X_2\ (0) = ... = X_n\ (0) = 0$

$Y_1\ (0) = Y_2\ (0) = ...Y_n\ (0) = 0$

The solution of the equation is

at $k_1 = k_2 = k$

$$X_n(t) = A_1^{\,n}\left(\sum_{i=0}^{n-1} B_{ni}t^i\right)\exp(-r_1 t) + A_2^{\,n}\left(\sum_{i=0}^{n-1} B_{ni}t^i\right)\exp(-r_2 t)$$

where,

$$B_{ni} = n/i! \sum_{m=0}^{n-1-i} \frac{(-1)^{n-1-i-m}(2n-2-i-m)!}{m!(n-m)!(n-1-i-m)!(v_3-r_1)^m (r_2-r_1)} n-1-i-m$$

$$r_{1'}\, r_2 = \frac{(v_1+v_2+v_3)+(v_1+v_2+v_3)^2 - 4v,v_3}{2}$$

$A_1 = v_1(v_3 - v_1)/(r_2 - r_1)$

$A_2 = v_1(v_3 - r_2)/(r_2 - r_1)$

$v_1 = Q/[(1+v_{2i}/vi)vi]$

$v_2 = k/((1+v_{2i}/v_i)v_i)$

$v_3 = k/v_{2i}$

1 2 i n

Q

v_{11} X_1 — v_{12} X_2 — v_{11} X_i — v_{in} X_n

k_1 k_2 k_1 k_2 k_1 k_2 k_1 k_2

v_{21} Y_1 — v_{22} Y_2 — v_{2i} Y_i — v_{2n} Y_n

(H) The Cascaded Reactor Model with Back Streaming : It is typical of sectionalized packed tray type towers in which backmixing takes place. The applicable equation is

$(\Delta V/Q_0)\,(dC_1/dt) + (1 + f)\,C_1 - fC_2 = C_o$

$(\Delta V/Q_0)\,(dC_2/dt) + (1 + 2f)\,C_2 - fC_3 = (1 + f)C_1$

$(\Delta V/Q_0)\,(dC_i/dt) + (1 + 2f)\,C_i - fC_{i+1} = (1 + f)C_{i-1}$

$(\Delta V/Q_0)\,(dC_{n-1}/dt) + (1+2f)\,C_{n-1} - fC_n = (1 + f)C_{n-2}$

$(\Delta V/Q_0)\,(dC_n/dt) + (1 + f)C_n = (1 + f)C_{n-1}$

The initial and boundary conditions are as follows

$t = 0;\ C_o = C_o\,\delta(0);\ C_1 = C_2 = \ldots\ldots C_n = 0$

The solution of the equation is

$$W(p) = \overline{C}_n(p)/\overline{C}_o(p) = \frac{!}{\lambda_n p^n + \lambda_{n-1}p^{n-1} \ldots + \lambda_j p^i + \lambda_1 p + 1}$$

Where,

$$\lambda_j = \frac{1}{(\xi^{n-1} - n^j)} \left[\sum_{k=0}^{n-j} \frac{k}{\xi} \, {}^{[(n-f)-k]} f \; \frac{(k+j)!((n-2)-k)!}{k!j!((n-j)-k)!\;(j-2)!} \right]$$

$\xi = 1 + f$

$f = \theta_1 Q_0$

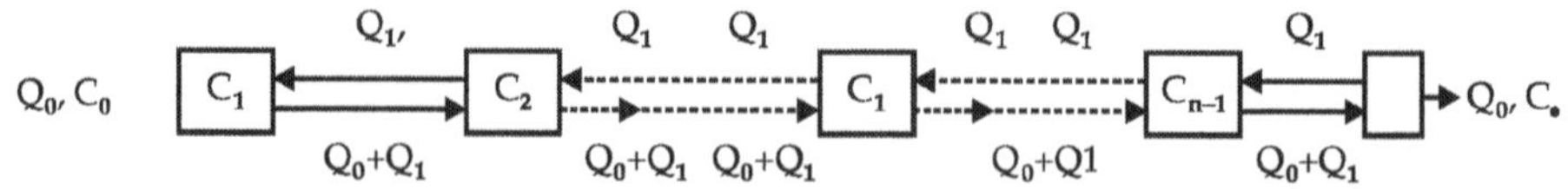

Table 8.2 : Combination Models of Flow Patterns

(A) Model

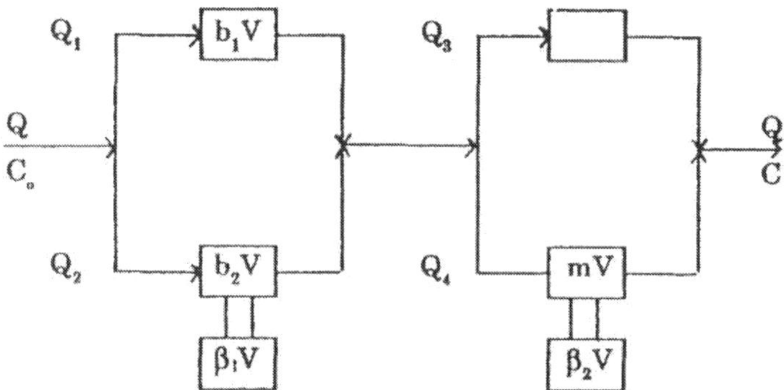

This model applies to packed columns, fixed-bed and fluidized bed reactors, extraction columns and well stirred tank reactors.

The equation of the model is

$$W(p) = \sum_{i=1}^{2} \frac{(Q_iQ_4/Q^2)\exp(-Qb_i\bar{t}p/Q_i)}{1+pmQ\bar{t}/Q_4}$$

$$+ (Q_iQ_3/Q^2) \exp [-(b_i/Q_1 + b_3/Q_3Qtp)]$$

The solution of the equation has the form

$$C(\theta) = \sum_{i=1}^{2} (\delta Q_iQ_3/Q^2(\theta - b_iQ/Q_i - b_3Q/Q_3)$$

$$+ (Q_iQ_4^2/mQ^3 \exp[-(Q_4/mQ)$$
$$(\theta - b_iQ/Q_i] \delta(\theta - bQ)Q_i)$$

$$F(\theta) = \sum_{i=1}^{2} [(\eta Q_iQ_3/Q^2)(\theta - b_iQ/Q_i - b_3Q/Q_3)]$$

$$+ (Q_iQ_4/Q^2) [1 - \exp(-(Q_4/mQ)(\theta - b_iQ/Q_i)] \times \eta (\theta - b_iQ/Q_4)$$

(B) Model

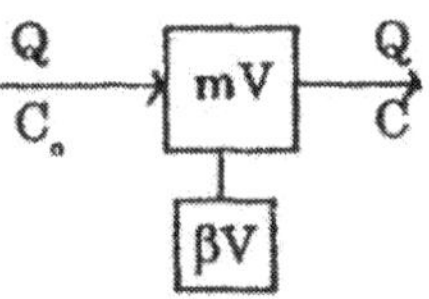

This model holds for mixing at low rotational speeds of the mixer in a reactor. Its equation has the form

$$\frac{dc}{dt} = (1/m\bar{t})(C_o - C)$$

Its transfer function is

$$W(p) = 1/(m\bar{t}p + 1)$$

Its solution takes the form

$C(\theta) = 1/m \exp(-\theta/m)$

$F(\theta) = 1 - \exp(-\theta/m)$

(C) Model

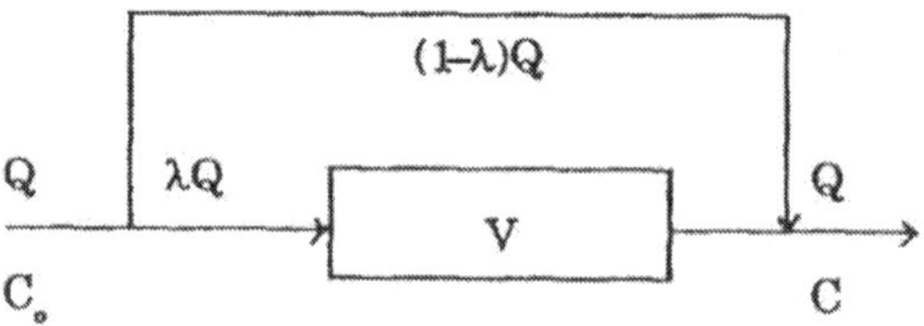

This model is applicable to well mixed stirred tank with a high rate of feed input and closely spaced inlet and exit. The model is described by an equation of the form

$dC/dt = (\lambda^2/\bar{t})(C_0 - C/\lambda)$

Its transfer function is

$W(p) = (1 - \lambda) + \lambda/(p\bar{t}/\lambda + 1)$

The Solution are

$C(\theta) = (1 - \lambda)\,\delta(\theta) + \lambda^2 \exp(-\lambda\theta)$

$F(\theta) = (1-\lambda)\,\eta(\theta) + \lambda\,[1 - \exp(-\lambda\theta)]$

(D) Model

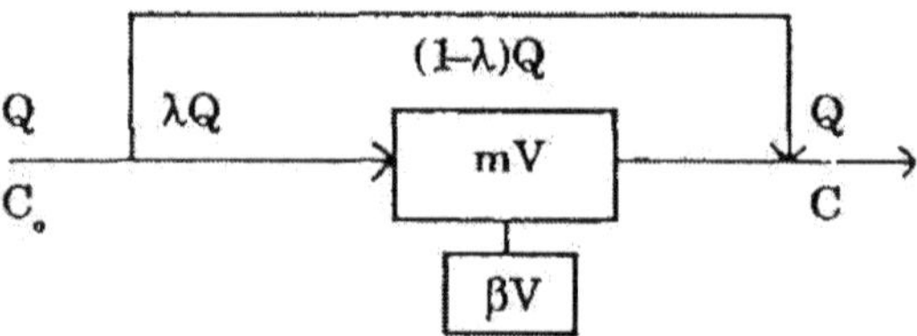

This model applies to same cases as the previous ones. Its equation is an follows

$dC/dt = (\lambda^2\,mt)(C_0 - C/\lambda)$

and its transfer function is

$W(p) = (1 - \lambda) + \lambda/(mtp/\lambda + 1)$

The solutions are

$C(\theta) = (1 - \lambda)\,\delta(\theta) + (\lambda^2/m) \exp(-\lambda\theta/m)$

$F(\theta) = (1 - \lambda)\,\eta(\theta) + \lambda\,[1 - \exp(-\lambda\theta/m)]$

(D) Model

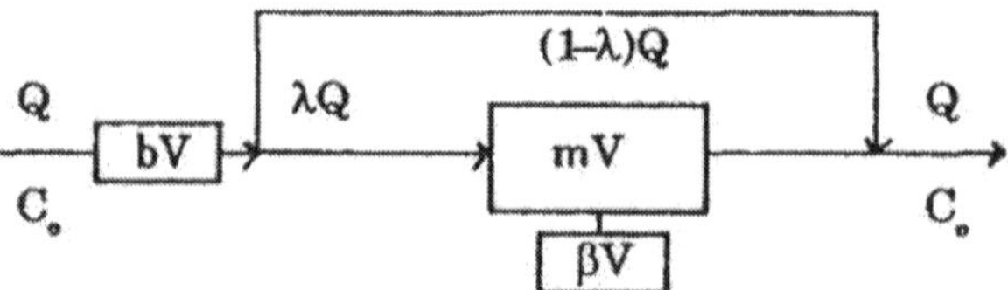

It is applicable to well stirred tubular, fluidized bed and fixed bed reactors and absorption towers. It's equations take the form

$$\frac{dC_b}{dt} = -(Q/b)(dC_b/dV)$$

$$\frac{dC}{dt} = (\lambda^2/m\bar{t})(C_b - C/\lambda)$$

and its transfer function is

$W(p) = \exp(-b\,\bar{t}\,p)\,[(1 - \lambda) + \lambda/(m\,\bar{t}\,p/\lambda + 1)]$

The solutions are as follows

$C(\theta) = (1 - \lambda)\,\delta(\theta - b) + (\lambda^2/m)\exp[(-\lambda(\theta - b)/m)]\delta(\theta - b)$

$F(\theta) = (1 - \lambda)\,\eta(\theta - b) + \lambda\,[1 - \exp(-\lambda(\theta - b)/m)]\eta(\theta - b)$

(E) Model

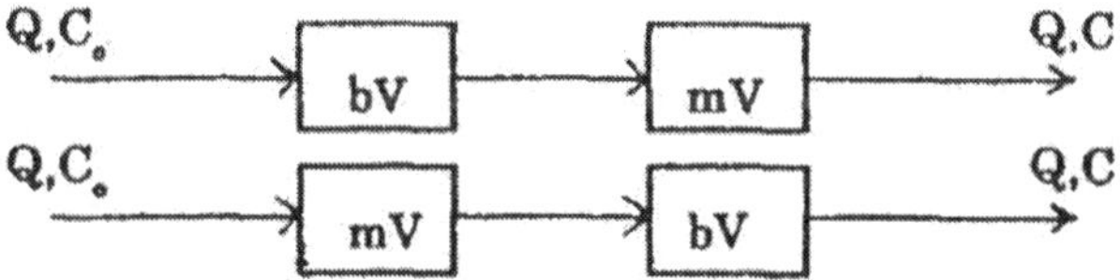

It is applicable to packed bed, tubular, fluidized bed and well stirred tank reactors. The transfer function of the model is

$W(p) = \exp(-b\,\bar{t}p)/(m\,\bar{t}p + 1)$

The solutions are as follows:

$C(\theta) = (1/m)\exp[-(\theta - b)/m)]\delta(\theta - b)$

$F(\theta) = [1 - \exp(-(\theta - b)/m)]\,\eta(\theta - b)$

(F) Model

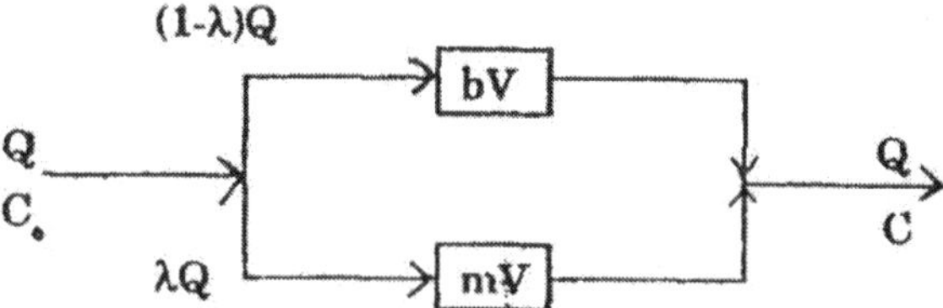

It is valid for fluidized and fixed bed reactors and well stirred tank reactors. Its transfer function is

$W(p) = (1 - \lambda)\exp[(-b\,\bar{t}\,p/1 - \lambda)] + \lambda\,(m\,\bar{t}\,p/\lambda + 1)$

The solutions are as follows

$C(\theta) = (1 - \lambda)\,\delta[\theta - b/(1 - \lambda)] + (\lambda^2/m)\exp(-\lambda\theta/m)$

$F(\theta) = (1 - \lambda)\,\eta\,[\theta - b/(1 - \lambda)] + \lambda[1 - \exp(-\lambda\theta/m)]$

(G) Model

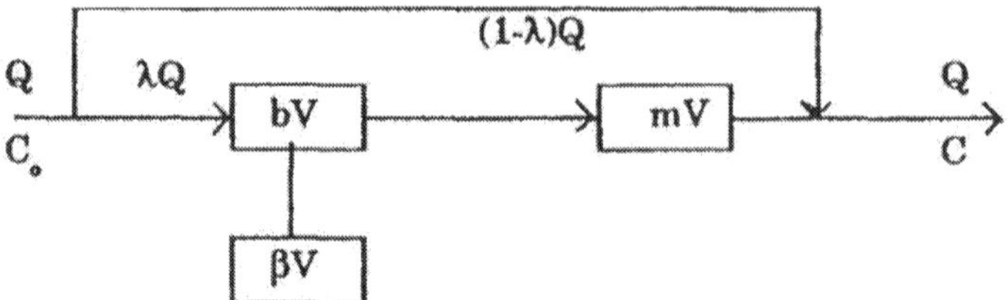

It is useful in calculating the flows in plate type columns, fluidized bed and fixed bed reactors, and well stirred tank reactors. Its transfer function takes the forms

$W(p) = (1 - \lambda) + \lambda \exp[(-b\bar{t}p/\lambda)]/(1 + m\bar{t}p/\lambda)$

Its solutions are as follows

$C(\theta) = (1 - \lambda)\ \delta(\theta) + (\lambda^2/m) \exp[-\lambda(\theta - b/\lambda)/m]\ \delta(\theta - b/\lambda)$

$F(\theta) = (1 - \lambda)\ \eta(\theta) + \lambda[1 - \exp(-\lambda\ (\theta - b/\lambda)/m]\eta(\theta - b/\lambda)$

(H) Model

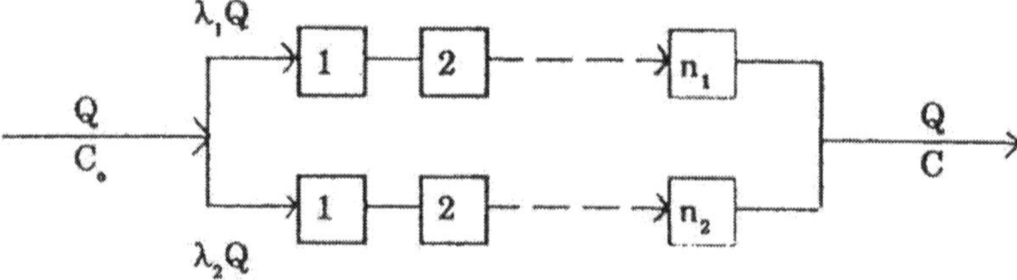

It is applicable to bubble type column, well stirred reactors and fountains bed reactors. The transfer function is

$$W(p) = \frac{\lambda_1}{(1+\bar{t}p/n\lambda_1)^{n1}} + \frac{\lambda_2}{(1+\bar{t}p/n\lambda_2)^{n2}}$$

Such that

$\lambda_1 + \lambda_2 = 1$ and $n_1 + n_2 = n$

The solutions are as follows

$$C(\theta) = \sum_{i=1}^{2} \frac{\lambda_i^{ni+1} n^{ni} \theta^{ni-1}}{(n_i - 1)!} \exp(-n\lambda_1\theta)$$

$$F(\theta) = \sum_{i=1}^{2} \left\{ \lambda_i [1 - \exp(-n\lambda_i\theta)] \sum_{J=1}^{ni} \left(\frac{1}{J!}\right) (n\lambda_i\theta)^j \right\}$$

8.1.5 Determination of Model Variables for Multiphase Flows by a Direct Method

When it comes to a multiphase system or a system in which the structure shows a pronounced in homogeneity and the manner in which the total volume is shared between the phases or discontinuities is not known, the tracer technique of flow-structure analysis is not at all easy to apply. For one thing, the response to a particular test input is accompanied by interfering factors such as molecular diffusion into the pores and capillaries of the particles, the films and pockets between the particles, convective diffusion in stagnant zones, adsorption and desorption of the tracer on the particles and the walls bounding the stream, etc.

In such cases, it is preferable to inject a change in terms of the flow rate of the disperse (or discontinuous) phase rather than in terms of tracer concentration. From an analysis of an appropriate response, it is an easy matter to determine the model variables for the effective (flow) part of the system. Responses to fluid-flow disturbances can then be statistically analyzed in much the same fashion as responses to standard disturbances in trace concentration. This is discussed at length by Kafarov *et al*[1].

The flow patterns most typical of chemical engineering which can be used to calculate operations based not only of fluid mechanics, but also on heat and mass transfer, chemical conversion, and biochemical processes are given in Tables 8.1 and 8.2. From them, it is also possible to determine the time to the completion of a given operation.

Many of the analytical solutions are presented in the tables as the respective Laplace transforms and are usually called transfer functions. A detailed mathematical formalization of the models, their analytical solutions, and their Laplace transformation are given by Kafarov, *et al*[1-3].

An operation takes the shortest time to completion in plug flow tubular units (linear system), and the longest in well-stirred tank units. It is relevant to note that in the courses on chemical engineering operations, processes and reactors, it is usual to proceed from the plug-flow tubular apparatus. In such cases, the average residence time is given by

$$\bar{t} = t - V/Q = L/v \qquad \text{...(8.15)}$$

where, V is the volume of the flow zone in m^3, Q is the volumetic throughput in $m^3\ s^{-1}$, L is the length of the flow zone, and v is the linear flow rate over the total cross section of the reactor.

As has been noted, the term C(t)dt defines the fraction of the tracer in the exit stream at ages t and t + dt. The average concentration of a particular material in the stream leaving a real reactor, $\overline{C}_A$ can be found if we multiply C(t)dt by the concentration of the same material, C_A, remaining in the stream at the age t + dt :

$$\overline{C}_A = \int_{t=0}^{t=\infty} C_A C(t)dt \qquad \text{...... (8.16)}$$

If we find C_A from independent equations will allowance for the kinetic relations, we can then establish the connection between the distribution function, or the stochastic component, and the deterministic component of a given operation or process.

8.1.5.1 Model Testing and Adjustment

The first step in testing a model is to see if the flow pattern chosen fits the flow(s) in the actual reactor. If the experimental response curve found with an impulse, step or sinusoidal test signal is the same as the plot of the analytical solution, this is an indication that the adopted model is adequate. The experimental response curves are obtained on a pilot geometrically similar to the projected commercial plant.

In most cases, stagnation, short-circulating and backstreaming can be observed visually or from an examination of the internal layout of the reactor and placement of its inlet and outlet.

Where a well-stirred (perfectly mixed) model is involved, the adequacy of the model can best be ascertained by plotting the response to a step disturbance (the C-diagram), that is, ln $(1 - c_i/c_o)$ versus time, if the plot yields a straight line. Its slope and its intercepts on the coordinate axes will give the parameters of the well-stirred (perfect mixing) model or of its combination. This stems directly from inspection of the perfect mixing model:

$$dc_i/dt = (c_o - c_i)/t_p \qquad ....(8.17)$$

Hence, on separating the variables and integrating, we get

$$\ln (1 - c_i/c_o) = t/t_p, \text{ or} \qquad .... (8.18)$$

$$c_i/c_o = 1 - \exp(-t/t_p) \qquad .... (8.19)$$

To state this a bit differently, a plot of ln $(1 - c_i/c_o)$ against time for a perfect mixing model disturbed by a step input should produce a straight line whose slope is inversely proportional to the residence time in the mixing zone, 1/tp, or accordingly, 1/m, where m is the volume of the mixing zone.

For an evaluation of the response curve applicable to a particular reactor model, it is convenient to use a plot of the rate function $\lambda(t)$. Physically, the rate function is a measure of the probability that the reactor will discharge the fluid element that has resided in it for a time t and left it during the interval (t, t + dt). The rate function is given by

$$\lambda(t) = -\frac{d}{dt}\ln\left\{\bar{t}[1-F(t)]\right\} \qquad ....(8.20)$$

For the perfect mixing model, $\lambda(t)$ should be a constant, because the probability of a fluid element leaving the reactor does not depend on its past history. For a plug flow the fact that all feed elements leave the reactor at time t = V/Q is a certainly. Accordingly, the plot of the rate function is a straight-line segment parallel to the axis of ordinates and drawn from a point $\theta = 1$ where $\theta = \bar{t}/t$ on the axis of abscissae, Fig. 8.1.

Plots of the rate function for arbitrary flows which do not show any particular variations in the average age (residence time) distribution are located, as can be seen from Fig. 8.2, between two mutually perpendicular lines each corresponding to the rate function of one of the two extreme cases (the perfect mixing case and the plug flow case). The fact that these functions are rising is an indication that the longer a fluid element resides in the reactor, the higher the probability that it will leave the reactor.

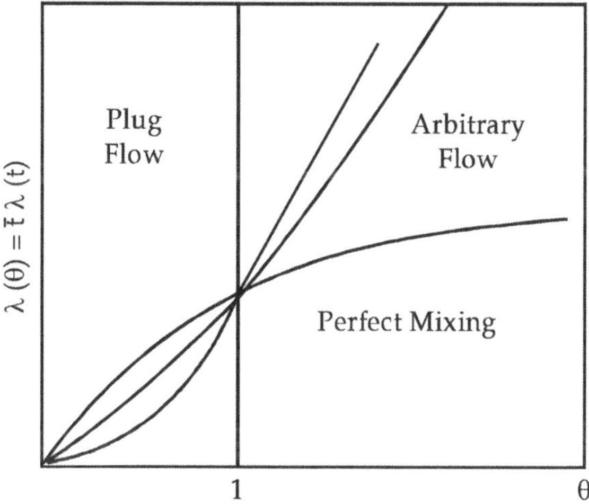

Fig. 8.1 : Rate Function Versus Dimensionless Time or Several Flow Models

For the same reason, a system with stagnation will show a rise in the λ-function just as the main body of the stream leaves the reactor. After that the probability for the remaining fluid elements to leave the reactor decreases, because most of the them are located in the stagnation zones. Now, the rate function will not rise without bound. Instead on passing through a peak, it will decrease, Fig. 8.2. Eventually, the fluid elements residing in the stagnation zones will leave the reactor as well. Again their probability of leaving the reactor will increase with time. To state this differently, on passing through a minimum the rate function will again rise without bound.

Variations in the rate functions for flows with short-circuiting can be explained similarly. The only difference from the previous case is that the flow (short-circuiting) zone and the stagnation region (now the main portion of the stream) change places within the system. This can readily be seen from the plot of the rate function in Fig. 8.2.

As we have seen, a major advantage of rate functions is that they bring out the existence of stagnation in a simple and straightforward manner.

For reactor which can be represented by models with one or several recycles and well-stirred tanks, the residence time distribution-function can best be found by means of Markov analysis. Their method yields a stochastic transition probability matrix which gives a complete description of the model.

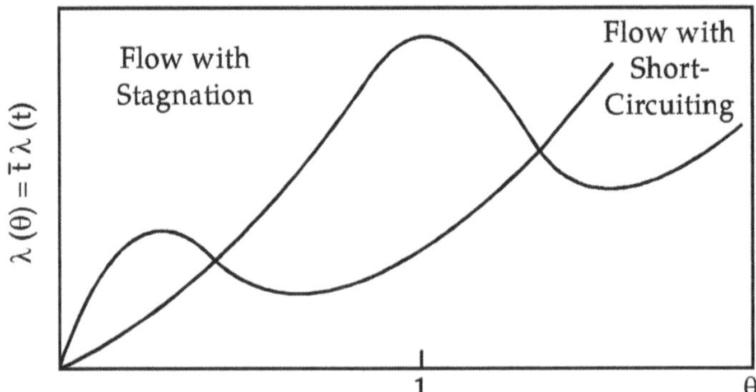

Fig. 8.2 : Rate Function Versus Dimensionless Time for Flows with Stagnation and Short-Circuiting.

For the purpose of analysis, the events taking place in chemical reactors may be divided into those occurring at microlevel (the microkinetics of an operation) and those occurring at a microlevel (the macrokinetics of an operation)[1]. The factors falling in the microkinetic class are the set of physico-chemical effects governing the rate of physical or chemical conversions at the molecular (atomic) level and within a local space of the reactor. The macrokinetics of an operation is concerned with the behaviour of a physico chemical system as an entity. Now, the operations based on fluid mechanics, heat transfer, and mass transfer and occurring on a large scale are superimposed on the effects at the micro level. The structure of the large-scale operations is decided by the reactor design, the type of agitator used, input of external energy, etc.

A sharp delineation between the events occurring at the micro level and the macro level is difficult to draw. Because of this, the investigator has to introduce effects occurring at intermediate levels as a link between the two extreme levels. This results in a complex, five-level hierarchical structure for the effects taking place in a physico-chemical system.

In regard to a constrained flow, these five levels are as follows : (1) the atomic molecular level; (2) the level of supermolecular or globular structures; (3) physio-chemical events related to the flow of a single element of the disperse phases, with allowance for chemical reactions and mass transfer between the phase; (4) physico-chemical events in an ensemble of disperse-phase elements constrained to move in a layer of the continuous phase; and (5) events controlling the fluid-mechanic macropattern in the reactor as a whole.

Leaving out the first three levels primarily determined by physico-chemical kinetics let us examine the fourth and fifth levels of the hierarchy.

Events at the fourth level control the flow pattern within a local space of the reactor, so it is natural to use the term local fluid mechanics. It is at work within a fluid element, but its size is such that it contains a sufficiently large amount of the disperse phase.

The basic quantitative characteristics of a physico-chemical at this hierarchical level are normal and tangential (shearing) stresses, strain, rates, viscosity, diffusivity, thermal conductivity, phase changes, etc.

As a rule, the semiempirical concept of local behaviour calls for knowledge of the ratio between the coefficient of turbulent transport and the kinematic viscosity. The manner in which the integration region is separated and the coefficient of turbulent transfer at the interface is found define a particular theory invoked to explain the interphase transfer.

The factors at the first, second, third and fourth levels are closely interrelated and form a set of what may be called fluid-mechanic microfactors. They control material transport in a multiphase, multicomponent system.

The fifth hierarchical level refers to events which define what we have referred to as the fluid-mechanic macropattern in a reactor, that is, the flow pattern in a reactor as a whole, and not in a local space within it.

The key factor at the fifth level is the design of the reactor involved. This includes its geometry, the type of agitator(s) and heat exchanger(s) used, placement of inlets and exits, provision (or otherwise) of baffles, diffusers, plates and the like. Also, the design determines the way in which external mechanical energy is applied to produce agitation in the system, input and removal of heat, and also fluid-mechanic, concentration and heat disturbances introduced by the feed streams of reactants.

Acting jointly, mechanical agitation, the disturbances introduced by the feed streams, and the reactor geometry produce a particular flow topology in the reactor. More abstractly the flow topology within a reactor (or the fluid-mechanic flow pattern) is controlled by the nature and location of macrodiscontinuities, stagnation regions, short-circuiting, laminar and turbulent flow zones, recycling, and so on.

Changes in the flow pattern in a reactor as a whole bring about changes in its basic quantitative characteristics, namely the distribution of continuous-phase and disperse phase elements by flow paths and residence time, the reactor hold-up for the continuous and disperse phase, the distribution of the disperse phase by size, etc.

The flow pattern and the reactor geometry control the formation of concentration and temperature fields in the reactor as a whole. These fields are further affected by the disturbances produced by the feed streams, heat disturbances associated with the feed streams, and also the manner of heat input to the reactor.

The basis for describing the factors at the first hierarchical level is provided by the phenomenological and statistical methods of physico-chemical kinetics and chemical thermodynamics. At this level, the objectives are to decode the mechanisms of the complex chemical reaction, to carry out stoichiometric analysis, to set up reaction rate equations, and to determine the kinetic constants involved.

At the second hierarchical level, the data amassed at the first level are enriched and interpreted in view of what is known about the degree of segregation of the system and the structure of Supermolecular groups. The working tools at this level are mathematical models for flow segregation, and also various theories of heterophase processes.

The basis for describing the factors at the third level is provided by the mechanics of small-scale streams around disperse-phase elements, the thermodynamics of surface phenomena, the techniques used in describing the equilibrium of multicomponent system, and various theories of interphase transport.

In describing the events at the fourth hierarchical level one may resort to the statistical mechanics of suspensions, fluid-mechanical models based on the interpenetration of multivelocity continuous media, the mechanics of suspended, fluidized disperse systems, models based on mathematical methods from the kinetic theory of gases, etc. Notably, for small parameter systems (those with low pressure, velocity, temperature, etc.), what happens in polydisperse media can best be described by extending Gibbs' statistical ensembles to the collections of macroelements (solid particles, drops, bubbles) of the disperse phase. Any stoichastic description of the system, supplemented by deterministic models for mass, energy and momentum transfer within the individual phases finally leads to a general mathematical model which reflects both the deterministic and stoichastic aspects of the fourth hierarchical level.

The equations from the first, second, third and fourth hierarchical levels make up a mathematical description of the events at the fifth level as the descriptions of the respective subsystems combined into a single system embracing the reactor as a whole. As practice has shown, this description must above all be simple and convenient to use. Therefore, data coming from the lower hierarchial levels must be "compressed" to the utmost and fed to the top hierarchical level in a sufficiently simple and compact form. As part of this "data compression", the quantities entering the lower-level equations are ranked in value, the most important (controlling) factors are identified, rigorous relations are replaced by the simpler "model" structures and the simpler mathematical formalisms, etc.

The intensity of mixing is determined by dispersion number, D/UL which contains two components, *viz.* geometric factor and intensity of dispersion.

$$\frac{D}{UL} = \text{(Intensity of dispersion (Geometric factor)}$$

$$= \frac{D}{Ud} \frac{d}{L} \qquad .....(8.21)$$

Where, d is the characteristic length of the bioreactor. The degree of mixing can be known if the variance of the trace is known and is calculated by using the following expression.

$$\sigma^2 = 2D/UL - 2(D/UL)^2 (1 - \exp(- UL/D)) \qquad .....(8.22)$$

The variance (σ^2) is known, therefore, by fit and trial method D/UL can be estimated. The degree of mixing levels in terms of dispersion coefficient are given as :

Amount of Dispersion	D/UL value
• Small amount of dispersion	≤ 0.002
• Intermediate amount of dispersion	0.025
• Large scale dispersion	>0.2

The number of reactors (n) required to satisfy the dispersion levels are given by the equation as follows :

$$\sigma^2 = \frac{1}{(n-1)!} \qquad ...(8.23a)$$

For a given number of tanks, D/UL can be easily estimated and some of the values are :

$n = \infty \quad D/UL = 0$

$n = 10 \quad = 0.055$

$n = 5 \quad = 0.011$

$n = 2 \quad = 0.33$

$n = 1 \quad = \infty$

8.2 Biofilm Kinetics Models

Lawrence and McCarty[4] presented an unified model that described the behaviour of suspended growth bacteria in wastewater treatment processes when the substrate utilization rate followed a Monod kinetics. The same was done by Rittman[5] for attached bacteria, following a method developed by Rittman and McCarty[6]. Both analyses were performed at steady state. Rittman, *et al.*[7-8] here explained different data using their model and it was also used satisfactorily to model secondary substrate utilization[9]. Saez[10-11] generalized the model presented by Lawrence and McCarty for suspended growth bacteria in order to apply it to system where substrate utilization rate can be represented by zero-order, first order, Monod or Contois Models. The purpose of this section is to generalize the biofilm model proposed by Rittman and McCarty[6] for comparing the responses of once through, fixed film reactors with different substrate utilization rate models, at steady state. The models include zero order, first order, Monod and Contois Kinetics.

Fig. 8.3 shows the block diagram for various process items required for developing a mathematical model. Fig. 8.4(a) illustrate and idealized biofilm that has uniform cell density of X_f ($M_x L^{-3}$, where M_x and L are mass of bacteria and length respectively) and a locally uniform thickness $L_f(L)^5$. Substrate concentration within the biofilm changes only in the z-direction, which is normal to the surface of the biofilm. It is assumed that all required nutrients are in excess concentration, except one which is called the rate limiting substrate, or simply the substrate. Fig. 8.4(b) shows two characteristic substrate concentration profiles. A 'deep' biofilm is one in which the substrate concentration decreases asymptotically to zero. The deep biofilm has the maximum possible substrate flux for a given S_e (MsL^{-3} where Ms is mass of substrate), the substrate concentration is liquid-biofilm interface. 'Shallow' biofilms are those in which the substrate concentration at the biofilm-solid support interface is greater than zero. For both type of biofilms, the substrate-concentration gradient becomes zero at or before the solid-surface boundary. With respect to Fig. 8.4, L(L) is equivalent depth of liquid through which the actual mass transport can be described by molecular diffusion alone, S and S_f(MsL^{-3}) are the substrate concentration in bulk liquid and in the biofilm respectively, X(L) is the longitudinal direction along the reactor.

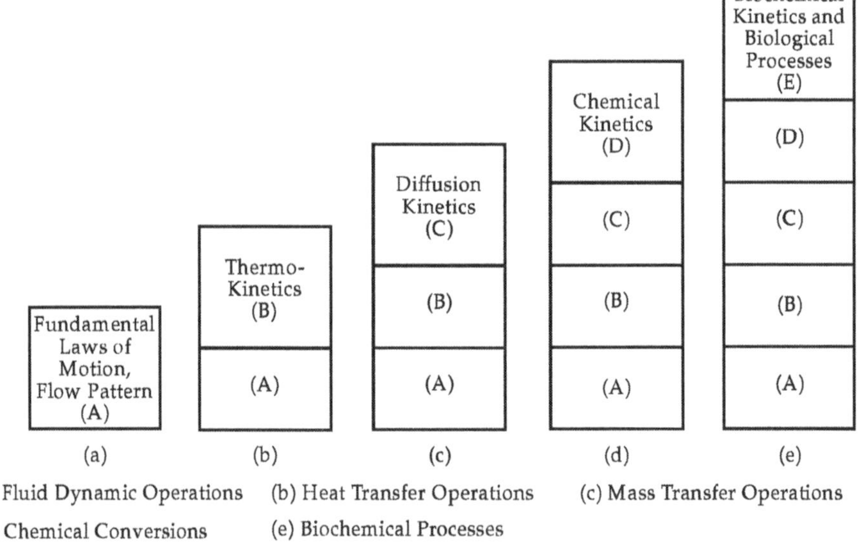

Fig. 8.3 : Schematic Diagram of Various Unit Operations and Unit Processes Required for a Biochemical Reactor

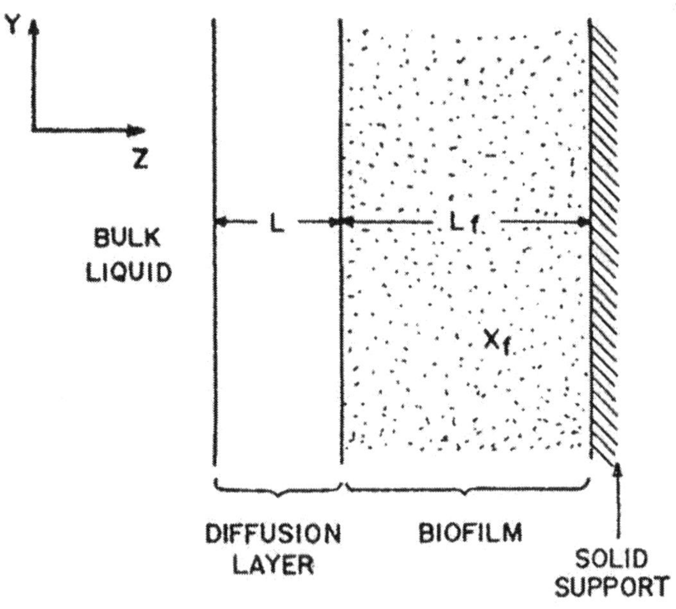

(a) Physical Concepts

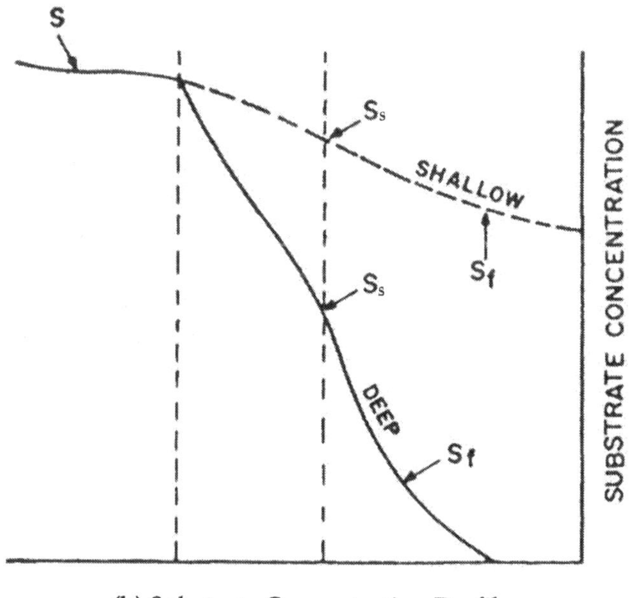

(b) Substrate Concentration Profiles

Fig. 8.4 : Conceptual Basis for Biofilm Model

8.2.1 Generalized Model

8.2.1.1. Substrate Mass Balance

In order to establish the substrate balance, consider an element of once through fixed bed reactor with total volume $dV(L^3)$ as shown in Fig. 8.5. It is

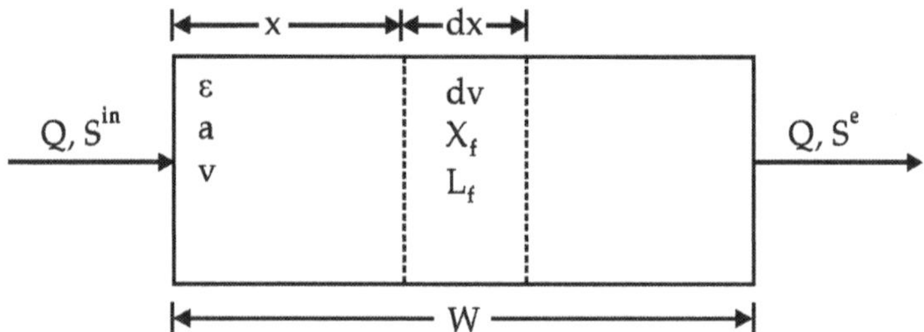

Fig. 8.5 : Schematic Representation of Once Through Fixed Film Reactor.

assumed that there is no back mixing in the reactor (plug flow). The porosity of the medium, Σ(dimensionless), and specific biofilm surface area or area per total volume, $a(L^{-1})$ are supposed to be constant through the entire reactor. The substrate mass balance in the element of reactor can be written as

$$\varepsilon dV\frac{\delta S}{\delta t} = -Qdx\frac{\delta S}{\delta x} + D_H dV\frac{d^2S}{dx^2} - (-\gamma_a)\,\varepsilon\, dv \quad - \quad aJdV \qquad .....(8.23)$$

(Accumulation) =(Convection) + (Dispersion) – (Removal by suspended bacteria) (Removal by biofilm)

Where
- t = time
- Q = liquid flow rate
- D_H = hydraulic diffusivity
- $(-\gamma_a)$ = substrate removal by suspended bacteria
- J = substrate flux into biofilm

Dividing equation 8.23 by dV, it forms the governing partial differential equation of substrate concentration in a control volume.

$$\varepsilon\frac{\delta S}{\delta t} = -v\frac{\delta S}{\delta x} + D_H\frac{\delta^2 S}{\delta x^2} - \varepsilon(-\gamma_a) - aJ \qquad .....(8.24)$$

Where, v is the superficial or empty bed flow velocity. Finally, at steady state equation 8.24 is transferred into

$$O = -v\frac{\delta S}{\delta x} + D_H\frac{\delta^2 S}{\delta x^2} - \varepsilon(-\gamma_a) - aJ \qquad .....(8.25)$$

The two boundary conditions are :

$$vs^{in} = vs - D_H \frac{dS}{dx} \text{at } x = 0 \qquad ...(8.26)$$

and

$$\frac{dS}{dx} = 0 \text{ at } x = w \qquad ... (8.27)$$

Where, $S^{in}(MsL^{-3})$ is the influent substrate concentration and W(L) is the length of the reactor. The solution of equation 8.25 gives steady state concentration of substrate at any point in the reactor including the effluent concentration $S^e(MsL^{-3})$. In order to integrate equation 8.25, it is required to know (–rs) and J as function of S and or X. Different substrate utilization models produce different expressions for $(-\gamma_a)$ and J. When dispersion is negligible, equation 8.25 takes the following form :

$$O = -V\frac{dS}{dx} - \sigma(-\gamma_a) - aJ \qquad ....(8.28)$$

For this case, the appropriate boundary condition is

$$S = S^{in} \text{ at } x = 0 \qquad ....(8.29)$$

8.2.1.1.1 Substrate Utilization Rate by Suspended Bacteria

The substrate utilization rate expression most commonly used is that proposed by Monod[12]. This expression relate the rate both to the concentration of active suspended micro-organisms $X_a(MxL^{-3})$ and to the concentration of substrate in bulk liquid S :

$$(-\gamma_a) = \frac{kX_aS}{K_a + S} \qquad ....(8.30)$$

Where, k = maximum rate of substrate utilization per unit of micro-organisms.

K_a = half velocity constant (M_aL^{-3}) or substrate concentration when $(-\gamma_a)$ X_a = 1/2 k.

The Monod model has the following limiting cases :

(*a*) for very low substrate concentration ($S<<K_a$), the substrate utilization rate is

$$(-\gamma_a) = \frac{kX_a}{k}S \qquad ....(8.31)$$

or kinetics of first order in S and

(*b*) for very high substrate concentration ($S>>K_a$), the rate is :

$$(-\gamma_a) = kX_a \text{ or a zero order kinetics} \qquad ...(8.32)$$

Contois[13] suggested that the expression for substrate utilization was given by equation (8.34)

$$(-\gamma_a) = \frac{kX_aS}{BX_a + S} \quad \text{....(8.33)}$$

where, B = kinetic coefficient ($MsMx^{-1}$) which represents the ratio between the substrate concentration and the active organism concentration when $(-\gamma_a)/X_a$ = 1/2k.

The limiting cases for the Contois model are equal to those for the Monod model if K_a is substituted by BX_a.

Two main differences exist between the Monod and Contois model. First, while the half velocity constant for the Monod model is constant, it is a linear function of X_a for the Contois model second, the specific growth rate, defined as the rate of active bacterial growth per unit of active organisms, for the Contois model is not independent of active organisms concentration as in the Monod case.

Equation 8.25 will be solved when substrate utilization rate by suspended bacteria can be represented by zero order (equation 8.32), first order (equation 8.31), Monod (equation 8.30), and Contois (equation 8.33), models. Since, X_a has been taken as a fixed constant, the expression for $(-\gamma_a)$ for the Monod case is the same as the Contois case. When X_a is not constant through the entire reactor the substrate utilization rate for the Contois model differs from the applicable Monod model.

8.2.1.1.2 Substrate Consumption by Biofilm

The substrate mass balance within differential section of biofilm when the mass transport is only by molecular diffusion can be written as

$$\frac{\delta S_f}{\delta t} = D_f \frac{\delta^2 S_f}{\delta z^2} - (-\gamma_{sf}) \quad \text{...(8.34)}$$

Where, D_f = molecular diffusivity of substrate within the biofilm

(γ_{sf}) = substrate utilization rate by the biofilm ($MsL^{-3}T^{-1}$).

Different expression are used to evaluate (γ_{sf}). These include zero order, first order, Monod and Contois kinetics. Equations 8.30 to 8.33 are applicable if S and X_a are replaced by S_f and X_f respectively. It is assumed that the kinetic coefficients within biofilm are the same than in the bulk liquid. At steady equation 8.34 is transformed into :

$$D_f \frac{\delta^2 S_f}{\delta z^2} = -(\gamma_{sf}) \quad \text{.... (8.35)}$$

The two boundary conditions are :

$$S_f = S_s \text{ at } z = 0 \quad \text{....(8.36)}$$

$$\frac{dSf}{dz} = 0 \text{ at } z = L_f \text{ (Shallow film)} \qquad ....(8.37)$$

$$S_f = 0 \text{ at } z = L_f \text{ (Deep film)} \qquad .....(8.38)$$

It should be pointed out that a deep film is only a particular case of a shallow biofilm when S is high enough. Also, equation 8.35 can be integrated analytically only for some $(-\gamma_{sf})$.

Since S_s and L_f are not known a priori, other equations must be developed. Across the diffusion layer, L (Fig. 8.4), Fick's first law is applied for one dimension to give :

$$J = \frac{D}{L}(S - S_s) \qquad ..... (8.39)$$

Where, D = molecular diffusivity of the substrate in the liquid.

Also, according to Fick's first law, the flux J can be determined by using.

$$J = -D_f \frac{dS_f}{dz} \quad \text{at } z = 0 \qquad ..... (8.40)$$

The last relationship needed to determine J as a function of S is obtained by making a mass balance on attached biomass in the element of reactor of Fig. 8.5.

$$adV \frac{\delta}{\delta t}(X_f L_f) = YJadV - bX_f L_f adV + \gamma_{dep}\, adV - \gamma_{shear} adv \qquad ....(8.41)$$

Where, Y = true yield of bacterial mass per unit of substrate mass utilized $(M_x M_s^{-f})$

b = decay Coefficient

γ_{dep} = deposition rate or rate at which the suspended bacteria are deposited in biofilm $(M_x L^{-2} T^{-1})$,

γ_{shear} = Shear loss rate, or rate at which biofilm is transferred to suspended media by shear stress. $(M_x L^{-2} T^{-1})$

Rittman[14] used the data of the experiments of Trulear and Characklis[15] to demonstrate that the shear loss rate may be approximated by equation 8.43

$$\gamma_{shear} = AX_f L_f \qquad ...(8.42)$$

Where, A is a function of L_f and shear stress.

At steady state, when the deposition rate is negligible and shear loss can be approximated by equation 8.42, equation 8.41 takes the form of :

$$L_f = \frac{JY}{b' x_f} \qquad ....(8.43)$$

Where, b′ = b + A (loss coefficient which includes decay and shear losses).

Rittman and McCarty[6] have defined a steady state biofilm represented by equation 8.43, as one having no net growth or loss (decay and shear losses) for the entire biofilm depth. For that condition, the growth due to substrate flux (JY) is equal to the total losses from the entire biofilm (bX_fL_f). At steady state, the biofilm thickness is assumed constant but different sections of the biofilm can be growing or decaying.

The following procedure can be used to obtained J as a function of S. First, S_f as a function of S_s and L_f can be obtained by solving equation 8.35, 8.36, 8.37 or 8.38. Equation 8.37 is used for shallow biofilm and equation 8.38 for deep biofilm. Second, the relationship just found can be used in conjunction with equation 8.39, 8.40 and 8.44 to obtain J, L_f and S_s as function of S and S_f as a function of S and Z.

A key concept of the steady state solution is the existence of a threshold substrate, $S_{min}(MsL^{-3})$, below which no steady state biofilm activity occurs[6]. The reason that S_{min} occurs is that the net rate of cell growth at a concentration less than S_{min} is always negative. In other wards, the biofilm is continually decaying and cannot exist in a steady state situation when S is less than S_{min}. At concentration above S_{min}, the biofilm will have a net positive growth rate and the biofilm can grow until it reaches its steady state thickness. If the deposition rate is negligible, the mass balance on attached biomass in a differential section of biofilm can be written as :

$$a\,dV\frac{\delta}{\delta t}(X_f dz = Y(-\gamma_{af})a\,dV\,dz - b'\,a\,dVX_f dz \qquad .....(8.44)$$

At $S = S_{min}$, it can be assumed that only a monolayer of bacteria forms the biofilm. Since there is essentially no diffusion limitation in a monolayer, $S_f = S_s$ for all Z. Solving equation 8.44 at steady state for S_s, it is obtained the minimum surface concentration S_{min}. Although the value of S_{min} cannot be determined rigorous without the knowledge of substrate flux and mass transport effects, it can be approximated by relation $S_{min} \approx Ss_{min}$, because the flux is very low. Then, S_{min} is evaluated by solving.

$$(-\gamma_{af}) = \frac{b'X_f}{Y} \text{at } S_f = S_{min} \qquad ....(8.45)$$

8.2.1.1.3 Substrate Flux Predicted by Different Kinetic Models

Table 8.3 shows the minimum substrate concentration in the bulk liquid for different kinetic models. The results have been obtained by solving equation 8.45. There is no S_{min} for zero-order model. For this case, the biofilm will be at steady-state only when $b' = kY$. If b' is greater kY the biofilm will disappear because losses are higher than the growth, the biofilm will grow continuously when kY is greater than b'. S_{min} for the first order model is slightly lesser than S_{min} for Monod and Contois cases. S_{min} depends upon X_f only for the Contois model.

Table 8.3 : Minimum Substrate Concentration in the Bulk Liquid

Sr. No.	*Kinetic Model*	S_{min}, MsL^{-3}
1.	Zero order	Zero (biofilm at steady-state only when b′=kY)
2.	First order	Ks b′/kY
3.	Monod	$Ks\frac{b'}{kY - b'}$
4.	Contois	$BX_f\frac{b'}{kY - b'}$

For zero order model, there is not a single solution for J when the biofilm is shallow, because one of the 'supposed' independent of others. However, for deep biofilms at steady-state (b′ = kY), the following equation give substrate flux, biofilm thickness, substrate concentration at the liquid-biofilm interface, and substrate concentration within the biofilm :

$$J = (\alpha^2 + \frac{2D}{L}\alpha S)^{1/2} - \alpha \qquad \text{... (8.46)}$$

$$L_f = \frac{D_f L}{D\alpha} J \qquad \text{... (8.47)}$$

$$S_s = S - JL/D$$

$$S_f = S_s(1 - Z/L_f) - \left[\frac{\alpha D}{2L\, D^2_f}(L_f Z - Z^2)\right] \qquad \text{... (8.48)}$$

Where, $\alpha = D_f k X_f L/D(M_s L^{-2} T^{-1})$. For high values of the substrate coefficient in the bulk liquid, the substrate flux is proportional to $S^{0.5}$.

Table 8.4 shows the response of the system with first order model for both shallow and deep biofilms. The substrate flux for both shallow and deep biofilms is found by trial and error, because the expression for J is implict. Since tan h (5.4) = cot h (5.4) = 1.0000, the expression to evaluate J for both shallow and deep biofilms is reduced to equation 8.49 when $(r_2 q_2 J) > 5.4$:

$$J = p_2 S \text{ for } (r_2 q_2 J) > 5.4 \qquad \text{...(8.49)}$$

Where, $p_2 = D_f\, r_2/(1 + L\, D_f\, r_2\, D)$. The condition $(r_2\, q_2\, J) > 5.4$ is equivalent to $S > 5.4/(p_2\, q_2\, r_2)$. Then, the substrate concentration in the bulk liquid at which a shallow first order biofilm is transformed into a deep one is :

$$S_{deep} = \frac{5.4}{p_2 q_2 r_2} \qquad \text{..... (8.50)}$$

Table 8.4 : Substrate Flux, Biofilm Thickness, Substrate Concentration at Liquid-biofilm Interface and Substrate Concentration within Biofilm for First Order Kinetics

Sr. No	*Biofilm*	*Expression*
I	**Shallow Biofilms**	
1.	Substrate Flux*	$J = (S - JL/D)D_f r_2 \tan h(r_2 q_2 \alpha J)$
2.	Biofilm thickness**	$L_f = q_2 J$
3.	Substrate concentration at liquid biofilm interface	$S_s = S - JL/D$
4.	Substrate concentration within biofilm	$S_f = S_s \dfrac{\cos h(r_2(L_f - Z)}{\cos h(r_2 L_f)}$
II	**Deep Biofilms**	
1.	Substrate Flux	$J = (S - JL/D)D_f\, r_2 \cot h(r_2 q_2 J)$
2.	Biofilm thickness	$L_f = q_2 J$
3.	Substrate concentration at liquid-biofilm interface	$S_s = S - JL/D$
4.	Substrate concentration within the biofilm	$S_f = S_s \dfrac{S_{inh}(r_2(L_f - Z)}{S_{inh}(r_2 L_f)}$

* $r_2 = (kX_f/K_a D_f)^{1/2}$ ** $q_2 = Y/(bX_f)$, $(Ms^{-1}TL^3)$

It can be concluded that for $S > S_{deep}$, the first order model predicts that the substrate flux, biofilm thickness and substrate concentration at liquid biofilm interface are all proportional to S. The substrate concentration in the bulk liquid at which substrate flux predicted by zero-order model is equal to the substrate flux predicted by the first order model, S_{eq} (MsL^{-3}), can be easily determined when $S_{eq} > S_{deep}$. For these cases :

$$S_{eq} = \frac{2\alpha(D/L - p_2)}{p_2^{\ 2}}; \quad (\text{if } S_{eq} > S_{deep}) \qquad \text{.... (8.51)}$$

Williamson and Chung[16] found an analytical expression to determine the substrate flux into a deep biofilm when the substrate utilization rate can be modeled by Model. The expression is given by equation 8.52.

$$J = \left[2\,k\,X_f D_f (S - JL/D - Ks \ln \left(\frac{Ks + S - JL/D^{1/2}}{Ks} \right) \right] \qquad \text{....(8.52)}$$

Again, equation 8.52 can be solved by trial and error procedure because it is implicit. However, there is a special limiting case for the last equation. When

$F_1 = (S - JL/D)$ is much greater than

$F_2 = Ks \ln((Ks + S - JL/D)/Ks)$, equation 8.52 can be simplified to equation 8.53

$$J = (\alpha^2 + 2D/L\ \alpha S)^{1/2} - \alpha; \ (\text{if } F_1 >> F_2). \qquad \text{... (8.53)}$$

This shows that for very high S values, the substrate flux into a deep Monod biofilm is equal to the substrate flux into a deep zero-order biofilm. The condition $F_1 > 100\ F_2$ is equivalent to $S_s > 648\ K_a$, and since S is always greater than or equal to S_s equation 8.53 can be used in an almost completely correct form when $S > 648\ K_a$. Since the model assumes that X_f is constant, equations 8.53 and 8.53 can be used with deep Contois biofilms; the only adjustment is to change K_a by BX_f.

There is no analytical solution for shallow Monod biofilms at steady state. The problem has been solved numerically. A model developed by Rittman and McCarty[6] presents an algebraic method to estimate the substrate flux as a function of S. This model was tested against a numerical integration of equation 8.35, based on Crank-Nicolson method. Saez *et al.*[10-11] have given other numerical solution Equations 8.35, 8.36, 8.37, 8.39 and 8.43 that give the behaviour of shallow Monod biofilms, were Non-dimensionalised giving :

$$\frac{d^2 S_f^*}{dz^{*2}} = \frac{2S_f^*}{1+S_f^*} \quad ...(8.54)$$

with the following boundary conditions :

$$S^*_f = S^*_s \text{ at } Z^* = 0 \quad ...(8.55)$$

and

$$\frac{dS_f^*}{dz^*} = 0 \text{ at } z^* = L_f^* \quad ...(8.56)$$

The non-dimensionalised forms of equations 8.37, 8.39 and 8.43 are respectively :

$$J^* = \frac{S^* - S_s^*}{L^*} \quad ....(8.57)$$

$$J^* = -D_f^* \frac{dS_f^*}{dZ^*} \quad \text{at } Z^* = 0 \text{ and} \quad ...(8.58)$$

$$L_f^* = \frac{J^*}{D_f^* \theta^*} \quad ...(8.59)$$

Where

$S^*_f = S_f/K_s$; $Z^* = Z/\tau$; $C = (2K_s D_f/kX_f)^{1/2}$;

$S_s^* = S_s/K_s$; $L^*_f = L_f/\tau$; $J^* = J\tau/DK_s$

$D^*_f = D_f/D$ and $\theta^* = 2b'/kY$

A finite-difference technique was used to solve the equations system. The finite difference approximation of equation 8.54 at node i is :

$$\frac{S^*_f(i+1) - 2S^*_f(i) + S^*_f(i-1)}{(\Delta Z^*)^2} = \frac{2S^*_f(i)}{1+S^*_f(i)} \quad ....(8.60)$$

Where ΔZ^* is the node spacing and it is equal to L^*_f/N where N is the number of nodes. On the other hand equation 8.60 is transformed into :

$$\Delta Z^* = \frac{J^*}{D^*_f \, \theta^* N} \quad .....(8.61)$$

The combination of equations 8.60 and 8.61 reduces to

$$\frac{(D_f{}^*\theta^* N)^2}{(J^*)^2}\left[S_f{}^*(i+1) - 2S_f{}^*(1) + S_f{}^*(i-1)\right]\left[1 + S_f^*(1)\right] - 2S_f{}^* = 0 \quad ....(8.62)$$

Since at $Z^* = 0$; $S_f^* = S_s^*$, equation 8.62 first mode 1 is

$$\frac{(D_f{}^*\theta^* N)^2}{(J^*)^2}\left[(S_f{}^*(2) - 2S_f^*(1) + S_a^*\right]\left[1 + S_f{}^*(1)\right] - 2S_f{}^*(1) = 0 \quad ... (8.63)$$

Since at $Z^* = L^*_f$; $\dfrac{dS_f{}^*}{dz} = 0$ or $S_f^*(N + 1) = S_f^*(N-1)$, equation 8.62 for node N is:

$$\frac{(D_f{}^*\theta^* N)^2}{J^{*2}}\left[2(S_f{}^*(N-1) - 2S_f{}^*(N)\right]\left[1 + S_f{}^*(N)\right] - 2S_f{}^*(N) = 0 \quad .....(8.64)$$

The flux J^* was evaluated from equations 8.54 and 8.58

$$J^* = \int_0^{t^*_f} 2D^*_f\left(\frac{S^*_f}{1+S^*_f}\right)dz^* \quad ....(8.65)$$

and using Simpson's 1/3 rule[15]

$$J^* = 2D_f{}^*\frac{L^*_f}{3N}\left[\frac{S_s{}^*}{1+S_s{}^*} + 4\frac{S_f{}^*(1)}{1+S_f{}^*(1)} + 2\frac{S_f{}^*(2)}{1+S_f{}^*(2)} \cdots\cdots \frac{S_f{}^*(N)}{1+S_f{}^*(N)}\right] \quad ...(8.66)$$

and using 8.59

$$J^* = \frac{2D_f{}^*}{3N}\frac{J^*}{D_f{}^*\theta^*}\left[\frac{S_s{}^*}{1+S_s{}^*} + 4\frac{S_f{}^*(1)}{1+S_f{}^*(1)} + 2\frac{S_f{}^*(2)}{1+S_f{}^*(2)} \cdots \frac{S_f{}^*(N)}{1+S_f{}^*(N)}\right] \quad ...(8.67)$$

or finally

$$-1 + \frac{2}{3N\theta^*}\left[\frac{S_s{}^*}{1+S_s{}^*} + 4\frac{S_f^*(1)}{1+S_f^*(1)} + 2\frac{S_f^*(2)}{1+S_f^*(2)} \cdots \frac{S_f^*(N)}{1+S_f^*(N)}\right] = 0 \quad ...(8.68)$$

Equations 8.57, 8.62, 8.63, 8.64 and 8.68 are a non-linear systems of (N+2) equations with (N+2) unknowns. The unknowns are $S_f^*(1)$, $S_f^*(2)$ ....$S_f^*(N)$; S_s^* and J^*. Since the system is non-linear, the generalized Newton method was used to solve it[17]. Once that the system is solved for a given S value, L^*_f can be obtained from equation 8.59. By using the definition of dimensionless variables, the dimensionless solution can be transformed into a physical domain. The per cent difference between algebraic solution and numerical method varies from 0.2 to 44.9% for a substrate concentration in the range of 0.005 to 1.0 mg/cm^3. Only for dillute substrates, the variation is large otherwise the agreement between the two methods is quite close.

8.2.2 Responses of Once Through Fixed Bed Reactors

The response of once through fixed bed reactors is given by the solution of equation 8.25 which gives the substrate concentration of any point on the reactor. That equation can be solved analytically only with shallow first order by biofilm when $S > S_{deep} = 5.4/(p_2q_2r_2)$. Introducing equations 8.30 and 8.51 into equation 8.25 and integrating;

$$S = C_1 \exp(r_3X) + C_2 \exp(r_4X) \quad ....(8.69)$$

Where,

$$C_1 = \frac{vS^{in}}{v(1 - r_3 / r_4 \exp(w(r_3 - r_4))) - D_H r_3(1 - \exp(w(r_3 - r_4)))} \quad .... (8.70)$$

$$C_2 = \frac{vS^{in}}{v(1 - r_4 / r_3 \exp(w(r_4 - r_3))) - D_H r_4(1 - \exp(w(r_4 - r_3)))} \quad ... (8.71)$$

$$r_3 = \frac{V + (v^2 + 4D_H T)^{0.5}}{2D_H} \quad ... (8.72)$$

$$r_4 = \frac{V - (v^2 + 4D_H T)^{0.5}}{2D_H} \quad ... (8.73)$$

$$T = ap_2 + \frac{\varepsilon k X_a}{Ks} \quad ... (8.74)$$

When dispersion is negligible and $S > S_{deep}$, the solution for shallow first order biofilm is simplified to

$$S = S^{in} \exp(-TX/v) \quad ... (8.75)$$

An analytical solution can be obtained with deep zero-order biofilm when dispersion is negligible. For this case, the solution is found by integrating equation 8.28 and with boundary condition equations 8.32 and 8.46 for $(-r_0)$ and J respectively. The solution is:

$$X = v(I_o - I(S)) \quad ... (8.76)$$

Where,

$$I(S) = P_o(q_o - r_o \ln q_o) \quad ... (8.77)$$

$$I_o = I\,S^{in} \quad \text{... (8.78)}$$

$$P_o = \frac{L}{Da^2\alpha} \quad \text{...(8.79)}$$

$$q_o(s) = a(\alpha^2 + 2D/L\ \alpha S)^{0.5} + r_o \quad \text{...(8.80)}$$

$$r_o = -\,a\alpha + \varepsilon kXa \quad \text{...(8.81)}$$

The solution procedure is the following :

1. first evaluate r_o from equation 8.81 and P_o form equation 8.79.
2. compute q_o (S^{in}) from equation 8.80,
3. evaluate I_o from equations 8.77 and 8.78,
4. compute $q_o(S)$ for a given value of S from equation, 8.80,
5. evaluation I(S) from equation 8.77,
6. compute X from equation 8.76, with this procedure it is possible to obtain a curve relating substrate concentration in the bulk liquid to the distance along the reactor.

A numerical technique is required with shallow first order biofilms when $S < 5.4/(p_2q_2r_2)$ and with shallow Monod biofilms when dispersion is not negligible. When dispersion is negligible, the solution can be obtained by using the subroutine RKGS, which solves a system of first order ordinary differential equations with given initial values. The sub-routine RKGS, implemented in FORTRAN, uses the fourth order Runge Kutta Formulae in the modification due to Gill[18].

8.2.3 Effect of Process Loading

The accumulation of attached biomass is critical for successful operation of any fixed film process. Based on the original work of Rittman and McCarty[6] the differential equation for bio-accumulation can be written as:

$$d/dt\,(X_fL_f) = YJ - b'\,X_fL_f + k_aX\,\varepsilon/a \quad \text{...(8.82)}$$

where k_a = first order attachment rate

X = suspended micro-organism in the liquid

ε = reactor liquid hold up

a = biofilm surface area per total reactor volume

b′ = shear loss rate due to decay (b) and shear (bs).

The shear loss (bs) depends on the hydrodynamic regime of the reactor and surface properties of attachment medium. High flow rates and other factors which increase turbulence and shear stress increase the rate of biofilm loss. Strong adhesion forces and sheltering are the surface properties which reduce biofilm loss rates. Good quantitative understanding of factors affecting shear loss are not yet available.

A portion of reactor continually exposed to a concentration less than S_{min} would grow or sustain no biofilm. S_{min} can be calculated from the following expression[6,21] :

$$S_{min} = K_a \frac{b'}{Yk - b'} \quad ....(8.83)$$

The shear loss rate can be defined in terms of shear stress

$$bs = 8.84 \times 10^{-2}\, \sigma^{0.58},\ day^{-1} \quad ...(8.84)$$

$$\sigma = \frac{200\mu v(1-\varepsilon)^2}{d^2 p \varepsilon^3 A_v F} \text{(Spherical packing), dgree / cm}^2 \quad ...(8.85)$$

Where μ, Av and F refer to viscosity, specific area and conversion factor respectively.

The relative performance of different reactor types depends on the process loading. Comparison among reactor types can change when loading is high instead of low. For biofilm processes, the term loading is most relevantly defined according to the reactor's ability to achieve effluent concentrations approaching S_{min}. For completely mixed and fixed medium biofilm processes, a low loading means that the effluent concentration (S°) is equal to or slightly greater than S_{min}. For fluidized bed biofilm process, low loading is evidenced by effluent concentration less than S_{min}. Conversely, highly loaded reactors produce effluent concentrations significantly greater than S_{min}. High or low loads donot necessarily indicate high or low percentage removals; high percentage removals, e.g. can be associated with high loads and with low loads.

Knowing how to select a biofilm process design that represents a high load or a low load is valuable. Completely-mixed or fixed-medium reactors with low loads are selected when requirements are for an effluent concentrations near S_{min}. Fluidized beds having low loads are selected for effluents concentrations below S_{min}. If the effluent concentration may be considerably greater than S_{min}, a high load is appropriate.

An approximate criterion for dividing the high and low load regions can be derived from basis of steady-state-biofilm kinetics. As a first approximation, a process with low load will be shallow biofilm and an effluent concentration approaching S_{min}, while a highly loaded process will have a deep biofilm. Rittman and McCarty[6] showed that the lowest substrate concentration that gives a deep biofilm, S_{deep} is estimated by :

$$S_{deep} = [4.6\, S_{min}\, (1 + (2D_f L^*)^{0.5}] \quad ...(8.86)$$

where, $$D_f^* = D_f/D \quad ...(8.87)$$

$$L^* = L/\tau \quad ....(8.88)$$

$$C = (2Ks\ Df/kX_f)^{0.5} \quad ...(8.89)$$

J_{deep}, the flux when the substrate concentration in S_{deep} is calculated with variable order for deep biofilm :

$$J_{deep} = \frac{K_a D}{\tau} C^*(S^*_{deep})^q \quad ...(8.90)$$

where,

$$S^*_{deep} = S_{deep}/Ks$$

$$C^* = \frac{2D_f^*\left[(2)^{0.5} + 2(L^*)D^*_f\right]^{1-2q}}{1+0.54\left[1+0.0121\ln(1+2L^*)\right]\left[1-8.325\,(\ln q/0.707)^2\right]} \quad ...(8.91)$$

$$q = 0.75 - 0.25\tan h[0.477A] \quad ...(8.91a)$$

$$A = \ln S^*_{deep} - \ln\left[2+\frac{\ln D_f^*}{2.303}\right] - 1.8\ln\left[1+2D_f^*L^*\right] + 0.353 \quad ...(8.91b)$$

For low loads, the actual flux, J is much less than J_{deep}; for high loads, J is much greater than J_{deep}. The actual flux for a steady state reactor is determined as :

$$J = \left(\frac{S^o - S^o}{a\theta}\right)\varepsilon \quad ...(8.92)$$

Where S°, S°, ε a, θ (= ε v/θ), V and Q refer to influent concentration, effluent concentration, liquid hold up, specific area, HRT, total reactor volume and feed flow respectively.

The dividing criterion between high and load occurs when $J = J_{deep}$. Substituting into equation 8.92. J_{deep} for J and S_{min} for S* yields a relationship that gives S* and θ values that divide high load from low loads.

$$S^\circ = S_{min} + a\theta J_{deep}/\varepsilon \quad ...(8.93)$$

For practice, the three load ranges are defined as :

(*a*) Low load; $S^\circ \le 0.33\ (S_{min} + a\theta J_{deep}/\varepsilon)$ ...(8.94)

(*b*) Intermediate load; $0.33\ (S_{min} + a\theta J_{deep}/\varepsilon) < S^\circ < 3.0(S_{min} + a\theta J_{deep}/\varepsilon)$ ...(8.95)

(*c*) High load; $S^\circ \ge 3.0\ (S_{min} + a\theta J_{deep}/\varepsilon)$ ...(8.96)

Within low load region, fluidized bed reactors have a significant advantage for attaining low effluent concentrations. In the high load region, effluent concentrations become similar for all process types, as substrate flux is independent of biofilm thickness. A decrease in biofilm loss rate is an effective means for reducing effluent concentration when the process is in the low load region. However, enhanced biofilm accumulation does not offer a large advantage when the process have a high load.

8.2.4 Young's Model[22]

The substrate concentration within the biofilm varies with respect to its depth and is always smaller in magnitude compared to the bulk substrate concentration. Since the substrate consumption rate by micro-organisms in the biofilm is dependent on the substrate concentration in their immediate vicinity, it is evident that overall substrate consumption by the entire biofilm will be dependent on effective substrate concentrate ($\bar{S}$), and not on bulk substrate concentration. In effect, $\bar{S}$, or the effective substrate concentration is that which would result in the same rate of substrate removal per unit of biological mass if the mass in the entire biofilm were completely mixed with the substrate with uniform substrate concentration everywhere. A safety factor relating the effective substrate concentration to the bulk substrate concentration, may then be defined as :

$$\bar{S} = \frac{S_B}{S_F} \qquad ...(8.97)$$

Since, in plug flow systems, effective substrate concentration is always less than bulk substrate concentration, it is anticipated that safety factor (SF) will always be higher than unity. Thus, SF signifies the extent of mixing in the system and its value approaches unity for completely mixed systems.

In plug flow systems, Young[22] assumes following relationship to ascertain the value of safety factor :

$$SF = 1 + (SF_\circ - 1)\exp(-k_s S_b) \qquad ...(8.98)$$

Thus, by definition of effective substrate concentration, the expression for substrate utilization could be written as :

$$\frac{dS}{dt} = -\frac{k\bar{S}X}{K_s + \bar{S}} \qquad ...(8.99)$$

Substituting equation 8.97 in equation 8.99 yields :

$$\frac{dS}{dt} = -\frac{kS_B X}{K_s(SF) + S_B} \qquad ...(8.100)$$

The 'effective half velocity constant' given by K_s(SF) would then be equal to or higher than K_s value measured in completely mixed systems.

Annachature, *et al.*[23-26] extended the Young's model to determine HRT(t), area available for substrate transfer per unit reactor volume (A), substrate flux (J) and liquid layer thickness (L_w) and are given by equation 8.103 and 8.104

$$t^* = \frac{bKs}{kX} \int_{S_f^*}^{S_f^*} \frac{\frac{dS_B{}^*}{S_B{}^*}}{(SF + S^*{}_B)} \qquad ...(8.101)$$

$$t^* = b.t \qquad ...(8.102)$$

$$-dS_B = J.A.\ dt \qquad ...(8.103)$$

$$t^* = \frac{b.Ks}{kX_f} \int_{S^*_C}^{S^*_F} \frac{dS_B{}^*}{J^*A^*} \qquad ...(8.104)$$

$$A^* = A\left(\frac{D_f K_s}{k X_f}\right)^{0.5} \qquad ...(8.105)$$

However, estimation of t by equation 8.104 demands a relationship between non-dimensional substrate flux $J^* = \left(\frac{J}{kX_f D_f K_s}\right)^{0.5}$ and non-dimensional bulk substrate concentration, S_B which is provided by equations 8.106 and 8.110 alongwith following the relationship[26] describing non-dimensional diffusive substrate flux through a stagnant liquid layer covering the biofilm.

$$L^*_f = \frac{kY}{b} J^*_s \qquad ...(8.106)$$

If acetic acid is predominant in reactor feed and other acids are negligible, the basic differential equation reduces to

$$\frac{\delta S}{\delta t} = D_f \frac{\delta^2 S}{\delta z^2} - \frac{k_a X S}{K_s + S}, \qquad ...(8.107)$$

Non-dimentional form of equation 8.107 could be written as :

$$\frac{bKs}{kX} \frac{\delta S^*}{\delta t} = \frac{\delta^2 S^*}{\delta z^{*2}} - \frac{S^*}{1+S^*} \qquad ...(8.108)$$

Under steady state, equation 8.108 simplifies to equation 8.109 :

$$\frac{\delta^2 S^*}{\delta z^{*2}} = \frac{S^*}{1+S^*} \qquad ...(8.109)$$

For deep biofilms, Suidan and Wang[27] proposed a non-dimensional relationship between J*, S_s* and L_f* and is given as

$$L_f^* = J^* + \tan h^{-1}\left(\frac{0.5 + J^*\left[1+(J^*/3.4)^{1.19}\right]^{-0.61}}{S_s^*}\right) \qquad ...(8.102)$$

$$J^* = \frac{D_W}{L_W^*}\left(S_B^* - S_s^*\right) \qquad ...(8.103)$$

$$L_w^* = L\left(\frac{kX_f}{D_f K_s}\right)^{0.5} \qquad ...(8.104)$$

Where, D_w and D_f refer to diffusivities of substrate in water and biofilm respectively. L^*_w and L are non dimensional liquid layer thickness and thickness of stagnant liquid layer respectively.

Annachatre and Khanna[25,26] presented an electrical analogy approach for estimation of liquid layer resistance and biofilm resistance through simple algebraic equations and also derived a relationship between the rate of biofilm growth and substrate flux across the biofilm to estimate biofilm thickness and maturation time.

8.2.5 Important Comments

Based on general knowledge of anaerobic reaction kinetics, it is possible to discuss the likelyhood of diffusional resistance being of significance to anaerobic processes. The following parameters have been used for evaluation :

$X = 50\ kg/m^3$;

$K_s = 0.2 kg\ COD/m^3$;

$Y = 0.1 kg\ VSS/kg\ COD$;

$\gamma x = 2 kg\ COD/kg\ VSS.d$; and

$D = 1 \times 10^{-4}\ m^2/d$

Transition to 1-order reaction occurs when $S/K_s = 2$ which gives 0.4 kg COD/m^3. The diffusional resistance starts to influence the efficiency of the bulk reaction in the 1-order range for :

$$\left(\frac{\gamma x X L^2}{K_s D}\right)^{1/2} = 2 \qquad ...(8.105)$$

which occurs for a film thickness of 1mm in the whole range of 1 order reaction Transition from bulk 0 - order or 1/2-order occurs for

$$\left(\frac{2DS}{\gamma_x X L^2}\right)^{1/2} = 1 \qquad ...(8.106)$$

which gives the borderline

$L^2 = 2 \times 10^{-6}\ S.$

For a bulk concentration of $0.4 kg\ COD/m^3$, the transition occurs for a film thickness close to 1mm. The final result is shown in Fig. 8.6. The indication is that diffusional resistance does not become of significance until the biofilm thickness reaches values above approximately 1mm for $K_s = 0.2$ kg COD/m^3

(and approximately 0.3 mm for K_s = 0.02kg COD/m^3). This indicates diffusional resistance is not of practical significance. The above conclusion is attributable entirely to high K_s value. However, this does not put the mind to rest, because it must be remembered that gradual transition from 1- to 0-order before the recognition of the diffusional effect was interpreted as an artificially high K_s–value, in the order of 100mg/L for glucose instead of 10. There is a need for the models described above presume the biofilm to be composed of a single microbial species growing or a single rate-limiting substrate. However, in an anaerobic fixed film reactor, the biofilm develop from a consortium of organisms, each requiring a different substrate for growth. Further, the metabolic product of one organism is often the substract for another and there is a spatial and temporal variation of biofilm composition within the reactor depending on operational conditions.

The competition of several microbial species for space and substrate within a biofilm has been described Wanner and Gujer[28]. The author have developed a model for steady state Substrate removal by antotrophic and hectrotrophic organisms competing verification of K_s-values for anaerobic processes in distinctly dispersed bacterial cultures. One of the interesting verification of the diffusional resistance concept is an analysis of the out diffusions of products of processes. The dominating product is inorganic carbon. Invariably do the diffusional resistance cause build up of the inorganic carbon species in rear of the biofilm. This may cause very significant increases of pH in the biofilm for an alkalinity producing reaction like denitrification. It may even increase the pH to such values that chemical precipitation occurs in the film, which could not be explained when considering the bulk liquid only. Nitrification is an acidity producing process. Accordingly the out diffusion of inorganic carbon causes decrease in pH inside the biofilm. That may give rise to toxicity in the biofilm without that being released when considering the bulk liquid alone.

Out diffusion of gaseous products of low solubility may give rise to bubble formation in the biofilm, which is experienced in the case of denitrification. There is little room for build up between the concentration in the bulk liquid and saturation. That phenomenon can significantly alter kinetics of the film and can give rise to sloughing off. The out diffusion in anarobic biofilms is interesting with respect to methane, as the methane produced in any biofilm has to move out of the biofilm. In case of low bulk methane concentration, the transport out of the biofilm will take place by diffusion, for which a build up of methane concentration in the biofilm in required. In case of strong diffusional resistance that may give rise to supersaturation and bubble formation in the rear of the biofilm, in analogy to build up of free nitrogen in denitrifying biofilms. In case of supersaturation in bulk liquid, the methane can move out only as bubbles. It is a fundamental, as yet unanswered question, how this formation and out movement of bubbles affects the performance of the biofilm. In case of fully penetrated biofilm, the effect may not be of great significance. But in cases of thick biofilms with significant diffusional resistances one might visualise that the out diffusion of bubbles may cause such disturbances that the whole diffusional pattern is brokeup. Micro currents generated by bulk movements may cause an artificial increase in the apparent diffusion rate with the result that the diffusional resistance becomes insignificant.

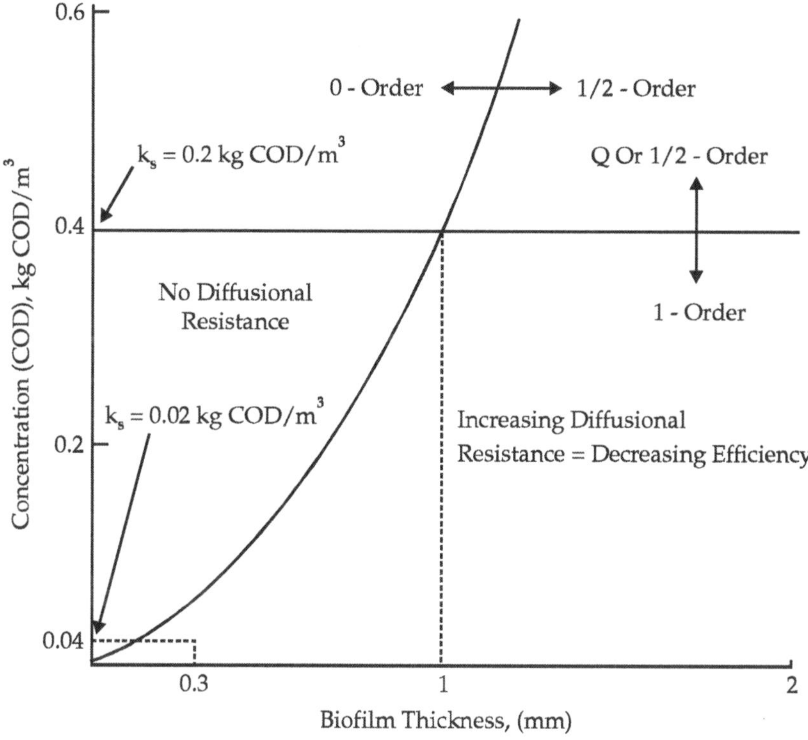

Fig. 8.6 : Diffusional Resistance in an Anaerobic Biofilm

For dilute wastewaters presents another problem with respect to performance, because the soluble COD concentration in the effluent is increased by the COD of the methane, which for a saturated solution amounts to approximately 80g COD/m^3.

For long, it has been part of the conventional thinking that adsorption and hydrolysis play a significant role in biological waste treatment. Adsorption is an important process in that it may provide the primary removal mechanism. It is also important to the growth mechanism of the biofilm. Suspended solids will be incorporated into the biofilm matrix through adsorption. Only a fraction of organic particulates can be hydrolysed to soluble matter. The inert matter left will occupy space in biofilm. In analogy to a similar concept for activated sludge. The following fractions have to be dealt with :

- active biomass consisting of acid phase 'bacteria and methanogen' :

- adsorbed hydrolysable organic matter
- adsorbed inert material, either inorganic and inert organic matter derived from influent or inert residue of lysed bacteria.

Growth kinetics considering these different fractions in biofilm are considered to be of importance because active bacteria are diluted by other fractions, which will influence the removal rate significantly. This concept is expected to be of particular importance to anaerobic processes due to very low yields. High concentration of inert material in the influent may even give rise to failure because the growth rate of the biofilm and sloughing off rate may dilute the methanogen and leave so little time for their slow growth. According by, removal of acetic acid and production of methane may be severely hampered. There is a great need for analysis of the growth of biofilm and the mechanisms that govern the bacterial densities.

It may also be mentioned that the attachment and sloughing off mechanisms of anaerobic biofilms are very important to process performance, that very little is known about it and that what is known is hardly applicable in an engineering context.

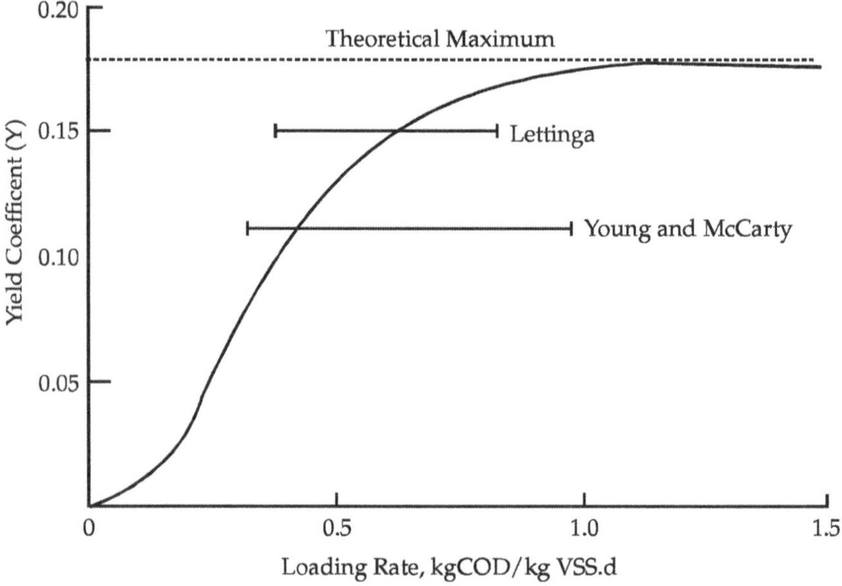

Fig. 8.7 : Yield Coefficient Versus Loading Rate

The stoichiometry of the anaerobic processes is of interest in relation to methane gas production, alkalinity changes and to nutrient requirements. Methane gas production accounts for the major part of the organic matter degraded during a stable anaerobic process. Unstable and stressed processes may result in concentrations of volatile fatty acids in the effluents of magnitude, which account for a considerable amount of the degraded organic for dilute wastes. Due to low yield coefficients, methane production in general surpasses 90 per cent of the COD processed. The maximum theoretical methane yield (with bacterial yield coefficent of zero) is 0.370 m^3 STP/kg COD_r. Low methane yield reported in many experiments may have been caused by leaks. It is rather uninteresting, but often seen that methane production is compared, to the influent COD as some part of the COD might be biodegradable. Fig. 8.7 shows the relationship between loading rate and yield coefficent for common nutrient within the biofilm.

The head loss through a fixed bed can be calculated by a well known Leva's equation :

$$\frac{\Delta P}{H} = 2fU^2 / d\left[(1-\varepsilon)^{3-n} / \gamma^{3-n}\varepsilon^3\right] \qquad ...(8.107)$$

In the operating velocity range, the Reynolds numbers, Re = $\rho dU/\mu$ is less than 10. The flow is laminar : n = 1, f = 100/Re. The shape factor (γ) is determined by the following relation:

$$Y = 4.87\ V^{2.3}/W \qquad ...(8.108)$$

Where, V, W and U refer to volume of grain and specific area of the particles and fluid velocity in the empty reactor respectively. The variation in void factor with biomass can be estimated if the concentration of bacteria is between 0 and 2 mg VSS per gram of media. Under these limits a linear relationship between void factor (ε) and biomass, mg VSS few gm of media (X) can be assumed according to the following formula :

$$\varepsilon = 0.035\ X + \varepsilon_o \qquad ...(8.109)$$

The detention period (θ) in a fixed bed without biomass is given by :

$$\bar{\theta} = (\varepsilon_o)\ (V/Q) \qquad ...(8.110)$$

Fixed film reactors often operate on porous support materials, such as activated carbon, which are also good adsorbents. Andreurs and Tien[29] reveal that in such cases, the bacteria and activated carbon complement each other in the removal of organics from the wastestream. For example, aromatic compounds have a high affinity for carbon but are more resistant to biodegradation, whereas common bacterial substrates are soluble and therefore adsorb poorly on carbon. Further, the bacteria are too large to enter the pores of the carbon and reduce the concentration of biodegradable organics at the carbon surface. Any reversibly-adsorbed biodegradable molecule therefore, tend to desorb and diffuse out of the carbon to the bacteria.

Based on these considerations, Andrew and Tien[29] have developed a mathematical model for predicting the overall removal of organic matter in a system where bacteria grow as a film on the carbon surface. The model is based on the assumption that organic substrate is soluble and adsorbs reversibly on activated carbon; the film density does not vary with thickness; the organic substrate limits bacterial growth and its uptake follows first order kinetics and the mass transfer resistance of external liquid film in negligible. Two main feature of interaction between adsorption and microbial growth have been considered in the model, *viz.* extra mass transfer resistance to adsorption resulting from the microbial film and the bio-regeneration of saturated adsorbent. The model can be applied to real waste systems whose exact composition may not be known.

The state of the art of biofilm kinetics models has been reviewed. The models have some important limitations. First, they are only applicable at steady-state and non-steady-state operation is rather common. Second they apply only when a single substrate limits the kinetics. Third, they assume that deposition rate is negligible and they rest on a poor correlation to estimate the shear loss rate. Finally, clogging is not taken into account, and it can result in short-circulating and loss of pore volume, both of which reduce treatment efficiency.

The properties like diffusivity (D, D_H and D_f) Y, k, b, X_f size (d_p), specific area (a) shear stress (σ), biofilm thickness (L_f), liquid film thickness (L), etc. need experimental methods to determine their exact values. The properties of biomass in the bulk solution cannot be assumed to be same as in the biofilm.

In addition, investigations have to be carried out to determine the bio-chemical pathways whereby by all higher fatty acids can be converted quickly to acetic acid for further biochemical reactions. So far this step could not be achieved and research is required urgently in this area.

It is equally important to shorten the period of growth of biofilm on support media. Investigation in this direction will result in maturation of fixed film reactors at an early date and it will be able to compete with aerobic systems.

In most of the models, isothermal conditions are assumed but in practice it does not happen. In addition, the hydraulic regime is assumed to be ideal plug flow or otherwise but it rarely happens. Therefore, hydraulic conditions must be incorporated into the models. Diffusion processes for biofilms have not been studied throughly and most often catalyst type equations are used which may not be true. Distribution of biomass is considered uniform which does not happen in practice. In addition to growth associated kinetics it is also necessary to include structural associated growth into the main set of equations. In addition, steady state has been assumed in most cases, it is not possible because the micro-organism are always growing. This effect has also to be incorporated into the basic equations.

8.3 Fixed Film Fixed Bad Reactor

8.3.1 Hines Model[30] and Atkin Sow Model[31]

The results from a laboratory scale study were used to examine the feasibility of using a fixed film bioreaction model to describe a packed bed bioreactor. The model was developed using a mass balance and film penetration procedure.

mass rate of substrate approaching the biofilm	+	mass rate of substrate penetrating the biofilm surface	=	mass rate of substrate leaving the biofilm surface

The model assumes that substrate is utilized by the biofilm where uptake occurs. A biofilm model is illustrated in Fig. 8.8. The following equation is used to describe the bio-utilization of substrate, *viz.* :

$$QS_Z - N_Y W\Delta Z = Q\left(S_z + \frac{\delta S}{\delta z}dz\right) \qquad ...(8.111)$$

Where Q, S, N_y, W, at Z refer to flow, substrate, penetration flux, width of media surface section, and depth of media surface section (direction of flow) respectively. Divinding equation 8.111 by z

$$\frac{Q\left(Sz - (Sz + (dS/dz)\,dz\right)}{z} = N_Y W \qquad ...(8.112)$$

By taking definition of a limit

$$-Q\ dS/dz = N_Y\ W \qquad ...(8.113)$$

The penetration flux (N_Y) can be described at steady state as equal to bioutilization. An approximate mathematical form can be substituted into this equation. A modified Monod form has often been applied to fixed film reactors, *viz.*:

$$N_Y = \frac{fhk_o S^2}{K_m + S} \qquad ...(8.114)$$

The linearized form (where $S >> K_M$) is :

$$N_Y = fhk_o S \qquad ...(8.115)$$

Therefore, the overall equation becomes :

$$-Q\frac{dS}{dz} = fhk_o S \qquad ...(8.116)$$

Integrating equation 8.116, results in

$$S_o/S_o = \exp\ (-\ fh\ WZ\ k_o/Q) \qquad ...(8.117)$$

Where, f and h are effectiveness factor and biofilm thickness respectively.

Eckenfelder consolidated the coefficients as follows :

$$K = fh\ wk_o \qquad ...(8.118)$$

$$D = Z \qquad ...(8.119)$$

$$S_o/S_o = \exp\ (-\ KD/Q^n) \qquad ...(8.120)$$

Where, K, n and D refer to treatability factor, characteristics constant of the filter and anaerobic reactor depth respectively. The above equation is

$$2.3 \log (S_o/S_o) = -KDQ^{-n} \quad ...(8.121)$$

$$\text{Slope} = -(K\ Q^{-n})/2.3 \quad ...(8.122)$$

$$\text{Log}(2.3 \times \text{Slope}) = -(\log K + (-n) \log Q)$$

$$= -\log K + n \log Q \quad ...(8.123)$$

Another expression developed by Eckenfelder is given as :

$$S_o/S_o = \exp(-K_Z S_a^m (A/Q)^n) \quad ...(8.124)$$

Where, K, S_a and A refer to observed reaction rate constant, specific area = (media surface area, (A_s/V) and cross sectional area of the fixed bed respectively.

If m = n = q, then :

$$\frac{S_a ZA}{Q} = \frac{A_s}{V}\frac{ZA}{Q} = \frac{As}{Q} = \frac{WZ}{Q} \quad ...(8.125)$$

and equation 8.124 reduces to equation (8.117). The other model used by Oleszkiewicz[32] has been found to fit experimental data quite well and is expressed as :

$$S_e/S_i = \exp(-K/L) \quad ...(8.126)$$

Equation 8.126 relates directly food/micro-organism (F/M) ratio,

$$S_e/S_i = \exp(-KAD/QS_i) \text{ or} \quad ...(8.127)$$

$$S_e/S_i = \exp[(-K(F/M))] \quad ...(8.128)$$

Where, h and L refer to biofilm thickness and loading rate respectively.

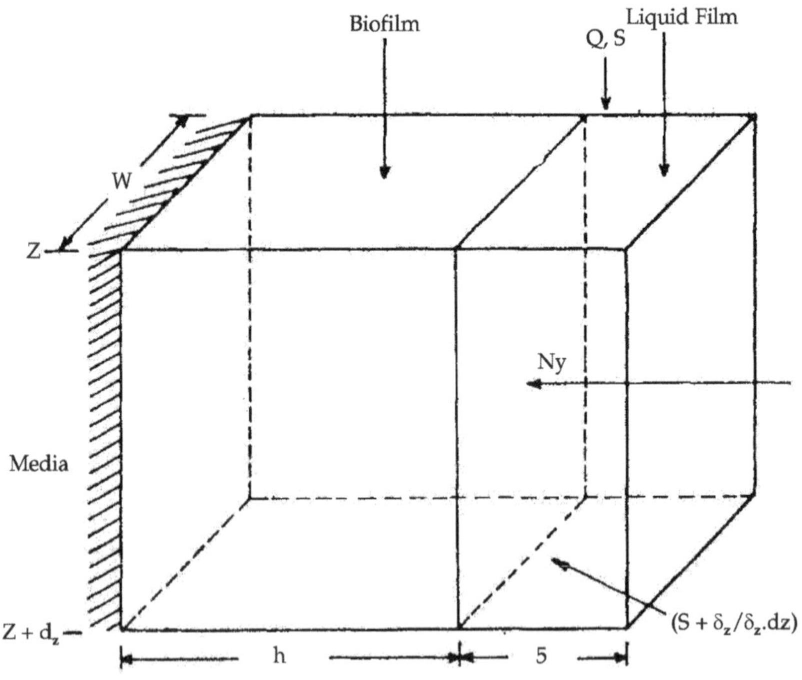

Fig. 8.8 : Model of Biofilm Utilization

8.3.2 DeWalle and Chian Model[33]

In a biological film reactor at steady state, a mass balance within the film can be made by equating :

$$\text{Input} - \text{Output} = \text{Uptake} \qquad ...(8.129)$$

According to Fick's law of molecular diffusion, the mass transfer rate ($\delta F/\delta t$) through a biofilm surface area (A) is proportional to the concentration gradient of substrate, S at the interface :

$$\frac{\delta F}{\delta t} = -Ad_s \frac{\delta S}{\delta z} \qquad ...(8.130)$$

Where, D_s $\delta S/\delta z$ and Z refer to diffusion coefficient substrate gradient and depth of biofilm starting from liquid film interface respectively. The biological uptake within the biological layer can be expressed as

$$\frac{dS_z}{dt} = \frac{US_zX}{K_s + S_z} \qquad ...(8.131)$$

Where, U, S_z and X refer to maximum substrate removal rate, substrate concentration at depth, z and biomass concentration respectively.

By making a material balance between a differential distance within biolayer and using equations 8.129 to 8.131, the results in the following equations for a unit-cross sectional area

$$\frac{d^2S_z}{dZ^2} = \frac{1}{Dx}\frac{US_zX}{K_s + S_z} \qquad ...(8.132)$$

Which, states that the derivative of the substrate concentration with respect to z is dependent on substrate and biomass. When axial dispersion becomes important, the term v $\delta s/\delta z$ will have to be added to right hand side of equation 8.132.

Equation 8.132 is a second order non-linear differential equation. As such it does not have a simple solution. However, the equation can be solved when S_z is much larger or much smaller than K_s. When S_z is much larger than K_s, equation 8.132 reduces to zero order kinetic equation :

$$\frac{d^2S_z}{dz^2} = \frac{UX}{D_s} \qquad ...(8.133)$$

Which can be integrated for 0a unit cross-sectional area and the mass transfer rate at the interface becomes

$$dF/dt = UXh \qquad ...(8.134)$$

Where, h refers to thickness of biofilm. Since U and X are not expected to vary greatly, equation 8.134 would indicate the rate is independent of substrate concentration but directly proportional to the thickness of the biolayer. Since the equation is derived for a unit cross-sectional area, the mass transfer rate is also proportional to the specific area of the solid medium within the unit. In such a situation the decrease of substrate within the biofilm is small compared to the concentration at the interface so that the rate of substrate utilization per unit of biomass is approximately the same as if the solids were evenly dispersed throughout the liquid. According to Pirt[34] and Saunder and Bazin[36], the thickness of biolayer(h) can be approximated by :

$$h=\left(\frac{2D_sS_b}{UX}\right)^{0.5} \qquad ...(8.135)$$

Where, S_b refers to substrate concentration in bulk liquid. However, equation 8.135 was derived based on the boundary conditions of zero substrate concentration at the interface of the biolayer and the solid support; it, therefore, violates the constraints that S_z is much larger than K_s and is, therefore, only an approximation. Since D_s, U and X are not expected to vary to a great extent for a given substrate equation 8.134 can, therefore, be further modified to indicate that the rate of substrate removal per unit volume is related to specific surface area (A/V) and the square root of the bulk substrate concentration.

$$\frac{1}{V}\frac{dF}{dt}=k_1(A/V)S_b \qquad ...(8.136)$$

Where, k_1 and V refer to coefficient based on zero order kinetics and reactor volume respectively.

When S_z is much smaller than K_s, equation 8.132 reduces to a first order kinetic equation:

$$\frac{d^2S_z}{dz^2}=\frac{UXS_z}{D_sK_s} \qquad ...(8.137)$$

Since D_s U, and K_s are not expected to vary greatly with a given substrate equation 8.137 reduces to

$$\frac{1}{V}\frac{dF}{dt}=k_2(A/V)S_b \qquad ...(8.138)$$

Where, k_2 refers to coefficient based on first order kinetics. The equation indicates that the rate of substrate removal per unit reactor volume is related to the specific area of the biofilm and the bulk substrate concentration, but is independent of film thickness. A similar equation has been used by Atkinson and Daoud[36] for a large film thickness. Equation 8.36 and 8.138 should be used in small successive sections in plug flow anaerobic reactors. However, since such detailed data are generally not available both equations can be used with respect to the whole depth of the reactor while S_b is substituted by the effluent concentration of anaerobic reactor.

Additional resistance to mass transfer from bulk liquid to the liquid-biofilm interface is caused by the liquid film adjacent to the biofilm. The thickness of this layer can be described by Nasselt[37] equation :

$$\delta = (3\mu Q/\rho g w)^{0.33} \qquad ...(8.139)$$

Where, Q, μ, ρ, g, W and δ refer to flow rate, viscosity, density, gravitational constant, width of biofilm and thick of liquid film respectively. Since mass transfer rate decreases with increasing liquid film thickness in laminar sublayer, the flux of the substrate would also be proportional to the flow rate to the power 1/3. This agrees with the analysis of Takeshi[32] that mass transfer coefficient k_L for laminar flow can be expressed as:

$$k_L = KQ^{1/3};\ N_{Re} < 10 \qquad ...(8.140)$$

$$k_L = KQ^{1/2};\ N_{Re} > 10 \qquad ...(8.141)$$

Both equation, 8.136 and 8.138 illustrate the importance of specific surface area (A/V) of the fixed film, respectively. Based on similar considerations Ames,[39] stated that the substrate removal was proportional to the specific area of the medium. Lamb and Owen[40] came to the identical conclusion.

8.3.3 Kornegay Model[41]

The substrate material balance at steady state with recycle is given as :

$$\text{Input} = \text{Output} + \text{Microbial consumption} \qquad ...(8.142)$$

$$Q\,(1+\gamma)\,(S+ds) = Q\,(1+\gamma)\,S - \left(\frac{\mu}{Y}M_o\right) \qquad ...(8.143)$$

Where Q, γ, S, μ, M_o and Y refer to hydraulic flow rate, hydraulic recycle ratio, concentration of rate controlling nutrient, specific growth rate, active mass of organisms in differential element and yield coefficient respectively.

It has been shown by Kornegay and Andrew[41] that the active microbial mass may be expressed as the product of the differential depth (dz), the cross sectional area (H), the specific area of the fixed bed (a), the unit mass of biological film (X) and the thickness of active layer (d). When this relation is substituted into equation 8.143 one obtains :

$$Q\,(1+\gamma)\,(S+ds) = Q\,(1+\gamma)\,S - \frac{\mu}{Y}(a)(X)(H)dZ \qquad ...(8.144)$$

The resulting expression for substrate removal with depth is then

$$-\frac{dS}{dZ} = \frac{\mu(a)(X)(d)(H)}{YQ\,(1+Y)} \qquad ...(8.145)$$

When the specific growth rate is expressed using Monod model, *i.e.* :

$$\mu = \mu_{max}\frac{S}{K_s + S} \qquad ...(8.146)$$

Equation 8.145 becomes :

$$-\frac{dS}{dZ} = \frac{\mu_{max}(a)(d)(H)(X)}{YQ(1+\gamma)} \frac{(S)}{(K_s + S)} \quad ...(8.147)$$

This expression may be integrated when S goes from S_i to S_e as the depth goes from 0 to z to produce the expression.

$$K_s \ln (S_i/S_e) + (S_i - S_e) = \frac{\mu_{max}}{Q(1+\gamma)}(a)(X)(d)(H)Z \quad ...(8.148)$$

It may be shown from mass balance on recycle system that the influent substrate concentration (S_i), may be determined by the equation :

$$S_i = \frac{(S_e\gamma + S_o)}{(1+\gamma)} \text{ and} \quad ...(8.149)$$

Equation 8.148 then becomes :

$$K_s \ln \left[\frac{S_o + \gamma S_e}{S_o(1-\gamma)}\right] + \left[\frac{S_o + \gamma S}{1+\gamma}\right] - S_o = \frac{\mu_{max}}{QY(Hr)}(a)(X)(d)(h)(Z) \quad ....(8.150)$$

when there is no recycle (r = 0), and

$$K_s \ln \frac{S_o}{S_o} + (S_o - S_o) = \frac{\mu_{max}}{QY}(a)(X)(d)(h)(Z) \quad ...(8.151)$$

Although equations 8.150 and 8.151 are mathematically correct they contain a number of elusive kinetic parameters which lead to either misuse or disuse by the practicing engineer. However, such parameters as yield, maximum specific growth rate, unit mass of biological film and active film thickness need not be determined on an individual basis for design application. These terms may be combined into a single term for simplification, but their presence and significance should be throughly understood.

Since the product [1/Y μ_{max} (X)(d)] may be evaluated from pilot plant studies, it would suggest that a single term, P be substituted into equations 8.151 such that:

$$P = \frac{1}{Y}(\mu_{max})(X)(d) \quad ...(8.152)$$

Equation 8.150 would then become :

$$K_s \ln \left[\frac{S_o + S_o}{S_o(1+\gamma)}\right] + \left[\frac{S_o + \gamma S_o}{1+\gamma}\right] - S_o = \frac{P}{Q(1+\gamma)}(a)(H)(Z) \quad ...(8.153)$$

and equation 8.151 becomes :

$$K_s \ln \left[\frac{S_o}{S_o}\right] + (S_o - S_o) = \frac{P}{Q}(a)(H)(Z) \quad ...(8.154)$$

The parameters included in the term P, should be constant for a given set of environmental factors and may be determined for the region of interest. With the value of P known the design engineer may then select the flow rate (Q), the depth (z), cross sectional area (H) and the specific area (a) for the anaerobic fixed bed reactor, which will most efficiently reduce the influent concentrations (S_0) to the desired effluent concentration (S_0). It must be remembered that K_s varies with velocity and must be evaluated at each flow rate. In addition, the relation between the flow rate (Q) and the saturation constant (K_s) must be known to optimize the reactor design.

The values of the constant, P may be evaluated in several ways. To obtain a rapid estimate, apply the waste to be treated to a pilot plant having a known filter depth (z) and cross-sectional area (H). The specific area (a) of the media must be known and recycle is not practiced. Whenever possible, increase the influent concentration (S_0) or decrease the depth (z) until the ratio S_0/S_0 approaches one. As this occurs, the value of K_s ln (S_0/S_0) will become insignificant. Therefore, the following expression can be used.

$$(S_0 - S_0) \cong \frac{P}{Q}(a)(H)(Z) \quad ...(8.155)$$

From the knowledge of influent and effluent substrate concentration, flow rate and reactor dimension, the value of P may be calculated from the approximate expression

$$P \cong \frac{(S_0 - S_0)Q}{(a)(H)(Z)} \quad ...(8.156)$$

The process can be move easily understood from an examination of differential form of the mathematical mode. It can be seen that when S is much greater than K_0, the equation

$$-\frac{dS}{dz} = \frac{\mu_{max}}{YQ}(a)(X)(d)(H)\frac{S}{(Ks+S)} \text{ and this reduce to :} \quad ...(8.157)$$

$$-\frac{dS}{dz} = \frac{\mu_{max}}{YQ}(a)(X)(d)(H) \quad ...(8.158)$$

By proper substitution, equation 8.158 becomes

$$-\frac{dS}{dz} = \frac{P}{Q}(a)(h) \quad ...(8.159)$$

Here, it is obvious that the rate of substrate removal with depth is constant and P may be estimated from expression:

$$P = \frac{\Delta SQ}{\Delta Z(a)(H)} \quad ...(8.160)$$

With the value of P known, the saturation constant (K_s) may be calculated for the flow employed. The above method only provides estimates of P and K_s but more accurate techniques of evaluation are available. Equation 8.154 can be written

$$(S_o - S_o) = \frac{P}{Q}(a)(H)(z) - K_s \, In(S_o - S_e) \qquad ...(8.161)$$

Which is similar in form to Y = C + mX. Therefore, a plot of $(S_o - S_e)$ versus ln (S_o/S_e) should produce a line with slope K_s and intercept (P/Q.a.H.z) at $S_o/S_e = 1$. The product (P/Q) (a)(H)(Z) represents maximum substrate removal that may be expected from an anerobic reactor having a depth (z), cross sectional area (H), media with a specific surface area (a). It may, therefore, be considered the maximum bioreactor capacity for a flow rate, Q. The term, P would then represent the capacity constant for a unit area of reactor media.

8.3.4 Kong Model[42]

In an anerobic fixed bed reactor, the reactor unit can be subdivided into 'n' infinitiesmal equal sections such that both the microbial mass and the substrate concentration in each section can be considered homogeneous. The cell yield coefficient is assumed to be equal for each section and it is taken as the mean cell yield coefficient.

By the above assumptions of subdivision the material balance can be applied to each section. When J^{th} subdivision is considered, the material balance can be expressed as follows :

$$v \frac{dS}{dt} = QS_{j-1} - FS_j - v \;\; \mu_j x_j / Y \qquad ...(8.162)$$

Where μ_g refers to the specific growth rate of micro-organism on J^{th} division. At steady state, the above equation becomes :

$$S_j = S_{j-1} - (v/Q)(\mu_j X_j/Y) \qquad ...(8.163)$$

$$S_1 = S_0 - (v/Q)(\mu_1 X_1/Y)$$

$$S_2 = S_1 - (v/Q)(\mu_2 X_2/Y)$$

$$\vdots \quad \vdots \quad \vdots$$

$$S_n = S_{n-1} - (v/Q)(\mu_n X_n/Y)$$

Summing up all S_j, they will give

$$\sum_{j=1}^{n} S_j = S_o + \sum_{j-1}^{n-1} S_j - \frac{1}{QY}\sum_{j-1}^{n} \mu_j X_j v \text{ or}$$

$$\sum_{j=1}^{n} S_j = S_o + \sum_{j-1}^{n-1} S_j - S_n - \frac{1}{QY}\sum_{j-1}^{n} \mu_j X_j v \text{ or}$$

$$S_0 - S_n = \frac{1}{QY}\sum_{j-1}^{n} \mu_j X_j v \text{ or} \qquad ...(8.164)$$

The cell mass balance for Jth subdivision will give :

$$v\frac{dx}{dt} = \mu_j X_j v + QX_{wj-1} - QX_{wj} - bX_j v \qquad ...(8.165)$$

At steady state, the above equation becomes :

$$v\mu_j X_j = Q(X_{wj} - X_{wj-1}) + bX_j v \qquad ...(8.166)$$

Substituting equation 8.145 into equation 8.164 will give

$$S_o - S_n = \frac{1}{QY}\sum_{j-1}^{n}\left[Q(X_{wj} - X_{wj-1}) + bX_j v\right] \qquad ...(8.167)$$

$$= \frac{1}{Y}\sum_{j-1}^{n}(X_{wj} - X_{wj-1}) + \frac{b}{QY}\sum_{j=1}^{n} X_j v$$

$$S_o - S_n = \frac{X_{w,n}}{Y} + \frac{b}{QY} X_t \qquad ...(8.168)$$

where,

$$X_{w,\,o} = 0, X_t = \sum_{j=1}^{n} X_j V \text{ (Total cell mass in the fixed bed reactor)}$$

Since, $X = X_t/V$ which is the total cell mass retained per unit reactor volume, then

$$S_o - S_n = (V/QY)\left(\frac{X_{w,n}Q}{V} = b\bar{X}\right)$$

$$= \left(\bar{X}V/QY\right)\left(QX_{w,n}/V\bar{X} + b\right)$$

$$S_o - S_n = (\bar{X}V/QY)(1/SRT + b) \qquad ....(8.169)$$

Where, $\quad SRT = \dfrac{\text{total cell mass retained/reactor volume}}{\text{sludge wasted/day/reactor volume}}$

Rearranging equation 8.169 gives

$$\frac{1}{STR} = \left(\frac{S_o - S_n}{\bar{X}}\right)\frac{Q}{V}Y - b \qquad ...(8.170)$$

where S_o and S_n refer to influent and effluent concentration respectively. The concept of solids retention time has been applied to fixed bed reactor system. This equation is analogous to activated sludge process. The present model does include both microbial growth kinetic concepts which can be more practical and meaningful for the design of an anaerobic fixed bed reactor system. The relationship between SRT and influent COD(S) is determined by plotting SRT versus reciprocal of influent COD can concentration at various hydraulic loading rates. The relationship is given as follows :

$$SRT = Slope(x)/S + Constant$$

The slope(x) is plotted against hydraulic loading rates and a relationship can be obtained as follows :

$$X = Constant - d.H$$

Therefore, the SRT versus hydraulic loading rate (H) can be expressed as

$$SRT = (a - dH)/S + C \quad ...(8.171)$$

$$SRT = (a - dH)H/L + C \quad ...(8.172)$$

Where a, d and c are obtained from the plots mentioned above.

The SRT for the fixed bed can be obtained from total mass retained and daily sludge wasted. The SRT for the reactor can also be estimated using independent variables, *viz.* : hydraulic loading (H), organic loading (L) or influent substrate concentration (S) using equations 8.171 and 8.172. The reactor performance can be expressed as a function of SRT and basic microbial growth kinetic concepts.

8.3.5 *Stover and Rankness Model*[43]

When considering a volume of fixed bed reactor, a mass balance of the substrate into and out of that volume at steady state can be made and is given as follows :

$$QS_i = QS_o + (dS/dtA)_G\,A \quad ...(8.175)$$

$$\begin{bmatrix}\text{mass of substrate}\\ \text{into the volume}\end{bmatrix} = \begin{bmatrix}\text{mass of substrate}\\ \text{out of the volume}\end{bmatrix} + \begin{bmatrix}\text{mass of substrate}\\ \text{consumed}\end{bmatrix}$$

By plugging the following substrate utilization rate developed by Stover and Kincannon[44,45],

$$\left(\frac{dS}{dtA}\right)_G = \left(\frac{U_{max}\,QS_i/A}{K_B + QS_i/A}\right) \quad ...(8.174)$$

Into equation 8.173, the resulting expression can be solved for required surface area (A) for design, as follows:

$$A = \frac{QS_i}{\left[\dfrac{U_{max}S_i}{S_i - S_e}\right] - K_B} \quad ...(8.175)$$

Where, U_{max}, K_B and QS_i/A refer to maximum specific substrate removal rate (kg/day/1000m^3), proportionately constant (kg/day/1000m^2) and substrate loading rate (kg/day/1000^2) respectively.

The expression can also be solved for effluent substrate concentration that would correspond to specific design loading and amount of surface available, as follows :

$$S_o = S_i - \frac{U_{max}\, S_i}{K_B + (QS_i / A)} \qquad ...(8.176)$$

8.3.6 Dispersion Model[46]

The complex pattern of substrate degradation is considered as a set of pseudo reaction, as follows :

$$S \rightarrow P \text{ (Product)}$$

$$\underset{\text{(Substrate)}}{S} \rightarrow \underset{\text{(Viable biomass)}}{X} \rightarrow \underset{\text{(dead biomass)}}{X_D}$$

Following the made assumptions, a set of rate expression can be written as follows :

$$\gamma_s = -kS_iX \qquad ...(8.177)$$

$$\gamma_x = Y\,k\,S_i - bX \qquad ...(8.178)$$

$$\gamma_{xo} = bX \qquad ...(8.179)$$

in which case the expressions represent the rate of change of substrate, viable cell and dead biomass. Here, the effectiveness factor has been lumped together with the intrinsic rate constant, k (= Ek_o).

Following the approach of the dispersion model for fixed film reactor, the following conservation equation at steady state can be used :

$$D_s \frac{d^2S}{dz^2} - v_o \frac{ds}{dz} - kS_i\, X = 0 \qquad ...(8.180)$$

Here, both S and X vary along the reactor. Furthermore, S_i depends on S (bulk liquid) in a certain way according to the mass transfer resistance. S_i is the concentration at the liquid film layer.

First, an expression for X will be derived. It is convenient to define the mass fraction of viable cells at a certain position of the reactor according to the following expression :

$$f_x = \frac{X}{X_o} \qquad ...(8.181)$$

where, $x_o = A_v \delta \rho_o$

The erosion of dead and living micro-organism follows a complex pattern. The situation can be substantially simplified by assuming that the composition of the eroded material, represents the average composition of the fixed film at the particular position at which erosion occurs. This is a so called tank assumption. Under these conditions, the total production rate of biomass, including dead and viable cells can be written as

$$R_r = \gamma_s + \gamma_{SD} = YkS_iX \quad ...(8.182)$$

Furthermore, the production rate of viable cells is :

$$R_v = \gamma_w = YkS_iX - bX \quad ...(8.183)$$

At steady state, with complete balance between erosion and growth rate, the following expression holds :

$$f_x R_r = R_v \quad ...(8.184)$$

Inserting the expression for R_r and R_v respecting, leads to is

$$f_x = 1 - \left(\frac{b}{YkS_i} \right) \quad ...(8.185)$$

In order to find an expression for S_i in a conventional form, the Fick's law is used :

$$N_a = k_s (S - S_i) = -\frac{\gamma_a}{A_v} \quad ...(8.186)$$

By utilizing the expression for γ_s, and f_x as well as definition of f_x, S_i can be expressed explicity :

$$S_i = \left(S + \frac{X_o b}{K_s A_v Y} \right) / \left(1 + \frac{kX_o}{k_s A_v} \right) \quad ...(8.187)$$

Eliminating S_i and X in the conservation equation and using a dimensionless parameter $\xi = S - \frac{b}{Yk}$, the equation can be expressed as :

$$D_s \frac{d^2\xi}{dz^2} - v_o \frac{d\xi}{dz} - K\xi = 0 \quad ...(8.188)$$

where $$\frac{1}{K} = \frac{1}{kX_o} + \frac{1}{k_s A_v} \quad ...(8.189)$$

The boundary conditions are :

$$z = 0 \rightarrow \xi = \xi_o; \; -D_s \frac{d\xi}{dz} = (\xi_o - \xi) v_o$$

$$z = L \rightarrow \xi = \xi; \ \frac{d\xi}{dz} = 0$$

The solution to differential equation is :

$$\frac{\xi}{\xi_o} = \frac{4\,\psi\,\exp(1/2\gamma)}{(1+\psi^2)\exp(\psi/2\gamma)-(1-\psi^2)\exp(-\psi/2\gamma)} \quad ...(8.190)$$

where,

$$\psi = \left[1+4\gamma\frac{L}{v_o}\left(\frac{1}{k}\frac{1}{X_o}+\frac{1}{k_s X_o}\right)^{-1}\right]^{0-5} \text{ and} \quad ...(8.191)$$

$$\gamma = \left(\frac{Ds}{v_o L}\right)$$

The solution can be substantially simplified by considering that the second term of the denominator can be neglected. Furthermore is substantial dispersion prevails, $\psi>>1$, the following simplification can be obtained under such conditions.

$$-\ln(\xi/\xi_o) = \ln((1+\psi)^2/4\psi) + 4\,(K^{1/2}/D_s) \quad ...(8.192)$$

Furthermore, when $\psi<<1$, the axial dispersion can be neglected and the solution reduces to :

$$-\ln(\xi/\xi_o) = KLv_o^{-1} \quad ...(8.193)$$

8.3.6.1 Hydrodynamics

The nature of impact of K on substrate degradation in a fixed film reactor depends in the governing mechanism as well as the hydrodynamic conditions. The hydrodynamics may be taken into account for different design and operating conditions by carefully selecting pertinents correlations. A set of correlations were selected for down-flow conditions, in order to illustrate the principle. The mass transfer coefficient can be estimated by data correlated by van Krevelen and Krekels[47].

$$\frac{k_s}{A_g D} = 1.8\left(\frac{\rho v}{A_f \mu}\right)^{0.5}\left(\frac{\mu}{\rho D}\right)^{0.33} \quad ...(8.194)$$

The liquid holdup correlation is estimated by using Otake and Okada[48] correlation :

$$h_L = 15.1\left(\frac{\rho v dp}{\mu}\right)^{2/3}\left(\frac{dp^3 g \rho^2}{\mu^2}\right)^{-0.4} \quad ...(8.195)$$

The effective wetted surface area is estimated by means of equation suggested by Puranik and Vogelpohl[49]

$$\left(\frac{Av}{Ag}\right) = 1.05\ (v_o dp/\mu)^{0.041}\ (v^2_o dp/\rho\sigma)^{0.133} \qquad ...(8.196)$$

Finally dispersion may simply be expressed according to Hofmann[50]

$$\left(\frac{D_s h_L}{v_o dp}\right) \approx 2 \qquad ...(8.197)$$

A simplified design equation can now be derived from equations 8.192 and 8.194, by taking the hydrodynamic conditions into consideration. After considerable algebraic manipulations, the following general form can be derived

$$-\ln(\xi/\xi_o) = \alpha + \beta\ L\ Ag^m\ v_o^{\ n} \qquad ...(8.198)$$

In general, the ratio (ξ/ξ_o) is set to S/S_o when evaluating experimental data, *i.e.*, critical conditions are neglected. Furthermore, α is considered zero. However, if plotting the left and side of equation 8.198 as a function of L, α attains a value greater than unity if axial dispersion prevails. It should be noted that 'α' depends on v_o while. β is a constant that only depends on physical properties of the system. The powers of m, and n will attain different values depending on governing mechanism. Table 8.5 shows the actual values obtained when utilizing the correlations presented earlier. The nomenclature for various parameters are give in Table 8.6.

Table 8.5 : The values of Exponents m and n is Equation 8.198*

Sr. No	*Mechanism*	*Liquid downflow*		*Liquid upflow*	
		m	*n*	*m*	*n*
1.	Kinetics	0.83	–0.69	1.00	–1.00
2.	Mass transfers	1.06	–0.19	1.50	–0.50
3.	Dispersion and kinetics	1.05	–0.10	1.00	–0.50
4.	Dispersion and mass transfer	1.16	+0.15	1.25	–0.25

* Source – Karlson, H.T. on Modelling on Fixed Film Reactors; 5th Int. Symp. (May 22/26 1988), Bologna, Italy.

8.4 Fixed Film Fluidized Bed Reactor

Expanded/fluidized bed reactor have much larger surface area perunit reactor volume, which increases the reactor micro-organisms concentration. The larger specific area allows shorter HRTs or lowering temperatures for the same degree of treatment is a given volume. Switzenbaum[51] used anaerobic fixed film expanded bed process and found it effective for the treatment of low strength soluble organic wastes at reduced temperature, at short retention time and at high organic loading rates. Hickey and Owen[52] used, a similar process and have shown this to be the effective for simultaneous generation of methane gas and stabilization of high strength wastewater including dairy, chemical, food processing, soft drink bottling and heat transfer liquids.

Table 8.6 : Nomenclature for Dispersion Model

Sr. No	*Parameter*	*Significance*
1.	A	surface area
2.	b	decay coefficient
3.	D	diffusion coefficient
4.	D	dispersion coefficient
5.	f	fraction
6.	g	gravitational constant
7.	h	hold up
8.	k	kinetic rate constant; mass transfer coefficient
9.	K	lumped rate constant
10.	L	reactor length
11.	m	power with respect to area
12.	n	power with respect to velocity
13.	N	mass transfer rate
14.	r	reaction rate
15.	R	rate of change of species
16.	S	substrate concentration
17.	V	velocity
18.	X	cell concentration
19.	Y	yield coefficient for growth
20.	Z	distance along reactor

Greek letter : α,β-constants; γ-dummy variable; δ-biofilm thickness μ-dynamic viscosity; ξ-reduced substrate concentration.

Subscripts : O-lumped (k); superficial velocity (v), viable and dead (x), inlet (S,ξ) g-geometric; i-interfacial; L-liquid; S-Substrate; V-Volume; x-viable cells; X_0-dead cells.

In an expanded bed, the particles remain in stationary contact whist in a filmdized bed the particles are in free motion. When the liquid is passed upward through an unrestrained bed of particles, the bed will initially expand slightly to take up a loose packed arrangement. If the flow is increased, the pressure drop across the bed increases as shown by the line-OA in Fig. 8.9. Eventually the pressure drop equals the gravity (corrected for buoyancy in the liquid) on the particles and the grains begin to move. This is point 'A' in the figure. During this period, the porosity increases and the pressure drops more slowly than before due to the net effect of increased porosity and velocity. When point 'B' is reached, the bed is in extremely loose condition with grains still in contact. Between points 'A' and 'B' the bed is unstable, the particles begin to losse contact and then adjust their position to present as little resistance to the flow as possible.

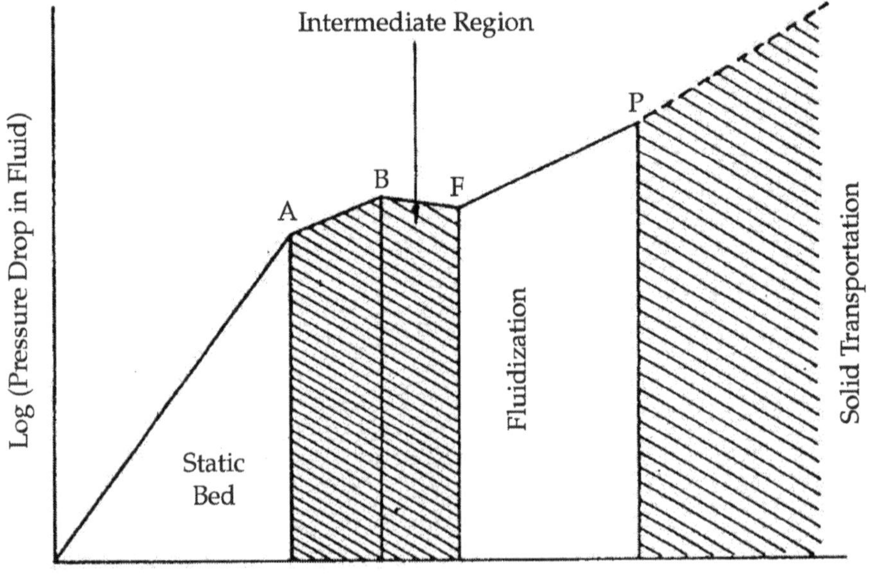

Fig. 8.9 : Pressure Drops in Fluidized Bed

As the velocity is further increased, the grains separate and time fluidization begins. This is the point, 'F' on the figure. By the time this point is reached, all particles are in motion and beyond that point, the bed continues to expand as the velocity is increased and maintains a uniform character. As the fluid velocity is increased further, the porosity increases, the bed of solids expands and eventually at point 'P' on the figure, all of the particles have been entrained in the fluid. The porosity approaches one and the bed ceases to exist. From this point on, there exists the simultaneous flow of two phases. At point 'P', the superficial velocity is approximately equal to terminal settling velocity of the particles.

The system as applied to wastewater treatment, consists of inert particles in a column which expanded with upward flow of waste through the column. The inert particles act as support surface for the growth of attached organisms. Fluidization allows the entire surface area of each particle to become available for biological attachment and subsequent reaction. The biomass hold up is achieved by allowing natural mechanisms of flocculation and adhesion to take place in a low shear environment and then to remove excess growth by particle/ particle or particle wall contact. In anaerobic systems, both expanded and fluidized beds operate at less than full fluidization.

8.4.1 Wang, Suidan and Rittman Model[53]

This model incorporates the concepts of liquid layer mass transport, biofilm molecular diffusion and Monod kinetics. That is, the substrate is transported, by molecular diffusion the bulk liquid to biofilm through a liquid layer of thickness (L). Within the biofilm, substrate utilization by micro-organisms and molecular diffusion occur simultaneously. The following assumption are made in developing the single stage substrate conversion biofilm model fluidized bed reactor.

- The biofilm is assumed to be attached to a flat plate of infinite length and width. Scanning electro micro-graphs of bioactive activated carbon media with drawn from anaerobic reactor revealed that the biofilm thicknesses are extremely thin, thus curvature effects were small, and the carbon surface is modelled as though it is a small flat plate.
- The biofilm has a uniform thickness (L_f) and a uniform cell density (X_{af}), that is assumed constant with depth.
- Substrate concentration within biofilm changes only in Z-direction, which is normal to the surface of the biofilm.
- The liquid and solids phases within reactor are completely mixed.
- The rate of reaction is limited by a single substrate.

The steady state material balance over the reactor volume of the completely mixed fluidized bed anaerobic process can be expressed as :

$$J_a = Q(S_{ao} - S_{ab}) = \frac{D_{aw}}{L} A(S_{ab} - S_{as}) \quad ...(8.199)$$

Where, J, Q, S_{ao}, S_{ab}, D_{aw}, L, A and S_{as} refer to acetate flux (mg COD/ cm^2.d), feed flow rate (cm^3/d), influent acetate concentration (mg COD/cm^3), acetate concentration is bulk liquid (mg COD/cm^3), molecular diffusivity of acetate in water (cm^2/d), liquid layer thickness (cm), total biofilm area normal to Z-direction (cm^2) and acetate concentration at the biofilm liquid interface (mg COD/cm^3) respectively.

Incorporating the Monod expression into a steady substrate material balance across the biofilm gives a second order differential equation, *viz.* :

$$\frac{d^2S_{af}}{dz^2} = \frac{k_a S_{af} X_{af}}{D_{af}(K_{sa} + S_{af})} \quad ...(8.200)$$

Where, S_{af} k_a, K_{sa}, X_{af} and D_{af} refer to acetate concentration within biofilm, maximum utilization rate of acetate, Monod half velocity coefficient, density of methanogens within biofilm and molecular diffusivity of acetate in biofilm respectively.

Two boundary conditions can be obtained at the biofilm-liquid interface, *i.e.*, at z = 0

$$S_{af} = S_{sa} \quad ...(8.201)$$

$$\frac{dS_{af}}{dz} \text{ at } z = 0 = \frac{-J_a}{D_{af}} \qquad ...(8.202)$$

Since the depth of active biomass is limited by substrate transport, the thickness of the active biofilm is given by the value of the film depth parameter, z, when the substrate concentration gradient becomes zero; *i.e.*

$$L_f = z \text{ at} \frac{dS_{af}}{dz} = 0 \qquad ...(8.203)$$

By defining two normalizing parameters, *viz.* :

$$S^*_{af} = \frac{S_{af}}{K_{sa}} \text{ and} \qquad ...(8.204)$$

$$Z^* = Z\left(\frac{k_a X_{af}}{D_{af} K_{sa}}\right)^{0.5} = Z\Phi_a \qquad ...(8.205)$$

Equations 8.200–8.204 can be transformed to dimensionless form :

$$\frac{d^2 S^*_{af}}{dz^{*2}} = \frac{S^*_{af}}{1 + S^*_{af}} \qquad ...(8.206)$$

$$S^*_{af} = S^*_{as} = \frac{S_{as}}{K_{as}} \quad \text{at } z^* = 0 \qquad ...(8.207)$$

$$\frac{dS^*_{af}}{dz^*} \text{ at } Z^* = 0 = -J^*_a = -\left(\frac{J_a \Phi_s}{k_a X_{af}}\right)^{0.5} \text{ at } z^* = 0 \qquad ...(8.208)$$

The dimentionless film thickness (L^*_f) is given by

$$L^*_f = Z^* \text{ at} \frac{dS^*_{af}}{dz^*} = 0 \qquad ...(8.209)$$

in which $L^*_f = L_f \, \Phi_a$

Equation 8.206 is a second order, non-linear ordinary differential equation that has no known analytical solution. Consequently, the single stage substrate conversion biofilm mathematical model defined by equations 8.206 to 8.208 can be solved numerically using a Symmetric Quardratures Technique. The bulk substrate concentration (S_{ab}) can be taken as the log mean of the COD of the influent and influent acetate concentration and is given as :

$$S_{ab} = \frac{S_{af} - S_{ae}}{\ln (S_{af} / S_{ae})} \qquad ...(8.211)$$

Where, S_{af} and S_{ae} refer to inlet acetate concentration and effluent acetate concentration from the reactor. S_{af} can be calculated as :

$$S_{af} = \frac{QS_{ao} + Q_R S_{ao}}{Q + Q_R} \quad \text{...(8.212)}$$

Where, Q and Q_R refer to feed flow rate and recycle flow rate respectively.

The solution to equation 8.206 is started at the biofilm-liquid interface, $Z^* = 0$ and is terminated when the substrate gradient changes sign (or $dS^*_{af}/dz^* = 0$). The value of Z^* at termination of the solution give L^*_f. The total rate of substrate conversion within biofilm is equal to J_a and is given by :

$$J_a = \int_0^{L_f} \frac{k_a S_{af} X_{af}}{(K_{as} + S_{af})} dz \quad \text{...(8.213)}$$

Stated in dimentionless form this becomes

$$J_a^* = \int_0^{L_f} \frac{S^*_{af}}{(1 + S^*_{af})} dz^* \quad \text{...(8.214)}$$

Equation 8.214 can be solved numerically as part of the model solution and its value is compared to the boundary value J_a, which is determined experimentally. This comparison is used to assess convergence and solution stability. If the calculated J_a did not approximate the measured J_a, the step size used in the integration is varied (normally reduced) until convergence is obtained.

Use of the single stage substrate conversion model requires as input the following parameters :

- Liquid layer thickness (L)

$$L = \frac{dp}{RJ_D S_C^{1/3}} \quad \text{...(8.215)}$$

Where, R, S_c, J_D, v, μ, ρ, ε and dp refer to the Reynolds number (= ρ v dp/μ), Schmidt number (= μ/ρ D_{aw}), the generalized Colburn, J, factor for mass transfer (= $0.010 + 0.863/R^{0.58} – 0.483/\rho$), superficial or empty bed velocity through the reactor, absolute viscosity of the liquid, density of liquid, bed porosity and diameter of the spherical particles respectively.

- Molecular Diffusivity of Acetate in Water (D_{aw})

This uses the Wilke and Chang Model[54], *viz.*,

$$D_{aw} = \frac{(7.4 \times 10^{-8})(\Phi M_B)^{0.5} T}{\mu V_a^{0.6}} \qquad ...(8.216)$$

Where M_B, T, μ, V_a, and Φ refer to molecular weight, temperature (°K), solution viscosity, solute molal volume at normal boiling point and association factor for solvent (= 2.6 for water as the solvent) respectively.

- Molecular Diffusivity of Acetate in Biofilm (D_{af})

The molecular diffusivity of acetate within biofilm (D_{af}) is less than D_{aw}, because a significant part of the biofilm is made of bacteria and their extracellular polymers. A ratio of D_{af}/D_{aw} = 0.8 was found by Williamson and McCarty[55].

A fully penetrated biofilm is a special case of biofilm kinetics for which the substrate concentration does not vary with the depth of the biofilm. Because substrate concentration is not reduced by diffusion in the biofilm, the flux is defined simply by

$$J_a = \frac{k_a S_{as}}{S_{sa} + K_{sa}} X_{af} L_f \qquad ...(8.216a)$$

S_{as} is the acetate concentration at the biofilm-liquid interface and is constant with the depth of biofilm. This equation can be rearranged as :

$$X_{af} L_f = \frac{J_a (S_{as} + K_{sa})}{k_a S_{as}} \qquad ...(8.217)$$

If external mass transport resistance is not significant, S_{as} can be approximated by S_{ab}. Equation 8.216 can also be arranged as :

$$\frac{J_a}{S_{af} L_f} = u = \frac{k_a S_{ab}}{S_{ab} + K_{sa}} \qquad ...(8.218)$$

The left hand side of this equation represents the acetate utilization per unit weight biomass (u).

8.4.2 Shieh Model[56]

Virtually all the bimass in an FFBR is a part of biofilm, substrate must be transported into the biofilm before they can be metabolized. The gelatinuous biofilm tends to retard substrate transport and thus causes the substrate concentration surrounding the micro-organism to be less than bulk liquid. Because the rates of microbial processes are determined by the substrate concentration surrounding the micro-organisms, it is essential that combined effects of microbial rate processes and physical mass transport phenomena are considered.

Another important consideration in modelling fluidized bed reactors (FFBR) is the mechanics of fluidization. For a given set of operating conditions, as analysis of the fluidization mechanics within FFBR yields two critical parameters, *viz.* :

- The equilibrium biofilm thickness.
- The bed porosity.

This information can, in turn, be used to calculate biomass concentration. The approach is quite similar to that being commonly employed for heterogeneous catalytic processes.

8.4.2.1 Effectiveness Factor–Mass Transfer Limitations

Substrate conversion in a heterogenous bioreactor such as an FBBR can be described by the following steps[57] :

- Transport of substrate from bulk liquid to the liquid-biofilm interface (external mass transfer).
- Transport of substrate within the biofilm (internal mass transfer).
- Substrate conversion reactions within biofilm.

Steps 2 and 3 take place simultaneously and thus neither can be said to control, while step-1 occurs in series with steps 2 and 3. For intrinsic reaction rate with positive dependence on substrate concentration (first order, Monod, etc.) the gradients established by steps 1 and 2 decrease the observed reaction rate by decreasing local (intra biofilm) substrate concentration. For intrinsic zero order kinetics, steps 1 and 2 can reduce the observed reaction rate by limiting substrate penetration into the biofilm.

8.4.2.1.1 External Mass Transfer

The significance of external mass transfer in FBBR can be assessed using a correlation developed for fluidized beds[58]

$$k_e = \left(\frac{0.81}{\varepsilon}\right)\left(\frac{D^{4/3}U\rho_1}{\mu^{1/3}dp}\right)^{1/2} \qquad \text{...(8.219)}$$

Where k_e, ε, D, U, ρ_1, μ and dP refer to external mass transfer coefficient, bed porosity, diffusivity of substrate in the bulk liquid, superficial upflow velocity, dynamic viscosity of liquid and bioparticle diameter respectively.

8.4.2.1.2 Effectiveness Factor Expressions

The effect of mass transfer on the intrinsic reaction rate is commonly quantified by an effectiveness factor[57]. Under the assumption of negligible external mass transfer, the effectiveness factor (η) is defined as

$$\eta = \frac{\text{observed reaction rate}}{\text{intrinsic reaction rate at bulk liquid condition}} \qquad \text{...(8.220)}$$

Mathematical expression for a bioparticle effectiveness factor are developed subject to the following constraints :

- Homogeneous biofilm of uniform thickness.
- Spherical media of uniform size.
- Negligible external mass transfer resistance.
- Internal mass transfer described by Fick's first law.
- Single limiting soluble substrate.
- Steady state conditions.

The bioparticle continuity equation for limiting substrate is given by :

$$\frac{D_e}{\gamma^2}\frac{d}{d\gamma}\left(\gamma^2\frac{dS}{d\gamma}\right)=R \quad ...(8.221)$$

Where D_e, γ, S and R refer to the effective diffusivity of substrate in the biofilm, radial distance measured from the bioparticle center, intrabiofilm substrate concentration and the observed substrate conversion rate per unit biofilm respectively.

For intrinsic zero order kinetics, the following reaction rate and boundary conditions apply, Fig. 8.10, *viz.*:

$$R = \rho k_o \quad ...(8.222)$$

$$S = S_b \qquad \text{at } \gamma = \gamma_p \quad ...(8.223)$$

$$\frac{dS}{d\gamma}=S=0 \ \text{ at } \gamma=\gamma_c \quad ...(8.224)$$

Where ρ, k_o, S_b, γ_p and γ_c refer to biofilm dry density, intrinsic zero order rate constant, bulk-liquid substrate concentration, bioparticle radius and substrate penetration depth respectively.

Integration of equations 8.221 and 8.222 subject to above boundary conditions yields

$$\left(\frac{\gamma_c^{\,3}}{\gamma_p}\right)-0.5\left(\frac{\gamma_c}{\gamma_p}\right)^2+\left(\frac{1}{2}-\frac{3}{\Phi^2_{\ o}}\right)=0 \quad ...(8.225)$$

Where $\Phi_o=\gamma_p\left(\frac{\rho k_o}{D_c S_b}\right)^{1/2}$, is the conventional zero order Thiele modulus.

The effectiveness factor for the intrinsic zero order reaction (η_o) is the ratio of biofilm volume with substrate concentration greater than zero to the total biofilm volume[59], *viz.*:

$$\eta_0 = \frac{1-(\gamma c/\gamma p)^3}{1-(\gamma m/\gamma p)^3} \qquad ...(8.226)$$

The functional relationship can be drawn by plotting η_0 versus ϕ_0 for various $(\gamma m/\gamma p)$. The functional relationship can be simplified by replacing the bioparticle radius of the conventional zero order Thiele modulus with a characteristic radius ($\bar{\gamma}$), defined as :

$$\bar{\gamma} = \frac{\text{biofilm volume}}{\text{biofilm exterior surface area}} \qquad ...(8.227)$$

$$\bar{\gamma} = \frac{\gamma p^3 - \gamma m^3}{3\gamma p^2} \qquad ...(8.228)$$

A modified zero order Thiele modulus (ϕ_{0m}) in then defined as :

$$\phi_{0m} = \bar{\gamma}\left(\frac{\rho k_o}{D_e S_b}\right)^{0.5} \qquad ...(8.229)$$

Numerical solution of equations 8.221, 8.227 and 8.229 yields a linear log-log relationship and is well described by the following empirical equation :

$$\eta_0 = 1.2712\ \phi^{-1.0}_{0m}$$

$$= \frac{3.814}{\left[1-(\gamma m/\gamma p)^3\right]}\left(\frac{D_0}{\rho k_0 \gamma_p^{\ 2}}\right)^{0.5} S_b^{0.5} \qquad ...(8.230)$$

An explicit relationship is thus established for the intrinsic zero order reaction in partially penetrated bioparticles which shows the effectiveness factor to be propositional to the bulk-liquid substrate concentration to the 0.50 power. If $\eta_0 = 1.0$, then $\phi_{0m} = 1.27$, the value at which transition occurs from full to partial substrate penetration of the biofilm. For $\phi_{0m} \leq 1.27$, the whole biofilm is active and intrinsic zero-order kinetics are observed. For $\phi_{0m} \geq 1.27$, the inner portion of the biofilm is inactive and the observed reaction rate is proportional to $S_b^{\ 0.50}$.

The intrinsic first order reaction rate and boundary conditions for equation 8.221 are as follows :

$$R = pk, S \qquad ...(8.231)$$

$$S = S_b \text{ at } \gamma = \gamma_p \qquad ...(8.232)$$

$$\frac{dS}{d\gamma} = 0 \text{ at } \gamma = \gamma_m \qquad ...(8.233)$$

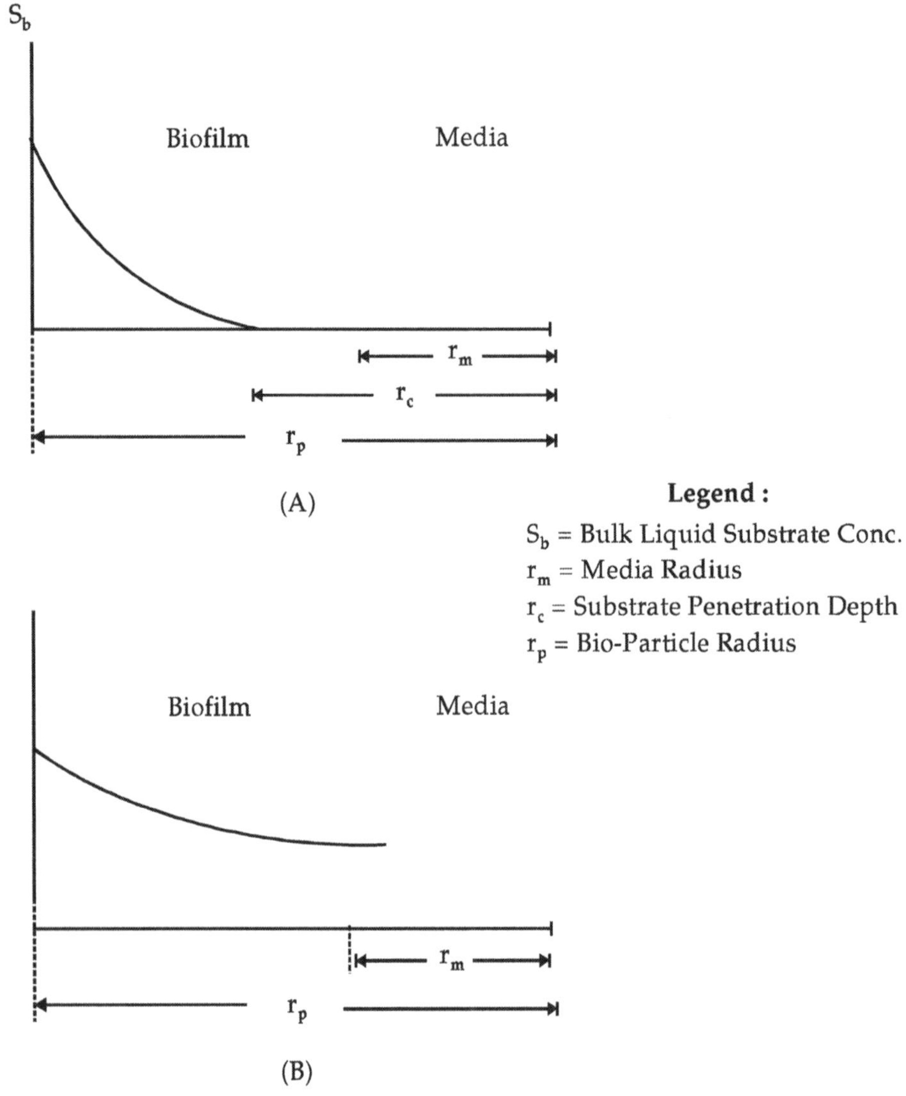

S_b

Biofilm Media

r_m

r_c

r_p

(A)

Legend :

S_b = Bulk Liquid Substrate Conc.

r_m = Media Radius

r_c = Substrate Penetration Depth

r_p = Bio-Particle Radius

Biofilm Media

r_m

r_p

(B)

Fig. 8.10 : Boundary Conditions for Intrinsic Zero Order Kinetics (A) and Boundary Conditions for Intrinsic First Order Kinetics (B).

Where k_1 refers to the intrinsic first order rate constant. Integration of equations 8.221 and 8.231 subject to boundary conditions yields :

$$S = \frac{S_b \gamma_p}{\gamma}\left[\frac{(\alpha\gamma_m + 1)\exp\{\alpha(\gamma - \gamma_m)\} + (\alpha\gamma_m - 1)\exp\{-\alpha(\gamma - \gamma_m)\}}{(\alpha\gamma_m + 1)\exp(\alpha\delta) + (\alpha\gamma_m - 1)\exp(-\alpha\delta)}\right] \qquad ...(8.234)$$

Where $\alpha = (Pk_1/D_e)^{0.5}$; and $\delta = (\gamma_p - \gamma_m)$, the biofilm thickness. Under steady state conditions, the observed substrate conversion rate by a bioparticle is equal to the mass transfer rate of substrate across the liquid-biofilm interface, ($W\gamma$ at γ_p), *i.e.*

$$W\gamma \text{ at } \gamma = \gamma_p = -4\Pi\, \gamma^2{}_p D_e \frac{dS}{dr} \text{ at } \gamma = \gamma_p$$

$$W\gamma \text{ at } \gamma = \gamma_p = -4\Pi\, S_b \gamma_p D_p$$

$$\left\{\frac{(\phi_1 - 1)[1 + \phi_1(\gamma_m / \gamma_p)\exp(\alpha\delta)] + (\phi_1 + 1)[1 - \phi_1(\gamma_m / \gamma_p)\exp(-\alpha\delta)]}{[\phi_1(\gamma_m / \gamma_p + 1)\exp(\alpha\delta)] + [\phi_1(\gamma_m / \gamma_p - 1)\exp(-\alpha\delta)]}\right\} \quad ..(8.235)$$

Where, $\phi_1 = \alpha\gamma_p = \gamma_p\,(\gamma k_1/D_e)^{0.5}$, is the conventional first order Thiele modulus. The bioparticle first order effectiveness factor (η_1) is defined as :

$$\eta_1 = \frac{W\gamma \text{ at } \gamma = \gamma_p}{-4/3\pi(\gamma_p{}^3 - \gamma_m{}^3)\rho k_1 S_b}$$

$$= \frac{\phi_1{}^2}{3[1 - (\gamma_m / \gamma_p)^3]}$$

$$\left\{\frac{(\phi_1 - 1)[1 + \phi_1(\gamma_m / \gamma_p)]\exp(\alpha\delta) + (\phi_1 + 1)[1 - \phi_1(\gamma_m / \gamma_p)]\exp(-\alpha\delta)}{[\phi_1(\gamma_m / \gamma_p + 1)]\exp(-\alpha\delta) + [\phi_1(\gamma_m / \gamma_p) - 1]\exp(\alpha\delta)}\right\} \quad ...(8.236)$$

The functional relationship between ϕ_1 and ϕ_2 can be made for various (γ_m/γ_p). Following the approach used previously to obtain a single explicit expression for the effectiveness factor, a first order Thiele Modulus (ϕ_{1m}) is defined in terms of the characteristics radius ($\bar{\gamma}$) as follows :

$$\phi_{1m} = (\gamma/\gamma_p)\phi_1 \quad ...(8.237)$$

Numerical solution of equations 8.236 and 8.237 for different (γ_m/γ_p) values is well described by first order, non-spherical form advanced by Aris[60] for homogenous reaction media :

$$\eta_1 = \frac{\coth(3\phi_{1m})}{\phi_{1m}} - \frac{1}{3\phi_{1m}{}^2} \quad ... (8.238)$$

If $\eta_1 = 1.0$, $\phi_{1m} = 0.11$, value at which transition occurs from observable intrinsic kinetics to a regime in which the observed reaction rate is limited by internal mass transfer.

8.4.2.2 Fluidization Mechanics

The main difference between conventional biofilm reactors and FFBRs is the free movement of bioparticles in the latter. In addition, the growth of biofilm changes the overall density of the bioparticles and, therefore, the expansion of the fluidized bed. Thus, the number and size of bio particles (and therefore, biofilm) per unit fluidized bed provide two critical pieces of information for the estimation FFBR biomass concentration. It is appropriate to define the fluidization mechanic in terms of measurable parameters. This provides the basis for calculating equilibrium biofilm thickness (δ), bed porosity (ε) and biomass concentration (X) which correspond to a given set of operating conditions[61-63].

There are several solid-liquid fluidization correlation, which link particle concentration in the fluidized state to the physical characteristics of a fluidized bed reactor[64-67]. Among them the Richardson-Zaki correlation is widely used, *i.e.,* :

$$U/Ut = \varepsilon^n \qquad ...(8.239)$$

Where, Ut and n refer to bio particle terminal settling velocity and expansion index respectively.

For smooth spherical particle, the bioparticle terminal settling velocity may be calculated using Newton's law:

$$U_t = \left[\frac{0.75(\rho_a - \rho_1)g\, d_p}{C_D \rho_1}\right]^{1/2} \qquad ...(8.240)$$

Where ρ_a, g and C_D refer to bioparticle density, gravitational acceleration and Newton's drag coefficient respectively. The relationship for drag coefficient and expansion index coefficient is defined by Mulcahy and LaMotta[58] and are given as follows :

$$C_D = 36.67\ N_{Ret}{}^{0.67};\ 40 < N_{Re} < 90 \qquad ...(8.241)$$

$$\eta = 10.35\ N_{Ret}{}^{0.18};\ 40 < N_{Re} < 90 \qquad ...(8.242)$$

Where, $N_{Ret} = [d_p \rho_1 U_t/\mu]$ defines the terminal Reynold number

The bio particle density (ρ_a) is :

$$\rho_a = \rho = (\gamma_m/\gamma_p)^3 + [\rho/1 - \rho]\ [1(\gamma_m/\gamma_p)^3 \qquad ...(8.243)$$

Where, ρ_m and ρ refer to media density and biofilm moisture content.

Using equations 8.239 to 8.243, the biomass concentration in a FFBR can be calculated as :

$$X = \rho\ (1-\varepsilon)\ [1 - (\gamma_m/\gamma_p)^3] = \rho V_m/AH_B\ [(\gamma_p/\gamma_m)^3-1] \qquad ...(8.243a)$$

Where, X, V_m, A and H_B refer to biomass concentration, media volume, cross-sectional area of the reactor and expanded bed height respectively.

Prediction of biomass concentration is a FFBR utilizing the correlation described above can be facilitated via the iterative procedure. For specific application, an estimated value of biofilm thickness (δ) and relevant information (U, A, V_m, etc.) are substituted into the fluidization correlation step wise. From these correlations, an estimated expanded bed height (H_B) is calculated. If height equals H_B, the estimated biofilm thickness (δ) is the desired design and operating value and the corresponding biomass can be calculated. Otherwise, a new δ is chosen to calculated another H_B value.

8.4.2.3 Overall Rates of Substrate Conversion

Recycle of reactor effluent is sometimes employed in the operation of FBBR to ensure uniform fluidization and an adequate loading rate, the overall rate expression for substrate conversion are strongly affected by FBBR hydraulic characteristics. It has been experimentally demonstrated via tracer studies that FBBR can be treated as a plug flow reactor if recycle ratio is less than 2, otherwise, a completely mixed flow model must be used[69].

8.4.2.3.1 FFBR Under Plug Flow Conditions, Recycle Ratio Less Than Two

The material balance on bulk-liquid concentration across the FBBR yields :

$$U\frac{dS_b}{dz} + R_v = 0 \qquad ...(8.244)$$

with boundary conditions :

$$S_b = S_{bi} \quad \text{at } z = 0 \qquad ...(8.245)$$

Where R_v, S_{bi} and Z refer to observed substrate conversion per unit fluidized bed volume, influent substrate concentration at z = 0 and the axial position respectively. The reaction term (R_v) is normally defined as :

$$R_v = RX \qquad ...(8.246)$$

For intrinsic zero-order kinetics :

$$R_v = RX = \eta_0 k_0 X$$

$$= \frac{3.814}{[1-(\gamma_m/\gamma_p)^3]}\left(\frac{k_0 D_e}{\rho \gamma_p^2}\right)^{0.5} X\, S_b^{0.5} \qquad ...(8.247)$$

Integrating equation 8.244 subject to equation 8.245 and 8.247 yields

$$S_b^{0.5} = S_{bi}^{0.5} - K_0 Z \qquad ...(8.248)$$

Where,

$$K_0 = \frac{1.907}{[1-(\gamma_m/\gamma_p)^3]}\left(\frac{k_0 D_e}{\rho \gamma_p^2}\right)^{0.5} (X/U),$$

an observed zero order rate constant.

The substrate concentration profile through an intrinsic zero order FBBR limited by internal mass transfer is described by half order equation. The value of K_0 will vary, depending on the characteristics of both substrate and biofilm and on prevailing operating conditions.

For intrinsic first order kinetics :

$$R_x = \eta_1\ k_1\ S_b\ X$$

$$= \left\{ \frac{\cot h\left[3\bar{\gamma}\left(\frac{\rho k_1}{D_e}\right)\right]^{0.5}}{\bar{\gamma}\,\rho\left(\frac{\rho k_1}{D_e}\right)^{0.5}} - \frac{1}{\left(\frac{3\rho\bar{\gamma}k_1}{D_e}\right)^{0.5}} \right\} k_1 S_b X \qquad ...(8.249)$$

Integration of equation 8.244 subject to equations 8.245 and 8.249 yields

$$S_b = S_{bi}\ \exp\ (-K_1 z) \qquad ...(8.250)$$

Where,

$$K_1 = \frac{X}{\bar{\gamma}\,\rho U}\left\{(\rho k_1 D_e)^{0.5}\ \cot h\left[\left(3\bar{\gamma}\frac{\rho k_1}{D_e}\right)\right]^{0.5} - \frac{D_e}{3}\right\}, \qquad ...(8.251)$$

an observed first order rate constant.

Therefore, the substrate concentration profile through an intrinsic first order FFBR limited by internal mass transfer is described by an exponential equation. The value of K_1 will vary as a function of substrate and biofilm as well as existing operating conditions.

8.4.2.3.2 FFBR Under Completely Mixed Conditions, Recycle Ratio Greater Than Two

A mass balance across FFBR an bulk liquid concentration yields :

$$\frac{S_{bi} - S_b}{\theta} = RX \qquad ...(8.252)$$

Where, $\theta = H_B/U$; the hydraulic retention time

For intrinsic zero-order kinetics :

$$\frac{S_{bi} - S_b}{\theta} = \frac{3.814}{\left[1-(\gamma_m/\gamma_p)^3\right]}\left(\frac{K_o D_e}{p\gamma_p{}^2}\right)^{0.5} X\,S_b{}^{0.5} \qquad ...(8.253)$$

Alternately,

$$S_b = S_{bi} + 2\,(K_0H_B)^2 - 2\,(K_0H_B)\,S_{bi}^{0.5} + (K_0H_B)^2 \qquad ...(8.254)$$

For intrinsic first order kinetics :

$$\frac{S_{bi} - S_b}{\theta} = \left\{ \frac{\cot h\left[3\bar{\gamma}(\rho k_1 / D_e)^{0.5}\right]}{\bar{\gamma}(\rho k_1/D_e)^{0.5}} - \frac{1}{(3\bar{\gamma}\rho k_1 / D_e)} \right\} k_1 S_b X \qquad ...(8.255)$$

Alternately,

$$S_b = \frac{S_{bi}}{(1 + K_1 H_B)} \qquad ...(8.256)$$

8.4.2.4 Limiting Bulk-Liquid Concentration

The effectiveness factor expressions defined above provide a convenient way of examining the effects of bulk-liquid substrate concentration on the observed substrate conversion rate of FFBR. Effectiveness factor expressions shown in equations 8.230 and 8.238 indicate that the observed substrate conversion rate is only effected by the bulk liquid substrate concentration in an intrinsic zero order FFBR. It has been observed that the limiting bulk-liquid substrate concentration which induces significant internal mass transfer effects in an intrinsic zero-order FFBR (η_0 = 0.90) is directly proportional to biofilm thickness. Higher bulk-liquid substrate concentration is required to penetrate into thick biofilm to ensure, that a majority of the biofilm is not substrate starved.

8.4.2.5 Critical Biofilm Thickness

A parameter which is indirectly controllable is the operation of an FFBR is the equilibrium upflow velocity. For a given medium, equilibrium biofilm thickness is dependent on superficial upflow velocity, expanded bed height and media volume. Biofilm thickness is in turn, the single most important parameter affecting biomass effectiveness and thus the overall performance of an FFBR[58].

Excessive accumulation of biofilm on the fluidized medium leads to a greater bed height increase which reduces the bed stability against hydraulic load variations. In extreme cases, wash-out off the bio-particles from the reactor could occur. It has been observed that the nature of curve between biomass concentration (X) and biofilm thickness (δ) is by and large quite close a parabola. The parabolic shape of this curve indicates that, to the left of the maximum, the rate of biomass increase due to bio film accumulation is faster than the rate of biomass decrease due to reduction in bioparticle per unit fluidized volume caused by bed expansion. To the right, opposite holds. If the objective is simply to maximize total biomass concentration, then the FFBR should be operated to control biofilm thickness at approximately 100 microns. Because of internal mass transfer limitations, however, this simplistic approach may not optimize overall

FFBR performance. The examine this possibility, it is necessary to determine the effect of biofilm thickness on the biofilm effectiveness factor.

A concept which helps in understanding FFBR performance characteristics is that of effective biomass concentration (X_a). The effective biomass concentration is less than X because of internal mass transfer limitations :

$$X_a = \eta X \qquad ...(8.257)$$

The effective biomass concentration is, in turn, related to the overall reaction term R_v as follows :

$$R_v = k_0 X_a \text{ (Zero-order kinetics) or} \qquad ...(8.258)$$

$$R_v = k_1 X_a S_b \text{ (first order kinetics)} \qquad ...(8.259)$$

Therefore, the observed substrate conversion rate at any point in FFBR is directly proportional to the effective biomass concentration at the point. Maximization of effective biomass will therefore maximize the observed substrate conversion rate in FFBR.

8.4.2.6 Critical Media Size and Density

The effect of medium characteristics on FFBR bed expansion is very important. Too small media size cause excessive bed expansion and a decrease in biomass concentration and thus in the rate of substrate removal. Too large media size decrease the available surface areas and thus the biomass concentration. FFBR bed stability to variations in superficial upflow velocity increase with increasing medium size and density. This is extremely important with regard to flow equalization requirements for a system utilising FFBR. Nevertheless, any such advantages must be weighed against the higher energy requirements for fluidization of dense and/or large media.

The effect of varying media density shows that increasing media density enhances the substrate removal rate. It is to be expected that for a given upflow velocity, less bed expansion and, therefore, greater biomass will result when a denser media is used.

It may also be mentioned that for a given set of operating parameters there exists a bio film thickness that affords optimal removal rate. The fact that in a FFBR system the biofilm thickness will affect particle size, specific density and thus flotation properties renders findings compatible with packed bed observation that substrate removal rate is independent of biofilm thickness[68]. There is a relationship between biofilm dry density and biofilm thickness :

$$\rho = 65 \text{ mg/mL}; \; 0 \le \delta < 300 \text{ microns}$$

$$\rho = 97 - 0.106\delta \; /300 < \delta \le 630 \text{ microns}$$

$$\rho = 30 \text{ mg/mL}; \; \delta > 630 \text{ microns}$$

8.4.2.7 Parameter Estimation

The intrinsic constants K_0 and D_e can be determined experimentally by a series of runs in a single FFBR or preferably by parallel runs in a series of FFBRs. In either case, the quantity to be varied from run to run is the equilibrium biofilm thickness. This is accomplished by varying media volume in FFBRs while maintaining constant superficial upflow velocities and expanded bed heights. The objective is to achieve biofilms thin enough to allow the intrinsic rate constant (K_0) to be measured directly. Intrinsic kinetics (negligible mass transfer limitations) are observed when K_0 calculated by equations 8.248 or 8.253 does not increase with increasing biofilm thickness. The effective diffusivity (D_e) can then be determined for the rest of the experimental runs (those in which observed K_0 decreases with increasing biofilm thickness) by applying equation 8.248 with 8.253 with K_0 and k_0 known. Because the effective diffusivity is a constant, averaging can be used to obtain a representative D_e value[68].

As with the zero-order case, a series of FFBR experimental runs are performed at various biofilm thickness. Intrinsic kinetics is observed when no increase in quantities, k_1 and η_1, is observed with decreased biofilm thickness. Under these conditions η_1 is unity and k_1 can be calculated directly by equation 8.250 or 8.255. The remaining runs with greater biofilm thicknesses are used to determine the effective diffusivity. From the slope of either $\ln(S_b)$ versus lnz plot or $(S_{bt}-S_b)/\theta$ versus S_b plot, η, k_1 can be calculated. The corresponding η_1 values are obtained by factoring out k_1. The effective diffusivity obtained stepwise by first calculated φ_{1m} at a given η_1 by trial and error solution of equation 8.238. Then D_e is calculated at given φ_{1m} and k_1 by equation 8.237.

For each experimental run, before superficial upflow velocity variation is begun, it is necessary to establish the basis for calculation of bed porosity in terms of easily observable parameter, bed height. Be definition of bed porosity, the following may be written

$$\varepsilon = 1 - (V_s/H_B A) \qquad \text{...(8.260)}$$

Where V_s refers to the total bioparticle volume in the fluidized volume ($H_B A$) and is determined as follows :

At a given superficial upflow velocity, a small portion of bed materials is withdrawn from the reactor and the change in expanded bed height (ΔH_B) is noted. The number of bioparticles in the sample (ΔN) is calculated as :

$$\Delta N = (W_m)/\rho_m \, (\pi d_m^{-3}/6) \qquad \text{...(8.261)}$$

Where W_m and d_m refers to weight of media in the sample and average bioparticle diameter. The total volume of bioparticles remaining in the expanded bed could then be calculated as :

$$V_s = \frac{(H_{B-2})(\Delta N)(\Pi \bar{d}_p^{\,-3}/6)}{\Delta H_B} \qquad \text{...(8.262)}$$

Where, H_{B-2} refers to the new expanded bed height after the sample was withdrawn from the reactor.

8.4.2.8 Engineering Applications

The only variables that can be controlled by the design engineer are media characteristics (size and density) as well as column height and allowable expansion. These parameters will in tern, determine upflow velocity, porosity and biofilm thickness. Equations defined earlier can be rearranged to determine HRT (θ) :

$$\theta = \frac{S_i^{0.55} - S_o^{0.55}}{(0.6216)X\left[3\gamma_p^2 / \rho^{0.5}(\gamma_p^2 - \gamma_m^2)\right]^{0.9} k^{0.55} D_e^{0.45}} = \frac{H_B}{v} \qquad ...(8.263)$$

Where θ and S_o refer to HRT based on expanded bed and substrate concentration at $z = H_B$ respectively. The biomass concentration is given by :

$$X = \frac{\rho V_m}{AH_B}\left[\left(\frac{\gamma_p}{\gamma_m}\right)^3\right] - 1 \qquad ...(8.264)$$

$$V_m = \frac{XAH_B}{\rho\left[(\gamma_p / \gamma_m)^3 - 1\right]} \qquad ...(8.265)$$

or

$$= 0.6035\, Q\left[\frac{\gamma_m^3}{\gamma_p^{2.1}}\right]\left[\frac{S_i^{0.55} - S_0^{0.55}}{(\rho k)^{0.55} D_e^{0.45}}\right] \qquad ...(8.266)$$

$$V_a = \left(\frac{V_m}{4\Pi / 3\gamma_m^3} \cdot \frac{4}{3}\pi\gamma_p^3\right) = V_m\left(\frac{\gamma_p^3}{\gamma_m}\right) \qquad ...(8.267)$$

Total value of biofilm V_b is then

$$V_b = V_a - V_m \qquad ...(8.268)$$

Therefore, the appropriate dimension of the reactor can be determined as follows :

$$AH_B = \frac{\rho V_M}{X}\left[(\gamma_p / \gamma_m)^3 - 1\right] \qquad ...(8.269)$$

$$= V_m/V_b\ [(\gamma_p/\gamma_m)^3 - 1] \qquad ...(8.270)$$

The procedure can be repeated with a different biofilm thickness or different media size or both. The only uncertainty will be in the kinetic parameters, ρ, k and D_e. Until sufficient range of values is published in the literature, the safest procedure will be to determine them experimentally.

A system of pipe lateral is used in large FFBR to introduce wastewater downwards into a flow distribution device located at the reactor bottom, which consists of number of inverted conical shaped structures. The openings of orifice or nozzles of the lateral can be calculated as follows :

$$Q = 19.63\, C\, d_1^{\,2} (h)^{0.51} \left[\frac{1}{1 - (d_1 / d_2)^4} \right]^{0.5} ; \ \frac{d_1}{d_2} > 0.3 \qquad ...(8.272)$$

$$Q = 19.63\, C d_1^2 (h)^{0.5} ; \ \frac{d_1}{d_2} < 0.3 \qquad ...(8.273)$$

Where Q, d_1, h, d_2, and C refer to flow (gpm), diameter of orifice (inches), differential head at orifice (ft of liquid), diameter of lateral (inches) and discharge coefficient respectively.

The minimum fluidization velocity is calculated by equation developed by Wen and Yu[70] :

$$U_{mf} = \frac{0.5\mu}{\rho_1 \gamma_m} \left[(33.7)^2 + 0.326 \frac{\rho_1 \gamma_m^{\,3} (\rho_m - \rho_1) g}{\mu^2} \right]^{0.5} - 33.7 \qquad ...(8.273)$$

Ergun[71] equation is used to estimate the head loss during fluidization :

$$\frac{H}{L} = 150 \frac{\mu}{\rho g} \frac{(1-\varepsilon)^2}{\varepsilon^3} \left(\frac{S_v}{6} \right)^2 v + 1.75 \frac{(1-\varepsilon)}{\varepsilon} \left(\frac{S_v}{6} \right) - \frac{v^2}{g}$$

for N_{Re} from 1 to 2000 ...(8.274)

Where H, L, g, S_v, and N_{Re} refer to head loss in depth of bed, height of bed acceleration of gravity, specific area (= grain surface area per unit grain volume = 6/deq for spheres and 6/ψ deq for irregular grains, ψ – Sphericity, *i.e.*, ratio of surface area of an equal volume sphere to the actual surface area of grain), and Reynolds number (= deq $\upsilon\rho/\mu$; deq-grain diameter of sphere of equal volume, v-superficial velocity) respectively.

In tapered fluidized bed configuration, the cross-sectional area gradually increases from bottom to the top of the reactor. This configuration provides the flow patterns a minimal back mixing, especially at the feed entry point and prevents plugging by maintaining a high inlet velocity. Relatively stable flow through the reactor can be achieved when entry cross section is sufficiently small and expansion is gradual (an angle of few degrees). Since the fluid velocity decreases with reactor, the height of the column allows a wide range of flow rates without the loss of bed material. Therefore, the tapered fluidized bed can effectively operate a wide range of feed flow rates. This system can maintain a stable operation over a wide range of volumetric flow rates than the conventional fluidized bed. Mathematical models for this system are yet to be developed.

8.5 Fixed Film Rotating Bed

As the rotating contractor system is a relatively new concept in biological wastewater treatment, few models have been proposed to described the process.

8.5.1 Kornegay Model[72]

Depending upon the method of operation the carbonaceous waste removal in rotating biological contactors may be obtained by suspended growth as well as that attached to the media. Therefore, the formulation of a mathematical model for the system is an extension of both fixed film and slurry reactor kinetics.

Any substrate removal experienced in rotating biological contactor must be due to the suspended organisms occupying the liquid volume, V or the microbial film attached to the wetted surface area. The substrate balance across the reactor is then :

Accumulation = (Inflow) – (Outflow) – (Consumption by Attached Growth) – (Consumption by Suspended Growth)

$$V\left(\frac{dS}{dt}\right) = QS_o - QS_e - \frac{\mu_g}{Y_g}M_o - \frac{\mu_s}{Y_s}X_S V \qquad ...(8.275)$$

This is true when :

- complete mixing is achieved in the liquid volume;
- organisms decay is neglected;
- maintenance energy is not included in explicitly terms.

where μ_g, Y_g, μ_s, Y_s, M_o and X_s refer to specific growth rate of the fixed film, apparent yield coefficient of fixed film organism, specific growth rate of suspended organisms, apparent yield of the suspended organisms, active biomass of fixed film organisms and concentration of suspended organisms respectively.

The active mass of attached organism (M_0) is equal to the product of the wetted area (A), the active depth (d) and the unit mass of the fixed microbial film (X_f). Furthermore, it can be determined that the wetted area may be expressed as :

$$A = 2\, N\, \Pi\, (\gamma_0^2 - \gamma_u^2)$$

Where N, γ_0 and γ_u and refer to number of discs, total disc radius and unsubmerged disc radius respectively.

$$V\frac{dS}{dt} = QS_o - QS_e - \frac{\mu_g}{Y_g} 2\, \Pi\, N\left(\gamma_0^2 - \gamma_u^2\right)X_r d - \frac{\mu_s}{Y_a} X_a V \qquad ...(8.277)$$

Using Monod's equation, along with equation 8.277

$$V\frac{dS}{dt} = QS_o - QS_e - \frac{\mu_{max}}{Y_g}2\Pi\left(\gamma_0{}^2 - \gamma_u{}^2\right)X_f d\left(\frac{\mu_s}{K_g + S}\right) - V\frac{\mu_{max}}{Y}X_a\left(\frac{S}{K_a + S}\right) \quad ....(8.278)$$

Equation 8.278 is a general equation describing the dynamic performance of a completely mixed rotating biological contactor in which removal by suspended and attached growth is significant. At steady state, the equation expressing system performance is :

$$Q(S_o - S_e) = \frac{\mu_{max}}{Y_g}2\pi(\gamma_o{}^2 - \gamma_u{}^2)X_f d\left(\frac{S_1}{K_g + S_1}\right) + \left(\frac{\mu_{max}}{Y_a}\right)X_a V\frac{S_1}{K_a + S_1} \quad ...(8.279)$$

In most cases, rotating biological contactors are operated at very short retention times and suspended organisms are insignificant in comparison to the attached growth. Under these condition X_a would essentially equal to zero or contribution for conversion of substrate is negligible and equation 8.279 would becomes :

$$Q(S_o - S_e) = \frac{2\mu_{max}}{Y_g}\pi\, N\, X_f d\left(\gamma_o{}^2 - \gamma_u{}^2\right)\left(\frac{S_1}{K_g + S_1}\right) \quad ...(8.280)$$

When reactors are operated in series the total, R_T, is the sum of removal in each reactor or

$$R_T = R_1 + R_2 + — R_{n-1} + R_n \quad ...(8.281)$$

If the reactors are indentical, the substrate concentration, S is the only variable from reactor to reactor and the total removal in a series is then :

$$Q(S_o - S_n) = \frac{2\mu_{max}}{Y_g}\Pi\, N\, X_f d\,(\gamma_o{}^2 - \gamma_u{}^2)\sum_1^n\left(\frac{S}{K_g + S}\right) \quad ...(8.282)$$

It is apparent that equations 8.280 and 8.282 contain the same kinetic parameters that are consolidated into the area capacity constant P as has been done for fixed bed reactors :

$$P = \frac{\mu_{max}}{Y_g}X_f\, d \quad ...\ (8.283)$$

Substituting the value of P in equation 8.280

$$Q\ (S_o - S_u) = 2PN\Pi\ (\gamma_o{}^2 - \gamma_u{}^2)\ (S/K_g+S) \quad ...(8.284)$$

The equation for a multi-stage system is :

$$Q(S_o - S_e) = 2P\Pi N(\gamma_o^2 - \gamma_u^2)\sum_{1}^{n}(S/K_g + S) \qquad \text{...(8.285)}$$

Under extreme loading conditions $S/(K_g + S)$ approaches unity (very high loading conditions), equation 8.284 becomes :

$$Q(S_o - S_e) = 2P\Pi N(\gamma_o^2 - \gamma_u^2) \qquad \text{... (8.286)}$$

While the equation for n-series reactors become

$$Q(S_o - S_e) = 2nP\Pi N(\gamma_1^2 - \gamma_u^2) \qquad \text{....(8.287)}$$

If the loading is very light, then it would be anticipated that the substrate concentration would be very low, and

$$\frac{S}{K_g + S} \approx \frac{S}{K_g} \qquad \text{.... (8.288)}$$

At these low substrate concentration, equation 8.284 would be :

$$Q(S_o - S_o) = 2P\Pi N(\gamma_o^2 - \gamma_u^2)\, S/K_g \qquad \text{... (8.289)}$$

and the fraction of soluble organics remaining in the first reactor would be

$$\frac{S_e}{S_o} = \frac{1}{[1 + 2PN\Pi / Qk_g(\gamma_o^2 - \gamma_u^2)]} \qquad \text{....(8.290)}$$

The soluble organic fraction remaining in the nth reactor operating at very low substrate concentrations would be :

$$\frac{S_n}{S_o} = \left(\frac{S_1}{S_o}\right)\left(\frac{S_2}{S_1}\right)\cdots\cdots\left(\frac{S_{n-1}}{S_{n-2}}\right)\left(\frac{S_n}{S_{n-1}}\right) \text{or} \qquad \text{...(8.291)}$$

$$\frac{S_n}{S_o} = \frac{1}{[1 + (2P\Pi N / Qk_g)(\gamma_o^2 - \gamma_u^2)]_1} \times \frac{1}{[1 + (2P\Pi N / Qk_g)(\gamma_o^2 - \gamma_u^2)]} \text{ etc.} \qquad \text{...(8.292)}$$

For identical reactors and operating conditions, this would be

$$\frac{S_n}{S_o} = \frac{1}{[1 + (2P\Pi N / Qk_g)(\gamma_0^2 - \gamma_u^2)]^n} \qquad \text{...(8.293)}$$

The kinetic parameters for suspended and attached micro-organisms must be determined in separate tests for rotating biological contractors, in which both contribute to the substrate removal. In the first series of tests, the retention time should be decreased until washout occurs. Removal by suspended growth is then negligible compared to the removal by fixed film organisms and the steady state performance may be expressed by the equation.

$$Q(S_o - S_e) = 2P\Pi N(\gamma_o^2 - \gamma_u^2)(S/K_g + S) \qquad ...(8.294)$$

This equation can be rearranged as :

$$\frac{K_g}{P}\frac{1}{S_o} + \frac{1}{P} = \frac{2N\Pi(\gamma_0^2 - \gamma_u^2)}{Q(S_0 - S_e)} \qquad ...(8.295)$$

Which plots as a straight line with a slope of K_g/P and intercept $1/P$ when S_o^{-4} is plotted against the terms on the right hand side of equation (8.295). This provides a method of evaluating K_g and P for fixed film organisms. If desired, the active depth (d), unit weight of biological film (X_f) and the yield (Y) coefficient, may be measured and μ_{max} calculated from the area capacity constant (P). If the reactor is to be operated at low retention times then removal by suspended organisms should be insignificant and this would be the only data required for design. However, if the reactor is to operate at the retention times greater than the reciprocal of specific growth rate, the kinetic constants for suspended organisms must also be obtained. A rotating biological contactor containing both suspended and attached organisms may be described by the equation :

$$Q(S_o - S_e) = 2PN\Pi N(\gamma_o^2 - \gamma_u^2)\left(\frac{S_o}{K_s + S_e}\right) + \frac{\mu_{max}}{Y_s} X_a V\left(\frac{S_e}{K_s + S_e}\right) \qquad ..(8.296)$$

The substrate removal be suspended growth is the difference between the total removal and that due to the fixed-film organisms or

$$Q(S_o - S_e) = 2P\Pi N(\gamma_o^2 - \gamma_u^2)\left(\frac{S_o}{K_s + S_e}\right) = \frac{\mu_{max}}{Y_s} X_s V\left(\frac{S_o}{K_s + S_e}\right) \qquad ...(8.297)$$

This equation can be rearranged as :

$$\frac{X_s V}{Y_s [Q(S_o - S_e) - 2P\Pi N(\gamma_o^2 - \gamma_u^2(S_e / K_g + S)]} = \frac{K_g}{\mu_{max}}\frac{1}{S_e} + \frac{1}{\mu_{max}} \qquad ...(8.298)$$

Therefore, a plot of $1/S_e$, versus the terms on left hand side of equation 8.298 should produce a line with a slope of K_g/μ_{max} and intercept of $1/\mu_{max}$. This will provide the data required to describe the removal by the suspended growth.

8.5.2 Andreadakis and Cailas Model[72]

The material balance equation is given by equation 8.299.

$$V\,dS/dt = QS_o - QS - \gamma AdX \qquad ...(8.300)$$

Where V, Q, S_o, S, A, d, X and γ refer to reactor volume, flow rate, influent and effluent substrate concentration, media area, biofilm thickness and substrate removal rate respectively.

Further development of the model depends on the adoption of a suitable expression for substrate removal rate (γ). For first order kinetics ($\gamma = K_1S$) and assuming steady state expression equation 8.300 becomes :

$$E = \frac{k_1 X d}{(Q/A + k_1 X d)} \quad ...(8.301)$$

Where $E = (S_o - S)/S_e$

For steady state conditions, adoption of the Grau kinetic expression ($\gamma = K_2 S/S_o$), yields :

$$E = \frac{k_2 X d}{QS_o / A + k_2 X d} \quad ...(8.302)$$

Adoption of the Monod expression yields :

$$\frac{A}{QES_0} = \frac{K_s}{k_3 X d}\frac{1}{S} + \frac{1}{k_3 X d} \quad ...(8.303)$$

Expression 8.303 is very similar to the equation developed by Clark and others[73] and Kornegay and Andrews[72] and likewise is based an mass balance of substrate in complete mixing reactors with consumption of substrate determined by applying a saturation function adopting Clark's definitions of the area capacity constant (P) and removal coefficient (R):

$$P = k_3 X d \quad ...(8.304)$$

$$R = QES_o/A \quad ...(8.305)$$

Equation 8.303 can be written as :

$$\frac{1}{R} = \left(\frac{K_4}{P}\right)\frac{1}{S} + \frac{1}{P} \quad ...(8.306)$$

This is a generalised equation for RBCs and presents as method for determining k_3 and P for the attached biomass on the media.

Assuming, Xd = constant (C), equation 8.301 and 8.302 can be written as :

$$E = \frac{k_1 C}{Q/A + k_1 C} = \frac{C'}{Q/A + C'}, \quad ...(8.307)$$

where, $C' = k_1 C$

and $$E = \frac{k_2 C}{QS_o / A + k_2 C} = \frac{C''}{QS_o / A + C'} \quad ...(8.308)$$

where, $C'' = k_2C$

Equations 8.307 and 8.308 are in essence the mathematical expressions for a number of empirically derived design graphs which have been proposed and widely and satisfactorily used for design purposes[74].

It should be noted that in RBC no practical means exist to control the amount of biomass on the media. The amount and type of biomass that develops will be that best suited to treating the wastewater, therefore, control is neither necessary nor desirable.

If is assumed that the biomass produced due to substrate utilization is controlled by micro-organisms decay (k_d), the depth of biofilm can be determined from the following expression :

$$d = \frac{YEQS_0}{Ak_dX} \qquad ...(8.309)$$

However, except possibly in the case of very low loading, the thickness of the biofilm is controlled by rotation of the media which provides constant shear force and causes continual sloughing of culture, therefore, maintaining a more or less constant film thickness of the order of few millimeters. It has also been observed that growth of the biofilm in excess of 70-200 microns does not result in any significant increase in substrate utilization[72]. This is attributed to the importance of the active portion of the biomass with respect to removal in contrast to the total active and non active mass. The concept of active biofilm has been usually related to mass transfer limitations.

Regardless of the cause, the existence of an active biofilm responsible for substrate removal, which is not proportional to the total biofilm thickness, but more or less constant is a fairly established fact. Therefore, the adoption of constant value for the product (Xd) in above equations, leads to volumetric loading rates for RBC seems valid.

8.5.3. *Diffusion Kinetics Theory*[75]

Fig. 8.11 shows a conceptual biological fixed film which two possible substrate concentration profiles :

- The substrate fully penetrates the biofilm.
- The substrate generates only partial penetration.

As it will be seen later on, the only difference that penetration could make occurs when the intrinsic rate is pseudo zero-order with respect to the substrate.

The assumptions involved for the development of this model are :

- That bacteria are uniformly distributed within a biofilm.
- That biofilm thickness is uniform.

- That the biofilm is at steady state. That is the density of the biofilm does not change with any experimental run (or for any reasonable time).
- That the suspended solids in the bulk fluid can be neglected.
- That the biofilm is considered to be an equiaccessible surface, this has also been called as 'quasi-stationary' method by Frank-kamenetskii.[76]
- That there is a single substrate limiting growth.

8.5.3.1. Monod Kinetics

Material balance on substrate at steady state within a differential element of thickness (dy), one can write :

$$\begin{bmatrix}\text{Mass in by}\\ \text{diffusion}\end{bmatrix} - \begin{bmatrix}\text{Mass out by}\\ \text{diffusion}\end{bmatrix} - \begin{bmatrix}\text{Consumption by}\\ \text{reaction}\end{bmatrix} = 0$$

Where the only transport of material is by diffusion in the negative Y-direction.

$$-AD_e\frac{dS}{dy} - \left[-AD_e\left(\frac{dS}{dy} + \frac{d^2S}{dy^2}dy\right)\right] - \frac{k_{max}S}{K_a + S}\ .\ A = 0 \qquad \text{...(8.310)}$$

Where D_e and A refer to effective diffusivity coefficient of S and area of the biofilm normal to the direction of the diffusional flux. Using dimensionless analysis, the physical meaning of some parameters will surface :

$$Z = y/L \qquad \text{...(8.311)}$$

$$\bar{S} = S/S_g \qquad \text{...(8.312)}$$

$$\zeta = K_a/S_a \qquad \text{...(8.313)}$$

Substituting the above dimensionless set of numbers in equation 8.310 with some manipulations, the following relationship is obtained :

$$\frac{d^2\bar{S}}{dz^2} = \phi^2\frac{\bar{S}}{1+(\bar{S}/\xi)} \qquad \text{...(8.314)}$$

The required boundary conditions are

$$\bar{S} = 1 \quad \text{at } z = 1 \qquad \text{...(8.315)}$$

$$\frac{d\bar{S}}{dz} = 0 \ \text{at } z = 0 \qquad \text{...(8.316)}$$

Where, $\phi^2 = \left[\frac{k_{max}}{K_a}\right]\frac{L}{D_e}$ ...(8.317)

ϕ is called the Thiele modulus. This parameter has an important meaning. Physically the square of ϕ is the ratio of the kinetic-reaction of interphase diffusion rate[76]. In this case if $\phi > 1$, a diffusion limited regime prevails and if $\phi < 1$, a kinetic limited regime prevails.

ξ indicates the degree of deviation from Monod kinetics, if any. Because of its non-linearity, equation 8.260 with boundary conditions 8.315 and 8.316 does not have an explicit solution. A numerical solution using Runge Kutta[77,78] should be used.

It is important to determine the internal effectiveness factor (η_i). This is also known as phase-utilization factor and is defined as:

$$\eta_i = \frac{\text{actual Rate in Presence of Defussion}}{\text{theoretical ratio without internal diffusion resistance } (S = S_s)} \quad ...(8.318)$$

Mathematically,

$$\eta_i = \left(\frac{\xi+1}{\xi}\right)\frac{1}{\phi^2}\frac{d\bar{S}}{dz} \text{ at } z = 1 \quad — (8.319)$$

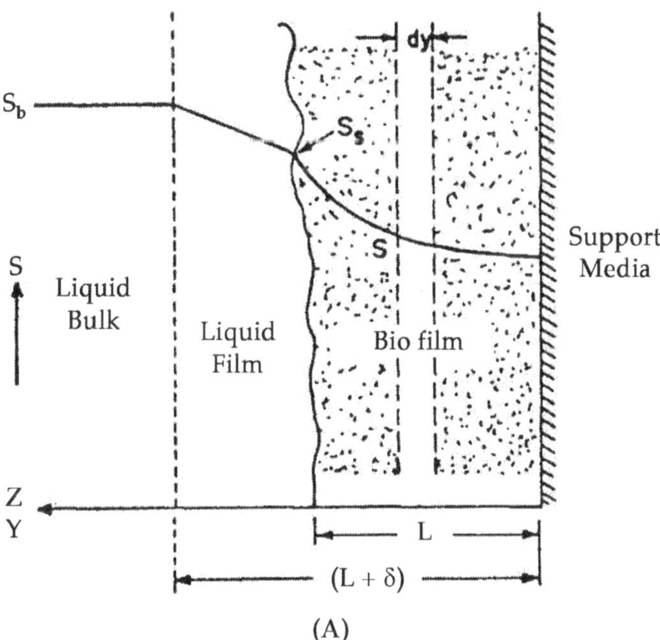

(A)

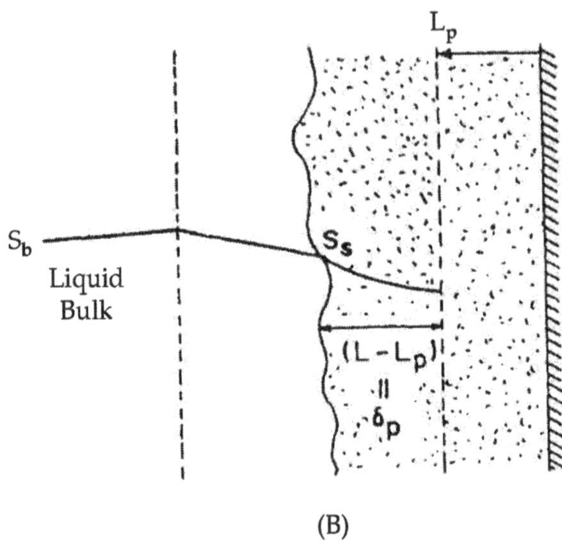

(B)

Fig. 8.11 : A Conceptual Representation of RBC Biofilm Showing the Concentration Profile for (A) Full and (B) Partial Penetration

Where, the substrate gradient evaluated at the biofilm interface can be found by a numerical approach, that is from the solution of equation 8.314 with two boundary conditions. S is a function of z, ϕ and ξ. So then $d\bar{S}/dz$ at $z = 1$ is a function of ϕ and ξ. Accordingly, η_i is a function of ϕ and ξ. Manipulation equations 8.317 and 8.318 the following is derived

$$\phi^2 = \frac{1}{\eta_i}\left[\frac{L}{D_e}\right]\left[\frac{1+\xi}{\xi}\right]\frac{R}{S_s} \qquad ...(8.320)$$

Where, R refer to observed rate.

Defining $\bar{\phi}$ as

$$\bar{\phi} = \eta_i\ \phi^2 \qquad ...(8.321)$$

Then from equation 8.320, it follows that :

$$\bar{\phi} = \frac{L}{D_e}\left(\frac{1+\xi}{\xi}\right)\frac{R}{S_s} \qquad ...(8.322)$$

Where ϕ is a modified Thiele modolus for this case. Ultimately, the functional relationship between η_i and ϕ or ϕ must be represented graphically to simplify the solution. It is seen that this functional relationship is implicity and not explicit. One method might be to plot η_i over ϕ with ξ as a parameter. Another method would be an asymptotic solution[79,80]. As defined by Salterfield[81] an asymptotic solution would be incorporating all intrinsic kinetic information into the function in such a way to cause all the asymptotes to superimpose, thereby bringing the family of curves closer together and replacing them by a single curve at high values of ϕ. However, this seeming simplification is purchased at the cost of complicating the modulus, and it becomes difficult to interpret the modulus in terms of physically reality. Bischoff[82] has presented a mathematical method (elliptic integrals of the first and second kind) for a general asymptotic solution for an adsorption reaction rate from (hetrogeneous mechanism) that leads to a definition of a general modulus somewhat different than the standard one. Paralleling the work of Bischoff[82], one can write a modified general Thiele modulus for the case of Monod kinetics as follows :

$$\phi_m = L\left(\frac{k_{max}/Ks}{2D_e}\right)^{0.5}\left(\frac{\xi^{-1}}{1+\xi^{-1}}\right)\left[\xi^{-1} - \ln(1+\xi^{-1})\right]^{0.5} \quad \text{.... (8.323)}$$

It is easy to plot ϕ_m versus η with ξ as parameter[83].

8.5.3.2 Pseudo First Order Kinetics

Depending on the relative values of Ks and S, Monod kinetics become pseudo first order if Ks >> S. Therefore, equation 8.314 becomes :

$$d^2\bar{S}/dz^2 = \phi^2\bar{S} \text{ and boundary conditions remain the same} \quad \text{...(8.324)}$$

Thiele modulus (ϕ) for the pseudo first order kinetics is given by

$$\phi^2 = \left[\frac{k_{max}}{K_s}\right]\frac{L}{D_e} \quad \text{...(8.325)}$$

Equation 8.324 is a second order differential equation, linear and homogenous. It possesses an explicity analytical solution with same boundary conditions. It can be shown that the solution[84,85] is:

$$\bar{S} = \frac{S}{S_o} = \frac{\cos h(\phi z)}{\cos h\,\phi} \quad \text{...(8.326)}$$

It gives the substrate concentration profile within a bio film of thickness L. It should be noted that at high ϕ values (internal diffusion regime), the film is not fully penetrated.

Keeping this inview it is possible to derive the phase utilization factor (internal effectiveness factor η_i) and is given by :

$$\eta_i = \frac{D_e}{L}\left[\frac{Ks}{k_{max}}\right]\frac{dS}{dz} \text{ at } z = 1 \quad \text{...(8.327)}$$

Substrate gradient should be evaluated at z = 1. This can be achieved by differentiating equation 8.326 and evaluating it at z = 1. Finally, it can be shown that :

$$\eta_i = \frac{\tan h\phi}{\phi} \qquad ...(8.328)$$

which is the required relationship. A plot of η_i versus ϕ can be drawn. It is seen that the drastic change in the slope occurs at $\phi = 1$. Hence, for $\phi < 1$, the reaction is kinetically controlled; while for $\phi > 1$, the reaction is interphase diffusion controlled. It can also be observed that at high ϕ values, η_i is small and hence the biofilm is not fully used because it is not fully penetrated. It is important to study some extreme conditions for large values of ϕ:

As ϕ tends to infinity in equation 8.328, tan h tends to unity and η_i tends to $1/\phi$, *i.e.* :

For large $\phi \rightarrow \eta_i \rightarrow \phi^{-1}$ ...(8.329)

Which is called an asymptotic solution.

The observed rate (R) is given by :

$$R = \frac{k_{max}}{K_s} S_s \, \eta_i \qquad ...(8.330)$$

For large ϕ values (>3 – 4), the observed rate becomes:

$$R = \frac{k_{max}}{K_s} S_s \, \phi^{-1} \qquad ...(8.331)$$

Substituting the value of φ form equation 8.325 into equation 8.331, the following is found :

$$R = ((k_{max}/K_s)\,(D_e/L))^{0.5}.\ S_s \qquad ...(8.332)$$

and setting

$$k_o = ((k_{max}/K_s)\,(D_e/L))^{0.5} \qquad ...(8.333)$$

Where k_o can be called the observed rate coefficient for this case ($\phi > 3 - 4$) of strong intraphase diffusion influence. Under such a case, R becomes:

$$R = k_o S_s \qquad ...\ (8.334)$$

where it can be seen that within this domain, the observed reaction rate is first over when time (intrinsic rate is likewise first order. It can be shown that for any intrinsic reaction rate order *n* that :

$$R = \frac{k_{max}}{K_s} S_s^{\,n} \, \eta_i;\ \text{and} \qquad ...\ (8.335)$$

$$\phi = L\left(\frac{k_{max}}{K_s}\frac{S_s^{\,n-1}}{D_e}\right)^{0.5} \qquad ...(8.336)$$

For large ϕ values, $\eta_i \rightarrow 1/\phi$ therefore R becomes :

$$R = \frac{1}{L}\left(\frac{k_{max}}{K_s}D_e\right)^{0.5} S_s^{(n+1)/2} \qquad ...(8.337)$$

Where it can be seen that the observed reaction rate order becomes (n+1)/2, when intrinsic rate order is n.

If the intrinsic rate is preudo-first order (n = 1), then the observed rate is still first order (n + 1)/2 = 1 even under a strong intraphase diffusion regime; that is, even under partial substrate penetration, and for pseudo-first order intrinsic kinetics, the observed rate order will not change. But this is not true for the case of pseudo-zero order intrinsic kinetics. Thus, under a serious intraphase diffusion regime and for n = 0, the observed rate will be equal to 1/2. It can be shown that the rate constant k_0 is proportional to $L^{0.5}$. That is, the relationship between k_0 and L is not linear for a pseudo-zero order intrinsic rate at a high ϕ values.

8.5.3.3 Pseudo-Zero Order Kinetics

If $K_a << S$ then Monod kinetics can be approximated by a pseudo zero-order in the substrate concentration. Under such a case, equation 8.314 becomes:

$$\frac{d^2\overline{S}}{dz^2} = \phi_0^{\,2} \qquad ...(8.338)$$

Boundary conditions have to be split into two categories, *viz.*, fully penetrated biofilm and partly penetrated biofilm.

8.5.3.3.1 Fully Penetrated Biofilm

The boundary conditions are the same

$$\overline{S} = 1 \quad \text{at } z = 1$$

$$\frac{d\overline{S}}{dz} = 0 \text{ at } z = 0$$

Thiele modulus (ϕ_0) for pseudo-zero order kinetics becomes :

$$\phi_0^{\,2} = \frac{k_{max}}{S_s}\frac{L}{D_e} \qquad ...(8.339)$$

Equation 8.338 with two boundary conditions may be solved to yield the following

$$\frac{d\bar{S}}{dz} = \phi_0^2 z \qquad ...(8.340)$$

$$\bar{S} = \frac{\phi_0^2}{2}(z^2 - 1) + 1 \qquad ...(8.341)$$

Where equations 8.340 and 8.342 give the substrate gradient and profile respectively. These equation are also valid for cases where $\phi_0 \leq (2)^{0.5}$ otherwise the situation will define a partial penetration case. The effectiveness factor is derived as before and is given by:

$$\eta_1^o = 1 \qquad ...(8.342)$$

When intrinsic reaction rate is pseudso zero-order and the film is fully penetrated, then the internal effectiveness factor is constant and equals unity. This type of result should be expected since the reaction rate is independent of S and the film is fully penetrated. If $\phi > (2)^{0.5}$ then the film is under partial penetration otherwise for $\phi \leq (2)^{0.5}$, the film is fully penetrated. The observed reaction rate (R) becomes equal to substrate flux at the bio film interphase. Therefore, the following can be found:

$$R = k_{max} \qquad ...(8.343)$$

From which a very important result surfaces : when the intrinsic reaction rate is pseudo-zero order, the observed rate is zero order too with respect to S. This conclusion is in agreement with LaMotta[86]. Under these conditions it can be written :

$$R = k_m(S_b - S_a) \qquad ...(8.344)$$

Solving for $S_a = S_b - k_{max}/k_m$ ...(8.345)

Under a very low resistance toward external mass transfer, S_a is approximately equal to S_b which is in agreement with Atasi and Borchardt[75].

8.5.3.3.2 *Partly Penetrated Bio-film*

In this case equation 8.338 remains the same

$$\frac{d^2\bar{S}}{dz^2} = \phi_0^2$$

Boundary conditions are defined as

$$\bar{S} = 1 \text{ at } z = 1$$

$$\frac{d\bar{S}}{dz} = \bar{S} = 0 \text{ at } z = z_p = L_p/L \qquad ...(8.346)$$

L_p is the distance from support media to the point where substrate flux stops. Therefore, the depth of penetration is given by :

$$\delta_p = (L - L_p) = (1 - Z_p) \quad ...(8.347)$$

Solving the basic differential equation with two boundary conditions, the following relationships are derived :

$$\frac{d\bar{S}}{dz} = \phi_0^2 (Z - Z_p) \quad ...(8.348)$$

$$\bar{S} = \frac{\phi_0^2}{2}(Z - Z_p)^2 \quad ...(8.349)$$

Where, $$\phi_0 = \left(\frac{k_{max}}{S_a}\frac{L}{D_e}\right)^{0.5} \quad ...(8.350)$$

The depth of penetration (Z_p) is obtained by substituting first boundary condition into equation 8.349.

$$Z_p = 1 - (2)^{0.5}/\phi_0 \text{ or} \quad ...(8.351)$$

$$\delta = L(2)^{0.5}/\phi_0 \quad ...(8.352)$$

If $Z_p > 0$, then $\phi_0 > (2)^{0.5}$ for partial penetration to occur :

$$\phi_0 > (2)^{0.5} \text{ partial penetration} \quad ...(8.353)$$

$$\phi_0 \leq (2)^{0.5} \text{ full penetration} \quad ...(8.354)$$

If $Z_p = 1$, then $L_p = L$, when ϕ_0 is infinite. It means that a surface penetration might exist only under severe intraphase diffusion. Substitute equation 8.351 into equation 8.346, the following is obtained :

$$S = \phi_0^2 (2)^{0.5} (Z + (2)^{0.5}/\phi_0 - 1)^2 \quad ...(8.355)$$

As ϕ_0 increases, the depth of penetration decreases until a condition exists where $Z_p = 1$ that is $L_p = L$. This will take place when $(2)^{0.5}/\phi_0 = 0$ ($\phi_0 \rightarrow$ is very large).

It is necessary to determine the internal effectiveness factor under partial penetration situation. Using the same procedure as given before :

$$\eta^o_{i,p} = 1 - Z_p \quad ...(8.356)$$

$$\eta^o_{i,p} = (2)^{0.5}/\phi_p \quad ...(8.357)$$

It becomes clear that functionality between fully penetrated and partically penetrated changes at the point where $\phi_0 = (2)^{0.5}$.

Same procedure is used to calculated the observed reaction rate, R. By evaluating the substrate concentration gradient at z = L, the following is obtained:

$$R = (2k_{max}\ D_e/L\ S_a^{0.5}) \qquad ...(8.358)$$

It was mentioned earlier that when the intrinsic reaction rate is pseudo zero order w.r.t the substrate concentration, then the observed reaction rate will show a half order dependence on S under partial bio film penetration and this is in agreement with LaMotta[86]. The experimentalist is bound to observe half order rather than zero order (this idea has been often misunderstood).

Observed reaction rate can be defined in terms of observable measurable quantities. S_a is difficult to determine. Therefore, R has been defined in terms of bulk substrate concentration, S_b.

$$R = \left(2k_{max}\frac{D_e}{L}\right)^{0.5}\left(\frac{1}{2k_{max}}\right)\left[\left(2k_{max}\frac{D_e}{L} + 4k^2{}_m S_b\right)^{0.5}\left(2K_m\frac{D_e}{L}\right)^{0.5}\right] \qquad ...(8.359)$$

Observed rate R can also be written in terms of the depth of penetration as follows :

$$R = k_{max}\frac{\delta_p}{L} \qquad ...\ (8.360)$$

This shows that observed rate is directly proportional to the partial depth of penetration for case of pseudo-zero order kinetics in S.

8.5.3.4 Interphase-Intraphase Diffusion and Reaction

The procedure adopted in this section can be used for any fixed-film biological reactor as all kinetic processes within the intraphase-interplace have been considered. However, there might be some specific difference involved, with certain special type of reactor systems. Such problems would require experimental determinations. Only linear kinetics has been considered as Atasi and Borchart[75] have demonstrated this mechanism experimentally.

At steady state, within biofilm, the following boundary condition is defined :

$$R = k_m\ (S_b - S_a) = D_e\ \frac{dS}{dz}\ \text{at } z = 1 \qquad ....\ (8.361)$$

Solving for S_a results in :

$$S_a = \frac{1}{(1 + \phi \tanh \phi / B_i)} \qquad ....(8.363)$$

Where, $B_i = k_m\ L/D_e$ (Biot number)[87] and ...(8.363)

$$\phi^2 = (k_{max}/K_s)(L/D_e) \qquad ....(8.364)$$

Biot number measures the ratio between intraphase and interphase gradients. The overall observed rate becomes :

$$R = \left[\frac{1}{k_m} + \frac{1}{k_m / k_s . \eta_i}\right]^{-1} . S_b \qquad \text{... (8.365)}$$

Defining the following

$$1/K_{ov} = 1/k_m + (1/(k_m/K_s . \eta_i)) \qquad \text{.... (8.366)}$$

where K_{ov} refers to overall first order observed reaction rate constant and η_i is the internal effectiveness factor defined by equation (η_i = tan h ϕ/ϕ). Therefore, the observed reaction rate becomes :

$$R = K_{ov} . S_b \qquad \text{.... (8.367)}$$

The overall observed reaction rate (including all transport phenomenon and reaction) is first order also when the intrinsic rate is pseudo-first order. It implies that the experimentalist will tend to observe first order kinetics in the bulk measurable substrate concentration only if the intrinsic rate is also pseudo-first order. It can be seen that under such conditions, the first order mechanism is additive and is composed of mass transfer (inter-phase, $1/k_m$) and the diffusion and reaction term (intraphase = $1/(k_{max}K_s).\eta_i$). If no intraphase resistance is present (η_i tends to unity), then :

$$\frac{1}{K_{ov}} = \frac{1}{k_m} + \frac{1}{(k_m / K_s)} \qquad \text{.... (8.368)}$$

If no interphase and intraphase diffusion prevails (η_i – 1; k_m is indefinitely large) then the reaction mechanism becomes homogeneous or

$$K_{ov} = k_{max}/K_s \qquad \text{... (8.369)}$$

The observed overall reaction rate in terms of an overall effectiveness factor η_i becomes :

$$R = \frac{k_{max}}{K_s} . S_b . \eta_o \qquad \text{... (8.370)}$$

Where, $$\frac{1}{\eta_o} = \frac{1}{\eta_i} + \frac{k_{max} / K_s}{k_m} \qquad \text{.... (8.371)}$$

This is a desired analytical relationship that includes all relevant parameters of the interphase-intraphase diffusion and reaction. It also reveals that it is additive in terms of linear kinetics. The term η_i^{-1} represents intraphase resistance, k^{-1}_m the interphase resistance and finally the intrinsic kinetic constants are represented by k_{max} and K_s. Equation 8.371 can be written in terms of Biot number (B_i).

$$\eta_o = \frac{\tanh \phi}{\phi \left[1 + \frac{\tanh \phi}{B_i}\right]} \qquad \text{... (8.372)}$$

This shows the functionality of η_o in terms of Thiele modulus and Biot number.

Some of the terms are not observable, therefore, a modification can be carried out as suggested by Wheeler[88] for intraphase diffusion problem :

$$\phi = \eta_o \phi^2 = R\frac{L}{D_e S_b} \quad ...(8.373)$$

Where refers to modified Thiele modulus which is measurable because it is a function of measurable parameters such as R, L, D_e and S_b. Relationship between ϕ and η_o is not explicit (implicit). A graphical representation can be made for overall effectiveness factor (η_o) and observable Thiele modulus with Biot number as a parameter.

8.5.3.4.1. Extreme Cases Related to Effectiveness Factor

This aspect includes both phases and utilization (external and internal).

For large values of ϕ and D_s, η_i and η_o will approach the following values :

$$\eta_o \rightarrow 1/D_a \quad ... (8.374)$$

$$\eta_i \rightarrow 1/\phi \quad ... (8.375)$$

These are called asymptotic solutions and D_a stands for Damkohler number[76].

For small values of ϕ and D_a (weak interphase and intraphase diffusion resistance), η_o and η_i approach the following limit :

$$\eta_o \rightarrow 1 \quad ...(8.376)$$

$$\eta_i \rightarrow 1 \quad ... (8.377)$$

Comparing the pseudo-zero and pseudo-first order mechanisms, it can be seen that the higher the reaction rate order, higher the influence of diffusion phenomenon, the other parameters being the same.

In equation 8.371, for large ϕ values, the following is true :

$$\eta_o = \frac{1}{\phi_i(1+\phi / B_i)} \quad ...(8.378)$$

Therefore, under this condition of strong diffusion resistance and finite B values, the following conditions hold :

$$R = \frac{k_{max}}{K_s}\frac{S_b}{\phi(1+\phi / B_i)} \quad ...(8.379)$$

If $\phi/B_i >> 1$; then :

$$R = \frac{k_{max}}{K_s}\frac{B_i}{\phi^2} S_b$$

Substituting the value of ϕ^2 from equation 8.325, the above equation reduces to :

$$R = k_m S_b \qquad \text{....(8.380)}$$

To summarise, when the intrinsic reaction rate is pseudo first order, the observed reaction rate is first order too in the substrate bulk concentration even under a strong intraphase diffusion regime. Under this situation, the depth of substrate penetration within the biofilm will not have an effect on the observed rate order.

When the intrinsic reaction rate is pseudo-zero order and with substrate's full penetration, the observed rate is zero order too in the substrate bulk concentration.

When the intrinsic reaction is pseudo-zero order w.r.t the substrate concentration; then the observed reaction rate will show a half order dependence on the bulk substrate concentration under conditions of partially penetrated biofilm.

The rate expression, in terms of the measurable bulk concentration, will not exhibit a Monod type mechanism, even when the intrinsic rate is so assumed.

8.5.4 Diffusion and Kinetic Model[89]

Two models of substrate kinetics for RBCs are being discussed in this section. The single substrate limiting model consists of mathematical equations on a mass balance and of the limiting substrate in a liquid film, a biofilm and a bulk liquid. The single substrate limiting model assumes one substrate is limiting over the entire depth of a biofilm and reaction is assumed to follow Monod kinetics. The dual substrate limiting model assumes rather an electron donor or an electron acceptor controls and rate of reaction. The limiting substrate of a reaction is determined from the relative concentration of both substrates. Both models account for the external mass transfer and internal diffusion alongwith bio-chemical reaction.

8.5.4.1 Substrate Removal in a RBC Process

The resistance to mass transport is believed to be concentrated at the boundary, layer and the rate of mass transfer is generally expressed as follows :

$$N_s = k'_a (S_b - S_a) \qquad \text{...(8.381)}$$

where N_s, k'_a, S_b and S_a refer to flux of S into biofilm, mass transfer coefficient of S, concentration of S in the bulk liquid and biofilm surface respectively. The mass transfer coefficient normally varies with the property of bulk liquid, the diffusional characteristics of S in the bulk liquid, and the extent of mixing near the biofilm surface. Such effects are incorporated into dimensionless form and the mass transfer coefficient is expressed as follows :

$$k'_a = a\, S_c^{m} R_e^{b} \qquad ...(8.382)$$

Where S_c, R_e, and (a, m and n) refer to Schmidt number, Reynolds number and coefficients (a, m and n) respectively. There are some theoretical and empirical relationship of mass transfer coefficients for specific geometry and flow regimes. In fixed film process modeling, investigators have simply adopted an empirical equation developed with non-porous solids media. Whether such a relationship holds in a semisolid media with high porosity is questionable. Additional research is needed to better determine the mass transfer coefficients in a fixed process.

There is some degree of dripping and tubulence, especially at high rotational speeds, which helps mixing. However, to avoid complexity the liquid film is assumed to be stagnant layer with a constant thickness. The rate of mass transfer from the liquid film to the biofilm is expressed as follows:

$$N_a = k''_a (S_L - S_a) \qquad ...(8.383)$$

Where k''_a and S_L refer to mass transfer coefficient of S and concentration of S in the liquid film. Assuming that only diffusion takes place through the liquid film; the mass transfer coefficient in this case varies with the liquid film thickness and based on this theory the mass transfer coefficient can be expressed as follows:

$$k''_a = D_a/\delta/2 \qquad ...(8.384)$$

where D_a and δ refer to diffusion coefficient of S in water and liquid film thickness respectively. Because S_1, is an average concentration in the liquid film, half the liquid film thickness is used in the above equation. The liquid film thickness appears to vary with rotational speed, however, because of lack of an adequate measurement technique reliable results have not been reported.

Under a steady state substrate concentration the biofilm shows a unique profile of substrate versus depth. The following mass balance equation with Monod kinetics is used:

$$D'_a \frac{d^2S}{dz^2} = k_a \times \frac{S}{(K_a + S)} \qquad ...(8.385)$$

Where D_a, k_a, X and K_a refer to diffusion coefficient of substrate within the biofilm, maximum utilization rate constant, organism concentration and saturation constant respectively. The two boundary conditions are :

$$S = S_a \text{ at } z = 0 \qquad ...(8.386)$$

$$\frac{dS}{dz} = 0 \text{ at } Z = Z_c \qquad ...(8.387)$$

Where Z_c is a limiting thickness. The limiting thickness is a result of limitation of the substrate transport within the bio film. It is defined as a thickness beyond which no further substrate removal occurs. However, limiting thickness cannot be measured and in many cases it varies because the biofilm is not in a true steady state. For example, in a RBC process even though the substrate concentration in the bulk liquid is maintained at steady state, substrate concentration in the biofilm varies because of the rotational speed. Therefore, the limiting thickness varies with the position of the element in space. Instead of equation 8.387, a boundary condition representing the substrate flux at the biofilm media interface is used in these models.

$$dS/dz = 0 \text{ at } Z_c = L \qquad \text{— (8.388)}$$

Where L refers to biofilm thickness. This does not imply that the limiting thickness is equal to the biofilm thickness.

8.5.4.2. Single Substrate Limiting Model

Generally, the limiting substrate of bio-chemical reaction is not restricted to an electron donor. Depending upon the relative concentration of electoron acceptor, an electron acceptor may become the limiting substrate. The rate of reaction in single substrate limiting model (SSLM) depends upon the concentration of one substrate throughout the depth of a biofilm. This is true if the other substrate is provided in excess with the bio film. The biofilm thickness is assumed to be uniform and biomass gain due to growth is assumed to be equal to biomass loss due to respiration and detachment.

From the geometry of RBC system, the time required for one revolution of the moving bed is expressed as :

$$t^* = 2\Pi/W \qquad \text{— (8.389)}$$

where t^* and W refer to time required for one revolution of moving bed and angular rotational speed respectively.

During one evolution the entire surface spends a fraction of time in the waste and the remainder in the upper atmosphere. The various fractions are simply expressed as follows :

$$f_w = \frac{A_w}{A_a + A_w} \qquad \text{— (8.390)}$$

$$f_a = 1 - f_w \qquad \text{— (8.391)}$$

where f_w, f_a, A_w and A_a refer to fraction of time spent in wastewater, fraction of time spent in upper atmosphere, submerged area, and area exposed to the upper atmosphere respectively.

Mass transfer of limiting substrate in liquid film may be expressed as follows :

$$\delta \frac{dS_L}{dt} = -k''_a (S_L - S_{sa}) \quad \text{— (8.392)}$$

where S_{sa} is the substrate concentration at the surface of the biofilm as exposed to the upper atmosphere (unsubmerged portion). The initial condition of the above equation is as follows :

$$S_L = S_b \text{ at } t = 0 \quad \text{— (8.393)}$$

Because two different equation are used to describe the external mass-transport, two equations with different boundary conditions are required to describe the substrate concentration within the biofilm. For an unsubmerged sector, the mass balance is expressed as follows:

$$\frac{\delta S_a}{\delta t} = D'_a \frac{\delta^2 S_a}{\delta z^2} - \frac{k_a \times S_a}{(K_a + S_a)} \quad \text{— (8.394)}$$

where S_a refers to the substrate concentration within biofilm exposed to the unsubmerged sector. The initial and boundary conditions are expressed as follows:

$$S_a (t = 0) = S_w (t = f_w t^*) \quad \text{— (8.395)}$$

$$\delta S_a / \delta z = 0 \text{ at } Z = L \quad \text{— (8.396)}$$

where S_w refers to the substrate concentration within the biofilm to the submerged sector. For a submerged sector the mass balance may be expressed as :

$$\frac{\delta S_w}{\delta t} = \frac{\delta^2 S_w}{\delta z^2} - \frac{k_a \times S_w}{K_a + S_w} \quad \text{— (8.397)}$$

The initial and boundary conditions are expressed as follows :

$$S_w (t = 0) = S_a (t = f_a t^*) \quad \text{— (8.398)}$$

$$\delta S_a / \delta z = 0 \text{ at } Z = L \quad \text{— (8.399)}$$

Equations 8.392 and equation 8.394 describe mass balance of the unsubmerged sector for a period of $0 \le t \le f_a t^*$, while the mass balance of the submerged sector as described by equation 8.397 is defined at $0 \le t \le f_w t^*$.

The substrate concentration in the bulk liquid is determined from a mass balance, which includes substrate carried by the bulk flow and the liquid film in addition to the mass transported to the biofilm of the submerged sector. The mass balance of the substrate in the bulk liquid is expressed as follows :

$$V \frac{dS_b}{dt} = Q (S_i - S_b) + \frac{A\delta}{t^*} (S_L (f_a t^*) - S_b) - k'_a f_w A \int_0^1 (S - S_{ws}) dY \quad . \text{(8.400)}$$

where Q, S_i, A and Y refer to flow rate, influent substrate concentration, total surface area and a dummy variable respectively.

The above equation is used for describing the dynamic behaviour of the RBC process with appropriate initial and boundary conditions. The initial and boundary conditions can be obtained from the result at a steady state. Equation 8.399 reduces to the following under steady state :

$$S_b = \frac{QS_i + (A\delta / t^*) S_L(f_a t^*) + k'_a f_w A \int_0^1 S_{ws} dy}{Q + (A\delta / t^*) + k'_a f_w A} \quad \text{... (8.401)}$$

The above equation involves one ordinary differential equation and two partial differential equations alongwith an algebraic equation. As shown in equations 8.393, 8.395 and 8.398, the initial and boundary conditions are expressed as constraints. And because of the non-linear nature of Monod equation, the model equation cannot be solved analytically. In order to solve this equation, finite difference is to be formed and a numerical iteration technique can be used. The computation begins with assumed value of S_b and S_a (t = 0). The computed results are compared with the assumed ones at the end of each interaction. The iteration continues until the assumed values are equal to the computation value.

8.5.4.3 Dual Substrate Model

May studies on the reaction kinetics have simply assumed an electron donor to the limiting substrate if the concentration of the electron acceptor is sufficiently high enough not to limit the reaction rate, this assumption can be justified. However, in general, the rate of reaction is not controlled by the electron donor. For example, in an activated sludge process the organic removal rate may not be independent of oxygen concentration in the mixed liquor. The effect of oxygen may be higher for a waste of high organic content. The electron acceptor becomes limiting substrate when its concentration is below critical values. Some researchers assume both substrates affect the reaction rate, and the rate of reaction is expressed as a function of both substrates, usually a product of Michaelis Menten equations[90-92]. While, other researchers assume that the reaction rate is limited by one substrate at a time and the limiting substrate is determined by the relative concentration of both substrates[93-94]. It is not clear whether both substrates limit simultaneously or either one limits separately. The rate of reaction is assumed to be limited by one substrate at a time in the dual substrate model (DSLM). For anaerobic reactors, sulphates, nitrates, CO_2, etc., play the same role as oxygen does in the aerobic systems.

A biochemical reaction in which an electron donor and an electron acceptor may be expressed as follows :

$$sS + cC = pP \quad \text{...(8.402)}$$

Where a, c and p refer to an electron donor, an electron acceptor and an end product respectively. s, c and p are stoichiometric constants respectively. Considering only substrate used for the reaction, a utilization ratio (f) can be defined as follows :

$$f = \frac{sMs}{cMc} \qquad \text{... (8.403)}$$

where Ms and Mc refer to molecular weights of electron donor and electron acceptor respectively.

The substrate utilization rate of the electron donor and the electron acceptor are related to the utilization ratio as follows :

$$f = \gamma_s / \gamma_c \qquad \text{...(8.404)}$$

where γ_s and γ_c refer to utilization rates of the electron donor and the electron acceptor respectively. When both substrates are much larger than the saturation constants, the substrate utilization rate follows zero-order kinetics and the above equations are reduced to:

$$f = k_s / k_c \qquad \text{.... (8.405)}$$

where k_s and k_c refer to maximum utilization rate constants for S and C respectively.

Either an electron donor or an electron acceptor can be a limiting substrate depending on their concentrations. Williamson and McCarty[93] developed a criteria to determine the limiting substrate. When an electron acceptor is a limiting substrate, the following inequality holds :

$$C < R_c / R_s - S \qquad \text{....(8.406)}$$

where k_c and k_a refer to saturation constants of C and S respectively.

If the above relationship does not hold, the electron donor is a limiting substrate. When S is a limiting substrate, the substrate utilization rate of S and C may be expressed as follows :

$$-\gamma_s = -\frac{k_a \times S}{(K_s + S)} \qquad \text{...(8.407)}$$

$$-\gamma_c = \frac{1}{f}(-\gamma_s) = -\frac{k_c \times S}{K_s + S} \qquad \text{...(8.408)}$$

On the other hand when C is limiting, the substrate utilization rate of C and S may be expressed as follows :

$$-\gamma_c = -\frac{k_c \times C}{(K_s + C)} \qquad \text{...(8.409)}$$

$$-\gamma_s = \frac{1}{f}(-\gamma_c) = -\frac{k_s \times C}{K_c + C} \quad ...(8.410)$$

In a fixed film process, the substrate concentration within the biofilm is dependent not only upon the reaction rate but also upon the external mass transport and the internal diffusion. The limiting substrate may be altered within a bio film and both substrates may limit in different zones. Four possible cases are shown in Fig. 8.12. The solid line represents the limiting substrate and non limiting substrate is shown with broken lines. Cases 1 and 2 are identical to the SSLM, while the other two represent dual limiting cases, which may be likely to occur in deep biofilms.

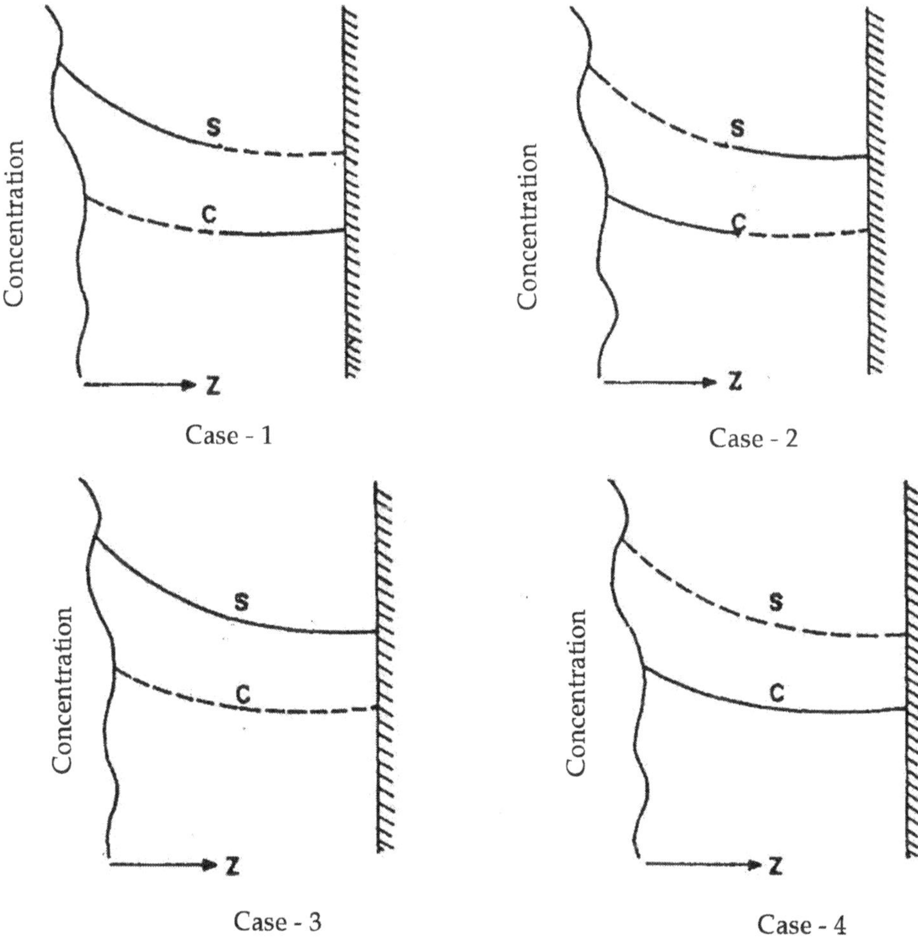

Fig. 812 : Possible Substrate Profiles of an Electron Donor and an Electron Acceptor within a Biofilm.

Model equations in DSLM are similar to those SSLM. Equation describing the external mass transport of the electron donar are identical to those presented in SSLM. Mass balance of S and C within biofilm may be written similarly to the equations described in SSLM. They can be expressed as follows :

$$\frac{dS}{dt} = D'_c \frac{d^2S}{dz^2} - \rho_s \quad \text{... (8.411)}$$

$$\frac{dC}{dt} = D'_c \frac{d^2C}{dz^2} - \gamma_c - R_m X \quad \text{... (8.412)}$$

where D'_c and R_m refer to diffusion coefficient of C within a biofilm and are endogeneous respiration rate respectively. In the above equation γ_s and γ_c determined after limiting substrate is defined. Equations 8.407 to 8.410 may be used.

The mass balance of S in the bulk liquid presented in SSLM can be used in the DSLM. When the substrate concentration within the biofilm is computed, the limiting substrate criterion must be checked at each step and appropriate equation for the rate of reaction is used.

8.5.5 Other RBCs Model

Joost[95] introduced a conceptual model to describe BOD reduction per stage and is given as follows :

$$\frac{\text{BOD reduction}}{\text{Stage}} = KC^aT^bS^ct^eR^dQ^fD^g \quad \text{...(8.413)}$$

Where K, C, T, S, R, Q, t and (a, b, c, d, e, f, and g) refer to treatability factor of the waste material, substrate concentration, wastewater temperature, reactor residence time, physical configuration constant, flower, submerged media depth, and partial regression coefficients respectively.

Weng and Molof[96] simplified the above model and is given as :

$$F = K\, C^aT^bS^c(WA)^d\, t^e\, Q^f\, D^g \quad \text{...(8.414)}$$

where F refers to fraction of the influent loading remaining in the effluent.

Antonie and Welch[97] suggested RBC system performance could be predicted with the equation :

$$\%\ \text{COD removal} = KL_o^{[(a+1)\ n-1]}\ (T^b\ S^c\ B^d)^{[1-(a+1)]N\ [1-(a+1)]} \quad \text{...(8.415)}$$

where, B, L_o and N refer to rotational speed, influent wastewater concentration and stage number respectively.

Wu[98] proposed a model of the form :

$$F = (K/N)L_o^{\,a}\ T^b\ S^c\ B^d\ A^e\ D^f\ Q^g \quad \text{.... (8.416)}$$

The speed of rotation, liquid detention time and submergence depth of the media are not considered to be the major controlling factors in RBC systems. A first order relationship between stage number β = log (K/N) is expressed :

$$\beta = K \exp(-K'N) \qquad ....(8.417)$$

where K and K′ refer to constants for a particular type of waste and RBC configuration including media and also type of reaction (aerobic/anaerobic).

Tait and Friedman[99] modelled stage soluble TOC remaining by the equation :

$$\ln C = \ln C_i - K_f\, t \qquad ....(8.418)$$

where, C, C_i, K_f and t refer to soluble TOC remaining, influent soluble TOC, pseudo first order rate constant HRT respectively.

The rate constant is found to be a function of influent TOC :

$$K_f = 0.3744 - 7.96,\ 10^{-4} C_i \qquad ...(8.419)$$

Friedman *et al.*[100] described RBC by the conventional plugflow first order model based on apparent first order reaction kinetics :

$$C_a = C_b \exp(-k_o t) \qquad ...(8.420)$$

where t and k_o refer to HRT and rate constant of the system respectively.

Schroeder[101] proposed a model based an mass transfer concepts to predict removal of organics in an RBC unit which was modified by Friedman, *i.e.*,

$$Mz = \frac{f\, h\, K_o\, A_s\, C_b^{\,2}}{(K_m + C_b)} \qquad ....(8.421)$$

Where M_a, f, h, A_s, K_m and K_o refer to mass of substrate removed per unit time, proportionality factor, effective biomass depth, submerged surface area per stage, half saturation constant and maximum area removal constant respectively. If f, h and K_o are constant for a particular substrate, the model reduces to the following :

$$Mz' = \frac{K''\, C_b^{\,2}}{K_m + C_b} \qquad .... (8.422)$$

where Mz′, K″ and A_s refer to substrate removal per unit total media surface area, K″ = (f h $K_o A_s$) and As is a constant for a given reactor media and depth of submergence respectively. Using the analogy of plug flow reactor, the following equation can be used :

$$\frac{K''\theta}{V} = K_m\left(\frac{1}{C_b} - \frac{1}{C_{bi}}\right) + \ln\left(\frac{C_{bi}}{C_b}\right) = \frac{fkK_oA_s}{V}\,\theta \qquad ...(8.423)$$

where V, θ and C_{bi} refer to reactor volume, HRT and influent substrate concentration respectively.

The value of f, h, K_o and Km are dependent on the substrate nature and biomass as the media. The variables A_s, θ and V are either controlled or selected by the designer. The value of K″ was empirically related by the relation :

$K'' = 6956\ (\theta C)^{0.664}$, where θ C is the system loading factor. ...(8.424)

Grieves[102] suggested the following set of equations based on mass transfer resistances to RBC process using both first order and Monod kinetics. Two models were developed, (*a*) a pseudo homogeneous and (*b*) a heterogeneous model. The first model involved external mass transfer resistance (an attached liquid film), while the latter involved external and internal mass transfer resistance. For first order Monod kinetics, the mode for an *n* stage reactor is :

$$\frac{C_b}{C_o} = \frac{1}{1 + N/Q[P_1 As + P_2 W(1 - \exp(-P_1 A\alpha / P_2 W)]} \quad ...(8.425)$$

where C_b, C_o, N, Q, A_s and W refer to effluent, influent substrate concentration, total number of media particle per stage, flow rate, submerged surface area of the media and rotational speed (rpm) respectively.

$$P_1 = \frac{K_L K_1}{1 + K_1} \quad .... (8.426)$$

$$K_1 = \frac{\mu_m X_a\, d(TF)^{n-1}}{Y K_s\, \eta\, K_L} \quad ... (8.427)$$

$$P_2 = K_2\, A\, \Delta L \quad .... (8.428)$$

where $K_{1.}$, TF, n, A, ΔL, η, d, X_a, μ_m, Y and K_a refer to mass transfer coefficient, treatability factor, number of stages, total media surface area, liquid film thickness, effectiveness factor active depth of biofilm, active biomass, maximum specific growth rate, yield coefficient and saturation constant respectively.

Pano and Middlebrooks[103] suggested a model which incorporates the first stage and remaining stages. They considered first stage separately since these stages receive raw wastewater while the influent to the other stages contain sloughed biomass from the proceeding stages with unconsumed substrate. The kinetics of substrate removal in stages (except for stage one) can be expressed by a common model :

$$Q(S_1 - \bar{S}) = \sum_{i=2}^{4} A_i (K_L)^{20}\, \theta^{\tau-20}\, S_i^{\,n}. \quad ....(8.429)$$

where S_i, $\bar{S}$, A_i, $(K_L)_{20}$, θ, T and n refer to first stage substrate concentration, mean substrate concentration, total available surface area/stage, reaction rate constant at 20°C, temperature factor and an apparent reaction order respectively.

Watanabe and Ishiguro[104] described the kinetic equation for steady state and completely mixed flow reactor :

$$\frac{C_o}{C_e} = 1 + \frac{K^* A}{Q} \quad ....(8.430)$$

where C_o, and C_e refer to influent and effluent substrate concentration respectively.

$$K^* = \left(\frac{1}{K} + \frac{1}{K_s}\right)^{-1} \quad ....(8.431)$$

where A, K and K_s refer total surface area, diffusion term and reaction term respectively.

As reaction term (K_s) becomes large in comparison to 1/K, it can be neglected. This implies, system is diffusion controlled of $1/K << 1/K_s$, then the system is controlled by a reaction kinetics.

Eckenfeldar and Vandevenne[105] developed a first order kinetic Model for an RBC treating industrial waste water, *viz.* :

$$\frac{S_n}{S_o} = \frac{1}{(1 + KAQ)^n} \quad .... (8.432)$$

where S_n, S_o, A, K and n refer to influent and effluent substrate concentration, effective surface area, proportionality factor and number of stages respectively.

Benjes[106] used plug flow reactor analogy to propose the following equation:

$$\frac{S_e}{S_o} = \exp[-K(V/Q)^{1/2}] \quad ....(8.433)$$

$$\frac{S_e}{S_o} = \exp[-S(A_s/Q)^{1/2}] \quad ... (8.434)$$

where V, K and S and A_s refer to reactor volume, empirical constants, and surface area respectively.

Kincannon and Groves[107] used a four stage RBC system. The performance can be predicted using first order kinetics plus incorporation of suspended biomass :

$$\frac{(S_o - S_e)Q}{(X_m + VX_a)} = K\,S_o \qquad \text{.... (8.435)}$$

where X_m, V, X_a and K refer to active biomass, reactor volume in a stage, suspended growth concentration and proportionality constant respectively.

Opatken[108] applied Levenspiel's equation for stage reactors following second order kinetics to calculate the soluble organic concentration at any stage is given by :

$$C_n = \frac{-1 + (1 + 4\,K\,\theta\,C_{n-1})^{0.5}}{2\,K\,\theta} \qquad \text{.... (8.436)}$$

where C_n and C_{n-1}, K and θ refer to n^{th} and $(n - 1)^{th}$ stage substrate concentration, second order reaction rate constant and HRT respectively.

Fujie[109] presented a method to estimate power economy (defined as the amount of BOD removed poor unit power consumption of RBC). Due to daily fluctuations in influent flow rate (Q) and influent substrate concentration (S_o), the daily mean power economy is given by :

$$\frac{W}{P_w} = \frac{\int Q\,(S_o - S_n)\,d\theta}{\int P_w d\theta} = \frac{\int QS_oX_n d\theta}{\int P d\theta} \qquad \text{.... (8.437)}$$

where S_n, X_n, θ, P_w and W refer to effluent substrate concentration, fractional substrate removal, HRT, requisite power, and rate of substrate removal per unit power consumption respectively.

For constant P_w, Q, S_o and X_n, this equation reduces to :

$$\frac{W}{P_w} = (QS_oX_n)/P_w \qquad \text{... (8.438)}$$

The fractional substrate removal in the effluent from n^{th} RBC, is described by:

$$X_n = 1 - \frac{1}{(1 + \phi\,KA_1/Q)(1 + \phi\,KA_2/Q)......(1 + \phi\,KA_w/Q)} \qquad \text{..(8.439)}$$

where A_i, K and ϕ refer to total surface area in the ith stage, first order reaction rate constant and rotational factor indicating effect of rotational speed on the rate of BOD removal respectively.

Since the surface area of the media is usually equal in each stage, the above equation can be simplified to :

$$X_n = 1 - \frac{1}{(1 + \phi(A/nQ))^n} \qquad \text{... (8.440)}$$

Power required is determined based on the analogy made between RBC unit and a CSTR in terms of the power number, Np. The requisite power per unit surface area of the media can be calculated using the equation :

$$\frac{P_W}{A} = \lambda_2 N^3 D^3 \quad \text{.... (8.441)}$$

where λ, D and N refer to an empirical constant, effective diameter of the reactor and rotational speed respectively.

Williams[110] used the following equation for finding out the required RBC media area requirements, *viz.* :

$$\frac{S_e}{S_o} = \exp[-K(A_s / 695Q)^{1/2}] \quad \text{... (8.442)}$$

where K, A_s and Q refer to treatability factor, total media area (ft^2) and flow (mgd) respectively.

Hsieh[111] suggested that following relationship to correlate the per cent COD reduction with some parameters of RBC :

$$P = [1 - \exp(-bN\theta)] \quad \text{.....(8.443)}$$

$$N = \frac{V_t V_L}{A_L n} \quad \text{... (8.444)}$$

where P, b, θ, N, V_t, V_L, A_L and n refer to fractional COD reduction, constant, HRT, media number, tank volume per stage, volumetric loading, areal loading and surface area per unit biomedia support respectively.

Ching-San[112] pseudo-homogeneous surface reaction mode is applied to an RBC system, the mass of reduction of substrate per unit media area is :

$$M = \gamma\, dNA \quad \text{..... (8.445)}$$

where M, γ, dN and A refer to mass of reduction of substrate, reaction rate, liquid film thickness and media area/total media respectively. The mass balance for each media surface of an RBC can be expressed using zero-order reaction model as :

$$V\frac{dS}{dt} = M \quad \text{..... (8.446)}$$

The liquid film thickness (dN) can be defined as :

$$dN = K\left(\frac{\mu NR}{\rho g}\right)^{0.5} \quad \text{..... (8.447)}$$

where N, K and R refer to rotational speed (rpm), constant and radius of the media respectively. The liquid film thickness can be described as :

$dN = K_2(N)^{1/2}$; where K_2 refers to a constant. ..... (8.448)

The final equation becomes :

$V (S_o - S_e) = K_3 N^{1/2} At$ ..... (8.449)

where K_3, V, and t refer to constant (= $K_1 . K_2$), liquid volume of a RBC and HRT = V/Q) respectively. Alternatively,

$Q (S_o - S_e)/A = K_3 N^{1/2}$ ...(8.450)

and this refers to substrate removal rate as a function of square root of rotational speed.

8.6 Other Types of Reactor Systems

8.6.1. CSTR Without Recycle[113,114]

CSTR system without solids recycle, the rates of cell mass and substrate concentration changes are given by the following equations :

$$\frac{dM}{dt} = \mu M - M/\theta; \text{ and} \quad ...(8.451)$$

$$\frac{dS}{dt} = -F + \frac{S_o - S}{\theta} \quad ...(8.452)$$

where, M, μ, F, S, S_o, θ and t refer to cell mass concentration, specific growth rate of micro-organism, volumetric substrate utilization rate, reactor and effluent substrate concentration, HRT and time respectively. The hydraulic retention time (HRT) equals solids retention period.

$$\mu = \left(\frac{Y}{M}\right)F \quad ...(8.453)$$

Under steady state the following is obtained :

$\mu = 1/\theta$ ...(8.454)

$F = (S_o - S)/\theta$; and ...(8.455)

$M = Y (S_o - S)$ ...(8.456)

Contois equation for specific growth of micro-organism is given by

$$\mu = \frac{\mu_m S}{(BM + S)} \quad ...(8.457)$$

where μ_m and B refer to maximum specific growth rate of micro-organism and kinetic parameter respectively. Rearranging this equation alongwith equation 8.451 results in the following :

$$\frac{\mu}{\mu_m} = \frac{S/S_o}{K + (1-K)S/S_o} \quad \text{...(8.458)}$$

Combining equations 8.54 and 8.458 result in the following :

$$\frac{S}{S_o} = \frac{K}{(\mu_m \theta - 1 + K)} \quad \text{...(8.459)}$$

Where K refers to dimensionless kinetic parameter (=YB). If S_o is the influent biodegradable COD (or VS concentration), S is the effluent biodegradable COD (or VS), S_{TO} is the influent total COD (or VS) and S_T is the effluent total COD (or VS), then equation 8.459 can be modified as follows :

$$\frac{S_T}{S_{TO}} = (S + S_r)/S_{TO} = R + (1 - R)K/(\mu_m \theta - 1 + K) \quad \text{...(8.460)}$$

where S_r refers to COD (or VS) concentration of non-biodegradable material in can be obtained from equation 8.464

$$\overline{F} = \frac{F}{S_0/\theta_m} = \frac{F}{(1-R)\,S_{TO}/\theta_m} \quad \text{...(8.461)}$$

The maximum substrate utilization rate is determined by taking a derivative of F (equation 8.464) with respect to θ and equating to zero (with second derivative to be negative).

$$F_{max} = (1-R)\,\mu_m\, S_{TO}/(1+K^{0.5})^2 \quad \text{...(8.466)}$$

Which occurs at $\theta = (1 + K^{0.5})/\mu_m$ ...(8.467)

The above analysis is now given in terms of HRT (θ); viz. :

$$\theta_m = 1/\mu_m;\ (\mu \rightarrow \mu_m \text{ as } S \rightarrow S_o) \quad \text{...(8.468)}$$

Therefore, for a completely mixed, continuous flow system under steady state the following expression can be derived:

$$\theta = \frac{1}{\mu_m} + \frac{K}{\mu_m}\left(\frac{S_o - S}{S}\right) \quad \text{...(8.469)}$$

Thus, the value of μ_m and K can be graphically known.

When μ_m and K are determined for a particular substrate, S can be predicted by :

$$\frac{S}{S_o} = \frac{K}{(\theta/\theta_m - 1 + K)} \quad \text{...(8.470)}$$

It is a known fact that a reduction of 1g COD is equivalent to production of 0.35 litres of methane at STP. Knowing the COD loading to the anaerobic bioreactor and volume of methane produced, the remaining COD in the digester can be calculated.

If C denotes the litres of CH_4 at STP produced/gCOD added to the reactor, and C_o is the litre of CH_4 at STP per g COD produced at infinite retention time, the biodegradable COD in the fermentor will be directly proportional to $(C - C_o)$ and C_o will be directly proportional to the biodegradable COD.

Using equation 8.470, the following is obtained

$$\left(\frac{C_c - C}{C_o}\right) = \frac{K}{(\theta/\theta_m - 1 + K)} \quad \text{or} \qquad ...(8.471)$$

$$C = C_o\left[1 - \frac{K}{(\theta/\theta_m - 1 + K)}\right] \qquad ...(8.472)$$

Equation 8.472 shows that if θ/θ_m is greater than [1 – k], the plot of C versus $1/\theta$ should be a straight line with $C \rightarrow C_o$ as $\theta \rightarrow \infty$. This method can be used to determine C_o per each residue studied. Note that, when K equals unity, the inequality, θ/θ_m is greater than [1 – K] is always satisfied. Since C_o is the litres of CH_4 produced per gram of COD added after an infinite reaction time, it is finally attainable methane production per gram of COD of the given substrate. Equation 8.471 can be rearranged to give:

$$\theta = \theta_m + \theta_m K\,[C/(C_o - C)] \qquad ...(8.473)$$

From equation 8.473 the plot θ versus $(C/C_o - C)$ yields a straight line with intercept equal to θ_m and slope equal to $K\theta_m$. Therefore, this equation can be used to determine the kinetic constants, μ_m and QK.

Since C is the methane per gm of COD added, the volumetric methane production rate (G) equals C times the loading rate.

$$G = \frac{CS_{TO}}{\theta} = \frac{C_oS_{TO}}{\theta}\left[1 - \frac{K}{(\theta/\theta_m - 1 + K)}\right] \qquad ...(8.474)$$

The maximum, volumetric methane production rate (G_{max}) is obtained by taking the derivative of G w.r.t θ and equating it equal to zero. Therefore,

$$G_{max} = \frac{C_oS_{TO}}{\theta} = \left[\frac{1 - K/\left(K + K^{1/2}\right)}{1 + K^{1/2}}\right] \qquad ...(8.475)$$

Which occurs at $\theta = \theta_m\,(1 + K^{1/2})$ ...(8.476)

Temperature has been shown to be the primary factor affecting μ_m. Substrate and substrate concentration has no detectable effect on μ_m. The effect of temperature can be described by the following empirical relationship.

$$\mu_m = 0.013T - 0.129 \quad ...(8.477)$$

8.6.2 CSTR System with Recycle[115]

Modification of the single stage process involving physical segregation of biochemically diverse acid and methane bacteria and microbial or chemical pretreatment of feed have shown to achieve better substrate conversion. Several researchers[116-119] have demonstrated the advantages in terms of improved gas yields, conversion efficiency and stability as well as reduction in HRT achieved through segregation of phases in anaerobic digestion. HRT in the methane phase could be reduced further in anaerobic activated sludge system, with concomitant maintenance of a high mean cell residence time (MCRT), by recirculation of sludge facilitating higher substrate conversion efficiencies and cost effective biogas designs.

Two phase CSTR system comprised of (*a*) acid phase, and (*b*) methane phase respectively.

Acid phase is described as follows :

The optimum hydraulic retention time in the acid phase of diphasic system has been evaluated from the maximum product formation criterion through the following steps :

- Steady state concentration of substrate and micro-organisms at various HRTs have been used to estimate the kinetic coefficients as defined by Metcalf and Eddy Inc.[120]
- The theoretical effluent substrate concentration(s) can be calculated from the equation as given by Metcalf and Eddy Inc.

$$S = \frac{Ks\,(1+\theta k_d)}{\theta\,(Yk-k_d)-1} \quad ...(8.478)$$

- The product output rate (P) can be calculated from the equation developed by Ghosh and Klass[117]

$$P = \frac{\alpha(U_e + m\theta)\,(S_o - S)}{\theta(U_p + U_e + m\theta)} \quad ...(8.479)$$

where α, U_e, U_p, m, k, and k_d refer to time product yield constant (mg/mg), substrate utilization coefficient for energy, substrate utilization coefficient for growth, maintenance coefficient (d^{-1}), maximum rate of substrate utilization per unit mass of micro-organisms and endogenous decay coefficient respectively.

Methane phase of CSTR is highlighted as follows :

- Steady state concentration of substrate volatile acids and micro-organisms at various HRTs and MCRTs can be used to estimate the kinetic coefficients.
- Minimum mean cell residence time (θ_c^m) can be estimated by using the conventional CSTR model equation.

$$\frac{1}{\theta_c^{\,m}} = \frac{YkS_o}{K_a + S_o} - k_d \qquad ...(8.480)$$

- A minimum safety factor (SF) of about 2.4 can be applied to θ_c^m for estimating θ_c design (θ_c^d) by using the conventional CSTR equation

$$\theta_c^{\,d} = \theta_c^{\,m} \times SF \qquad ...(8.481)$$

- The average values of $X\gamma$ and X can be plugged into the following conventional CSTR equation

$$\theta = \theta_c\,(1 + \gamma - (X\gamma/X)\gamma) \qquad ...(8.482)$$

where $X\gamma$, X, θ_c, θ and γ refer to biomass concentration in recycle stream, biomass concentration in reactor, MCRT, HRT and recycle ratio respectively.

- Value of sludge density (w), total head (H), hydraulic flow (Q) and efficiency (η) can be used to calculate break horse power (BHP) for various values of γ

$$BHP = \frac{WQ\gamma H}{75\eta} \qquad ...(8.483)$$

- Total methane production rate is equal to the product of specific biogas yield and total substrate removed. The specific biogas yield is assumed to be constant.

8.6.3 Down flow Stationary Film (DSF) Anaerobic Attached Growth Reactor[121]

This section examines basic reactor operating parameters to be used in DSF design with an optimum surface area to volume ratio, determine kinetic constants and predict reactor performance. Monod kinetics successfully applied to CSTR process have been applied to DSF reactors. The empirical steady state model is based on the following assumptions :

- Anaerobic digestion is a process characterized by a heterogenous microbial population which metabolises components of complex wastes. The rate limiting substrate are soluble organics are determined by chemical oxygen demand (COD) analysis.

- Substrate completely penetrates the biofilm and no substrate gradient exists across the bio-film depth, L_f (Thiele Modulus = 0).
- Biofilm accumulation rate is small and can be neglected.
- Concentration of non-biodegradable organic material in the reactor is small and can be neglected.
- Reactor is completely mixed in terms of soluble substrate.
- The surface area to reactor volume ratio is in the optimum range.

Based on the previous assumptions a steady state mass balance on substrate and biomass in the reactor results in equations 8.484 and 8.485 :

$$X = \frac{Y(S_o - S)\,Q/V}{\mu} \qquad ...(8.484)$$

$$\mu = \frac{QX_L}{VX} + k_d \qquad ...(8.485)$$

where X, X_L and V refer to reactor biomass concentration (g – volatile suspended solids and attached film solids), suspended biomass concentration and reactor volume respectively. The distribution of cell ages in a biological waste treatment process is such that significant number of cell are not in the log growth phase. Equation 8.485 incorporates decay rate to account for endogenous material respiration.

SRT(θ_s) can be defined as :

$$\theta_s = \frac{VX}{QX_L} \qquad ...(8.486)$$

Equation 8.430 can be re-arranged as :

$$\mu = 1/\theta_s + k_d \text{ or} \qquad ...(8.487)$$

$$\mu = \mu_0 + k_d \qquad ...(8.488)$$

where μ_o refers to net specific growth rate.

If a model is to describe DSF reactor performance, it is important to know the relationship between HRT and SRT. In CSTR biological process μ_0 and 1/SRT are equal to 1/HRT. This indicates that biomass growth and removal are dependent on HRT. In DSF reactors biomass is attached to and growing on support media. In this case μ_o and 1/SRT are equal to 1/HRT multiplied by a washout factor. It is difficult to measure SRT in DSF reactors; however, determination of washout factor for wastes will enable reactor performance to be predicted as a function of easily measured parameters such as HRT of IR.

If the biofilm accumulation rate is small and is neglected, the μ_o for each steady state can be described by equation 8.489.

$$\mu_o = \frac{1}{\theta_s} - \frac{f}{\theta_n} \qquad ...(8.489)$$

where f and θ_n refer washout factor and HRT respectively.

Implicit in the calculation of kinetic parameters in the assumption that the system is at steady state and biomass concentration remains constant during the evaluation period. With the large biomass concentration in DSF reactors and low growth rates of anaerobic bacteria, biomass accumulation rate will be insignificant. Specific growth rate of bacteria can be described by equation 8.490.

$$\mu = \frac{\mu_m S}{Ks + S} \qquad ...(8.490)$$

Microbial substrate used is related to bacterial growth by equation 8.491:

$$U = \frac{-kS}{(Ks + S)} \qquad ...(8.491)$$

where, U refers to specific utilization rate (gCOD/gVSFS-d).

Equation 8.491 can be linearised using the Hanes method[122] which enhances the kinetic parameters, k and ks to be determined from routine, reactor measurements. Equations 8.488 and 8.491 can be combined and rearranged to given a linear expression for μ_0

$$\mu_0 = -\ YU - k_d \qquad ...(8.492)$$

This equation indicates that net specific growth rate (μ_0) is directly related to specific substrate use rate (U). A plot of θ_0 versus U allows Y and k_d to be determined. The maximum specific growth rate (μ_m) can be determined as follows:

$$\mu_m = Yk \qquad ...(8.493)$$

These constants are valuable for kinetic analysis and enable determination of amount of solids that accumulate in the reactor. Combining equations 8.484, 8.488 and 8.490 COD removal efficiency (E) and reactor biomass concentration (X) can be predicted as a function of HRT :

$$E = \left[1 - \frac{K_s((f/\theta_n) + k_d)}{S_0(\mu_m - (f/\theta_n) - k_d}\right] 100 \qquad ...(8.494)$$

$$X = \frac{Y(S_0 - S)}{(f/\theta_n + k_d)\theta_n} \qquad ...(8.495)$$

'X' describes the total biomass concentration in the reactor. Biofilm biomass can be determined by subtracting the suspended biomass from the total biomass concentration (X).

The organic carbon balance for steady state operating conditions is given as :

$$COD_i = sCOD_e + gCOD_p + mCOD_e + is\ COD_e + bCOD \quad ...(8.496)$$

where, COD_i, $sCOD_e$, $gCOD_p$, $mCOD_e$, is COD_e, and bCOD refers to total influent COD, soluble effluent COD, gaseous methane produced (COD equivalent), methane in effluent (COD equivalent), insoluble COD in the effluent and biomass accumulated in biofilm (COD equivalent) respectively.

Total methane production rate is product of specific biogas yield and total substrate removal rate. It is assumed that specific biogas yield is constant.

8.6.4 Anaerobic Baffled Reactor (ABR)[123]

This new process uses a series of vertical baffles to enforce wastewater flow under and over them as it passes from the influent to the effluent. The bacteria within the reactor tend to rise and settle with gas production, but move horizontally at a relatively slow rate. The wastewater can, therefore, come into contact with large active biological mass as it passes through the ABR and the effluent is relatively free of biological solids.

The fixed film model provides a better agreement between fundamental parameters and reactor performance. The apparent reason is that the sludge particles within the reactor sludge blanket act as fluidized spheres with surface area through which the solute must diffuse for consumption. The deep fixed film model used was based upon that described by Williamson and McCarty, *i.e.*, :

$$D_f \frac{\delta^2 S_f}{\delta z^2} = \frac{k\, S_f X_f}{(K_a + X_f)} \quad ...(8.497)$$

and as modified by Rittman and McCarty to variable order model :

$$\gamma_{ai} = C_i S_i^{qi} \quad ...(8.498)$$

Where, γ_{ai}, C, q, S and i refer to specific reaction rate for substrate consumption per unit surface area of biofilm, variable order reaction coefficient, which is function of D_f, K_a, k, X_f and the molecular diffusivity in bulk liquid (Dw), variable reaction order, which is a function of S, the length of effective diffusion layer (L) and the parameter associated with C and location identification.

Equation 8.498 is specifically derived for a flat plate with a biofilm of sufficient depth so that the substrate concentration approaches zero within biofilm. Application to more nearly spherical as in the ABR is satisfactory if the particle diameter relative to its effective biofilm depth is sufficiently large. The ABR can be thought to consist of a series of upflow reactors connected by means of baffles. Recycle can be employed to reduce the extent of organic acid production in the first chamber that might cause a depressed pH. Recycle, ineffect made the plug flow system act more like a completely mixed system.

Each chamber with the ABR is considered to be completely mixed as a result of rapid gas evolution at high loadings. With specific flow rate, Q_i (flow rate per unit volume of chamber, i, including recycle flow) and a specific biofilm surface area, a_i (surface area per unit chamber, i, liquid volume), a mass balance on substrate give the change of substrate concentration with time in chamber, i.:

$$\frac{dS_i}{dt} = -a_iC_iS_i^{qi} + Q_iS_{i-1} - Q_iS_i \qquad ...(8.499)$$

where S_i and S_{i-1} refer to bulk liquid substrate concentration in chamber, i and the influent the chamber i respectively. At steady state, equations 8.498 and 8.499 can be combined and solved for S_i :

$$S_i = S_{i-1} - (a_i/Q_i)\, C_iS_i^{qi} \qquad ...(8.500)$$

Equation 8.500 can be solved iteratively to determine S_i for each chamber with a knowledge of the appropriate kinetic coefficients, bio film surface area within each chamber, influent wastewater substrate concentration and flow rate and recycle flow rate. A mass balance of influent and recycle streams can be made to estimate the substrate concentration of the stream entering the reactor. An initial value can be assumed for recycle stream concentration, and then refined through successive iterations around the reactor.

There are group of parameters that are specific for a given reactor and need to be measured or specified for or given system. These are Q_o, Q_r, S_o, T, a_i and L. The diffusion layer depth (L) can be determined from the liquid phase mass transfer coefficient (k_L) within the sludge bed as estimated from the relationship of Willson and Geankopolis[124] for a packed bed at very low Reynolds numbers :

$$k_L = \frac{1.09\,(Re)^{2/3}\,(Sc)^{2/3})u}{\varepsilon} \qquad ...(8.501)$$

where ε refers to liquid void fraction (liquid volume/liquid plus biomass volume). This relationship is valid for Re is the range of 0.0016 to 55 and void fraction (ε) in the range of 0.35 to 0.75.

The parameter L is related to liquid phase mass transfer coefficient (k_L) according to :

$$L = Dw/k_L \qquad ...(8.502)$$

It may perhaps be better to use the calculated mass transfer coefficient (k_L) directly but the model as originally developed is based upon the concept of a diffusion layer with depth, L. In either case the results are same.

It is apparent that a biological sludge blanket, is hardly an ideal packed bed system. It is assumed that the bioflocs are spherical with dismeter d and the flocs touch each other. Another assumption is made that there exists no gas liquid interphase inside a floc and the gas bubbles grow at the outside surface of the floc. Gas production will increase the mixing of the bulk liquid phase, a phenomena not considered by in the relationship of Wilson and Geankopolis. On the other hand, within dense sludge phase, the mass transfer is actually decreased. Therefore, gas production may not be considered in calculating k_L. Further, it is assumed that dead spaces within reactor have no negative impacts.

Another important parameter to be evaluated is the fraction of VSS that represents active methanogenic acetate using micro-organisms, the value of which is termed as f_{ai}.

This reactor is simple in design and requires no special gas or sludge separation system. The over and under liquid flow reduces washout and does not require unusual settling properties per microbial culture. The ABR can be operated for long periods of time on soluble waste without sludge wasting.

The most appropriate organic wastewater for this system is a completely soluble one. Small amounts of degradable suspended solids could be accepted without significantly diluting the biological solids within the chambers. Colloidal suspensions such as starches, and dilute milk wastes should be treatable. Poorly bio-degradable suspended material would dilute the microbial solids and therefore, are likely to reduce the maximum volumetric rates that are possible.

The biofilm model developed appears to be applicable to sludge blanket reactors. It is not yet feasible to predict the optimum design for an operational ABR scale up factors are in general difficult to predict. Mixing by gas evolution should be more important in large reactors due to the fact that gas is produced throughout the whole column height. This should lead to greater evolution of gas per unit horizontal cross-sectional area in full scale system, leading to more complete mixing throughout the chamber. It should also result in greater turbulence and resulting in higher mass transfer rates. This is likely to lead to better efficiencies in large scale reactors. However, other scale factors such as liquid upflow velocities may decrease these advantages.

Further, research is needed on the effect of ABR performance through recirculation, frequency of biological solids wasting and temperature. Also requiring evaluation are feasibility of intermitte and several operation and system economics. Most important, a field scale evaluation is needed to better evaluate the ABR's potential.

8.6.5 Anaerobic Batch and Semi-Continuous CSTR System[126]

This section presents a model of microbial growth, substrate utilization and biogas production by a set of non-linear differential equations.

Batch system equations are :

$$\frac{dX}{dt} = \mu_{max} \frac{SX}{(Ks + S)} \quad \text{and} \qquad ...(8.503)$$

$$\frac{dS}{dt} = -\frac{\mu_{max}}{Y}\left(\frac{S}{(Ks+S)}\right) \quad ...(8.504)$$

Where, Ks, S and S_0 refer to half velocity saturation constant, substrate concentration and initial substrate concentration respectively. The interacting differential equations describing the dynamic behaviour of the continuous system are :

$$\frac{dS}{dt} = -\frac{\mu_{max} S(X/Y)}{(Ks+S)} + (S_0 - S) D \quad ...(8.505)$$

$$\frac{dX}{dt} = -\frac{\mu_{max} SX}{(Ks+S)} + (X_0 - X) D \quad ...(8.506)$$

where, S_0 and X_0 refer to influent substrate concentration and initial microbial concentration respectively. In dimensionless forms the above equations take the forms as:

$$\frac{dX_n}{dt_n} = \frac{X_n S_n}{b+S_n} \quad ...(8.507)$$

$$\frac{dS_n}{dt_n} = -\frac{X_n S_n}{b+S_n} \quad ...(8.508)$$

where, $X_n = (X/Y)S_0$; $S_n = S/S_0$ and $t_n = t\ \mu_{max}$ respectively:

$$\frac{dX_n}{dt_n} = \frac{X_n S_n}{b+S_n} + (X_{on} - X_n)D_n \quad ...(8.509)$$

$$\frac{dS_n}{dt_n} = -\frac{X_n S_n}{b+S_n} + (1 - S_n)D_n \quad ...(8.510)$$

Where $X_{on} = X_0/S_0$; and $D_n = \mu/\mu_{max}$ respectively :

Biogas production rate (Q_G) in nondimentional form is :

$$Q_G = D_n (1 - S_n) \quad ...(8.511)$$

8.6.6 Anaerobic Batch CSTR System with Inhibitor[125]

This model considers two steps, *i.e.*, acidogenesis corresponding to the conversion of easily fermentable sugar to acetic acid and methanogenesis corresponding to the decarboxylation of this acid into methane and carbon dioxide. The metabolic production (acid or methane) are not as proportional to the respective production of biomass and each kinetic is represented by a different equation. The unionized acid concentration inhibits the micro-organisms growth (acidogenic and methanogenic) and the production rate of methane.

Microbial growth is represented by :

$$dX/dt = D(X_o - X) - k_d\, x + \mu X \qquad ...(8.512)$$

$$\mu = \mu_{max} \frac{1}{(1 + K/SUB + AH/K_i)} \qquad ...(8.513)$$

where, SUB, K, AH and K_i refer to acidogenic or methanogenic substrate concentration, constant, unionized acetic acid concentration and inhibition coefficient respectively. Mass balance during acidification and methanization are expressed by the following relationships for the substrate:

Acidogenesis

$$\frac{dS}{dt} = D\,(S_o - S) - \left[\frac{dS}{dt}\right]_{x_1} - \left[\frac{dS}{dt}\right]_a \qquad ...(8.514)$$

Methanogenesis

$$\frac{dA}{dt} = D(A_o - A) + \left[\frac{dA}{dt}\right]_p - \left[\frac{dA}{dt}\right]_{x_2} - \left[\frac{dA}{dt}\right]_a \qquad ...(8.515)$$

Expressions of Rates

Acidogenesis

Conversion of substrate to acidogenic biomass and acetic acid are expressed as follows :

$$\left[\frac{dS}{dt}\right]_{x_1} = \frac{\mu_1 X_1}{Y_{15}} \qquad ...(8.516)$$

$$\left[\frac{dS}{dt}\right]_a = Ksx_1 \,.\, X_1 + Knx_1\, X_1 \frac{S}{(K_{sn} + S)} \qquad ...(8.517)$$

The latter equation reflects the energy conservation during substrate catabolism. It includes two terms corresponding to an energy consumption for growth and for maintenance. Therefore, the rate of acetate production is :

$$\left[\frac{dA}{dt}\right]_p = Y_{sa} \left[\frac{dS}{dt}\right]_a \qquad — (8.518)$$

Methanogenesis

The rate of acetate consumption for methanogenic biomass production $(dA/dt)X_2$ and decarboxylation into methane and carbon dioxide $(dA/dt)_m$ are :

$$\left[\frac{dA}{dt}\right]_{x_2} = \frac{\mu_2 X_2}{Yx_2 a} \quad \text{and} \qquad ...(8.519)$$

$$\left[\frac{dA}{dt}\right]_m = \left[\frac{dCH_4}{dt}\right]\frac{1}{Y_{ma}} \quad ...(8.520)$$

The unionized acid concentration (AH) is calculated from the total acetate acid dissociation concentration as follows :

$$(AH) = \frac{(A)}{1+(K_e/H^+)} \quad ...(8.521)$$

Methane Production

The rate of methane production is described by an equation derived from that of Yarovenko and Nakhmanovich[126] which is as follows :

$$\left(\frac{dCH_4}{dt}\right) = V_m^{max}(X_2)\frac{(AH)}{AH+K_m}\cdot\frac{K_{im}}{K_{in}+AH} \quad ...(8.522)$$

The above equation require a numerical analysis and analytical solutions are not possible. The nomenclature for various mathematical terms are defined in Table 8.7.

Table 8.7 : Nomenclature for Various Mathematical Terms Used in Model 5.6.6.

S.N.	*Term*	*Definition*
1.	dA/dt	variation rate of total acetic acid concentration ($kg/m^3.d$)
2.	dCH_4/dt	methane production rate ($kg/m^3.d$)
3.	dS/dt	variation rate of conversion of acidogenic phase substrate ($kg/m^3.d$)
4.	dX/dt	variation rate of micro-organisms concentration ($kg/m^3.d$)
5.	A	total acetic acid concentration (kg/m^3)
6.	AH	unionized acetic acid (kg/m^3)
7.	D	dilution rate (d^{-1})
8.	H*	hydrogen ion concentration ($mol/10^{-3}m^3$)
9.	K	constant
10.	S	substrate concentration (kg/m^3)
11.	SUB	acidogenic or methanogenic substrate concentration (kg/m^3)
12.	V	specific production rate (kg CH_4/m^3/kg of methanogenic bacteria)
13.	a	relative to acetic acid
14.	d	relative to micro-organism death
15.	e	relative to acid dissociation
16.	m	relative to methane
17.	n	relative to maintenance energy
18.	p	produced by micro-organisms
19.	S	relative to substrate
20.	x	relative to micro-organism
21.	θ	relative to influent
22.	1	relative to acidogenic phase
23.	2	relative to methanogenic phase

8.6.7. Another Approach to Anaerobic CSTR System[127]

For the design industrial biogas reactor, the dependence of biogas productivity (Q_G/V_R) or the digestion tank loading rate (B_R) must be known. This can be established experimentally. Process modelling for continuous reactor involving back mixing without recycle is proposed as follows :

From Monod kinetics it follows that the product is formed according to :

$$\gamma_p = k_p \frac{C_{kom}}{K_a + C_{kom}} C_x \quad \text{and with simplification} \qquad ...(8.523)$$

$$\gamma_p = k'_p \, C_{kom} \, C_x \qquad ...(8.524)$$

It follows that biogas production in the single-stage stirred tank reactor proceeds according to :

$$\frac{Q_G}{V_R} = \frac{C^o{}_{max} - C_{kom}}{t} \frac{1}{\rho_G} = k_p C_{kom} C_x \qquad ...(8.525)$$

Assuming that the growth of micro-organisms is self limiting :

$$\mu_{obs} = \mu_{max}\left(1 - \frac{C_x}{C_{x\,max}}\right) \qquad ...(8.526)$$

It follows from material balance for micro-organisms in the single-stage CSTR that :

$$C_x = C_{x\,max}\left(1 - \frac{1}{\bar{t}\,\mu_{max}}\right) \qquad ...(8.527)$$

$$= C_{x\,max}\left(1 - \frac{\bar{t}*}{\bar{t}}\right) \qquad ...(8.528)$$

From the experimental data it can be proved that :

$$Y = Y_{max}\left(1 - \frac{\bar{t}*}{\bar{t}}\right) = Y_{max}\left(1 - \frac{B_R}{B_{R\,max}}\right) \qquad ...(8.529)$$

Therefore, it follows that :

$$\frac{Q_G}{V_R} = Y_{max}\left(1 - \frac{\bar{t}*}{\bar{t}}\right) B_R \qquad ...(8.530)$$

Equation 8.530 can be used to design industrial biogas reactors after the model constants ($Y_{max}, \bar{t}*$) have been determined.

where B_R, C_i, k, k', K_a, Q_G, γ, $\bar{t}$, $\bar{t}^*$, V_R, Y, ρ_G, O, k_{om}, X, P, and max refer to mass flow rate per volume (M_{OM} $L^{-3}T^{-1}$), total concentration of material i(M_iL^{-3}), reaction rate constant (T^{-1}), pseudo reaction rate constant ($M^{-1}L^3T^{-1}$), saturation constant (ML^{-3}), gas volume flow rate (L^3T^{-1}), reaction rate ($ML^{-3}T^{-1}$), mean HRT (T), maximum mean HRT (T), reaction gas volume (L^3), gas yield coefficient ($L^3M_{om}^{-1}$), gas density (ML^{-3}), initial value, convertible organic mass, biomass, product and maximal value respectively.

8.7 Upflow Anaerobic Sluge Blanket Reactor

The knowledge about the flow pattern in any type of reactor system is essential. UASB is a new concept, therefore, it is essential to have the basics of these flow pattern in this type of reactor system. Residence time distribution studies can be performed by means of injection of LiCl solution as a tracer in the influent of the reactor and measurement of the response of this stimulus on several locations in the reactor and in the influent. The dimensionless concentration time curve for the model's effluent is obtained by shifting the calculated $C_d/C_o - \theta$ curve horizontally over the reduced residence time of plug flow region ($\theta_{pf} = V_{pf}/V_R$). The value of the region of dead space in the reactor can be estimated by means of the measured F-curve, because

$$\int_0^\infty [1 - F(\theta)]\, d\theta = 1 \qquad ...(8.531)$$

and the integral is equal to the area between F curve and line $C/C_o = 1$ in the $C/C_o - \theta$ diagram. If in the experimental results this area turns to be smaller than unity, the reactor volume occupied by the flowing liquid must be smaller than the real volume of the reactor. The difference between these volumes is the volume of the region of dead space. Two flow patterns have been suggested and the results are similar[138].

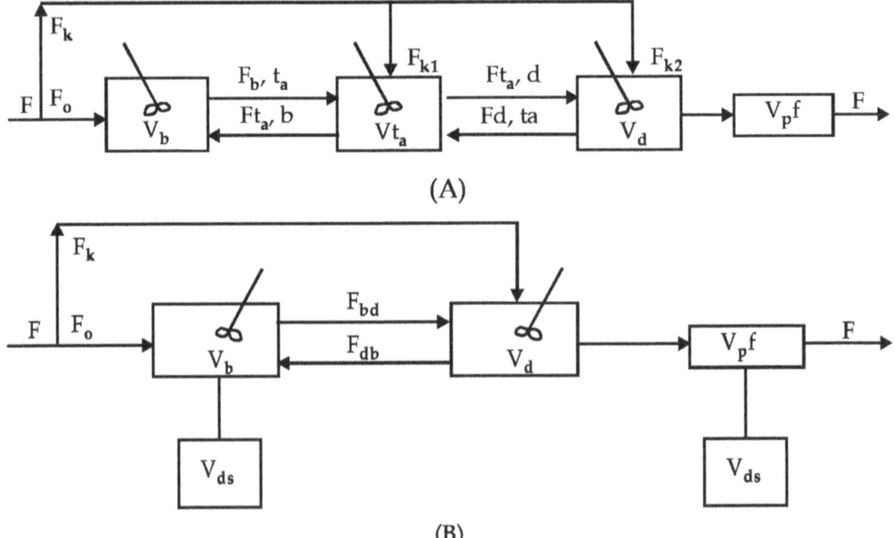

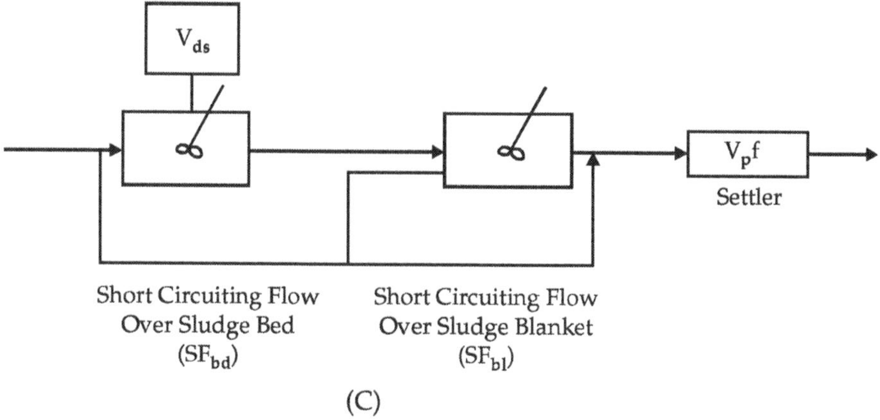

(C)

Fig. 8.13 : Block Diagram of Fluid Flow Pattern in UASB Reactor

First flow pattern is shown in Fig. 8.13 and the necessary equations are :

$$V_R = V_b + V_{ta} + V_d + V_{pf} \text{ [three mixing zone + plug flow + bypass]} \quad ...(8.532)$$

$$F = F_o + F_k \quad ...(8.533)$$

$$F_k = F_{k1} + F_{k2} \quad ...(8.534)$$

$$F_{b'ta} = F_o + F_{ta,b} \quad ...(8.535)$$

$$F_{ta,d} = F_{b'ta} + F_{k1} + F_{d'ta} - F_{ta'b} \quad ...(8.536)$$

$$\frac{dC_b}{dt} = [F_oC_o + F_{ta}\,C_{ta} - F_{b'ta}\,C_b]/V_b \quad ...(8.537)$$

$$\frac{dC_{ta}}{dt} = [F_{k1}C_o + F_{b'ta}\,C_b + F_{d'ta}C_d - F_{ta'b}C_{ta}]/V_{ta} \quad ...(8.538)$$

$$\frac{dC_d}{dt} = [F_{k2}C_o + F_{ta'd}\,C_{ta} - F_{d'ta}C_d - FC_d]/V_d \quad ...(8.539)$$

Second flow is also shown in Fig 8.13 and the necessary equations are :

$$V_R = V_b + V_d\,V_{pf} + V_{ds} \text{ [two mixing zone + plug flow + bypass + dead space]} \quad ...(8.540)$$

$$F = F_o + F_k \quad ...(8.541)$$

$$F_{bd} = F_o + F_{db} \quad ...(8.542)$$

$$\frac{dC_b}{dt} = [F_oC_o + F_{db}C_d - F_{bd}C_b]/V_b \quad ...(8.543)$$

$$\frac{dC_d}{dt} = [F_kC_o + F_{bd}C_b - F_{db} - FC_d]/V_d \quad ...(8.544)$$

Where, F F, V, b, bd, d, db, ds, R, ta and pf refer to flow rate, response on an up-step stimulus (dimensionless) volume, sludge bed, from bed to blanket, form blanket to bed, dead space, reactor, area of transition, and plug flow respectively.

The above equation are coupled and can be solved with the aid of an computer. The following conclusion can be derived :

- At the influent concentration and liquid residence time applied (0.50-1.43 kg C/m^3 and 4.4-9.8 h respectively), the gas produced by the anaerobic bacteria provides good mixing of the fluid inside the reactor. In order to prevent turbulence in the settler on the top of the reactor, this region has to be shielded effectively from the region in the reactor where anaerobic conversion take place. This could be effected by means of plates placed at an inclination of 45°.
- High concentration of mud in the blanket (C_{md} greater than 30 kg/SS/m^3) may cause extra thickening of the suspended solids in the bed, resulting in occurrence of dead space.
- Upflow reactor for anaerobic treatment of wastewater should contain so much sludge that the height of the sludge bed is 1.5-2.5m. At bed heights of 0.4m most of the influent bypasses the bed.
- Scaling up of upflow reactors in horizontal direction is no problem if the influent distribution has been given special attention. With the present knowledge the optimum reactor height seems to lie between 4 and 6m.

If an set up stimulus response experiment is simulated, the following mass balance can be derived for the tracer in sludge bed and sludge blanket :

$$V_{bd}\frac{dC_{bd}}{dt} = (1 - SF_{bd})\phi_v\, C_{inf} - (1 - SF_{bd})\phi_v\, C_{bd} \qquad ...(8.545)$$

$$= (- SF_{bd})\phi_v(C_{inf} - C_{bd})$$

$$V_{bi}\frac{dC_{bi}}{dt} = (- SF_{bi})\ \phi_v\ C_{bd} + (SF_{bi} - SF_{be})\phi_v\ C_{inf} \qquad ...(8.546)$$

$$- (1 - SF_{bi})\phi_v C_{bi}$$

$$C_{eff/t} = SF_{bi}\, C_{inf} + (1 - SF_{bi})C_{bi}\ 1\ t\text{-}\tau\text{-settler} \qquad ...(8.547)$$

The ratio of the short circuiting flow over sludge blanket and the short circuiting over the sludge bed is normally is a function of height of the sludge blanket. Because of the dependence of SR_{bi} on SF_{bd}, the quotient SF_{bi}/SF_{bd} is taken to avoid influences on changes is SF_{bd} on SF_{bi}. It is experimentally known that the short circuiting flow over the sludge blanket decreases with increasing height of the sludge blanket. The height of the sludge blanket is defined as[129]:

$$H_{bi} - H_{re} - H_{bd} - H_{se} = \text{Height of Sludge Blanket} \qquad ...(8.548)$$

Where, H_{bi}, H_{re}, and H_{se} refer to height of sludge blanket, reactor, sludge bed and settler respectively. An empirical formula has been developed to describe the flow model:

$$SF_{bd} = (-0.25\ H_{bd} + 0.95)\ (0.42\ v_{SG} + 0.44) \quad ...(8.549)$$

$$SF_{bi} = (0.16H^2_{ib} - 1.24H_{bi} + 2.5)\ (-0.16v^2_{SG} + 1.6v_{SG})\ (SF_{bd})...(8.550)$$

The definition of various terms is defined in Table 8.8. Where SF_{bd}, SF_{bd} H_{bd}, H_{bi}, and V_{sa} refer to short circuting flow over the sludge bed, short circulating flow over the sludge blanket, height of the sludge bed, height of sludge blanket and overall superficial gas velocity respectively.

Table 8.8 : Definition of Various Terms in Model 8.7

Sr.No.	*Terms*	*Definition*
1.	SF	short circulating flow as a fraction of influent flow
2.	C	concentration
3.	H	height
4.	V	reactor volume
5.	v	velocity
6.	ϕ_v	volumetric flow rate
7.	bd	bed
8.	bl	blanket
9.	eff	effluent
10.	G	gas phase
11.	inf	influent
12.	S	superficial

The optimal height of the sludge bed can be defined as the height for which the short circuiting flows are as small as possible. It can be concluded that for normal superficial velocity (1–1.5 m/h), a height between 3.5 and 4 m is sufficient. This is in contradiction to the results of former investigators[130-132] who found a minimum in the short circuiting flow over the sludge bed as a function of bed height for a height of ≈ 2.5m in the same system. Because the short circuiting flows are strongly affected by the superficial gas velocity, and some further investigations are desirable. Two adjustable parameters are necessary and sufficient to describe the fluid flow in a well-functioning (no dead space in the reactor) USAB reactor, *i.e.*, the short circuiting flow over the sludge blanket.

The absence of dead space under normal operating conditions indicate that adsorption of L* on materials with ion exchange capabilities can be neglected: the amount of tracer leaving the reactor equals the amount of tracer supplied with the influent.

8.7.1 Model for UASB[133]

The UASB reactor can be divided in three compartments : the sludge blanket and the sludge bed, both perfectly mixed w.r.t the liquid phase, and a plug flow region, assumed to describe the internal settler.

The sludge bed and sludge blanket donot have a constant volume. Therefore, the balance equations have to apply to mass instead of concentration since mass is a conservative state variable. However, dry weight measurements of micro-organisms in the sludge bed over longer periods showed a constant concentration of 80-90 kg dry weight/m^3, independent of the place in the sludge bed, Fig. 8.13(C) :

$$\frac{dM_{x'bd}}{dt} = \frac{d(V_{bd}C_{x'bd})}{dt} = V_{bd}\frac{d\varepsilon_{x'bd}}{dt} + C_{x'bd}\frac{dV}{dt} \quad ...(8.551)$$

$$dC_{x'bd}/dt = 0 \quad ...(8.552)$$

Therefore, it follows

$$\frac{dM_{x'bd}}{dt} = C_{x'bd}\frac{d(V_{bd})}{dt} \quad ...(8553)$$

The sludge bed volume is calculated from the actual sludge mass in the bed and constant concentration. This sludge bed volume is substrated from the total reactor volume minus the constant settler volume, the remaining volume is the sludge blanket. In combination with the actual sludge mass in the blanket the sludge concentration results.

- **Sludge Bed :** Liquid phase–substrate balance in the sludge is :

  $$dM_{a'bd}/dt = (\text{net transport}) - (\text{consumption}) \quad ...(8.554)$$

  The transport terms depends on the structure of the fluid flow model

  $$T_{a'bd} = (1 - SF_{bd})\phi_{in}(C_{S'in} - C_{S'bd}) \quad ...(8.555)$$

  The consumption of substrate is :

  $$\gamma_{a'bd} = \frac{\mu_{bd}}{Y_{a'x}}C_{x'bd} \quad ...(8.556)$$

- **Solid Phase :** The micro-organisms balance in the sludge is more complex:

  $$dM_{x'bd}/dt = \text{transport} + \text{growth} - \text{consumption including decay} \quad ...(8.557)$$

  Concerning the sludge transport, it holds in a presudo-steady state that Biomass flow from bed to blanket (dragging) = biomass flow from blanket to bed (settling) ... (8.558)

  Dragging flow = $X\phi_{g'bd}C_{x'bd}$, where X is a dragging constant, ...(8.559)

$$\text{Settling flow} = \frac{M_{x'bi}}{\text{average settling time}} \quad ...(8.560)$$

$$M_{x'bi} = h_b\, A\, C_{x'bi} \quad ...(8.561)$$

$$\text{Average settling time} = h_{bi}/v_s \quad ...(8.562)$$

The particles are assumed to fall over the height of the sludge blanket, equation 8.557 results in :

$$X\, \phi_{g'bd}\, C_{x'bd} = v_s\, A\, C_{bi} \quad ...(8.563)$$

The transport of sludge from be to blanket is due to gas produced in and on the sludge particles. From the impulse balance it follows :

$$\text{upward forces} = \text{gravitational forces} \quad ...(8.564)$$

$$(\rho_L - \rho_G) = X'\, (\rho_s - \rho_L) \quad ...(8.565)$$

Where X′ is a newly defined drag coefficient :

$$X' = \frac{\rho_L - \rho_G}{\rho_S - \rho_L} = \frac{1000 - 1}{1070 - 1000} = 14 \quad ...(8.566)$$

Every single gas bubble has the ability to carry 14 times its own volume biomass upward.

- **Production :** The micro-organisms use fatty acid to reproduce themselves. This growth rate can be described by :

$$\gamma_{x'bd} = Y_{a'x}\, \gamma_{a'bd} \quad ...(8.567)$$

multiplied by the sludge bed volume, this gives the biomass production per unit time.

- **Endogenous Respiration and Decay :** This aspect is neglected because the possibility to store sludge over long periods of time indicates low maintenance energy requirement.

- **Gas Phase :** Solubility of CH_4 in water is negligible, so amount of CH_4 formed in the sludge bed and sludge blanket can be calculated. The yield of CH_4 on substrate, Y_{s,CH_4} follows from the stoichiometric equation :

$$CH_3COOH \rightarrow 0.04 \text{ biomass} + 0.94\, CO_2 + 0.94\, CH_4 \quad ...(8.568)$$

$Y'_{S,CH_4} = 0.94$ mole CH_4/mole substrate or

$Y'_{S,CH_4} = 0.25$ kg CH_4/kg substrate

The CH_4 flow evolving from sludge bed is :

$$\phi_{S,CH_{4,bd}} = \overline{V}\, V_{bd}\, Y_{s,CH_4}\, \gamma_{s,bd}/MM_{CH_4},(m^3/h) \quad ...(8.569)$$

- **Sludge Bed :** The substrate balance over the sludge blanket is :

$$\frac{dM_{x,bi}}{dt} = (1\text{–}SF_{bi})\phi_m C_{s'bd} + (SF_{bd} - SF_{bi})\phi_m\ C_{s'in}$$

$$- (1 - SF_{bi})\phi_{in'}\ C_{s'bi} - \gamma s_{'bi}\ V_{bl} \quad ...(8.570)$$

The balance for micro-organisms over the sludge blanket is :

$$\frac{dM_{x,bl}}{dt} = \eta_{dr}\ x'\phi_{g'bd} - C_{s'bi}\ v_o A + \gamma_{x'bl} + (1 - \eta)(1\text{–}SF_{bi})\phi_{in} C_{x'bl} \quad ...(8.571)$$

The last terms represents mass flow of sludge that cannot be retained by settler and washout; η is the efficiency of the settler and is defined as:

$$\eta = \frac{1 - C_{x'eff}}{C_{x'bl}} \quad ...(8.572)$$

Table 8.8 : Definition of Various Terms for Model 8.7.1

Sr.No.	*Terms*	*Definition*
1.	P	pressure (bar)
2.	m	molar
3.	g/G	gas
4.	ϕ	volumetric flow rate
5.	γ	conversion rate
6.	dr	drag coefficient
7.	X	biomass
8.	L	liquid
9.	X′	drag coefficient based on biomass base
10.	T	transport
11.	s	substrate
12.	S	solid
13.	$\overline{V}$	molar volume (m^3/mol)
14.	V	volume
15.	MM	molar mass (kg/kg mol)
16.	M	mass
17.	A	area of reactor
18.	h	height

The total CH_4 flow leaving the sludge blanket is

$$\phi_{g'bi} = \phi_{CH_4'bi} + \phi_{g'bd}\frac{P_{bd}}{P_{bi}} \quad ...(8.573)$$

- **Settler :** The gas is separated from the liquid and particles before settler. Dissolved materials are retarded as long as the retention time of the settler compartment. This does not hold for the gas. The nomenclature of various mathematical terms are give in table 8.8.

8.7.2 Model for USAB[134]

The flow model is presented in Fig. 8.13(A) and the mass balance for the substrate over the bed is as follows :

$$V_b \frac{dC_{fa'b}}{dt} = F_o\, C_{fa'o} + F_{db}\, C_{fa'd} - q_b\, C_{mb} V_b \qquad ...(8.574)$$

The sludge concentration in the bed (C_{mb}) has been taken as constant, furthermore $V_b = h_b\, A_R$

For specific conversion rate of substrate, the following equation hold

$$q_b = q_{max,fa} \frac{C_{fa'b}}{K_{fa} + C_{fa'b}} \qquad ...(8.575)$$

At steady state equation 8.574 can be simplified to :

$$F_o\, C_{fa'o} + F_{db}\, C_{fa'd} - F_{bd}\, C_{fa'b} - q_b\, C_{mb}\, V_b = 0 \qquad ...(8.576)$$

Substrate conversion in the blanket can be described by means of mass balance that is analogous to equation 8.574, *i.e.* :

$$V_d \frac{dC_{fa'd}}{dt} = F_k\, C_{fa'o} + F_{bd}\, C_{fa'b} - F_{db} C_{fa'd} - FC_{fa'd} - q_d\, C_{md}\, V_d \qquad ...(8.577)$$

$$V_d = h_d A_R \text{ and} \qquad ...(8.578)$$

$$q_d = q_{max,fa} \frac{C_{fa'd}}{(K_{fa} + C_{fa'd})} \qquad ...(8.579)$$

At steady state 8.577 simplifies to :

$$F_k\, C_{fa'o} + F_{bd} C_{fa'b} - F_{db}\, C_{fa'd} - FC_{fa'd} - q_d\, C_{md}\, V_d = 0 \qquad ..(8.580)$$

The sludge concentration (C_{md}) can be calculated using the following formulation :

$$C_{md} = \frac{\phi_{gb}' K_{tb}\, C_{mb}}{\phi_{gb}' K_{td} + (V_s - V_L) A_R}, \qquad ...(8.581)$$

where $$\phi_{gb'} = \phi_{gb} \frac{10}{(10 + h_a + h_{st})} \qquad ...(8.582)$$

It is assumed that the volume of backmixing stream (F_{ab}) is equal to the volume of gas that escapes per unit time out of the bed (experimental evidence), *i.e.* :

$$F_{ab} = 1.0\ \phi'_{g,b} \qquad ...(8.583)$$

Experimental data revealed that in UASB, the bypassing flow rate is dependent on the height of the sludge bed and is related as follows:

$$= (0.13 + 0.17h_b)F \qquad ...(8.584)$$

The two mass balance in combined with other equations presented can be used to calculate the quantitative values of $C_{fa'b}$ and $C_{fa'd}$:

Gas production in the bed can calculated from the following relationship:

$$\phi_{gb} = \frac{\gamma_{me'b}\ V_b}{C_{me}} \qquad ...(8.585)$$

where $$\gamma_{me'b} = me_{max'fa}\left[\frac{C_{fa'b}\ C_{mb}}{K_{me} + C_{fa'b}}\right] \qquad ...(8.586)$$

Gas production in the blanket can be calculated from the following relationship.

$$\phi_{gd} = \frac{\gamma_{me'd} V_d}{C_{me}} \qquad ...(8.587)$$

where, $$\gamma_{me'b} = me_{max'fa}\frac{C_{fa'd}\ C_{md}}{(K_{me} + C_{fa'd})} \qquad ...\ (8.588)$$

Total gas production in the system is :

$$\phi_g = \phi_{gb} + \phi_{gd} = \frac{\gamma_{me'b}\ V_b + \gamma_{me'd}\ V_d}{C_{me}} \qquad ...\ (8.589)$$

Mass balance for the macro-organisms in the bed is given :

$$V_b \frac{dC_{ap'b}}{dt} = F_{db}C_{ap'd} - F_{bd}C_{ap'b} + \gamma_{ab'b}V_b \qquad ...\ (8.590)$$

Where, $\gamma_{ap'b} = K_{ap}(C_{fa's} - C_{fa'b})$ ... (8.591)

Bacterial mass balance in the blanket is given by :

$$V_d \frac{dC_{ap'd}}{dt} = F_{db}C_{ap'b} - F_{db}C_{ap'd} - FC_{ap'd} - \gamma_{ab'b}V_d \qquad ...(8.592)$$

Where, $\gamma_{ap'd} = K_{ap}(C_{fa'o} - C_{fa'd})$ .... (8.593)

At steady state both above equation are equal to zero and the bacterial waste products in the effluent can be estimated :

$$C_{ap'd} = \frac{(\gamma_{ap'b} V_b + \gamma_{ap'd} V_d)}{F}, \text{ As} \qquad ...(8.594)$$

$$V_b = A_R h_b;\ V_d = A_R h_d \text{ and } F = V_L A_R$$

Therefore,

$$C_{ap'd} = \frac{(\gamma_{ap'b} h_b + r_{ap'd} h_d)}{V_L} \qquad ...(8.595)$$

Substrate conversion can be estimated, because

- $C_{a,e} = C_{fa'd} + C_{ap'd}$ and the overall conversion can be calculated by : .... (8.596)

- $$\xi_a = \frac{C_{fa,o}\ C_{s,e}}{C_{fa,o}} \qquad ... (8.597)$$

The specific organic loading rate (kg/Cm^3.d) that is converted in the reactor is given by

- $$\phi'''_s = \frac{F\xi_a\ C_{fa,o}}{V_R} \qquad ...(8.598)$$

The above analysis is a complete scenario even under strongly loaded conditions. In practice overloading should be avoided. In such a situation, the fatty acid concentrations in the bed and the blanket are small and almost equal which also means that the specific sludge loads in both reactor parts are nearly equal. The simplification of basic material balances results in the following :

$$V_b \frac{C_{fa'b}}{dt} = F_o (C_{fa'o} - C_{fa}) - q\, C_{mb} V_b \qquad ...(8.599)$$

$$V_d \frac{dC_{fa'd}}{dt} = F_R (C_{fa'o} - C_{fa}) - q\, C_{md} V_d \qquad ...(8.600)$$

Under steady state

$$F_o(C_{fa'o} - C_{fa}) = q\ C_{mb}\ V_b \qquad ...(8.601)$$

$$F_k(C_{fa'o} - C_{fa}) = q\ C_{mb}\ V_b \qquad ...(8.602)$$

The bacterial waste products are given by :

$$F_{db}\ C_{ap} - F_{bd}\ C_{ap} + \gamma_{ap'b}\ V_b = 0 \qquad ...(8.603)$$

$$F_{bd}\, C_{ap} - F_{db}\, C_{ap} - F\, C_{ap} + \gamma_{ap,d}\, V_d = 0 \quad ...(8.604)$$

These two equations can be solved and the result is :

$$FC_{ap} = \gamma_{ap'b}\, V_b + \gamma_{ap'd}\, V_d \quad ...(8.605)$$

Table 8.9 : Definition of Various Mathematical Terms in Model 8.7.2.

Sr.No.	*Term*	*Definition*
1.	a, K, k	constant
2.	A	surface area
3.	C	concentration
4.	F	fluid flow
5.	f	fraction
6.	h	height
7.	me	specific methane production rate
8.	v	velocity
9.	q	specific substrate conversion rate
10.	γ	production or conversion rate
11.	V	volume
12.	X	amount of sludge
13.	ϕ_g''	volume of gas/time
14.	ϕ_g'''	volume of gas per unit area per unit time
15.	ϕ_g'''	volume of gas per unit volume per unit time
16.	ϕ_g	specific conversion of substrate (kg/m^3.d)
17.	ap	unidentified products
18.	b	sludge bed
19.	bd	from bed to blanket
20.	d	blanket
21.	db	from blanket to bed
22.	e	effluent
23.	fa	fatty acids
24.	g	gas
25.	gb	gas produced in the bed
26.	gb'	gas produced at the top of the bed
27.	gd	gas produced in the blanket
28.	k	bypassing
29.	m	sludge
30.	mb	sludge from bed
31.	md	sludge from blanket
32.	me	methane
33.	tb	transport of sludge from bed
34.	td	transport of sludge from blanket

Finally,

$$C_{ap} = \frac{K_{ap}}{F}(C_{af'o} - C_{af})(V_b + V_d) \text{ or} \quad ...(8.606)$$

$$C_{ap} = \frac{K_{ap}}{v_L}(C_{af'o} - C_{af})(h_b + h_d) \quad ...(8.607)$$

The methne production rate is given by :

$$\phi_{g,me} = K_{sg}\, F(C_{s,o} - C_{s,e}) \text{ or} \quad ...(8.608)$$

$$\phi''_{g,me} = K_{sg}\, v_L\, (C_{s,o} - C_{s,o}) \quad ...(8.609)$$

The total biogas production can be known if the fraction of methane content in the gas is known :

$$\phi''_{g,me} = \frac{\phi''_{g,me}}{f_{me}} \quad ...(8.610)$$

The amount of sludge in the reactor is given by:

$$X_R = C_{mb}\, V_v + C_{md}\, V_d \quad ...(8.611)$$

The amount of substrate that can be converted is given by

$$\phi_g = \phi'''_{g,}\, V_R \quad ...(8.612)$$

The nomenclature for various mathematical terms is given in Table 8.9.

8.8 General Models

To determine the maximum loading rate (L_m) that can be applied to the any type of reactor system is given by DeWalle and Chian[33]:

$$\frac{1}{V}\frac{dF}{dt} = L = \frac{L_{max} S_o}{K_S + S_e} \quad ...(8.613)$$

where L and K_s refer to loading rate and inhibition cofficients respectively. Stover *et al.*[135] developed an expression based on monomolecular kinetics relating substrate utilization as a function of surface area of the media and is given by :

$$U = \frac{U_{max}(QS_i/A)}{[K_B + (QS_i/A)]} \quad ...(8.614)$$

Where U, U_{max}, K_B and A refer to specific substrate utilization, maximum specific substrate utilization, an inhibition coefficient and media surface area respectively.

The above expression can be substituted in the mass balance equation for substrate entering and leaving a particular void volume of a reactor system as:

Input – output – utilization = 0

$$QS_i = QS_e + dS/dt\ A \quad ...(8.615)$$

Substituting equation 8.614 into 8.615 gives :

$$QS_o = S_e + [(U_{max}\ QS_i/A)(K_B + QS_i/A)] \quad ...(8.616)$$

The expression facilitates the evaluation of the effluent substrate concentration achievable or media surface required for a specific substrate utilization rate as :

$$S_o = S_i - [U_{max}\ S_i/\ (K_B + (QS_i/A)] \quad ...(8.617)$$

$$A = QS_i/[U_{max}S_i/\ ((S_i - S_e) - K_B)] \quad ...(8.618)$$

Data collected by Young[136] and reported by Young and Dahab[137] indicated a linear relationship between efficiency and the inverse of HRT in the voids within the rock filled reactors. This can be represented as :

$$E = 100\ (1 - a/HRT) \quad ...(8.619)$$

where E, and a refer to efficiency and a proportionality constant respectively 'a' also indicates the threshold HRT beyond which the efficiency will be reduced to zero. The above equation can be modified as it is not possible to achieve 100 percent efficiency, *i.e.*

$$E = E_m\ (1 - a/HRT) \quad ...(8.620)$$

where E_m refers to maximum removal efficiency.

Morris *et al.*[138] referred refractory organic material as that portion of organic substrate which is resistant to biological degradation and remain undegraded at the time when the rate of stabilization has decreased so as to be insignificant and is given by the expression :

$$S\gamma = R\ .\ S_i \quad ...(8.621)$$

where $S\gamma$, S_i and R refer to refractory and total concentration and refractory fraction respectively. Taking into consideration the refractory fraction, the basic expression developed from Monod kinetics for evaluating the influent substrate concentration (S_e) is given by :

$$S_e = [K_s(\mu_{max}.\ SRT - 1)] + RS_i \quad ...(8.622)$$

A plot of S_o/S_i as a function of $(S_i\ .\ SRT)^{-1}$ yields an intercept, R on the ordinate axis as SRI approaches infinity. Most of the reactor can be categorised as a pseudo mixed system and therefore, following set of equation may be applied (keeping in mind that they can be considered only empirical in nature).

ln + Produced = out + Accumulation

$$Q\ S_0 + dS/dt\ V = Q.\ S + 0 \quad ...(8.623)$$

Rearranging this equation results in :

$$S = \frac{S_0}{1 + K\,\theta_n} \text{ or} \qquad ...(8.624)$$

$$E = \frac{S_0 - S}{S_0} = K\theta_n \cdot \frac{S}{S_0} \qquad ...(87.625)$$

If there are voids, then in terms of void fraction (ε), the equation is:

$$S = \frac{S_0}{1 + (K\,\varepsilon\,\theta_n / \varepsilon)} \text{ or} \qquad ...(8.626)$$

$$S = \frac{S_o}{1 + (K'\theta_n{}'}; K' = K\varepsilon \text{ and } \theta'_n = \theta_n / \varepsilon, \text{ or} \qquad ...(8.626)$$

$$E = \left(1 - \frac{1}{1 + K'\theta'_n}\right) \qquad ...(8.627)$$

The sludge production rate (P_x) can be estimated by :

$$P_x = \frac{YQ\,(ES_o)\,(10^3 g / kg)^{-1}}{1 + k_d \text{ HRT}} \qquad ...(8.628)$$

Another expression for removal efficiency is given by :

$$E + \left[\frac{S_o - \dfrac{kg(1 + k_d - SRT)}{(SRT \,.\, Ks) - (1 \mid k_d \,.\, SRT)}}{S_o}\right] \qquad ...(8.629)$$

where k_g, k_d and K_s refer to kinetic constants.

Smith *et al.*[139] developed a model for determination of specific growth rate of the microorganisms for fixed film system after extensive experimentation. This model is given as:

$$E = \left(\frac{S_o - S}{S_o}\right) = K_\mu \theta^\mu \qquad ...(8.630)$$

Where K_μ, θ and μ refer to constant, HRT or (SRT) and specific growth rate constant respectively.

Oh and Yang[140] developed as empirical relationship between efficiency and HRT (θ) and is as follows :

$$E = E_{max}\,(1 - a/\theta^n) \qquad ...(8.631)$$

Where *a* and *n* refer to threshold HRT and a constant respectively. This expression is similar to that reported by Young and McCarty. The difference might be due to reactor configuration and media.

For UASB system, Oh and Yang[140] developed the following relationship :

$$F/M(\text{BOD based}) = 2.11.10^7 \exp(-0.2015\ E) \quad ...(8.632)$$

$$F/M(\text{COD based}) = 4.86.10^6 \exp(-0.1935\ E) \quad ...(8.633)$$

A general relationship relating biogas yield with substrate loading rate is given by the following equation :

$$Y(m^3CH_4/m^3.d) = a\ X\ (\text{kg COD}/m^3.d) + \text{const.} \quad ...(8.634)$$

where Y, a, and X refers to biogas yield, specific biogas yield (m^3. CH_4/kg COD) and loading rate respectively.

Stover, *et al.*[135] developed an expression relating total gas production as a function of mass substrate loading rate by the following expression :

$$G = G_{max}\ (QS_i/A/[G_B + QS_i/A] \quad ...(8.635)$$

where G, G_{max}, A and G_B refer to total specific gas production rate (m^3/m^2.d), maximum specific gas production, area and inhibition coefficient respectively.

Similar expression can be used for methane in place of total gas production.

McCarty[141] has given the following expression for estimating the quantity of CH_4 production, *i.e.* :

$$QCH_4\ (m^3/d) = 0.35m^3/kg\ [EQS_o\ (10^3 g/kg)^{-1} - 1.42\ Px] \quad ...(8.636)$$

where Q, E, 1.42, 0.35 and Px refer to flow rate, substrate efficiency conversion factor for cell tissue to BOD_L, theoretical conversion factor for the amount of methane produced from conversion of one kg of BOD_L and net mass of cell tissues produced per day respectively.

Pearson *et al.*[142] developed a semiempirical relation for methane production. The relative ultimate biogas or methane production, referred to as actively is defined as :

$$A = \frac{v}{V} \quad ...(8.637)$$

Where A, v and V refer to the relative ultimate gas or methane production, ultimate gas (methange) production in the reactor containing toxicant and ultimate gas (methane) production obtained in the control reactor (no toxicant added) respectively. The relationship of relative gas production A versus toxicant concentration (T) produces inverse S-Curves that could be modelled by two parameter function

$$A = [1 + (T/T_{50})^a]^{-1} \quad ...(8.638)$$

where T_{50} and *a* refer to half kill dose or toxicant dose, resulting in a 50 per cent reduction of ultimate gas production and a sensitivity exponent defined as the rate of change of relative gas production at a given dose and half kill dose. Using linear regression method, the above equation simplifies to

$$\log \quad (1/A - 1) = a \log (T) - a [\log T_{50}] \quad ...(8.639)$$

Effect of temperature is most pronounced in anaerobic systems. Kennedy *et al.*[143] proposed relationship between temperature and loading rate and temperature and methane production rate expressed by :

$$Gp = 0.167 \, T - 0.692 \quad ...(8.640)$$

$$LR = 0.557 \, T - 1.546 \quad ...(8.641)$$

where Gp, LR and T refer to methane production (m^3/d), loading rate (kgCOD/m^3.d) and temperature (°C) respectively.

REFERENCES

1. Kafarov, V, V, and Dorokhov, I. N., System Analysis of Operation in Chemical Engineering, Book-I, Fundamentals of Strategy, Moscow, Nauka (1976).
2. Album of Mathematical Description and Control of Algorithms of Unit Operations of Chemical Eng., Issues 1-5 Moscow, (1965-1975).
3. Kafarov, V. V. and Vetokhin, V.N., Operating Systems in Chemical Engineering, Nauka, Moscow (1980).
4. Lawerence, A.W and McCarty, P. L., Unified Basis for Biological Treatment Design and Operation, Journal of Sanitary Engg. Div. ASCE, 96(SA-3), p757-778(1970).
5. Rittman, B.E. Comparative Performance of Biofilm Reactor Types, Biotech Bio-Engg. 24, p.1341-1370 (1982).
6. Rittman, B.E. and McCarty, P.L., Model Steady State Biofilm Kinetics, Biotech. Bio-Engg. 22, p2343-2357 (1980).
7. Rittman, B.E. and McCarty, P.L., Design of Fixed Film Process with Steady State Biofilm Model, Prog. Wat. Tech. (Toronoto) 12, p271-281 (1980).
8. Rittman, B.E. and McCarty, P.L., Evaluation of Steady State Biofilm Kinetics. Biotech. Bio-Engg., 22, p2359-2373 (1980).
9. Namkung, E. and Rittman, B.E. Predicting Removal of Trace Organic Compounds by Biofilms, JWPCF, 55, p1366-372 (1983).
10. Saez, P.B. Cinetica de los, Processes Biologicos Usados enal Tratamiento de Aguas Resoduales, Presenteds at the X-Inter-American Conference of Chemical Engg., held at Santiago, Chile, (Nov. 6-11, 1983).
11. Saez, P.B. Cinetica de los Proceses Anaerobicos, Apuntes de Ingenieria 14, p97-215 (1984).
12. Monod, J. La Tecnique de Culture Continue, Theorie de Applications, Annals Institute Pasteur 79, p390-410 (1950).
13. Contois, D.E., Kinetics of Baterial Growth : Relationship between Population and Specific Growth Rate of Continuous Cultures, 21, p40-50 (1959).

14. Rittman, B.E, The Effect of Shear on Biofilm Loss Rate, Biotech. Bioengg, 24, p501-506 (1982).
15. Truelear, M.G. and Characklis, W.G., Dynamics of Biofilm Process, Presented at the 53rd Annual conf., of Wat. Poll. Cont. Fed. Held at Las Vegas, USA, (September 28-October 3, 1980).
16. Williamson, K.J. and Chung, T.H. Dual Limitation and Substrate Utilization Kinetics within Baterial Films, Presented at the 49th National Meeting of the American Institute of Chemical Engineers, held at Houston (USA), (March 19, 1975).
17. Gerald, C.F., Applied Numerical Analysis, 2nd Ed. Addison-Wesely Publishing Company, California (1980).
18. Ralston, A., Mathematical Methods for Digital Computers, John Wiley and Sons, N.Y. (1977).
19. Rittman, B.E., Comparative Performance of Biofilm Reactor, Biotech, Bio. Engg. 24, p1341 (1982).
20. Rittman, B.E., The Effect of Shear stress on Biofilm Loss Rate, Bio.Tech. Bio. Engg. 24, p501 (1982).
21. Rittman. B.E., McCarty, P.L., Variable Order Model of Bacterial Film Kinetics; J. Environ, Engg. Div., ASCE 104 (EE.5) p889 (1978).
22. Young, J.G., The Anaerobic Filter For Wastewater Treatment, Ph.D Thesis, Stanford Univ. (1968).
23. Annachatre, A.P. and Khanna, P. Fixed Film Biomethanation Modelling J. Envn., Engg. ASCE 116, p49-60 (1990).
24. Annachatre, A.P. and Khanna P., Unsteady State Biofilm Kinetics, J. Envn. Engg. (ASCE), 113(2), p429-433 (1987).
25. Annachatre, A.P., and Khanna, P., Unified Basis of Biofilm Kinetics, J. Envn. Engg. (ASCE), 113(5), p116-1169 (1987).
26. Annachatre A.P. and Khanna, P. Methane Recovery From Water Hyacinth Through Whole Cell Immobilization Technology, Bio-Tech. Bio-Engg., 29(7), p805-818 (1987).
27. Suidan, M.T. and Wang,. Y.T., Unified Analysis of BioKinetics, J. Envn. Engg. (ASCE), 11(5), p634-646 (1985).
28. Wanner, O. and Gujer, W., Competition in Biofilms, Wat. Sci., Technol., 17(1), p. 27-44 (1985).
29. Andrews, G.F. and Tien, C., Bacterial Film Growth in Adsorbent Surfaces, AICHE, J. 27 (3), p396-403 (1981).
30. Hines, D.C.M. and Genung, R. K., Production of Ammonia in a Packed Bed. Anaerobic Upflow (ANFLOW) Bioreator, AICHE, Symposium Series 209, Vol. 77, p. 286-294 (1981).
31. Atkinson, B, The Overall Rate of Substrate Uptake by Microbial Films, Parts I and II, Trans, Inst. Chem. Engrs (1974).
32. Oleszkiewicz, J.A. Low Temperature Anaerobic Biofiltration, JWPCF, p1465-1475 (1982).
33. DeWalle, F.B. and Chian, E.S. K, Kinetics of Substrate Removed in a Completely Mixed Anaerobic Filter, Bio.Tech. Bio.Engg. 27, p1275-1295 (1976).
34. Port, S.J., J. Appl. Chem. Bio.Tech. 23, p. 389 (1973).
35. Saunder, P. T. and Bazin, M.J., Appl. Chem, Biotech. 23, p847 (1973).

36. Atkinson, B. and Daoud, I.S., Trans. Inst. Chem. Eng., 46, p19 (1968).
37. Nusselt, W., Z, Verh. Dent. Ing. 60, p569 (1916).
38. Takeshi, K. J., Chem. Engg, Japan, S, p132 (1972).
39. Ames, W.F., J. Sanit, Eng. Div (ASCE), 88(SA-3), p21 (1962).
40. Lamb, R. and Owen, S.G.S., Wat, Poll. Cont., 69 p209 (1970).
41. Kornegay, B.H., Kinetics of Fixed Film Biological Reactors, JWPCF, 40, pR-460 (1968).
42. Kong, W.T. Kinetics of Trickling filters, Biotech. Bio. Engg, 25, p603 (1983).
43. Stover, R.L. and Rankness, Process Evaluation of Fixed Film Biological Treatment Plants, Proc., 2nd Intl. Conf. Fixed Film Biol. Processes, Vol. 1, p814-830 (1984).
44. Kincannon, D.F. and Stover, R.L., Use of R.B.C. on Meat Industry Wastewaters, Proc. 5th Natl. Symp. Food Processing Wastes, EPA Technology Series, EPA-660/2-74-058 (1974).
45. Kincannon, D. F., and Stover, R.L., Design Methodology for Fixed Film Reactors–RBC and Biological Towers, Civil Engrs. Practicing and Design Engineers 2, p,107 (1982).
46. Karlsson, H.T. and Nilsson, B.K. On Modelling of Fixed Film Reactors, Proc. 5th Intl. Symp. on Anaerobic Digestion, Bologna (Italy), p195-198 (1988).
47. van Krevelen, D.W. and Kvekeles J.T.C., Recueil Trans. Chem. Pays-Bas 67, p512 (1948).
48. Otake, T. and Oada, K., Kogaku Kogaku 17, p176 (1963).
49. Puranik, S.S. and Vogelphol, A., Chem. Eng. Sci, 29, p51 (1974).
50. Hofmann, H. Intl. Chem. Engg., 17, p19 (1977).
51. Swetzenbaum, M.S. The Anaerobic Attached Fixed Film Expanded Bed Reactor For the Treatment of Dilute Organic Wastes. Ph. D. Thesis, Cornel Univ. (1978).
52. Hickey, R.F. and Owens, R.W., Methane Conversion from High Strength Industrial Wastes with Anaerobic Biological Fluidized Bed. 3rd Symp. on Biotech Energy Production and Conservation, John Wiley and Sons, Inc., N.Y. (1981).
53. Wang, Y.T. Suidan, M.T. and Rittman, B.E., Kinetics of Methanogens in and Expanded Bed Reactors, J. Envn. Engg (ASCE), 112 (1), p155-170 (1985).
54. Wilke, C.R. and Chang, R. Correlation of Diffusion Coefficients in Dilute Solutions of ACHE, J. I, p264 (1955).
55. Williamson, K. and McCarty, P.L., JWPCF 48 p.9-24 (1976).
56. Shieh, W. K., A Suggested Kinetic Model For Fluidized Bed Biofilm Reactor (FBBR), Biotech, Bioeng, 22, p667 (1980).
57. LaMotta, E.J., Evaluation of Diffusional Resistances in Substrate Utilization by Biological Film, Ph. D. Thesis, University of North Carolina at Chapel Hall (1976).
58. Mulcahy, L.T. and LaMotta, E.J., Mathematical Model of the Fluidized Bed Biofilm Reactor, Report No. Inv. E. 59-78-2. Dept of Civil Engg. Univ. of Massachusetts, Amherst (1978).
59. Shieh, W.K., Mulahy, L.T. and LaMotta, E.J., Trans, Instn., Chem. Engg., 59, p129 (1981).
60. Aris, R., Elementary Chemical Reactor Analysis, Englewood Cliff, N.J. Prentice Hall (1969).
61. Hermanowiez, S.W. and Ganczaryk, J.J., Biotech/Bioengg. Res. Des. 61, p125 (1983).

62. Webb, C, Black, G. M and Atkinson, B., Chem. Engg. Res. Des. 61, p125 (1983).
63. Shieh, W.K. and Chem, C.Y., Chem Engg. Res. Des. 62, p133 (1984).
64. Richardson, J.F. and Zaki, W.N., Trans. Instn., Engg. 32, p35 (1954).
65. Lewis, E.W. and Bowerman, E.W., Chem. Eng. Prog. 48, p605 (1952).
66. Wen, C.Y. and Yu, Y.H., Chem. Eng. Prog. Symp. Series, 62, p100 (1962).
67. Valsalos, I.A, and Tjatjopoulos, G.T., AICHE. J., 28(2), p346 (1982).
68. Mulcahy, L.T. Shich, W.K. and LaMotta, E.J., Water-1980, AICHE, Symp. Ser 209, 77, 273, (1981).
69. Li, A. Personal Communication (1992).
70. Wen., C.Y., Yu, Y., H., Chem., Engg., Prog., Symp. Series 62, p100 (1962).
71. Ergum, S. Chemical Engineering Progress, 48, p.89-94 (1952).
72. Kornegay, B.H. and Andrew J.F., Kinetics of Fixed Film Biological Reactors. JWPCF 40, p(R) 460 (1968).
72a. Andreadakais, A.D. And Cailas, M., Evaluation of a simple Mode for Organic Carbon Removal by a RBC, 2nd Int. Cong. On Fixed Film Biological Processes, Vol l. p664-680 (1984), Arlington, Wirginia.
73. Clark, J.H., Moseng, E.M. And Asano, T., Performance RBC Under Varying Wastewater Flow, JWPCF, 50 p896 (1978).
73a. Antonie, R.L. Fixed Biological Surface-Wastewater Treatment, CRC Press, (1976).
74. Grady, C.P.L. And Lim, H.C., Biological Wastewater Treatment-Theory and Application. Marcal Dekker, Inc, New York (1980).
75. Atasi, K.Z. and Borchardt, J.A. Kinetics of the Rotating Biological Contactor, Proc. Nat Cong., Environ. Eng. ASCE, Boulder, CO (1983).
76. Frank-Kamenetakii, D.A., Diffusion and Heat Transfer in Chemical Kinetics. Plenum Press (1969).
77. Aris, R., Introduction to the Analysis of Chemical Reactors, Prentice-Hall Inc (165).
78. Levenspiel, O., Chemical Reaction Engg., J. Wiley and Sons, Inc., N.Y. (1972).
79. Burden, R.L. And Reynolds. A.C., Numerical Analysis, Prindle, Weber and Schmidt, Inc. Boston (1978).
80. Carnahan, B, Wilkes, J.O., Applied Numerical Methods, J. Wiley and Sons Inc., (1969).
81. Satterfield, C.N., Mass Transfer In Hetrogenous Catalysis, M.I.T. Press (1970).
82. Bischoff, K.B., Effectiveness Factor for General Reaction Rate Forms, AIch E.J., 11, p351 (1965).
83. Pitcher, W. H. Engineering of Immobiolized Enzyme Systems, Catal. Rev.-Sci.Engg. 12, p37 (1975).
84. Spiegel, M.R., Applied Differential Equation, Prentice Hall, Inc., (1981).
85. Wylie, C.R., Advanced Engineering Mathematics; McGraw Hill Book Co. Inc (1951).
86. LaMotta, E.J., Evaluation of Diffusional Resistances in Substrate Utilization by Biological Films; Ph.D. Thesis, Univ. North Carolina, Chapel Hill (1974).
87. Aris, R. Mathematical Theory of Diffusion and Reaction in Permeable Catalyst, Vol.1, Oxford Univ. Press (1975).

88. Wheeler, A. and Emmet, P.H., Catalysis, Vol.2. Reinhold (1955).
89. Cheng, T.H. and Borchardt, J.A; Investigation of Biological Oxidation Using RBCs; Ph. D. Thesis, The Univ. of Michigan (1978).
90. Famuler, J., Application of Mass Transfer to Rotating Biological Contractors, JWPCF, 50 p. 653 (1978).
91. Mueller, J.A., Nitrification in RBCs, JWPCF, 52 p688 (1980).
92. Film Reactor; Water Research 10, p935 (1976).
93. Williamson, K.J. and McCarty, P.L., A Model of Substrate Utilization by Bacterial Films, JWPCF, 48, p9 (1976).
94. Williamson K.J. and Chung, T.H. Dual Limitation of Substrate Utilization Kinetics within Bacterial Films, Paper Presented at 49th National Meeting of AIChE, Houston, Texas (1975).
95. Joost, R.H., System Action in Using Biological Surface Waste Treatment Process, Proc. 24th Purdue Industrial Waste Conf. 24, p365-373 (1969).
96. Weng, C.N. and Molof, A.H., Nitrification in Biological Fixed Film RBC., JWPCF, 46(7), p1674-1685 (1974).
97. Antonie, R.L. and Welch, F.M., Preliminary Results of a Novel Biological Process for Treating Dairy Wastes, Proc. 24th Purdue Industrial Waste Conference, 24, p115-126 (1969).
98. Wu, Y.C., Modelling of RBCs. Biotech, Bio-Engg., 22, p2055-2064 (1980).
99. Tait, S.J. and Friedman, A.A., Anaerobic Rotating Contactor for Carbonaceous Wastewater, JWPCF, (52(8), p2257-2269 (1980).
100. Friedman, A.A. Woods, R. C., and Wilkey R.C., Kinetic Response of Rotating Biological Contractor Proc. 31st Purdue Industrial Waste Conf., 31, p420-433 (1976).
101. Schroeder, E.D., Water and Wastewater Treatment, McGraw Hill, N.Y. (1976).
102. Grieves, C.G., Dynamic and Steady State Models For RBCs Ph.D. Thesis, Clemson Univ., Clemson, SC(1972).
103. Pano, A. and Middlebrooks, E.J., Kinetics of RBC Proc., Ist Int. Conf on Fixed film Biological Processes, Kings Island, OH, I, p261–308 (1982).
104. Watanbe, Y and Ishiguro, M., Kinetics of a Submerged RBC. Prog. Wat. Tech. 10(5/6), p187–195(1978).
105. Eckenfelder, W.W. and Vanedevenne, L., A Design Approach for RBC Treating Industrial Wastewaters, Proc. First Natl. Symposium/Workshop on RBC Technology, Champion, PA, II, p1065–1075(1980).
106. Benjes, H.H., Small Communities Wastewater Treatment Facilities–Biological Treatment System in Design, Seminar Handout Small Wastewater Treatment Facilities, USEPA, Technology, Transfer, Jan. (1978).
107. Kincannon, D.F. and Groves, S., Role of Suspended Solids in the Kinetics of RBC System Proc. Ist Natl. Symp/Workshop on RBC Technology, Champion, P.A, I, p433–448(1980).
108. Opatken, E.J., RBC – Second Order Kinetics, Proc. Intl. Conf. Fixed Film Biological Processes, Kings Island, OH, I, p210–232 (1982).
109. Fujie, K., Operational Design and Power Economy of RBC, Water Res., 17(9), p1153–1162 (1983).

110. Williams, R.B. Process Selection, Design, Startup, and Operation of an RBC; Proc., Ist Intl. Conf. on Fixed Film Biological Processes, Arlington, Virginia, p1791–1838 (1984).
111. Hsieh, C.N., Variables affecting the Performance of Biological Disk Filtration units., Masters Thesis, A.I.T., Bangkok, Thailand (1972).
112. Ching–San, H., Nitrification Kinetics and its RBC Application, J. Amer Soc. Civil. Engr. Envn. Engg. Dn.(6), p473–487 (1982).
113. Chen, Y.R. and Hashimoto, A.G., Biotechnol. Bioengg., Symp 10, p1020–1028 (1980).
114. Chen Y.R. and Hashimoto, A.G. Biotechnol Bioengg., Symp. 8, p269–279 (1978).
115. Saraswat, N and Khanna, P., Methane Recovery From Water Hyacinth Through Anaerobic Activated Sludge Process, Biotech. Bioengg., XXVII, p240–246 (1986).
116. Ghosh S. and Pohland, F.G., JWPCF, 46, p. 748 (1974).
117. Ghosh S. and Klass, D.L., JWPCF, 47, p31 (1975).
118. Claussen, C.C., and Gaddy, J.L. AICHE, 74 p. 56 (1978).
119. Massey, M.L. and Pohland, I.G., J. San. Engg., Div., ASCE 96, p 2204 (1978).
120. Metcalf and Eddy Inc., Wastewater Engineering Treatment, Disposal and Reuse, Tata–McGraw Hill Publishing Company Ltd. N. Delhi, (1979).
121. Kennedy, K.J. and Droste, R.L., Kinetics of Downflow Anaerobic Attached Growth Reactor, JWPCF, 59(4), p212–221 (1987).
122. Grady, C.P.L., and Lim, H., Biological Wastewater Treatment Theory and Application, Marcel Dekker, Inc., N.Y. 28, p1713(1986).
123. Bachamann, A; Beard, V.L., and McCarty, P.L. Performance characteristics of the Anaerobic Baffled Reactor, Water Res. 19(1) p99–106 (1985).
124. Wilson, E.J. and Geankopolis, C.J., Liquid Mass Transfer at Very Low Reynolds Numbers is Packed Beds, Ind. Engg. Chem. Fund., 5, p9–14(1966).
125. Moletta, R. and Albaganae, Dynamic Modelling of Anaerobic Digestion, Water, Res. 20(4) p427–434 (1986).
126. Yarovenko, V.L. and Nakhamanovich, B.M., Kinetics Product Synthesis in Continuous Alcoholic Fermentation, Pure Appl. Chem. 36, p397–405 (1973).
127. Neumann, W. and Ruckauf, H. Modelling and Scale up of Biogas Reactors. 5th Intl. Symp. Anaerobic Digester, Bologna, Itlay, p305–311(1988).
128. Heertjes, P.M. and van der Meer, R.R., Fluid Flow Pattern in Upflow Reactors For Anaerobic Treatment of Beet Sugar Factory Wastewater, Biotech. Bioengg. XXIV, p443–459 (1982).
129. Bolle, W.L. and Zoetemeyer, Modelling the Liquid Flow in UASB Reactors, Biotech. Bioengg., XXVIII, p1615–1620 (1986).
130. Heertjes, P.M. and van der Meer, R.R. Biotech. Bioengg. 20, p1577–1594 (1978).
131. Walpot, J.I. Biotechnol. Group Delft Univ. Technol., The Netherlands, Intern., Ref. (1980).
132. Van dar Meer, R.R., Ph.D Thesis, Delft Univ. Technol., The Netherland (1980).
133. Bolle, W.L. and van Gil, W., An Integral Dynamic Model for the UASB Reactor; Biotech. Bioengg., XXVIII, p1621–1636 (1986).
134. van den. Meer R.R. and Heertges P.M., Mathematical Description of Anaerobic Treatment of UASB, Biotech, Bioengg., 25(11), p2531(1983).

135. Stover, E.L. and Gonzalez, R., Anaerobic Fixed Film Biological Treatment Kinetics of Fuel Alcohol Production Wastewater, 2nd Intl. Conf. Fixed Film Biol. Process, p1625 (1984).
136. Young, J.C., The Anaerobic Filters For Wastewaters Treatment, Ph.D Thesis, Stanford University, Stanford, California (1968).
137. Young, J.C. and Dahab, M.F. Effect of Media on the Performance of Fixed Bed Anaerobic Reactors, Proc. Specialized Seminar of IAWPRC held in Copenhagen, Denmark (June, 1982), Water Sci. Technol. 15(8/9), 1982.
138. Morris, G.R., Jewell, W.J. and Loehr, R.C., Anaerobic Fermentation of Animal Waste, A kinetic Design Evaluation, Proc. 32nd Purdue Industrial Conf, Lafayette, Indiana, p689(1977).
139. Smith R.E., Reed, M.J. and K.I. Ker, J.T., Two Phase Digestion of Swine Waste, J Envn. Engg. Div. (ASCE), Trans, p1123–1128 (June 1977).
140. Oh, Y.M. and Yang, B.S., Anaerobic Wastewater Treatment using Floating Media (Development of the Upflow Floating Anaerobic Filter Process), Wat. Sci. Tech. 18, p. 225–237 (1986).
141. McCarty, P.L., Anaerobic Waste Treatment Fundamentals, Chemistry and Microbiology, Public Works, 95, p107(1964).
142. Pearrson, F. and Chang, S.C., Toxic Inhibition of Anaerobic Bio-oxidation, JWPCF, 52(3), p472–482 (1980).
143. Kennedy, K.J. and van den Berg, L., Effect of Temperature and Overloading on the Performance of Anaerobic Fixed Film Reactors, Proc. 36th Purdue Industrial Waste Conference, Purdue Univ. Ann. Arbor, Michigan (1981).

ADDITIONAL REFERENCES

1. Evaluation of Design Parameters for Anaerobic Rotating Biological Drum Contactor Treating Paper Mill Effluent, S.N. Kaul, T. Nandy and A.G. Bhole, Jr. of Environmental Science and Health, Vol. A27, No. 7, p1589–1618, (1992).
2. Pilot Plant Studies on Fixed Bed Reactor System for Biomethanation of Distillery Spent Waste, T. Nandy, S.N. Kaul, P.P. Pathe and C.V. Deshpande Intern. Jr. Environmental Studies, Vol. 41, p87–107, (1992).
3. Denitrification of Secondary Effluent from a Municipal Wastewater Treatment Unit using Anaerobic Fixed Film Moving Bed Reactor, S.N. Kaul, A.G. Bhole, Y.V.V. Satyanarayana and T. Nandy, Indian Jr. of Environmental Protection, Vol. 12, No., 11 p801–808, (1992).
4. Anaerobic Upflow Bioreactor for Pharmaceutical Wastetreatment, Mukti Deb Roy., T. Nandy and S.N. Kaul Indian Jr. of Environmental Protection, Vol. 13, No. 1, p. 41–46, (1993).
5. Evaluation of Kinetic Constants of Anaerobic Fixed Film Fixed Bed Reactor System Treating Herbal Pharmaceutical Wastewater, T. Nandy and S.N. Kaul, Jr. of Environmental Science and Engineering, Vol. A26, No. 5, p689–709, (1991).
6. Energy Recovery from Waste Paper Based Pulp and Paper Mill Wastewater, H.M. Rudriah, A.G. Bhole, S.N. Kaul and T. Nandy Jr. of Indian Association for Environmental Management, Vol. 17, No. 1, p6–14, (1990).

7. Upflow Anaerobic Sludge Blanket Reactor for Wastewater Treatment–An introduction P.P. Pathe, T. Nandy and S.N. Kaul, Ind. Jr. of Environmental Protection, Vol. 10, No. 7, p493–501, (1990).
8. Anaerobic Rotating Biological Drum Contactor for Distillery Spent Wash, T. Nandy, P.P. Pathe and S.N. Kaul, Asian Environment, Vol. II, No., p54–65, (1989).
9. Energy Recovery from Cattle Dung using Packed Bed Reactor System, P.P. Pathe, T. Nandy, C.V. Deshpande, S.N. Kaul and S.D. Badrinath, Asian Environment, 9(3), p4, (1987).
10. Package Wastewater Treatment (Anaerobic) for Domestic Sewage, P.P. Pathe, T. Nandy and S.N. Kaul Ind. J. Envrn. Prot., 9(3) p197, (1988).
11. Use of Whole Cell Immobilized Reactor for Treatment of Partially Treated Sewage, P.P. Pathe, T. Nandy and S.N. Kaul, J. Chem. Engg. World, (In Press).
12. Attached Film Fixed Bed Reactor for Treatment of Dairy Wastewater, S. Shanta, Jayshree Venkatraman, S. Saoji and S.N. Kaul, Ind. J Envn. Prot., 8(1), p39, (1988).
13. Anaerobic Rotating Biological Drum Contactor for the Treatment of Dairy Wastes, S. Shanta, K. Khakar, S.N. Kaul, S.D. Badrinath and N.G. Swarnkar J. Ind. Chemical Engineer, XXIX(3), (1987), (Special Feature).
14. Non–conventional energy sources-A perspective under Indian conditions, S.N. Kaul, S. Shanta, C.V. Deshpande, P.P. Pathe and T. Nandy J. Chem. Engg. World, XXII (7-8), p66, (1987).
15. Anaerobic Treatment of Human Wastes S. Shanta, S.N. Kaul, S.D Badrinath and S.K. Gadkari, Ind. J. Envn. Prot., 7(1) p. 39, (1987).
16. Upflow Anaerobic Suspended Bed and Fixed Bed Fixed Film Reactor Systems for Sewage Treatment, P.P. Pathe, T. Nandy and S.N. Kaul, Proc. 2nd National Conv. Engg. Institution of Engineers, Lucknow, p82, (April, 1987).
17. Biological Denitrification Using Fluidized Bed Reactor System, S.K. Gadkari, B. Chakladhar, S.N. Kaul and S.D. Badrinath, J. Asian Envn., 8(3), p4, (1986).
18. Biogas from Industrial Wastewaters, S.N. Kaul, S.K., Gadkari and V.P. Deshpande J. Inst. Pub. Helt. Engrs., 3, p5(1986).
19. Packed Bed Anaerobic Reactor for Treatment of Meat Wastes K.L. Saxena, S.N. Kaul, M.Z. Hassan, S.D. Badrinath and S.K. Gadkari, J. Asian Environment Vol. 8(2), p.20 (1986).
20. Biogas from Anaerobic Digestion of Human Wastes S. Shanta, S.N. Kaul, S.D. Badrinath, S.K. Gadkari Proc. of the National Conference of Consortium for Rural Technology, New Delhi, (August 1986).
21. Treatment of Synthetic Milk Waste by Anaerobic Upflow Contactor Filter, S. Shanta, Seema Saoji, S.N. Kaul and S.D. Badrinath, Paper Presented at the 39th Annual Meeting of the Institution of Indian Chemical Engineers at Hyderabad, (December 1986).
22. Evaluation of Kinetics Constants for Various Wastewater using Anaerobic Fixed Film Fixed Bed Reactor, S. Shanta, S.N. Kaul, S.D. Badrinath, S.K. Gadkari and V.P. Deshpande, IAWPC, Tech. Annual-XII, 8, (1985).
23. Anaerobic Bio–oxidation of Strong Soluble Organic Wastewater, S.D. Badrinath, S.N. Kaul, S. Shanta and A.G. Bhole, J. Asian Environment, Vol. 7.1, (June 1985).
24. Two Stage Fixed Film Reactor for Treatment of Distillery Wastewater, S.D. Badrinath and S.N. Kaul, IAWPC-Technical Annual XI, p61 (1984).

25. Treatment of Distillery Spent Wash using Anaerobic Packed Bed Reactor System, S.N. Kaul, S.D. Badrinath, Ph.D. Thesis, Nagpur University, (1986).
26. Studies on Fluidised Bed System for the Wastewater Treatment, S.N. Kaul, S.K. Gadkari, Ph.D. Thesis, Nagpur University, (1988).
27. Studies on Treatment of Domestic Sewage using Anaerobic Rotating Biological Drum Contactor, S.N. Kaul, V.P. Deshpande, Ph.D. Thesis, Nagpur University, (1990).
28. Studies on Anaerobic Fixed Film Fixed Bed Reactor System for the Treatment of Pharmaceutical Wastewater, S.N. Kaul, T. Nandy, Ph.D. Thesis, Nagpur University, (1992).
29. Resource Recovery from Tannery Wastewater using Fixed Film Fixed Bed Bio–Reactor, S.N. Kaul and P.K. Mukhopadhyay, Ph.D. Thesis, Devi Ahilya Vishwavidyalaya, Indore, (1993).
30. Anaerobic Rotating Biological Drum Contactor for Pharmaceutical Wastewater, S.N. Kaul, T.A. Singh, M. Tech. Dissertation, Nagpur University, (1988).
31. Treatment of Wastewater Based Pulp and Paper Mill Effluent by Anaerobic Rotating Biological Drum Contactor, S.N. Kaul, H.M. Rudraiah, M. Tech. Dissertation, Nagpur University, (1989).
32. Denitrification of Secondary Effluent using Anaerobic Fixed Film Rotating Bed Reactor S.N. Kaul, and L.A. Malik, M. Tech. Thesis, Nagpur University (1990).
33. Anaerobic Treatment of Food Processing Waste by Packed Bed, S.N. Kaul, and V. Jayshree, M. Tech. Dissertation, Nagpur University, (1991).
34. Denitrification of Secondary Effluent from a Domestic Wastewater Treatment Plant using UASB Reactor, B. Sahu, M. Tech. Dissertation, Nagpur University (1992).
35. Anaerobic Rotating Biological Drum Contactor for Treatment of Tannery Wastewater. S.N. Kaul and A.K. Dhage, M. Tech. Dissertation, University of Baroda, (1992).
36. Biological Sulphate Reduction using Upflow Anaerobic Sludge Blanket Reactor System, S.N. Kaul and Kanishka Desarkar, B. Tech. Project, Nagpur University, (1993).
37. Performance of Full Scale Wastewater Treatment Facility for Finished Leather Industry, S.N. Kaul, P.K. Mukherjee and T. Nandy, J. Envn. Sci. and Engg., Vol. A28 (6), p1277 (1993).
38. An Evaluation of Simple Kinetic Mode for Treatment of Domestic Sewage by using AnRBC, S.N. Kaul, V.P. Deshpande and C.V. Dehspande, J. Bio-resource Tech., 38, p31 (1991).
39. An Overview of Waste Management in Sugar Industry, S.N. Kaul, P.P. Pathe and T. Nandy, J. Ind. Ass. of Envn. Management, Vol. 17(2) (1990).
40. Determination of Kinetic Constants for Two Stage Anaerobic Packed Bed Reactor for Dairy Waste Treatment, V. Jayshree, S.N. Kaul and S. Shanta, Bio-resource Tech. 40, p253 (1992).
41. Some Salient Mathematical Consideration for Various Types of Anaerobic Bio-reactors, V. Jayshree and S.N. Kaul, J. Asian Envn., 13, 036 (1991).
42. Experience with Full Scale Aanerobic Fixed Film Reactors, R.A. Daryapurkar and S.N. Kaul, J. Envn. Sci. and Engg. Vol. A26 (3), p317 (1991).
43. Effluent of Cage Diameter and Rotational Speed on the Performance of Anaerobic Moving Bed Reactor Treating Sewage, V.P. Deshpande S.N. Kaul, J. Asian Envn. 12, p2 (1990).

44. Energy Recovery from Wastewater Based Pulp and Paper Mill Effluent, H.M. Rudriah, A.G. Bhole, S.N. Kaul and T. Nandy, J. IAEM, Vol. 10(6), June (1990).
45. Rotating Biological Contractor Treatment Kinetics for Dairy Wastewater, V. Jayshree. S.N. Kaul and S. Shantos, J. Ind. Public Heath Vol. (April 2, 1990).
46. Domestic Sewage Treatment using Anaerobic Moving Bed Reactor, V.P. Deshpande, S.N. Kaul and C.V. Deshpande, Ind. J. Envn. Prot. Vol. 10(6), (June 1990).
47. Upflow Anaerobic Sludge Blanket Reactor for Wastewater Treatment, P.P. Pathe, T. Nandy and S.N. Kaul, Ind. J. Envn. Prot. Vol. 10(7) (1990).
49. Kinetics and Model Description for UASB Reactor and for Various Wastewaters, P.P. Pathe, T. Nandy and S.N. Kaul, Ind. J. Envn. Prot. Vol. 10(8) (1990).
49. Moving Bed Biological Reactor for Wastewater-Some Considerations, S.N. Kaul, P. P. Pathe and T. Nandy, J. Chem. Engg. World, Vol. XXIV (12), (Dec. 1989).
50. Fluidized Bed Reactor for Wastewater Treatment, S.N. Kaul, and S.K. Gadkari, J. Chem. Engg. World, Vol. XXV(2), (1990).
51. LTC Wastewater Treatment by Anaerobic Methods, S. Shanta, S. N. Kaul and R. A. Daryapurkar, J. Asian Envn. Vol. 13(2), p47, (1991).
52. Three Stage Anaerobic Upflow Packed Bed Reactor System for Treating Basic Drugs Industrial Wastewater, K. Misra, T.A. Sihorwala and S.N. Kaul, Ind. J. Envn. Prot., 12(10), p751 (1992).
53. Anaerobic Fixed Film Fluidized Bed Reactor for High Strength Wastewater, S.N. Kaul, V. Jayshree and T. Nandy, IAWPRC (Australia) Proc. Vol. 1, p210 (1991).
54. Attached Film Fixed Bed Reactor for Treatment of Dairy Wastewater, V. Jayshree, S. Shanta and S. N. Kaul, J. Chem. Engg. World, 24(5), p54 (1989).
55. Anaerobic Biochemical Reactor Choice for Wastewater Treatment, S.N. Kaul, J. Asian Envn. 10(1), p42 (1988).
56. Effect of Depth on Reactor Performance of Anaerobic Packed Bed Reactor for Dairy Wastewater, V. Jayshree and S.N. Kaul, Ind, J. Envn. Prot. Vol. 8(10) p743 (1988).
57. Performance Evaluation of Two Stage Anaerobic Fixed Film Fixed Bed Reactor Treating Dairy Wastewater, using Special Media, S. Shanta, V. Jayshree and S.N. Kaul, Ind. J. Envn, Prot. Vol. 8(12), p919 (1988).
58. Anaerobic Bio-oxidation of Strong Soluble Organic Wastewater, S.N. Kaul, S.D. Badrinath and S. Shanto, J. Asian Envn., Vol. 7, p1 (June 1985).
59. Energy Recovery from High Strength Wastewaters, S.N. Kaul, S.D. Badrinath and B.B. Sundaresan, Proc. Bio-energy Society, p135 (1984/85).
60. Two Stage Fixed Film Reactor for Treatment of Distillery Wastewater, S.N. Kaul and S.D. Badrinath, IAWPC-Tech. Annual XI, p61 (1984).
61. Studies Related to Support Matrix for the Anaerobic Fixed Film Fixed Fixed Bed Reactor. S. Shanta, S.N. Kaul, A.S. Jawarkar and K.K. Singh, Accepted for Publication in the J. Envn. Science and Engg. (USA).

CHAPTER 9

SCALE UP FORMULATIONS FOR ANAEROBIC SYSTEMS

9.0 Introduction

The ultimate purpose of all pilot-plant and model experiments is crystallized in a phrase by L.H. Baekeland[1] that has become famous : commit your blunders on a small scale and make your profit on large scale. David[2] author of the world's first handbook of chemical engineering, emphasized the value of the experiments on a scale intermediate, between that of the laboratory and full scale production. A small experiment made upon a few grammes of material in the laboratory will not be of much use in guiding to erection of a large scale works, but there is no doubt that an experiment based upon few kilogram will give nearly all the data required.

Small scale plant is employed in chemical engineering for two main purposes. The first is as a forerunner to a full-sized production unit that is not yet built. In this case small scale equipment is called a pilot plant and its principal function is to provide design data for the ultimate large one, although it may also be required to produce small quantities of a new product for trial. The second purpose is to study the behavior of an existing plant of which the small unit is an reproduction. In this case, the small-scale equipment is what is ordinarily called a model and its chief function is to exhibit the effect of change in shape in operating conditions more quickly and economically than would be possible by experiments on a full-sized prototype. The functions of a pilot plant belong to the sphere of process development, those of a model of an existing plant to the sphere of process study. For the present purpose, it is not important whether the small scale unit is the forerunner from which a full sized counterpart will ultimately be scaled up, as in the case of a pilot plant, or whether the small scale unit is itself a scaled down model of an existing piece of plant.

Design data required for each piece of equipment can then be classified under six heads :

- data available from past experience ;
- data given in the laboratory reports or which can be derived from the laboratory results;

- data available in the literature;
- data which can be approximated sufficiently for design purpose by means of thermodynamic relations, the theorem of corresponding status or some of the many semi-empirical correlations that have appeared in recent years;
- data which could be obtained by further research in the laboratory;
- other practical data felt to be necessary for design purposes. Information falling under this head can be determined only in a pilot plant.

A pilot plant is unlikely to yield the maximum possible amount of information unless the 'Critical' components atleast are designed and operated in accordance with model theory. The first step is to derive the similarity criteria which govern the operations or processes to be studied on small scale. They may be obtained either by dimensional analysis or from fundamental differential equations of the process. A study of similarity criteria will reveal the conditions under which the model should be tested in order that the results may simulate those obtained under given conditions on the large scale. It will also show whether there are likely to be appreciable scale effects, in which case it will be necessary to apply corrections to the pilot plant results before they are used for full scale design. Finally, such a study will reveal those awkward cases which two or more similarity criteria are incompatible and where, consequently, it is not possible to emulate large scale results in a small apparatus with any certainly. Even in these cases, experiments with small-scale units over a sufficiently wide range of conditions can give valuable information for the design of full-sized plant.

The safest method of evaluating scale effects and allowing for incompatibility of similarly criteria is to scale a process up by easy stages through two or three pilot plants or increasing size. The effects, if any, of chance of scale on rate parameters and yield can then be observed directly and extrapolated to the full scale. A graduated succession of pilot plants is part of the traditional technique of process development, a technique that is both costly and slow. The principal aim of model theory is to allow of larger steps in the development and reduce their number and duration, and it is to be hoped that in future the need for more than one pilot stage between laboratory and the full stage will become increasingly rare. Two important factors are to be considered. *viz.*

- process study;
- elimination of avoidable errors.

9.0.1 Process Study

This aspect implies the examination and improvement of existing full-scale process plant. The principal object of using models is to determine the effects of modifications in design or major changes in operating conditions without increasing the expense or risk of making these changes on the large scale. For example, to establish the effect of making of altering the shape or position of a baffle wall in a furnance by means of a full scale trial, it will be necessary to take the furnance out of service, allow it to cool down, break out the old baffle wall, build the new one, and heat sufficiently slowly to avoid damaging the brick work. The required information could probably be obtained by means of a model in a few days and without interrupting production.

Similarly major change in operating conditions might conceivably damage a production plant, can be tested first on a small scale provided that model apparatus is available which can be relied upon to reproduce the behaviour of the prototype. Model theory indicates the necessary conditions for such reproducibility. It is, therefore, necessary to keep a pilot plant commissioning long after its original process development function has been commissioned. As the detailed design and construction of a full scale plant proceed, the pilot plant is then available to solve last minute problems. In case of difficulty or dangerous process, it can provide a training good for production plant operatives.

9.0.2 Elimination of Avoidable Errors

It is necessary to examine the possible sources of errors in the alternative procedure of calculating plant dimensions from laboratory data and theoretical or empirical design equations, and to evaluate the gain in accuracy and dependability of data to be expected from pilot plant model studies. The basic design equation is the rate equation which gives the rate at which a physical or chemical process takes place and hence the size of the equipment. The components which form a part of the basic rate equation are :

- numerical factor;
- physical and chemical properties;
- operating variables;
- dimensional and time factors.

The numerical factor is a dimensionless factor which is needed to balance the rate equation when the latter is expressed in dimensionless form. It is normally a shape factor which varies with the geometrical shape of the equipment but is independent of its size.

Physical and chemical properties represent the usual form in which the laboratory data are fed into the design calculation. They may be determined in the laboratory for this purpose, taken from literature or reference handbook or approximated by means of some theoretical or empirical rate.

Operating variables include temperature, pressure, concentrations, rates of flow and residence times. In general, they constitute the independent variables which are under plant operator's control.

The dimensional factor highlights the manner in which the total quantity of change per hour depends upon the dimensions of the equipment whether, for example, it varies with the length, surface or reactor volume of the reactor. The time factor shows the variation of rate with time. At constant rate, both dimensional and time factors are incorporated in a rate coefficient. For rates which vary with time, time factor enters as a differential.

Each of the above classes of quantity represents a possible source of error when introduced into the basic rate equation without experiment. In a correctly planned pilot plant or model experiment, most of the above sources of error can be avoided.

9.1 The Principle of Similarity

The Principal of similarity is concerned with the relations between physical systems of different sizes, and therefore, it is fundamental to scaling up or down of physical and chemical processes. The principal of similarity is usually coupled and not infrequently confused with the method of dimensional analysis. Although, historically the two have been linked together, yet logically they are quite distinct. The principle of similarity is a general principle of nature, dimensional analysis is only one of the technique by which the principle may be applied to specific cases, the other technique being to start from the generalised equations of motion of the system. Physical systems and material objects are characterized by :

- size
- slope
- composition

All three are independent variables. The principle of similarity is more concerned with the general concept of shape as applied to complex system with the implication of the fact that the shape is independent of size and composition. This principle states that the spatial and temporal configuration of a physical system is determined by the ratio of magnitude within the system itself and does not depend upon the size or nature of units in which these magnitudes are measured.

Similarity can be defined in two ways, by specifying the ratios either of different measurements in the same body or of corresponding measurement in different bodies.

The second method has practical advantages that a single scale ratio is substituted for a number of shape factors. Four similarity states are important in biochemical engineering, *viz.* :

- geometrical similarity
- mechanical similarity

- thermal similarity
- chemical similarity.

Each of the above states necessitates all previous one. For example, complete chemical similarity would require thermal, mechanical and geometrical similarity. All actual cases of similarity in fact contain an element of approximation because of disturbing factors always present which prevent ideal similarity from being attained. Often the effects of such departures from ideal similarity are negligible. When not negligible, they give rise to scale effects and a correction of some kind has to be applied when experimental results are scaled up and down. Sometimes the requirements for similarity with respect to two important factors are totally incompatible, giving rise to the difficult case of mixed regime in which even an approximation to similarity may be impossible to achieve without drastic changes in the process.

Two bodies are geometrical by similar when to every point in one body there exists a corresponding point in the other. Mechanical similarity comprises of static or static force similarity, kinematic similarity and dynamic similarity, *viz.* :

- geometrically similar bodies are statically similar when under constant stress their relative deformations are such that they remain geometrically similar;
- geometrically similar moving systems are kinematically similar when corresponding particle trace out geometrically similar paths in corresponding interval of time;
- geometrically similar moving systems are dynamically similar when the ratios of all corresponding forces are equal.

Geometrically similar systems are thermally similar when corresponding temperature differences bear a constant ratio to one another and when the system, if moving are kinematically similar. Geometrically and thermally similar systems are chemically similar when corresponding concentration differences bear a constant ratio to one another and when the systems if moving are kinematically similar. The intrinsic ratios or criteria which define chemical similarity in addition to those required for kinematic and thermal similarity, are :

- ratio of rate of chemical formation to rate of bulk flow;
- ratio of rate of chemical formation to rate of molecular diffusion. Normally this ratio may be neglected in comparison to the first one in most cases.

Generally, the rate of a chemical reaction may be independently varied by changing the temperature. In practice both the chemical equilibrium and the relative rates of unwanted side reactions vary with temperature and there is usually a narrow temperature range within which the reaction must proceed on both the small and large scale in order to ensure maximum yield. In both model and prototype, the reaction time will be of the same order and this requirement fixes the relative velocities in continuous – flow systems. These velocities are incompatible with the velocities necessary for kinematic similarity except at very low or very high velocities. Therefore, in scaling up a continuous chemical reaction and especially where there is an optimum reaction time beyond which the yield or quality is reduced, it is advantageous to operate both model and prototype either in the streamline region or with a high degree or tubulence. If neither condition is feasible, there will be an unpredictable scale effect and it would be prudent either to scale up in several stages or to allow ample factor of safety in design.

Mechanical, thermal or chemical similarity between geometrically similar systems, can be specified in terms of criteria which are intrinsic ratio of measurements, forces or rate within each system. Since these criteria are ratio of like quantities, they are dimensionless and there are two general methods of arriving at them. Where the differential equations that govern the behaviour of the system are unknown, but provided one knows all the variables which would enter into the differential equations, it is possible to derive similarity criteria by means of dimensional analysis. When differential equations are known but cannot be integrated, the similarity criteria can be obtained from differential from. When differential equations are known and can be integrated, there, is in general no need for either similarity criteria or model experiments, since the behaviour of the large scale system can be directly calculated.

9.2 Π – Buckingham Theorem

This Theorem falls into two parts, *viz.* :

- the solution of every homogenous physical equation has the form :

  $$\phi_1 (\Pi_1, \Pi_2, \Pi_3 \ldots\ldots \Pi_n) = 0 \qquad \ldots(9.1)$$

  where Π_1, Π_2, ... Π_n represent a complete set of dimensionless groups of various variables and dimensional constant in the equation

- if an equation contain 'n' separate variables and dimensional constants and these are given dimensional formulae in terms of 'm'–primary quantities, then the number of dimensionless groups in a complete set is (n – m)

Every complete physical equation is either dimensionally homogeneous or capable of being resolved into two or more separate equations that are dimensionally homogenous and a dimensional constant is any constant appearing in a physical equation which changes in value when the units of measurements of primary quantities are changed. Some of the important dimensionless groups are :

- Pressure coefficient ($F/\rho v^3L^2$)

- Reynolds number ($\rho vL/\mu$)
- Weber number ($\rho v^3L/\sigma$)
- Froude number (v^2/Lg)
- Grashof number ($\beta g\Delta TL^3\rho^2/\mu^2$)
- Prandtl group ($C\mu/k$)
- Nusselt number (hL/k)
- Schmidt number ($\mu/\rho D$)
- Arrhenius number (E/RT)
- Power number ($Pgc/\rho N^3L^5$)
- Sherword number (k_mL/D)
- Radition number ($rC_pv/\sigma eT^3$)

where C, D, E, F, g, g_c, h, k, k_m, L, P, R, T, v, β, Δ, μ, e and σ refer to specific heat, diffusivity, activation energy, force, acceleration of gravity, Newton's law conversion factor, heat transfer coefficient, mass transfer coefficient, length, power consumption, gas constant, linear velocity, coefficient of cubical expansion, difference, viscosity, density and surface tension respectively.

9.3 Similarity Criteria and Scale up Equations

9.3.1 Flow of Fluids

The basic differential equation for isothermal flow of a Newtonian viscous fluid are Navier-Stokes equations

$$\underset{\text{I}}{\rho\frac{\delta u}{\delta t}}+\underset{\text{II}}{\rho\left(u\frac{\delta u}{\delta x}+v\frac{\delta u}{\delta y}+w\frac{\delta u}{\delta z}\right)}=\underset{\text{III}}{\rho g\,\mathrm{Cos}\,\alpha_z}-\underset{\text{IV}}{\frac{\delta p.}{\delta x}}$$

$$+\underset{\text{V}}{\frac{1}{3}\mu\left(\frac{\delta u}{\delta x}+\frac{\delta v}{\delta y}+\frac{\delta w}{\delta z}\right)}+\underset{\text{VI}}{\mu\left(\frac{\delta^2 u}{\delta^2 x}+\frac{\delta^2 v}{\delta^2 y}+\frac{\delta^2 w}{\delta^2 z}\right)} \quad \text{...(9.2)}$$

Where u, v, and w, ρ, and μ, p, t and g refer to velocities in x, y and z directions, fluid density and viscosity, pressure, time and accleration due to gravity acting at an angle α_x to x-direction respectively.

The generalized dimensional equation may be written

$$\underset{\text{I}}{\left(\frac{\rho v}{t}\right)}+\underset{\text{II}}{\left(\frac{\rho v^2}{L}\right)}=\underset{\text{III}}{(\rho g)}-\underset{\text{IV}}{\left(\frac{\Delta p}{L}\right)}+\underset{\text{V \& VI}}{\left(\frac{\mu v}{L^2}\right)} \quad \text{...(9.3)}$$

I and II are dimensionally equivalent and, hence, there will be three independent dimensionless groups. Therefore, the dimensionless equation can be written as :

$$\phi\left(\frac{\rho vL}{\mu},\phi\frac{v^2}{Lg},\frac{\delta p}{\rho v^2}\right)=\text{constant or} \qquad ...(9.4)$$

$$\frac{\Delta p}{\rho v^2}=\phi'\left(\frac{\rho vL}{\mu},\frac{v^2}{Lg}\right) \qquad ...(9.5)$$

If the effect of the surface tension is to be incorporated, an additional dimensionless number (Weber number) is to be added, *i.e.*

$$\frac{\Delta p}{\rho v^2}=\phi'\left(\frac{\rho vL}{\mu},\frac{v^2}{Lg},\frac{\rho v^2 L}{\sigma}\right) \qquad ...(9.6)$$

It may be mentioned that variations in viscosity with a shear rate, changes in viscosity and density due to the frictional heating and compression-wave phenomena have not been considered because these aspects play a minor role in anaerobic biological reactor system.

9.3.2 Thermal Processes

The basic differential equations of heat flow by forced convection in a moving fluid are the Navier-Stokes equation (9.2) alongwith heat transfer equation (9.2) *i.e.*,

$$\underset{\text{I}}{\rho C_p\left(u\frac{\delta T}{\delta x}+v\frac{\delta T}{\delta y}+w\frac{\delta T}{\delta z}\right)}+\underset{\text{II}}{k\left(\frac{\delta^2 T}{\delta x^2}+\frac{\delta^2 T}{\delta y^2}+\frac{\delta^2 T}{\delta z^2}\right)}=\underset{\text{III}}{-\rho C_p\frac{\delta T}{\delta t}} \qquad ...(9.7)$$

The dimensional form of equation (9.7) is as follows :

$$\underset{\text{I}}{\left[\frac{\rho C_p vT}{L}\right]}+\underset{\text{II}}{\left(\frac{kT}{L^2}\right)}=-\underset{\text{III}}{\left[\frac{LC_pT}{t}\right]} \qquad ...(9.8)$$

Rearranging equation (9.8) to incorporate heat transfer coefficient (h) results in

$$\underset{\text{I}}{\left[\rho C_p\ vT/L\right]}+\underset{\text{II}}{\left[kT/L^2\right]}=-\underset{\text{III}}{\left[hT/L\right]} \qquad ...(9.9)$$

Rearranging equation (9.8) results in the following equation, *i.e.*

$$\phi\left(\frac{\rho C_p vL}{k}, \frac{hL}{k}\right) = \text{constant} \qquad ...(9.10)$$

Equality of Peclet number (ρC_p vL/k) in model and prototype systems ensure, that the ratio of heat flows by conduction and forced convection across corresponding surfaces shall be equal which is one of the conditions for thermal similarity. The other condition for thermal similarity is that of fluid flow pattern (kinematic similarity). The generalized dimensionless equation for heat transfer by forced convection is given as follows :

$$\phi\left(\frac{\rho vL}{\mu}, \frac{v^2}{Lg}, \frac{\rho C_p vL}{k}, \frac{hL}{k}\right) = \text{constant} \qquad ...(9.11)$$

In model and prototype systems where both Reynolds and Prandl numbers are equal, it follows algebraically that Peclet numbers are also equal. If gravitational effects are negligible, *i.e.* Froude number (v^2/Lg) can be neglected, the resultant dimensionless equation is given as follows :

$$hL/k = \phi\ (\rho vL/u,\ C_p\mu/k) \qquad ...(9.12)$$

where C_p, k, T and h refer to specific heat at constant pressure, thermal conductivity temperature and overall heat transfer coefficient respectively.

9.3.3 Radiation Effects

This aspect may have to be included in an anaerobic biological reactors. If the order of magnitude is small compared to conduction and convection, it may be neglected. The basic differential equation is as follows :

$$H = \frac{dQ}{dt} = \sigma e\left(T_1^{\,4} - T_2^{\,4}\right)L^2 \qquad ...(9.13)$$

where H, Q, T_1 and T_2, e, σ and L^2 refer to rate of heat transfer, net quantity of heat transferred, absolute temperatures of hot and cold surfaces, combined emissivities of hot and cold surfaces, Stefan Boltzman constant and radiating surfaces respectively. Equation (9.13) in dimensional is given as follows :

$$\left[\frac{Q}{kLT}\right] = \left(\sigma e T_1^{\,4}\right)\left[\frac{T_1^{\,4} - T_2^{\,4}}{T_1^{\,4}}\right] \qquad ...(9.14)$$

Radiation of heat is generally accompanied by conduction to and from the radiating surfaces; so the dimensionless group of conduction must be included, *viz.*:

$$\phi\left(\frac{H}{\sigma e L^2 T}, \frac{T_1}{T_2}, \frac{\rho C_p L^2}{kT}\right) = \text{constant} \quad \text{... (9.15)}$$

If L/t is considered as velocity, then the basic dimensionless equation becomes :

$$\phi\left(\frac{\rho C_p v}{\sigma e T^3}, \frac{T_1}{T_2}, \frac{\rho C_p L v}{k}, \frac{H}{kLT}\right) = \text{constant} \quad \text{...(9.16)}$$

9.3.4 *Diffusional Processes*

The basic equations for mass and heat transfer are quite analogous. The differential equations for mass transfer with forced convection are the Navier-Stokes equation (9.2) alongwith mass transfer equation, *i.e.* :

$$\underset{\text{I}}{\left(u\frac{\delta C}{\delta x} + v\frac{\delta C}{\delta y} + w\frac{\delta C}{\delta z}\right)} + \underset{\text{II}}{D\left(\frac{\delta^2 C}{\delta x^2} + \frac{\delta^2 C}{\delta y^2} + \frac{\delta^2 C}{\delta z^2}\right)} = \underset{\text{III}}{-\frac{\delta C}{\delta t}} \quad \text{...(9.17)}$$

where C and D refer to concentration of diffusing substance for unit volume and diffusion coefficient of diffusing substances respectively. The dimensional form of equation, (9.17) is as follows :

$$\underset{\text{I}}{\left(\frac{vC}{L}\right)} + \underset{\text{II}}{\left(\frac{CD}{L^2}\right)} = \underset{\text{III}}{-\left(\frac{C}{t}\right)} \quad \text{...(9.18)}$$

Normally terms I and II are dimensionally equivalent therefore, there is one dimensionless number. For steady and unsteady state conditions, the dimensionless equation can be written as :

$$\phi(vL/D) = \text{constant (steady state)} \quad \text{...(9.19)}$$

$$\phi(L^2/Dt) = \text{constant (unsteady state)} \quad \text{...(9.20)}$$

Using Navier-stokes equation alongwith diffusional equation, the resultant dimensionless equation is as follows :

$$\phi\left(\frac{\rho v L}{\mu}, \frac{v^2}{Lg}, \frac{\Delta p}{\rho v^2}, \frac{vL}{D}\right) = \text{constant} \quad \text{...(9.21)}$$

This equation on simplification yields :

$$\phi\left(\frac{\rho v L}{\mu}, \frac{v^2}{Lg}, \frac{\Delta p}{\rho v^2}, \frac{\mu}{\rho D}\right) = \text{constant} \quad \text{...(9.22)}$$

Generally, Froude number is not important in an anaerobic biological reactor system and can be ignored.

9.3.5. Chemical Processes

9.3.5.1 Homogenous Reactor

The basic equations of importance are :

- Navier stokes equation (9.2).
- Heat transfer equation with convection and conduction.
- Mathematical equation connecting mass transfer rate with consumption of substrate or formation of products.
- All these equations are coupled. Heat transfer equation is given as follows:

$$\rho C_p\left(u\frac{\delta T}{\delta x}+v\frac{\delta T}{\delta y}+w\frac{\delta T}{\delta z}\right)+k\left(\frac{\delta^2 T}{\delta z^2}+\frac{\delta^2 T}{\delta y^2}+\frac{\delta^2 T}{\delta z^2}\right)+\rho C_p\frac{\delta T}{\delta t}U=qU \quad ...(9.23)$$

where q and U refer to heat of reaction per unit mass of product and chemical reaction rate expressed as mass of product formed per unit volume and time respectively.

The above equation can be symbolically represented in the dimensional form as follows:

$$\left[\frac{\rho C_p vT}{L}\right]+\left[\frac{kT}{L^2}\right]+\left[\frac{\rho C_p vT}{t}\right]=[qU] \quad ...(9.24)$$

The resultant dimensionless equation is as follows :

$$\phi\left(\frac{\rho C_p vL}{k},\frac{qUL}{\rho C_p vT}\right)=\text{constant} \quad ...(9.25)$$

Similarly the generalized mass transfer equation is given by :

$$\left(u\frac{\delta C}{\delta x}+v\frac{\delta C}{\delta y}+w\frac{\delta C}{\delta z}\right)+D\left(\frac{\delta^2 C}{\delta x^2}+\frac{\delta^2 C}{\delta y^2}+\frac{\delta^2 C}{\delta z^2}\right)+\rho Cp\frac{\delta T}{\delta t}=qU \quad ...(9.26)$$

The dimensionless representation of this equation is as follows :

$$\phi\left(\frac{vL}{D},\frac{UL}{Cv}\right)=\text{constant} \quad ...(9.27)$$

Radiation effects cannot be neglected and, therefore, the generalized dimensionless equation for a homogenous chemical reaction taking place in a moving – fluid is as follows :

$$\phi\left(\frac{\rho vL}{\mu},\frac{v^2}{Lg},\frac{\Delta p}{\rho v^2},\frac{\rho C_p vL}{k},\frac{qUL}{\rho C_p vT},\frac{vL}{CU},\frac{Q}{\sigma eL^2T^4},\frac{T_1}{T_2}\right)=\text{Constant} \quad ...(9.28)$$

Gravitational and frictional effects are to be ignored in chemically reacting system. Therefore, the generalized basic equation for mass transfer (homogenous reaction) becomes:

$$\frac{UL}{Cv}=\phi\left(\frac{\rho vL}{\mu},\frac{C_p\mu}{k},\frac{\mu}{\rho D},\frac{qC}{\rho C_p T},\frac{qCv}{\sigma eT^4},\frac{T_1}{T_2}\right) \quad ...(9.29)$$

If radiation effects are negligible in comparison to other important parameters, the resultant equation is reduced to the following form which was derived by Damkohler for a continuous reactor system[3] :

$$\frac{UL}{Cv}=\phi\left(\frac{\rho vL}{\mu},\frac{C_p\mu}{k},\frac{\mu}{\rho D},\frac{qC}{\rho C_p T}\right) \quad ...(9.30)$$

9.3.5.1.1 Temperature Effects on Rate of Reaction

The reaction rate depends upon two important variables, *viz.* :

- temperature (T)
- concentration (C)

The generalized equation for rate of a homogenous-chemical reaction is given by

$$U = K_n\ \phi\ (C_1, C_2, \ldots\ldots C_n) \quad ...(9.31)$$

where K_n, C_1, C_2 ...... C_n and ϕ refer to reaction rate constant of nth order, concentration of reactors and dimensionless kinetic factor which incorporates activity coefficients of reactants and products. To incorporate the effect of temperature, it is necessary to use the modified Arrhenius equation, *i.e.*

$$K_n = AT^{1/2} \exp\ (-\ E/RT) \quad ...(9.32)$$

Where A, E and R refer to a constant, activation energy and universal gas constant respectively.

The generalized dimensionless equation becomes:

$$\frac{K\,C^{2n-2}\,L^2T}{v^2}=\phi\left(\frac{\rho vL}{\mu},\frac{C_p\mu}{k},\frac{\mu}{\rho D},\frac{qCR,}{E\rho C_p},\frac{qC}{\sigma eT^4},\frac{T_1}{T_2}\right) \quad ...(9.33)$$

If $KC^{2n-2}L^2T/v^2$ is replaced by $KC^{2n-2}\,t^2\,T$, equation (9.33) will be useful for batch reactors.

The limitations of the above equation are :

- useful where there is one reversible reaction;
- useful incase of simultaneous reaction, where they are all of the same order;
- if several reactions of different order occur simultaneously, then L.H.S group would be replaced by several groups, one for each order of reaction

All these and together with the fact that order of some or all of the constituent reactions may be unknown, at present limits the practical utility of equation (9.33).

9.3.5.2 Heterogenous Reactors

This aspect is of utmost utility for anaerobic fixed film reactors. The rate of heterogenous biological/biochemical reaction depends upon the interfacial area between phases utilizing the concept of catalysis by making use of specific area or interfacial areas per unit volumes and alongwith the kinetics of heterogenous reaction rate (U) which is expressed as mass of substrate or product formed per unit interfacial area per unit time.

Therefore UL/Cv can be replaced by U'sL/Cv hetergenous reactions. In addition s is inversely related to some references length (L) for a geometrically similar system. Therefore, U'sL/Cv becomes U'/Cv

Again, reaction rate U' is related to temperature and concentration, *i.e.*

$$U' = \phi[K_n'\ X\ \alpha\ (C_1, C_2, \ldots.C_n)] \qquad \ldots(9.34)$$

where ϕ, K'_n, X and α refer to unknown function, reaction rate constant, kinetic factor and dimensionless factor proportional to catalytic activity of interface.

The basic dimensionless equation for anaerobic fixed film reactor under steady state condition is given as follows :

$$\left[\frac{K'\alpha\, s\, C^{2n-2}L^2T}{v^2}\right] = \phi\left(\frac{\rho v L}{\mu}, \frac{C_p\mu}{k}, \frac{\mu}{\rho D}, \frac{qCR}{E\rho C_p}, \frac{qC}{\sigma e T^4}, \frac{T^1}{T^2}\right) \qquad \ldots(9.35)$$

If the system operates under unsteady state conditions, then $\left[\frac{K'\alpha\, s\, C^{2n-2}L^2T}{v^2}\right]$ can be replaced by $(K', \alpha, s, C^{2n-2}\,t^2, T)$.

9.4 Effect of Several Dimensionless Numbers

The dimensionless similarity criteria are ratio of physical quantities which are functions of various forces or resistances which control the reaction rate as discussed in earlier section. If there are several controlling factors of different kind, there will be several dimensionless criteria. An example can be cited:

Resistance to the motion of a fluid may be due to viscous drag, gravitational forces or surface tension, the corresponding criteria being the Reynold, Froude and Weber numbers respectively. These criteria are mutually incompatible for a homogenous system of different absolute magnitudes, *i.e.*

$v \alpha L^{-1}$ (Reynolds number being equal) ...(9.36)

$v \alpha L^{1/2}$ (Froude number being equal) ...(9.37)

$v \alpha L^{-1/2}$ (Weber number being equal) ...(9.38)

It is possible to satisfy two criteria at a time by adjusting fluids with different physical properties for a system. For the reliable scaling up or down of a complex physical or chemical process, two conditions are necessary, *viz.*:

- the regime should be relatively pure (the reaction rate should depend chiefly upon few dimensionless numbers);
- the regime should be of the same type on both the small and prototype units.

The regime refers to particular force, flow, or resistance factor which controls the overall rate of change rather than the rate determining process in a system. In planning a series of pilot plant or model experiments, corresponding conditions in the large scale prototype must be constantly born in mind. The danger of change in regime occurs chiefly when it is necessary to extrapolate the similarity relation. The dimensionless equations developed in section 9.3 are based on Fundamental Principles and same relationship will emerge when extended principle of similarity is used based on Π – Buckingham's theorem and any number of dimensionless groups can be used.

9.5 Effect of Temperature and Other Consideration

Increase in temperature tends to increase the rate of physical, chemical, biochemical and biological reactions by decreasing the resistance factor in the generalised rate expression. The properties that are most affected are :

$k = A \exp(-E/Rt)$; [Rate constant] ...(9.39)

$1/\mu = B \exp(-Ev/RT)$; [Viscosity effect] ...(9.40)

$D = C \exp(-E_d/RT)$; [Diffusivity] ...(9.41)

where

A, B and C refer to constants which are temperature dependent, *i.e.*

$A = a, T^{1/2}$; $B = b$

$T^{1/2}$ and $C = c\, T^{1/2}$ and a, b and c are true constants.

The prevailing regime where both chemical (biochemical/biological) and dynamic processes take place depend upon the experimental determination of two parameters, *viz.*

- the 10° temperature coefficient;
- the Reynolds index (the Reynolds number exponent of the prevailing fluid regime) Therefore, the general rules can be summarised based on the above two facts, *viz.;*
- a 10° temperature coefficient greater than 2 characterizes a chemical regime; a coefficient below 1.5 characterizes a dynamic regime;
- a Reynolds index approximating to zero characterizes a chemical regime or a stream line dynamic regime; an index between 0.5 and 0.8 characterizes a turbulent dynamic regime with fixed interface; and index between 3 and 5 characterizes a dynamic regime with free interface, *i.e.* a two phase liquid-liquid or gas-liquid or liquid solid;
- where the 10° temperature coefficient is greater than 1.5 and the Reynolds index is below 0.5, a mixed chemical-dynamic regime is suggested.

There may arise at the boundary surface departures from similarity which are termed boundary or wall effects and which may, unless controlled, render it almost impossible to predict large scale performance from model experiments. A fixed film fixed bed reactor element in which reactor diameter is only two or three the packing material would giva little information about behaviour of a full scale reactor owing to the predominant wall effect in the element. Boundary effects cannot be eliminated by increasing the extent of the system under control. The boundary effects may have four different kinds of effects upon a physical, chemical (biochemical/biological) reaction proceeding within it, *viz.* :

- the fluid flow pattern and frictional resistance is influenced;
- the transfer of heat into or out of the system is possible;
- the absorption of matter from or release of matter to the fluid stream;
- can positively or negatively catalyse a chemical reaction (this aspect is most difficult to assess).

For chemical process industry, it is generally accepted rule that the ratio of prototype size to bench scale unit can be between 3 to 5. However, in wastewater systems the reactor sizing can be extended upon 10. Beyond this ratio, it is difficult to maintain flow, thermal and chemical regime under the control of the engineer.

9.6 Scale up Formulation for Fixed Film Fixed Bed Reactor

Scale up is defined as the successful start up and operation of a full scale unit whose design is in part based upon experimentation and demonstration at a smaller scale of operation. Experience of actual workers in this field suggests that there are recurring scale up problems in the development of many process studies. Differences that may be of particular importance in moving from laboratory scale to full scale plant are :

- slope which can lead to differences in turbulence, fluid short circuiting or stagnation zones;
- mode (and scale) of operation resulting in different residence time distributions;
- surface area to volume ratio, flow patterns, and geometry which result in significantly different gradients of concentrations and temperatures;
- size and shape of media used for biomass attachment in the reactor;
- flow stability;
- heat losses and temperature maintenance;
- wall, edge and end effects.

9.6.1. *Applications of Π-Buckingham Theorem for Development of Scale up Equation for Fixed Film Fixed Bed Reactor*[A]

The basic variables for the determination of substrate conversion, biogas production and biomass yield is given in Table 9.1. The total number of basic variables are 15 and there are three dimensions (length, mass and time; L, M and S). Therefore, the total number of basic dimensionless numbers should be 12 according to Π-Buckingham theorem.

Table 9.1 : Basic Variables for Development of Scale up Equations

Sr.No.	*Variable*	*Symbol*	*Unit in M(L)S, (MLT)*
1.	Influent substrate concentration	S_i	ML^{-3}
2.	Attached biomass concentration	X	ML^{-3}
3.	Substrate reaction rate constant	k	S^{-1}
4.	Hydraulic detention period	θ	S
5.	Yield coefficient	Y	MM^{-1}
6.	Media particle-diameter	dp	L
7.	Diffusivity	D	L^2S^{-1}
8.	Kinematic viscosity	υ	L^2S^{-1}
9.	Specific growth rate	μ_{max}	S^{-1}
10.	Water density	ρ	ML^{-3}
11.	Half substrate velocity constant	K_a	ML^{-3}
12.	Biomass retention period	θ_c	S^{-1}
13.	Diameter of the bioreactor	D_l	L
14.	Height of the bioreacter	H	L
15.	Biomass decay coefficent	k_d	S^{-1}

The fixed film reactor is characterized by three basic factors, *viz.* :

- substrate conversion;
- biomass production;
- biogas generation.

It is logical that there should be three basic equations for the above basic factors. As an example, calculation for one dimensionless group is shown using the basic principle and the others can be computed on similar basis.

$$\Pi_1 = d_p^{x_1}\theta^{y_1}\rho^{z_1}v^{-1} = M^o\ L^o\ S^o \qquad ...(9.42)$$

$$= L^{x1}\ S^{y1}\ M^{x1}\ L^{3z1}\ SL^{-2} = M^o\ L^o\ S^o$$

$$\Sigma L = x_1 - 3z_1 - 2 = 0 \qquad ...(9.43)$$

$$\Sigma M = z_1 = 0 \qquad ...(9.44)$$

$$\Sigma S = y_1 + 1 = 0 \qquad ...(9.45)$$

$x_1 = 2$, $y_1 = -1$ and $z_1 = 0$ and, therefore,

$\Pi_1 = \frac{d_p^{\ 2}}{v\theta}$; if $\theta = \frac{d_p}{v}$, then Π_1 can be rearranged as follows :

$$\Pi_1 = \frac{d_p^{\ 2}}{\phi v}.\frac{v}{d_p} = \frac{d_p.v}{\upsilon}, = \frac{\rho d_p v}{\mu} \qquad ...(9.46)$$

Similarly, other dimensionless numbers can be calculated. The three basic scale up equations based on Π-Buckingham theorem are summarised in equations 9.47 to 9.49 for substrate conversion, biomass production and biogas generations. Basically there are 11 dimensionless numbers [(μ_m d_p/v and k_d d_p/v) are independent dimensionless numbers. Therefore, ratio of these two numbers k_d/μ_m does not define any new dimensionless numbers)].

- **Substrate Conversion**

$$\left(\frac{S_i - S}{K\,\theta\,S_i}\right) = \frac{K_1 N_{pe}^{a}\,N_{Re}^{b}}{I}$$

$$\frac{N_{th}{}^{c}(X/K_s)^d\ \mu_m(d/v)^e\ \theta_c(V/d_p)^f\ k_d(d_p/v)^g(k_d/\mu_m)^h}{II}$$

$$\frac{k(d_p/v)^i}{III}\ \frac{(D_i/d_p)^j\,(h/d_p)^k}{IV} \qquad ...(9.47)$$

- **Biomass Production**

$$\frac{M^*}{Q(S_i - S)} = \frac{K_2 N^{a'}{}_{Pe} N^{b'}{}_{Re}}{I}$$

$$\frac{N^{c'}{}_{th}(X/Ks)^{d'}\ \mu_m(d_p/v)^{e'}\theta_c(v/d_p)^{f'}K_d(d_p/v)^{g'}(k_d/\mu_m)^{h'}}{II}$$

$$\frac{K(d_p/v)^{i'}}{III}\ \frac{(D_i/d_p)^{j'}(h/d_p)^{R'}}{IV} \qquad ...(9.48)$$

- **Biogas Generation**

$$\frac{M^{**}}{Q(S_i - S)} = \frac{K_3 N_{Pe}{}^{a'} N_{Re}{}^{b'}}{I}$$

$$\frac{N_{th}{}^{c'}(X/Ks)^{d'}\ \mu_m(d_p/v)^{e'}\theta_c(v/d_p)^{f'}K_d(d_p/v)^{g'}(k_d/\mu_m)^{h'}}{II}$$

$$\frac{k(d_p/v)^{i'}}{III}\ \frac{(D_i/d_p)^{j'}(h/d_p)^{R'}}{IV} \qquad ...(9.49)$$

where

(*a*) M^* and M^{**} refer to rate of mass production of biomass and biogas respectively.

(*b*) N_{pe} (Peclet number), N_{Re} (Reynolds number) and N_{th} (Thiele number) refer to vd_p/D, vd_p/v and $d_p\ (\mu_m/DYL_a)^{1/2}$ respectively.

(*c*) Terms I, II, III and IV refer to respectively dimensionless number based on hydraulics in the system, dimensionless number based on properties of microorganisms, dimensionless number based on properties of substrate and dimensionless number based on bioreactor system.

(*d*) a, b, c, d, e, f, g, h, i, j, and k are exponents in the above equations.

(*e*) K_1, K_2 and K_3 are constants in the above equations.

Equations 9.47 to 9.49 are coupled equations. However, for all practical purpose the specific biogas yield and biomass yield is constant for a specific type of reactor and substrate. Therefore, the scale up equations reduces to one equation *viz.* 9.47. It must be mentioned that the parameters in this equation must be adjusted to some standard temperature and pH conditions, it is assumed that other physico-chemical parameters are by and large quite constant and will not hamper the biological activity. The expression for effect of temperature and pH must be developed for all the rate constants for substrate removal (k), biomass growth and decay (μ, k_d, K_s and Y), and biogas yield. The hydraulic properties must also be known.

It is necessary to simplify the basic equation (9.47) for practical application. Velocity of wastewater inside the reactor can also be expressed in terms of θ and is numerically equivalent to a ratio of total media height to HRT (ht/θ) as explained below :

$$\text{Velocity} \quad (v) = \text{flow/Area} \qquad \text{...(9.50)}$$

$$Q = \frac{\text{Void volume } (V_v)}{\theta} \qquad \text{...(9.51)}$$

$$v = \frac{V_v}{\text{Area} \cdot \theta} \qquad \text{...(9.52)}$$

It is assumed that the porosity is same throughout the media height or in other words, at any cross section in the media, the total cross-sectional area of the openings in the media is same,

$$V_v = \text{Area, ht; ht - total media height} \qquad \text{...(9.53)}$$

$$\text{Area} = V_v/\text{ht}$$

Parameters like, velocity in terms θ, diffusivity (D), viscosity (υ) and specific gravity (ρ) clubbed together (D.ρ.θ/v) reflect hydraulic properties of the reactor system.

Similarly, parameters like maximum specific growth yield (μ_m), biomass (X), yield coefficient (Y) and decay coefficent (k_d) comprise of (μ_{max}.X.Y/K_d) represent properties of biomass.

Substrate utilization rate (reaction rate constant, k), influent substrate concentration (S_i) and half velocity substrate concentration (K_S) represent the properties of substrate and interaction of substrate with biomass.

The dimensionless terms D_i/ht, ht/dp and h/ht represent physical configuration of the reactor.

Equation 9.47 can be modified as follows :

$$\left(\frac{S_i - S}{S_i} k \cdot \theta\right) = K_1 \left[\frac{D}{\vartheta} \frac{\mu_m}{k_d} \frac{X}{K_a} \frac{\rho}{S_i} . Y . K . \theta\right]^a \left[\frac{D_t}{ht}\right]^b \left[\frac{ht}{dp}\right]^c \left[\frac{h}{ht}\right]^d \qquad \text{...(9.54)}$$

$$\left(\frac{S_i - S}{S_i} k \cdot \theta\right) = K_1 \left[\frac{D}{\vartheta} \rho.\theta \frac{\mu_m}{k_d} .X.Y \frac{k}{S_1 Ks}\right]^n \left[\frac{D_t}{ht}\right]^b \left[\frac{ht}{dp}\right]^c \left[\frac{h}{ht}\right]^d \qquad \text{...(9.54a)}$$

I II III IV

Terms I, II, III and IV refer to hydraulic, micro-organisms, substrate and physical configuration respectively. Equation 9.51 can be further simplified as follows :

$$\theta = \frac{V_v}{Q} = \frac{p\,(\Pi/4)\,D_1^{\,2}h}{Q} \qquad ...(9.55)$$

where p and V_v refer to voids ratio and void volume respectively

$$\left(\frac{S_i - S}{S_i}\right) = \frac{K_1}{k}\left[\frac{D.\rho.\mu_m\, XYk}{\vartheta\, k_d\, K_s}\right]^a \left[\frac{(\theta)^{a-1}}{S_i^{\,a}}\right]\left[\frac{Di^b\, h^d}{d^c{}_p\, ht^{(b+d-c)}}\right] \qquad ..(9.56)$$

$$= \left(\frac{K_1 . K_2}{k\,(h_t)^{b+d-c}}\right)\left(\frac{\theta^{a-1}}{S_1^{\,a}}\right)\left(\frac{Di^b}{d_p^{\,c}}\right)\left(h^d\right) \qquad ...(9.57)$$

Where

$$K_2 = \left[\frac{D.\rho.\mu_m .X.Y.k}{\vartheta\; k_d\;\; K_s}\right]^a$$

$$\frac{S_i - S,}{S_i} = \frac{K_1\, K_2}{K\,(ht)^{b+d-c}}\left(\frac{p\Pi D_1^{\,2}h}{4Q}\right)^{a-1} \frac{D_i^{\,b} h^d}{d_p^{\,c}\, S_1^{\,a}} \qquad ...(9.58)$$

$$\eta = \frac{K_1 K_2}{k(ht)^{b+d-c}}\left(\frac{p\Pi^{a-1}}{4}\right)\frac{D_1^{\,2}(^{a-1})\, D_1^{\,b}\, h^{a-1}\, h^d}{Q^a\, D_p^{\,c}\, S_i^{\,a}}.Q \qquad ...(9.59)$$

$$\eta = K_3 \frac{D_i^{(2a+b-2)}\; h^{(a+d-1)}}{(QS_i)^a\; (d_p)^c} Q \qquad ...(9.60)$$

where $$K_3 = \frac{K_1 K_2}{k(ht)^{I(b+d-c)}}\left[\frac{p\Pi}{4}\right]^{a-1}$$

and the final equation for scale up can be defined as:

$$\eta = \frac{S_i - S}{S_i} = K_3 \frac{D_i^{(2a+d-2)}\; h^{(a+d-1)}}{(QS_i)^{(a-1)}\; d_p^{\,c}}.Q \qquad ...(9.61)$$

where a, b, c and d are exponents and Q, S_1, S and η refer to influent flow and concentration, effluent concentration and efficiency of substrate removal respectively.

Equation 9.61 represents a simplified equation for scale up purpose. In any design problem Q, S_i and desired efficiency (η) will be known before hand. Media particle size (dp) is to be pre-decided based on media selection studies, past experience and availability of the media of that size. D_i and h are the only unknowns in the equation. To calculate bioreactor diameter (D_i) media, height (h) can be assumed based on one's experience and using equation (9.62) D_i can be estimated.

Alternatively, if there are any constraints on available area (bioreactor diameter $-D_i$), then D_i can be pre-determined and media height (h) can be estimated using equation (9.61). The various exponents and constants have to be determined by conducting extensive bench scale studies. Similar expressions can be developed for other types of fixed films reactors.

*9.6.2 Scaling up Procedure for Biofilm Processes**

The scaling procedure is limited to situations having the following characteristics :

- the same reactions and micro-organism types occur in the prototype and the scaled process;
- the same biofilm reactor configuration is used in both processes;
- the same effluent concentration of the same rate limiting substrate is expected from both processes;
- the same temperature occurs for both processes.

Three steps are necessary for the development and experimental evaluation for scaling a biofilm process, *viz.* :

- discussion on the key criteria for similitude;
- development of a method to apply the criteria for similitude;
- application of the method for some type of system.

*9.6.2.1 Criteria for Similitude of Biofilm Reactors**

Given the restriction stated earlier (same reaction, micro-organism types, reactor configuration, effluent concentration and temperature), three key criteria define similitude for biofilm reactors, *viz.*, biofilm concentration of substrate, biofilm shear-loss rate and mass balance.

The typical representation of a biofilm and its kinetics, presented in Fig. 9.1 was proposed by Williamson and McCarty[5]. Simultaneous reaction and diffusion of the substrate occur within biofilm. Diffusion is described by Fick's second law, while the reaction is described by Monod function. The transport of substrate for bulk liquid to the biofilm is described by Fick's first law.

* Adapted from scaling procedure for Biofilm process by J.A. Mamen and B.E. Rittman, Wat. Sci. Tech. 22 (1-2), p. 329-346 (1990).

Because the intrinsic reaction parameters (K_a and k) and the diffusion parameters (D and D_f) are the same at a constant temperature for the same substrate reaction and micro-organisms, the kinetic parameters which can change between the prototype and scaled process are L, X_f and L_f

9.6.2.1.1 *Surface Concentration and External Mass Transport Resistance (First Criterion of Similitude)*

Rittman and McCarty[6,7] demonstrated that the flux into a steady-state biofilm depends on the surface concentration (S_a). If the biofilm loss rate does not change, the substrate flux and biofilm accumulation per unit surface area (X_fL_f), are fixed when S_a is held constant. Therefore, the first fundamental criterion for maintaining the same biofilm mass accumulation and substrate flux is to keep S_a constant. The first similitude criterion, Ω_1 is S_a. S_a is related to the bulk-liquid concentration, S, through Fick's first law can be solved for S to give :

$$S = S_a + J\,L/D$$

Maintaining the same value when S_s, J and D also are held constant requires that L is the same for the prototype and scaled reactors. L represents external mass transport resistance and is related to an external mass transport coefficient (k_m) by :

$$L = D/k_m$$

Numerical estimate for L are often obtained from empirical correlation[8, 9]. One correlation that is applied successfully to biofilm reactors was presented by Wilson et.al[10].

$$L = \frac{D\varepsilon}{1.09v} R_e^{2/3}\, S_c^{2/3}; 0.0016 < R_e < 55 \qquad \text{...(9.62)}$$

where $v\ (= Q/Acs)$, Q, Acs, Sc, μ, ρ, $R_e \left(= \frac{d_p v \rho}{\mu}\right)$, ε,

and d_p refer to superficial velocity, total flow rate into the reactor, cross-sectional surface area Schmidt number, water viscosity, water density, Reynolds number, medium porosity, and particle diameter, respectively.

For use in similitude analysis here, μ, ρ, D, and Sc are fixed because temperature is constant. Thus the ratio of L′ values for the scaled process and the prototype is :

$$\frac{L'}{L} = \left(\frac{R_e'^{0.66}}{R_e}\right)\left(\frac{v}{v'}\right)\left(\frac{\varepsilon'}{\varepsilon}\right) \qquad \text{...(9.63)}$$

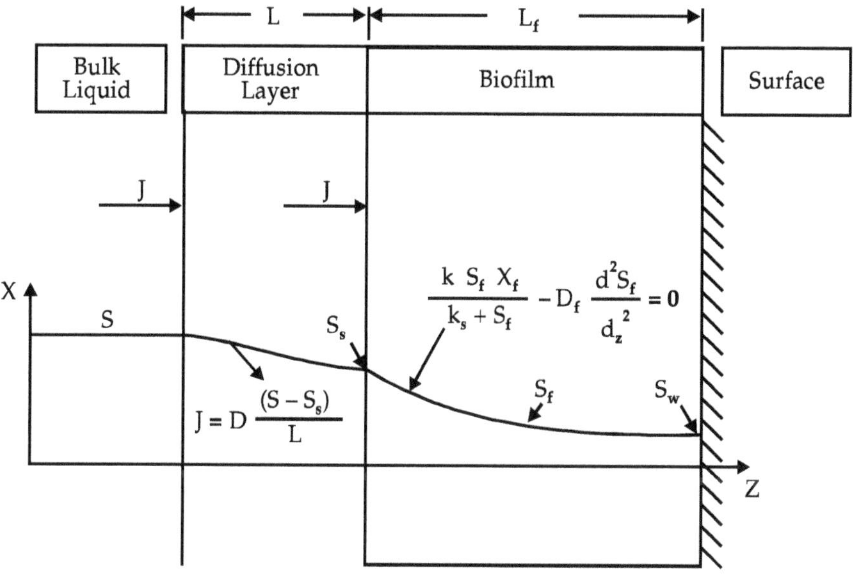

Legend

S – Substrate Concentration in the bulk Liquid

S_s – Substrate Concentration at the Interface of the Biofilm and the Bulk Liquid

k – Maximum Specific Substrate Utilization Rate

K_s – Monod Half Maximum Rate Concentration

X_f – Density of Micro-organism within the Biofilm

L_f – Thickness of Biofilm

J – Substrate Flux

D – Molecular Diffusion Coefficient in Water

D_f – Molecular Diffusion Coefficient within Biofilm

L – Depth of Effective Diffusion Layer

Fig. 9.1 : Conceptual Basis for the Biofilm Model

where prime (/) stands for the scaled process. Both processes have the same mass transport resistance when L′/L = 1.0. Substituting the definition of R_e into equation 9.62 yields :

$$\left(\frac{d_p^{\ 0.66}}{v'^{0.33}}\right)\left(\frac{v^{0.33}}{d'_p{}^{\ 0.66}}\right)\frac{\varepsilon'}{\varepsilon} = 1 \qquad ...(9.64)$$

Equation 9.64 provides a revised first scaling criterion (ϕ^*_1) which is correct when S = S′ at the same time Ss = S′s. f^*_1 is stated directly in equation 9.65:

$$\phi^*_1 = \frac{d_p^{\ 0.67}}{v^{0.33}}\varepsilon \qquad ...(9.65)$$

The scaled process and prototype have the same external mass transport resistance and S_a value when

$$\phi^*_1 = \phi^*_2 \qquad ...(9.66)$$

9.6.2.1.2 Rate of Biofilm Shear Loss (Second Criterion of Similitude)

The accumulation of attached biomass is a critical determinant of process performance, especially when loading and specific growth rate are low[11, 12]. The total attached biomass (X_fL_f) which has units of biomass per unit area, contains two of the three independent variables identified above. The accumulation of attached biomass depends upon the balancing of the biofilm growth and loss. Rittman[8] and McCarty[6] showed that a steady-state biofilm is described by :

$$\int_{X=v}^{X=L_f} \left(Yk \frac{S_r}{K_a + S_r} - k_d - b_a \right) X_f \, dz = 0 \qquad ...(9.67)$$

where Y, and b_s refer to true yield coefficient and first order shear loss coefficient respectively.

Equation 9.67 can be integrated and simplified to give the steady-state value of the attached biomass[6] (X_fL_f).

$$X_fL_f = \frac{JY}{k_d + b_a'} \qquad — (9.68)$$

In order to have X_fL_f' for scaled process the same as $X_f L_f$ for the prototype,

$$\frac{JY}{K_d + b_a} = \frac{JY}{k'_d + b'_a} \qquad — (9.69)$$

Since reaction and micro-organisms are similar $Y' = Y$ and $K'_d = k_d$ and the total removal rate must be similar $J' = J$; therefore, the similitude occurs when $b'_a = b$.

Rittmann[11] presented a relationship for b_a when $L_f \leq 30\mu m$.

$$b_a = 8.42 \times 10^{-2} \sigma^{0.58}$$

Where σ refers to shear stress on biofilm in dynes/cm^2. Shear stress can be calculated for spherical packed media[11, 13].

$$\sigma = 200 \frac{\mu v (1-\varepsilon)^1}{d^2{}_p \varepsilon^2 a \, 7.46 \times 10^9} \qquad — (9.70)$$

where 'a' refers to specific area of the reactor and 7.46×10^9 converts days^2 to second^2.

The specific surface area for quasi-spherical medium is given by equation 9.71.

$$a = \frac{6(1-\varepsilon)}{\psi\, d_p} \qquad ...(9.71)$$

where ψ refers to shape factor (1.0 for spheres and less than 1.0 for irregular surfaces).

For fluidized beds, shear stress can be found from Rittmann[11]:

$$\sigma = \frac{(\rho_p - \rho)(1-\varepsilon)\, g}{a} \qquad ...(9.72)$$

where ρ_p refers to density of particles

In equations (9.70 – 9.72), μ, ρ, and g can be considered constants. On the other hand, v, ε, d_p and ψ could change from the prototype to the scaled process. For fixed bed systems combining equations 9.71 and 9.72 and keeping $\sigma^1 = \sigma$ will result in the following :

$$\frac{v'(1-\varepsilon')\psi'}{d_p' \varepsilon'^3} = \frac{v(1-\varepsilon)\psi}{d_p \varepsilon^3} \qquad ...(9.73)$$

Therefore, the second criterion for similitude, ϕ_2 for fixed films is

$$\phi_2 = \frac{v(1-\varepsilon)\psi}{d_p \varepsilon^3} \qquad ...(9.74)$$

For fixed bed systems, often $\varepsilon^1 = \varepsilon$ in which case ϕ_2 becomes :

$$\phi_2 = \frac{v\psi}{d_p} \qquad ...(9.75)$$

For fluidized bed reactors, $\sigma^1 = \sigma$ when

$$(\rho'_p - \rho)\, \psi'\, d'_p = (\rho_p - \rho)\, \psi\, d_p \qquad ...(9.76)$$

Therefore, ϕ_2 for fludized bed is :

$$\phi_2 = (\rho_p - \rho)\psi\, d_p \qquad ...(9.77)$$

The similitude with respect with respect to shear stress assumes that the shear stress loss rate (b_a) are the same in the prototype and the scaled process and it occurs when $\phi'_2 = \phi_2$.

9.6.2.1.3 Mass Balance (Third Criterion of Similitude)

The third criterion for similitude (ϕ_3) is that the same substrate mass balance exists for each system. A simple steady state balance on substrate in a control volume is given by :

$$Q(S_o - S_e) - JaV = 0 \qquad ...(9.78)$$

where S_o and S_e and V refer to influent and effluent substrate concentrate and control volume respectively.

For the conditions of similitude, $S'_e = S_e$ and $J' = J$, and the variable parameters are S_o, V, a and Q. If the detention time (θ) is defined as V/Q, the number of variables is reduced to three, so, θ and a. Rearranging equation (9.78) gives :

$$a\theta = \frac{S_o - S}{J} \qquad ...(9.79)$$

When all the terms in RHS of equation (9.79) are fixed (i.e. $S'_o = S_e$), the third criterion of similitude (ϕ_3) is

$$\phi_3 = a\theta \qquad ...(9.80)$$

Substitution of the definition of 'a' and elimination of constants transforms ϕ_3 to

$$\phi_s = \frac{(1-\varepsilon)\phi}{d_p\psi} \qquad ...(9.81)$$

When S_o is not held constant, equation (9.83) must be rearranged so that all the fixed values are on RHS, *i.e.* :

$$\frac{a\theta J - S_o}{J} = -\frac{S_e}{J} \qquad ...(9.82)$$

Thus, when S'_o does not necessarily equal S_o, criterion ϕ_3 is equal to

$$\phi'_3 = \phi_3 = (S_o - aJ\theta) = (S_o' - a'J'\theta') \qquad ...(9.83)$$

For a completely mixed reactor $S_e = S$ and ϕ_3 defines a family of (S_o, Q) pairs. For other reactor, S_o is specified by setting S equal to some appropriate average concentration and ϕ_3 specifies θ. Table 9.2 summarises the three criteria for fixed bed and fuidized bed biofilm reactors. The criteria are given for the general condition and for the common special case where $Y = Y'$, $\varepsilon = \varepsilon'$ and $\rho_p = \rho'_p$

9.6.2.2 *Using the Similitude Criteria*

It is clear from Table 9.2 that all three criteria cannot be satisfied simultaneously when S_o is held constant, since the functional relationship between v and d_p are different for each criterion. In some cases, however, either ϕ_1^* or ϕ_1 can be selected as a primary similitude parameter, while ϕ_2 can deviate from exact similitude (*i.e.* $\phi_2 = \phi'_2$). In other case, ϕ_2 must be primary, while ϕ_1^* or ϕ_1 can deviate. Choice of which criterion should be primary is based upon which mechanism exerts greater control over the process performance. If S_o is allowed to change, all three criteria can be satisfied simultaneously.

9.6.2.2.1 *Determining which Mechanisms Control*

External mass transport control is of prime importance when it causes the interfacial substrate concentration (S_o) to be significantly lower than S. The difference between S and S_s is calculate for the prototype with Fick's first law:

$$S - Ss = JL/D \qquad ...(9.84)$$

If ($S - S_s$) is small in comparison to S, external mass transport resistance is not important and ϕ_1 or ϕ_1^* can be allowed to vary. On the other hand, when ($S - S_s$) represents a significant fraction of S (more than 10%), ϕ_1 or ϕ_1^* must be considered a primary criterion.

Table 9.2 : Summary of Similitude Criteria

	Criterion	Bed type	*Value of Criterion*	
			General	*Special case***
1.	**When** $S_o = S'_o$			
	ϕ_1^*	Both	$\frac{d_p^{0.67}}{v^{0.33}}\varepsilon$	$\frac{d_p^{0.67}}{v^{0.33}}$
	ϕ_2	Fixed	$\frac{v(1-\varepsilon)\psi}{d_p\varepsilon^3}$	$\frac{v}{d_p}$
	ϕ_3^*	Both	$\frac{(1-\varepsilon)\theta}{d_p\psi}$	$\frac{\theta}{d_p}$
2.	**For any** S_o			
	ϕ_1	Both	S_a	S_s
	ϕ_2	fixed	$\frac{v(1-\varepsilon)\psi}{d_p\varepsilon^3}$	$\frac{v}{d_p}$
		fluidized	$(\rho_p - \rho)\psi\ d_p$	d_p
	f_3	Both	$S_o - (6(1-\varepsilon)\theta J/d_p\psi)$	$S_o-(6(1-\varepsilon)\theta J/d_p\psi)$

** $\varepsilon' = \varepsilon$, $\rho'_p = \rho_p$ and $\psi' = \psi$

The importance of shear loss rate is determined by two factors. The first factor is the degree to which shear loss contributes to the total biofilm loss rate. Rittman[II] showed that the total biofilm loss rate (b) is made up of the maintenance loss rate (k_d) and the shear loss rate (k_d) and the shear loss rate (b_s), *i.e.*

$$b = k_d + b_s \qquad (9.85)$$

Clearly, when b_s is vanishingly small fraction of b, shear loss does not need to be an important scaling criterion.

Another situation in which shear loss rate is not very important occurs when the biofilm accumulation (X_fL_f) is sufficiently great that the biofilm is deep[7], in which case the substrate flux is independent of X_fL_f. A deep steady-state biofilm exists when S is sufficiently greater than S_{min}, the minimum substrate concentration capable of sustaining a steady state biofilm,[6,7] Smin is given by :

$$S_{min} = \frac{K_s b}{\mu - b} \qquad ...(9.86)$$

If $S > 10\ S_{min}$, the usually is sufficiently deep that small deviations in shearing are unimportant.

Once the primary of mass transport resistance or biofilm loss rate is determined, the proper use of the similitude criteria to scale up or scale down biofilm process involves four steps, *viz.* :

- decide whether S should be held constant;
- use ϕ_1 and ϕ_2 or use ϕ_1^* or ϕ_2 as appropriate to determine feasible value of v′ for desired values of d_p, ε′, Y′ and ρ_p';
- determine the reactor detention time (θ) from ϕ_3 or ϕ_3^*;
- calculate the reactor length (L_c) after selection of a cross-sectional area (Acs).

9.6.2.2.2 *Scaling Procedure when ϕ_1 and ϕ_2 are Equally Important*

In some situations, neither external mass transport mass transport (ϕ_1), nor the biofilm shear loss rate (ϕ_2) can be ignored and, therefore, an alternate scaling procedure must be found. It consists in satisfying the criterion ϕ_2 first. Then, a bulk substrate concentrate for the scaled process is chosen to given same S_s in both the prototype and the scaled process (criterion - ϕ_1, using equations 9.88 and 9.65). Finally, similitude with respect to ϕ_3 gives the detention time for the scaled process.

Table 9.3 : Possible Combinations of Criteria Primacy and Selection Criteria

	Significant Mechanism						
			Criteria to be used				
Situation	*External mass transport*	*Biofilm shear loss*	*So (variable)*			*$S_o = S'_o$*	
1.	No	No	ϕ_1,	ϕ_3		ϕ_3^*	
2.	Yes	No	ϕ_1,	ϕ_3		ϕ_1^*,	ϕ_3^*
3.	No	Yes	ϕ_1,	ϕ_2,	ϕ_3	ϕ_2,	ϕ_3^*
4.	Yes	Yes	ϕ_1,	ϕ_2,	ϕ_3	Impossible	

Table 9.3 lists the possible combinations involving the similitude criteria. It also shows the that none, one or both of the criteria involving shear loss and external mass transport can be primary, when neither is primary, only criterion ϕ_3 or ϕ_3^* is needed. When only external mass transport has primary, ϕ_2 and ϕ_3 or ϕ_2 and ϕ_3^* must be met. Finally; if external mass transport and shear loss are

important, criteria, f_1, f_2 and f_3 must be considered and S_o' cannot be maintained equal to S_o. The value of S can be estimated as :

$$S = \frac{S_o - S_e}{In(S_o / S_e)} \qquad ...(9.87)$$

9.7 Scale Up Using Biochemical Process Simulator

Biochemical reactor come in a variety of configuration, using either immobilized or free and employing many possible reaction kinetic expressions.

In order to scale up laboratory data, laboratory reactor must be stimulated to determine kinetic constants, and to verify the applicability of certain assumptions as to mixing, particle effectiveness factor and kinetic expressions.

Since the modelling process involves repeated solution of many different trial models, the task of programming to solve the many simultaneous nonlinear differential equations can be time consuming, especially if recycle streams are involved. For this reason, a Biochemical Process Simulator (BPS) was developed for use on IBM-PC to allow the user to select his mixing model, kinetic expression, and reactor configuration from a series of graphics driven menus.

9.7.1 Programme Description

A flow sheet summarising the screen and the data or decision to be made on this is given in Fig. 9.2. The main screen is entered first, which consists of a series of menu choices : file specify, configure, run, results and exit. When file specify is chosen, the bottom of the screen becomes active allowing the user to specify drive (A, B, C) and eight character name with the BPS automatically added. When the configure cell is chosen, a second screen appears allowing the user to specify one of the following : free cell, activated sludge and immobilized. The activated sludge is included separately because it is an important biological reactor used in wastewater processing and employs solid-settlers and solids recycle both of which are common in biochemical reactor design. Once a configuration has been chosen, a screen will appear allowing for the specification of the inlet stream concentrations, nutrients, etc. as well as reactor kinetic expression, type of reactor hydraulics (CSTR, plug flow, etc.) and type of settler model if solids recycle is used. Once these selections are made, the appropriate kinetics screen will appear allowing input of values for kinetic constants or choice default values from a menu of common substrate. After the kinetics selections are made, the appropriate reactor/settler screen will appear allowing data input for such variables as recycle ratio, number of tanks in series (if needed). Peclet number and reactor length (for axial dispersion), sludge solids settling constant (if needed for flux model), reactor residence time and reactor volume (for batch reactor) and volumetric flow rate.

Once this data is entered all parameters are fixed and solution can be generated by returning to Main and choosing the cell marked RUN. A screen will appear, giving the status of solution and will tell when the solution is complete. The results are obtained by either going to Main Menu choosing cell marked RESULTS or by exiting the program and calling up the file with the chosen file name and extension (PRN). Custom tailored graphs can be created by importing this (PRN) file into standard spread sheet program such as Lotus 123R. If any changes need to be made in one of the parameters and the solution repeated, the chosen variables can be changed and the solution re-run without re-specifying all remaining parameters.

9.7.2 Kinetic Models

The choices of kinetic expressions for substrate removal ($-\gamma_a$) under free cell, activated sludge, or immobilized configuration are given in Table 9.4. The net rate of cell growth (γ_z) is taken to be :

$$\gamma_x = (-\gamma_s)\, Y_x - k_d X \qquad ..(9.88)$$

The rate of product formation kinetics is taken to be :

$$\gamma_p = (-\gamma_s)\, \gamma_p \qquad ...(9.89)$$

The units associated with values of the kinetic constants fed into the program determine the units associated with output variables. The units become very critical in immobilized reactors, where enzyme or cell concentrations are quoted as dry mass per unit volume of particle per volume reactors.

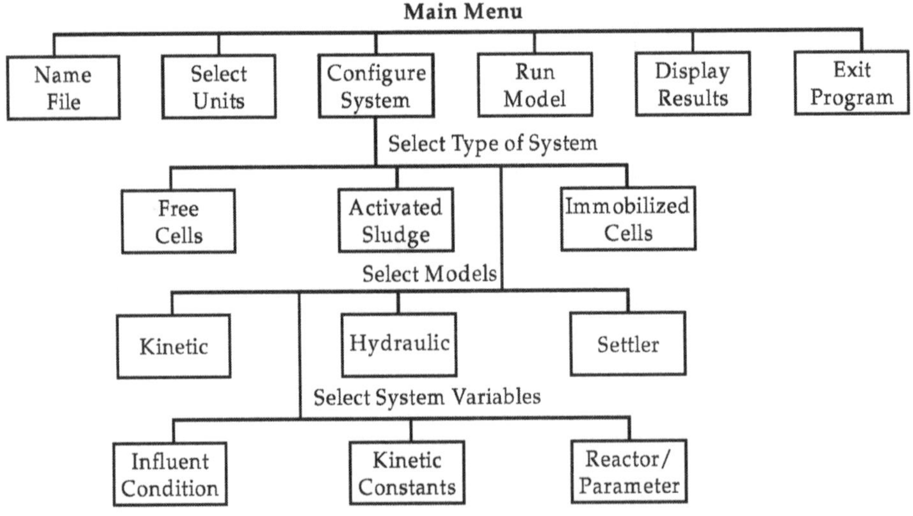

Fig. 9.2 : Menu Flow Sheet

**Source :* Dr. Alberto Bertucco, Universita di Padova, Padova Itlay

Table 9.4 : Kinetic-Model Equations For $(-\gamma_a)$

A Free cell or Activated Sludge	
1. Michaelis Menten/Monod	$= k_o X S/Y_x(K_m + S)$
2. Competitive Inhibition	$= k_o X S/Y_x(K_m + S + I K_m/K_I)$
3. Non-Competitive Inhibition	$= k_o X S/Y_x(K_m + S) (1 + I/K_I)$
4. Product Inhibition (not a choice under ASP)	$= [k_o X S/Y_x (K_m + S)]. [(K_p/(K_p + P)]$
5. Substrate Inhibition	$= [k'_o X S/Y_x(K_m + S + S^2/K_I)]$
B Immobilized Cells	
1. Michaelis Menten	$= [k_o ES/K_m + S]; E= E_o \exp(-k_d t)$
2. Substrate Inhibition	$= [k_o E S/ (K_m + S + S^2/K_I)]; E = E_o \exp(-k_d t)$
3. Product Inhibition	$= [k_o E S/(K_m + S)]. [1/(1+P/K_p)]; E = E_o \exp(-k_d t)$
4. Equilibrium	$= [k_o' E(S-S_o)]/[K'_m + (S-S_o)]$

where E, k'_a, k'_m and S_a , refer to g cell per reactor volume, elementary kinetic constant (g-substrate per unit time, constant for a given value of S_a (g-substrate per reactor volume) and equilibrium substrate concentration respectively.

9.7.3 Effectiveness Factor

In immobilized cell system as well as in cell flocs, substrate and product diffusion within the particle can be the rate limiting step, resulting rates less than predicted based on only bulk liquid concentration[13,14]. An additional diffusion resistance can also occur across a stagnant external liquid film across the particles, but this resistance is taken to be negligible compared to the resistance to diffusion within the particle.

The rate of consumption for substrate $(-\gamma_s)$ under conditions of intraparticle substrate diffusion is represented by

$$(-\gamma_s) = \eta \times \text{(reaction rate under bulk liquid concentrations)} \quad ...(9.90)$$

where η = effectiveness factor

$$= \frac{\text{observed reaction rate}}{\text{reaction rate under bulk liquid concentration}} \quad ...(9.91)$$

At steady state, observed reaction rate, which is the numerator of equation (9.91) is equal to the diffusion rate within the particle when evaluated at the particle surface. Therefore, if a substrate concentrate profile could be calculated within the particle, and its slope evaluated at the surface, then η and the reaction rate could be calculated in the reactor.

In order to evaluate the concentration profile within the particle, a steady state balance on substrate and product in a spherical shell of the particle is written and yields

$$D_a \frac{d^2S}{dr^2} + \frac{2}{r}\frac{ds}{dr} + \gamma_s = 0 \quad ...(9.92)$$

$$D_p \frac{d^2P}{dr^2} + \frac{2}{r}\frac{dP}{dr} = \gamma_p = 0 \quad ...(9.93)$$

The boundary conditions are :

at $r = 0$; $ds/dr = 0$; $dP/dr = 0$ ...(9.94)

at $r = R$; $k_L (S_o - S_R) = D_o\, ds/dr$ at $r = R$ ...(9.95)

where k_L and S_R refer to external mass transfer coefficients and substrate concentration at the particle surface respectively.

This system of differential equations and boundary conditions can be reduced into a single differential equation by adding the two differential equations, together with the equations (9.88) and (9.89) to yield :

$$P = P_o + (S_o - S)\, Y_p\, D_r/D_p \quad ...(9.96)$$

Therefore, only the first differential equation needs to be solved which can be rewritten in terms of a dimensionless radius, $\bar{t} = r/R$:

$$\frac{d^2S}{dr^2} + \frac{2}{\bar{r}}\frac{dS}{d\bar{r}} + \frac{R^2\gamma_a}{Ds} = 0 \quad ...(9.97)$$

Equation 9.97 with boundary conditions (9.94) and (9.95) are solved using the method of orthogonal collocation[15] which converges fast for this system. The constants in collocation equations are solved by Newton-Raphson on technique to give point values of substrate within the particle as a function of dimensionless radius. The derivative at $\bar{t} = 1$ is evaluated and substituted into equation (9.95) to allow the calculation of η for a given value of external substrate concentration (S_o). The solution is repeated many times for different values of S_o ranging from the feed concentration to near Zero, thus giving a table of values of h as a function of external particle substrate concentration. This generated table allows the effectiveness factor to change along the length of the reactor as the bulk substrate concentration decreases. Equations for different reactor types are given in Table 9.5.

Table 9.5 : Different Reactor Types

1. **Batch**

$\frac{dS}{dt} + (-\gamma_a) = 0$	:	$\phi = \frac{\text{Reactor Volume}}{\text{Fresh Feed Rate}}$
$*\frac{dX}{dt} + (-\gamma_x) = 0$	:	$R = \text{Recycle Ratio} = \frac{Q_R}{Q_o}$
$\frac{dP}{dt} + (-\gamma_P) = 0;$	:	$X_R = \frac{\gamma - \text{biomass}}{\text{volume}}$

2. **CSTR**

$(S - S_o) + (-\gamma_g)\ \theta = 0$: θ_c = Sludge age

$*-(1 + R)\ X + X_o + RX_R + (-\gamma_x)\ \theta = 0$; i = reactor index (1—N)

$(P_o - P) + (\gamma_p)\ \theta = 0$: $\overline{X} = \frac{\Sigma_i X_i}{N}$ = Average biomass concentration

$$*\theta_e = \frac{\theta}{1+R\,(1-X_R\,/\,X)}$$

3. **N-CSTR in Series for i = 2 ....N**

$S_i - S_{i-1} + (-\gamma_s)\ \theta/(R+1) = 0$

$X_i - X_{i-1} + (-\gamma_x)\ \theta/(R=1) = 0$

$P_i - P_{i-1} + (-\gamma_p)\ \theta/(R+1) = 0$

For i = 1

$(1 + R)S_I - S_o - RS_N + (-\ \gamma_s)\theta = 0$

$*(1 + R)X_1 - X_o - RX_r\ (-\ \gamma_r)\theta = 0$

$(1 + R)P_1 - P_o - RP_N + (-\gamma_p)\theta = 0$

$$*\theta_c = \frac{\theta XN}{(R+1)X_N - RX_\gamma}$$

4. **Plug Flow**

$\frac{dS}{d\theta} + \frac{(-\gamma_s)}{(R+1)} = 0$: $Z' = Z/L$, L= Reactor length

$*\frac{dX}{d\theta} + \frac{(-\gamma_x)}{(R+1)} = 0$; Z = reactor axial co-ordinates

$\frac{dP}{d\theta} + \frac{(-\gamma_p)}{(R+1)} - 0$: or $\overline{X} = \int_0^1 XdZ'$

$\frac{dS}{dz'} + (-\gamma_s)\frac{\theta}{(R+1)} = 0$:

$*\frac{dX}{dz'} + (-\gamma_x)\frac{\theta}{(R+1)} = 0$:

$\frac{dP}{dz'} + (-\gamma p)\frac{\theta}{(R+1)} = 0$

5. **Axial dispersion**

$$-\frac{1}{P_{os}}\frac{d^2S}{dz'^2} + \frac{dS}{dz'} + (-\gamma_s)\frac{\theta}{(R+1)} = 0$$

$$*\frac{-1}{P_{ex}}\frac{d^2X}{dz'^2}+\frac{dX}{dz'}+(-\gamma_x)\frac{\theta}{(R+1)}=0$$

$$-\frac{-1}{P_{ep}}\frac{d^2P}{dz'^2}+\frac{dP}{dz'^2}+(-\gamma_x)\frac{\theta}{(R+1)}=0$$

$*\theta_c = q\overline{X}/[(R+1.........]$

$Z' - Z/L$; L-Reactor length; Z-Reactor axial co-ordinates

Pe-Peclet number (uL/D) = $L^2(1 + R)/Z$

$\overline{u}$ -Average axial velocity

$$\overline{X} = \int_0^1 X\, dz' \; X_{out} = X \text{ at } Z = L$$

* Equation not used in immobilized system

9.7.4 *Reactor Mixing Models*

The type of liquid mixing may be modeled according to batch, a single CSTR, N-equal volume CSTR in series, plug flow and dispersion equations given in Table 9.5. The biological reactor for free cells and activated sludge configurations, with the exception of batch system, can have continuous feed, cell setting and solids recycle.

For free cell flow reactors using no solids separation or recycle, the recycle rate (R) would be set to 0.0 and the settler system and recycle calculations are by passed. For immobilized systems, there are no solids leaving the reactor and hence no settler or solids recycle are needed. The immobilized reactor choices are : batch, CSTR, column in plug flow and column in axial dispersion flow. For immobilized systems, the biomass or enzyme concentration in the reactor is designated by replacing X with E in the product and substrate equations in Table 9.5. This change allows consistency with the kinetic equations.

In system utilizing solids recycle, an index of solids residence time must be specified. An approximately to a solids residence time, called sludge age is often used and designated as θc. Sludge age and its approximation to a mean solids residence time are discussed elsewhere[16]. By making a solids balance around the settler, the sludge age is related to other parameters by the relationships given in Table 9.5. The sludge age has significance in predicting solids settling characteristics and is one of the independent variables which must be specified.

9.8 Some Considerations

The Scaling up formulation has been described from three view points, *viz.* :

- Conventional - Π - Buckingham Theorem
- Fixed film system
- Modeling system.

From the foregoing discussion, it is still difficult to determine the size of the reactor by using fixed film system or modeling approach. However, general capacity of the reactor may be known. In addition, there are several assumptions involved in both fixed film and modeling approach which may or may not be suitable under field conditions. However, these approaches, do save time and the labour involved in extensive planned experimentation for scale up which is mandatory for extended principle of similarity using Π - Buckingham Theorem. However, by application of this theorem, one can definitely determine the total size of the reactor system based upon the extensive experimental studies an bench scale units. It may be emphasized that scale up formulation using fixed film or before, one can directly apply the similarity criteria developed from fixed film. Similar conclusions can be made for modeling system.

It is not necessary to develop scale up equations for all type of reactor, particularly for those process for which enough experience and expertise is available. Laboratory treatability studies coupled with determination of kinetic constants would suffice to design a large wastewater treatment facility directly from bench scale units. No intermediate scale units are necessary. However, similar situation does not exist for anaerobic bioreactors. Scanty information is available for substrate, biomass and biogas kinetics for several types of attached and unattached biomass systems. It is, therefore, necessary that detailed experimentation and modeling exercise are required to formulate are required to formulate the scale up equations for at least some of the importation anaerobic bioreactors, *viz.* :

- Upflow anaerobic sludge blanket reactor;
- Fixed film fixed reactor;
- Fixed film fluidized bed reactor;
- Fixed film moving bed reactor;
- CSTR system with recycled floc reactor.

REFERENCES

1. Baekeland, L.H.J., Ind. Eng. Chem. 8, p184(1916).
2. Davis, G.E., A Hand book of Chemical Engineering, Davis Bros., Manchester, England (1901).
3. Damkohler, G.Z, Elektrochem, 42, p846(1936).
4. Kaul, S.N. *et al.* Biogas Recovery from Low and Medium Strength Wastewater using Fixed Film of Reactor system, Final Report Submitted to Ministry of Non-Conventional energy Resources. N. Delhi (Oct. 1992).

5. Williamson, K. and McCarty, P.L.A., Model of Substrate Utilization by Bacterial Biofilm, JWPCF, 48(1), P 9-24(1976).
6. Rittman, B. E and McCarty, P.L. Design of Fixed Film Processes with Steady State Biofilm Model, Biotech and Bioengg 22, p. 2343-2357 (1980).
7. Rittman, B.E. and McCarty P.L. Evaluation of Steady state Biofilm Kinetics, Biotech and Bioengg 22, p 2359-2373 (1980).
8. Rittman, B.E. The Kinetics of Trace Organics Utilization by Bacterial Film, Ph.D. Thesis, Civil Engg. Dept, Standard Unity. California (1979).
9. Chang, T. and Rittman, B.E., Comparative Study of Biofilm Shear Loss on Different Adsorptive Media, JWPCF, 60, p. 362-368 (1988).
10. Wilson, E and Geankoplis, C., Ind. Engg. Chem. Fundamentals, 59(1966).
11. Rittman, B.E., The effect of Shear Stress on Biofilm Loss Rate, Biotech and Bio-engg. 24, p 501-506 (1982).
12. Rittman, B., Choosing and Determining Loading Types for Fixed Film. Reactors, Proc. 2nd Intl. Conf. Fixed Film Biological Processes, Washington, D.C. (1984).
13. Wadrak, D.T. and Carbonell, R.G. Effectiveness Factors for Substrate and Product Inhibition, Biotech and Bioengg 17, p 1761-1773 (1975).
14. Frouws, M.J., Combined External and Internal mass Transfer Effects in Heterogenous Catalysis (Enzyme), Biotech. Bioengg. 18, p 53-62 (1976).
15. Finlayson, B.A., The Method of Weighted Residuals and Variational Principles with Application in Fluid Mechanics and Mass Transfer, Academic Press, NY (1962).
16. Sundstrom, D.W., Wastewater Treatment, Prentice Hall, NY(1979).

General Review

1. Johnstone, R.E. and Thring, M.W., Pilot Plants Models and Scale up Methods, McGraw Hill, NY (1979).
2. Biotechnology Processes Scaleup and Mixing, Ed(s)-Chester, S.H. and Oldshue, J.Y., Amer. Inst. of Chemical Engineers, NY (1987).
3. Chemical Engg. (Vol. 1 and 2), Coulson, J.M., Richardson, J.F., Backhurst J.R. and Harkar, J.H., Pergamon Press, Oxford (1990).
4. S.N. Kaul Scale-up equations for Film Fixed Bed Reactor (Tannery), Ministry of Non-Conventional Energy Sources, N. Delhi (Oct. 1992).

CHAPTER 10

DESIGN, OPERATION AND MAINTENANCE

10.0 Introduction

The chapter highlights the general principles of design, operation and maintenance of anaerobic systems. Emphasis has been given to fixed film reactor systems and therefore no attempt has been made to cover conventional digesters (sludge and anaerobic CSTR) which are normally used and enough information is available about them. The following aspects will be considered in general terms :

- system design
- design criteria
- applicability
- problems associated
- advantages and disadvantages.

Design aspects have been covered in modeling, kinetics and scaleup formulation chapters. However, it was felt that low qualitative description be given for better understanding.

10.1 Anaerobic Stationary Fixed Film Treatment

10.1.1 Systems Design

The design of stationary fixed film reactor consists of wastewater distribution system, a biofilm support structure. Head space, effluent draw off and recycle facilities (if required).

The up or down flow made of operation dictate a stationary film support to maintain the film of micro-organisms in the reactor. Additionally, to prevent setting of suspended solids on parts of the film support surface, the stationary film support is arranged in more or less vertical channel and are made of potters clay, hard rock particles, ceramic raschig rings, plastic berl saddles, draintile clay, needles punched polyester or ployvinyl chloride, refractory bricks, activated carbon, etc. Reactor size and height have relatively little effect on the performance (when expressed in surface to volume ratio) but the reactor configuration and operation have marked effect on performance. Generally multichannel reactors have not performed as well as reactors with only few channels[1]. The design of distributor at the bottom or top is very important keeping in view uniform distribution of wastewater.

10.1.1.1 Recirculation

As quoted by van den Berg, *et al.*[1] recirculation generally improves the performance. With wastes containing large amounts of hard to digest suspended solids (*e.g.* piggery waste) recirculation helps to keep these in suspension and aid in their degradation[2].

10.1.2 Design Criteria

Support material affects the rate of start-up markedly. Reactor made from rigid foam polyvinyl chloride could not be started at all, while the glass reactors were slow to start up, presumably because bacteria had difficulty attaching themselves to smooth inert surface. Solid polyvinyl chloride (PVC, used extensively in biological wastewater treatment) is a good film support media. It has been found by experience that inert support media with a rough surface enhanced biofilm accumulation and reactor performance. Physical roughening of smooth plastic surface and addition of sawdust to clay support media before firing have shown to enhance start up and overall reactor performance[1].

10.1.2.1 Surface to Volume Ratio

The amount of retained biomass is stationary fixed film reactor depends on the surface to volume ratio and is therefore limited by support matrix area. The importance of surface to volume ratio on start up and ultimate loading rates for reactor of the same height cannot be overlooked. Reactors with larger surface to volume ratio achieved higher rates of methane gas production[3].

10.1.2.2 Organic Volumetric Loading Rate

Loading rates and organic removal efficiencies depended on the total amount of active biomass retained as well as on the type of waste. For a wide variety of industrial wastes, loading rates of 5 to 25 kg COD/m^3.d were readily obtained with 70-95% COD removal efficiency depending on loading rate and type of waste[1]. van den Berg *et al.*[4] with their experience with bean blanching waste, found that the small fixed film reactors could be loaded substantially higher than larger ones and this difference is not explained. The size of the reactor did not affect the COD removal efficiency. Maximum COD loading rates, using bean blanching waste (9,500 – 10,000 COD mg/L, mainly soluble) were as high as 20 kgCOD/m^3.d (depending an surface to volume to volume ratio and size) to achieve 86% removal efficiency, Table 10.1.

Table 10.1 : Performance Data for Anaerobic Fixed Film Reactors of Two Different Sizes

Sr. No.	*Parameters*	*Fixed Film Reactor Size* 0.7- Litre*	35 - Litre**
1.	Minimum hydraulic retention time, days	0.5	1
2.	Maximum COD loading rate+,	20++	10
3.	COD removal efficiency at maximum loading rate, %	86	86
4.	Suspended COD of the effluent		
*	%	0.09	0.09
**	% of the total effluent COD	65	65

* Surface to volume ratio, 140 m^2/m^3

** Surface to volume ration, 120 m^2/m^3 - reactor may not reached maximum loading rate in test run

+ kg COD/m^3.d; ++ Independent of waste strength (0.5-2.0% COD).

For simulated sewage sludge (55,000 COD mg/L, over 45,000 which is suspended) the maximum COD loading rate of 12 kgCOD/m^3.d with an efficiency of 70% removal is reported.

10.1.2.3 Temperature

Kennedy, *et al.*[5] reported the effect of temperature or the performance of fixed film reactors treating bean blanching waste and industry waste. The maximum loading rates decreased linearly with temperature. The fixed film reactor treating bean blanching waste, mainly containing soluble starch and protein (total COD = 10g/L) was reported to be capable of achieving high loading without substantial change in COD removal efficiency at a temperature range of 10°C to 35°C. They observed that a decrease in temperature from 35°C to 25°C, decreased the maximum steady-state loading rate by 37% while at 10°C the loading rate was reduced by 75% of the maximum loading rate at 35°C (18.4 kgCOD/m^3.d). The COD removal efficiency was independent of temperature and remained at 88 ± 3%.

Similar results were obtained for chemical industry waste (total COD = 14g/L). The maximum steady-state loading rate and the volumetric methane production rate decreased by less than 25% between 35°C and 25°C. As with the bean benching waste there was no appreciable change in the COD removal efficiency or in the digester gas composition with temperature. The effect of temperature on the performance of fixed film reactors digesting chemical industry waste at maximum loading rate is presented in Table 10.2.

They proposed relationships between temperature and loading rate and temperature and methane production rate expressed by:

$$GP = 0.167T - 0.692 \quad ...(10.1)$$

$$LR = 0.557T - 1.546 \quad ...(10.2)$$

where GP, LR and T refer to methane production rate ($m^3/m^3.d$), loading rate ($kgCOD/m^3.d$) and operating temperature respectively.

Table 10.2 : Effect of Temperature on Fixed Film Reactors

Sr. No.	*Parameters*	*Temperature, °C*	
		25	*35*
1.	Loading rate, $kgCOD/m^3.d$	14.0	17.9
2.	Film Surface loading rate, $kgCOD/m^2.d$	0.001	0.0128
3.	HRT, days	1.0	0.78
4.	% CODr	81	84
5.	% CH_4	54	55
6.	Biogas yield, $m^3CH_4/m^3.d$	3.7	4.9
7.	Biogas yield, $m^3CH_4/m^2.d$	0.026	0.039
8.	Volatile solids, mg/L		
	- acetic acid	280±20	180±20
	- propionic acid	170±20	180±20

The coefficients in these relationships presumably depend on the nature of the substrate and type of fixed film support material as well as support material configuration and surface to volume ratio. Van den Berg *et al.*[1] reported the studies conducted by Kennedy, *et al.*[2] which stated that the fixed film reported can be operated at thermophilic temperatures (55°C) but maximum loading rates and COD removals were similar to those at mesophilic temperatures.

10.1.2.4 Hydraulic Retention Time

Hydraulic retention time (HRT) is the ratio between void volume of the reactor and the volumetric flow rate. An intermediate HRT is desirable for some high strength wastewater due to poor conversion of wastes to methane at very low HRTs. With laboratory experience, it was found that the HRT varies from few hours to number of days depending on the waste characteristics and strength.

10.1.2 Applicability

Stationary fixed film reactors could be changed over from one waste to another with relatively little loss of capacity and could adapt readily to changes in temperature as low as 10°C. This is important for installations where character of the wastewater changes rapidly due to the season or production schedule. van den Berg, *et al.*[3] reported that the reactors could start up very quickly after a period of starvation (one or two days to reach maximum capacity after 3 weeks of starvation).

The reactor could be used to remove the treated wastewater intermittently than continuously and intermittent addition of wastes appears to be feasible

Intermittent loading increase the rate of methane production and hence the rate of conversion of COD, but decreases the COD removal. The latter may be caused by short HRT for the waste.

In stationary fixed film reactors COD removal depends on types of wastes and hydraulic retention times. Waste with hard to digest solids showed lower removals, particularly at short hydraulic retention time. Reactors could handle both low and high strength nitrogen wastes (pear picling waste, piggery waste). Further, this fixed film reactor with effluent removal from the bottom has been found to produce methane with high suspended solids.

Due to the self mixing feature of the fixed film reactors, they could treat (this mixing is produced by the rising gas bubbles which causes every channel to act as gas lift pumps) dilute and concentrated wastes equally well. The rapid self mixing distribute waste quickly through the reactor before local high concentration of volatile acids could develop.

The fixed film reactors could handle severe over loading and organic shock loads without serious problems and could be operated at temperature lower than optimum and still be loaded at high rates without affecting digester gas composition or COD removal efficiency.

Mesophillic fixed bed reactors tolerated sudden shock loads at constant hydraulic loading (caused by sudden increase in waste strength) and recovered normal performance within a few days, if the alkalinity was sufficiently high to maintain the pH above 6.2[3].

Kennedy, *et al.*[3] reported that for chemical industry waste (T-COD = 14g/L) the fixed bed reactor could be over loaded 8 times their normal rate for a 24 hour period and, recovery to be possible within 12-48 hours while still being loaded normally. COD removal decreased with increasing overloading rates and was temperature dependent. During overloading at a loading rate of 61 kgCOD/m^3.d (0.43 kgCOD/m^2.d), COD removal efficiencies at 25°C and 35°C were 44% and 61% respectively.

It is evident that the COD, removal during overloading decreased with increased rate of overloading while methane production rates increased. Further, Kennedy, *et al.*[5] stated that repeated overloading improve the reactors as they are more stable and could be loaded at higher steady state rates than before overloading. This may be due to activation of inactive film by the availability of substrate and nutrients[5]. It is reported that sloughing of the biofilm occurred during organic and hydraulic shock loadings. Fixed bed reactors are able to withstand large toxic shock loads[1].

In terms of biofilm thickness control, the fixed bed is need a special biofilm control device in the form of backwash[6], but it does not happen often.

10.1.4 Problems Associated with Stationary Fixed Film Reactor

10.1.4.1 Plugging of Reactor

Non-uniformity of biofilm thickness in fixed bed reactor occurs due to excessive growth of biofilm near the top of the reactor. Under certain conditions, this non-uniform growth can cause plugging at the top of packing and partial plugging of some channels.

Factors like, width of channels, smoothness of packing (will determine the smoothness of occasional sloughing), recirculation rate and the composition of waste determine whether or not plugging will occur.

vad den Berg[1] reported on several methods which have potential for maintaining reasonably thin biofilm. These are :

- organic and hydraulic shock loads - the effect in film sloughing will be greatest near the top of the reactor where load enters. This method will not be of much use for channels already blocked,
- recirculation of effluent - as already discussed, for wastes with a high suspended solids content, intermittent pumping of liquid from the bottom to the top of the reactor helps to maintain a uniform, relatively thin film,
- recirculation of gas - large gas bubbles rising in channels should help the sloughing process and may even open blocked channels,
- reactor configuration - a relatively thin layer of course packing on the top of the packing material may accumulate the excess biofilm and cope with it,
- improvement in the flow distribution system on top of the reactor to avoid too low liquid velocity in these channels,
- horizontal spacing of channels to improve mixing and reduce dead space.

10.1.4.2 Start-up of the Reactor

Rate of start-up depends on the type of inoculum, the type and strength of waste, level of volatile acids maintained and the characteristics of the material used. For example reactor were difficult to start-up with chemical industry (toxicity could not be demonstrated), while reactor started steadily on food processing wastes or sugar waste. Sewage digester sludge generally required longer period to adapt than inoculum from an active digester fed with food processing waste. The rate of startup was faster with sugar waste at 5000 CODmg/L than with a strength of 10,000 CODmg/L or higher. Also, reactors started up faster when volatile acid levels were maintained at about 1000 mg/L then with below 600 mg/L[3]. Several factors presumably play a role : concentration of critical types of bacteria, ecological relationships and how close the waste resembled the substrate to which the inoculum was accustomed.

Support material as well as the number of channels in a reactor affect the startup and ultimate loading rates[1]. This effect was caused by differences in mixing patterns because it affects the amount of dead space and short circuiting. Recirculation rate also improves mixing patterns and rate of startup. Horizontal spaces in banks of vertical channel also provide an improvement in the rate of startup by reducing the amount of dead space.

10.1.5 Advantages and Disadvantages

These are described as follows :

- elimination of mechanical mixing (mixing in the reactor is provided entirely by the action of rising bubbles);
- recycle not necessary;
- better stability at higher loading rates;
- simplicity in construction;
- low head loss - less than 15 cm is normal operation;
- clear, odour and nuisance free effluent;
- efficiency not affected by intermittent or transient nature of flows;
- lower quality in circumstances where the influent suspended solids concentration is high;
- requires more care in startingup of the reactor;
- depending on the type of packing a smaller or greater fraction of the reactor volume is lost for retaining – sludge.

10.2 Upflow Anerobic Sludge Blanket Treatment (UASB)

10.2.1 System Design

Like the other system discussed, proper influent distribution, head space, and effluent draw off facilities must be designed. The key to successful operation of the UASB is to keep the sludge within the system (*i.e.* maintaining the solids without any support material). This is accomplished with the internal gas-solids separator and by minimization of mechanical mixing and/or sludge recirculation for the sake of improving the ability of sludge to settle.

10.2.1.1 Gas-Solids Separator

The gas-solids separator (GSS) located in the upper part of the reactor is particularly important. Mainly it serves for retaining the anaerobic sludge within the anaerobic reactor. This is accomplished by separating as effectively as possible the gas bubbles from the system at approximately 2m beneath the effluent weir.

Thus, a quiescent zone is created in the upper part of the reactor where sludge flocs or particles can flocculate, settle out and/or can be entrapped in the sludge blanket present.

In order to combat buoyancy of the sludge, the dimensions of the GSS - device should be such that the gas liquid interface in the gas collector will be well mixed by the up - flowing gas bubbles. The surface area of the gas-liquid in the gas collector should not be too small, because it might lead to severe scumming[7].

Stringent design criteria for the GSS-device is not yet provided. Lettings *et al.*[7] provided the following guidelines :

- The first main objective of the GSS-device is to accomplish an effective separation of the gas. For this purpose proper baffle plates should be installed beneath the aperture between the gas collectors.
- The second objective is entrapping sludge flocs (granules) conveyed into the settler compartment with the upflowing solution and returning them back into the digestion compartment. For this purpose the inclined wall of the settler should be approximately 50°C from the horizontal. The surface load of the reactor should not exceed 2m/h in flocculent sludge bed systems.
- The height of the settler compartment should not be lower than 1.5-2m and perhaps even should exceed 2m.

10.2.1.2 Feed Inlet System

In a UASB reactor, short circuiting of the liquid flow could be reduced (if necessary) by increasing the number of the feed inlet points.

Acceptable results can be achieved in highly loaded processes (*i.e.* at greater than 1-2 kgCOD/m^3.d) with one feed inlet point per 5-10m^2, although a secondary start-up of the process may proceed rather slow in that case. For purpose of shortening a secondary start-up, a larger number of feed inlet points should be pursued. This applies particularly for low loaded systems. In that case one feed inlet point per m^2 may be sufficient particularly in granular sludge and/or dense flocculent sludge bed systems[7]. In rather voluminous flocculent sludge bed systems (10-15 kgVSS/m^3) quite satisfactory results have been obtained with only one feed inlet point per 2m^2 in treating raw sewage at liquid retention times of 8 hours at temperatures in the range of 15°C - 20°C[8]. Some rough guidelines for the required number of feed-inlet points are presented in Table 10.3[9].

Table 10.3 : Rough Guidelines for the Number of Feed-Inlet Nozzles Required in a UASB Reactor

Sr. No.	Type of Sludge	Area (m^2) per Nozzle
1.	Dense flocculent sludge (exceeding 40 kg DS/m^3)	One at loads less than 1 - 2 kgCOD/m^3.d
2.	Thin flocculent sludge (less than 40 kg DS/m^3)	Five at loads exceeding approximately 3 kgCOD/m^3.d
3.	Thick granular sludge	One at loads of approximately 1-2 kgCOD/m^3.d

10.2.1.3 The Hight-Area Ratio

The amount of wastewater to be treated, the design capacity of the reactor and the maximum permissible surface load (tentatively 1-1.5 m^3/m^2.h for flocculent sludge systems) roughly dictates the height-area ratio of the reactor.

10.2.1.4 Recycling

With a sophisticated feed distribution system installed in a UASB reactor, effluent recycle (to fluidize the sludge bed) is not necessary as sufficient contact between wastewater and sludge is guaranteed even at low organic loads[7]. So far, effluent recycle is not generally applied in UASB reactors.

However results of a 12.5 liter (2 m height) UASB reactor with high organic and hydraulic loads applied indicate that effluent recycle certainly will result in a high treatment efficiency[7].

10.2.1.5 Mechanical Mixing

In order to prevent difficulty with buoying sludge, the gas liquid interface in the gas collector has to be kept well stirred.

At high organic loading rates the agitation caused by gas evolution will usually be sufficient to prevent an excessive buoying of sludge. However, at low organic loading rates - depending upon the design of the gas-solids separator and the type of waste treated, some form of mechanical agitation of the gas - liquid interface in the gas collector may be required. This applies in particular to wastes that can readily form a scum layer, *i.e.* diary wastes and raw sewage[10].

10.2.2 Design Criteria

10.2.2.1 Organic Loading

None of the full scale plants built so far has been designed for space loads exceeding 15kgCOD/m^2.d at 19°C, despite the fact that considerably higher loads have been applied in small and large pilot plants. The conservative design criteria are chosen due to the lack of all scale experience of UASB operating reactors[7].

10.2.2.2 *Temperature*

A treatment temperature of 30-35°C is being used in most of the current operating full scale plants. Lettinga *et al.*[9] reported that recent experimental results conducted by Hulshoff Pol (unpublished work) indicate that the rate of start-up can be maintained significantly by increasing the operational temperature from 30°C to 38°C.

Experiments were conducted to study the effect of temperature on the specific activity (g COD/g VSS.d) using two types of feeds, a mixture of volatile fatty acid (VFA's) and dry solids from potato[11]. Temperature exceeding 45°C were found to be detrimental which results in a sharp reduction of the specific activity amounting to 50-60% of the activity over the whole temperature range 29-40°C. Otherwise the systems adapts very rapidly to temperature changes in the 10-40°C range.

Guidelines for the design capacity of UASB plants treating mainly soluble wastes in relation to temperatures are presented in Table 10.4[7].

Table 10.4 : Tentative Rough Design-Capacities for UASB-Reactors Treating mainly Liquid Wastes in Relation to the Temperature (Conservative Figures)

Temperature (°C)	*Design - capacity kgCOD/m³.d*
40	15 - 25
30	10 - 15
20	5 - 10
15	2 - 5
10	1 - 3

10.2.2.3 *Hydraulic Retention Time*

Hydraulic retention times (HRT) are in the range of 4 to 24 hrs. The HRT depends upon the effluent characteristics and the treatments objectives. Very short hydraulic retention times (3-8 hrs) could be applied with medium concentrated wastewater (1-3 kg/m^3 of soluble COD). With more concentrated wastewater (10-50 kg/m^3 of soluble COD), longer hydraulic retention times (approximately 1 day) have to be applied. In both cases, however, a reduction of 80-98% of a soluble COD load at approximately 15 kgCOD/m^3.d could be obtained.

10.2.2.4 *Sludge Bed Height*

So far no sound arguments have been provided to restrict the total amount of granular sludge in a UASB reactor. Experiments carried out in a 6 m^3 pilot plant in which a dense-mainly granular - sludge bed occupied the lower 1-2 m of the reactor (with a more flocculent sludge blanket above it) have shown that space loads up to 45 kgCOD/m^3.d with potato processing waste and 30 kgCOD/m^3.d with sugar best waste can be well accomodated[7]. Any evidence that a maximum should be set for the height of the granular bed was not obtained from these experiments.

The basic design criteria for reactors of 30 m^3 and 800 m^3, treating liquid sugar wastes and beet sugar wastes respectively are given in Table 10.5.

10.2.3 Applicability

Although the UASB process was originally developed for the treatment of mainly soluble low- and medium-strength wastewater, satisfactory results have already been achieved with complex wastes, at optimal and suboptimal mesophilic temperatures.

The potential of the UASB concept for treating mainly soluble liquid wastes has been demonstrated in both full-scale and pilot UASB reactors, as well as in numerous bench-scale UASB experiments with various types of wastes. *e.g.*, from sugar beat (soured as well as unsoured), bean blanching, sauekraut, alcoholic fermentations, potato processing, as well composite VFA wastes etc.

Table 10.5 : Basic Design Features of the Anaerobic Treatment Plants in Operation in 1979[12]

Particulars	*30m³ Plant*	*800m³ Plant*
Tank configuration	Cylindrical	Rectangular
Building material	Steel	Concrete
Height (m)	6	4.5
Bottom surface (m^2)	5	178
Depth of digesting zone (m)	4.9	3.3
Depth of settling zone (m)	1.1	1.2
Type of wastewater	Liquid sugar	Beet sugar
Organic loading (kgCOD/d)	400	13,000
(kgCOD/m^3.d)	13.3	16.25
Influent concentration (mg COD/L)	17,000	3,000
Purification efficiency (%)	94	88
Average hydraulic flow (m^3/h)	1	180
Peak flow (m^3/h)	1.5	240
Hydraulic settler surface load (m^3/m^3.h)	0.5	1.5
Gas Production (m^3/m^2.h)	1.7	1.2

In treating mainly soluble wastes, generally very high loading rates can be applied. Some results achieved with VFA substrates, and alcoholic waste are summarized in Table 10.6.

Table 10.6 : Results Obtained with UASB Reactors Using Granular or Mainly Granular Seed Sludge, and VFA Solutions, Alcoholic Wastewater and Potato-Processing Waste as Feed[9]

Substrate Characteristics				*Experimental Conditions*					*COD Reduction*	
Type	*COD (mg/L)*	*Soured (%)*	*Medium used for growth of seedsludge inoculum*	*Reactor Volume*	*Volume of Sludge Bed*	*COD Sludge Load (kg/kg VSS-d)*	*Temp. (°C)*	*COD Load (kg/m³.d)*	*Total (%)*	*Filtered Effluent (%)*
VFA	1000C_2 1000C_3	100	VFA	30L	15-20L	2-3	30	62	-	80-90
Alcoholic	-	0	Sugarbeet waste	2.7L	~1.5	30	22	-	>95	
				28L	~10	0.7	80	14	-	85-90
Potato	2.5-4.2		Digested	6m³	<2m³	0.27	19	3-5	88	95
Processing	3.3-5.0	6-16	sewage	6m³	<2m³	0.65	26	10-15	86	95
	3.5-4.5		sludge	6m³	<2m³	0.97	30	15-18	83	95
	3.5-7.1			6m³	~4m³	1.45	35	25-45	84	93

* Methanol 51%, Ethanol 27%; Propanol 12%, Butanol 10%

Table 10.7 : Results Obtained with Three of Complex Wastes Using Flocculant-Sludge UASB Reactor[9]

Waste (mg/L)	*Influent COD**		*Temperature Applied (°C)*	*COD Load Applied (kg/m³.d)*	*COD Reduction Filtered (%)*	*Achieved** Unfiltered (%)*	*Voluem of Reactor*
	Total (mg/L)	*Soluble*					
Domestic sewage	322-950	235-460	15-20	max. 2.0	30-80		6m³
			9-12	max. 2.0			
Calf-fattening	9500	3800	30	4	93	90	25L
			25	2	90	85	25L
Slaughterhouse	1500-2200	600-1100	30	2.5-3.0	75-85	65-80	30m³
			20	1.5-2.5	78-85	55-75	

* COD values for domestic sewage comprise average values of 5-15 composite daily samples.

** Filtered : COD reduction based on filtered effluent and raw influent;
Unfiltered : COD reduction based on raw effluent and influent samples.

In applying very high sludge loads, appropriate adjustments have to be made to the design of the GSS because, under these conditions, a considerable fraction of the sludge granules will be redispersed in the liquid medium above the sludge bed, because of the marked turbulence about by rising gas bubbles, as well as the increasing tendency of the granules to float[9].

In treating complex (*i.e.* partially insoluble) wastes, generally low loading rates are applied. High loading rates are possible with complex wastes only when employing granular sludge-bed reactors. In flocculent sludge-bed UASB reactors the presence of poorly or non-biodegradable suspended matter in the wastewater will result irrevocably in a sharp drop in the specific methanogenic activity, because the dispersed solids will be trapped in the sludge. Moreover, any significant granulation will not occur under these conditions. The maximum loading potential of such a flocculent sludge-bed system is in the range of 1-4 kgCOD/m^3.d[9]. This applies particularly to low-strength wastes in which the insoluble fraction is less than about 50%, but is also true for medium- and high-strength wastes which after hydrolysis (and acidogensis), do now allow an easy separation of the remaining solids from the liquid. Some relevant results for complex wastes with flocculent and granular sludge-bed UASB reactors are presented in Tables 10.7 and 10.8 respectively.

The performance of the UASB system with respect to the removal of SS is fairly poor, particularly at low temperatures. This presumably is due to channelling in the sludge bed, resulting from the very low gas production at these low temperatures. At higher temperatures, finally dispersed matter present in the raw sewage is removed considerably more efficiently, allowing the system to be exposed to higher hydraulic and organic loads. The maximum loading potential of flocculent sludge bed reactors for raw sewage has not bee established for temperatures exceeding 20°C.

Significantly higher loading rates can be accommodated in granular-sludge UASB reactors compared with flocculent-sludge bed reactors, and it is recognized that the SS of the considerable turbulence resulting from the vigorous gas evolution. A primary or secondary settler therefore has to be installed in line with the anaerobic reactor.

The UASB Reactor was found to perform well at an intermittent feeding. Start-up difficulties were not reported for interruptions of even for few weeks. It is of practical importance for situations where large daily and/or weekly variations occur in the pollution load; as a result installation of an equalization tank can be avoided or the size of the tank could be reduced significantly.

The process accommodates fairly well to hydraulic and organic shock loads, temperature fluctuations, and low influent pH values, provided that the reactor pH remains well above 6.0 and that the sludge load applied is below the maximum specific COD removal rate of the sludge at the temperature prevailing in the reactor.

Table 10.8 : Results Obtained with Three Types of Complex Wastes Using Granular Sludge Reactor[9]

Origin	*Waste Water*		*Reactor*						
	COD (g/L)	*Fraction Dissolved (%)*	*Volume (L)*	*Height (m)*	*Amount of Seed Sludge (g)*	*COD Load (kg/m³.d)*	*Temp. °C*	*COD Reduction Filtered Effluent (%)*	*Unfiltered Effluent (%)*
Rendering wastes	5500	70	60	2	1500	27	30	94	64
	3000	85	60	2	1500	63	30	80	63
Slaughterhouse	1.5-2.2	50	30	1.3	1000	10	30	87	-
	1.5-2.2				1000	6	20	91	-
Raw sewage	0.2-0.9	5-35	120	2	3300	0.7-2.7	8-20	60-89	54-72

* Based on filtered effluent samples and unfiltered influent samples.

The UASB reactor (as in the case of an anaerobic filter) seems to be relatively insensitive to inhibition by sulphide due to its high sludge retention time[13]. In the UASB system, an attractive volumetric loading rate could be achieved when the methanogenic activity of the sludge is still very low. In UASB reactor treating yeast-production waste, loading rates up to 14 kgCOD/m^3.d could be applied at hydrogen sulphide concentrations of approximately 90 mg/L. COD removal efficiency of 60-80% could still be reached under these circumstances[9].

The UASB reactor is able to tolerate extremely high organic and hydraulic loading rates up to 30 kgCOD/m^3.d and 8 m^3/m^3.d respectively once the sludge has been adapted to the waste none of the full scale plants built so far has been designed for space loads exceeding 15 kgCOD/m^3.d at 30°C.

Results of a few pilot scale experiments conducted indicate the feasibility of the UASB reactor's use for denitrification and acid fermentation. As in the case of other anaerobic wastewater treatment systems, the UASB treatment may require further post treatment in the form of cascade aeration or the effluent could be discharge into the sewer system, for further treatment.

10.2.4 Problems Associated with Anaerobic Upflow Sludge Blanket Reactor

10.2.4.1 Start-up

To achieve high treatment efficiencies at high loading the formation of a highly settleable and active sludge in the UASB reactor is of utmost importance. Granular type of sludge is reported to have these properties[11].

It has been found that space loads exceeding 5 kgCOD/m^3.d can already be accommodated within 6-7 weeks in using VFA's (Volatile fatty acids) mixtures as substrate, provided the start-up of the process is managed well[14-15].

The guidelines to cultivate a granular sludge during the start-up of the process are given in Table 10.9[9]. Following the guidelines given in Table 10.9, it is possible to cultivate a granular sludge on VFA feeds. The first granules have been observed 4-6 weeks after the start of the experiment, but as pointed out by Hulshoff Pol *et al.*[15], various factors are involved in the granulation process.

Lettinga *et al.*[9] reported that the rate of start-up can be enhanced significantly by supplying a suitable carrier material to the seed sludge, *e.g.* anthracite particles (Hulshoff Pol, unpublished work).

Table 10.9 : Tentative Guidelines for the First Start-Up of a UASB Plant Using Digested Sewage Sludge as Seed

1.	Amount of seed sludge: 10-15 kg VSS/m^3
2.	Initial sludge load: 0.05-0.1 kgCOD/kgVSS/day
3.	No increase of the sludge load unless all VFA's are more than 80% degraded
4.	Permit the wash - out of voluminous (poorly settling) sludge
5.	Retain the heavy part of the sludge

10.2.4.2 Factors Affecting the Granulation Process

As bacterial granulation must be primarily governed by bacterial growth, the granulation process will be affected by the following[15]:

(*a*) Environmental conditions, such as :

- the availability of essential nutrients, because growth conditions should be optimal;
- the temperature, since the specific activity of methanogenic sludge is highly temperature dependent;
- the pH, which should be in the optimal range (6.5-7.8);
- the type of wastewater, with regard to the composition of the waste, the biodegradability of the organic matter, the presence of finely dispersed non-biodegradable organic and inorganic matter, the ionic-composition (concentration of uni- and divalent-cations) and the presence of inhibitory compounds.

(*b*) The type of the seed sludge, *i.e.* with respect to its specific activity, its settleability and the nature of the inert fraction.

(*c*) The process conditions applied during the start-up, such as :

- the procedure followed in increasing the loading rate, *e.g.* the extent of overloading and the allowed wash-out of suspended solids;
- the amount of seed sludge used.

Distillery waste and corn-starch wastes are reported to have problems with the formation of granular sludge although ultimate granulation of the sludge occurred.

Experiments were conducted to verify the influence of sludge loading rate, the addition of small amount of granular sludge to the seed sludge, and to investigate the effect of NH_4^+ and Ca^{++} concentration in the influent on the pelletization process of sludge granules[15]. Further investigations are necessary to establish criteria.

10.2.4.3 Sludge Washout

The results of all start-up experiments carried out with UASB reactors reveal a very significant wash-out of sludge during the initial start-up phase of the process and consequently a deep depression is developed in the retained (heavier) amount of the seed sludge. Due to the wash-out of the sludge-ingredients, a considerable fraction of the net bacterial growth during the initial phase of the start-up is also lost with the effluent.

To prevent the formation of a deep depression in the retained sludge, the following measures are proposed[15] :

- The selection of the right type of seed sludge.
- There is evidence that a thick, relatively inactive seed sludge is beneficial to a good sludge retention[14].
- Adjusting the wastewater composition.

  The addition of nutrients or vitamins if necessary to assure optimal growth conditions.
- Proper management.

During the start-up, long periods of over and under-loading must be prevented. Under-loading leads to the development of voluminous sludge. Overloading is detrimental because of the gas production that will occur in the gas-solids separator (GSS), which will hamper the settlement of the sludge in the GSS.

Sludge wash-out during the operation of the process is closely connected to the finely dispersed sludge present in the reactor. When the top of the sludge blanket remains well below the effluent weir of the reactor, the sludge washout is considerably less than the sludge accretion from growth. In this case the sludge lost in the effluent may consist of a considerable amount of suspended solids which may originate from the suspended solids in the influent. When the sludge blanket reaches the effluent weir under steady loading conditions, the wash-out of sludge and the accretion of sludge by growth will range over a similar order of magnitude per unit time[11].

Temporary drastic wash-out of the sludge may occur under an excessive expansion of the sludge blanket due to the result of a shock loading or due to suddenly deteriorating conditions (*i.e.* nutrient, high concentration of finely flocculating matter, etc.) This may last until a new steady bed could be established.

Some relevant data concerning the wash-out of sludge are summarized and are presented in Table 10.10.

Table : 10.10 : Suspended Solids Washout as Found in $6m^3$ Pilot-Plant and Full-Scale Experiments with the UASB Process[11]

Period of Experiment (day)	*COD Space Load ($kg/m^3.d$)*	*HRT (Hour)*	*SS Washout* Total (mg SS-COD/g COD_{infl})*	*SS Washout* After 30 min. Settling (mg SS-COD/g COD_{infl})*	*Total Amount of Sludge in Reactor* (kg TS)	*Total Amount of Sludge in Reactor* (kg TVS)
1. Potato waste ($6m^3$ pilot plant)[b]						
408-413	10-13	6.3	70-120	40-80	110-120	89-97
423-427	15-17	6.0	70-90	-	125-130	110-114
429-434	22-30	3.9	60-90	30-50	145-155	127-136
436-440	33-43	3.5	40-110	30-50	200-210	174-183

Day	*COD Space Load (kg/m^3-d)*	*HRT (hour)*	*SS Washout (,\gSS-COD/g COD_{infl}*)*	*Total Amount of Sludge in Reactor* (kg TS)	*Total Amount of Sludge in Reactor* (kg VTS)
2. Sugar-beet waste (200 m^3 full-scale plant)[c]					
54	11.2	6.7	100	6500	2280
61	13.9	7.6	60	6650	2530
68[d]	12.9	7.5	200	6650	2530
75	14.0	8.0	74	6500	-

a SS COD-content in potato wastes varied between 100-150 mg COD/g $COD_{infl.}$ and in the sugar beet waste between 70-150 mg COD/g $COD_{infl.}$ during the experimental periods considered.

b Sludge growth; 0.180-0.20 g sludge-COD/g$COD_{infl.}$(30°C); 0.26-0.30 g sludge-COD/gCODinfl. (20°C)

c Sludge growth; 0.08-0.16 sludge-COD/g $COD_{infl.}$

d Only 20% of COD was present as VFA-COD, whereas this was 80-85% during the other days.

10.2.4.4 Foaming

Excessive foaming has been encountered in the event of relatively poor treatment efficiencies, due to overloading or nutrient deficiency, and at very high gas production rates. Difficulties may be expected in the water that are relatively rich in proteins, such as potato starch wastewater[11]. Foaming could be effectively depressed by adding anti-foaming agents to the feed solution.

10.2.5 Advantages and Disadvantages

Advantages

- Simple construction
- No need for any form of mechanical mixing except, if applied, at low organic loads (2-4 kgCOD/m^3.d), at high hydraulic loads (3-4 m^3/m^2.d) and in treating wastes containing a significant amount of undissolved solids.
- Comparatively less investment when compared to an anaerobic filter or a fluidized bed system.
- Higher loading capacities and treatment efficiencies compared to that of anaerobic filter process.
- No real need for applying effluent recycle.

Disadvantages

- High wash out of suspended matter.
- Significant wash-out of sludge during the intial phase of the process and consequently a deep depression is formed in the retained amount of seed sludge.
- Necessity to develop dense and preferably a granular type of sludge in the reactor on the wastewater submitted to the reactor.
- Requirement of sufficient amount of granular seed sludge when the formation of dense, and granular sludge is not possible.

10.3 Anaerobic Expanded/Fluidized Bed Treatment

10.3.1 System Design

Like the anaerobic filter, the design of an expanded/fluidized bed consists of a wastewater distributor, a medium support structure, medium, head space, effluent draw off and recycle facilities. For high strength wastes, a device for separating the excess biomass from the support medium and subsequent wasting of this excess growth is generally incorporated into the design.

10.3.1.1 Recycling

Since the medium should be kept in the fluidized state, upflow velocities must be high enough to keep the particles in suspension, and thus effluent recycle is practised.

In addition to this, a certain amount of recycle is useful as it can :

- held neutralize the pH of the incoming wastewater;

- reduce the amount of alkalinity required;
- reduce the effect of toxic biodegradable compounds;
- minimize the effect of shock loadings;
- compensate for veriability of influent flow rate.

Increasing recycle in effect allows the process to tend towards the results and operational characteristics of a completely mixed system.

10.3.1.2 Separation Equipment

In fluidized beds, some sort of separating equipment of other biomass retention measure is incorporated in the design as process failure could result in total of biomass within 15 minutes. Expanded bed reactors thus have some inherent risk.

10.3.2 Design Criteria

10.3.2.1 General

The expanded/fluidized bed reactor can be designed using either he organic volumetric loading rate (OVLR) or solid retention time (SRT) approach. The kinetics of substrate removal in the fluidized bed reactor will determine the actual SRT required for a given degree of treatment efficiency. Once the SRT is established together with values for the kinetics parameters, the kinetics equations can be used together with the information from fluidization mechanics to establish values for other design parameters (hydraulic retention time, fluidization velocity, recycle ratio, etc.)

A less rigorous and strictly empirical approach will be to use the organic volumctric loading ratc (OVLR) to achieve a given degree of treatment.

10.3.2.2 Solid Retention Time

The solid retention time (SRT) is the average retention time of organisms in the system. In the expanded/fluidized bed process the SRT is normally defined as :

$$SRT = \frac{\text{volatiles suspended solids (VSS) in the reactor}}{\text{volatile solids lost in the effluent or int entionally wasted / day}}$$

In a biological reactor the organism specific growth rate is equal to the reciprocal of the solid retention times of the system

The organism specific growth rate is expressed according to the kinetic equation given:

$$\mu = 1/X \, . \, dX/dt = YK - b$$

where; μ = organisms specific growth rate, $m^3/m^3.h$

X = organisms concentration, kgVSS/m^3

Y = organisms specific yield coefficient

K = specific substrate utilization rate, kgCOD/kgVSS.h

b = organisms decay coefficient, h^{-1}

thus

$$1/SRT = YK - b$$

10.3.2.3 Organic Volumetric Loading Rate (OVLR)

Numerous authors[16,17,18] have used this parameter to illustrate the efficiency of fluidized bed reactors in comparison to other systems in treating wastewaters. The OVLR is used when it is difficult to determine the reactor biomass concentration.

The organic volumetric loading rate (OVLR) to the system is defined as :

$$OVLR = \frac{QS_0}{V} = \frac{S_0}{T}$$

where, Q = influent flow rate, m^3/d

S = influent substrate concentrate kg/m^3

V = reactor volume, m^3

T = reactor hydraulic retention time, d

Numerous pilot scale tests have shown high COD removal of more than or equal of 80% at COD loading of 10-20 kg/m^3.d for variety of industrial waste.

It has been found that the biomass concentration was affected by variation in the organic loading, influent substrate concentration, and hydraulic retention time[9]. Similarly COD loading appears to be the major determinant of biomass concentration[20].

10.3.2.4 Hydraulic Retention Time (HRT)

Hydraulic retention time (HRT) is calculated on the basis of expanded/ fluidized bed volume. The HRT is the ratio between the expanded/fluidized bed volume and the influent flow rate of the wastewater.

In the case of wastewater treatment, due to the relative insensitivity of the process performance to HRT, the system should be designed at a low HRT (of the order of several hours). The actual design HRT depend on wastewater organic strength[21].

Laboratory and pilot-scale experiments have been conducted with HRT values varying from a low of the five minutes to several days for various industrial wastes. COD loadings have ranged from 0.65 to 60 kg/m^3.d[22].

The organic removal efficiency decreased with decreasing HRT at a constant loading rate has been reported by number of authors[19-20]. Optimum HRT for mesophilic (35°C) anaerobic fluidized bed reactors lie within the range of 6-13 hours[19].

10.3.2.5 Recycle Ratio

Expanded and fluidized beds operate with very high recycle rates. For concentrated wastes, a very high degree of recycle is needed in order to keep the bed particles in suspension and at the same time to dilute the organic materials present in high concentration. For dilute wastes, like municipal wastewater, the recycle ratio is reduced to reasonably low values.

The recycle ratio (α) is given by the ratio between the influent flow rate (Q) and the recycle flow rate (Q_r)

Thus, recycle ratio (α) = Q/Q_r

Based on the design values of substrate removal rate, v_x= 0.01 kg.COD/kg.VSS.h; biomass concentration, X = 20 kg.VSS/m^3; hydraulic loading rate, Q = 10 m^3/m^2.h; COD removal efficiency, E = 80%; and total particle density, p = 1.02×10^3 kg/m^3, a plot as presented in Fig. 10.1 is developed to show the effect of influent wastewater concentration. Coupon the degree of recycle required in expanded and fluidized bed reactors[23].

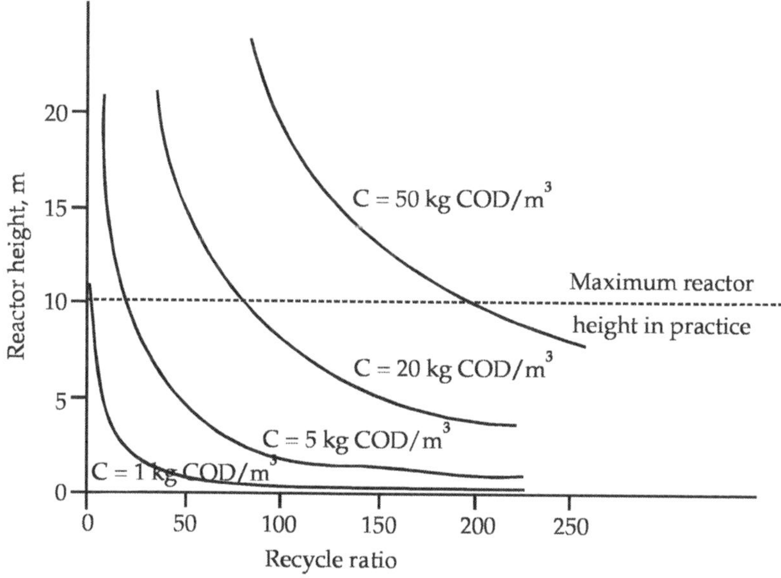

Fig. 10.1 : Recycle Rates for Expanded/Fluidized Beds

However, the assumed upflow superficial velocity of 10 m^3/m^2.h is an estimation based on an overall bed particle specific density identical for pure biological flocs. For very thin biofilms on high density media, the particle size will be of great importance in considerations regarding the needed superficial upflow velocity[24].

10.3.2.6 Filter Media

The medium used for biofilm attachment is comprised of small diameter inert particles, such as sand, anthracite, or granular activated carbon, which are maintained in the fluidized state.

Various support media have been tested, including sand, PVC particles, granular activated carbon, and diatomaceous earth. A range of particle sizes and densities have been examined. There are trade-offs between size and density of particles and stabililty of operation with these systems. Smaller particles provide greater specific surface area-to-volume ratio and thus provide greater surfaces for attached biofilms. In addition, lighter particles can be fluidized at lower upflow velocities which reduce the recycle rate necessary to achieve a given HRT.

In the design of an anaerobic expanded/fluidized bed, small and light particles which are easy to fluidize and provide a large specific surface are being used. There would exist however, a minimal particle size and/or density in order to prevent carry-over.

If a high shear stage is to be used to disengage the biomass and the medium then it is likely that a tough medium, such as sand, would be chosen in preference to a more fragile one like activated carbon[25].

The technical data based on available literature are given in Table 10.11[26].

Table 10.11 : Available Technical Data

Particulars	*Expanded Bed*	*Fluidized Bed*
Reactor media		
Inert material type	sand/gravel/plastic	granite/sand/carbon
Inert material diameter (mm)	0.3-3	0.2-1
Inert material submergence (%)	100	100
Bed expansion (%)	20-40	30-100
Specific surface area, m^2/m^3	1000-3000	1000-2500
Depth of reactor (m)	2-4	4-8
Radius of reactor (m)	2-3	2-3
Vertical velocity, empty bed (including recycle) (m^3/h)	2-10	6-20
Recycle ratio	2-100	5-500

10.3.2.7 Temperature

Temperature was found to be an important variable affecting process efficiency, but the process was found to compensate well for changes in temperature. However, the optimum temperature for the treatment is around 35°C.

High rate conversions of soluble organics with anaerobic fluidized bed reactors at thermophilic (55°C) temperatures, after a 5 month accumulation of biomass has been reported[10]. A mature microbial attached film was developed successfully at 55°C over a short time period and high solids concentrations and film depth (60 g/L and 170 cm respectively) were achieved with these thermophilic films. Medium strength (1.5 to 3 g/L COD) and high strength (5-16 g/L COD) soluble wastes were treated with a 70% removal efficiency at a volumetric loading rate of 30 gCOD/L.d.

Experimental results indicate that thermophilic anaerobic fluidized bed reactors are inferior to mesophilic reactors in several ways. The most important was the inferior organic removal efficiencies achieved by thermophilic anaerobic fluidized bed reactors under a number of operating conditions[19].

10.3.2.8 Process Design

Since anaerobic fluidized bed treatment technology is relatively a new approach, process design criteria is not available at present time.

Pilot plant studies were completed by Dorr-Oliver$^{T.M.}$ of U.S.A for industrial wastewater treatment as reported[27] in order to derive information for process design of single phase and two phase anaerobic fluidized bed systems. The authors presented the range of operating conditions and the range of process design parameter values for single and two phase fluidized bed systems as given in Tables 10.12 through 10.14. The design values presented for single phase and fluidized bed reactors are summarized in Table 10.15.

Table 10.12 : Operating Condition for Fluidized Bed Reactor of Single Phase System

Characteristic	*Reactor Value*
Mean sand size, mm	0.5
Hydraulic loading rate, m^3/m^2.h	25-33
Controlled bed expansion, %*	90-110
pH range 6.7-7.2	
Temperature, °C	30-35

* Percent expansion of settled sand bed.

Table 10.13 : Operating Conditions for Fluidized Bed Reactors of Two Phase System

Characteristic	*First Stage Reactor Value*	*Second Stage Reactor Value*
Mean sand size, mm	0.5	0.5
Hydraulic loading rate, m^3/m^2-h	25-33	25-33
Controlling bed expansion, %	40-60	90-110
pH range	5.7-6.2	6.7-7.2
Temperature, °C	30-35	30-35

* Percent expansion of settled sand bed.

Table 10.14 : Design Parameter Values for Single Phase and Two Phase Fluidized Bed Reactors Operated at 30°C to 35°C

Phase	*Reactor Biomass Concentration gVSS/L*	*Reactor Organic Loading Rate kgCOD/kgVSS.day*	*Reactor Volumetric Loading Rate kgCOD/m³.d*
Single Phase Organic to methane	15-25	0.4-1.0	10-20
Two Phase Organics to acetic acid	15-20	2.0-4.0	30-40
Acetic acid to methane	8-15	2.0-4.0	25-35

Table 10.15 : Single and Two Phase Fluidized Bed Design Values

Design Characteristic	*Single Phase Reactor Design Value*	*Two Phases Design Values*	
		First Stage	*Second Stage*
Volumetric loading rate, kgCOD/m³.d	15	40	30
Controlled fluidized bed height, m	10.5	10.5	10.5
Hydraulic loading rate, m/h	24	24	17
Reactor area, m²	127	48	51
Fluidized bed volume, m³	1333	504	536

* A 20% COD reduction is assumed in the first stage reactor of the two-phase system.

10.3.3 Applicability

Shock loadings (in terms of temperature and loading strength) had relatively little influence on the process[28]. Expanded bed reactor was found to be efficient for the treatment of particulate waste[29].

The anaerobic expanded bed process has been demonstrated to be effective for treating low strength waste (COD less than 600 mg/L) at short retention times (several hours) and at high organic loading rates (up to 8 kgCOD/m^3.day) even at low temperatures (10°C, 20°C)[30].

Due to the relative insensitivity of the process performance to hydraulic retention time, the system could be designed at a low HRT (on the order of several hours). The actual HRT will depend on wastewater organic strength. Little gas is produced when treating low strength wastes and as a result anaerobic treatment of sanitary wastewater cannot be regarded as a larger energy producer.

However, for high strength industrial wastes (COD range 5,000-50,000 mg/L) BOD and COD removal varies between 60-95 and 65-85 percent respectively in 0.3 to 4.9 days HRT over a wide range of organic loading (4-25 kgCOD/m^3.day) with a considerable amount of methane production.

Experiments carried out on anaerobic fluidized beds to study the effect of shock loads, showed that the process was unaffected by large instantaneous fluctuations in temperature, flow rate, organic concentrations, and organic loading rate[30].

However, the optimum temperature for the treatment is around 35°C and for lower temperatures, reduced removal efficiencies are observed. The energy produced could be used to heat the influent wastewater depending on its temperature. Typically 0.4 litres CH_4 were produced per gram of COD removed at 35°C. Methane content averages approximately 70% of the biogas with a range between 65-75% as presented in Table 10.16[31].

Table 10.16 : Summary of Methane Produced

Wastes	*COD gm/L*	*$LCH_4/gCOD_r$*	*% CH_4*	*$kgCOD/m^3.d$*
Food process	7-10	0.40	70	3.5-24.1
Chemical	12	0.41	82	3.5-5.7
Soft drink	4-18	0.41	60	–
Zimpro supernatant	7.8*	–	72	3.4-16.7*

* Based on ultimate BOD, 35 days.

Expanded/fluidized beds are able to withstand severe hydraulic overloading. They are less suited for organic overloading compared to fixed film reactors.

Laboratory scale experiments carried out for the treatment of black liquor evaporator condensate from a kraft mill, showed that the process in the expanded bed was upset after exposure to the toxic wastes while the fluidized bed showed no signs of toxicity influence even at high volumetric loads[32]. The fixed bed reactor survived the toxicity effects. Generally, expanded/fluidized beds are believed to have a relatively poor ability to withstand the effect of toxic compounds and other environmental factors compared to fixed bed reactor[26].

Results of a few pilot-plant scale experiments conducted indicated the feasibility of using expanded/fluidized bed reactor for denitrification.

10.3.4 Problems Associated with Expanded/Fluidized Bed Reactors[26]

10.3.4.1 Non-attached Biomass

In fluidized bed reactors, the high vertical velocity will take the free swimming organisms with the flow and deteriorate the effluent quality.

10.3.4.2 Suspended Organics

Expanded and fluidized beds have their biomass structure damaged by significant amount of suspended organics in the influent.

10.3.4.3 Foaming

Foaming problems were reported in expand/fluidized beds.

10.3.4.4 Gas Bubbles

Problems with gas bubbles are frequently met with in expanded/fluidized beds as in fixed beds and sludge blanket reactors. Gas bubbles may adhere to flocs/bed particles and cause these to rise in the reactor, and may result in wash-out of biomass or deterioration of the effluent quality

10.3.4.5 Start-up

Although all anaerobic reactors have start-up, expanded and fluidized bed reactors are believed to be the most troublesome in this aspect.

10.3.5 Advantages and Disadvantages

Advantages

All the advantages claimed for the anaerobic expand/fluidized bed reactors are derived directly or indirectly from the high concentration of biomass. Generally 10-40 kg/m^3 of volatile solids loading can be achieved.

Expanded and fluidized beds have several important advantages over anaerobic filter. These include the following[16] :

- no danger of clogging;
- small headloss;
- easier removal or addition of active material;
- better hydraulic circulation (avoidance of short circuiting);
- ability to operate at lower required detention times and/or higher organic loading rates;
- greater surface are available per unit of reactor volume;
- greater efficiency;
- better mitigation against toxic shocks and dilution of high strength wastes.

Further, the following advantages for biological fluidized bed systems have been claimed[25] :

- The high biomass concentration achievable leads to a small plant design which could reduce the land area requirement by up to 80 per cent;
- a large reduction of capital cost is achieved due to the greatly reduced reactor volumes.

Disadvantages

One disadvantage of this system is that recycling of effluent may be necessary to achieve bed expansion, and the system is more complex.

10.4 Anaerobic Rotating Biological Contactors

10.4.1 System Design

The major components of the equipment are the central shaft along with the stub, shaft bearings, media support structures, media drives and covers. In addition to the equipment, consideration should be given to the geometry of the tankage, the hydraulic regime through the system, control of solids build-up, monitoring and servicing of the equipment. Each component is discussed in greater depth in the subsequent. Each manufacturers's design offers certain features that are unique to its manufacturing process.

10.4.1.1 General Shafts

The current shaft design has been restricted to an overall length of approximately 27-feet of which 25-feet is devoted to media support. The shapes and sizes of the shafts of the major RBC manufacturer's are given in Table 10.17.

Table 10.17 : Shaft Characteristics

Manufacturer	*Shape*	*Size (in.)*	*Thickness (in.)*	*Section Modulus (in.)*	*Ref.*
Autotrol	Square	16 × 16	1.00	282	33
Clow	Round	30	0.625	415	34
Crane-Cochrane	Round	30	0.75	492	35
Lyco	Round	28	0.75	426	36
Lyco Hormel	Octagonal	24	0.75	344	37
Walker Process	Round	30	0.75	492	38

Although the RBC shafts rotate at a speed of only 1.5 RPM, the shaft undergoes constant reversing stresses. The weakest link in the shaft design are the welds, particularly the vertical welds on the shafts. The problem is compounded by human errors during the manual welding process. Poor welding practices have resulted in several shaft failures. Stringent quality control requirements during manufacturing would help alleviate this problem. It is recommended that on small size projects, the specifications should require the manufacturer to test all shafts by X-ray and ultra-sound and furnish certified copies of the test results to the Engineer. On intermediate size projects, require a certain number (10-20 percent) of shafts, to be picked are random by the Engineer, to be X-rayed and ultra-sound tested. On large projects, in addition to certain shafts being X-rayed and ultra-sounded, a couple of shafts should be picked at random (after all shafts are fabricated) and tested under full load conditions to the number of rotations equal to the service life of the project (normally 20-25 years)[39].

Another major cause of shaft failures has been the overloading of the shafts due to excessive biofilm growth. Several options are available to alleviate this problem[40]. These should be evaluated by the design engineer and the ones pertinent to the particular design should be incorporated in the design of the facilities.

The simplest, most fool proof method is to design all shafts to carry the maximum biofilm load that the shaft would ever be subjected to. Either ample safety factor on biofilm thickness or provisions to slough excessive growth should be incorporated in the design. Depending upon the implications of a shaft failure on the overall performance to the plant, stringent design and testing procedures may be specified.

The United State Environment Protection Agency's Checklist for a Trouble-free Facility[41] recommends that :

"...biofilm growths of 0.18 inches have been observed. Therefore, the engineer should specify a load bearing capacity for each shaft that considers the maximum anticipated biofilm growth, the capacity to strip excess biofilm, and an adequate margin of safety.

The engineer should require the manufacturer to provide adequate assurance that the shaft and media support structure are protected from structural failure for the design life of the facility. All fabrication during construction should conform to quality control standards."

Excessive biofilm thicknesses a detrimental to RBC's process performance as well as the structural integrity of the shafts. The EPA MREL Checklist[41] commends that :

"A means for removing excess biofilm growth should be provided such as air or water stripping, chemical additives, rotational speed control/reversal, etc."

The structural shaft design should be specified to be based standards. The design life (number of cycle) and the load at which the shaft will operate should also be specified.

10.4.1.2 Media Support Structure

Some of the RBC manufactures design uses a steel or stainless steel radial media support structure. Others do not in which case the media also acts as a structural member. The support structure too, has undergone several generation of modifications. The modifications strive to make the units more price competitive and also more energy efficient by reducing the dead weight of the unit. This may lead to compromise in the structural integrity of the support structure. The dilemma the engineer is faced with is that with each generation of changes he has less long term operating information to evaluate the particular manufactures equipment. It is therefore, important for the engineer to evaluate the manufacturer's design carefully before specifying it.

The main purpose of the support structure are :

(*a*) Support the plastic media, thereby eliminate the media having to act as a structural member.

(*b*) In the case of one manufacturer, the support structure holds the media pack such that it does not rest on the centre shaft, thereby providing an annular space between the shaft and media for free passage of air and sloughed solids in and out of the media.

(*c*) Facilitate easy removal and replacement of media pack in the field.

The first two items are discussed under the subsection entitled 'Media'. For the last item, the engineer should bear in mind that the covers have to be removed first and adequate provision for this should be made. Also, provision for removing the media, *i.e.* adequate working space and clearances remove the rods and media pack should also be provided. The type and material of all fasteners should be such that it would not corrode or be difficult to unfasten after several years of operation.

The material of the radial support structure should be evaluated carefully. The effects of any significant industrial wastes or in-plant recycle streams on the support structure should be evaluated. Generally, radial support structure made from galvanized or coal tar coated carbon steel, stainless steel or plastic (for rods) in furnished by the manufacturers.

As mentioned earlier, since the support structure design is in the stage of flux, it is recommended that warranties in excess of the standard one year should be specified by the engineer.

10.4.1.3 Shaft Bearings

Most of the manufacturers will furnish either tappered or spherical roller bearings in 4-15/16 inch or 7 inch diameters. There is not a significant difference in the cost of the two types of bearings of the same size[39]. The bearings should have the capability of withstanding shaft misalignments, and should have positive seal to prevent dirt and moisture from entering the bearing race. In general, spherical roller bearings offer greater shaft misalignment capability than tappered roller bearings[42]. One method of minimizing bearing failure due to excessive misalignment is to specify "shaft centre line out of alignment" not to exceed 0.0015 inches total indicator reading (TIR) per inch of shaft length during shaft fabrication[39]. The inner race of either type of bearing has two Allen-key type set screws which hold the race onto the shaft. The proper torquing of these set screws is important for the proper service life of the bearings. It is recommended that either the bearing be fully mounted and lubricated prior to shipment of the RBC to the job site or alternately, a qualified manufacturer's representative should do the final bearing installation and adjustments in the field.

One of the problems with the bearings is freeting. This could lead to the inner ring freezing onto the shaft, and make removal of the race at a later date very difficult. In some instances, operators have had to cut the ring off the shaft. It is recommended that the specifications require the shaft to be coated with "Never Seez" or equivalent compound to minimize freezing problem.

Life of the bearings should be based on the maximum shaft weight specified rather than the maximum expected weight of the shaft under operating conditions. The RBC bearings are generally fitted with mechanical type seal. Some manufacturers offer bearings with a set of two seals on each side. The space between the seals is purged with grease which act as a vapour barrier. The grease fittings should be provided with extension tubes which are fastened securely at a height which allows greasing the bearings with ease. On larger facilities, drum mounted grease guns are more economical than cartidge type grease guns. On such facilities, provisions should be made to allow carting the grease drum around. Ramps to get onto the RBC tanks should be provided where necessary. On older plants, bearing failures occurred due to corrosion and lack of proper sealing[43]. Present design requires greasing once every three months instead of twice a week for the older RBC's. A three month cycle should be specified.

10.4.1.4 Media

Media discs are vacuum formed from 40-50 mil. high density polyethylene (HDPE's) co-polymer sheets. Various media configurations or corrogation patterns have been selected by the manufacturers, each with its own claimed advantages. The corrogations increase the surface area of the sheet by up to 20 percent, given structural rigidity to the sheets, and act as spacers between sheets. Media may be mounted on the shaft in pie-shaped sectors, held in place by the radial support structure, wrapped around the shaft like a jelly roll, or thermo-welded media may be 'keyed' on to a square shaft.

The engineer should be cognizent of the potential ramifications associated with media replacement. RBC's with radial support arm assemblies combined with pie-shaped media wedges are conducive to field replacement of media without removal of the shaft from the RBC tank. Normally for the self supporting type media, the shaft may have to be taken out of the tank and in the case of one manufacturer, may require sending it to the manufacturer's facilities to rewrap the media. Appropriate arrangements to tackle this problem must be incorporated into the design.

Media failure can occur directly due either to degradation of the HDPE from prolonged exposure to heat, concentrated organic solvents, or ultraviolet (UV) radiation or breakage resulting from the excessive weight of heavy biofilm growth[44]. Media failure can also occur due to excessive shifting in the radial support structure. These problems have been well understood by the manufacturers and corrective design actions have been taken. However, a serious and little understood cause of media failure is breakage due to stress cracking. Media failures due to stress cracking have occurred as little as 6 to 9 months following start-up[44]. Stress cracking is an internal or external rupture in the plastic caused by tensile stresses of lesser magnitude than the short-term mechanical strength of the material. It is reported that HDPE can withstand many chemical when not under stress, but that under stress, the same materials may have a severe deterious effect. It therefore appears that RBC's that utilize media that is subjected to stresses (used as a structural member) may be subject to failure due to environmental stress cracking.

Therefore, the RBC and plastic manufacturers should be consulted before specifying the percent of these compounds to be added to the plastic before extruding media sheets.

10.4.1.5 Drives

The purpose of the discussion is to point out design features which may be included in the design to improve the O and M of the facility.

RBC models experienced problems with leaky oil bath chain casings, chain alignment and belt adjustments[43]. The current shaft mounted design eliminates these problems. On some of the older facilities, the drives were housed inside the covers to protect them from the elements. However, this created other problems, such as lack of adequate head room and working clearances, darkness, humid conditions, flies and mal-odours for the operators to work in. The current thinking is to locate the drive on the outside of the covers.

The drives need a semi-annual oil change. In warmer climates, synthetic oils may be used to reduce oil change frequently to once per annum.

The speed of the mechanical drives may be varied by changing sheaves or installing an electronic variable speed controller. Extra sheaves and belts should be stocked at the facility if speed is to be changed by using interchangeable sheaves. In general, the RBC mechanical drives have been reliable and have exhibited very few problems.

10.4.1.6 Covers

The first individual RBC covers were modified acryllic-vacuum formed, rigidized with spray up fiberglass[43]. There units were brittle and cracked over a period of time. They also had a tendency to sag. The present day covers are constructed of fiberglass reinforced plastic (FRP). These are hand layered on molds and each penal is manually constructed. It is therefore, recommended that stringent quality control requirements be specified and the general contractor and field engineer be required to inspect each cover for manufacturing deficiencies. Covers are also available with foam sandwich construction when additional insulation is required. The covers should be specified to handle expected wind loads either when fully assembled or partly assembled. The cover anchoring holes should allow for field adjustment such that the covers are not stressed when installing. The hardware should be stainless steel which would allow for future removal and reinstallation of the covers.

Access doors should have ERP ribbing to minimize buckling and the hinges should be continuous piano hinge type. Riviting on RFP should not be allowed and all bolts should have either backup plates or large diameter washers to reduce stress concentrations at bolt holes.

10.4.1.7 RBC Design

The prime factors of concern at the outset of the RBC design are :

(*a*) Structural integrity of the media, media support and media attachment;

(*b*) Structural integrity of the shaft including fatigue and manufacturing;

(*c*) Reliability and efficiency of drive system;

(*d*) Ability to remove shafts in future;

(*e*) Basin baffling;

(*f*) Means of monitoring shaft loads;

(*g*) Ability to take basins in and out of service.

10.4.2 Design Criteria

10.4.2.1 Hydraulic Loading

RBC trains are normally arranged in parallel modules. Flow distribution is achieved by fixed weirs. The lack of positive flow measurement and control to each train may lead to overloading or underloading of some trains. It is recommended that these features be incorporated in the design. Any plant side streams such as supernatant should be introduced upstream of the RBC trains to assure complete mix and even loading to the trains. Easily accessible inter-stage baffling and influent step feeding provisions should be made. Where flow equalization is not provided and significant diurnal flow variations are expected, provision for recirculating RBC effluent should be made.

The tank geometry should maintain minimum scour velocitites to limit suspended solids settling in the tanks. Also, as recommended by EPA MERL Checklist, primary sedimentation will minimize the solids settling problem. Some manufacturers recommend installation of fillets at the inter stage walls of the tanks to minimize solids settling in dead zones.

10.4.2.2 Load Cells

Load cells provide a better indication of the biofilm thickness than visual examinations. It is a recommended that load cells be furnished at least on all first and second stage shafts of all trains. Load cells at other locations may be desirable. It is recommended that the load cell components be made of corrosion resistant materials or are housed in corrosion resistant enclosures. The electronic load cells are more expensive than hydraulic type, but they can be wired to give an early warning of impending shaft overload problems and corrective action may be taken sooner.

10.4.2.3 Process Flexibility

In order to allow the operator to operate the RBC facility successfully, the EPA MERL Checklist recommends that :

Adequate flexibility in process operation should be provided by considering one or more of the following :

- Variable rotational speeds in first and second stages.
- Multiple treatment trains.
- Removable baffles between stages.
- Positive influent flow control to each unit or flow train.
- Positively controlled alternate flow distribution systems such as step feed.
- Recirculation of secondary clarifier effluent.

10.5 Process Control

For anaerobic systems to function properly, the operator must monitor certain parameters such as pH, alkalinity, volatile acids, COD, BOD, sulphides, temperature, gas and make adjustments accordingly. This chapter will discuss in detail the parameters that have to be monitored and provide some guidelines for these parameters. Because these parameters will vary with different types of wastes, it is up to the operator to monitor, chart and ultimately determine what is the best range for each parameter for a plant to produce a well treated effluent.

10.5.1 Test Methods and Sampling

Various analyses are used to determine the condition, progress, and efficiency of an operating anaerobic reactor. Tests, such as pH, alkalinity, volatile acids, COD, BOD, sulphides, SS, TS, SVS, gas production and analysis, are part of the analysis process. All of these tests should be performed according to the latest edition of "Standard Methods for the Examination of Wastewater," which is prepared and published jointly by the American Public Health Association, American Water Works Association, and the Water Pollution Control Federation.

Laboratory analysis results are only as good as the sampling technique followed. In analyses, the results of single grab samples are rarely reliable. Analyses of raw and treated effluent should be made on composite samples, if possible, which are comprised of a reasonable number of properly preserved single samples. Common sense dictates that recirculation samples should be taken only after a reasonable recirculation period. pH should be measured as soon as possible after collection to ensure valid results. Determine volatile acids within 48 hours after sample collection provided the sample is properly stored at 4°C. If the analysis must be delayed past 48 hours the sample can be preserved with formaldehyde or other microbial limiting substances. Some of the volatile acids will be destroyed using this techniques; however, acetic and propionic acids, the major constituents of the volatile acids, should not be altered.

Preservative techniques and maximum holding periods for some selected parameters are shown in Table 10.18. The operator must maintain control of the anaerobic process and to assist in quality control, Table 10.19 lists typical sampling frequencies.

Table 10.18 : Sample Preservation Techniques

Parameter	*Preservation*	*Maximum Holding Time*
Acidity/Alkalinity	Refrigeration at 4°C	24 hrs.
BOD	Refrigeration at 4°C	6 hrs.
COD	2ml/L Conc. H_2SO_4	7 days
Cloride	None required	7 days
pH	On site	None
Solids	None available	7 days
Ammonia Nitrogen	40 mg/L $HgCl_2$ -4°C or 0.8ml/L Conc. H_2SO_4	7 days
Kjeldhal Nitrogen	40 mg/L $HgCl_2$ -4°C or 0.8ml/L Conc. H_2SO_4	Unstable
Nitrate Nitrogen	40 mg/L $HgCl_2$ -4°C or 0.8ml/L Conc. H_2SO_4	7 days
Nitrate Nitrogen	40 mg/L $HgCl_2$ -4°C	1-2 days
Organic Carbon	2ml/L H_2SO_4 (pH-2.0)	7 days
Phosphorus	40 mg/L $HgCl_2$ -4°C or 0.8ml/L Conc. H_2SO_4	7 days
Sulphate	Refrigeration at 4°C	7 days
Sulphide	2 ml/L Zinc Acetate	7 days
Specific Conductance	None required	7 days
Metals, Total	5 ml/L HNO_3	6 months
Metal, Dissolved	Filtrate : 3ml/L 1:1 HNO_3	6 months
Phenolics	1.0 gm $CuSO_4$+H_3PO_4 to pH 4.0 at 4°C	24 hrs.

10.5.2 Control Parameters

pH : A reactor pH is used extensively to judge the general conditions of the anaerobic reaction. In most anaerobic reactors a neutral condition, as indicated by an average pH of 6.8 to 7.2, is considered normal and represents, to a certain extent, a proper balance between the materials and leaving the reactor. A pH in this range also indicates the presence of certain buffering substances. One step in anaerobic treatment, methane fermentation, produces a large quantity of carbon dioxide. A samples's unusually high pH may result from carbon dioxide released by excessive agitation of the sample before analysis. Because of reactor alkalinity, pH changes very slowly; therefore it must be used in conjuction with other analysis for proper process control. A reactor can become completely upset the pH ever changes.

Table 10.19 : Minimum Sampling Frequencies

Parameter	*Minimum Frequency*
Feed Inflow/Effluent	
Flow rate	Daily
pH	Daily
Temperature	Daily
Alkalinity	Daily
Volatile Acids	Biweekly
COD	3days/week
BOD	Weekly
Solids	Weekly
Nitrogen-NH_3-N	Biweekly
Total Kjeldhal Nitrogen	Biweekly
Total Phosphorus	Biweekly
Sulphide	Biweekly
Metals	Monthly
REACTOR CONTENT	
Temperature	Daily
pH	Daily
COD	Weekly
Solids	Weekly
Volatile Acid Fractions	Biweekly
Gas Production	Daily
Methane	Weekly
Carbon Dioxide	Weekly
Hydrogen Sulphide	Weekly

Alkanity : Alkalinity also aids in laboratory control of a anaerobic reactors. Calcium, magnesium, and ammonium bicarbonates are examples of buffering substances normally found. The anaerobic process ammonium bicarbonate and the others are contained in raw waste. Alkalinity and pH in a reactor contents may vary directly with the raw waste and with detention time. Alkalinity results must be interpreted carefully because high alkalinities can parallel high volatile acids during an acid phase operation.

Volatile Acids : One of the best anaerobic reactor control parameters is volatile acids concentration. Volatile acids are intermediate digestion products.

Proper operating levels are determined by volatile acids raw waste characteristics, design load and detention time, the degree of digestion, and testing method. When comparing volatile acids to alkalinity, the volatile acids/alkalinity ratio can indicate process conditions. Although individual values reveal current conditions, process control is best accomplished by closely monitoring the ratio's rate of change. If the ratio exceeds 0.8, pH depression and inhibition of methane production occurs and the process is failing. Increases above 0.3 to 0.4 indicate upset and the need for corrective action. Typical volatile acids concentrations range from 50 to 300 mg/L.

Gas : Other effective process control tools are recordings of gas production and gas analyses. In reactors equipped with gas collection and gas metering systems, maintain accurate records and correlate the results with routine control analyses. Increases in CO_2 concentrations in gas forecast difficulties in the process so long as it is not associated with incremental feeding. This because CO_2 increases after feeding. Continuous feeding helps to limit this variable. An increase in CO_2 content may result from a shift in the reactors chemical balance and is normally accompanied by increased volatile acids or reduced pH or both.

Of the two major components of gas, CO_2 is easier to analyze. The sum of CO_2 and CH_4 should be approximately 95%. Therefore, if the CO_2 content is determined, the amount of CH_4 can be calculated by subtracting the CO_2 value from 95. Typical CO_2 values in a well operated reactor range between 30 and 35%. When the percent of CO_2 is greater than 35%, trouble may be on the way.

Loss of gas production is also an obvious sign of process difficulty if there has been no interruption in the feed. Analysis for CO_2 and CH_4 combined with a plot of gas production versus loading rate can afford excellent control parameters. Under proper anaerobic digestion, the CO_2 and CH_4 in the has should range from about 30 to 35% and 65 to 70%, respectively. It is recommended that the CO_2 to CH_4 ratio and the volume of gas produced per kilogram of COD removed he used as a standard parameter of reactor condition. The CO_2 to CH_4 ratio will change as swiftly as the volatile acids content changes in digestion.

Temperature : Temperature strongly influences biological reactions. Because anaerobic treatment is a biological process, the reaction rate is decidedly influenced by temperature. Reaction almost ceases at approximately 10°C (50°F). Most anaerobic reactors operate in the mesophilic temperature range of 27°C to 38°C (80°C to 100°F). The optimum temperature range is 35°-40°C. If at all possible, the temperature should never increase or decrease more than 1°C (2°F) from the optimum temperature range.

10.5.3 Record Keeping

Anaerobic treatment is essentially a biochemical process and a rather involved one. Its effective management and control require observation of the raw, intermediate, and end products.

To decide what measurement and tests are essential and desirable, consider the variability of the quantity and quality of the raw wastewater, the kind and number of units in the system, and the acceptable range of quality of the treated effluent.

10.6 Process Failure and Remedies

As with any biological process, anaerobic reactor systems are subject to failures resulting from unstable operation or toxic materials influent to the process. This section outlines causes and effects of process failures and provide control strategies for remedying upsets.

10.6.1 Causes of Failure

The three basic causes of the anaerobic reactor instability or failure are hydraulic overload, organic overload, and toxic overload[45]. A brief description of each overload situation follows.

Hydraulic Overload : Hydraulic overload occurs when the effective retention time is reduced to a point at which the organisms cannot reproduce fast enough to avoid washout. Hydraulic overload can result from overpumping a dilute feed, sludge production exceeding reactor capacity, or from the reduction of effectives reactor volume by grit deposition, scum formation, or poor mixing. Hydraulic overload can also cause alkalinity washout, which results in a poorly buffered system.

Organic Overload : Organic overload occurs when the organic feed rate exceeds the rate which the process microorganisms can consume the organics under balanced conditions. Organic overload results from a sudden increase in feed rate, a sudden increase in feed organic concentration, too rapid start-up, excessive loading on an infrequent basis, or a feed which is too high in carbohydrates.

Toxic Overload : The anaerobic treatment process to certain compounds including sulphides, volatile acids, heavy metals, calcium, sodium, potassium, dissolved oxygen, ammonia and chlorinated organic compounds[46]. The *inhibitory concentration* of a substance depends on many variables, including pH, organic loading, temperature, hydraulic loading, the presence of other materials, and the ratio of the toxic substance concentration to the biomass concentration[47]. Tables 10.20 through 10.22 summarize inhibitory levels of several compounds[48].

Effects of antaginism and synergism are important, particularly concerning cation inhibition. Antagonism is the reduction of the inhibitory effect of one substance by the presence of another. Synergism is the increase of the inhibitory effect of one substance by the presence of another. Table 10.23 illustrates the antagonistic and synergistic behavior of certain cations. Often, a biological system acclimates to a toxic material. When acclimating, process organisms adjust the material's adverse effects are felt less severely[49].

Table 10.20 : Effect of Ammonia Nitrogen on Anaerobic Process

Ammonia Concentration, as N, mg/L	*Effect*
50-200	Beneficial
200-1000	No adverse effect
1500-3000	Inhibitory at pH over 7.4-7.6
Above 3000	Toxic

10.6.2 Failure Indicators

Instability in the anaerobic process occurs when the series of microbiological reactions become uncoupled. Acid formers outproduce acid consumers (methane formers) and a sharp rise in volatile acids results. The uncoupling may be a result of methane-forming organism inhibition; organic overload, which allows the faster growing acid formers to outproduce the methane formers ability to consume acids; or washout of the slow growing methane formers. Table 10.24 summarizes the physical condition of anaerobic reactor upset.

When volatile acids accumulate, the following chemical reaction occurs :

$$\underset{\text{Bicarbonate}}{HCO_3^- + HVA} \longrightarrow \underset{\text{Volatile acid}}{VA^-} + \underset{\text{Water}}{H_2O} + \underset{\text{Carbon dioxide}}{CO_2}$$

Bicarbonate alkalinity is consumed, carbon dioxide production increases, and eventually the pH falls. As the pH falls, the equilibrium between ionized and un-ionized volatile acids shifts toward the un-ionized state. Therefore, by increasing the amount of unionized volatile acid, pH depression increases the potential toxicity of high volatile acid concentration[50].'

Table 10.21 : Total Concentration of Individual Metals Required to Severely Inhibit Anaerobic Process

	Concentration, mg/L		
Metal	*Percent Dry Solids*	*Moles Metal/ kg Dry Solids*	*Soluble Metal, mg/L*
Copper	0.93	150	0.5
Cadium	1.08	100	–
Zinc	0.97	150	1.0
Iron	9.56	1710	–
Chromium			
+6	2.20	420	3.0
+3	2.60	500	–
Nickel	–	–	2.0

Table 10.22 : Stimulating and Inhibitory Concentrations of Light Metal Cations

	Concentration, mg/L		
Cation	*Stimulatory*	*Moderately Inhibitory*	*Strongly Inhibitory*
Calcium	100-200	2500-4500	8000
Magnesium	75-150	1000-1500	3000
Potassium	200-400	2500-4500	12000
Sodium	100-200	3500-5500	8000

During an upset, volatile acids may begin to rise well before bicarbonate alkalinity is consumed. pH depression does not occur until alkalinity is depleted. By the time pH depression is detected, however, the process may be well on its way to failure. Also, because the normal alkalinity test includes a portion of the volatile acids, calculation of bicarbonate alkalinity (BA) is important in ensuring adequate buffer capacity. To calculate BA,

$$BA = \text{Total alkalinity} - 0.71 \text{ Volatile Acids}$$

The value 0.71 is the product of two factors-one of the factors 0.83, converts volatile acids as acetic to volatile acid alkalinity, and the other factors, 0.85, accounts for the fact that titration to pH 4.0 for total alkalinity measures only 85% of the volatile acids[51].

Gas production rate and composition are excellent indicators of reactor health. Therefore it is important to monitor both to predict an upset. During process upset methane production declines. The carbon dioxide fraction of the gas may increase. The total gas production rate may remain unchanged despite falling methane production rates because of increased CO_2 production.

The rate of change in gas production is also determining digestor health. During a hydraulic or organic overload, the methane production rate gradually declines or, during an organic overload, may first increase, then slowly decline[45].

10.6.3 Control Strategy

During a upset situation take the following steps[46] :

1. Maintain near neutral pH.
2. Determine cause of the imbalance.
3. Correct the cause of the imbalance.
4. Provide pH control until treatment returns to normal.

Table 10.23 : Synergistic and Antagonistic Cation Combinations

Toxic Cations	*Synergistic Cations*	*Antagonistic Cations*
Ammonium	Calcium, Magnesium, Potassium	Sodium
Calcium	Ammonium, Magnesium	Potassium, Sodium
Magnesium	Ammonium, Calcium	Potassium, Sodium
Potassium	–	Ammonium, Calcium, Magnesium
Sodium	Ammonium, Calcium, Magnesium	Potassium

Table 10.24 : Summary of Process Failure

Indicators

Volatile acids concentration increases
Bicarbonate alkalinity drops
pH falls
Gas production rate drops
Percentage of CO_2 in gas increases

Typical causes of process failure

Hydraulic overload
increased freed flow
excessive sludge production
grit and scum accumulation
alkalinity washout

Organic overload

increase in sludge production
increase in feed concentration
change in feed characteristics
too rapid startup
infrequent feeding

Toxic overload

heavy metals
detergents
chlorinated organics
oxygen
cations
sulphides

Solutions

Adjust alkalinity using a supplement
Adjust feed schedule
Industrial pretreatment
Clean reactor
Restart reactor

10.6.3.1 pH Control

The key to controlling the reactor of pH is to provide bicarbonate alkalinity to react with acids and to buffer the system near pH 7. Bicarbonate can be added as bicarbonate itself or as a base which reacts with dissolved carbon dioxide to produce bicarbonate. Chemicals commonly used include lime, sodium bicarbonate, sodium carbonate, ammonium hydroxide, sodium hydroxide, and gaseous ammonia.

Lime Addition. Lime increases alkalinity as follows :

Lime (500 to 1000 mg/L)

$$\underset{\text{Lime}}{Ca(OH)_2} + \underset{\text{Carbon dioxide}}{2CO_2} \longrightarrow \underset{\text{Calcium bicarbonate}}{Ca(HCO_3)_2}$$

Lime (1000mg/L)

$$\underset{\text{Lime}}{Ca(OH)_2} + \underset{\text{Carbon dioxide}}{1CO_2} \longrightarrow \underset{\text{Calcium carbonate}}{CaCO_2} + \underset{\text{Water}}{H_2O}$$

These reaction have several serious implications. First, consumption of dissolved carbon causes CO_2 from the overlying gas to enter solution. Removal of CO_2 can cause a vacuum to develop within the tank. This can draw air into the reactor or even tip a floating cover[52].

Lime produces calcium bicarbonate only up to the point of solubility for calcium bicarbonate, or approximately 1000mg/L. Beyond that point calcium carbonate, which is relatively insoluble and readily precipitates is produced. Lime will continue to react with CO_2 without producing additional alkalinity. The pH may then increase to a point above the desirable pH range. Also, if the CO_2 fraction is below 10% of the overlying gas, pH during lime addition will suddenly and uncontrollably increase. Other bases react similarly[53].

Sodium bicarbonate addition. Sodium bicarbonate directly adds bicarbonate alkalinity. It does not react with CO_2 and therefore, eliminates the potential for vacuum development. Care must be taken to avoid sodium toxicity from overdosing. If only bicarbonate is being added, the pH will not increase beyond 8.3.

Chemical dosage for pH control. Required chemical dosage for pH control can be calculated using several methods. Using the titration method, a representative sample is titrated with the chosen chemical to the target pH (6.7 to 6.8 for lime, 7.0 or slightly higher for other bases). The dosage for the entire system is then calculated using a direct ratio between tank volume and sample size.

Dosage can also be calculated based on volatile acid concentration using the following procedure :

1. Determine the excess alkalinity desired (500 to 1000 mg/L).

2. Determine the alkalinity required to neutralize the volatile acids present (0.833 × VA).
3. Calculate the necessary alkalinity :

   Excess Alkalinity + Calculated Alkalinity – Measured Alkalinity = Needed Alkalinity.
4. Calculate the amount of chemicals needed by converting from $CaCO_3$ basis to chemical basis, adjust for commercial purity.

Dosage can also be calculated using Table 10.25 which provides the quantities of alkalies required to neutralize a given amount of volatile acids[52].

Table 10.25 : Quantities of Various Alkalies Required to Neutralize Volatile Acids

Actual kg acid/ 1000 L	*NH_3 anhydrous ammonia, kg*	*NH_4OH aqua ammonia, L*	*Na_2CO_3 anhydrous soda, ash, kg*	*NaOH liquid caustic soda, kg*	*NaOH flake caustic soda, kg*
1.00	0.11	0.75	0.33	0.50	0.25
2.00	0.21	0.82	0.67	1.01	0.50
2.99	0.32	1.22	1.00	1.51	0.75
4.00	0.43	1.63	1.33	2.01	1.01
4.99	0.54	2.03	1.67	2.51	1.26
5.99	0.64	2.44	2.01	3.03	1.51
6.99	0.75	2.84	2.33	3.52	1.76
7.99	0.86	3.25	2.67	4.03	2.01
8.99	0.96	3.65	3.00	4.52	2.26
9.99	1.07	4.05	3.34	5.03	2.52
20.01	2.15	8.14	6.68	10.09	5.04
30.05	3.22	12.23	10.05	15.16	7.58
40.12	4.30	16.33	13.41	20.23	10.11
50.24	5.39	20.45	16.71	25.32	12.66
60.37	6.47	24.58	20.17	30.47	15.21
101.17	10.85	41.17	33.81	50.98	25.49
205.09	22.00	83.52	68.56	103.38	51.69

When applying the chemical, dose the calculated amount over a 3 or 4 day period, mix well, monitor VA, pH, and alkalinity frequently. Make sure the vacuum relief device is operable. It is also important to avoid toxicity from the cations associated with the alkalies.

10.6.3.2 Other Control Methods

If possible, an ailing reactor's feed rate should be reduced until VA fall to a normal concentration. Reseeding the reactor with actively digesting sludge can significantly accelerate recovery time. The pH can be adjusted without chemical variation by gas scrubbing and recycle. Gas is scrubbed or CO_2, then recirculated through the tank. The CO_2 leaves the solution and enters the gas phase causing pH to increase[54].

To alleviate heavy metal toxicity, add sodium sulphide or ferrice or ferrous sulphate. Because toxic heavy metal sulfides have low solubility, and are less soluble than ferric sulphide, the toxic metals precipitate as sulphides if present[45]. Ferric chloride addition can similarly control sulphide concentration by displacing toxic high sulphides with chloride and removing the sulphides through precipitation as ferric sulphide[45].

10.7 Safety

With any process there are some special and highly recognized safety problems. This section is a basic guideline which covers the major safety aspects of anaerobic treatment. The main hazard associated with anaerobic treatment is the recovered gas, composed of methane and carbon dioxide. Such gas is both explosive and asphyxiating. Following discussion will cover gas handling, storage and monitoring.

10.7.1 Biogas

Biogas is a by-product of anaerobic digestion. It is composed primarily of methane (65 to 70%) and carbon dioxide (25 to 30%), with trace amounts of nitrogen, hydrogen, hydrogen sulphide, and oxygen. Table 10.26 summarizes the characteristics of common gases common to the wastewater industry. Methane is lighter than air and usually can be trapped in high places. Carbon dioxide, on the other hand, is heavier than air and is usually found in low places. Both displace air, which will cause asphyxiation because of lack of oxygen. These gases are colourless and odourless. An anaerobic reactor should be equipped to monitor for combustibles and oxygen deficiency.

10.7.1.1 Handling and Storage

This section will deal with the hazards involved in the safe handling and storage of biogas and the equipment used to accomplish this. Three key safety elements at anaerobic facilities are : providing adequate ventilation; maintaining an explosion-proof facility; and performing frequent inspections and preventive maintenance on all anaerobic systems safety and monitoring devices.

Biogas handling systems are kept under positive pressure to prevent leakage of atmospheric air into the system. In order for an explosion or a fire to occur, a certain amount of oxygen must be mixed with the combustible gas. The explosive range of air to biogas is approximately 20:1 to 5:1. Avoid mixing particularly in the ratio indicated. Maintain pressures in all gas collection systems, and instruct all personnel directly involved in operation on the workings of system mechanisms. Accidentally draining anaerobic reactor below the operating range of the floating cover can introduce air into the digestor.

Closely monitor liquid levels and take corrective action if the level drops and air is drawn into the reactor.

Gas leaks from the gas collection/storage system produce hazards similar to those previously defined in that they may result in an atmosphere conducive to fire or explosion, or both. Gas leaks are detected more easily than air leaks because a gas leak may register on a gas detector, or may noticed by smell. The nose, however, should not be relied on as an instrument for specific gas detection. Periodically check the gas main and appurtenances with a portable gas detector. Above all, there must be no smoking, sparks, or open flames in areas where gas may be present from leaks or from exposed reactor. Also, electrical installations must be well-maintained and explosion-proof. An installation is no longer explosion-proof if even one element, such as a light switch, is not properly installed.

The most important precaution against gas build-up is good mechanical ventilation throughout the reactor facility. Positive ventilation also provides a means for the removal of carbon monoxide, or unburned hydrocarbons, that could conceivably be produced in boilers, gas engines, or other equipment powered by sludge or natural gas. Stacks from such equipment should be sufficiently extended, located, and designed to avoid short-circuiting of exhaust gases back through the ventilation system.

Safety equipment associated with biogas system which requires regular or periodic inspection and maintenance includes flame arresters, pressure regulating valves, automatic control valves, automatic pilot valves, gas burner and controls, gas compressors and controls, and gas condensate traps. Flame arresters protect against flashback from waste gas burners, gas engines, and gas-fired boilers. They must be checked and cleaned on a scheduled basis to prevent possible blockage. Pressure relief devices, which provide relief if excessive pressure develops in the sludge gas systems, should always be vented on the outside. Inspect relief valves and pressure regulating valves on a scheduled basis to ensure that they relieve or regulate at the designated pressures. Check automatic gas and pilot valves preceding gas burners engines to ensure that they open and seat properly on demand of the gas unit. Clean gas burners on a scheduled basis and adjust their controls as necessary. Inspect and overhaul gas compressors on a scheduled basis, and eliminate all elements that produce or add to overheating. Drain gas condensate traps and replace seals as necessary.

10.7.1.2 Safety Considerations of Operation

Maintain tank pressures through such devices as cover position, liquid level indicators, and gas pressure. Avoid creation of vacuum conditions. Post "NO SMOKING" signs and enforce them.

Maintain all safely devices in operable condition including monitoring according to manufacturer's recommendations or at least monthly. Check all pipe joints for gas leakage on weekly basis.

Table 10.26 : Characteristics of Common to the Wastewater Industry

Gas and Chemical Formula	*Specific Gravity*	*Explosive Limit LEL UEL*	*Max.Safe 15-min Limit Exposure (% by vol in air)*[a]	*Max. Safe 8-h Exposure (% by vol in air)*[a]	*Common Properties*	*Physiological Effects*	*Location of Highest Concentration*	*Most Common Sources*	*Simplest and Safest Testing Method*
Carbon Dioxide CO_2	1.5	Non Flammable	1.5	0.5	Colourless, odourless, nonflammable; may cause acid taste in large quantities	Acts on respiratory nerves; 5% cannot be endured for more than a few minutes	Down low but may rise if heated	Sludge, sewer gas, combustion carbon and its compounds	Oxygen deficiency indicator
Carbon Monoxide CO	0.97	12.5 74.2	0.04	0.005	Colourless, odourless, tasteless, nonirritating flammable, explosive, poisonous	Combines with hemoglobin of blood causing oxygen starvation; fatal in 1h at 0.1%; unconsciousness in 30 min at 0.25% and causes headaches in a few hours at 0.02%	Up high specifically if in presence of illuminating gas	Manufactured fuel gas, flue gas, combustion and fires	CO indicator
Hydrogen Sulfide H_2S	1.19	4.3 46	0.0015	0.001	Rotten egg odour in small concentrations, colourless, flammable, and explosive	Paralyzes the respiratory system; lessens the sense of small as concentration increases; rapidly fatal at 0.05%	Down low, can be higher if air is hot and humid	Coal gas, petroleum sewer gas and sludge gas	Lead acetate paper, lead acetate ampoules, H_2S detector

Contd...

1	2	3	4	5	6	7	8	9	10	11
Methane CH_4	0.55	5 15	No limit providing sufficient oxygen (at least 19%) is present		Colourless, odourless, tasteless, explosive, flammable;	Deprives tissues of oxygen; dose not support life	At top, increasing to certain depth	Digestion of sludge, anaerobic reactors	Combustible gas indicator oxygen deficiency indicator	
Nitrogen N_2	0.97	Non-flammable	—	—	Colourless, tasteless, odourless, and nonflammable	In very high concentrations reduces oxygen intake; dose not support life	Up high and sometimes in low areas	Sewer and sludge gas	Oxygen deficiency indicator	
Sludge	Gas varies	5.3 19.3	Varies with composition	—	Flammable, practically odourless, and colourless	Will not support life	Up high	Digestion of sludge	Combustible gas indicator oxygen deficiency indicator	

[a] Conforms to "Threshold Limit Values for Chemical Substances and Physical Agents in the Work Environment and Biological Exposure Indices"

Source : WPCF Manual of Practice No. 1 "Safety and Health in Wastewater Systems" (1983).

To Shutdown, follow procedures that will :

- Plan the operation carefully;
- Relieve pressures gradually;
- Vent gases to the atmosphere;
- Isolate reactor for 30 days;
- Purge gas system and reactor;
- Ventilate reactor;
- Follow all applicable procedures for entering a "confined space."

10.7.2 Taking an Anaerobic Reactor Out of Service

Anytime an anaerobic reactor is taken out of service, or personnel enter and work inside a reactor, danger is involved. Proper procedures and equipment are necessary to perform the tasks safely. Each plant should review their operation and prepare readily available procedures.

When a reactor is taken out of service and its contents emptied a dangerous situation arises because biogas may mix with air and an explosion could result. To avoid this problem, inject carbon dioxide under the cover, before the reactor seal is broken, to displace the gas. Once the seal is broken, continuously vent the area with explosion-proof fans.

When entering an anaerobic reactor, the danger again is gas. As a minimum, before entering the reactor, test the atmosphere for oxygen deficiency and combustibles. Also continuously purge with face to bring in fresh air.

Working in an anaerobic reactor is dangerous because it is dark, slippery, gas may be present, and various equipment may be inside. Lock out power to all equipment. To ensure safety; supply sufficient light and ventilation and provide combustible and oxygen deficiency equipment to monitor for the gas. All of the safety equipment that is listed below should be available.

- Explosion-proof lighting and extension cords.
- Explosion-proof blowers or fans.
- Self-contained air breathing apparatus.
- Safety harness and rope.
- Combustible and oxygen deficiency meter.
- Hold card or tags.
- Spark-proof tools.
- Rubber gloves and boots.
- "NO SMOKING" signs.
- Universal tester with carbon monoxide and hydrogen sulphide ampules.

Preparing for Entry – The following steps should be considered :

- Shut down operations. Stop feed but continue heating and mixing for at least 30 days, or until gas is no longer being produced. This is eliminate further methane production during dewatering.
- On floating covers, withdraw enough supernant until the cover is firmly resting on support brackets. On fixed covers, raise the liquid level to its highest level.
- Close all inlet valves and gas valves to the anaerobic reactor. These valves shall be tagged off or chain locked.
- A checklist shall be posted at entrances to the confined space and an emergency telephone easily accessible. A safety sign reading: "No Smoking", or "No Spark-producing," or "No Open Flames" shall be prominently displayed in the area.
- Inject an inert gas, such as CO_2, or nitrogen, into the gas area and vent until a reading of below 2.5% combustible gas is detected. If this is not practical, purging with air may be used if all sources of ignition are excluded and the site is carefully monitored.
- All electrical disconnects for mixing equipment, pumps and so on, must be locked out and tagged. All inlet and gas piping leading into the reactor must be blanked off, valved, and tagged.
- Self-contained positive pressure air breathing equipment, rope, and a safety harness must be stationed at the entrance way.
- Providing that a telephone is easily accessible, there must be at least one observer for each employee stationed at the entrance or two at the top of the plant. The observer should be capable of monitoring the assigned employee and must be equipped and capable of removing the employee in an emergency.
- Side hatches should be used if available.
- All employee working in or outside of the reactor must leave all smoking materials (cigars, cigarettes, pipes, matches, lighters) in their lockers.
- Hard hats with a chin strap must be worn.
- The condition of built-in-rungs must be checked every time an employee ascends or descends. A ladder is preferred even when built-in-rungs are present.
- All personnel involved in the cleaning operation must be thoroughly trained in the hazards of biogas, and qualified to perform first aid including artificial respiration.

A successful and safe reactor cleaning operation involves planning in advance. The job supervisor should use a daily checklist of the safety procedures involved. The storage or lack of necessary materials, or the failure to follow the correct procedures, could result in a fatality. When working with digester gas, assume no margin of error.

Entry : The following are general steps describing entry in an anaerobic reactor :

- Monitor atmosphere. Oxygen level must be 19.5% or greater; combustible level must be 20% or less. Test all possible areas of the reactor to make sure gas is not pocketed. Fans should be kept in continuous operation. The tank must be continuously monitored for oxygen and methane while employees are in the tank. A means of communicating with the employee inside the tank should exist, so as to notify the employee of a hazardous condition. A horn is recommended. In an emergency, when the reactor must be entered, and the gas is either in the explosive range or oxygen-deficient, the employees must be equipped with a self-contained pressure air breathing apparatus. The employees must wear a safety harness with a life line, and be attended by two observers outside the entrance. For the most part, however, entrance under these conditions should not even be considered.
- Open the cover and install two fans, one to exhaust gas, and one to intake air. Remember, use only explosion-proof motors and spark-proof tools.
- Remove as much reactor content as possible.
- Water is generally needed to break up biomass. Extreme care should be taken when using high pressure hoses. Compressed air is not recommended because it tends to release gas that has been trapped.

REFERENCES

1. van den Berg, L., Kennedy, K.J. and Samson, R., Anaerobic Downflow Stationary Fixed Film Reactor: Performance Under Steady-State and Non-Steady Conditions. Water Sci. Technol., 17(1): 89-102, (1985).
2. Kennedy, K.J. and van den Berg, L., Stability and Performance of Anaerobic Fixed-Film Reactors During Hydraulic Overloading at 10-35°C. Water Res., 16(9): 1391-1393, (1982).
3. Kennedy, J.L. and Droste, R.L., Effect of influent Concentration on the Start-up of Anaerobic Downflow Stationary Fixed Film (DSFF) Reactors. Proceedings of the 38th Industrial Waste Conference, Purdue University, Ann Arbor Science, Ann Arbor, Michigan, (1984).
4. van den Berg, L., Lentz., C.P. and Amstrong, D.W., Anaerobic Waste Treatment Efficiency Comparisons Between Fixed Film Reactors, Contact Digestors and Fully Mixed Continuously Fed Digestors. Proceedings of the 35th Industrial Waste Conference, Purdue University. Ann Arbor Science, Ann Arbor, Michigan, (1980).

5. Kennedy, K.J. and van den Berg. L., Effects of Temperature and Overloading on the Performance of Anaerobic Fixed Film Reactors. Proceedings of the 36th Purdue Industrial Waste Conference, Purdue University. Ann Arbor, Michigan, (1981).
6. Switzenbaum, M.S., A Comparison of the Anaerobic Expanded/Fluidized Bed Process. Proceedings of a Specialised Seminar of the IWAPRC, held in Copenhegan, Demark, 16-18 June 1982. Water Sci. Technol., 15(8/9), (1982).
7. Lettinga, G. *et al.*, Design, Operation and Economy of Anaerobic Treatment. Proceedings of a specialised Seminar of the IAWPRC, held in 16-18 June 1982, Copenhagen, Denmark. Water Sci. Technol., 15(8/9): 177-195, (1982).
8. Grin, D., Roersma, R and Lettinga, G., Anaerobic Treatment of sewage in UASB Reactors. 2nd International Symposium on Anaerobic Digestion, Travemunde, 6-11 September, Travemunde, Poster Session, (1981).
9. Lettinga, G. *et al.*, High-Rate Anaerobic Waste-Water Treatment Using the UASB Reactor Under a Wide Range of Temperature Conditions. Biotechnol. Genetic Engg. Rev., 2 (Oct.): 252-284, (1984).
10. Lettinga, G. *et al.*, Use of the Upflow Sludge Blanket (USB)-Reactor Concept for Biological Wastewater Treatment, Especially for Anaerobic Treatment. Biotechnol. Bioeng., 22(4): p699-734, (1980).
11. Lettinga, G and Vinken, J.N., Feasibility of the Upflow Anaerobic Sludge Blanket (UASB) Process for the Treatment of Low Strength Wastes. Proceedings of the 35th Industrial Waste Conference, Purdue University, 1980. Ann Arbor Science, (1980).
12. Pette, K.C. *et al.*, Full Scale Anaerobic Treatment of Beet-Sugar Wastewater. Proceedings of the 35th Industrial Waste Conference, Purdue University, 1981. Ann Arbor Science, Ann Arbor, Michigan, (1980).
13. Parkin, G.F. and Speece, R.E., Attached versus Suspended Growth Anaerobic Reactors: Response to Toxic Substances. Proceedings of a Specialised Seminar of the IAWPRC held in Copenhagen, Denmark, 16-18 June 1982, Water Sci. Technol., 15(8/9): 261-289, (1982).
14. De Zeeuw, W. and Lettinga, G., Acclimation of Digested Sewage Sludge during Start-up of an UASB-reactor. Proceedings of the 35th Industrial Waste Conference, Purdue University, 1981. Ann Arbor Science, Ann Arbor, Michigan, (1980).
15. Hulshoff Pol, L. *et al.*, Granulation in UASB Reactors. Proceedings of a specialised Seminar of the IAWPRC held in Copenhagen, Denmark, 16018 June 1982. Water Sci. Technol., 15(8/9), 1983.
16. Stephenson, J.P. and Murphy, K.L., Kinetics of Biological Fluidized Bed Wastewater Dentrification. Prog. Water Tech., 12: 159, (1980).
17. Sutton, P.M. and *et al.*, Dorr-Oliver's Oxitron System Fluidized Bed Treatment Water and Wastewater Treatment Process. Biological Fluidized Bed Treatment of Water and Wastewater. Cooper, P.F. and Atkinson, B (eds.), Elis Horwood, Chichester, England, (1981).
18. Jewell, W.J., Development of the Attached Microbial Film Expanded-Bed Process for Aerobic and Anaerobic Waste Treatment. Biological Fluidized Bed Treatment of Water and Wastewater. Cooper, P.F. and Atkinson (eds.) Ellis Horwood, Chichester, England, (1981).
19. Rudd, T., Hicks, S.J. and Lester, J.N,. Comparison of a Synthetic Meat Waste by Mesophillic and Thermophilic Anaerobic Fluidized Bed Reactors. J. Environ Technol. Letters, 6(5): 209-224, (1985).

20. Schraa, G. and Jewell, W.J., High Rate Conversions of Soluble Organics with a Thermophilic Anaerobic Attached Film Expanded Bed. J. Water Pollut. Control Fed., 56(3): 226 232, (1984).
21. Switzenbaum, M.S., Sheehan, K.C. and Hickey, R.F., Anaerobic Treatment of Primary Effluent. Environ. Technol. Letters, 5: 189-200.
22. Switzenbaum, M.S., Anaerobic Fixed Film Waste Treatment, Review. Enzyme and Microbial Technol., 5(4): 242-250, (1983).
23. Vigeneswaram, S., Balasuriya, B.L.N., and Viraragharam, T., Anaerobic Wastewater Treatment. Attached Growth and Sludge Blanket Process, Environmental Sanitation Reviews, No. 19/20, 51-53, (1986).
24. Jewell, W.J., Anaerobic Attached Film Expanded Bed Fundamentals. Paper presented at First International Conference on Fixed Film Biological Processes, 20-23 April 1982, Kings Island, Ohio, (1982).
25. Cooper, P.F. and Wheeldon, D.H.V., Fluidized and Expanded Bed Reactors for Wastewater Treatment. J. Water Pollut. Control, 79)2): 286-301, (1980).
26. Henz. M. and Harremoes, P., Anaerobic Treatment of Wastewater in Fixed Film Reactors, A Literature Review. Proceedings of a specialized Seminar of the IAWPRC held in Copenhagen, Denmark, 16018 June 1982. Water Science and Technology, 15 (8/9): 1-101.
27. Sutton, P.M. and Li, A., Single and Two Phase Anaerobic Stabilization in Fluidized Bed Reactors. Proceedings of a Specialised Seminar of the IAWPRC, 16-18 June 1982, Copenhagen, Denmark. Water Sci. Technil., 15(8/9): 333-344, (1982).
28. Jewell, W.J., Switzenbaum, M.S. and Morris, J.W. Municipal Wastewater Treatment with the Anaerobic Attached Microbial Film Expanded Process. J. Water Pollut. Control, 53(4): 482-490, (1981).
29. Morris, J.W. and Jewell, W.J., Organic Particulate Removal with the Anaerobic Attached Film Expanded Bed Process. Proceedings of the 36th Industrial Waste Conference, Purdue, University, May 1981. Ann Arbor Science, Ann Arbor, Michigan, (1982).
30. Jewell, W.J. and Switzenbaum, M.S., Anaerobic Attached Film Expanded Bed Reactor Treatment J. Water Pollut. Control, 52(1): 1953-1965, (1980).
31. Jeris, J.S., Industrial Wastewater Treatment Using Anaerobic Fluidized Bed Reactor, Proceedings of a Specialised Seminar of the IAWPRC, 16-18 June 1982, Copenhagen, Denmark. Water Sci. Technol., 15(8/9): 169-176, (1982).
32. Norrman, J., Treatment of a Black Liquor Condensate from Pulp and Paper Industries. Proceedings of a Specialized Seminar of the IAWPRC, 16-18 June 1982, Copenhagen, Denmark. Water Sci. Technol., 15(8/9): 247-259, (1982).
33. Autotrol Wastewater Treatment Systems – Design Manual. Autotrol Corporation, Bio-Systems Division, Milwaukee, Wisconsin, (1979).
34. Clow Envirodisc Division, Florence, Kentucky, (1981).
35. Crane Cochrane Rotating Biological Contractor Bulletin No. 32.01. King of Prussia, Pennsylvania, (1981).
36. LYCO Rotating Biological Surface. Wastewater Equipment Catalog. Marlboro, New Jersey, (1983).
37. EDS Hormel, Environmental System Division, - Catalog coon Rapids, Minnesota, (1979).

38. Walker Process Corporation, RBC Presumbittal Package, Aurora, Illinois, (1984).
39. Personal Communication with Sib Banerjee of Clow Corporation, Florence, Kentucky, June 12, (1984).
40. Chestner, W.H. and J. Iannone, Review of Current RBC Performance and Design Procedures Report prepared for USEPA, MERL, Cincinnati, Ohio by Roy F. Weston, Inc. (Publication Pending).
41. USEPA, Rotating Biological Contractors (RBCs) Checklist for a Trouble-Free Facility, Washington, District of Columbia, Revised May, (1984).
42. Personal Communications with D. Murrill of Rexnord, Mechanical Power Division, Atlanta, Georgia, July 3, (1984).
43. Personal Communication with R. Davie of Envirex Corporation, Milwaukee, Wisconsin, June 14, (1984).
44. Brenner, R.C., Etal., Design Information on Rotating Biological Contractors, USEPA, MERL, Cincinnati, Ohio, Publication Pending.
45. Graef, S.P., and Andrwes, J.F., Stability and Control of Anaerobic Digestion. J. Water Pollut. Control Fed. 46, 666 (1974).
46. McCarty, P.L., Anaerobic Waste Treatment Fundamentals Public Works, 95, (1964).
47. Kugelman, I.J., and McCarty, P.L. Cation Toxicity and Stimulation in Anaerobic Waste Treatment. J. Water Pollut. Control Fed. 37, 97 (1965).
48. MERL EPA, Process Design Manual Sludge Treatment and Disposal (1979).
49. Kugelman, I.J., and Chin, K.K., Toxicity, Synergism and Antagonism in Anaerobic Waste treatment. Anaerobic Biological Treatment Process. Advances, in Chemistry Series, 105. Amer. Chem. Soc. (1971).
50. McCarty, P.L., and McKinney, R.E. Volatile Acid Toxicity in Anaerobic Digestion. J. Water Pollut. Control Fed. 33, 223 (1961).
51. Brovko, N., and Chen, K.Y., Optimizing Gas Production, Methane Content and Buffer Capacity in Digester Operation. Water and Sew. Works, 54 (1977).
52. Municipal Operations Branch EPA 430/9-76-001, Anaerobic Sludge Digestion. (1976).
53. Barber, N.R., Lime/Sodium bicarbonate Treatment Increases Sludge Digester Efficiency. J. Environ. Sci., 2 28 (1978).
54. Andrews, J.F., Control Strategies for the Anaerobic Digestion Process. Water and Sew. Works, Parts I, II 122, 62, (1975).

CHAPTER 11

CASE STUDIES ABROAD AND IN INDIA

11.0 Introduction

The technology of anaerobic digestion has developed significantly in recent years. It has gained increased credibility from successful applications in industry as well as other sectors.

Anaerobic biological waste treatment offers advantages over aerobic systems in terms of lower energy requirements, less biological sludge production and the potential for energy recovery in the form of methane gas. The development of innovative reactor designs, based on the optimization of growth and retention of micro-organisms, has created an impetus to re-evaluate the anaerobic treatability of many industrial waste streams. The comparison of anaerobic and aerobic biotechnology per metric ton COD destroyed is given Table 11.1

Table 11.1 : Comparison of Anaerobic and Aerobic Biotechnologies

Parameter	*Anaerobic*	*Aerobic*
Electricity	-	1100Kwh
Methane	1.1×10^7 BTU	-
Net cell production	20-50 kg	400-600 kg

The most recent and significant, advances in anaerobic digestion are related to the technology's ability to accommodate relatively high rates of organic loading. Companies are also interested in using anaerobic digestion for the biodestruction of organic materials that are not removed in conventional aerobic treatment. As application of anaerobic technology to various process streams increase, more success is inevitable, resulting in industries that are more economically competitive because of their more judicious use of natural resources.

Companies adopting anaerobic technologies fall into the following categories :

- Companies scheduled to expand their production process capacity and whose treatment facilities are already at capacity.
- Companies that discharge to publicly owned treatment works whose surcharge for treatment has increased substantially.

- Companies that have relatively monotonous, highly concentrated waste streams contribution a major portion of total waste load.
- Companies in areas where extremely high land costs make conventional aerobic digestion to expensive.
- Clusters of small and medium scale units can join to treat their wastewater through common effluent treatment plants using anaerobic route if possible.

In addition to companies that adopt digestion as the treatment method of choice, several companies maintain active research programme to evaluate new methods, including anaerobic digestion, for treating new and existing waste streams. Two important reasons for such programs are :

- Anticipated changes in environmental statues.
- The need to develop more cost-effective treatment techniques.

In India, the companies seldom sustain their R and D programme in waste treatment and waste minimization. Therefore, MNES is funding several R and D programmes and also encouraging to step up demonstration units, Table 11.2.

Although the technology is not without faults, it often represents an alternative to wastewater treatment that can provide substantial savings to the industry. Many aspects of technology contribute to such savings :

- Land cost saved by installing an anaerobic reactor.
- Reduced sludge volume.
- Decreased energy requirements.
- Energy credits from methane in the bioconversion, traditionally cited as sources of savings.

The state of art in anaerobic treatability is relatively simple. An anaerobic culture is mixed with substrate in the presence of micro-nutrients. Samples to be tested must be maintained in an oxygen free environment. Subsequently, the activity of the bioprocess is monitored by measuring the quantity and quality of biogas produced. Like any biological process, anaerobic organisms may require an acclimation period (depending on the origin of culture). Because the technology has been applied to a relatively limited array of waste-types, caution should be exercised in drawing conclusions from results of a negative assay. The appropriate source of a culture often makes the difference between successful assay and a failure to produce methane.

Lack of sophisticated water management within a facility is one of the largest obstacles to the adoption of anaerobic digestion. The more characteristics that are known about waste stream, the better the chances for successful treatment. Segregation and testing of individual streams are commensurate with a successful testing programme. Inorganic compounds such as cyanides, heavy metals, sulphates, and nitrates can be removed through pre-treatment and render an otherwise non-treatable sample biologically degradable.

11.1 Case Study — Anaerobic Treatment of Pharmaceutical Fermentation Wastewater[1]

US, EPA was able to subcategorie the pharmaceutical industry based on unit manufacturing processes and some common wastewater characteristics.[2] The subcategories selected are :

- fermentation
- biological and natural extraction
- chemical synthesis
- formulations
- research.

US, EPA survey determined that fermentation wastewaters are typically characterized by high BOD, COD and TSS, large flows and a pH range of about 4.0 to 8.0. Such wastewaters contain small amounts of organic extraction solvents and significant food stuffs such as grain flours and oils, sugar, starches, protein and macro- and micro-nutrients. Fermentation production throughout the pharmaceutical industry is generally continuous and results in the generation of wastewater with relatively consistent and predictable characteristics.

Table 11.2 : Statement of Financial Assistance for R and D Programmes by MNES, N. Delhi*

Year	BE/RE (a)	R and D (b)	BE/RE (c)	Demonstration (d)	Total BE/RE (a+c)	Total Exp. (b+d)
1985-86	1.600	1.200	-	0.370	1.600	1.490
1986-87	1.950	1.150	0.800	1.700	2.750	2.850
1987-88	1.500	1.630	1.000	1.600	2.500	3.230
1988-89	1.500	1.840	0.800	0.800	2.300	2.610
1989-90	0.900	1.120	0.700	0.670	1.600	1.790
VII-Plan	7.450	6.860	3.300	5.110	10.750	11.970
1990-91	1.200	1.280	1.080	1.660	2.280	2.940
1991-92	1.050	2.215	1.245	1.376	2.295	3.526
VII-Plan						
1992-93	1.340	1.340 (0.625)	3.260	3.260	4.600	4.600 (1.931)
1993-94	4.60	1.65	-	2.50	4.60	4.15

* *Source* : MNES, N.Delhi

Chemical synthesis production, on the other hand is usually conducted batchwise and in 'campaigns', resulting in generation of a complex wastewater which can very considerably in the constituents and flow rate and which can sometimes be incompatible with biological treatment systems. It is generally known that anaerobic treatment process function most reliably on waste systems which do not exhibit sudden changes in component makeup. Therefore, performance results have been given for fermentation processes only. The wastewater also contains residual levels of antibiotics which did not affect performance of ASP, but which were of potential concern in anaerobic treatment considering the slower organism growth rates and the balance among acid formers, acetogens and methanogens which must be maintained. The general wastewater characteristics are given in Table 11.3.

Table 11.3 : General Fermentation Wastewater Characteristics

Sr. No.	*Parameter*		*Value*
1.	Flow,	m^3/d	1730
2.	COD,	kg/d	15000
3.	BOD,	kg/d	7150
4.	TSS,	kg/d	2880
5.	TKN,	kg/d	430
6.	NH_3-N,	kg/d	280
7.	Total-P,	kg/d	80
8.	Sulphate,	kg/d	350

A preliminary economic analysis was first prepared based on hypothetical system design. Although the analysis included many assumptions, the projected payback period was in the range of 1.3 to 2.7 years which encouraged the initiation of bench and pilot plant studies. Four anaerobic processes were short listed, *viz.*

- a down flow anaerobic filter,
- a combination of downflow/upflow anaerobic filter,
- an UASB, and
- a low rate anaerobic filter.

11.1.1 Anaerobic Filters

These were selected first for pilot testing, since attached growth systems offer a potentially long SRT. These systems are generally regarded as tolerant of load swings or brief toxic slugs, since only the surface layer of the biomass attached to the support media is directly exposed to wastewaters. The downflow/upflow filter was constructed by placing two identical columns in series so that liquid flow was downward in the first column and upward in the second. Because the short HRT did not allow sufficient solubilization of the matter to occur, it was concluded that the observed COD removal was not primarily attributable to the conversion of COD to methane as desired. This conclusion was further supported by the unstable and very low rates of gas generation, Table 11.4.

Table 11.4 : Performance of Upflow-Downflow Filters

Sr. No.	*Parameters*		*Reactors Type*		
			Down-Flow		*Downflow/Upflow*
1.	Media depth, m		1.1		2.4
2.	HRT, d		2.5		2.5
3.	Temperature, °C		35		35
4.	CH_4 yield, m^3/kg COD		unstable		unstable
5.	*Loading, kg COD/m³.d*	*Percent Removal*			
		COD	TSS	COD	TSS
	1.15	0	87	-	-
	1.38	55	79	-	-
	1.54	48	90	-	-
	1.60 (System Failed)	69	96	-	-
	1.0	-	-	58	86
	2.1	-	-	47	94
	3.9	-	-	62	85
	5.3	-	-	2	90
	5.5 (System Failed)	-	-	10	68

11.1.2 An UASB Reactor

Downflow/Upflow filter performance revealed that an active sludge blanket had formed in upflow reactor and that biomass formation on upflow media was incomplete, suggesting that improved performance of the reactor compared to downflow reactor may have been distributable to the presence of sludge blanket rather than to the presence of additional upflow media. Therefore, UASB reactor system was tried and the performance is given in Table 11.5.

11.1.3 Low Rate Anaerobic Reactor (An CSTR)

High suspended solids coupled with high rate anaerobic reactor system which were evaluated did not satisfactorily degrade these solids in the reactor time available. Low rate systems implies that the reactor operate at a slightly lower temperature, degrade the waste at a lower rate and are loaded much less heavily than the more common high rate systems. Therefore, reactor size is large and SRT is of the order of 1-2 years. The performance data are given in Table 11.6.

Table 11.5 : Performance of UASB Reactor System

Sr. No.	*Parameter*	*Value*	
1.	Reactor Depth, m	1.2	
2.	HRT, d	0.48-0.52	
3.	Temperature, °C	35-37	
4.	CH_4 yield, m^3/kg COD	0.18-0.30	
5.	*Loading, kg COD/m^3.d*	*Per cent Removal*	
		COD	*TSS*
	1	24	30
	3	61	51
	5 upto 5 kgCOD/m^3.d-O.K.	73	75
	8	74	85
	10 (System failed-reduction in sludge blanket because of excessive wash-rate of biomass)	68	0

11.1.4 Full Scale Implementation

Low rate anaerobic reactor performance has been found to be stable and can handle raw waste solids as well as return activated sludge. The fermentation raw waste stream is separately pre-treated through two low rate reactors operating in parallel before joining the chemical, synthesis raw waste stream. The combined wastewater is further treated in the existing ASP. The excess sludge from ASP is added to low rate anaerobic reactor system for anaerobic digestion. The performance data are provided in Table 11.7.

Table 11.6 : Performance Data of Low Rate Anaerobic Reactor

Sr. No.	*Parameter*	*Value*	
1.	Reactor height, m	5	
2.	Reactor diameter, m	10	
3.	HRT, d	10	
4.	Temperature, °C	15	
5.	CH_4 yield, m^3/kg COD	0.30	
6.	*Loading, kg COD/m^3.d*	*Per cent Removal*	
		COD	*TSS*
	0.09	62	85
	0.20	71	85
	0.38	72	80
	0.65	59	88
	0.84 (Includes RAS)*	69	94
	1.03 (do)	70	93
	1.10 (do)	70	83

* RAS – addition of return activated sludge (aerobic)

11.2 Case Study – Anaerobic Treatment for Pulp Paper Wastewaters[3]

Basic research completed in Scandinavia suggests that anaerobic processes will provide enhanced degradation of chlorinated organics from bleach plants as well as improved removal of sulphur compounds from pulping and evaporators condensates. General information required for implementation of anaerobic treatment in pulp and paper industry is as follows :

- General suitability of process effluents (strength, flow, temperature and suspended solids).
- Biodegradability of wastewater organics (treatment efficiency, methane production).
- Presence and effects of toxic and inhibitors compounds.
- Process selection, design and economics.

Table 11.7 : Performance Data of Full Scale Plant

Sr. No.	*Parameter*	*Value*		Overall % Removal		
(A) Chemical Stream				*COD*	*BOD*	*TSS*
1.	Flow, mgd	300		—	—	—
2.	Equalization basin mg	1000		—	—	—
3.	Aerobic System* mg	600		—	—	—
4.	Settler, mg	450		90	95	80
(B) Fermentation System						
5.	Flow, mgd	500		—	—	—
6.	2-Anaerobic reactor (parallel), mg	4000	each	65	65	80
7.	Sludge thickening and filter press	NA		—	—	—
8.	Total effluent dischange, mgd	825		—	—	—

* Addition of polymer after ASP

11.2.1 *Suitability, Biodegradability and Toxicity of Pulp and Paper Wastewaters*

The screening program involved chemical characterization using conventional methods as well as biological testing using an anaerobic serum bottle technique[4] which provides a rapid assessment of several anaerobic treatability characteristics such as COD_r, NOD_r, biogas generation and the inhibitory properties of wastewater on anaerobic micro-organisms. The cumulative methane production pattern show no initial lag periods and there is consistent increase in methane production with increasing wastewater concentrations tested. The results show that the non-sulphur pulping liquor should be readily treatable anaerobically without inhibition of the anaerobic biomass. Sulphur based pulping liquor demonstrated an inhibitory effect which may be attributed to the presence of significant quantities of organic and inorganic sulphur. The lag periods were longer. However, once methane production began, the rate of accumulation was similar to that of the control, indicating that acclimation to inhibitory component is possible. The complex nature of methane production curve for this wastewater suggest sequential removal of several biodegradable constituents. Summary of mill and inplant streams for anaerobic treatability screening is given in Table 11.8.

Table 11.8 : Constituents of Pulp and Paper Mill Effluents in Various Sections

Sr. No	*P and P Technology*		*In-Plant Streams*
1.	**Kraft Pulping**		
	• Unbleached kraft	-	Combined mill effluent
	• Bleached kraft	-	Woodroom effluent Weak black liquor Bleach plant caustic extract Bleach plant chlorination filtrate Evaporator and digester condensates
2.	**Sulphite Pulping**		
	• Sulphite and groundwood		Spent liquor
	• Ammonia base sulphite		Combined mill effluent
	• High yield sulphite		Spent liquor
	• Very high yield sulphite		Spent liquor
	• Neutral semi-chemical		Spent liquor
3.	**Mechanical Pulping**		
	• CTMP and NSSC		Combined mill effluent (CME)
	• Peroxide bleached TCMP		-do-
	• CMP		Spent liquor
	• TMP and CTMP		CME
	• TMP		Chip wash effluent Combined mill effluent
4.	**Miscellaneous Pulping**		
	• Non-sulphur pulping		Combined mill effluent White water recycle
	• Masonite hard board		Combined mill effluent

11.2.2 Process Selection and Design

Three processes were studied, *viz.*

- an UASB,
- a fludized bed,
- a hybrid sludge blanket/anaerobic filter.

Several of the pilot plants were operated successfully at organic loading as high as 25 kg COD/m^3.d corresponding to detention period less than one day. The performance summary of these reactor systems[5] is highlighted in Table 11.9.

Table 11.9 : Comparison of Various Anaerobic Treatment Options

Sr. No.	*Reactor*	*COD, mg/L*	*Loading Rate, kg COD/m³.d*	*Per cent Removal*	
				BOD	*COD*
1.	Hybrid	22100	16.2	75	54
2.	UASB	12700	14.5	85	56
3.	UASB	19500	15.3	91	58
4.	Fluidized bed	12700	13.2	81	54
5.	Fluidized bed	18940	16.2	90	58

Among all these reactor UASB reactor system was found to be technically and costwise quite attractive. The economics of the system is given as follows:

- Reactor volume 13,000 m^3 (6-reactors)
- COD loading 130,000 kg/d
- COD loading rate 10 kg COD/m^3.d
- HRT 2 days
- Treatment efficiency 50% COD
  >75% BOD
- Biogas production 20,000 m^3/d
- Capital cost C–$ 6,000,000
  Annual biogas benefits C–$ 900,000
  Annual operating cost C–$ 500,000 (N and P to be supplemented)

The design loading rate of 10 kg COD/m^3.d was reached after one of the operation by the addition of active biomass imported from the Netherland. The loading rate has been increased to 20 kg COD/m^3.d with BOD removal efficiency of about 80 per cent.

11.2.3 Research Issues in the Pulp and Paper Industry

The pulp and paper effluents contain variable amounts of materials that are toxic or inhibitory in a biological treatment plant. The effects of these time-varying factors can be mediated partially by altering process design to include the following :

- equalization,
- recycle and bypass capability,
- operating strategies can also be optimized with on-line monitoring and control that can respond to process stress by manipulating a number of system variables.

The anaerobic-aerobic process conditions required to remove resins and fatty acids, chlorinated organics and total reduced sulphur compounds are being examined. In order to deal effectively with wastewaters from chlorine and peroxide bleacheries, traditional anaerobic treatment will need to be changed to detoxification prior to methane production.

11.3 Case Study - Anaerobic Wastewater Treatment of a Fuel Ethanol Facility[7]

A pilot plant study on a 6m^3 UASB was conducted to determine the efficiency of the system. Twenty one week pilot study demonstrated that the highly variable wastewater generated at ethanol facility could be effectively pre-treated using an anaerobic system. The effluent is being further treated in a two stage pilot trickling filter. The performance of the pilot plant studies is given in Table 11.10 :

Table 11.10 : Results Obtained from UASB Reactor Systems for Fuel Ethanol Effluent

Sr. No.	*Parameter*	*Influent*		*Effluent*	*%Removal*
1.	T-COD, mg/L	3267		874	76
2.	S-COD, mg/L	2889		416	86
3.	T-BOD, mg/L	2441		288	88
4.	S-BOD, mg/L	1910		181	90
5.	TSS, mg/L	414		330	-
6.	2-Stage TF after an UASB, mg/L	-		41 (BOD-1 stage) 19 (BOD-2nd stage)	
7.	Volumetric Loading	9.3	kg T-COD/m^3.d		
8.	HRT	9.4	hr		
9.	Methane Content	83%			
10.	Methane yield	0.33	m^3/kg TCOD^r		
11.	Flow rate	4-20	m^3/d		
12.	Chemical Requirement				
	• Caustic	0.080	kg NaOH/kg T-CODi		
	• Ammonia	0.005	kg N/kg T-CODi		
	• Phosphoric acid	0.001	kg P/kg T-CODi		

11.3.1 Full Scale Plant[5]

The wastewater from various sources is collected in 7600 m^3 surge tank. Wastewater is pumped from surge tants through heat exchanges to cool the water to an acceptable temperature and then entered the suction side of the recycle pumps of first stage TF. After biological treatment in first stage and intermediated clarification, transfer pumps carry the water to the second stage TF. Second stage effluent enters the final clarifier and is discharged into the Ohio river.

From pilot plant study it was known that additional equalization, more reliable temperature control and improved pH control would be required for trouble free operation of the full scale facility. In addition to UASB, several other unit processes were added to WTP area, *viz.* :

- gas holder;
- equalization basin (7600m^3);
- primary clarifier was added between surge and equalization tank to reduce oil and grease and incoming solids;
- a 205m^3 surplus tank was added to store an on-site supply of acclimated sludge (re-seeding of UASB if need arises);
- a sludge thickner (primary, secondary intermediate clarification sludge and excess sludge from UASB were concentrated before dewatering); and
- UASB reactor consists of 47m^3 conditioning tank and 2050m^3 UASB. These are inserted between equalization basin and Ist stage TF.

The performance data and the cost-benefit analysis are given in Table 11.11.

Table 11.11 : Comparison of Pilot Plant and Full Scale Unit

Sr. No.	*Parameter*	*Pilot Plant*	*Full Scale*
1.	T-CODi, mg/L	3627	5655
2.	T-CODe, mg/L	847	646
3.	% Removal	76	89
4.	S-CODin, mg/L	2889	5348
5.	S-CODe, mg/L	419	2819
6.	% Removal	86	95
7.	Volumetric load, kg T-COD/^{3}d	9.3	12.3
8.	HRT, h	9.4	11.0
9.	Biogas yield, m^3/kg CODr	0.33	NA
10.	Dollars/kg S-COD	-	$0.16
11.	Cost category	Cost Contribution	
	• Capital cost depreciation	-	60%
	• Chemicals	-	15.5%
	• Operation and supervision	-	13%
	• Maintenance		6%
	• Lab and analysis facilities		4%
	• Utilities		1.5%

11.4 Case Study – Treatment of Variety of Substrate[9]

11.4.1 Whey Wastewater (USA)

The raw effluent of the milk plant consists of whey mixed with wash and cleaning water. The composition of whey and the total waste flow is given in Table 11.12.

Table 11.12 : Composition of Whey Wastewater

Sr. No.	*Parameter*	*Whey*	*Total flow*
1.	Flow, m^3/d	200	420
2.	COD, g/L	55	12
3.	BOD, g/L	33	19
4.	TKN, mg/L	1100	580
5.	T-Phosphorus, mg/L	650	310
6.	pH	4.5-5.5	5.0-10.6
7.	Protein, %	0.6-0.9	-
8.	Lactose, %	4-5	-
9.	Fat, %	0.04	-

This wastewater has excellent characteristics for anaerobic option and, therefore, UASB reactor system has been in operation. The performance data obtained from the UASB system is given in Table 11.13.

Table 11.13 : Performance Data of UASB System for Whey Wastewater

Sr. No	*Parameter*	*Influent*	*Effluent*	*% Reduction*
1.	COD, mg/L	33,000	5,600	86
2.	BOD, mg/L	14,500	680	96
3.	Other Details			
	• HRT, hr	16		
	• Biogas production	0.50, m^3/kg COD		
	• Biogas composition	64% CH_4		
	• Sludge settling	25-54 m/h		
	• Overall BOD_r, %	99 (included anaerobic + aerobic RBC System)		

The treatment scheme is as follows :

- Whey and other wastewater are combined together in a mixing tank (lime addition if required).
- An UASB reactor system (heat exchanger outside if required).
- Gas can be flared or used for power generation.
- Effluent for an UASB is routed through RBC system (lime addition if required).
- Settling tank followed by RBC.
- A storage tank for anaerobic UASB sludge (excess-sludge stored and used whenever there are upsets).

11.4.2 Wheat Industries (Ireland)

The characteristics of wheat starch wastewater are indicated in Table 11.14.

Table 11.14 : Characteristics of Wheat Starch Wastewater

Sr. No.	*Parameter*	*Value*
1.	Flow, m^3/L	35
2.	COD, g/L	20.2
3.	BOD, g/L	12.2
4.	SS, g/L	1.5-3.0
5.	TS, g/L	10-15
6.	TKN, mg/L	45
7.	Phosphorus (P), mg/L	4
8.	Temperature, °C	20
9.	pH	5.0-6.5

An UASB system was selected and the complete treatment system scheme is identified as follows :

- treatment with stream if required;
- equalization tank (NaOH added if needed);
- an UASB reactor with gas holders, flow system and steam generation unit; and
- equalization tank for mixing the effluent from UASB and other process wastewater before final discharge.

The loading rate of about 7.5 kg $COD/m^3.d$ has been selected because the wastewater contains large amounts of suspended solids. Only a minimal of $150m^3$ of sludge was imported from Holland to start up this $2400m^3$-plant. The addition of trace elements helped in increasing the efficiency of the UASB system.

11.4.3 Maize Starch Industry (The Netherland)

The maize factory produces starch, glucose and other starch derivatives : The composition of wastewater from various sources is given in Table 11.15.

Table 11.15 : Characteristics of Maize Starch Industry

Sr. No.	*Parameter*	*Starch*	*Glucose*	*Derivates*	*Total*
1.	Flow, m^3/h	40	5	4	40-50
2.	COD, g/L	9	1.5	30	5-26
3.	TKN, mg/L	-	-	-	50
4.	P-total, mg/L	-	-	-	2
5.	pH	-	-	-	5-6

As the production is batchwise, large fluctuation in salt, COD and sulphite concentrations may occur. Therefore, equalization basin is necessary. The flow scheme is as follows :

- Equalization basin with a mixer (dosing of N and P and also lime whenever required).

- UASB system with effluent passing through a cascade aerator. Provision has been made for gas storage, flare and steam generation.
- The effluent from passing through cascade reactor is let into a settling basin before final discharge.
- There is a storage tank for maintaining excess UASB sludge generated (this is used whenever there are upsets in the treatment system).

The system is working at present at a loading rate of 11kg/m^3.d and the COD removal efficiency is about 90-95 per cent. During peak loads (lasting 2-3 hours), the loading rates can be as high as 30 kg COD/m^3.d The amount of biogas produced is more than 8,00,000 m^3/year with a methane content of about 70 per cent. The treatment plant is shut down during weekends. When the reactor is started up again, the plant is at full capacity, in a few hours the gas production rate reaches its maximum (50m^3/h).

11.4.4 Leachate Treatment (The Netherland)

The quality of leachate varies depending on the age of the waste through which it percolates. In relatively, young landfill sites, the leachate can be highly polluted with low-molecular weight organic acids and by ammonia, chloride, sulphate and heavy metals. After a period of 2-4 year the metabolic activity of the landfill itself will increase, resulting a drop of COD of the leachate from 25,000 to 2000/ 5000 mg/L. The COD/BOD ratio increases and the organic pollutants now mainly consist of long chain carbohydrates which are difficult to degrade. For treatment of this wastewater, non biological processes such as reverse osmosis are suitable. The reactor (UASB) was operated at an average loading rate of 7.3 kg COD/ m^3.d and at a temperature of 27°C. The biogas produced contained 85 per cent methane. The characteristics of raw and treated effluent are given in Table 11.16.

Table 11.16 : Characteristics of Wastewater and Performance of UASB System

Sr. No.	*Parameter, mg/L*	*Raw Influent*	*Treated Effluent*	*% Reduction*
1.	COD	4600	720	-
	(COD/BOD-6 to 9)	-	-	81
2.	TKN	190	175	13
3.	Fatty acids	1930	380	77
4.	P-Total	6.8	1.6	67
5.	**Heavy Metals**			
	• Cr	0.13	0.06	63
	• Ni	0.12	0.05	59
	• Cu	0.08	0.06	54
	• Zn	1.14	0.19	80
	• Pb	0.02	0.006	67
	• Cd	0.004	0.003	67
	• Fe	155	36	69

At Baval, the leachate is collected in or drainage system over a water-imperable foil. The leachate is re-circulated over the solid waste. A two stage wastewater treatment has been designed incorporating a UASB pretreatment and a post treatment according to reverse osmosis process. The wastewater characteristics and discharge requirement at Baval are given in Table 11.17.

Table 11.17 : Performance of UASB System for Leachate Treatment

Sr. No.	*Parameter, mg/L*	*Leachate (young)*	*Discharge Requirements,*	*Required % Reduction*
1.	T-COD	26,000	75	99.7
2.	TKN	1330	4	99.6
3.	BOD	-	5	-
4.	NH-N	1300	3	99.8
5.	Fatty acid (COD)	18000	-	-
6.	pH	6.6-6.8	-	-
7.	SS	100	-	-

A $60m^3$ UASB reactor with reverse osmosis ($2m^3/h$) facility was installed. The UASB reactor operates at a loading rate of 15 kg $COD/m^3.d$. The 80-96 per cent COD reduction resulted in $400m^3$ of biogas with 72 per cent CH_4 content. The performance of the combined treatment scheme is given in Table 11.18.

Table 11.18 : Performance of UASB System and Reverse Osmosis for Leachate

Sr. No.	*Parameter, mg/L*	*Leachate*	*After UASB*	*After Reverse Osmosis*
1.	COD	26,000	4,000	6
2.	TKN	1,330	1270	15
3.	NH_4^+-N	1,300	1,260	13
4.	Fatty acid(COD)	19000	2,300	-
5.	SO_4^{-2}	380	97	-
6.	**Heavy Metals**			
	• Cd	0.02	0.004	<0.1
	• Cu	0.03	0.020	<0.1
	• Hg	0.02	0.002	<0.1
	• Pb	0.11	0.060	<0.1

11.45 Liquorice Factory (The Netherland)

The design of the full scale plant was based on limited laboratory research as the company did not want to have pilot studies. The characteristics of liquorice are given in Table 11.19.

Table 11.19 : Characteristics of Liquorice Syrup

Sr. No.	*Parameter, mg/L*	*Liquorice Syrup*	*Remarks*
1.	Flow, m^3/d	50	
2.	COD	12,000	
3.	BOD	5800	
4.	TKN	260	
5.	Protein, per cent	3.9	The protein contributes
6.	Starch, per cent	31.0	15-13 per cent of total COD
7.	Sugar, per cent	31	

To prevent coagulation of the proteins in the system, the residence time of wastewater in the sewerage system was reduced from an average of 3 hours to 30 minutes. It is generally known that protein causes problems relating to sludge flotation and build up of scum layers in UASB. The three phase separator was designed to counter these problems. The wastewater is produced in an eight-hour day, five days a week but since continuous operation increases the stability of the anaerobic process and reduces the size of the required reactor, a buffer tank of $120m^3$ was incorporated. Additional advantages of this buffer tank are :

- Reduction of final discharge taxes because of discharge is continuous.
- Stable reactor performance due to buffering of large fluctuations in COD load brought about, for instance by cleaning activities.

32 per cent of COD could be converted mainly into acetic acid and propionic acid by acidifying bacteria. Caustic dosing was needed to control pH in this buffer tank. By recirculating the anaerobic effluent over the buffer tank, a considerable reduction of this dosing can be achieved. The treatment scheme is as follows :

- Buffering tank with a mixer (equalization basin, $120m^3$ with NaOH dosing facility).
- UASB, $50m^3$.
- Recirculation of treated effluent for UASB into Buffering tank.
- Storage tank for storing UASB sludge.
- Supply of certain trace metals becomes necessary after some time (6 months period).

11.5 Case Study – Yeast Fermentation Wastewater (UA)[10]

The anaerobic system (UASB) is used as a primary step in the treatment of wastewater from several different fermentation streams. At one facility the anaerobic effluent is discharged directly into municipal sewers (Dixie Yeast Factory), while at the other facility the anaerobic effluent is first polished for ammonia removal in a secondary activated sludge system and then discharged into municipality (Bush industrial Products). The performance summary of the treatment of these two plants is highlighted in Table 11.20.

Table 11.20 : Comparison of Design and Operating Data for UASB System

		Plant			
		Dixie Yeast		*Bush Industrial Products*	
Sr. No.	*Parameter*	*Design*	*Operating*	*Design*	*Operating*
1.	Digester volume, m^3	1750	1750	5000	5000
2.	Flow, mgd	0.23	0.32	0.40	0.50
3.	COD_i, mg/L	21000	15000	33500	25000
4.	CODr per cent	62	65	60	65
5.	BODr, per cent	85	95	80	85
6.	Methane content, per cent	70	75	70	65
7.	Loading rate, kg COD/m^3.d	10.9	10.7	10.3	9.5
8.	Biogas production, m^3/d	5068	5097	14600	16000
9.	Methane yield, m^3/kg COD	0.30	0.32	0.33	0.34

11.6 Case Study – Coal Conversion Wastewaters[11]

This investigation addresses to attempt development of an operating strategy and basis of an anaerobic sequencing batch reactor (AnSBR) biological treatment of coal conversion wastewaters. These anaerobic reactors operated in a fill and draw mode. The reactor cycle begins with fill, a period of time when wastewater is pumped into the reactor which contains the biomass from previous cycles. Mixing is provided to promote contact between the biomass and influent organics. At the end of reaction (RECT) mixing is discontinued and settling of the biomass is provided. The treated supernatant is removed during draw. Idle is the period between the end of draw and the beginning of the next fill period.

The screening tests were performed using serum bottle technique. The three serum bottles were each filled with an anaerobic sludge (collected from different sources including a mixture of sludges collected from 10 different municipal anaerobic digesters), a vitamin and mineral solution and one of the coal conversion wastewater constituents. The gas production was monitored and compared to serum bottle controls. The results were recorded after 60-90 days of operation and are highlighted in Table 11.21.

Six-2L AnSBR were being monitored. These have been started using different sludge source mixtures and start up policies. The reactors were fed with a synthetically prepared coal conversion wastewater (CCW). The composition of wastewater is given in Table 11.22. Data on acclimation studies from six different reactors is presented in Table 11.23.

Table 11.21 : Characteristics of Coal Conversion Wastewater

Sr.No.	*Constituents*	*No. of Triplicates*	*Initial Concentration mg/L (Range)*	*Days to Complete Degradation (Range)*	*Possible Inhibition at Conc., mg/L*
1.	Phenol	8	200-1000	10-40	1000
2.	*o*-Cresol	6	200-500	*	200
3.	*m*-Cresol	7	100-500	20-70	-
4.	*p*-Cresol	5	200-500	10-19	-
5.	Aniline	7	100-250	*	-
6.	2,3-Xylenol	7	100-250	*	200
7.	2,4-Xylenol	2	100-250	*	250
8.	2,6-Xylenol	2	100-250	*	250
9.	3,4-Xylenol	7	100-250	*	200
10.	3,5-Xylenol	7	100-250	*	200
11.	Catechol	4	200-1000	45-66	1000
12.	Resorcinol	4	200-1000	24-50	-
13.	Hexanoic acid	2	250-500	<1	-
14.	Benzoic acid	3	250-500	5-7	-
15.	Pyridene	2	150-250	28	-
16.	Guaiacol	2	150-250	14-20	-
17.	4-Methyl catechol	2	150-250	45-48	-
18.	2,3,5 Trimethyl Phenol	3	100-1000	**	250
19.	2,4,6 Trimethyl Phenol	2	100-250	**	250
20.	3,4,5 Trimethyl Phenol	2	100-250	**	250

* - none in 90 days; ** - none after 60 days

There has been no evidence of degradation of aniline or xylenols. Cresols are being degraded, but at different rates and after different periods of time.

11.7 Case Study—Fludized Bed Reactor for Different Types of Wastewater[12]

The fluidized bed reactor has been used for treatment of various industrial wastewaters using sand as a support media for biomass attachment. The correlation between pilot plant and full scale performance on Soya Processing wastewater (Grain Processing Corporation) is summarised in Table 11.24.

Table 11.22 : Composition of Wastewater for An-SBR System

Sr. No.	*Constituent*	*Concentration, mg/L*	*Equivalent TOC, mg/L*
1.	Phenol	40000	3060
2.	*o*-Cresol	500	385
3.	*p*-Cresol	500	385
4.	p-Cresol	500	385
5.	Aniline	50	39
6.	3,4-Xylenol	50	39
7.	3,5-Xylenol	50	39
8.	Acetic Acid	400	160
9.	Propionic acid	125	61
10.	*n*-Butyric acid	50	28
11.	Total TOC		4581

Table 11.23 : Acclimation Studies Data Obtained from 6 An-SBR System

Sr. No.	*Operating Parameter*	*Reactor (R)*					
		R_1	R_2	R_3	R_4	R_5	R_6
1.	Influent TOC, mg/L	5.1	4.9	1.1	2.3	2.3	4.6
2.	Weekend feed, ml	100	75	100	75	75	38
3.	Source of TOC						
	% CCW	22	46	100	100	100	100
	% Glucose	78	54	–	–	–	–
	H_2O dilution	*	*	1:4	1:2	1.2	*
4.	Reactor LSS, g/L	8.7	6.3	2.5	2.9	3.8	4.1

* - none

Table 11.24 : Comparison of Pilot Plant and Full Scale Units

Sr. No.	*Parameter*	*Pilot Plant*	*Full Scale*
1.	Mean Volumetric Loading, kg COD/m^3.d	13.0	15.6
2.	Mean Feed Values, mg/L		
	COD	10,914	10,556
	BOD	6,556	7,960
3.	Removal Efficiencies, %		
	COD	-	76
	BOD	83	83

Performance data on high volumetric loadings on pulp and paper wastewater is given in Table 11.25.

Table 11.25 : Performance Data for Fluidized Bed Reactor for Pulp and Paper Wastewater

Sr. No.	*Parameter*	*Test Period (Days)*		
		97-124	*168-179*	*193-198*
1.	Loading rate, kg COD/m^3.d	24	35	102
2.	HRT, days	0.52	0.29	0.10
3.	% COD_r	92	86	69
4.	% BOD_r	93	81	74
5.	Gas			
	• Biogas yield, m^3/kg COD_r	0.48	0.50	0.46
	• % CH_4	69	69	62
6.	Influent, mg/L			
	• COD	14,500	13,900	13,300
	• BOD	10,800	11,100	11,400
	• VOA	9,410	9,296	8,600
	• SS	73	29	106
7.	Effluent, mg/L			
	• COD	1,150	1,900	4,180
	• BOD	740	1,840	3,650
	• VOA	820	1,120	2,394
	• SS	140	124	147

The performance data for two phase fluidized bed for corn processing wastewater is summarized in Table 11.26.

Table 11.26 : Performance Data for Two Phase Fluidized Bed for Corn Processing

Sr. No.	*Parameter*	*Corn Processing Wastewater*		*Corn Processing Wastewater With Waste Activated Sludge*	
		I	*II*	*I*	*II*
1.	Loading Rate, kg COD/m^3.d	16.0	18.7	20.2	25.4
2.	Feed values, mg/L				
	• COD	8840	9310	10193	8430
	• BOD	5770	5196	6211	5307
3.	Effluent values, mg/L				
	• S-COD	772	1072	1077	1163
	• S-BOD	397	504	673	903
4.	% Removal				
	COD	91.3	88.5	89.4	86.2
	BOD	93.1	91.7	89.2	83.4

11.8 Membrane Anaerobic Reactor System (MARS)[12]

A suspended growth or contact reactor is coupled with ultrafiltration membrane models which serve to completely retain system biomass and provide an effluent that is essentially free of suspended solids. Candidate waters included industrial streams with COD greater than 20,000 mg/L (such as cheese whey or permeate) and low flow wastewaters where conventional aerobic technology is being considered. Performance data of this system is given in Table 11.27.

Table 11.27 : Performance Data for Membrane Anaerobic Reactor System for Various Types of Wastewater

Sr. No.	*Waste (Location)*	*Influent, mg/L*			*Effluent, mg/L*		
		COD	*BOD*	*TSS*	*COD*	*BOD*	*TSS*
1.	Wheat Starch (Ontario)	11,240	4,200	2,267	194	30	<10
2.	Wheat Starch (Minnesota)	35,175	15,463	13,300	270	74	<10
3.	Sweet Whey Permeate (Minnesota)	59,790	28,417	512	305	-	<10
4.	Acid Whey Permeate (New York)	55,107	23,502	107	457	27	<10

11.9 Case Study–Thermal Conditioning Liquors[13]

The thermal conditioning liquors (TCL) are generated as a by-product of the sludge dewatering operations. As part of the thermal conditioning process, biological cells in the sludge are hydrolysed and oxidized. A significant portion of the particular solids matter is converted into dissolved solids. After the conditioned solids have been separated from the liquid by settling and dewatering, the liquid phase contains high concentration of pollutants.

Because of the nature of the waste, and because of its obvious potential for anaerobic treatment, the Region of Peel and the Ontario Ministry of the Environment Commissioned Gore and Storrie Limited to conduct a laboratory scale investigation into anaerobic biological treatment of TCLs generated at Lakeview. Subsequent studies tested a variety of anaerobic processes and configurations including sludge blanket, fluidized bed and filter reactors. None of the anaerobic processes available were found to be suitable. Therefore, a hybrid configuration reactor combining anaerobic suspended growth and fixed film filter technology was found suitable. Such a design configuration has never been researched before or used elsewhere. The concentrations of different constituents of wastewater component are not given. However laboratory studies revealed that it was possible to remove 70 per cent of S-COD, and 80 per cent of S-BOD in the hybrid reactor. Reactor loadings were above 6.3 kg COD/m^3.d with a HRT of about 38 hours.

11.9.1 Full Scale Application

The hybrid process has consistently removed 80-85% of BOD pollutant load from waste stream. At this rate plant treatment cost was reduced by $ 150,000 per year. Additional benefits from biogas has been estimated to be around $ 310,000 per year.

The reactor facility was constructed by modifying two existing 27m diameter digesters. The liquid volume of each reactor is 2910m^3. The $1.15 million cost included all related constructions and control building rehabilitation. It is a compact reactor system. The bottom section of the reactor acts as a suspended growth zone. The top portion of the reactor was changed into filter zone. The biological growth and conversion takes place in the suspended growth zone, producing gas and biomass solids. Equalized feed distribution and mixing of the suspended growth are achieved by recirculation of the lower zone contents. Recirculation flow rate adjustment can be made without causing solids to rise in the filter zone where undesirable accumulation would occur. The upper reactor is based on anaerobic filter technology to take advantage of its high performance and resistance to shock loads. Gas rising from lower zone provides vertical mixing between the two reactor zones. The agitating action of the rising gas dislodges and returns solids to the zone, keeping the filter media clear.

The facility has consistently achieved 72 per cent COD and 80 per cent BOD reductions, while producing between 10,000 to 14,000 m^3 of fuel gas per day. The economics of aerobic and anaerobic (hybrid) system is given as follows :

- Annual electrical power costs to improve waste stream to discharge quality

- by aerobic system only	$	190,000
- by anaerobic + hybrid system	$	40,000
• Power cost savings	$	150,000
• Value of CH_4 produced	$	310,000
• Annual cost savings	$	460,000
• Hybrid construction cost	$	1,500,000
• Pay back period		39 months

The hybrid process was developed specifically for the treatment of TCL where solids growth, accumulation and precipitation during anaerobic treatment were of critical importance and concern. This process can be applied to any high strength wastewater, *e.g.*, landfill leachate, pharmaceutical, dairy, food processing, textile, petrochemicals and pulp and paper wastes. This process is useful where there is a possibility of high accumulation of inerth solids in fixed film reactor zone. In addition, this process can be applied where upflow sludge blanket process has had difficulty in maintaining biomass in the reactor. This process does not require a separate settling tank to capture and return biomass to the reactor.

11.10 Case Study—Celboric Process for Different Types of Industrial Wastewater[14]

This is an upflow randomly packed anaerobic bioreactor which has high process reliability and stability. Its longterm continuous performance relies on the ability to measure and control the quantity of solids that remain in the reactor. The company identified tracer chemicals which are extremely bio-refractory, but donot interact in any way with the micro-organisms in the reactor. Using one of these tracers, the volume within the reactor that is not occupied by solids can be measured accurately, thereby assisting in biomass control. Three full scale installations have been successfully operated in Texas and the performance data is given in Table 11.28.

Table 11.28 : Performance of Three Full Scale Plants at Texas (USA)

		Plant		
Sr.No.	*Parameter*	*I*	*II*	*III*
1.	Location	Vernon	Bishop	Pampa
2.	Major feedstocks	Guar beans	Natural gas, Propylene	Butane gas
3.	Major products	Water soluble polymers	Methanol, Formaldehyde Acetate esters Butylene glycol, Pentaery thritol Engineering platters	Acetic acid Methyl formate, Acetic anhybride, Ethyl Acetate Propionic acid, Methyl ethyl ketone, Formic acid Methyl and Ethyl acrylate
4.	Wastewater load	COD-20,000kg/d Flow-0.3mgd	50,000kg/d 1.0mgd	60.000kg/d 1.0mg
5.	System performance			
	• Equalization, mg	0.2	1.5 day	1.0 day
	• Reactor, mg	0.3	0.5	0.5mg
	• Loading, kgCOD/m^3.d	10.0	10.0	11.0
	• % CODr	65.0	80	90
	• CH_4 value, $/year	0.15×10^6	0.4×10^6	0.65×10^6

11.11 Case Study—Hospital Wastewater[15]

Biogas recovery from high as well as low strength wastewaters through anaerobic digestion is potentially one of the most attractive methods of solving the twin problem of energy production and pollution control resulting in cost-effective wastewater treatment package. Compared to conventional anaerobic treatment where the mean cell residence time (MCRT) is rarely greater than twice the hydraulic retention time (HRT), fixed film processes achieve MCRT of the order of 10 to 100 times the HRT. The research project funded by Ministry of Non-Conventional Energy Sources (MNES), N. Delhi to National Environmental Engineering Research Institute (NEERI), Nagpur has resulted in the demonstration of fixed film reactor technology for biomethanation of hospital wastewater.

11.11.1 Hospital Wastewater Characteristics

The wastewater is primarily constituted by sewage generated within the hospital and includes discharges from all sections of the hospital *viz.*, laboratories, operation theatres, kitchen, etc., and the characteristics of combined wastewater is given in Table 11.29.

Table 11.29 : Characteristics of Hospital Wastewater

		*Values**	
Sr. No.	*Parameter*	*Range*	*Average*
1.	pH	6-7	6.5
2.	Total suspended solids	200-400	230
3.	Volatile suspended solids	50-200	100
4.	COD	200-500	300
5.	BOD	100-250	150

* All values except pH are in mg/L

11.11.2 Process Development

Anaerobic fixed film process for biomethanation of hospital wastewater has been investigated on bench scale at NEERI, Nagpur for a wastewater flow of 80 liters/day. A pilot plant for the treatment of 50 Cum/day wastewater has been set up at Government Medical College, Nagpur. Based on these investigations, the design criteria for biogas generation from hospital wastewater have been evolved. Biochemical oxygen demand (BOD) reduction of 90-95 per cent, chemical oxygen demand (COD) reduction of 80-90 per cent and gas yield of 0.3 Cum/kg COD utilized have been achieved. The biogas contains about 60-65 per cent methane content. The reduction of pathogens in the system ranges between 94-98%. The effluent BOD is invariably less than 20 mg/L and, hence, the treated effluent, after disinfection, can be discharged directly on land or in receiving water bodies. The cost-benefit analysis for the treatment of 100, 500 and 1000 cum hospital wastewater/day is presented in Table 11.30.

Table 11.30 : Cost Benefit Analysis for Fixed Film Reactor Treating Hospital Wastewater

Sr. No.	*Plant Capacity (Cum/day)*	*Capital Cost (Rs.)*	*Annual Operation and Maintenance Cost (Rs.)*	*Annual Benefits, (Rs.)*
1.	100	4.5×10^5	1.35×10^5	0.08×10^5
2.	500	7.0×10^5	1.85×10^5	0.38×10^5
3.	1000	13.5×10^5	2.5×10^5	0.75×10^5

The details about the support media and pilot plant are given in Table 11.31.

Table 11.31 : Characteristics of Support Media

		Support Media		
Sr.No.	*Parameter*	*Nylon Rope*	*PVC Pieces*	*Brick Granules*
1.	Porosity, %	95	85	50
2.	Particle size, mm	0.3 720 length	10×10	4.76-9.51
3.	Bulk density, kg/m^3	-	200	900-950
4.	Surface area m^2/m^3 reactor vol.	250	1200	150

The design details of the fixed film reactor are given in Table 11.32.

Table 11.32 : Design Details of Pilot Plant

Sr. No.	*Name of the unit*	*Design HRT (hr)*	*Volume Required (m^3)*	*No. of Units and Size (mm)*	*Type of Construction*	*Platform and size in mm*
1.	Sump well	-	10.0	1 No 300 dia × 1500 ht	RCC and bricks work	-
2.	Wastewater storage tank	2.5	20.0	2 Nos. 2500 × 2000 × 2000 ht	MS	1 No. 5600 × 3250 × 300 ht
3.	Fixed film reactor	10.0	10.0 (each unit)	2 Nos. 2800 dia × 3823 ht	MS	2 Nos. 3250 dia × 300 ht
4.	Gas holder	24.0	5.0	1 No. 2500 dia × 1500 ht	MS/RCC and brick work	-

The performance data obtained in the laboratory and from the pilot plant are highlighted in Tables 11.33-11.38.

Table 11.33 : Fixed Film Reactor Studies : Laboratory Results

Reactor : Brick Granules

Sr. No.	HRT (hrs)	OLR (kg COD/ m^3/d)	TVA (mg/L)	Alkalinity	TVA/ Alk	COD (mg/L) Inf.	COD (mg/L) Eff.	COD (mg/L) % Red.	BOD (mg/L) Inf.	BOD (mg/L) Eff.	BOD (mg/L) % Red	Gas Yield (m^3 / kg CODr)
1.	24	0.3	25	230	0.109	300	40	86.7	–	–	–	0.192
2	12	0.6	20	194	0.103	300	45	85.0	143	9.5	94.8	0.172
3.	8	2.21	28	198	0.141	738	62	91.6	249	14.0	94.4	0.157
4.	6	2.67	47	251	0.187	667	151	77.4	384	31.0	91.9	-
5.	3.3	4.08	32	235	0.140	564	93	83.5	326	31.0	90.5	0.221
6.	2.5	4.00	24	245	0.100	412	64	84.5	209	18.0	91.4	0.265
7.	2.0	4.14	29	253	0.110	383	69	82.0	129	19.0	85.1	0.332

Table 11.34 : Fixed Film Reactor Studies : Laboratory Results

Reactor : PVC Pieces

Sr. No.	HRT (hrs)	OLR (kg COD/ m^3/d)	TVA (mg/L)	Alkalinity	TVA/ Alk	COD (mg/L) Inf.	COD (mg/L) Eff.	COD (mg/L) % Red	BOD (mg/L) Inf.	BOD (mg/L) Eff.	BOD (mg/L) % Red.	Gas Yield (m^3 / kg CODr)
1.	24	0.3	25	240	0.104	300	48	84.0	–	–	–	0.210
2.	12	0.6	20	198	0.101	30	37	87.7	143	8.3	94.2	0.245
3.	8	2.21	30	217	0.138	738	89	87.9	249	22.0	91.2	0.210
4.	6	2.67	50	229	0.128	667	103	84.6	384	34.0	91.2	0.126
5.	4.5	3.00	32	235	0.140	564	76	86.5	314	29.0	90.8	0.255
6.	3.4	2.91	24	245	0.100	412	59	85.7	209	19.3	90.8	0.256
7.	3.0	3.04	29	253	0.110	383	61	84.1	129	12.6	90.2	0.280

Table 11.35 : Fixed Film Reactor Studies : Pilot Plant Results

Reactor : Corrugated Plastic Rings

Sr. No.	Flow m^3/d	HRT (days) O*	HRT (days) O	OLR (kg COD m^3/d	pH Inf.	pH Eff.	COD (mg/L) Inf.	COD (mg/L) Eff.	COD (mg/L) % Red.	BOD (mg/L) Inf.	BOD (mg/L) Eff.	BOD (mg/L) Red	Gas Yield m^3/kg CODr	Gas Prod. (1/ day)	CO_2 %
1.	0.8	20.1	20.0	0.01	6.6	6.8	596	100	83	195	9	95	-	-	-
2.	1.6	10.4	10.0	0.02	6.5	6.8	392	50	87	157	9	94	0.73	400	8.2
3.	3.2	5.2	5.0	0.04	6.7	6.9	200	39	81	122	7	94	0.82	420	8.8
4.	6.4	2.6	2.5	0.17	6.4	6.6	449	73	84	135	7	95	0.34	810	8.6
5.	8.0	2.1	2.0	0.23	6.3	6.6	480	56	88	147	16	89	0.54	1822	9.6
6.	16.0	1.0	1.0	0.41	6.1	6.4	413	75	82	166	13	92	0.56	3024	8.8
7.	24.0	0.6	0.6	0.65	6.4	6.7	434	88	80	218	23	89	0.59	5104	8.7

* Based on empty bed volume.

Table 11.36 : Fixed Film Reactor Studies : Pilot Plant Results

Reactor : Brick Granules

		HRT (days)			*pH*		*COD (mg/L)*			*BOD (mg/L)*					
Sr. No.	*Flow m^3/d*	*O**	*O*	*OLR (kg COD m^3/d)*	*Inf.*	*Eff.*	*Inf.*	*Eff.*	*% Red.*	*Inf.*	*Eff.*	*Red*	*Gas Yield m^3/kg CODr*	*Gas Prod. (1/day)*	*CO_2 %*
1.	0.8	20.1	12	0.01	6.9	7.1	239	87	63	87	22	75	-	-	-
2.	1.6	10.4	6	0.02	6.9	1.3	167	62	63	70	12	83	-	-	-
3.	3.2	5.2	3	0.04	6.7	7.1	208	51	76	128	9	93	-	-	-
4.	6.4	2.6	1.5	0.17	6.4	6.6	449	102	77	75	10	87	-	-	-
5.	8.0	2.1	1.2	0.23	6.3	6.6	480	95	80	100	16	84	0.40	1232	8.5
6.	16.0	1.0	0.6	0.41	6.1	6.3	413	94	77	166	21	87	0.47	2424	9.3

* Based on empty bed volume.

Table 11.37 : Fixed Film Reactor Studies : Pilot Plant Results

Reactor : Brick Granules

		Total Coliforms			*Faecal Coliform*			*Faecal streptococci*		
Sr. No.	*HRT (Days)*	*Inf.*	*Eff.*	*% Red.*	*Ind.*	*Eff.*	*% Red.*	*Inf.*	*Eff.*	*% Red.*
1.	6.0	6.3×10^7	6.4×10^5	98.97	2.6×10^7	4.0×10^5	98.40	5.3×10^4	2.5×10^3	95.20
2.	3.0	3.5×10^8	7.8×10^6	97.74	1.3×10^7	3.9×10^5	97.69	9.1×10^4	4.6×10^3	95.00
3.	1.5	1.2×10^8	4.0×10^6	96.66	2.8×10^7	7.7×10^5	97.60	2.1×10^5	1.0×10^4	95.00
4.	1.2	1.3×10^8	5.0×10^6	96.15	7.5×10^6	2.0×10^5	97.20	1.5×10^5	9.3×10^3	93.09
5.	0.6	2.3×10^7	9.2×10^5	96.08	2.1×10^7	7.0×10^5	97.10	8.7×10^5	6.6×10^4	92.92
6.	0.4	2.5×10^7	1.2×10^6	95.20	2.8×10^7	1.2×10^6	96.25	8.0×10^5	6.0×10^4	92.50

Table 11.38 : Fixed Film Reactor Studies : Pilot Plant Results

Reactor : Brick Granules

		V. cholerae			*A. aeruginosa*			*S. aureus*		
Sr. No.	*HRT (Days)*	*Inf.*	*Eff.*	*% Red.*	*Ind.*	*Eff.*	*% Red.*	*Inf.*	*Eff.*	*% Red.*
1.	6.0	8.8×10^5	9.7×10^5	98.90	8.4×10^4	4.0×10^3	95.75	5.2×10^2	1.8×10^1	96.29
2.	3.0	7.4×10^5	1.0×10^4	98.60	1.3×10^4	5.0×10^2	95.41	1.0×10^3	4.1×10^1	95.80
3.	1.5	5.5×10^5	1.7×10^4	96.90	3.6×10^3	1.6×10^2	95.50	2.2×10^3	1.0×10^2	95.00
4.	1.2	4.5×10^4	1.8×10^3	96.00	2.2×10^3	1.1×10^2	95.00	1.6×10^3	8.3×10^1	94.70
5.	0.6	7.8×10^5	3.7×10^4	95.25	1.9×10^3	1.1×10^2	94.73	1.6×10^3	8.6×10^1	94.50
6.	0.4	7.2×10^5	3.6×10^4	95.00	1.5×10^3	1.0×10^2	94.00	1.4×10^3	8.2×10^1	94.10

11.11.3 Salient Features of the Technology

- The process ensures complete treatment and no secondary treatment, except disinfection, is required.
- The process is cost-effective as compared to the aerobic process involving high energy input.
- Biogas is generated as a by-product which can be used as a fuel or for generating electricity.
- The operation of the system is simple, and the operation and maintenance costs are lower compared to aerobic treatment.

11.12 Case Study—Distillery Spentwash Wastewater[16]

Biogas recovery from high as well as low strength wastewaters through anaerobic digestion is potentially one of the most attractive methods of solving the twin problem of energy production and pollution control resulting in cost-effective wastewater treatment package. Compared to conventional anaerobic treatment where the mean cell residence time (MCRT) is rarely greater than twich the hydraulic retention time (HRT), fixed film processes achieve MCRT of the order of 10 to 100 times the HRT. The research project funded by Ministry of Non-conventional Energy Sources (MNES), New Delhi to National Environmental Engineering Research Institute (NEERI), Nagpur has resulted in the development of fixed film reactor technology for biomethanation of distillery spentwash.

11.12.1 Distillery Spentwash Characteristics

With major constituents of carbohydrates and minerals, distillery spentwash from sugarcane molasses fermentation is characterize by high biochemical oxygen

Table 11.39 : Characteristics of Molasses-Based Distillery Spentwash

Sr. No	*Characteristics**	*Range*	*Average*
1.	pH	4.1-4.5	4.3
2.	Temperature	70-80	75.0
3.	Total Solids	60-80	72.0
4.	Volatile Solids	40.2-53.4	44.6
5.	Suspended Solids	5.5-9.4	7.3
6.	TVA as acetic acid	3.1-4.8	4.3
7.	Total Nitrogen	1.2-3.0	2.0
8.	Sodium as Na	1.2-2.5	1.9
9.	Potassium as K	9.1-12.0	11.2
10.	Calcium as Ca	1.0-2.5	1.9
11.	Phosphorus as P	0.043-029	0.15
12.	Sulphate as SO_4	2.5-4.0	3.4
13.	COD	60-80	71.5
14.	BOD	45-55	51.0

* In g/L except for pH and temperature

demand (BOD), chemical oxygen demand (COD), total organic carbon (TOC), low pH, high temperature, high dissolved solids and high concentration of inorganics especially sulphate and potassium, Table 11.39.

11.12.2 Process Development

Anaerobic diphasic fixed film reactor technology for biomethanation of distillery spentwash has been investigated on bench scale at NEERI, Nagpur. A pilot plant of 10 cum/day capacity has been set up at Western Maharashtra Development Corporation Distillery, Chitali, District Ahmednagar, Maharashtra. Based on these investigations, the design criteria for biogas generation from distillery spentwash have been evolved. Chemical oxygen demand (COD) reduction of 65-70 per cent and biogas yield of 0.4-0.45 cum/kg COD utilized have been achieved. The biogas contains 65-70 per cent methane. The cost-benefit analysis for the treatment 450 and 900 cum distillery spentwash/day is detailed in Table 11.40.

Table 11.40 : Cost Benefit Analysis for Biogas Generation From Distillery Spentwash for Various Plant Capacities

Plant Capacity (cum/day)	*Capital Cost (Rs) (Budgetary)*	*Annual Operation and Maintenance Cost (Rs)*	*Annual Benefit (Rs)*
450	150×10^5	42×10^5	84×10^5
900	355×10^5	64×10^5	168×10^5

11.12.3 Salient Features of the Technology Package

- The process is cost-effective as compared to other treatment process.
- Biogas is generated as a by-product which can be used as a fuel in the boiler.
- The coal equivalent of energy generated is 4.0-4.5 MT/100 cum of distillery spentwash.
- The pay-back period is about 3 years.
- The operation of the process is simple and the maintenance costs are low.

11.13 Case Study—A Diphasic System for Distillery Spentwash[17]

The technology developed is based on the concept of diphasic system of anaerobic digestion. The configuration of hybrid reactor design of methane digester, which combines the merits of upflow sludge blanket and packed bed reactor was employed. The other important features of the technology are, *viz.*, unique distribution, mixing and temperature maintenance system, open top of the methane digester—a built in safety system and an efficient microbial consortia isolated and developed in R and D Centre of Daurala Sugar Works.

11.13.1 Pilot Plant Details

The pilot plant to treat 100m^3/d has been designed on the basis of the data collected from 10m^3/d. The pilot plant has seven distinct stations which are as follows :

- spent wash storage tank.
- acid phase reactor
- methane phase reactor
- Settler (Conventional and Lamellae)
- gas holder
- gas flaring device
- hydrogen sulphide scrubbing tower with chemical tank

The composition of distillery effluents and its variations during the study period are given in Table 11.41.

Table 11.41 : Characteristics of Distillery Spentwash

Sr. No.	Parameter	Minimum	Maximum	Average
1.	pH	4.50	5.20	4.70
2.	Dry matter, %	7.00	12.00	9.50
3.	Ash, %	1.80	3.25	2.40
4.	VSS, %	5.2	10.00	7.50
5.	Nitrogen, mg/l	550	2,000	1,250
6.	Phosphorus, mg/L	25	45	30
7.	Sulphate, mg/L	7,000	12,000	10,000
8.	COD (BOD), mg/L	85000 (35,000)	1,20,000 (50,000)	1,00,000 (45,000)

The storage tank was designed for 8 hours retention period in order to achieve the primary settling of effluent. The material of construction was mild steel and to prevent it from corrosion, epoxy painting was applied on internal surface. The specifications of storage tank are as follows :

- diameter, 3m
- height, 6m
- capacity, 42.4m^3

The acid phase reactor was designed for a retention period of 40 hours. For the distribution of feed, nozzles were provided on a circular coil and the over flow was taken from the top of the tank. For mixing the contents of acid phase reactor, recirculation arrangements were provided. In the winter months, the reactor contents were recirculated through an appropriately designed shell and tube exchanger to maintain the temperature to 37°C using distillery effluent as heating medium. The design features of acid phase reactor are given as follows:

- height — 8.5m
- diameter — 5.5m
- total volume — $200m^3$
- working volume — $160m^3$
- HRT — 1.66 day
- feed distribution — eight nozzles packed at 90° C to each other
- internal surface coating — epoxy painting
- insulation — nil
- number of recirculation parts — 2
- recirculation rate — 2.5:1 to 12.5:1 (reactor : influent)
- material of construction — mild steel
- rate of feeding effluent to acid fermenter — $0\text{-}8m^3/h$

Methane phase reactor was designed an a liquid flow of $100m^3/d$ basis and having a HRT of 10 days. The recirculation arrangement was also provided with shell and tube heat exchanger for maintaining the temperature of the reactor to 37°C even during the peak winter. At the top of the reactor, a gas liquid separator was provided with gas collection hoods which in turn were joined with a header to collect the biogas. As per the characteristics of the hybrid reactor, a bed of packing material (bricks) of particular dimension was provided just below the gas collection traps which acted as an anaerobic filter. The design features of methane reactor are given as follows :

- design — hybrid type-combination of UASB and packed bed
- height — 12m
- diameter 12.8m
- total volume — $1300m^3$
- working volume — $1080m^3$
- packing width — 0.5m
- packing material — bricks
- feed distributions — through nozzles
- internal surface coating — nil
- insulation — nil
- number of recirculation ports — 2
- rate of recirculation — $0\text{-}30m^3/h$
- rate of feeding acidified effluent to methane reactor — $0\text{-}8^3/h$

The conventional settler was designed on the basis of liquid loading rate of $20m^3/m^2.d$. The settler was attached to the pumping station by which about 30-35 per cent of the under flow could be recycled was active biomass to methane reactor. The design features are given as follows :

- straight height of the settler 7m
- height of compaction zone 2.2m
- diameter 2.5m
- settler volume $38m^3$
- material of construction mild steel
- settler under flow line to recycle the biomass to methane phase $0\text{-}8m^3/h$

Lamelle settlers have been installed for a flow of $100m^3/d$.

Gas flow metre (Medco) was installed in between gas holder and gas compressor to measure the gas flow ($0\text{-}200m^3/h$). The acid and methane phase reactors were innoculated with a laboratory growth consortia of micro-organisms specially a screened and selected for fast degradation of distillery effluents. The acid and methane phase reactors were gradually stabilized by slow feeding of effluent and by monitoring carefully the process parameters which are responsible for fast growth of acidogens and methanogens.

The kinetic studies were carried out taking various process parameters such as organic loading rate, hydraulic retention period, COD removal efficiency, biomass formation, biogas and methane production, etc. During stabilization, the VSS content of the reactor increases steadily to about 7.0 kg/m^3 which was reflected in the increased biogas production. The VFA profile was found to increase substantially beyond the designed load of 10 kg $COD/m^3.d$ which was reflected in decreased COD removal and biogas production. Upto a VFA level of 2500mg/L, the efficiency of methane reactor in terms of COD removal, biogas production was found to be optimal but it got affected adversely beyond 3000mg L and at a level of 5000mg/L (COD loading $-12kg/m^3.d$), drastic reduction in biogas production was observed. AT a hydraulic retention time of 10 days, the maximum biogas production was achieved. The biomass was studied against the various loading rates of COD. Upto 10 kg $COD/m^3.d$, the VSS content was consistent in the reactor and the COD removal rate was also found to be maximum, ranging between 1.04-1.15 kg COD/kg VSS.d. The rate of biomass accumulation was found to be in the range of 0.07-0.1 which corresponds to RXa value of 64 mg VSS/L.d.

The effect of recirculation (mechanical mixing) was examined on the performance of methane reactor. Although, no mixing saves energy, but decreases the biogas production significantly. Further, it also causes lowering of pH at the bottom of the reactor. Moreover, recirculation is essentially required to maintain the reactor temperature. The recirculation of 10:1 to 12.5:1 was found to be optimum for the process.

The process of bio-methanation is carried out by several groups of micro-organisms and ultimately by several species of methanogens. Since the mesophilic temperature are in the range of 32-43°C, therefore, the effect of temperature on the performance of methane reactor was examined. A temperature of 36 to 38°C was found to be optimum for maximum gas production.

The beneficial effect of nickel and ferric chloride as trace nutrients were examined on methanogens/methane reactor performance. It was observed that the addition of 10 and 2 mg/L on nickel and ferric chloride respectively were responsible for enhancing methane generation and better BOD removal efficiency. The performance highlight of this treatment technology given as :

- BOD reduction 83%
- COD reduction 70%
- COD loading 9-11 kg COD/m^3.d
- Total HRT (Acid+methane Phase) 11-12 days
- Biogas production 33-35m^{3}m^3 of effluent
- Methane 63-65%
- Hydrogen sulphide less than 0.5%

11.14 Case Study—Varion Distillery Effluent Treatment and Golden Fish Culture

11.14.1 Introduction

VCDL's distillery is located at Vedanarayanapuram, near Chingleput, on the banks of Palar river. It is a molasses-based unit and produces about 33,000 litres of alcohol per day and discharges about 5 lakh litres of effluent. The effluent is characterised with on average BOD of about 45,000 and is dark brown in colour. The temperature of the effluent is about 80°C at the discharge point. While the temperature is brought down within a period of 12 to 24 hours by logooning, it has been found most difficult to get rid of the dark brown colour and to reduce the BOD to the permissible limit of 30 prescribed by the Pollution Control Board. The distillery effluent is considered to be the worst of all the Industrial effluents of the world, adversely affecting the environment in many ways. Several studies have been made in many countries to solve this problem, but most of them have been found be not quite adequate to bring down the BOD to the desired level. Over the past five years VCDL has been carrying out R and D works and has obtained valuable data which could solve the problem effectively.

11.14.2 Anaerobic Digestion

Though considered the worst, the effluent contains mostly organic substances, there being no heavy metal in it. The organics are readily digested by different groups of micro-organisms. A digester to anaerobically treat the effluent has been designed and is under operation. The technology for this is based on our studies on laboratory and pilot plant scale digester units. The effluent water after being cooled to about 35°C is let into the anaerobic digester and allowed to act by a microbiological process. Whereby methane producing bacteria. Methanomonas methanica produces methane gas, utilizing the rich organic matter contained in the effluent. In order to increase the efficiency of the conversion process the anaerobic digester tanks are suspended with biofilms. When the anaerobic digester tanks are suspended with biofilms. When the biofilms are suspended in the effluent, the surface area required for the organisms to stick on the multiply. Increases many hundred folds. It is estimated that in one cubic metre volume of effluent the suspended biofilms provide about $100m^2$ of surface area. This enhances the activity of the Mothanomonas bacteria and production of methane gas. While the methane gas is produced by the organism, the organic molecules of the effluent are broken down to more useful and less toxic substances. Since sulphur is a contaminant in the molasses used in the distillery, the effluent coming out of the distillery also contains small quantities of sulphur. In the anaerobic digestor system, the sulphur is acted upon by sulphur bacteria, Thiobacillus thio-oxidans, resulting in the formation of hydrogen sulphide. When the digestion is in full swing, after an initial maturity period of about 3 months, the gas which comes out is a mixture of methane, hydrogen-sulphide and carbon-di-oxide. This gas is utilized as a substitute for fuel in the boiler. It can replace coal and/or furnace oil used for the boiler in the distillery. It has been estimated that the total quantity of methane gas produced would be about 10,000 m° from out of 5 lakh litres of effluent per day and this could replace about 75 to 85 per cent of the fuel requirement of the distillery, valued at about Rs. 80 lakhs/year.

11.14.3 BOD Reduction

The holding capacity of the digestor is so designed to accommodated about 25 lakh litres of effluent, which is equivalent to five days' discharge from the distillery. In order to improve the efficiency of methane gas production, the contents of the digestor are continuously recycled until about 80 to 85 per cent reduction in the BOD level is achieved. Regular monitoring of the various factors influencing the digestion process such as the pH of the medium, temperature, etc. is carried. The digestion process is continuous and hence automatic device to regulate the inflow and outflow of effluent and various other controls and checks have been provided.

11.14.4 Bioconversion Ponds

The bacteria which produce the gases have a short life about 30 to 45 minutes. They multiply fast, utilizing the organic substance and also die very fast. Thus, a good portion of the organic matter is converted into a large mass of dead and live bacterial cells. The biomass is withdrawn regularly from the digestor and taken to bio-conversution tanks. These tanks are aerobic and extra oxygen to quicken the bio-conversion process is provided through floating aerotors. The original BOD of about 45,000 in the raw effluent gets reduced to about 4,500 in the anerobic digestor and when this digested effluent is taken to the aerobic bio-conversion tanks and treated for 24-48 hours the BOD comes down about 1000. At this BOD level, the digested effluent with the suspended microbial biomass becomes a rich medium for fish culture.

11.14.5 Selection of Fish Species

In our studies to find out the best way of utilizing the microbial biomass and to handle the effluent water, which continues to be dark-brown coloured even after the treatment under the anaerobic and aerobic systems, we found certain species of fish could carry out the cleansing operation very effectively. A comparative study of different species and strains of fish for this purpose revealed that the golden coloured Tilapia was comparatively the very best. Therefore, were concentrated out studies on developing a strains of Tilapia which would be nutritionally acceptable and economically viable. The common Tilapia which had found its way into India from the Africa countries has been condemned as one of the worst fish species which have contaminated Indian waters, the reason being that it multiplies very rapidly suppressing and smothering other economically important species in inland waters. It is considered a wild type and is gray coloured, bony, small-sized and not considered suitable for human consumption. The golden coloured mutant Tilapia on the other hand, is very acceptable in the market for its thick and white meat, good cooking qualities and absence of fishy smell white cooking. It has a taste of its own which has consumer preference.

11.14.6 Breeding of Golden Fish

We have collected different strains of the golden and red coloured Tilapia as well as other strains of fast growing and fatty, grey and brown coloured Tilapia from different sources. Through intensive cross breeding programme we have been able to develop a hybrid fish which we named as 'Golden Fish'. One of the parents of this hybrid is the original mutant with golden colour and the other is a grey coloured mutant. When these two are crossed, the golden colour being dominant, we obtain F-1's which are mostly golden coloured. After cross-breeding, the eggs are collected from the mouths of female parents and are hatched in special containers. The fries which come out of the eggs are pooled and taken to sex-reversal tanks. When these fries are fed with a harmone 'testosterone', at a particular concentration, the female reproductive structures are suppressed, and they behave like males. Thus over 98 per cent of population of fries in the total population behave like males and they are preferred for intensive aquaculture. It has been found that the males grow 30 to 40 per cent faster and physically fatter than the females, the energy utilized by the females for reproduction being cut off by the process of sex-reversal. The harmone is generally mixed with the feed given to the fry in the sex-reversal tanks and the process continues for about 3 weeks. By the end of this period, the 'all-male-population' is transferred to fingerling ponds.

11.14.7 *Fingerling Ponds*

The fingerling ponds are supplies with specially prepared feed so that the fingerlings could grow faster and attain a size of about 30 g in 6 weeks. Certain amounts of cannabolism among the fries has been noted and this is attributed to the fries seeking to get more of animal feed than vegetable feed. Also, variation in the size of the fries lead to the bigger ones eating away the smallest ones. By minimising the size variation in the fries and fingerling and by improving the quality of the feed, we are trying to reduce the rate of cannabolism. When once the fingerlings reach the size of about 30-50 gms, they are ready for transfer to fish grow-out ponds.

11.14.8 *Semi-intensive Fish Culture*

We have designed the fish grow-out ponds in such a manner that there is plenty of water surface exposed to sunlight which helps growth of micro-flora and fauna, which become rich food for fish. The design of the first pond with a central partition helps in many ways, particularly when the water is agitated with floating aerators, it moves continuously. The surface aerators help to improve the dissolved oxygen in the water, which is essential for better productivity of the pond. The two major factors which are important for a successful culture of the fish are plentiful biomass which become feed for the fish and adequate oxygen which would help the fish to grow faster. It is also necessary to provide extra oxygen on two other counts, *viz.*, to help the thick population of the fish to grow fast and in rapid multiplication of the microflora in the water, which helps conversion of the waste organic matter contained in the water into useful microbial biomass. During day time there is rich algal bloom in the pond, which results from the photosynthetic activity of the algae utilizing the carbon-di-oxide of the atmosphere and the sunlight. In the process, the algae release oxygen in the water, which helps fish and microbial growth. The alage becomes feed for the fish. While there is sufficient oxygen released into the water through algal growth during day time, the photosynthesis being cut off during might, the demand for oxygen increases in the night. Hence, continuous agitation through floating aerators in the night time becomes essential. Our studies have shown that the fish grow-out pond could be of optimum size of 140 mtrs. × 15 mts. with 1 mtr. depth of water. The dissolved oxygen required has been found to be optimum at about 5 ppm. The optimum population for an half-an-acre fish pond of standard size, with proper monitoring of the oxygen, ammonia, and pH levels, would be about 12,000 to 15,000. The fish grow from about 50 gms to 250 gms size within about four months and would be ready for harvest. The feeding of the fish has to be done at periodic intervals during day time and specially formulated floating fish feeds are being fed. The conversion ratio of feed into fish has been found to be in the range of 2:1.

11.14.9 *Fish Culture with Digested Effluent*

Having standardized the size of fish ponds and the culture techniques, we started utilizing the digested effluent water for fish culture. The effluent, which after successive anaerobic and aerobic digestions containing 1000 BOD is taken to the fish ponds and added every day to cover about 2 to 24 cm. height or about 10,000 to 15,000 litres per acre per pond per day. Our estimates of the evaporation rate of the water from the pond surface showed that it varied from about 10,000 to 15,000 litres per acre of waterspread. Our 'Golden Fish' Farm has been designed for a waterspread of about 50 acres so that the entire quantity of 5 lakh litres of digested effluent when distributed to the fish ponds evaporates every day. The Farm itself is about 75 acres in a pond, but the water surface in the ponds works out to about 50 acres. Therefore, the entire effluent water coming out every day is distributed in the 50 acre waterspread and gets evaporated. No water gets out since it is almost cent per cent a closed system, thereby we have in this system final BOD at zero level and therefore zero level of environmental pollution. At the present level of productivity we harvest on an average of about 15 tonnes of fish per acre through two crops in a year. We have the built-up capacity to produce 750 tonnes of fish from the 50 acres of grow-out ponds. We are marketing the fish in Madras City and several other places as 'live fish'. The fish on maturity, after about 5 to 6 months, is harvested and transferred in aerated water tanks to the sales point, where again they are let into aerated water tanks before being sold 'live' to the customers.

11.14.10 Increase Fish Productivity

After establishing the Golden Fish Farm for fully handling the 5 lakh litres of digested effluent every day, our R and D efforts were diverted to increase the productivity of fish from unit volume of water. The three important factors which was directly influence the productivity of Golden Fish are : (*i*) continuous replacement of the fish pond water with fresh water so as to improve its quality, (*ii*) daily removal of the sludge, *i.e.*, excreta from the thickly populated fish through provision of a drain at the base of the pond, and (*iii*) supply of protein-rich floating fish feed with good percentage of protein derived from animal source, as against sinking fish feed. Two different types of ponds were designed and tested for productivity and the results are as follows:

11.14.11 Intensive Fish Culture Ponds

The rectangular pond measure about half-an-acre and about 1.2 metres. deep, the side walls were lined with cement concrete slabs. The bottom was filled with a clay layer of about 15 cm. A central drain was provided with PVC pipes to withdraw the sludge and fish metabolites. The half acre pond was aerated with three 15 to 2 HP floating or injector type aerators. The stocking was done with 38,000 fingerlings of 15 to 20 g size. An extruder-made floating feed with 25 per cent protein content was provided. We could also add about 10,000 litres of digested effluent per acre in this pond. The result was, we could harvest in six months, about 15 tonnes of fish of 450 to 500 gm size and in one ha in a year about 120 tonnes. The feed conversion ratio ranged from 1.8:1 to 2.0:1. This we consider as an intensive culture system, whereas fish culture in the earthern ponds, described earlier, which yielded about 35 to 40 tonnes/ha as semi-intensive system. The sludge from the pond is rich in plant nutrients and is used as manure in agricultural fields.

11.14.12 Sulphur-intensive Fish Culture Ponds

In another super-intensive system a large 4.2 acre or 1.6 ha cement concrete pond with the side walls build of R.C.C. structure and the bottom lined with granite structure with central PVC drain was constructed. Along the periphery 36 tanks of about 130 m^2 were constructed. Provision has been made for regular inflow and outflow of fresh water so that every day about 5 to 10 of the pond water is replaced with fresh water. Aeration and feeding was done on the same basis as in intensive culture system. The stocking density was about 120 fingerlings per m^2 of waterspread and the depth of water in the ponds is about 1.2 m. The floating feed contained 25 to 30 per cent proteins, half of it being derived from animal source. The feed conversion ratio is about 1.8:1. The productivity per acre is about 250 to 300 tonnes per year, with three crops, each crop reaching the harvestable size of about 450 to 500 g in about four months. This works out to about 650 to 700 tonnes per ha per year.

11.14.13 Conclusion

In conclusion it may be stated that the wastewater, considered to be one of the worst industrial popullants, is economically utilized through anaerobic digestion to produce methane gas which replaces coal as fuel in the industrial boiler and the digested effluent is purified by intensive fish culture, the fish consuming the microbes and other organic molecules, providing almost zero BOD treatment system. The technology has been developed almost indigenously through our R and D efforts, there being a new design for anaerobic digestor and semi-intensive and intensive fish culture systems along with floating fish feed manufacture having been developed for the first time in the country. The potentials are great for this technology for being adopted to treat other industrial effluents and to substantially increase inland fish production to wipe out protein malnutrition in India.

11.15 Some Comments

This section has datailed the wastewaters treatment through recourse to biomethanation route. Several types of wastewater and reactor configurations and has been highlighted. In India, there are few plants working on these principles. It is, therefore, necessary, to undertake a project which determines the actual performance of these anaerobic reactor alongwith the problems that are associated with operating these anaerobic bioreactors, it will develop confidence for implementing more of these plants and also avoid duplication of R and D work pertaining to development of bioreactors. The time saved can be used directly to implement the successful plants designed in other areas and waste types.

REFERENCES

1. Data Provided by Abbot Laboratories, Illinois for Full Scale Plant at Abbot's North Chicago Facility (1987).
2. US, EPA, Development Document for Final Effluent Limitations Guidelines and Standards for Pharmaceutical Manufacturing Point Source Category, EPA-440/1-83/084, (September 1983).
3. Data provided by Environment Canada, Wastewater Technology Center, Burlington, Ontario, Canada.
4. Cornacchio, L. Modified Serum Bottle Testing Procedure for Industrial Wastewater, presented at 21st Cand., Symp. Wat. Poll. Res., Burlington, Ontaris (April 30, 1986).
5. Hall, E.R. And Prong, P.D. Evaluation of Anaerobic Treatment For NSSC Wastewater, presented at 1986 Environmental Conf. TAPPI, New Orleans (April 1986).
6. Welander, T. and Anderson, P.E. Anaerobic Treatment of Wastewater from the Production of Chemi-thermo-mechanical Pulp, Wat. Sci. Tech. 17 (1), p. 103 (1985).
7. Data obtained from Ashland Petroleum Company, Ashland, Kentucky, USA (1986).
8. Lanting, J. and Gross, R.L. Anaerobic Pre-treatment of Ethanol Production wastewater, Proc. 40th Ind. Waste conf. Purdue Univ. p. 905-914 (1985).
9. Data Obtained from Grontmig Consulting Engineers, DeHolle Bilt 22, De Bilt, The Netherlands (1990).
10. Data Obtained from Biothane Corporation, Camden, New Jersey, USA (1990).
11. Personal communication with Dr. James P. Earley, Department of Civil Engg. Univ. of Notre Dame, Notre Dame, Indiana.
12. Data Obtained from Dorr-Oliver, Inc, Stanford, Connecticut, USA.
13. Personal communication, Hein, W.D., G.S. Processes, Inc, Scottsdale, Arizona.
14. Personal communication with Mr. A.M. Sobkowicz, Badgar Engineer Inc, Cambridge, Massachusetts, USA.
15. Technology Process Package Developed by NEERI, Nagpur (1993).
16. Technology Process Package Developed by NEERI, Nagpur (1993).
17. Diphasic Anaerobic Digestion of Distillery Effluents in Scaled-up Pilot Plant at Daurala Sugar Works, Meerut, Ministry of Non-Conventional Energy Sources (MNES), N. Delhi (September 1992).
18. Personal Discussion with Dr. G. Rangaswamy, Chairman, R and D Centre Vorion Chemicals and Distilleries Ltd., Madras.

CHAPTER 12

FUTURE RESEARCH AREAS IN ANAEROBIC PROCESSES

12.0 Introduction

Prior to 1965, anaerobic wastewater treatment applications were primarily involved with stabilization of waste sludges. During the last two decades, laboratory research has led to the development of a new generation of anaerobic biofilm reactors for treating high strength organic waste. At the present time, processes such as the anaerobic (1) fixed bed (2) fluidized bed (3) expanded bed (4) rotating biological contactor (5) UASB and others are routinely being tested with a wide variety of high strength organic wastes. Pilot scale testing and commercial development have taken place with several of these processes. Accelerated development and commercial applications are anticipated.

At the end of each chapter of this book, some observations have been made regarding R and D required in that area. This Chapter summarises some of the details which require immediate attention.

12.1 Compounds Non-amenable to Biodegradation

Most organic compounds are biodegradable to an extent. While they may not be degraded to the extent one would like, nevertheless, there exist bacteria that are capable of carrying out slight changes in the nature of these molecules. There do exist, however, classes of extremely refractory organic compounds that because of their toxic, complex, or inert characteristics, are resistant to microbial attack. This condition may manifest itself in the individual molecules or occur only above certain threshold concentrations. The resistance encountered can be result of many factors, which include, solubility, molecular size, amount of tertiary branching, and the number, nature and position of the substituents. There exists general rule to determine the relative biodegradibility/non-biodegradibility of a compound[1, 2]. A few of them are as follows :

- materials that can pass through cell membranes are more readily available to the microbes and are degraded faster;
- microbes prefer non-aromatic or cyclic aromatics over aromatics. The presence of the substituents in the benzene ring usually increases biodegradability;

- soluble compounds are more easily degraded than insoluble;
- a high degree of branching imparts greater resistance to degradation;
- dispersed compounds provide more surface area for attack, and therefore, degrade better;
- compounds with unsaturated bonds are degraded more readily than saturated compounds;
- such materials are alcohols, aldehydes, acids, esters, amides, and amino acids are preferred over the corresponding alkanes, alkenes, ketones, dicarboxylic acids, nitriles amines and chloroalkanes.

These are some of the rules which may be useful while considering bio-oxidation through anaerobic and aerobic routes.

12.2 Research Needs – Anaerobic Biofilm Systems

Tables 12.1 and 12.2 list some of the often reported advantages and disadvantages given for employing anaerobic biofilm systems instead of conventional anaerobic processes for treating high strength waste. Most of the advantages are related to (1) the high concentration of biomass retained within the reactor and (2) lack of oxygen transfer constraints for anaerobic systems. At present our understanding of anaerobic biofilm processes is at a level comparable to our understanding of aerobic biofilm processes in 1965.

Both basic microbial research and applied research are needed if we are to take full advantage of the economic benefits potentially available from biofilm systems. Fig. 12.1 is a very simplified view of the organic carbon flow involved in waste stabilization. Complex interactions between three and perhaps five or more distinctly different types of facultative and anaerobic bacteria are involved in the conversion of complex organic materials to methane. Basic research conducted during the last two decades has provided insight and understanding of the latter two steps shown in Fig. 12.1 and the results of this research are currently being applied to the design and operation of anaerobic systems on routine basis. However, the microbial activities related to steps I and II are poorly understood at the present time. Many complex organic substrates, especially organic solids, appear to resist these initial steps and hence these activities limit both the rate and level of treatment attainable. Other organic substrates, such as humics, and organochlorides are attacked so slowly, if at all, that anaerobic treatment is ineffective or impractical. Better fundamental knowledge of enzymatic attack mechanisms involved, perhaps coupled with genetic engineering benefits, may unlock these rate limiting steps and permit the development of inexpensive pre-treatment or waste conditioning techniques for these recalcitrant wastes.

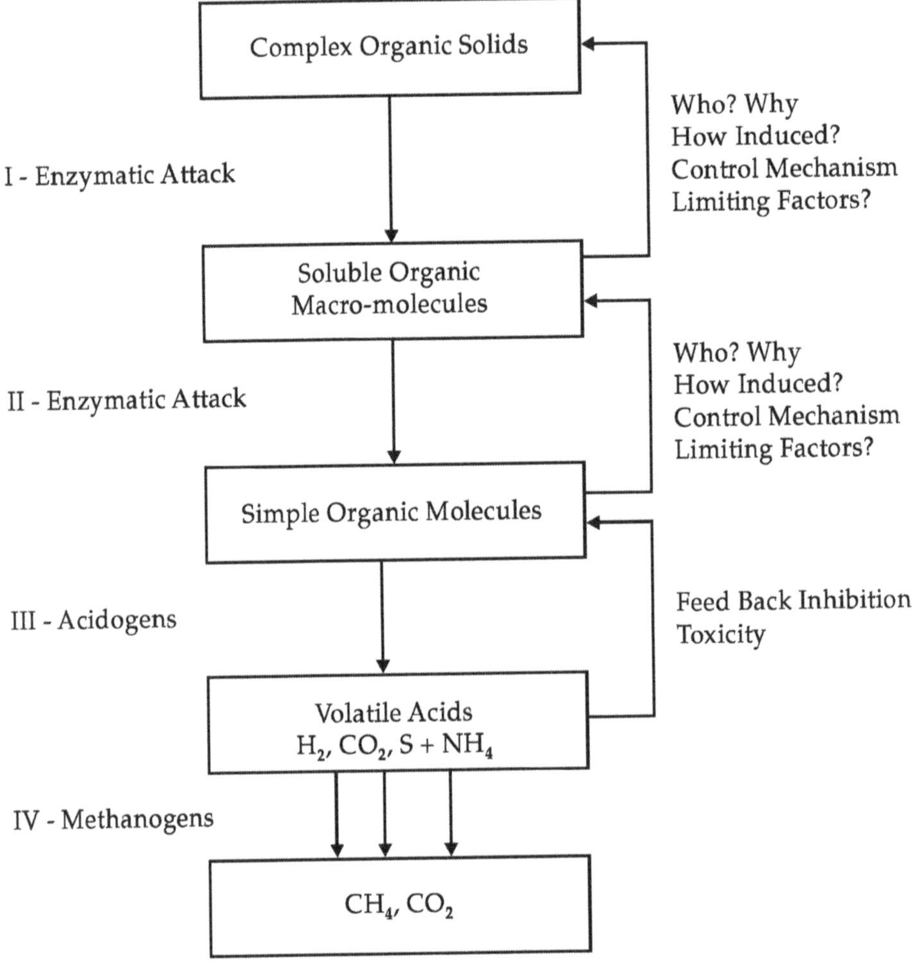

Fig. 12.1 : Simplified Overview of Carbon Flow in Methane Generation.

As indicated in Table 12.2, anaerobic biofilm systems require a long start-up period due to low micro-organisms growth rates. Full biomass development may require a year or longer. Seed sludges, usually obtained from nearby operating digesters, may or may not contain appropriate mixes of microbial populations for the wastes being treated. While initial attachment of bacteria to support media is quite rapid, subsequent growth and film development appears to take place rather slowly. In addition to the inherently low mass growth rates reported for most anaerobic systems several other possibilities exist that might limit growth and hence maturation of biofilm systems.

Well developed biofilms contain mixed population of acidogenic and methanogenic bacteria, with methanogens removing low molecular weight acid and H_2 as they are generated. Start-up procedures, however, present some unique problems. For example, if the initial attachment of floc particles to a clean surface occurs randomly, it is possible that much of the media surface will be coated with areas of reasonably pure microbial colonies. Should this occur, it is likely that increasing local intracolony concentrations of metabolic end products (volatile acids and H_2) would result in inhibition of the acidogens. Similar colonies of methanogens located in adjacent areas would be starved for acetate, H_2 and CO_2 and methanogenic activity would be limited. Externally, on a microscopic scale, the symptoms of 'Sour' digester, *e.g.* high volatile acids, low alkalinity, high CO_2 and low CH_4 would be perceived. These symptoms are frequently associated with start-up. Traditional start-up procedures include low organic loading rates, excess buffering and long periods for the development of mixed commensal population.

Table 12.1 : Potential Anaerobic Biofilm Process Advantages

Sr. No.	*Item*
1.	Biomass Attachment Yields long MCRTs
	* stability of biochemical system
	* resistance to washout
2.	High Active Biomass Concentration in Reactor
	* amenable to high strength influenece
	* oxygen transfer not limiting
	* better response to temperature fluctuations than anaerobic suspended culture system
	* small reactor volumes (low HRTs)
3.	Low Growth Rates
	* minimal solids production
	* low nutrient requirements.
4.	Production of Methane
5.	Especially Suitable for Low Cost Pre-treatment

Table 12.2 : Potential Anaerobic Biofilm Process Disadvantages

Sr. No.	*Item*
1.	Long Start Up Periods
2.	Toxicity Problems from Substrate Precursors
	* sulphides
	* ammonia
	* foaming
3.	Solids Deposition and Plugging
	* organic
	* inorganic
4.	Off-gas Scrubbing Requirements
5.	Poor Quality Effluents
	* solids that are difficult to settle
	* relatively high soluble organics

Several start-up procedures for improving start-up procedures are possible. Two separate seed sources, one rich in acidogens and other primarily consisting of methanogens may be passed across the support media *via* alternating pumping cycles to yield the layered situation. This would provide the opportunity for acidogenic end products to be rapidly removed from the local microscopic environment. Eventually, growth, death and decay would, balance the populations to yield desired situation (as in case of mixed growth). Alternatively, other procedures, such as reduced headspace pressures, could be used to the reduce the solubility of toxic or inhibitory gaseous end products during startup. Clearly, fundamental research concerning microbial attachment would be of benefit to all anaerobic biofilm systems.

Little detailed information is available concerning growth, death and decay rates of anaerobic biofilms. Following initial biomass attachment to support media or during recovery from extended influent toxic loads, it is desirable to encourage maximum growth, even at the expense of methane generation. Rapid microbial growth might be encouraged by augmenting normal substrates with simple sugars and/or acetate alongwith adequate buffering agents. Simple mechanical manipulation of off-gas system can enhance growth yields. Should these observation be substantiated by additional research, useful operational procedures can be readily implemented to encourage rapid microbial growth during critical periods.

Virtually nothing is known about death and decay in anaerobic biofilms. Fundamental knowledge concerning the mechanisms involved and the factors affecting death and decay are necessary if one wishes to model biofilm systems with the same degree of sophistication that is applied to aerobic suspended culture system. While major emphasis is being placed on the question of why and how bacteria stick to surface, little information is available for describing why sloughing occurs in anaerobic biofilms. It is very probable that microbial populations adjacent to support media undergo significant changes as growth proceeds outward towards bulk solution. As growth continues, local concentration of substrate become limiting near the support surface, and accumulation of metabolic end products forces change in metabolic processes and physiological responses. These factors undoubtedly affect sloughing and reactor performance. New knowledge concerning death and decay mechanisms and their rates, alongwith better description of biofilm characteristics as function of depth are required prior to the development of models that will permit the prediction of anaerobic biofilm performance.

At a different level, mechanistic studies of substance transport to and through porous biofilms are necessary to optimise reactor performance. Transport of substrate and metabolic end products to and from the biofilm surface can be described by models similar to those developed for other film systems. However, descriptions of rates of accumulation of the metabolic end products and their transport within porous biomass provides a challenge that will require complex experimentation and sophisticated modeling techniques before we can optimise reactor development. Similarly, the accumulation of micro bubbles of product gases within the filaments may act to restrict substrate diffusion between species and enzyme transport to the solution. In addition, the information of insoluble complexes and inorganic precipitates from sulphides and carbonates can be expected to affect substrate transport. The accumulation of these and organic solids over extended periods of operation can lead to plugging and sloughing problems that result in the measurable loss of substrate removal performance with time. A thorough understanding of the mechanisms and rates involved in solids formation should have significant impact on the engineering choice of biofilm reactor type or the pre-treatment techniques required for specific waste streams. The possible potential rate limiting mechanisms are as follows :

- bulk solution transport of substrate;
- diffusion of substrate across boundary layer and into biofilm;
- microbial reaction rate;
- micro-bubble blankets;
- diffusion of metabolic products through biofilm and boundary layer;
- removal of end products by convective transport in bulk solution.

It is necessary to know the amount of different kinds of fatty acids generated during anaerobic reactions. The possible solution lies in developing bacterial cultures which could convert all fatty acids to acetic acid and at a faster rate, so that the other fatty acids do not hamper or slow down the methaogenic activity.

Many large gaps exist in the application of anaerobic processing to the treatment of industrial wastes. Little or no work has been published involving the treatment of such groups as chlorinated alkanes, phthalate esters, or polychlorinated biphenyls by anaerobic processes. In addition the basic pathways for the production of methane from many organic substances is not yet well understood and needs to be developed further.

Much of the future work done on anaerobic systems will revolve around process medications. By manipulation of such factors as SRT, reactor types, and loading rates, process efficiencies will be increased towards the optimal performance level. Other chemical factors, especially those of temperature and pH, need to be carefully studies to maximize the growth rate of micro organisms involved in the biological degradation of organic compounds. Intensive experimentation in the above-mentioned areas will increase the level of technology in the treatment of industrial wastes by anaerobic processes to a much higher plateau than is realized today.

A promising area of future research that relies on the creation of new technology instead of modification of old technology is that of genetic engineering. Applications of this new innovative technology are opening new frontiers in many areas of science, including the wastewater treatment field. While it is highly probable that given a sufficient amount of time, bacteria will acquire to ability to degrade most of the new industrial pollutants on prolonged exposure, it may be possible to significantly speed up the evolutionary process. By selective use of naturally occurring breeding processes, the production of new characteristics in bacteria can be completed many times than if the change occurred by natural evolution. Using the techniques of true genetic engineering, the production of new characteristics in bacteria can be increased by an order of magnitude 10 times faster than natural evolution[3].

The genetic manipulation of organisms can be brought on by many techniques. Genetic material can be transferred from one organism to another *via* viruses, plasmids, bacteriophages, or even protoplasts. Geneticists are only beginning to realize the potential of genetic engineering using these methods. Some of the possibilities include membrane changes, adding new proteins to cells, and making additional enzyme systems operational. Conversely, properties can be deleted from the genetic makeup by careful introduction of mutagens.

As stated above, plasmid transfer is one possible method of changing the genetic makeup of a particular organism. Plasmids are small bits of double stranded DNA that contain specific information for the production of enzymes needed by the cells in cases of extreme stress. An average plasmid can contain approximately 20-30 genes. By a highly specialized transfer process, the strands of DNA from a donor plasmid can be incorporated into another organism's genetic inventory. The exchanged DNA is duplicated to form a double strand. Conjugation between two organisms can result in the exchanging of this new material, resulting in the eventual establishment of the new genes in the population[4].

This technique was the one employed by Chakrabarty and his colleagues at General Electric in the genetic engineering of a microbe that could degrade oil[4]. This was a particularly difficult feat since oil is composed of a variety of hydrocarbons and most microbes can degrade only one specific hydrocarbon. To produce an organism that could degrade oil, plasmids from a variety of organisms, each containing the genetic information for the production of enzymes necessary to degrade a different hydrocarbon, had to all be incorporated into one cell. Growth experiments were performed to test the success of the transplants and it was found that "new" organism grew much faster than single plasmid ones, indicating that it was using all the individual hydrocarbons contained in oil, simultaneously.

Another possible use of this techniques in wastewater treatment involves increasing organisms resistance to toxicants. A case of this type has already been demonstrated by Chakrabarty and his colleague[4]. By transforming the MER plasmid from cells of *Pseudomonas oleovorans* to *Pseudomonas putida*, the resistance of *P. putida* to mercury could be increased by a factor 25 to 35 times. It is thought that the MER plasmid increases the cell's ability to bind mercury with intracellular protein which is retained in the cell.

While genetic engineering is one solution to solving the problems of treating industrial pollutants it must be emphasized that it is not the only way. As stated before, process modifications while not as "trendy" as the area of genetic engineering, will play a large part in increasing the acceptability of anaerobic processes.

Recombinant DNA application to anaerobic fixed for methane biosynthesis should aim at isolation and screening of methanogenic bacteria containing plasmid(s), development of plasmid vector for gene cloning, as well as techniques for permeabilizing cells to the rDNA molecules. The availability of the plasmid vector will help in manipulation of target gene(s) for construction of more efficient methanogenic organisms.

Hungate's Roll Tube Technique may be used to isolate cultures from various sources, *e.g.* paddy field soil, compost pit, lake sediment, and hospital and distillery wastes. Cultures should be tested for their homogeneity using phase contrast microscopy, fluorescence microscopy and scanning electron microscopy. Isolates should be tested for their growth on carbon sources such as methanol, acetate, formate, $H_2 : CO_2$ and methylamines as well as for their sensitivity to certain drugs like puromycin, etc. Observations with enzymes like disaggregate and hyaluronidase indicate their efficacy to disaggregate cells of *Methanosarcina barkeri* and may thus serve as a tool to permeabilize cells to recombinant DNA molecules.

A shuttle vector may be designed using macro promotor to express and select for puromycin marker. The mobilization of shuttle vector to *M. barkeri* or others may be demonstrated by puromycin sensitivity, etc.

Anaerobic bacteria play a key role in anaerobic treatment of municipal and industrial wastes. Anaerobic digestion of industrial waste/wastewater is currently a treatment of choice as it is energy efficient and generates methane. The success of large scale industrial application of anaerobic digestion of waste depends upon better understanding of complex interdependent microbial activity. The anaerobic digestion broadly involves three trophic groups of bacteria. The first comprises the hydrolytic, commonly referred to as acidogens, because they initially hydrolyse the substrate into short chain organic acids. The second group is the heteroacetogens which produce acetic acid and hydrogen and the third, the methanogens, which produce methane at the cost of acetate and H_2 and CO_2. An important factor in heteroacid production is the concentration of hydrogen in the reactor which affects both pH and the redox potential. The excessive production of acetate and other acids like butyrate and propionate from alternative biochemical pathways lead to a 'sour' digester inhibiting methane production. Multiplex PCR will help in the assessment of presence and ratio of different organisms at various trophic levels so that the substrate dependent inter-species relationship can be maintained for maximum efficiency of treatment system.

Other Aspects

The other aspects that need immediate attention are :

- Choice of reactor system (with and without support media for biomass attachment). It is also desirable to develop hybrid reactor systems which could have merits of the original type of reactor systems minus its demerits.
- Determination of kinetic constants for different reactor types and wastewaters. Little information is available on these aspects. Determination of biofilm thickness, biomass concentration, etc., in a fixed film reactor system has to be rationally developed. Also the determination of methods to determine activity, toxicity, etc., of biofilm need some attention.
- Scale up formulations for different types of reactor systems have to be developed. Scanty information is available on these aspects.
- Mathematical modeling for optimization of process, size, cost effectiveness and all related aspects also needs special R and D support.
- Production of hydrogen and exploring the possibility of recovering sulphur must also be included in future R and D activities. This may also include removal of H_2S gas from biogas generated from the anaerobic reactors.
- Conversion of methane to methanol and other product can also be included, instead of trying to convert methane gas into heat and electricity. It is possible to obtain variety of chemicals from methane gas by reforming process as is being practised in petrochemical complexes. However, this application will apply to very large scale of biomethanation plants.
- Determination of methods for preserving 'seed sludge' for large duration of periods. This seed sludge can be used during the failure or upset of the plants or can be sold to other biomethanation plants for start-up of bioreactors.
- Thermodynamically, it is possible to convert excess CO_2 into methane by Shift reaction in presence of steam and H_2. R and D efforts are required to develop specialized cultures which can convert CO_2 into methane and, therefore, one can get enriched biogas.
- Biological methods are also required to remove H_2S from the biogas. Extensive research efforts are needed to improve the efficiency of this process and at a reasonable price.
- Conversion of methane directly into electricity is possible. However, intense research is required.

- Developing specialized cultures for thermophilic anaerobic biomethanation. These bacteria should not lyse at higher temperatures. If these cultures are developed, it is possible to reduce the HRT of the bioreactor considerably. However, the cultures should remain viable and active in the bioreactor.
- There is a general necessity to develop bacterial cultures which can work under cryophilic temperature range. These bugs can be utilized at places where the temperatures are quite low, particularly in extreme nothern and eastern part of India.
- To develop special biochemical compounds, through the agency of microbes to mark the odours emanating from malodorous compounds, or alternatively develop bacterial cultures which can inhibit the formation of malodorous compounds or utilise them for their synthesis and energy requirements.
- It is equally essential for their determine the exact biochemical pathways for various types of organic compounds and also to determine the rate limiting step(s) involved in this process.

REFERENCES

1. Painter, H.A., Biodegradability, Proc. Royal Soc. London, Ser. B. 185 (1079), p. 149-158 (1974).
2. SCS Engineers, Selected Biodegradation Techniques for Treatment and/or Ultimate Disposal of Organic Materials, Technical Completion Report, EPA-600/2-79-006.
3. Widdas, R. and Adult, C.R., Progress in Research Related to Genetic Engineering and Life Synthesis Internl. Rev. Cytology, 38(1974) (Academic Press).
4. Chakrabarty, A.M., which Way Genetic Engineering, Industrial Research, Jan. 45, (1976).

www.ingramcontent.com/pod-product-compliance
Ingram Content Group UK Ltd.
Pitfield, Milton Keynes, MK11 3LW, UK
UKHW040241300726
14061UKWH00001BD/77